Lecture Notes in Computer Science　　10848

Commenced Publication in 1973
Founding and Former Series Editors:
Gerhard Goos, Juris Hartmanis, and Jan van Leeuwen

Editorial Board

More information about this series at http://www.springer.com/series/7407

Willem-Jan van Hoeve (Ed.)

Integration of Constraint Programming, Artificial Intelligence, and Operations Research

15th International Conference, CPAIOR 2018
Delft, The Netherlands, June 26–29, 2018
Proceedings

 Springer

Editor
Willem-Jan van Hoeve
Carnegie Mellon University
Pittsburgh, PA
USA

ISSN 0302-9743   ISSN 1611-3349 (electronic)
Lecture Notes in Computer Science
ISBN 978-3-319-93030-5   ISBN 978-3-319-93031-2 (eBook)
https://doi.org/10.1007/978-3-319-93031-2

Library of Congress Control Number: 2018944423

LNCS Sublibrary: SL1 – Theoretical Computer Science and General Issues

Printed on acid-free paper

This Springer imprint is published by the registered company Springer International Publishing AG
part of Springer Nature
The registered company address is: Gewerbestrasse 11, 6330 Cham, Switzerland

Preface

This volume contains the papers that were presented at the 15th International Conference on the Integration of Constraint Programming, Artificial Intelligence, and Operations Research (CPAIOR 2018), held in Delft, The Netherlands, June 26–29, 2018. It was co-located with the 28th International Conference on Automated Planning and Scheduling (ICAPS 2018).

The conference received a total of 111 submissions, including 96 regular paper and 15 extended abstract submissions. The regular papers reflect original unpublished work, whereas the extended abstracts contain either original unpublished work or a summary of work that was published elsewhere. Each regular paper was reviewed by at least three Program Committee members, which was followed by an author response period and a general discussion by the Program Committee. The extended abstracts were reviewed for appropriateness for the conference. At the end of the reviewing period, 47 regular papers were accepted for presentation during the conference and publication in this volume, and nine abstracts were accepted for presentation at the conference. Three papers were published directly in the journal *Constraints* via a fast-track review process. The abstracts of these papers can be found in this volume. The EasyChair system was used to handle the submissions, reviews, discussion, and proceedings preparation.

In addition to the regular papers and extended abstracts, three invited talks were given, by Michela Milano (University of Bologna; joint invited talk with ICAPS), Thorsten Koch (Zuse Institute Berlin and Technische Universität Berlin), and Paul Shaw (IBM). The abstracts of the invited talks can also be found in this volume.

The conference program included a Master Class on the topic "Data Science Meets Combinatorial Optimization," with the following invited talks:

- Siegfried Nijssen (Université catholique de Louvain): Introduction to Machine Learning and Data Mining
- Tias Guns (Vrije Universiteit Brussel): Data Mining Using Constraint Programming
- Kate Smith-Miles (University of Melbourne): Instance Spaces for Objective Assessment of Algorithms and Benchmark Test Suites
- Bistra Dilkina (University of Southern California): Machine Learning for Branch and Bound
- Elias Khalil (Georgia Institute of Technology): Learning Combinatorial Optimization Algorithms over Graphs
- Barry O'Sullivan (University College Cork): Recent Applications of Data Science in Optimization and Constraint Programming

The organization of this conference would not have been possible without the help of many individuals. First, I would like to thank the Program Committee members and external reviewers for their hard work. Several Program Committee members deserve additional thanks because of their help with timely reviewing of fast-track papers, shepherding regular papers, or overseeing the discussion of papers for which I had a conflict of interest. I am also particularly thankful to David Bergman (Master Class Chair), Bistra Dilkina (Publicity Chair), and Joris Kinable (Sponsorship Chair) for their help in organizing this conference. Special thanks is reserved for the conference chair, Mathijs de Weerdt, who also acted as the liaison with Delft University and the organization of ICAPS. His support was instrumental in making this event a success.

Lastly, I want to thank all sponsors for their generous contributions. At the time of writing, these include: the *Artificial Intelligence* journal, Decision Brain, SAS, Springer, Delft University of Technology, the Association for Constraint Programming (ACP), AIMMS, Gurobi, GAMS, Pipple, the European Association for Artificial Intelligence (EurAI), and Cosling.

April 2018   Willem-Jan van Hoeve

Organization

Program Chair

Willem-Jan van Hoeve   Carnegie Mellon University, USA

Conference Chairs

Mathijs de Weerdt   Delft University of Technology, The Netherlands
Willem-Jan van Hoeve   Carnegie Mellon University, USA

Master Class Chair

David Bergman   University of Connecticut, USA

Publicity Chair

Bistra Dilkina   University of Southern California, USA

Sponsorship Chair

Joris Kinable   Eindhoven University of Technology, The Netherlands

Program Committee

Tobias Achterberg   Gurobi, Germany
Chris Beck   University of Toronto, Canada
Nicolas Beldiceanu   IMT Atlantique, France
David Bergman   University of Connecticut, USA
Timo Berthold   Fair Isaac Germany GmbH, Germany
Natashia Boland   Georgia Institute of Technology, USA
Andre Augusto Cire   University of Toronto, Canada
Mathijs de Weerdt   Delft University of Technology, The Netherlands
Bistra Dilkina   University of Southern California, USA
Ambros Gleixner   Zuse Institute Berlin, Germany
Carla Gomes   Cornell University, USA
Tias Guns   Vrije Universiteit Brussel, Belgium
Matti Järvisalo   University of Helsinki, Finland
Serdar Kadioglu   Oracle Corporation, USA
Joris Kinable   Eindhoven University of Technology, The Netherlands
Philippe Laborie   IBM, France
Jeff Linderoth   University of Wisconsin-Madison, USA
Andrea Lodi   Polytechnique Montréal, Canada

Michele Lombardi   University of Bologna, Italy
Laurent Michel   University of Connecticut, USA
Michela Milano   University of Bologna, Italy
Nina Narodytska   VMware Research, USA
Siegfried Nijssen   Université catholique de Louvain, Belgium
Laurent Perron   Google France, France
Gilles Pesant   Polytechnique Montréal, Canada
Jean-Charles Régin   Université Nice-Sophia Antipolis, France
Andrea Rendl   University of Klagenfurt, Austria
Louis-Martin Rousseau   Polytechnique Montréal, Canada
Ashish Sabharwal   Allen Institute for Artificial Intelligence (AI2), USA
Domenico Salvagnin   University of Padua, Italy
Pierre Schaus   Université catholique de Louvain, Belgium
Andreas Schutt   CSIRO and The University of Melbourne, Australia
Peter Stuckey   CSIRO and The University of Melbourne, Australia
Michael Trick   Carnegie Mellon University, USA
Charlotte Truchet   Université de Nantes, France
Pascal Van Hentenryck   University of Michigan, USA
Willem-Jan van Hoeve   Carnegie Mellon University, USA
Alessandro Zanarini   ABB Corporate Research, Switzerland

Additional Reviewers

Babaki, Behrouz   Gavanelli, Marco
Bai, Junwen   Gottwald, Robert Lion
Belov, Gleb   Grenouilleau, Florian
Björck, Johan   Gualandi, Stefano
Bliem, Bernhard   Hendel, Gregor
Booth, Kyle E. C.   Hojny, Christopher
Bridi, Thomas   Huang, Teng
Cherkaoui, Rachid   Khalil, Elias
Cohen, Eldan   Kimura, Ryo
Csizmadia, Zsolt   Kiziltan, Zeynep
Davarnia, Danial   Legrain, Antoine
De Filippo, Allegra   Lin, Qin
de Una, Diego   Lippi, Marco
Dey, Santanu   Lo Bianco, Giovanni
Duc Vu, Minh   Lozano Sanchez, Leonardo
Emadikhiav, Mohsen   Marinakis, Adamantios
Fages, Jean-Guillaume   McCreesh, Ciaran
Farias, Vivek   Mehrani, Saharnaz
Fischetti, Matteo   Miltenberger, Matthias
Flener, Pierre   Müller, Benjamin
Gagnon, Samuel   Nonato, Maddalena

Olivier, Philippe
Pecin, Diego
Perez, Guillaume
Pralet, Cédric
Prouvost, Antoine
Raghunathan, Arvind
Rehfeldt, Daniel
Roli, Andrea
Römer, Michael
Serra, Thiago
Serrano, Felipe

Shi, Qinru
Tanneau, Mathieu
Tesch, Alexander
Tubertini, Paolo
Turner, Mark
Urli, Tommaso
Van Cauwelaert, Sascha
van den Bogaerdt, Pim
Vigerske, Stefan
Witzig, Jakob

Extended Abstracts

The following extended abstracts were accepted for presentation at the conference:

- Magnus Björk, Pawel Pietrzak and Andriy Svynaryov: Modelling Real-World Strict Seniority Bidding Problems in Airline Crew Rostering
- Emir Demirović, Nicolas Schwind, Tenda Okimoto and Katsumi Inoue: Recoverable Team Formation: Building Teams Resilient to Change
- Andreas Ernst, Dhananjay Thiruvady, Davaatseren Baatar, Angus Kenny, Mohan Krishnamoorthy and Gaurav Singh: Mining, Matheuristics, and Merge-Search
- Alexandre Gondran and Laurent Moalic: Finding the Chromatic Number by Counting k-Colorings with a Randomized Heuristic
- Elias Khalil and Bistra Dilkina: Training Binary Neural Networks with Combinatorial Algorithms
- Varun Khandelwal: Solving Real-World Optimization Problems Using Artificial Intelligence
- Shiang-Tai Liu: Objective Bounds of Quadratic Programming with Interval Coefficients and Equality Constraints
- Günther Raidl, Elina Rönnberg, Matthias Horn and Johannes Maschler: An A*-Based Algorithm to Derive Relaxed Decision Diagrams for a Prize-Collecting Sequencing Problem
- Mark Wallace and Aldeida Aleti: Using CP to Prove Local Search Is Effective

Abstracts of Invited Talks

Same, Same, but Different: A Mostly Discrete Tour Through Optimization

Thorsten Koch[1,2]

[1] Zuse Institute Berlin, Takustr 7, 14195, Berlin, Germany
[2] Technische Universität Berlin, Str. des 17. Juni 136, 10623, Berlin, Germany
koch@zib.de
http://www.zib.de/koch

Abstract. This talk will give a short tour through selected topics in mathematical optimization. Though these topics are quite diverse, they also have a lot in common.

The tour will start at mixed-integer non-linear optimization (MINLP), proceed to mixed-integer optimization (MILP), it will then make short detour to linear programming (LP) and exact solutions, then proceed to algorithms, software, modelling, and parallel computing, jumping to gas networks as an application, from there visit Steiner tree problems, and finally arrive back at MILP.

On route, we will take the opportunity to point out a few challenges and open problems.

Empirical Model Learning: Boosting Optimization Through Machine Learning

Michela Milano

DISI, University of Bologna, Bologna, Italy
michela.milano@unibo.it

Abstract. One of the biggest challenges in the design of decision support and optimization tools for complex, real-world, systems is coming up with a good combinatorial model. The traditional way to craft a combinatorial model is through interaction with domain experts: this approach provides model components (objective functions, constraints), but with limited accuracy guarantees. Often enough, accurate predictive models (e.g. simulators) can be devised, but they are too complex or too slow to be employed in combinatorial optimization.

In this talk, we propose a methodology called Empirical Model Learning (EML) that relies on Machine Learning for obtaining decision model components that link decision variables and observables, using data either extracted from a predictive model or harvested from a real system. We show how to ground EML on a case study of thermal-aware workload allocation and scheduling. We show how to encapsulate different machine learning models in a number of optimization techniques.

We demonstrate the effectiveness of the EML approach by comparing our results with those obtained using expert-designed models.

Ten Years of CP Optimizer

Paul Shaw

IBM. 1681, route des Dolines, 06560 Valbonne, France
paul.shaw@fr.ibm.com

Abstract. CP Optimizer is the IBM constraint solving engine and part of CPLEX Optimization Studio. This talk takes a look at both the motivation and history of CP Optimizer, and the ten year journey from its beginnings until today.

At selected points, I will delve into the operation of different features of the engine, and the motivation behind them, together with how performance improvements in the automatic search were achieved.

From more recent history, I will concentrate on important developments such as the CP Optimizer file format, presolve, explanations for insolubility and backtrack, and lower bounds on the objective function.

Abstracts of Fast-Track
Journal Papers

Online Over Time Processing
of Combinatorial Problems

Robinson Duque[1], Alejandro Arbelaez[2], and Juan Francisco Díaz[1]

[1] Universidad del Valle, Cali, Colombia
{robinson.duque,juanfco.diazg}@correounivalle.edu.co
[2] Cork Institute of Technology, Cork, Ireland
alejandro.arbelaez@cit.ie

In an online environment, jobs arrive over time and there is no information in advance about how many jobs are going to be processed and what their processing times are going to be. We study the online scheduling of Boolean Satisfiability (SAT) and Mixed Integer Programming (MIP) instances that are well-known NP-complete problems. Typical online machine scheduling approaches assume that jobs are completed at some point to minimize functions related to completion time (e.g., makespan, minimum lateness, total weighted tardiness, etc).

In this work, we formalize and present an online over time problem where arriving instances are subject to waiting time constraints. To formalize our problem, we presented an extension of the Graham notation $(\alpha|\beta|\gamma)$ that allowed us to represent the necessary constraints. We also proposed an approach for online scheduling of combinatorial problems that consisted of three parts. Namely, training/testing models for processing time estimations; implementation of a hybrid scheduling policy using SJF and MIP; and usage of instance interruption heuristics to mitigate the impact of inaccurate predictions.

Unlike other approaches, we attempt to maximize the number of solved instances using single and multiple machine configurations. Our empirical evaluation with well-known SAT and MIP instances, suggest that our interruption heuristics can improve generic ordering policies to solve up to 21.6x and 12.2x more SAT and MIP instances. Additionally, our hybrid approach observed results that are close to a semi clairvoyant policy (SCP) featuring perfect estimations. We observed that with very limited data to train the models our approach reports scenarios with up to 90% of solved instances with respect to the SCP.

Finally, we experimented using models that were trained with different feature families and observed an interesting trade-off between the quality of the predictions and the computational cost to calculate such features. For instance, *Trivial* features are basically free to compute but they have impact on the quality of the models. On the other hand, *Cheap* features offer an interesting trade-off between prediction quality and computational cost. This abstract refers to the full paper [1].

Reference

1. Duque, R., Arbelaez, A., Díaz, J.F.: Online over time processing of combinatorial problems. In: Constraints Journal Fast Track of CPAIOR (2018)

Deep Neural Networks as 0-1 Mixed Integer Linear Programs: A Feasibility Study

Matteo Fischetti[1] and Jason Jo[2]

[1] Department of Information Engineering (DEI), University of Padova
matteo.fischetti@unipd.it

[2] Montreal Institute for Learning Algorithms (MILA) and Institute
for Data Valorization (IVADO), Montreal
jason.jo.research@gmail.com

Abstract. Deep Neural Networks (DNNs) are very popular these days, and are the subject of a very intense investigation. A DNN is made by layers of internal units (or neurons), each of which computes an affine combination of the output of the units in the previous layer, applies a nonlinear operator, and outputs the corresponding value (also known as activation). A commonly-used nonlinear operator is the so-called rectified linear unit (ReLU), whose output is just the maximum between its input value and zero. In this (and other similar cases like max pooling, where the max operation involves more than one input value), for fixed parameters one can model the DNN as a 0-1 Mixed Integer Linear Program (0-1 MILP) where the continuous variables correspond to the output values of each unit, and a binary variable is associated with each ReLU to model its yes/no nature. In this paper we discuss the peculiarity of this kind of 0-1 MILP models, and describe an effective bound-tightening technique intended to ease its solution. We also present possible applications of the 0-1 MILP model arising in feature visualization and in the construction of adversarial examples. Computational results are reported, aimed at investigating (on small DNNs) the computational performance of a state-of-the-art MILP solver when applied to a known test case, namely, hand-written digit recognition.

Intruder Alert! Optimization Models for Solving the Mobile Robot Graph-Clear Problem

Michael Morin[1,2], Margarita P. Castro[1], Kyle E. C. Booth[1], Tony T. Tran[1], Chang Liu[1], and J. Christopher Beck[1]

[1] Department of Mechanical and Industrial Engineering, University of Toronto, Toronto, ON, Canada
{mmorin,mpcastro,kbooth,tran,cliu,jcb}@mie.utoronto.ca
[2] Department of Operations and Decision Support Systems, Université Laval, Québec, QC, Canada
michael.morin@osd.ulaval.ca

We investigate optimization-based approaches and heuristic methods for the *graph-clear problem* (GCP), an $\mathcal{NP}$-Hard variant of the pursuit-evasion problem. The goal is to find a schedule that minimizes the total number of robots needed to "clear" possible intruders from a facility, represented as a graph. The team of robots can use *sweep* actions to remove intruders from *contaminated* nodes and *block* actions to prevent intruders from traveling between nodes. A solution to the GCP is a schedule of sweep and block actions that detects all potential intruders in the facility while minimizing the number of robots required. Solutions such that cleared vertices at each time step form a connected subgraph are termed *contiguous*, while those that prevent recontamination and, therefore, the need to sweep a node more than once, are called *progressive*.

We prove, via a counter-example, that enforcing contiguity may remove all optimal solutions and, conversely, that preventing two special forms of recontamination does not remove all optimal solutions. However, the completeness for the general case of progressive solutions remains open.

We then present *mixed-integer linear programming* (MILP) and *constraint programming* (CP) approaches, as well as new heuristic variants for solving the GCP and compare them to previously proposed heuristics. This is the first time that MILP and CP have been applied to the problem. Our experimental results indicate that our heuristic modifications improve upon the heuristics in the literature, that constraint programming finds better solutions than the heuristics in run-times reasonable for the application, and that mixed-integer linear programming is the superior approach for proving optimality. Nonetheless, for larger problem instances, the optimality gap for CP and MILP remains very large, indicating the need for future research and improvement. Given the performance of CP and MILP compared to the heuristic approaches, coupled with the appeal of the model-and-solve framework, we conclude that they are currently the most suitable approaches for the graph-clear problem.

Margarita P. Castro, Kyle E. C. Booth— Equally contributing authors.

Contents

Time-Bounded Query Generator
for Constraint Acquisition

Hajar Ait Addi[1], Christian Bessiere[2], Redouane Ezzahir[1],
and Nadjib Lazaar[2(✉)]

[1] LISTI/ENSA, University of Ibn Zohr, Agadir, Morocco
{hajar.aitaddi,r.ezzahir}@uiz.ac.ma
[2] LIRMM, University of Montpellier, CNRS, Montpellier, France
{bessiere,lazaar}@lirmm.fr

Abstract. QuAcq is a constraint acquisition algorithm that assists a
non-expert user to model her problem as a constraint network. QuAcq
generates queries as examples to be classified as positive or negative.
One of the drawbacks of QuAcq is that generating queries can be time-
consuming. In this paper we present Tq-gen, a time-bounded query
generator. Tq-gen is able to generate a query in a bounded amount
of time. We rewrite QuAcq to incorporate the Tq-gen generator. This
leads to a new algorithm called T-quacq. We propose several strategies
to make T-quacq efficient. Our experimental analysis shows that thanks
to the use of Tq-gen, T-quacq dramatically improves the basic QuAcq
in terms of time consumption, and sometimes also in terms of number
of queries.

1 Introduction

Constraint programming (CP) has made considerable progress over the last
forty years, becoming a powerful paradigm for modeling and solving combinatorial problems. However, modeling a problem as a constraint network still
remains a challenging task that requires some expertise in the field. Several
constraint acquisition systems have been introduced to support the uptake of
constraint technology by non-experts. Freuder and Wallace proposed the match-
maker agent [7]. This agent interacts with the user while solving her problem. The user explains why she considers a proposed solution as a wrong one.
Lallouet et al. proposed a system based on inductive logic programming with the
use of the structure of the problem as a background knowledge [10]. Beldiceanu
and Simonis have proposed MODELSEEKER, a system devoted to problems with
regular structures and based on the global constraint catalog [2]. Bessiere et al.
proposed CONACQ, which generates membership queries (i.e., complete examples) to be classified by the user [4,6]. Shchekotykhin and Friedrich have extended

This work was supported by Scholarship No. 7587 of the EU METALIC non redundant program.

CONACQ to allow the user to provide *arguments* as constraints to speed-up the convergence [12].

Bessiere et al. proposed QUACQ (for Quick Acquisition), an active learning system that is able to ask the user to classify partial queries [3,5]. QUACQ iteratively computes membership queries. If the user says *yes*, QUACQ reduces the search space by discarding all constraints violated by the positive example. When the answer is *no*, QUACQ finds the scope of one of the violated constraints of the target network in a number of queries logarithmic in the size of the example. This key component of QUACQ allows it to always converge on the target set of constraints in a polynomial number of queries. Arcangioli et al. have proposed the MULTIACQ system as an attempt to make QUACQ more efficient in practice in terms of number of queries [1]. Instead of finding the scope of one constraint, MULTIACQ reports all the scopes of constraints of the target network violated by the negative example. Despite the good theoretical bound of QUACQ-like approaches in terms of number of queries, generating a membership query is NP-hard. It can then be too time-consuming when interacting with a human user. For instance, QUACQ can take more than 20 min to generate a query during the acquisition process of the sudoku constraint network.

In this paper, we introduce TQ-GEN, a time-bounded query generator. TQ-GEN generates queries in an amount of time not exceeding a waiting time upper bound. We incorporate the TQ-GEN generator into the QUACQ algorithm to reduce the time complexity of generating queries. This leads to a new version called T-QUACQ. Our theoretical and experimental analyses show that the bounded waiting time between queries in T-QUACQ is at the risk of reaching a *premature convergence* state and asking more queries. We then propose strategies to make T-QUACQ efficient. We experimentally evaluate the benefit of these strategies on several benchmark problems. The results show that T-QUACQ combined with a good strategy dramatically improves QUACQ not only in terms of time needed to generate queries but also in number of queries, while achieving the convergence state in most cases.

The rest of this paper is organized as follows. Section 2 presents the necessary background on constraint acquisition. Section 3 presents the algorithm TQ-GEN for time-bounded query generation. Section 4 describes how we use TQ-GEN in QUACQ to get the T-QUACQ algorithm. Section 5 analyzes the correctness of the algorithm. Experimental results and strategies to make T-QUACQ efficient are reported in Sect. 6. Section 7 concludes the paper.

2 Background

The constraint acquisition process can be seen as an interplay between the user and the learner. User and learner need to share a *vocabulary* to communicate. We suppose this vocabulary is a set of n variables $X = \{x_1, \ldots, x_n\}$ and a domain $D = \{D(x_1), \ldots, D(x_n)\}$, where $D(x_i) \subset \mathbb{Z}$ is the finite set of values for x_i. A constraint c_Y is defined by a sequence Y of variables of X, called *the constraint scope*, and the relation c over D of arity $|Y|$. An assignment e_Y on a set of variables $Y \subseteq X$ *violates* a constraint c_Z (or e_Y is *rejected* by c_Z) if $Z \subseteq Y$ and the projection e_Z of e_Y on the variables in Z is not in c. A *constraint network* is a set C of constraints on the vocabulary (X, D). An assignment on X is a *solution* of C if and only if it does not violate any constraint in C. $sol(C)$ represents the set of solutions of C.

In addition to the vocabulary, the learner owns a *language* Γ of relations, from which it can build constraints on specified sets of variables. Adapting terms from machine learning, the *constraint bias*, denoted by B, is a set of constraints built from the constraint language Γ on the vocabulary (X, D), from which the learner builds the constraint network. We denote by $B[Y]$ the set of all constraints c_Z in B, where $Z \subseteq Y$. The *target network* is a network C_T such that for any example $e \in D^X = \Pi_{x_i \in X} D(x_i)$, e is a solution of C_T if and only if e is a solution of the problem that the user has in mind.

A *membership query* ASK(e) is a classification question asked to the user, where e is a *complete* assignment in D^X. The answer to ASK(e) is *yes* if and only if $e \in sol(C_T)$. A *partial query* ASK(e_Y), with $Y \subseteq X$, is a classification question asked to the user, where e_Y is a *partial* assignment in $D^Y = \Pi_{x_i \in Y} D(x_i)$. The answer to ASK($e_Y$) is *yes* if and only if e_Y does not violate any constraint in C_T. A classified assignment e_Y is called a positive or negative *example* depending on whether ASK(e_Y) is *yes* or *no*. For any assignment e_Y on Y, $\kappa_B(e_Y)$ denotes the set of all constraints in B rejecting e_Y.

We now define *convergence*, which is the constraint acquisition problem we are interested in. We are given a set E of (complete/partial) examples labeled by the user as positive or negative. We say that a constraint network C *agrees with* E if examples labeled as positive in E do not violate any constraint in C, and examples labeled as negative violate at least one constraint in C. The learning process has *converged* on the learned network $C_L \subseteq B$ if:

1. C_L agrees with E,
2. For any other network $C' \subseteq B$ agreeing with E, we have $sol(C') = sol(C_L)$.

We say that the learning process reached a *premature convergence* if only (1) is guaranteed. If there does not exist any $C_L \subseteq B$ such that C_L agrees with E, we say that we have *collapsed*. This can happen when $C_T \nsubseteq B$.

We finally define the class of biases that are *good* for a given time limit, that is, those biases on which bounding the query generation time does not hurt.

Definition 1 (τ-good). *Given a bias B on a vocabulary (X, D), given the maximum arity k of a constraint in B, and given τ a time limit, B is τ-good on (X, D) if and only if $\forall Y \subset X$ such that $|Y| = k$, $\forall C_i, C_j \subset B[Y]$, finding an assignment e on Y such that $e \in sol(C_i) \setminus sol(C_j)$, or proving that none exists, takes less than τ.*

3   Time-Bounded Query Generation

To be able to exhibit its nice complexity in number of queries, QUACQ must be able to generate *non redundant* queries. A query is *non redundant* if, whatever

---

Algorithm 1. TQ-GEN

---

1 **In** $\alpha, \tau,$ time_bound: parameters;
2 **InOut** ℓ: parameter; B: bias; C_L: learned network;
3 time $\leftarrow 0$;
4 **while** $B \neq \emptyset$ *and* time $<$ time_bound **do**
5      $\tau \leftarrow min(\tau,$ time_bound $-$ time$)$;
6      $\ell \leftarrow max(\ell,$ minArity$(B))$;
7   choose $Y \subseteq X$ s.t. $|Y| = \ell \wedge B[Y] \neq \emptyset$;
8      $e_Y \leftarrow solve(C_L[Y] \wedge \neg B[Y])$ in $t < \tau$;
9   **if** $e_Y \neq nil$ **then return** e_Y ;
10   **else**
11   **if** $t < \tau$ **then**
12             $\mid$ $C_L \leftarrow C_L \cup B[Y]; B \leftarrow B \setminus B[Y]$;
13   **else** $\ell \leftarrow \lfloor \alpha \cdot \ell \rfloor$;
14   time $\leftarrow$ time $+ t$;
15 **return** nil;

---

the user's answer, it allows us to reduce the learner's version space (i.e., the subset of 2^B currently agreeing with all already classified examples). In the context of QUACQ, a query ASK(e) is non redundant if e does not violate any constraint in the currently learned network C_L, and it violates at least one constraint of the current bias B in which we look for the missing constraints (i.e., $\kappa_B(e) \neq \emptyset$), We denote such an example e by $e \models (C_L \wedge \neg B)$. QUACQ has to solve an NP-hard problem to generate a non redundant query. Therefore, the user can be asked to wait a long time from a query to another.

We propose TQ-GEN, a query generator able to generate a query in a bounded amount of time (time_bound). We will see later that this bounded time is at the risk of reaching *premature convergence* and/or asking more queries than necessary. The idea behind TQ-GEN is that instead of looking for an assignment e on X such that $e \models (C_L \wedge \neg B)$, we look for a partial assignment e_Y such that $e_Y \models (C_L[Y] \wedge \neg B[Y])$, for some set $Y \subseteq X$.

3.1   Description of TQ-GEN

The algorithm TQ-GEN (see Algorithm 1) takes as input the set of variables X, a reduction factor $\alpha \in]0, 1[$, a solving timeout τ, a time limit to generate a query time_bound, an expected query size ℓ, a current bias of constraints B, and a current learned network C_L. We start by initializing the counter time of the time TQ-GEN has already consumed in its main loop. In line 5, we set τ so that the next execution of the main loop cannot exceed time_bound. In line 7, we choose a subset Y of size ℓ. To be able to generate a non redundant query on Y, it is required that $B[Y]$ is not empty. To guarantee that such an Y exists, we need ℓ to never be smaller than the smallest arity in B (line 6). In line 8, TQ-GEN tries to generate a query on Y of size ℓ in a time less than τ. If such a query is found in less than τ, we return it in line 9. Otherwise, either

$C_L[Y] \wedge \neg B[Y]$ is unsatisfiable and τ is sufficient to prove it, or $C_L[Y] \wedge \neg B[Y]$ is too hard to be solved in τ. If $C_L[Y] \wedge \neg B[Y]$ is unsatisfiable, the constraints in $B[Y]$ are redundant to C_L and they can be removed from B (line 12). These constraints have to be put in C_L to avoid generating later a query violating one of these redundant constraints, but they are useless in terms of the set of solutions represented by C_L. They can safely be removed from C_L at the end of the learning process. If $C_L[Y] \wedge \neg B[Y]$ is too hard, we reduce the expected query size ℓ with a factor α (line 13). In line 14, the time spent to try to generate a query is recorded in order to ensure that TQ-GEN will never exceed the allocated time `time_bound` (see line 4). The last attempt shall not exceed the remaining time (i.e., `time_bound` − `time`) (line 5).

4   Using the TQ-GEN Algorithm in QuAcq

In this section, we present T-QUACQ (Algorithm 2), an integration of TQ-GEN into QuAcq. T-QUACQ differs from the basic version presented in [3] at the shaded lines (i.e., lines 2, 4, 7, 10 and 15).[1]

T-QUACQ initializes the constraint network C_L to the empty set (line 1). In line 2, the parameters of TQ-GEN are initialized such that $\alpha \in\,]0..1[$ and $\ell \in [\mathtt{minArity}(B), |X|]$. In line 4, we call TQ-GEN to generate a query in bounded time. If no query exists (i.e., $B = \emptyset$), then the algorithm reaches a convergence state (line 6). If a query exists and TQ-GEN is not able to return it in the allocated time, T-QUACQ reaches a premature convergence (line 7). Otherwise, we propose the example e to the user, who will answer by *yes* or *no* (line 8). If the answer is *yes*, we can remove from B the set $\kappa_B(e)$ of all constraints in B that reject e (line 9). We can also adjust the expected size of the next query following a given strategy (line 10). This function is discussed later in Sect. 6.3. If the answer is *no*, we are sure that e violates at least one constraint of the target network C_T. We then act exactly as QuAcq by calling the function FindScope to discover the scope of one of these violated constraints and FindC to select which constraint with the given scope is violated by e (line 12). If a constraint c is returned, we know that it belongs to the target network C_T, we then add it to the learned network C_L (line 13). If no constraint is returned (line 14), this is a condition for collapsing as we could not find in B a constraint rejecting one of the negative examples. Functions FindScope and FindC are used exactly as they appear in [3]. When the answer is *no*, we can also adjust the expected size of the next query following a given strategy (line 15).

5   Theoretical Analysis

In this section we analyze the correctness of TQ-GEN and T-QUACQ. The role of TQ-GEN is to return a query that is non redundant with all queries already asked to the user.

---

[1] QuAcq also contains a line for returning "collapse" when detecting an inconsistent learned network. This line has been dropped from T-QUACQ because we allow it to learn a target network without solutions.

Algorithm 2. T-QUACQ

---

1 $C_L \leftarrow \emptyset$;

2 $initialize(\alpha, \tau, \texttt{time\_bound}, \ell)$;

3 **while** $true$ **do**

4 $\quad$ $e \leftarrow$ TQ-GEN $(\alpha, \tau, \texttt{time\_bound}, \ell, B, C_L)$;

5 $\quad$ **if** $e = nil$ **then**

6 $\quad\quad$ **if** $B = \emptyset$ **then return** "convergence on C_L" ;

7 $\quad\quad$ **return** "premature convergence on C_L" ;

8 $\quad$ **if** $Ask(e) =$ yes **then**

9 $\quad\quad$ $B \leftarrow B \setminus \kappa_B(e)$;

10 $\quad\quad$ $adjust(\ell, yes)$;

11 $\quad$ **else**

12 $\quad\quad$ $c \leftarrow$ FindC$(e,$ FindScope$(e, \emptyset, X, false))$;

13 $\quad\quad$ **if** $c \neq nil$ **then** $C_L \leftarrow C_L \cup \{c\}$;

14 $\quad\quad$ **else return** "collapse" ;

15 $\quad\quad$ $adjust(\ell, no)$;

Proposition 1 (Soundness). TQ-GEN *is sound.*

Proof. The only place where TQ-GEN returns a query is line 9. By construction, e_Y is an assignment which is solution of $C_L[Y]$ and that violates at least one constraint from $B[Y]$ (line 8). Thus, $\kappa_B(e_Y) \neq \emptyset$, and by definition e_Y is a non redundant query. $\qquad\square$

Proposition 2 (Termination). *Given a bias B on the vocabulary (X, D), if* $\texttt{time\_bound} < \infty$ *or if B is τ-good on (X, D), then* TQ-GEN *terminates.*

Proof. If $\texttt{time\_bound} < \infty$, it is trivial. Suppose now that B is τ-good on (X, D), $\texttt{time\_bound} = \infty$, and TQ-GEN never goes through line 9 (which would terminate TQ-GEN). At each execution of its main loop, TQ-GEN executes either line 12 or line 13. ℓ decreases strictly at each execution of line 13. Hence, after a finite number of times, ℓ will be less than or equal to the maximum arity in B. As B is τ-good, the cutoff τ will no longer be reached in line 8, and the next executions of the loop will all go through line 12. Thanks to line 6, ℓ cannot be less than the smallest arity in B. Thus, the set Y chosen in line 7 is guaranteed to have a non empty $B[Y]$. As a result, B strictly decreases in size at each execution of line 12, B will eventually be empty, and TQ-GEN will terminate. $\qquad\square$

We now show that under some conditions TQ-GEN cannot miss a non redundant query, if one exists.

Proposition 3 (Completeness). *If the bias B is τ-good, and* $\texttt{time\_bound} >$ $(|B| + \lceil log_\alpha(\frac{k}{n}) \rceil) \cdot \tau$, *with $n = |X|$ and k the maximum arity in B, then* TQ-GEN *is complete.*

Proof. TQ-GEN finishes by either returning a query in line 9 or *nil* in line 15. If a query is returned, we are done as TQ-GEN is sound (Proposition 1). Suppose *nil* is returned in line 15. According to the assumption on time_bound and the fact that each execution of the main loop of TQ-GEN takes at most τ seconds, we know that TQ-GEN has enough time to execute $|B| + \lceil log_\alpha(\frac{k}{n}) \rceil$ times its main loop before returning *nil* in line 15. In each of these executions, line 12 or line 13 is executed. Each time line 13 is executed, ℓ is reduced by multiplying it by the factor $\alpha \in]0..1[$. As ℓ cannot be greater than n when entering TQ-GEN, after $\lceil log_\alpha(\frac{k}{n}) \rceil$ executions, we are guaranteed that $\ell \leq n \cdot \alpha^{\lceil log_\alpha(\frac{k}{n}) \rceil} \leq n \cdot \frac{k}{n} = k$. As B is τ-good, TQ-GEN will be able to solve the formula $C_L[Y] \wedge \neg B[Y]$ in less than τ seconds for all Y, as $|Y| = \ell \leq k$. As a result, TQ-GEN has enough time for $|B|$ executions of the loop before reaching the time_bound limit. Thanks to line 7, we know that the set Y has a non empty $B[Y]$. Thus, line 12 removes at least one constraint from B, and B will be emptied before the time limit. Therefore, we have converged, and there does not exist any non redundant query. □

Theorem 1. *If $C_T \subseteq B$, T-QUACQ is guaranteed to reach (premature) convergence. If in addition B is τ-good and time_bound $> (|B| + \lceil log_\alpha(\frac{k}{n}) \rceil) \cdot \tau$, with $n = |X|$ and k the maximum arity in B, then T-QUACQ converges.*

Proof (Sketch). We first prove premature convergence. Let E be the set of all examples generated during the execution of T-QUACQ and C_L be the returned network. If C_L does not agree with E this means that there exists $e_Y \in E$ such that e_Y is positive and $e_Y \not\models C_L$, or e_Y is negative and $e_Y \models C_L$. As FindScope and FindC are sound, we only consider examples classified in line 8 of T-QUACQ. Suppose first that in line 8, e_Y is positive (e_Y^+). By construction, e_Y has been generated by satisfying $C_L[Y]$ (line 4), that is, $\not\exists c_Z \in C_L[Y] \mid e_Y \not\models c_Z$ at the time of generating e_Y. As line 9 removes from B all constraints rejecting e_Y, we are guaranteed that C_L agrees with $\{e_Y^+\}$ at the end of T-QUACQ. Suppose now that e_Y is negative (e_Y^-). As $C_T \subseteq B$, FindC returns a constraint c rejecting e_Y (line 12) and c is added to C_L in line 13. Thus, C_L agrees with $\{e_Y^-\}$ at the end.

We now prove that T-QUACQ converges when B is τ-good and time_bound $> (|B| + \lceil log_\alpha(\frac{k}{n}) \rceil) \cdot \tau$. By Proposition 3 we know that under this assumption, TQ-GEN always returns a non redundant query if one exists. As a result, TQ-GEN returns *nil* only when B has been emptied of all its redundant constraints in line 9, which means that T-QUACQ has converged on C_L. □

6 Experiments

In this section, we experimentally analyze our new algorithms. We first describe the benchmark instances. Second, we evaluate the validity of the time-bounded query generation by comparing a baseline version of T-QUACQ to the QUACQ algorithm. This baseline version allows us to observe the fundamental characteristics of the approach. Based on these observations, we discuss possible strategies and parameter settings that may make our approach more efficient. The only parameter we will keep fixed in all our experiments is time_bound, that we set to 1 s, as we consider it as an acceptable waiting time for a human user [9]. All tests were performed using the Choco solver[2] version 4.0.4 with a simulation run time cutoff of 3 h, 2 Gb of Java VM allowed memory on an Intel(R) Xeon(R) @ 3.40 GHz.

6.1 Benchmarks

We used four benchmarks from the original QuAcq paper [3] (Random, Sudoku, Golomb ruler, and Zebra), and two additional ones (Latin square, Graceful graphs).

Random. We generated binary random target networks with 50 variables, domains of size 10, and m binary constraints. The binary constraints are selected from the language $\Gamma = \{=, \neq, \leqslant, \geqslant, <, >\}$. We have launched our experiments with $m = 12$, and $m = 122$.

Sudoku. The sudoku logic puzzle with 9×9 grid must be filled with numbers from 1 to 9 in such a way that all the rows, all the columns, and the 9 non overlapping 3×3 squares contain the numbers 1 to 9. The target network has 81 variables with domains of size 9, and 810 binary $\neq$ constraints on rows, columns and squares. We use a bias of 19,440 binary constraints taken from the language $\Gamma = \{=, \neq \leqslant, \geqslant, <, >\}$.

Golomb Ruler (prob006 in [8]). The problem is to find a ruler where the distance between any two marks is different from that between any other two marks. Golomb ruler is encoded as a target network with n variables corresponding to the n marks. For our experiments, we selected the 8, 12, 16 and 20 marks ruler instances with bias of 660, 3,698, 12,552, and 32,150 constraints, respectively, generated using the language $\Gamma = \{=_0, \neq_0, \leqslant, \geqslant, <, >, \Vdash_{xy}^{zt}, \nVdash_{xy}^{zt}\}$ where $=_0$ and $\neq_0$ respectively denote the unary constraints "equal zero" and "not equal zero", and $\Vdash_{xy}^{zt}$ and $\nVdash_{xy}^{zt}$ respectively denote the distance constraints $|x - y| = |z - t|$ and $|x - y| \neq |z - t|$.

Latin Square. A Latin square is an $n \times n$ array filled with n different Latin letters, each occurring exactly once in each row and exactly once in each column. We have taken $n = 10$ and the target network is built with 900 binary $\neq$ constraints on rows and columns. We use a bias of 29,700 constraints built from the language $\Gamma = \{=, \neq, \leqslant, \geqslant, <, >\}$.

Zebra. Lewis Carroll's zebra problem has a single solution. The target network has 25 variables of domain size 5 with 5 cliques of "$\neq$" constraints and 14 additional constraints given in the description of the problem. We use a bias of 3,250 unary and binary constraints taken from a language with 20 basic arithmetic and distance constraint.

Graceful Graphs (prob053 in [8]). A labeling f of the n nodes of a graph with q edges is graceful if f assigns each node a unique label from $0, 1, \ldots, q$ and when

---

[2] www.choco-solver.org.

each edge (x, y) is labeled with $|f(x) - f(y)|$, the edge labels are all different. The target network has node-variables $x_1, x_2, \ldots x_n$, each with domain $\{0, 1, \ldots, q\}$, and edge-variables $e_1, e_2, \ldots e_q$, with domain $\{1, 2, \ldots, q\}$. The constraints are: $x_i \neq x_j$ for all pairs of nodes, $e_i \neq e_j$ for all pairs of edges, and $e_k = |x_i - x_j|$ if edge e_k joins nodes i and j. The constraints of B were built from the language $\Gamma = \{\neq, =, \parallel_{xy}^z, \nparallel_{xy}^z\}$ where $\parallel_{xy}^z$ and $\nparallel_{xy}^z$ denote respectively the distance constraints $z = |x - y|$ and $z \neq |x - y|$. We used three instances that accept a graceful labeling [11]: $GG(K_4 \times P_2)$, $GG(K_5 \times P_2)$, and $GG(K_4 \times P_3)$, whose number of variables is 24, 35, and 38 respectively, and bias size is 12,696, 40,460, and 52,022 respectively.

For all our benchmarks, the bias contains all the constraints that can be generated from the relations in the given language. That is, for a commutative relation c (resp. non-commutative relation c') of arity r, the bias contains all possible constraints c_Y (resp. c'_Y), where Y is a subset (resp. an ordered subset) of X of size r.

6.2   Baseline Version of T-QUACQ

The purpose of our first experiment is to validate the approach of time-bounded query generation and to understand the basics of its behavior. We defined a baseline version of T-QUACQ, called T-QUACQ.0, that we compare with QUACQ. T-QUACQ, presented in Algorithm 2, is parameterized with time_bound, α, τ and ℓ used by function *initialize*, and what function *adjust* does.

Once time_bound has been fixed, as said above, to 1 s, there remains to specify the other parameters and function *adjust*. In T-QUACQ.0, to remain as close as possible to the original QUACQ, we set ℓ to $|X|$ and the function *adjust* at lines 10 and 15 of Algorithm 2 simply resets ℓ to $|X|$. The impact of the parameters α and τ will be discussed later. For this first comparison between T-QUACQ.0 and QUACQ, we use two CSP instances: sudoku and $GG(K_5 \times P_2)$. They are good candidates for this analysis because QUACQ can be very time-consuming to generate queries on them.

Table 1 reports the comparison of QUACQ and our baseline version T-QUACQ.0 on the sudoku and $GG(K_5 \times P_2)$ instances. The performance of T-QUACQ.0 is averaged over ten runs on each instance. In this first experiment, we have arbitrarily set α to 0.5. $\#q$ denotes the total number of asked queries, $totT$ denotes the total time of the learning process, $\#Conv$ denotes the number of runs of T-QUACQ.0 in which it reached convergence, and $\%Conv$ denotes the average of the convergence rate over the ten runs. We estimated the convergence rate by the formula $100.\frac{|C_T| - \#missingto(C_L)}{|C_T|}$, where $\#missingto(C_L)$ is the number of constraints that have to be added to the learned network C_L to make it equivalent to the target network C_T.

From Table 1, we observe that when $\tau = 5$ ms, for both instances, T-QUACQ.0 is able to converge on the target network, as QUACQ (obviously) does. The interesting information is that T-QUACQ.0 does this in a total cpu time for generating all queries that is significantly lower than the time needed by QUACQ. QUACQ

Table 1. T-QUACQ.0 versus QUACQ (`time_bound` $= 1s$)

CSP	Algorithm	$(\alpha, \tau$ (in ms))	#q	$totT$ (in $seconds$)	#Conv	%Conv
Sudoku 9×9	QUACQ	-	9,053	2,810	-	100%
	T-QUACQ.0	(0.5, 0.001)	12	14	0	1%
		(0.5, 0.024)	9,132	37	10	100%
		(0.5, 5)	9,612	62	10	100%
		(0.5, 900)	9,557	41	5	94%
$GG(K_5 \times P_2)$	QUACQ	-	4,898	3,144	-	100%
	T-QUACQ.0	(0.5, 0.001)	11	62	0	1%
		(0.5, 0.024)	7,495	56	0	93%
		(0.5, 5)	5,610	43	10	100%
		(0.5, 900)	1,888	40	0	41%

needs 46 min to converge on the instance of sudoku and 52 min to converge on the instance of graceful graphs, whereas T-QUACQ.0 converges in 1 min or less on both instances.

Let us focus a bit more on how the two algorithms spend their time. Figure 1 reports the waiting time from one query to another needed by QUACQ and T-QUACQ.0 to learn $GG(K_5 \times P_2)$. We selected a fragment of 100 queries near to the end of the learning process for each algorithm. On the one hand, we see that T-QUACQ.0 never exceeds the bound of `time_bound` $= 1$ s between two queries, thanks to its TQ-GEN time bounded generator. On the other hand, we observe that in these 100 queries close to convergence, generating a query in QUACQ is time consuming because it requires solving a hard CSP. There are three negative queries (that is, queries followed by small queries of FindScope and FindC) requiring from 20 to 50 s to be generated, and many positive queries (that is, not followed by small queries) requiring from 20 to 200 s to be generated.

Once the approach has been validated by this first experiment, we tried to understand the behavior of T-QUACQ.0 when pushing τ to the limits of the range 0..`time_bound`. We instantiated τ to a very small value: 0.001 ms, and a very large value, close to `time_bound`: 900 ms. Results are reported in Table 1.

When τ takes the large value of 900 ms, T-QUACQ.0 fails to converge (the convergence rate is 94% in sudoku, and 41% in $GG(K_5 \times P_2)$). The explanation is that τ is so close to `time_bound` that if TQ-GEN fails to produce a query of size $|X|$ in 900 ms, there remains only 100 ms to produce a query of size $\alpha \cdot |X|$. In case TQ-GEN cannot make it, T-QUACQ.0 returns premature convergence because the time limit has been reached.

When τ takes its small value 0.001 ms, T-QUACQ.0 fails to converge (the convergence rate is 1% on both instances). The explanation in this case is that τ is so small that the bias is not τ-good. That is, the solver at line 8 of TQ-GEN fails to terminate even for the smallest sub-problems of two variables. Thus, TQ-GEN will spend time looping through lines 8, 13, and 6 until reaching `time_bound`.

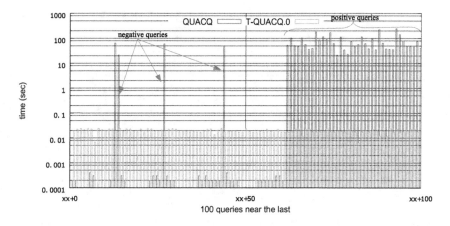

Fig. 1. Time to generate queries on $GG(K_5 \times P_2)$ (T-QUACQ.0 versus QUACQ).

After having tried these extreme values for τ, let us now use Theorem 1 to theoretically determine the values of τ that guarantee convergence. (Remember that α is set to 0.5.) According to Theorem 1, τ must be less than $1/(19440 + log_{0.5}(2/81)) = 0.05$ ms on sudoku and less than $1/(40, 460 + log_{0.5}(3/35)) = 0.0247$ ms on $GG(K_5 \times P_2)$. We launched an experiment with $\tau = 0.024$ ms, which meets the theoretical bound for both sudoku and $GG(K_5 \times P_2)$. The results are reported in Table 1. T-QUACQ.0 converges on sudoku but returns premature convergence on $GG(K_5 \times P_2)$ with a convergence rate equal to 93%. On sudoku, the bias is τ-good when $\tau = 0.024$ ms, so the two conditions for convergence of Theorem 1 are met. On $GG(K_5 \times P_2)$, $\tau = 0.024$ ms is too small for ensuring τ-goodness because the bias contains ternary constraints. Thus, the first condition for convergence of Theorem 1 is violated and T-QUACQ.0 fails to converge.

Our last observation on Table 1 is related to the number of queries. We consider only the cases where the learning process has converged, that is, sudoku with $(\alpha, \tau) = (0.5, 0.024)$ and $(\alpha, \tau) = (0.5, 5)$, and $GG(K_5 \times P_2)$ with $(\alpha, \tau) = (0.5, 5)$. We observe that T-QUACQ.0 respectively asks 1%, 4%, and 14% more queries than QUACQ.

To understand why T-QUACQ.0 asks more queries than QUACQ on $GG(K_5 \times P_2)$, we launched T-QUACQ.0 with different values of α. (τ is kept fixed to 5 ms.) Interestingly, we observed that the number of queries varies significantly with α. For α taking values 0.2, 0.5, and 0.8, T-QUACQ.0 requires respectively, 6, 654, 5, 610, and 5000 queries to converge. We then measured the size of queries in T-QUACQ.0 with these three values of α. Figure 2 reports the size of queries asked by T-QUACQ.0 to learn $GG(K_5 \times P_2)$ with α equal to 0.2, 0.5, and 0.8. We make a zoom on the $400th$ to $450th$ iterations of T-QUACQ.0. We observe that the larger α, the greater the size of the query returned by TQ-GEN and the smaller the number of queries. When α is small, this often leads to queries of very small size. For $\alpha = 0.2$, all queries have size $\lfloor \alpha \cdot 35 \rfloor = 7$ (because $GG(K_5 \times P_2)$ has

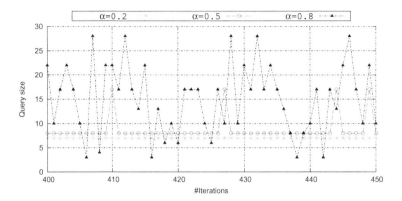

Fig. 2. Size of queries generated by TQ-GEN on $GG(K_5 \times P_2)$ with $\tau = 5$ ms.

35 variables). For $\alpha = 0.5$, a few queries have size $\lfloor \alpha \cdot 35 \rfloor = 17$ but almost all have size $\lfloor \alpha^2 \cdot 35 \rfloor = 8$. Generating queries of small size can be beneficial at the beginning of the learning process, when queries are often negative, because function FindScope will quickly find the right scope of the missing constraint inside a small subset Y. But at the end of the learning process, when most queries are positive, a short query leads to very few constraints removed from B in line 9 of Algorithm 2. Hence, convergence is slow in terms of number of queries. This is what happened in Table 1 on $GG(K_5 \times P_2)$ with $\alpha = 0.5$ and $\tau = 5$ ms. These observations led us to propose more flexible ways to adjust the query size during the learning process.

6.3   Strategies and Settings

Following our first observations on our baseline version T-QUACQ.0, we expect that there is room for improvement by making the use of the query size ℓ less brute-force (reset to $|X|$ after each query generation in T-QUACQ.0). We propose here to adjust it in a more smooth way, to let T-QUACQ concentrate on the size of query that is the most beneficial at a given point of the learning process.

We propose the following *adjust* function (see Algorithm 3). Given a query generated by TQ-GEN, if the answer is *yes*, *adjust* increases ℓ by a factor α, and if the answer is *no*, *adjust* decreases ℓ by a factor α. The intuition behind such adaptation of the query size is that when we are in a zone of many *no* answers (early learning stage), short negative queries lead to less queries needed by FindScope to find where the culprit constraint is, whereas in a zone of *yes* answers (late learning stage), larger positive queries lead to the removal of more constraints from B, and thus faster convergence. T-QUACQ using this version of the function *adjust* is called T-QUACQ.1 in the following.

We expect that the efficiency of T-QUACQ.1 will depend on the initialization of the parameters α and τ in function *initialize* (line 2 of Algorithm 2).

Algorithm 3. *adjust* function of T-QUACQ.1

---

1 **In** ℓ, *answer*, **InOut** ℓ
2 **if** *answer* = *yes* **then**
3 | $\ell \leftarrow min(\lceil \frac{\ell}{\alpha} \rceil, |X|)$;
4 **else**
5 | $\ell \leftarrow \lfloor \alpha \cdot \ell \rfloor$
6 **return** ℓ

Concerning the parameter ℓ, we observed that its initial value has negligible impact as it is used only once at the start of the learning process. We thus set function *initialize* to always initialize ℓ to $|X|$.

Concerning α and τ, to find the most promising values of these parameters, we made an experiment on graceful graphs. On these problems, our base version T-QUACQ.0 performed worse than QUACQ. The results of T-QUACQ.1 are shown in sub-figures (a), (b), and (c) of Fig. 3. The x-axis and the y-axis are respectively labeled by α ranging from 0.1 to 0.9, and $log_{10}(\tau)$ in μs (that is, each value of y corresponds to 10^y μs) ranging from 10 μs to 1 s. Darker color indicates higher number of queries. The number in each cell indicates the convergence rate %*Conv*.

Let us first analyze the convergence rate of T-QUACQ.1. We observe the same results as already seen with our baseline version, that is, premature convergence when τ is too small (10 μs) or very large (1 s), When τ does not take extreme values, we observe convergence in many cases. The range of values of τ that lead to frequent convergence (in fact convergence for all values of α except 0.9) goes from [1 ms, 100 ms] on the small instance to [10 ms, 100 ms] on the larger. Concerning α, we observe that its value does not have any impact on convergence except the very large value 0.9, which leads T-QUACQ.1 to return premature convergence on the harder instances even for values of τ that give convergence with all other values of α. (See $\alpha = 0.9$ in sub-figures (b) and (c) of Fig. 3.) This is explained by the fact that with such a large α, finding the right size ℓ of the query to generate can require too many tries (when τ is reached) and lead to exhaust the time bound.

Let us now study the impact of τ and α on the number of queries asked by T-QUACQ.1. We restrict our analysis to the cases where T-QUACQ.1 has converged. We observe that when T-QUACQ.1 converges, the larger τ, the lower number of query (see sub-figures (a), (b), and (c) of Fig. 3). Concerning the impact of α on the number of queries, we observe that, the greater α (except 0.9 which leads to premature convergence), the lower the number of queries. The reason is that a large α leads to a smooth adjustment of the size of the queries, depending on the computing time allowed by τ and the positive/negative classification of previous examples. This especially has the effect that T-QUACQ.1 generates large queries at the end of the learning process, which lead to faster convergence, as seen with T-QUACQ.0. In the following we choose 0.8 as a default value for α.

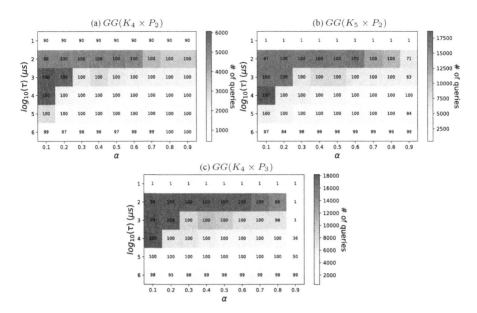

Fig. 3. Number of queries and convergence rate performed by T-QUACQ.1 on graceful graphs. Darker color in color bar indicates higher number of queries, and the number in each cell of the map indicates the convergence rate. (Color figure online)

To validate the observations made on graceful graphs, and in order to select the most promising value of τ, we extended our experimentation to Golomb rulers. Golomb rulers have the nice property that the basic model does not only contain binary constraints. It also contains ternary and quaternary constraints, which makes query generation more difficult. We used four instances of Golomb rulers of size $n = 8$, 12, 16, and 20. We set α to 0.8, and vary τ from $10\,\mu s$ to $1\,s$. We added the value $\tau = 50\,ms$ (that is, $log_{10}\,(50\,ms) = 4.7$) inside the interval $[10\,ms, 100\,ms]$ as these values were looking the most promising for convergence in our previous experiment. The results of T-QUACQ.1 on those problems are shown in Fig. 4, where the x-axis and the y-axis are respectively labeled by $log_{10}(\tau)$ in μs, and the problem size n.

We first analyze the impact of τ on the convergence rate. The results in Fig. 4 show us that the larger the problem size, the greater the value of τ for convergence. We observe that T-QUACQ.1 converges for no instance at $\tau = 10\,\mu s$, 1 instance at $100\,\mu s$, 2 instances at $1\,ms$, and 3 instances from $50\,ms$ to $1\,s$. For $n = 20$, convergence is never reached, but $\tau = 10\,ms$ and $\tau = 50\,ms$ give the best results. If we combine these results with those obtained on graceful graphs, it leads us to the conclusion that the best value for τ is $50\,ms$. Let us now analyze the impact of τ on the number of queries when T-QUACQ.1 converges. We observe that for all the instances, the number of queries required for convergence is almost the same regardless of the value of τ. In the following we set τ to $50\,ms$.

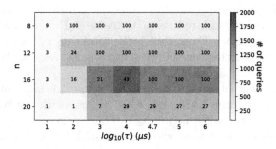

Fig. 4. Number of queries and convergence rate performed by T-QUACQ.1 on Golomb rulers.

Table 2. T-QUACQ.1 versus QUACQ, $\alpha = 0.8$, $\tau = 50\,\text{ms}$

Benchmark ($|X|, |D|, |C|$)	Algorithm	$totT$ (in *seconds*)	$MT(q)$(in *seconds*)	#q
Zebra (25, 5, 64)	QUACQ	1.29	0.13	706
	T-QUACQ.1	1.34	0.11	547
rand-50-10-12 (50, 10, 12)	QUACQ	204	5.01	253
	T-QUACQ.1	13	0.36	325
rand-50-10-122 (50, 10, 122)	QUACQ	88	1.68	1,217
	T-QUACQ.1	21	0.22	1,222
$GG(K_4 \times P_2)$ (24, 16, 164)	QUACQ	976	13	1,989
	T-QUACQ.1	47	0.39	1,273
$GG(K_5 \times P_2)$ (35, 25, 370)	QUACQ	3,144	512	4,898
	T-QUACQ.1	110	0.65	2,317
$GG(K_4 \times P_3)$ (38, 26, 417)	QUACQ	7,206	367	5,796
	T-QUACQ.1	150	0.89	2,883
Sudoku 9×9 (81, 9, 810)	QUACQ	2,810	1,355	9,053
	T-QUACQ.1	69	0.33	6,873
Latin-Square (100, 10, 900)	QUACQ	7,200	1,234	12,204
	T-QUACQ.1	120	0.56	7,711
Golomb-ruler-12 (12, 110, 2, 270)	QUACQ	11,972	2,808	2,445
	T-QUACQ.1	1,184	0.94	916

We finally validate this optimized version of T-QUACQ.1 on other benchmark problems. Table 2 reports the results of QUACQ and of T-QUACQ.1 with the parameters $\alpha = 0.8$ and $\tau = 50$ ms. $totT$ is the total time of the learning process, $MT(q)$ the maximum waiting time between two queries, and $\#q$ the total number of asked queries.

The first important observation is that T-QUACQ.1 has converged for all instances presented in Table 2. Second, what we saw with the baseline version T-QUACQ.0 remains true with T-QUACQ.1: Time to generate queries is short, almost always orders of magnitude shorter than with QUACQ. Finally, the good surprise comes from the number of queries. Compared to T-QUACQ.0, the number of queries in T-QUACQ.1 drops significantly thanks to the smooth adjustment of the size of the queries. The number of queries in T-QUACQ.1 is even smaller than the number of queries in QUACQ on all but two instances, despite QUACQ is free to use as much time as it needs to generate a query.

7   Conclusion

We have proposed TQ-GEN, a query generator that is able to generate a query in a bounded amount of time, and then to satisfy users tolerable waiting time. TQ-GEN is able to adjust the size of the query to generate so that the query can be generated within the time bound. We have also described T-QUACQ, a QUACQ-like algorithm that uses TQ-GEN to generate queries. Our theoretical analysis shows that the bounded waiting time between queries is at the risk of reaching a *premature convergence*. We have then proposed strategies to better adapt query size. Our experiments have shown that T-QUACQ combined with a good strategy dramatically improves QUACQ in terms of time needed to generate queries and also in number of queries, while still reaching convergence.

References

1. Arcangioli, R., Bessiere, C., Lazaar, N.: Multiple constraint acquisition. In: Proceedings of the Twenty-Fifth International Joint Conference on Artificial Intelligence, IJCAI 2016, New York, pp. 698–704 (2016)
2. Beldiceanu, N., Simonis, H.: A model seeker: extracting global constraint models from positive examples. In: Milano, M. (ed.) CP 2012. LNCS, pp. 141–157. Springer, Heidelberg (2012). https://doi.org/10.1007/978-3-642-33558-7_13
3. Bessiere, C., Coletta, R., Hebrard, E., Katsirelos, G., Lazaar, N., Narodytska, N., Quimper, C., Walsh, T.: Constraint acquisition via partial queries. In: Proceedings of the 23rd International Joint Conference on Artificial Intelligence, IJCAI 2013, Beijing, China, pp. 475–481 (2013)
4. Bessiere, C., Coletta, R., O'Sullivan, B., Paulin, M.: Query-driven constraint acquisition. In: Proceedings of the 20th International Joint Conference on Artificial Intelligence, IJCAI 2007, Hyderabad, India, pp. 50–55 (2007)
5. Bessiere, C., et al.: New approaches to constraint acquisition. In: Bessiere, C., De Raedt, L., Kotthoff, L., Nijssen, S., O'Sullivan, B., Pedreschi, D. (eds.) Data Mining and Constraint Programming. LNCS (LNAI), vol. 10101, pp. 51–76. Springer, Cham (2016). https://doi.org/10.1007/978-3-319-50137-6_3

6. Bessiere, C., Lazaar, N., Koriche, F., O'Sullivan, B.: Constraint acquisition. In: Artificial Intelligence (2017, in Press)
7. Freuder, E.C., Wallace, R.J.: Suggestion strategies for constraint-based match-maker agents. Int. J. Artif. Intell. Tools **11**(1), 3–18 (2002)
8. Jefferson, C., Akgun, O.: CSPLib: a problem library for constraints (1999). http://www.csplib.org
9. Lallemand, C., Gronier, G.: Enhancing user experience during waiting time in HCI: contributions of cognitive psychology. In: Proceedings of the Designing Interactive Systems Conference, DIS 2012, pp. 751–760. ACM, New York (2012). https://doi.org/10.1145/2317956.2318069
10. Lallouet, A., Lopez, M., Martin, L., Vrain, C.: On learning constraint problems. In: Proceedings of the 22nd IEEE International Conference on Tools with Artificial Intelligence, ICTAI 2010, Arras, France, pp. 45–52 (2010)
11. Petrie, K.E., Smith, B.M.: Symmetry breaking in graceful graphs. In: Rossi, F. (ed.) CP 2003. LNCS, vol. 2833, pp. 930–934. Springer, Heidelberg (2003). https://doi.org/10.1007/978-3-540-45193-8_81
12. Shchekotykhin, K.M., Friedrich, G.: Argumentation based constraint acquisition. In: Proceedings of the Ninth IEEE International Conference on Data Mining, ICDM 2009, Miami, FL, pp. 476–482 (2009)

Propagating LEX, FIND and REPLACE with Dashed Strings

Roberto Amadini[✉], Graeme Gange, and Peter J. Stuckey

Department of Computing and Information Systems,
The University of Melbourne, Melbourne, Australia
{roberto.amadini,gkgange,pstuckey}@unimelb.edu.au

Abstract. Dashed strings have been recently proposed in Constraint Programming to represent the domain of string variables when solving combinatorial problems over strings. This approach showed promising performance on some classes of string problems, involving constraints like string equality and concatenation. However, there are a number of string constraints for which no propagator has yet been defined. In this paper, we show how to propagate lexicographic ordering (lex), find and replace with dashed strings. All of these are fundamental string operations: lex is the natural total order over strings, while find and replace are frequently used in string manipulation. We show that these propagators, that we implemented in G-STRINGS solver, allows us to be competitive with state-of-the-art approaches.

1 Introduction

Constraint solving over strings is an important field, given the ubiquity of strings in different domains such as, e.g., software verification and testing [7,8], model checking [11], and web security [6,21].

Various approaches to string constraint solving have been proposed, falling in three rough families: automata-based [13,15,20], word-equation based [5,16] and unfolding-based (using either bit-vector solvers [14,17] or constraint programming (CP) [18]). Automaton and word-equation approaches allow reasoning about unbounded strings, but are limited to constraints supported by the corresponding calculus – they have particular problems combining string and integer constraints. These approaches both have scalability problems: from automata growth in the first case, and disjunctive case-splitting in the second. Unfolding approaches first select a length bound k, then substitute each string variable with a fixed-width vector (either by compiling down to integer/bit-vector constraints [2,14,17] or using dedicated propagators [18]). This adds flexibility but sacrifices high-level relationships between strings, and can become very expensive when the length bound is large – even if generated solutions are very short.

A recent CP approach [3,4] introduced the *dashed-string* representation for string variables, together with efficient propagators for dashed-string equality and related constraints (concatenation, reversal, substring). However, numerous other string operations arise in programs and constraint systems involving

ⓒ Springer International Publishing AG, part of Springer Nature 2018
W.-J. van Hoeve (Ed.): CPAIOR 2018, LNCS 10848, pp. 18–34, 2018.
https://doi.org/10.1007/978-3-319-93031-2_2

strings. In this paper, we focus on two: lexicographic ordering and find. Lexicographic ordering is the most common total ordering over strings, and is frequently used to break variable symmetries in combinatorial problems. Find (or *indexOf*) identifies the first occurrence of one query string in another. It appears frequently in problems arising from verification or symbolic execution, and is a convenient building-block for expressing other constraints (e.g., replacement and non-occurrence).

The original contributions of this paper are: *(i)* new algorithms for propagating lexicographic order, find and replace with dashed strings; *(ii)* the implementation of these algorithms in the G-STRINGS solver, and a performance evaluation against the state-of-the-art string solvers CVC4 [16] and Z3STR3 [5]. Empirical results show that our approach is highly competitive – it often outperforms these solvers in terms of solving time and solution quality.

Paper Structure. In Sect. 2 we give background notions about dashed strings. In Sect. 3 we show how we propagate lexicographic ordering. Section 4 explains find and replace propagators. In Sect. 5 we validate our approach with different experimental evaluations, before concluding in Sect. 6.

2   Preliminaries

Let us fix an alphabet Σ, a maximum string length $\ell \in \mathbb{N}$, and the universe $\mathbb{S} = \bigcup_{i=0}^{\ell} \Sigma^i$. A *dashed string* of *length* k is defined by a concatenation of $0 < k \leq \ell$ blocks $S_1^{l_1,u_1} S_2^{l_2,u_2} \cdots S_k^{l_k,u_k}$, where $S_i \subseteq \Sigma$ and $0 \leq l_i \leq u_i \leq \ell$ for $i = 1, \ldots, k$, and $\Sigma_{i=1}^k l_i \leq \ell$. For each block $S_i^{l_i,u_i}$, we call S_i the *base* and (l_i, u_i) the *cardinality*. For brevity we will sometimes write a block $\{a\}^{l,u}$ as $a^{l,u}$, and $\{a\}^{l,l}$ as a^l. The i-th block of a dashed string X is denoted by $X[i]$, and $|X|$ is the number of blocks of X. We do not distinguish blocks from dashed strings of unary length and we consider only *normalised* dashed strings, where the *null element* $\emptyset^{0,0}$ occurs at most once and adjacent blocks have distinct bases. For each pair $C = (l, u)$, we define $\mathsf{lb}(C) = l$ and $\mathsf{ub}(C) = u$. We indicate with $<$ the total order over characters of Σ, and with $\prec$ the lexicographic order over Σ^*.

Let $\gamma(S^{l,u}) = \{x \in S^* \mid l \leq |x| \leq u\}$ be the language denoted by block $S^{l,u}$. We extend γ to dashed strings: $\gamma(S_1^{l_1,u_1} \cdots S_k^{l_k,u_k}) = (\gamma(S_1^{l_1,u_1}) \cdots \gamma(S_k^{l_k,u_k})) \cap \mathbb{S}$ (intersection with $\mathbb{S}$ excludes the strings with length greater than ℓ). A dashed string X is *known* if it denotes a single string: $|\gamma(X)| = 1$. A block of the form $S^{0,u}$ is called *nullable*, i.e. $\epsilon \in \gamma(S^{0,u})$. There is no upper bound on the number of blocks in a dashed string since an arbitrary number of nullable blocks may occur. The *size* $\|S^{l,u}\|$ of a block is the number of concrete strings it denotes, i.e., $\|S^{l,u}\| = |\gamma(S^{l,u})|$. The size of dashed string $X = S_1^{l_1,u_1} S_2^{l_2,u_2} \cdots S_k^{l_k,u_k}$ is instead an overestimate of $|\gamma(X)|$, given by $\|X\| = \Pi_{i=1}^k \|S_i^{l_i,u_i}\|$.

Given two dashed strings X and Y we define the relation $X \sqsubseteq Y \iff \gamma(X) \subseteq \gamma(Y)$. Intuitively, $\sqsubseteq$ models the relation *"is more precise than"* between dashed strings. Unfortunately, the set of dashed strings does not form a lattice according to $\sqsubseteq$. This implies that some workarounds have to be used to determine a reasonable lower/upper bound of two dashed strings according to $\sqsubseteq$.

Intuitively, we can imagine each block $S_i^{l_i,u_i}$ of $X = S_1^{l_1,u_1} S_2^{l_2,u_2} \cdots S_k^{l_k,u_k}$ as a continuous segment of length l_i followed by a dashed segment of length $u_i - l_i$. The continuous segment indicates that exactly l_i characters of S_i *must* occur in each concrete string of $\gamma(X)$; the dashed segment indicates that n characters of S_i, with $0 \le n \le u_i - l_i$, *may* occur. Consider, for example, dashed string $X = \{$B,b$\}^{1,1}\{$o$\}^{2,4}\{$m$\}^{1,1}\{!\}^{0,3}$ illustrated in Fig. 1.

Fig. 1. Representation of $X = \{$B,b$\}^{1,1}\{$o$\}^{2,4}\{$m$\}^{1,1}\{!\}^{0,3}$. Each string of $\gamma(X)$ starts with B or b, followed by 2 to 4 os, one m, then 0 to 3 !s.

Given a dashed string X, we shall refer to *offset positions* (i,o), where i refers to block $X[i]$ and o is its *offset*, indicating how many characters from the beginning of $X[i]$ we are considering. Note that, while positions are 1-based, offset are 0-based to better represent the beginning (or dually the end) of a block. Positive offsets denote positions relative to the beginning of $X[i]$, and negative offsets are relative to the end of $X[i]$. For example, position $(2,3)$ of $X = a^{1,2}b^{0,4}c^3$ refers to 3 characters after the beginning of second block $b^{0,4}$; this position can be equivalently expressed as $(2,-1)$.

The *index* into a string $w \in \Sigma^*$ indicates a character position in range $1 \ldots |w|$, assuming that character positions are 1-based. For example, the index of the first occurrence of *"abc"* in *"dfababcdeabc"* is 5.

Converting between *indices* and *positions* is relatively straightforward (see the pseudo-code below), though this conversion might lose precision when dealing with blocks $S^{l,u}$ having non-fixed cardinality, i.e., having $l < u$. Indeed, we may have POS-TO-MIN-IND$(X,$ IND-TO-MIN-POS$(X,i)) < i$. Consider for example $X = a^{0,1}b^{0,2}$ and $i = 3$: we have IND-TO-MIN-POS$(X,i) = (2,1)$ and POS-TO-MIN-IND$(2,1) = 2$.

> **function** IND-TO-MIN-POS$(X = S_1^{l_1,u_1} \cdot \ldots \cdot S_n^{l_n,u_n}, idx)$
> $\quad i \leftarrow 1$
> $\quad o \leftarrow idx - 1$
> $\quad$ **while** $i \le n \wedge o \ge u_i$ **do**
> $\quad\quad (i,o) \leftarrow (i+1, o - u_i)$
> $\quad$ **end while**
> $\quad$ **return** (i,o)
> **end function**
> **function** POS-TO-MIN-IND$(X, (i,o))$
> $\quad$ **return** $1 + o + \sum_{j=1}^{i-1} l_j$
> **end function**

2.1 SWEEP Algorithm

Equating dashed strings X and Y requires determining two dashed strings X' and Y' such that: *(i)* $X' \sqsubseteq X, Y' \sqsubseteq Y$; and *(ii)* $\gamma(X') \cap \gamma(Y') = \gamma(X) \cap \gamma(Y)$. Informally, we can see this problem as a semantic unification where we want to find a refinement of X and Y including all the strings of $\gamma(X) \cap \gamma(Y)$ and removing the most values not belonging to $\gamma(X) \cap \gamma(Y)$ (note that there may not exist a greatest lower bound for X, Y according to $\sqsubseteq$).

In [3] the authors propose SWEEP, an algorithm for equating dashed strings. SWEEP works analogously to the sweep algorithm for timetable reasoning of CUMULATIVE [1]: to equate X and Y, for each block $X[i]$, we wish to find the earliest and latest positions in Y where $X[i]$ could be matched. Once these positions are computed, they are used to refine the block: roughly, $X[i]$ may only contain content between its earliest start and latest end, and any content between the *latest* start and *earliest* end must be included in $X[i]$. This process is repeated symmetrically to refine each block $Y[j]$.

2.2 G-STRINGS Solver

The SWEEP algorithm is implemented in G-STRINGS,[1] an extension of GECODE solver [12]. It implements the domain of every string variable X with a dashed string $dom(X)$, and defines a *propagator* for each string constraint. Propagators take advantage of SWEEP for refining the representations of the involved variables. For example, string equality $X = Y$ is simply propagated by equating $dom(X)$ and $dom(Y)$ with SWEEP; the propagator for $Z = X \cdot Y$ is implemented by equating $dom(Z)$ and $dom(X) \cdot dom(Y)$, where $dom(X) \cdot dom(Y)$ is the concatenation of the blocks of $dom(X)$ and $dom(Y)$, taking care of properly projecting the narrowing of $dom(X) \cdot dom(Y)$ to $dom(X)$ and $dom(Y)$.

G-STRINGS implements constraints like string (dis-)equality, (half-)reified equality, (iterated) concatenation, string domain, length, reverse, substring selection. Since propagation is in general not complete, G-STRINGS also defines strategies for branching on variables (e.g., the one with smallest domain size or having the domain with the minimum number of blocks) and domain values (by heuristically selecting first a block, and then a character of its base).

3 Lexicographic Ordering

The constraint $\text{LEX}_{\preceq}(X, Y)$ enforces a lexicographic ordering between X and Y. Propagating LEX on an unfolded sequence is largely straightforward: roughly speaking, we walk pointwise along X and Y to find the first possible difference, and impose the ordering on that element. An extension of this procedure which enforces domain consistency is given in [10]. Unfortunately, dashed strings represent sequences where we do not know the exact cardinality of each block.

---

[1] G-STRINGS is publicly available at https://bitbucket.org/robama/g-strings.

X	$\min_{\preceq}(D(X))$	$\max_{\preceq}(D(X))$
$c^{0,10} \cdot a^{0,10} \cdot d^{0,10}$	$c^0 a^0 d^0$	$c^0 a^0 d^{10}$
$c^{0,10} \cdot a^{1,10} \cdot d^{0,10}$	$c^0 a^1 d^0$	$c^{10} a^1 d^{10}$
$c^{0,10} \cdot a^{1,10} \cdot d^{1,10}$	$c^0 a^{10} d^1$	$c^{10} a^1 d^{10}$
$a^{1,10} \cdot c^{0,10} \cdot d^{1,10}$	$a^{10} c^{10} d^1$	$a^1 a^0 d^{10}$
$c^{1,10} \cdot d^{0,10} \cdot a^{1,10}$	$c^1 d^0 a^1$	$c^1 d^{10} a^{10}$
$a^{0,10} \cdot d^{0,10} \cdot a^{1,10} \cdot d^{0,10}$	$a^0 d^0 a^1 d^0$	$a^0 d^{10} a^1 d^{10}$
$a^{0,10} \cdot d^{0,10} \cdot a^{1,10} \cdot d^{1,10}$	$a^{10} d^0 a^{10} d^1$	$a^0 d^{10} a^1 d^{10}$

(a)

	$a^{0,10}$	$d^{0,10}$	$a^{1,10}$	$e^{1,10}$	
min. succ	$[\,a^+$	a^+	e^-	$\$$	$]$
lex. min.	$[\,a^{10}$	d^0	a^{10}	e^1	$]$
max. succ	$[\,d^-$	a^+	e^-	$\$$	$]$
lex. max.	$[\,a^0$	d^{10}	a^1	e^{10}	$]$

(b)

Fig. 2. (a) Lexicographic bounds for several dashed strings. (b) Least and greatest successors for each block of $a^{0,10} \cdot d^{0,10} \cdot a^{1,10} \cdot e^{1,10}$. The notation d^+ (resp. d^-) indicates the least/greatest successor of the block begins with a finite sequence of d's, followed by some character $x > d$ (resp. $x < d$).

3.1 Lexicographic Bounds on Dashed Strings

Propagation of $\textsc{lex}_{\prec}(X, Y)$ essentially reduces to propagating two unary constraints: $X \preceq \max_{\prec}(D(Y))$, and $\min_{\prec}(D(X)) \preceq Y$. However, the behaviour of max and min under $\preceq$ is perhaps counterintuitive. In the lexicographic minimum (resp. maximum), each block $S^{l,u}$ takes its least (greatest) character, and *either* its minimum cardinality l *or* its maximum cardinality u: but which cardinality is chosen depends on the following blocks, i.e., the *suffix* of the string.

Consider the string $a^n \cdot b \cdot X$, for some characters $a, b \in \Sigma$ and string $X \in \Sigma^*$. If $a < b$, increasing the number n of a's produces a smaller string under $\preceq$. If instead $a > b$, adding successive a's can only result in bigger strings under $\preceq$. If $a = b$, then we must recursively consider the prefix of X.

But we do not need to perform this recursion explicitly. All we need to know is *(i)* the first character of the suffix, and *(ii)* whether the sequence increases or decreases afterwards. In essence, we need to know whether the suffix is above or below the infinite sequence a^∞. We use the notation a^+ to denote a value infinitesimally greater than a^∞, but smaller than the successor $succ_<(a)$ of a in Σ under $\leq$; similarly, a^- denotes a value infinitesimally smaller than a but greater than $(pred_<(a))^\infty$, where $pred_<(a)$ is the predecessor of a in Σ under $\leq$.

Example 1. Figure 2(a) shows the behaviour of the lexicographic min- and max- for several dashed strings. The last two instances highlight a critical point: if the minimum value of the base of a block matches the minimum *immediate* successor, we must consult the character appearing *after* the contiguous run. Figure 2(b) illustrates the computation of least and greatest successors for a dashed string.

The computation processes the blocks in reverse order, updating the current suffix for each block. The suffix is initially $\$$, a special terminal character such that $\$ \prec x$ for each $x \in \Sigma^*$. The final block is non-empty, so the suffix is updated; and is greater than the current suffix, so becomes e^-. The next block is again non-empty, and below the current suffix, so it is updated to a^+. The next block is nullable, and its value d is greater than the current suffix. So, for computing the lex-min we omit this block, carrying the suffix backwards. We then reach $a^{0,10}$

Table 1. Calling LEX-STEP($\{d,e\}^{0,1}, a, Y_{max}, (1,0)$).

Call	Result	X'
LEX-STEP($\{d,e\}^{0,1}, a, Y_{max}, (1,0)$)	**EQ**$((1,0))$	ϵ
LEX-STEP($\{f,g\}^{0,4}, d, Y_{max}, (1,0)$)	**EQ**$((1,0))$	ϵ
LEX-STEP($\{d,e,f\}^{2,4}, a, Y_{max}, (1,0)$)	**EQ**$((1,2))$	$\{d\}^{2,2}$
LEX-STEP($\{a,b,c\}^{0,3}, a, Y_{max}, (1,2)$)	**LT**	$\{d\}^{2,2}\{a,b,c\}^{0,3}\{f,g\}^{1,4}$

with a successor of a^+. This gives us successors $[a^+, a^+, e^-, \$]$. We can then compute the lex-minimising value for each block by looking at its successor: the first block $a^{0,10}$ has successor a^+. As $a < a^+$, the block takes maximum cardinality a^{10}. The next block $d^{0,10}$ again has successor a^+. But this time, since $d > a^+$, the block takes its minimum cardinality 0. This process is repeated for each block to obtain the lexicographically minimum string. The pseudo-code for computing the minimum successor and the lex-min value is shown in Fig. 3.   □

function MIN-SUCC($X = S_1^{l_1,u_1} \cdot \ldots \cdot S_n^{l_n,u_n}$)
  $M_{succ} \leftarrow \emptyset$
  $suff \leftarrow \$$
  for $i \in n \ldots 1$ **do**
    $M_{succ}[i] \leftarrow suff$
    $c \leftarrow \min(S_i)$
    if $c < suff$ **then**
      $suff \leftarrow c^+$
    else if $l_i > 0$ **then**
      $suff \leftarrow c^-$
    end if
  end for
  return M_{succ}
end function

function LEX-MIN($S^{l,u}, c_{succ}$)
  $c \leftarrow \min(S)$
  if $c < c_{succ}$ **then**
    return c^u
  else
    return c^l
  end if
end function

Fig. 3. Computing the minimum successor for each block of X, and the lex-min value for block $S^{l,u}$ given the computed successor.

The LEX propagation for $X \preceq \max_{\prec}(D(Y))$ is implemented by LEX-X shown in Fig. 4. Essentially it walks across the blocks of X, by comparing them with $\max_{\prec}(D(Y))$. If it finds that the block must violate LEX, then it returns **UNSAT**. If the block may be strictly less than $\max_{\prec}(D(Y))$ we terminate, the constraint can be satisfied. If the lower bound of the block equates to $\max_{\prec}(D(Y))$ we force it to take its lower bound value, and continue processing the next block.

The *strict* parameter of LEX-X is a Boolean flag which is true if and only if we are propagating the strict ordering $\prec$. The propagation of $\prec$ is exactly the same of

function LEX-X$(X, Y, strict)$
    Let $X = X_1 \cdot \ldots \cdot X_n$.
    $X_{succ} := $ MIN-SUCC(X)
    $Y_{max} \leftarrow \mathbf{max}_{\preceq}(Y)$
    $pos \leftarrow \langle 1, 0 \rangle$
    $X' \leftarrow [\,]$
    for $j \in 1 \ldots n$ **do**
        match LEX-STEP$(X_j, X_{succ}[j], Y_{max}, pos)$ **with**
            case UNSAT $\Rightarrow$
                return UNSAT
            case LT $\Rightarrow$
                $X' \leftarrow X' \cdot X_j \cdot \ldots \cdot X_n$
                break
            case EQ$(pos') \Rightarrow$
                $X' \leftarrow X' \cdot$ LEX-MIN$(X_j, X_{succ}[j])$
                $pos \leftarrow pos'$
        end
    end for
    if $strict \wedge j = n$ **then**
        return UNSAT
    end if
    POST$(X \leftarrow X')$
    return SAT
end function

Fig. 4. Algorithm for propagating $X \preceq \max_{\preceq}(D(Y))$.

$\preceq$, with the only difference that if dashed string X is completely consumed after the loop then we raise a failure (this means that $\min_{\preceq}(D(X)) \geq \max_{\preceq}(D(Y))$).

Example 2. Let X be $\{d, e\}^{0,1}\{f, g\}^{0,4}\{d, e, f\}^{2,4}\{a, b, c\}^{0,3}\{f, g\}^{1,4}$ and $Y = \{a, b\}^{0,3}\{c, d\}^{0,3}\{a\}^{1,3}\{d, e\}^{1,3}$. For propagating $X \preceq Y$, we first compute X_{succ} as $[a, d, a, f, \$]$. Then, $Y_{max} = \max_{\preceq}(D(Y)) = dddaeee = \{d\}^{3,3}\{a\}^{1,1}\{e\}^{3,3}$. The calls to LEX-STEP, its returning values, and the progression of values for X' are shown in Table 1 (as a result of the propagation, X is replaced by X').

4   Find and Replace

Find and replace are important and widely used string operations, since much string manipulation is achieved using them. Formally, FIND(Q, X) returns the start index of the *first* occurrence of Q in X, or 0 if Q does not occur (assuming 1-based indexing). Similarly, REPLACE(Q, R, X) returns X with the first occurrence of Q replaced by R (returning X if Q does not occur in X).

    FIND is surprisingly difficult to express as a decomposition. An attempt might start with encoding the occurrence as $\exists a, b.\ X = a \cdot Q \cdot b \wedge |a| = idx - 1$. This ensures that Q occurs at index idx. However, it omits two important aspects:

function LEX-STEP$(X_j = S^{l,u}, succ, Y, \langle i, o \rangle)$
    Let $Y = \{a_1^{\alpha_1} \cdot \ldots \cdot a_m^{\alpha_m}\}$.   ▷ Greatest word in the domain of variable Y
    $c_S \leftarrow \min(S)$
    if $succ > c_S$ **then**   ▷ Better than our best successor, so saturate
        $cap \leftarrow u$
    else   ▷ Use only what we must.
        $cap \leftarrow l$
    end if
    while $cap > 0$ **do**
        if $i > m$ **then**   ▷ No more Y to match.
            return UNSAT
        end if
        $c_Y \leftarrow a_i$
        $card_Y \leftarrow \alpha_i - o$
        if $cap > 0 \wedge c_S > c_Y$ **then**   ▷ Greater than the next character in Y
            return UNSAT
        else if $cap > 0 \wedge c_S < c_Y$ **then**
            return LT   ▷ Globally satisfied
        else
            if $card_Y \leq cap$ **then**
                $cap \leftarrow cap - card_Y$
                $(i, o) \leftarrow (i + 1, 0)$
            else
                return EQ$(\langle i, o + cap \rangle)$ ▷ Bounds coincide so far.
            end if
        end if
    end while
    return EQ$(\langle i, o \rangle)$
end function

Fig. 5. Processing a single block X_j of the left operand of LEX$_\leq$.

(i) FIND(Q, X) identifies the *first* (rather than any) occurrence of Q; *(ii)* Q can be absent from X. Both problems arise from the same cause: the difficulty of encoding *non-occurrence*. To do so, we would essentially need to add a disequality constraint between Q and every $|Q|$-length substring of X (this problem will recur when we discuss encodings for unfolding-based solvers in Sect. 4.3). Thus, developing a specialised propagator for FIND appears prudent. As we shall see in Sect. 4.2, it also serves as useful primitive for implementing other constraints.

4.1 FIND

The constraint $I = $ FIND(X, Y) returns the index I of the character in Y where the string X appears first, or 0 if X is not a substring of Y. Note the use of

function PROP-FIND-MIN$(I, [X_1, \ldots, X_n], [Y_1, \ldots, Y_m])$
    $start \leftarrow$ IND-TO-MIN-POS$([Y_1, \ldots, Y_m], \min(D(I) - \{0\}))$
    $nochanges \leftarrow true$
    repeat
        for $i \in \{1, \ldots, n\}$ **do**   ▷ Scanning forwards
            if $\neg nochanges$ **then**
                $start \leftarrow est(X_i)$
            end if
            $start, end \leftarrow$ PUSH$^+(X_i, Y, start)$
            $est(X_i) \leftarrow start$
            $start \leftarrow end$
        end for
        if $end > (m, +\mathsf{ub}(Y_m))$ **then**
            $D(I) \leftarrow D(I) \cap \{0\}$ ▷ Prefix cannot fit
            return $D(I) \neq \emptyset$
        end if
        for $i \in \{n, n-1, \ldots, 1\}$ **do**   ▷ Scanning backwards
            $end \leftarrow$ STRETCH$^-(X_i, Y, end)$
            if $end > est(X_i)$ **then**
                $est(X_i) \leftarrow end$
                $nochanges \leftarrow false$
            end if
        end for
        $start \leftarrow end$
    until $nochanges$
    $D(I) \leftarrow D(I) \cap (\{0\} \cup \{$POS-TO-MIN-IND$([Y_1, \ldots, Y_m], end).. + \infty\})$
    return $D(I) \neq \emptyset$
end function

Fig. 6. Algorithm for propagating the earliest possible start position I of string X in the string Y.

1-based indexing, which is standard in maths (and mathematical programming systems), but not so standard in computer science.[2]

The propagator for $I = $ FIND(X, Y) is implemented using the PUSH and STRETCH algorithms used by the G-STRINGS solver to implement equality of dashed strings. PUSH$^+(B, Y, (i, o))$ attempts to find the earliest possible match of B after o characters into $Y[i]$ (the ith block of Y), while STRETCH$^+(B, Y, (i, o))$ finds the *latest* position B could finish, assuming B begins at most o characters into $Y[i]$. There are analogous versions PUSH$^-$ and STRETCH$^-$ which work backwards across the blocks.

The algorithm works by first converting the minimal index of the current domain of I into a earliest possible starting position in X. This is achieved by consuming characters from the upper bound of length of X blocks. We then use

---

[2] We use this indexing to comply with MiniZinc indexing [2]. However, translating between this and the corresponding 0-indexed operations (e.g., the Java `indexOf` method or the C++ `find` method) is trivial.

PUSH to find the earliest position in Y where X_i can occur, for each block X_i in turn, recording this in $est(X_i)$. If the earliest end position for this process is after the end of Y we know that no match is possible, we update the domain of I, and return $false$ if it becomes empty.

The PUSH procedure may have introduced gaps between blocks. Hence we STRETCH backwards to pull blocks forward to their actual earliest start position assuming all later blocks are matched. This may update the earliest start positions for each block Y_i. If we found any earliest start positions changed during the STRETCH operation, we repeat the whole PUSH/STRETCH loop until there are no changes. At the end of this, end holds the earliest match position for X in Y. We convert end to an index and update the domain of I to reflect this.

Example 3. Consider the constraint $I = \text{FIND}(X, Y)$ where $X = \{a\}^{1,1}\{b\}^{2,2}$ and $Y = \{b, c\}^{0,12}\{a\}^{3,3}\{d\}^{1,2}\{b, c\}^{2,4}\{a\}^{5,5}\{b\}^{3,3}\{a, c\}^{0,8}$, and $D(I) = \{0 \dots 38\}$. The starting position $start$ from index 1 is $(1, 0)$. We begin by PUSHing the block $\{a\}^{1,1}$ to position $(2, 0)$, we then PUSH the block $\{b\}^{2,2}$ to position $(4, 0)$. Starting from $end = (4, 2)$ we STRETCH the block $\{b\}^{2,2}$ to position $(4, 0)$, but STRETCHing $\{a\}^{1,1}$ we simply return $end = (4, 0)$ because the previous block is incompatible. We repeat the main loop by PUSHing the block $\{a\}^{1,1}$ starting from $(4, 0)$ to position $(5, 0)$, and PUSH the block $\{b\}^{2,2}$ to position $(6, 0)$. Starting from $end = (6, 2)$ we STRETCH the block $\{b\}^{2,2}$ to position $(6, 0)$, and STRETCHing $\{a\}^{1,1}$ to $(5, 4)$. We repeat the main loop this time finding no change. The earliest index for the match is hence POS-TO-MIN-IND$(Y, (5, 4)) = 0 + 3 + 1 + 2 + 5 = 11$. We thus set the domain of I to $\{0\} \cup \{11 \dots 35\}$. □

The algorithm for propagating the upper bound on the index value is similar. It uses PUSH⁻ to find the latest start position for each block, then STRETCH⁺ to improve these. This process continues until fixpoint. Finally the index of the latest start position of first block X_1 is used to update $D(I)$.

If we have determined $0 \notin D(I)$, from either a definite match or propagation on I, we know there is some occurrence of X in Y. In this case, we know that $Y = \Sigma^{\text{lb}(I)-1, \text{ub}(I)-1} X \Sigma^*$, and we propagate between X and Y accordingly.

We check for definite matches only if the string X is completely fixed. We build the fixed components of Y by replacing blocks $S^{l,u}$ having non-singleton character sets ($|S| > 1$) by special character $\$ \notin \Sigma$, and singleton character blocks where $S = \{a\}$ by a string $a^l \$ a^l$, unless $l = u$ in which case we use a^l. We do this because if a block is singleton character but has unfixed cardinality (e.g., $a^{3,6}$), then the block may participate in matches to either side, but cannot form matches *across* the block since its cardinality is not fixed. Thus it splits the component string, contributing its lower bound to either side (e.g. $a^3 \$ a^3$). We then do a substring search for X in this string. If there is a match we convert the resulting index of this string match into a position in Y, and then compute the latest possible index in Y corresponding to this position. We propagate this as an upper bound of I, and remove 0 from the domain of I.

Example 4. Consider the constraint $I = \text{FIND}(X, Y)$ where $X = \{a\}^{1,1}\{b\}^{2,2}$ and $Y = \{b, c\}^{0,12}\{a\}^{3,4}\{d\}^{1,2}\{b, c\}^{2,4}\{a\}^{5,5}\{b\}^{2,3}\{a, c\}^{0,8}$. X is fixed to string "*abb*", so we create the fixed components of Y as "$\$aaa\$aaad\$d\$aaaaabb\$bb\$$" and search for X in this string. We find a match at the 17th character, which corresponds to Y position $(5, 4)$. The last possible index for this position is $27 = 12 + 4 + 2 + 4 + 5$. We update the domain of I to $\{1 \ldots 27\}$. $\square$

4.2 REPLACE

The constraint $Y_2 = \text{REPLACE}(X_1, X_2, Y_1)$ requires that Y_2 is the string resulting from replacing the first occurrence of X_1 in Y_1 by X_2. If X_1 does not occur in Y_1, then $Y_2 = Y_1$. We encode this constraint by using concatenation and FIND as follows:

$$\exists n, A_1, A_2. \left(\begin{array}{l} n = \text{FIND}(X_1, Y_1) \\ \wedge\ |A_1| = \max(0, n - 1) \\ \wedge\ Y_1 = A_1 \cdot X_1^{(n>0)} \cdot A_2 \\ \wedge\ Y_2 = A_1 \cdot X_2^{(n>0)} \cdot A_2 \\ \wedge\ \text{FIND}(X_1, A_1) = (|X_1| = 0) \end{array} \right)$$

The encoding ensures that Y_2 is Y_1 with some part $Z_1 \in \{X_1, \epsilon\}$ replaced by $Z_2 \in \{X_2, \epsilon\}$. Index n encodes the position where we find X_1 in Y_1. If it does not occur we force $Z_1 = Z_2 = \epsilon$ which makes $Y_1 = Y_2 = A_1 \cdot A_2$. If it does occur we force $X_1 = Z_1$ and $X_2 = Z_2$, and the length of A_1 to be $n - 1$. The last constraint is redundant: it ensures X_1 appears in A_1 if and only if $X_1 = \epsilon$. It is included to strengthen propagation (due to the loss of information when converting between indices and positions).

The FIND constraint allows us to encode other sub-string like constraints, e.g., $\text{STARTSWITH}(Q, X) \iff (\text{FIND}(X, Q) = 1)$ and $\text{CONTAINS}(Q, X) \iff (\text{FIND}(X, Q) > 0)$. Note that because FIND is functional, these decompositions can be used in any context (e.g., in negated form).

4.3 Encoding FIND and REPLACE in Unfolding-Based Solvers

As discussed in Sect. 4.1, expressing the *non*-existence of target strings poses difficulties. In unfolding-based solvers (e.g., GECODE+S, or bit-vector solvers), this manifests in an encoding of $\text{FIND}(Q, R)$ which is of size $\text{ub}(|Q|) \times \text{ub}(|R|)$: essentially, we compute the set I of all the indices where Q and R align, then we return $\min(I)$ if $I \neq \emptyset$, otherwise we return 0:

$$\text{FINDAT}(Q, R, i) = \begin{cases} 1 & \text{if } \forall j = 1, \ldots, |Q| : Q[j] = R[i + j - 1] \\ 0 & \text{otherwise} \end{cases}$$

$$\text{FIND}(Q, R) = \begin{cases} \min(I) & \text{if } I \neq \emptyset \\ 0 & \text{otherwise} \end{cases}$$

where $I = \{i \in \{1, \ldots, |R|\} \mid \text{FINDAT}(Q, R, i) = 1\}$.

5   Experimental Evaluation

We have extended the G-STRINGS solver with the proposed filtering algorithms. We evaluated the propagators on three classes of problems and compared the performance of G-STRINGS against CVC4 [16] and Z3STR3 [5], two state-of-the-art SMT solvers supporting the theory of strings.[3]

First, we consider the PISA suite, consisting of instances arising from web-application analysis. However the PISA instances are rather small and, being heavily used SMT benchmarks, CVC4 and Z3STR3 are well tuned for these; they are included here largely as a sanity check. We then evaluate scalability with two sets of constructed combinatorial instances.

PISA. The PISA benchmark suite, described in [22], consists of a number of problems derived from web-application analysis. It contains find-related constraints like INDEXOF, LASTINDEXOF, REPLACE, CONTAINS, STARTSWITH, ENDSWITH. We compare CVC4 and Z3STR3 and G-STRINGS using different maximum string lengths $\ell \in \{500, 1000, 5000, 10000\}$; results are given in Table 2.

Table 2. Comparison of solvers in the PISA instances. Times are given in seconds. Satisfiable problems are marked ✓ while unsatisfiable are marked ×.

Instance	sat	Z3STR3	CVC4	G-STRINGS			
				500	1000	5000	10000
pisa-000	✓	0.02	0.12	0.00	0.00	0.00	0.00
pisa-001	✓	0.04	0.00	0.00	0.00	0.00	0.00
pisa-002	✓	0.02	0.01	0.00	0.00	0.00	0.00
pisa-003	×	0.01	0.00	0.00	0.00	0.00	0.00
pisa-004	×	0.01	0.56	0.05	0.11	0.55	1.07
pisa-005	✓	0.02	0.03	0.00	0.00	0.00	0.00
pisa-006	×	0.01	0.58	0.06	0.11	0.56	1.1
pisa-007	×	0.01	0.68	0.05	0.11	0.54	1.09
pisa-008	✓	0.03	0.01	1.44	1.44	1.45	1.44
pisa-009	✓	2.68	0.00	0.00	0.00	0.00	0.00
pisa-010	✓	0.01	0.00	0.00	0.00	0.00	0.00
pisa-011	✓	0.01	0.00	0.00	0.00	0.00	0.00
Total (sat)		2.83	**0.17**	1.44	1.44	1.45	1.44
Total (unsat)		**0.04**	1.82	0.16	0.33	1.65	3.26
Total		2.87	1.99	**1.60**	**1.77**	3.10	4.70

---

[3] We used Ubuntu 15.10 machines with 16 GB of RAM and 2.60 GHz Intel® i7 CPU. Experiments available at: https://bitbucket.org/robama/exp_cpaior_2018.

Table 3. Comparative results on the partial de Bruijn sequence problems.

Solver	opt	sat	unk	ttf	Time	score	borda	iborda
CVC4	0	**20**	0	14.2	600.0	10.0	9.6	11.5
Z3STR3	0	6	14	484.2	600.0	1.5	0.0	0.0
G-STRINGS	**6**	**20**	0	**0.0**	**446.0**	**15.0**	**22.4**	**20.5**

Table 4. Comparative results on the substring selection problems.

Solver	opt	sat	unk	ttf	Time	score	borda	iborda
Z3STR3	2	9	11	358.5	546.5	5.0	3.6	8.0
CVC4	2	18	2	60.1	564.4	8.0	4.6	9.0
G-STRINGS $\vert \cdot \vert \leq \vert \cdot \vert$	15	**20**	0	0.1	189.8	18.4	31.3	31.0
G-STRINGS $\preceq$	**18**	**20**	0	**0.05**	**80.5**	**19.5**	**42.5**	**34.0**

Unsurprisingly, the current versions of CVC4 and Z3STR3 solve the PISA instances near instantly. For G-STRINGS, performance on satisfiable instances is very efficient, and independent of maximum string length. Results on the unsatisfiable instances highlight the importance of search strategy. Using a naive input-order search strategy, several of these time out because the search branches first on variables irrelevant to the infeasibility; but by specifying the first decision variable, these instances terminate quickly. This suggests integrating dashed strings into a learning solver may be worthwhile, as the limitation is in the enumeration procedure, rather than the propagators themselves. It appears the unsatisfiable instances scale linearly with maximum string length for G-STRINGS.

Partial de Bruijn Sequences. The de Bruijn sequence $B(k, n)$ (see, e.g., [9]) is a cyclic sequence containing exactly one occurrence of each string of length n in an alphabet of size k. We consider instead the following variant: given a set S of fixed strings in the ASCII alphabet, what is the shortest string X containing at least one occurrence of each string in S:[4]

$$\underset{X}{\text{minimize}} \quad |X|$$
$$\text{subject to} \quad \text{CONTAINS}(X, s), \ s \in S.$$

We generated sets of n random strings having total length l, for $n \in \{5, 10, 15, 20\}$ and $l \in \{100, 250, 500, 750, 1000\}$ (one instance per pair of parameters). Results are given for G-STRINGS, CVC4 and Z3STR3. We also attempted an unfolding approach, using a MiniZinc implementation [2] of the decomposition of Sect. 4.3 to compile down to integer constraints; however the conversion failed due to memory exhaustion even for $l = 100$. The maximum string length for G-STRINGS was set to l, being an upper bound on the minimum sequence length.

---

[4] This model considers only non-cyclic sequences. For cyclic sequences, we need only to replace each occurrence of FIND(X, s) with FIND$(\text{CONCAT}(X, X), s)$.

The results are shown in Table 3. It shows for each solver the number of problems where an optimal solution is proven (opt), the number where a solution was found (sat), and the number where no solution was found (unk). It then gives the average time to first solution (ttf), and solve time (where 600 s is the timeout). The last three columns are comparative scores across the solvers: score gives 0 for finding no solution, 0.25 for finding the worst known solution, 0.75 for finding the best known solution, a linear scale value in (0.25, 0.75) for finding other solutions, and 1 for proving the optimal solution. borda is the MiniZinc Challenge score [19], which gives a Borda score where each pair of solvers is compared on each instance, the better solver gets 1, the weaker 0, and if they are tied the point is split inversely proportional to their solving time. iborda uses a similar border score but the comparison is just on the objective value (proving optimality is not considered better than finding the optimal solution – this is actually the incomplete score of MiniZinc challenge, devised to evaluate local search solvers).

Clearly the G-STRINGS solver significantly outperforms the other two solvers. It instantly finds a solution to all problems, and is the only solver capable of proving optimality. For these benchmarks Z3STR3 is dominated by the other two solvers (see the 0 score for Borda-based metrics).

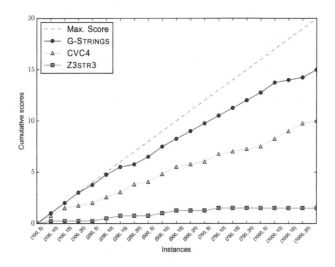

Fig. 7. Cumulative score for partial de Bruijn sequence problem.

Substring Selection. To test a typical application of LEX, we generated instances of a *substring selection* problem: given a set S of n strings, find the k-longest distinct substrings appearing in at least m original strings. LEX is used to break symmetries among selected substrings.

$$\underset{x_1,\ldots,x_k}{\text{maximize}} \quad |x_1| + \ldots + |x_k|$$

$$\text{subject to} \quad \sum_{s \in S} \text{CONTAINS}(s, x_i) \geq m, \ i \in 1, \ldots, k$$

$$x_{i-1} \preceq x_i, \ i \in 2, \ldots, k$$

$$|x_{i-1}| > 0 \rightarrow x_{i-1} \prec x_i, \ i \in 2, \ldots, k.$$

As for the partial de Bruijn problem, we selected a set S of n random strings, with $n \in \{5, 10, 15, 20\}$, and we tuned the total length $l \in \{100, 250, 500, 750, 1000\}$. We then fixed $k = \frac{|S|}{2}$, and selected uniformly in $m \in \left[\frac{|S|}{2}, |S| - 1\right]$. We give results for G-STRINGS, CVC4 and Z3STR3; for unfolding approaches, the conversion again failed due to memory exhaustion.

As the SMT string theory lacks terms for lexicographic ordering, we replace the LEX symmetry breaking using $\preceq$ (the last two lines of the model) by length symmetry breaking using $|x_{i-1}| \leq |x_i|$. We used pair-wise inequalities $\bigwedge_{1 \leq i < j \leq k} x_i \neq x_j$ to have distinct strings. For G-STRINGS, we report performance with symmetry breaking either using LEX, or just on string lengths.

The results shown in Table 4 clearly show that G-STRINGS is superior for these problems, and that using LEX for symmetry breaking is significantly better than simply breaking on length, particularly for proving unsatisfiability.

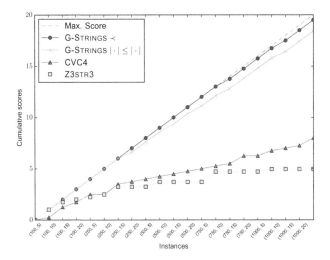

Fig. 8. Cumulative score for maximum substring selection problem.

Finally in Figs. 7 and 8 we show the cumulative score for each solver by sorting the instances lexicographically over (l, m) parameters. We can see that G-STRINGS is dominant, except for the larger de Bruijn sequence problems where CVC4 is quite competitive. On the maximum substring selection problem, the LEX constraints make the G-STRINGS performance almost perfect.

6 Conclusions

In this paper we have the continued work on dashed-string representation for string constraints, defining propagation algorithms for two constraints which lack compact decompositions: lexicographic ordering and substring search. On small verification instances, our approach is competitive with highly-tuned SMT solvers. On constructed combinatorial instances, our dashed-string based propagators substantially outperform current SMT-based string solvers. Results on the verification instances also highlight the importance of autonomous search. Future directions include therefore the study of suitable string search heuristics and the development of learning string solvers.

Acknowledgments. This work is supported by the Australian Research Council (ARC) through Linkage Project Grant LP140100437 and Discovery Early Career Researcher Award DE160100568.

References

1. Aggoun, A., Beldiceanu, N.: Extending CHIP in order to solve complex scheduling and placement problems. Math. Comput. Model. **17**(7), 57–73 (1993)
2. Amadini, R., Flener, P., Pearson, J., Scott, J.D., Stuckey, P.J., Tack, G.: Minizinc with strings. In: Logic-Based Program Synthesis and Transformation - 25th International Symposium, LOPSTR 2016 (2016). https://arxiv.org/abs/1608.03650
3. Amadini, R., Gange, G., Stuckey, P.J.: Sweep-based propagation for string cosntraint solving. In: AAAI 2018 (2018, to appear)
4. Amadini, R., Gange, G., Stuckey, P.J., Tack, G.: A novel approach to string constraint solving. In: Beck, J.C. (ed.) CP 2017. LNCS, vol. 10416, pp. 3–20. Springer, Cham (2017). https://doi.org/10.1007/978-3-319-66158-2_1
5. Berzish, M., Zheng, Y., Ganesh, V.: Z3str3: a string solver with theory-aware branching. CoRR abs/1704.07935 (2017). http://arxiv.org/abs/1704.07935
6. Bisht, P., Hinrichs, T.L., Skrupsky, N., Venkatakrishnan, V.N.: WAPTEC: whitebox analysis of web applications for parameter tampering exploit construction. In: Proceedings of ACM Conference on Computer and Communications Security, pp. 575–586. ACM (2011)
7. Bjørner, N., Tillmann, N., Voronkov, A.: Path feasibility analysis for string-manipulating programs. In: Kowalewski, S., Philippou, A. (eds.) TACAS 2009. LNCS, vol. 5505, pp. 307–321. Springer, Heidelberg (2009). https://doi.org/10.1007/978-3-642-00768-2_27
8. Emmi, M., Majumdar, R., Sen, K.: Dynamic test input generation for database applications. In: Proceedings of the ACM SIGSOFT International Symposium on Software Testing and Analysis (ISSTA), pp. 151–162. ACM (2007)
9. Fredricksen, H.: A survey of full length nonlinear shift register cycle algorithms. SIAM Rev. **24**(2), 195–221 (1982)
10. Frisch, A.M., Hnich, B., Kiziltan, Z., Miguel, I., Walsh, T.: Propagation algorithms for lexicographic ordering constraints. Artif. Intell. **170**(10), 803–834 (2006)
11. Gange, G., Navas, J.A., Stuckey, P.J., Søndergaard, H., Schachte, P.: Unbounded model-checking with interpolation for regular language constraints. In: Piterman, N., Smolka, S.A. (eds.) TACAS 2013. LNCS, vol. 7795, pp. 277–291. Springer, Heidelberg (2013). https://doi.org/10.1007/978-3-642-36742-7_20

12. Gecode Team: Gecode: generic constraint development environment (2016). http://www.gecode.org
13. Hooimeijer, P., Weimer, W.: StrSolve: solving string constraints lazily. Autom. Softw. Eng. **19**(4), 531–559 (2012)
14. Kiezun, A., Ganesh, V., Artzi, S., Guo, P.J., Hooimeijer, P., Ernst, M.D.: HAMPI: a solver for word equations over strings, regular expressions, and context-free grammars. ACM Trans. Softw. Eng. Methodol. **21**(4), Article No. 25 (2012)
15. Li, G., Ghosh, I.: PASS: string solving with parameterized array and interval automaton. In: Bertacco, V., Legay, A. (eds.) HVC 2013. LNCS, vol. 8244, pp. 15–31. Springer, Cham (2013). https://doi.org/10.1007/978-3-319-03077-7_2
16. Liang, T., Reynolds, A., Tinelli, C., Barrett, C., Deters, M.: A DPLL(T) theory solver for a theory of strings and regular expressions. In: Biere, A., Bloem, R. (eds.) CAV 2014. LNCS, vol. 8559, pp. 646–662. Springer, Cham (2014). https://doi.org/10.1007/978-3-319-08867-9_43
17. Saxena, P., Akhawe, D., Hanna, S., Mao, F., McCamant, S., Song, D.: A symbolic execution framework for JavaScript. In: S&P, pp. 513–528. IEEE Computer Society (2010)
18. Scott, J.D., Flener, P., Pearson, J., Schulte, C.: Design and implementation of bounded-length sequence variables. In: Salvagnin, D., Lombardi, M. (eds.) CPAIOR 2017. LNCS, vol. 10335, pp. 51–67. Springer, Cham (2017). https://doi.org/10.1007/978-3-319-59776-8_5
19. Stuckey, P.J., Feydy, T., Schutt, A., Tack, G., Fischer, J.: The miniZinc challenge 2008–2013. AI Mag. **2**, 55–60 (2014)
20. Tateishi, T., Pistoia, M., Tripp, O.: Path- and index-sensitive string analysis based on monadic second-order logic. ACM Trans. Softw. Eng. Methodol. **22**(4), 33 (2013)
21. Thomé, J., Shar, L.K., Bianculli, D., Briand, L.C.: Search-driven string constraint solving for vulnerability detection. In: ICSE 2017, Buenos Aires, Argentina, 20–28 May 2017, pp. 198–208 (2017)
22. Zheng, Y., Ganesh, V., Subramanian, S., Tripp, O., Berzish, M., Dolby, J., Zhang, X.: Z3str2: an efficient solver for strings, regular expressions, and length constraints. Formal Methods Syst. Des. **50**(2–3), 249–288 (2017)

Designing Fair, Efficient, and Interpretable Policies for Prioritizing Homeless Youth for Housing Resources

Mohammad Javad Azizi, Phebe Vayanos[✉], Bryan Wilder, Eric Rice, and Milind Tambe

CAIS Center for Artificial Intelligence in Society,
University of Southern California, Los Angeles, CA 90007, USA
{azizim,phebe.vayanos,bwilder,ericr,tambe}@usc.edu

Abstract. We consider the problem of designing fair, efficient, and interpretable policies for prioritizing heterogeneous homeless youth on a waiting list for scarce housing resources of different types. We focus on point-based policies that use features of the housing resources (e.g., permanent supportive housing, rapid rehousing) and the youth (e.g., age, history of substance use) to maximize the probability that the youth will have a safe and stable exit from the housing program. The policies can be used to prioritize waitlisted youth each time a housing resource is procured. Our framework provides the policy-maker the flexibility to select both their desired structure for the policy and their desired fairness requirements. Our approach can thus explicitly trade-off interpretability and efficiency while ensuring that fairness constraints are met. We propose a flexible data-driven mixed-integer optimization formulation for designing the policy, along with an approximate formulation which can be solved efficiently for broad classes of interpretable policies using Bender's decomposition. We evaluate our framework using real-world data from the United States homeless youth housing system. We show that our framework results in policies that are more fair than the current policy in place and than classical interpretable machine learning approaches while achieving a similar (or higher) level of overall efficiency.

1 Introduction

This paper addresses the problem of designing policies for prioritizing heterogeneous allocatees on a waiting list for scarce resources of different types so as to maximize overall efficiency. The allocatees differ in their intrinsic characteristics which, combined with the characteristics of their assigned resource, impact the efficiency of the policy. We consider a policy-maker who is able to enforce adoption of the computed policy. However, since the allocated resources are viewed as *common property*, i.e., as belonging to all members of the community, the policy should satisfy certain fairness requirements while also being interpretable, making it easy to explain why a particular assignment was made.

© Springer International Publishing AG, part of Springer Nature 2018
W.-J. van Hoeve (Ed.): CPAIOR 2018, LNCS 10848, pp. 35–51, 2018.
https://doi.org/10.1007/978-3-319-93031-2_3

We are particularly motivated by the problem of allocating housing to home-less youth. In the U.S., hundreds of thousands of homeless youth are forced to live in emergency shelters or on the streets, where they run a high risk of violence, substance abuse, and sexual exploitation [27]. To help support this vulnerable population, the U.S. government directs federal resources towards programs that assist homeless youth [21]. The Homeless Management Informa-tion System (HMIS) database collects information on these services. Analysis of the HMIS database has shown that providing housing to homeless individuals produces large gains in long-term health and stability [5,22]. Unfortunately, the number of homeless youth in the U.S. far exceeds the housing resources avail-able [9]. Moreover, once a house is procured, there are potentially hundreds of local homeless youth that are eligible for the resource.

Given the immense difference that housing programs can make for youth, policy-makers and communities must allocate these precious resources efficiently. Most communities employ a *Coordinated Entry System* (CES) in which organiza-tions within the same community pool both their housing resources and youth. When a housing resource becomes available, the waitlisted youth are ranked based on a set of *priority rules* and the house is allocated to the highest ranking individual [9].[1] The current prioritization tool, the TAY Triage Tool,[2] ranks the youth based on a *vulnerability score*, the Next Step Tool (NST) score, that relies on six key experiences that increase the risk of prolonged homelessness [23]. Thus, the current policy is not directly tied to outcomes (due mostly to lack of outcome data at the time of design). Instead, it is purely based on factors intrinsic to each youth that determine their ability to exit homelessness without supportive housing. The increasing availability of outcome data, combined with a strategic push to better coordinate housing resources [21], constitute a sig-nificant opportunity to improve the current policy to better match supply and demand for resources. We now summarize the main desiderata of such a policy:

(a) *Maximize Efficiency.* Given the scarcity of housing resources, it is critical to design an efficient policy explicitly tied to outcomes for allocating houses to the homeless youth. We thus seek to improve upon the efficiency of the current policy (which is not tied to outcomes) as measured in terms of the expected number of stably housed youth at the end of the intervention.

(b) *Ensure Fairness.* Housing resources constitute common property and can prove invaluable for the homeless youth. It is thus natural to seek an alloca-tion policy that is in some sense fair. Since there is no universally accepted measure of fairness, the proposed framework should afford the policy-maker the flexibility to select the fairness criteria that they wishes to enforce. For example, one could require that the probability of a stable exit for a youth in

---

[1] To date there is no regulation in place that enforces the current policy. However, previous analysis has shown that communities follow this policy in practice [20].

[2] Transition Age Youth (TAY) is a Service Prioritization Decision Assistance Tool that can be accessed at http://orgcode.nationbuilder.com/tools_you_can_use. It is incorporated work from the TAY Triage Tool of Rice [23] which can be accessed at http://www.csh.org/wp-content/uploads/2014/02/TAY_TriageTool_2014.pdf.

the system should be equal across different races, or independent of the vulnerability score of the youth so that, independently of their backgrounds and past experiences, youth are equally likely to transition into stable housing.

(c) *Customize Interpretability.* Currently, communities can decide to comply or not with a recommended allocation. Hence, policies should be interpretable: it should be easy to explain the structure of the policy and to justify a particular matching. For example, a policy which assigns priority based on a linear scoring rule may be viewed as interpretable. From our discussions with communities across the country, it appears that much of the success of the current policy can be attributed to its interpretability. Since interpretability is subjective, we allow the policy-maker to customize the policy structure.

Given the above desiderata, a natural question is: how to design classes of policies that conveniently trade-off between efficiency, fairness, and interpretability? We note that this question arises in many other contexts, e.g., in the design of policies for the U.S. Kidney Allocation System and the U.S. Public Housing Program. In this paper, we propose a framework for designing such policies that is applicable to all these contexts. We now summarize our main contributions:

(a) We introduce a data-driven framework for optimizing over interpretable policies, to which a policy-maker may add flexibly defined fairness constraints. We give a mixed-integer program for computing optimal policies.

(b) To enhance scalability for complex policy classes, we give an approximate solution approach which relaxes the problem to a form amenable to Bender's decomposition. This offers significant speedup and allows us to optimize over much more sophisticated policies (e.g., multi-level decision trees, compared to linear policies in previous work).

(c) We conduct an empirical evaluation using real-world data from homeless youth across the U.S. We compare to both the status-quo TAY prioritization as well an array of approaches from the literature. Our exact approach offers significant improvements in fairness compared to previous approaches for optimizing the same class of models, while our approximate approach allows us to improve fairness in a complementary way, by using a more expressive class. In both cases, we obtain efficiency comparable to the best (unfair) alternatives, and better than the status-quo TAY.

Literature Review. Allocation problems have been studied extensively in computer science and operations research. Much of this work considers incentives issues, where agents may misreport their true preferences to obtain a better match [1, 8, 13, 26], and the focus is on satisfying axiomatic properties (e.g. Pareto optimality or strategy-proofness). We do not consider strategic reporting since information reported by the youth can generally be verified. Instead, we focus on balancing efficiency, fairness, and interpretability. None of these goals are considered in this previous work, and all are crucial features of our domain. Another line of research considers nonstrategic online resource allocation, e.g., in the "Adwords" setting [4, 7, 19]. The focus here is on algorithms which provably approximate the optimal efficiency. By contrast, our goal is to find exactly

optimal policies out of a feasible set which is constrained by fairness and interpretability. Lastly, much previous work considers organ (e.g., kidney) allocation
[2,11,24,25]. Our paper is most closely related to that of Bertsimas et al. [2],
who optimize the U.S. Kidney Allocation System over a class of linear policies.
We improve upon their approach in several ways: *(i)* we propose an *exact* formulation of the allocation problem that enables us to guarantee fairness, while
Bertsimas et al. use a heuristic method that cannot guarantee fairness; *(ii)* our
model is *exact*, incorporating the order in which youth and housing resources
arrive, to provide accurate prioritization; *(iii)* we consider larger classes of interpretable policies (e.g., based on decision trees). These contributions translate
into substantial empirical improvement.

Our work is also related to recent applications of mixed-integer programming
to machine learning [3,17,18]. Our approach uses a mixed-integer program (MIP)
to optimize over classes of policies (linear models, decision trees) also used in
the machine learning literature. Previous work has shown the promise of using
MIPs in machine learning; however, we are not aware of any work using such
techniques to construct policies for resource allocation.

Lastly, our work is related to interpretable machine learning. Many interpretable models have been proposed, including decision rules [14,29], decision
sets [12] and generalized additive models [15,16]. In this work, motivated by the
policies currently used in the homeless youth housing system and U.S. Kidney
Allocation System, we build on decision trees, which have been used to create interpretable models in many contexts [6,10,28]. We make two contributions
compared to this previous work. First, we introduce two new model classes which
generalize decision trees to respectively allow more flexible branching structures
and the use of a linear scoring policy at each node of the tree (Examples 3 and 4
of Sect. 2.3). Second, we use these models to parameterize the allocation policy
itself rather than the learning system. Thus, the final policies produced by our
system are interpretable, not just the predicted success probabilities.

Notation. We denote sets (resp. random variables) using uppercase blackboard
bold (resp. uppercase script) font. We denote the indicator function with $\mathcal{I}(\cdot)$.

2   Model, Problem Statement, and Interpretable Policies

2.1   System Model

We model the homeless youth housing allocation system as an infinite stream
of housing resources indexed by $h \in \{1, \ldots, \infty\}$ that must be allocated to an
infinite stream of youth indexed by $y \in \{1, \ldots, \infty\}$. Associated with each housing
resource h is a random feature vector $\mathscr{F}_h \in \mathbb{R}^{n_h}$ which includes, without loss
of generality, the (random) arrival time $\mathscr{A}_h \in \mathbb{R}$ of the house in the system and
may also include e.g., the type of house (rapid rehousing, permanent supportive
housing, etc.). Accordingly, associated with each youth y is a random feature
vector $\mathscr{G}_y \in \mathbb{R}^{n_y}$ which includes the arrival time $\mathscr{T}_y \in \mathbb{R}$ of the youth in the
system and may also include e.g., the intrinsic characteristics of the youth (age,

history of abuse, history of substance use, etc.). Not all youth are eligible for all housing resources. Whether a youth is compatible with a particular house can be determined based on the features of the house and the youth. We let $\mathbb{M}(\mathscr{F}_h) \in \mathbb{R}^{n_y}$ denote the set of all youth feature vectors that are compatible with house h. For example, we may wish to enforce that $\mathbb{M}(\mathscr{F}_h) := \{\mathscr{G}_y : \mathscr{A}_h \geq \mathscr{T}_y\}$ so that a house must be allocated immediately upon arrival in the system. Thus, youth y is eligible for house h if and only if $\mathscr{G}_y \in \mathbb{M}(\mathscr{F}_h)$. The probability of a successful outcome (a safe and stable exit) when youth y is placed in house h is denoted by $p(\mathscr{G}_y, \mathscr{F}_h)$. The probability of a successful outcome if youth y is not offered a house is denoted by $\bar{p}(\mathscr{G}_y)$. We assume that both these quantities are perfectly known (can be estimated from data). In our numerical experiments, see Sect. 5, we will estimate these quantities using data from the HMIS database.

Our aim is to design interpretable parametric point-based policies that prioritize the youth for housing resources so as to maximize overall welfare. In particular, we consider parametric policies with parameter vector $\beta \in \mathbb{R}^n$ that map the features of the youth and the house to a score, see Sect. 2.3 for examples of such policies. We denote the score obtained for a youth y and house h for a given parameter choice β by $\pi_\beta(\mathscr{G}_y, \mathscr{F}_h)$. Then, youth y will have priority over youth y' if $\pi_\beta(\mathscr{G}_y, \mathscr{F}_h) > \pi_\beta(\mathscr{G}_{y'}, \mathscr{F}_h)$. We assume that ties are broken using a suitable tie-braking rule (e.g., at random). We thus let $\mathscr{R}$ be a permutation of the set $\{1, \ldots, \infty\}$ where the quantity $\mathscr{R}(y)$ denotes the tie-breaking score of youth y: when $\pi_\beta(\mathscr{G}_y, \mathscr{F}_h) = \pi_\beta(\mathscr{G}_{y'}, \mathscr{F}_h)$, youth y will be given priority over youth y' if and only if $\mathscr{R}(y) > \mathscr{R}(y')$.

Given a parameter vector β we now formalize the allocation process. For $t \in [0, \infty]$, we let $\mathbb{Y}(t)$ denote the set of youth that are available in the system at time t. We omit the dependence of $\mathbb{Y}(t)$ on β to minimize notational overhead. Thus, $\mathbb{Y}(0)$ denotes the initial state of the system. Suppose that a youth y arrives in the system at time t. Then $\mathbb{Y}(t+) = \mathbb{Y}(t) \cup \{y\}$. Suppose instead that house h arrives in the system at time t. Then, the house will be allocated, among all the compatible youth, to the one with the highest score (accounting for the tie breaking rule). In particular, it will be assigned to the youth[3]

$$y^\star = \underset{y}{\operatorname{argmax}} \left\{ \mathscr{R}(y) \; : \; y \in \operatorname{argmax}_y \left\{ \pi_\beta(\mathscr{G}_y, \mathscr{F}_h) \; : \; \mathscr{G}_y \in \mathbb{M}(\mathscr{F}_h), \; y \in \mathbb{Y}(t) \right\} \right\}.$$

Subsequently, youth $y^\star$ leaves the system, i.e., $\mathbb{Y}(t+) = \mathbb{Y}(t) \backslash \{y^\star\}$. Thus, given β, the allocation system generates: (i) an infinite random sequence $\{(\mathscr{Y}_i(\beta), \mathscr{H}_i)\}_{i=1}^\infty$ of matches, where $\mathscr{H}_i \in \{1, \ldots, \infty\}$ denotes the ith allocated house and $\mathscr{Y}_i(\beta)$ the youth to which the ith house is allocated under the policy with parameters β, and (ii) a set $\lim_{t \to \infty} \mathbb{Y}(t)$ of youth that will never receive a house.

2.2 Problem Statement

Given the model described in Sect. 2.1, the expected probability of a safe and stable exit across all youth is a complicated function of the parameters β and is expressible as

---

[3] By construction, there will be at most one youth in this set. Moreover, since there is a severe shortage of houses, we can assume w.l.o.g. that this set will never be empty.

$$\mathcal{P}(\beta) := \mathbb{E}\left[\lim_{N\to\infty}\frac{1}{N}\sum_{i=1}^{N}p(\mathcal{G}_{\mathcal{Y}_i(\beta)},\mathscr{F}_{\mathscr{H}_i}) + \lim_{t\to\infty}\frac{1}{|\mathbb{Y}(t)|}\sum_{\mathscr{Y}\in\mathbb{Y}(t)}\overline{p}(\mathcal{G}_{\mathscr{Y}})\right],$$

where the expectation is taken with respect to the distribution $\mathbb{P}$ of the random features of the houses and the youth, which include their arrival times and determine permissible matchings. The first (second) part in the expression above corresponds to the probability that a randomly chosen youth that received (did not receive) a house will have a safe and stable exit under the matching.

From the desiderata of the policy described in the Introduction, we wish to be able to enforce flexible fairness requirements. These requirements take the form of set-based constraints on the random sequence $\{(\mathcal{Y}_i(\beta),\mathscr{H}_i)\}_{i=1}^{\infty}$ of matches. For example, we may require that, almost surely, the proportion of all houses that provide permanent support that go to individuals with high vulnerability scores is greater than 40%. We denote by $\mathbb{F}$ the set in which the sequence of matchings is required to lie. Then, choices of β are restricted to lie in the set

$$\mathbb{S} := \{\beta\in\mathbb{B} \ : \ \{(\mathcal{Y}_i(\beta),\mathscr{H}_i)\}_{i=1}^{\infty}\in\mathbb{F}, \ \mathbb{P}\text{-a.s.}\},$$

where $\mathbb{B}\subseteq\mathbb{R}^n$ captures constraints that relate to interpretability of the policy (e.g., constraints on the maximum number of features employed, see Sect. 3.2).

The problem faced by the policy-maker can then be expressed compactly as

$$\text{maximize } \{\mathcal{P}(\beta) \ : \ \beta\in\mathbb{S}\}. \tag{1}$$

Unfortunately, Problem (1) is very challenging to solve since the relation between β and the random sequence $\{(\mathcal{Y}_i(\beta),\mathscr{H}_i)\}_{i=1}^{\infty}$ can be highly nonlinear, while the distribution of the features of the youth and the houses is unknown. In Sect. 3, we will propose a data-driven mixed-integer optimization approach for learning the parameters β of the policy in Problem (1).

2.3   Interpretable Policies

In what follows, we describe several policies that can be employed in our framework and that possess attractive interpretability properties.

Example 1 (Linear Scoring Policies). A natural choice of interpretable policy are linear (or affine) in the features of the houses and the youth. These are expressible as $\pi_\beta(\mathcal{G},\mathcal{F}) := \beta^\top(\mathcal{G},\mathcal{F})$, where one uses one-hot encoding to encode categorical features. The feature vector can naturally be augmented by nonlinear functions of the features available in the dataset as one would do in standard linear regression. To reinforce interpretability, one may wish to limit the number of permitted non-zero coefficients of β.

Example 2 (Decision-Tree-Based Scoring Policies). We refer to those policies that take the form of a tree-like structure (in the spirit of decision-trees in machine learning, see Introduction) as decision-tree-based scoring policies. In each internal node of the decision-tree, a "test" is performed on a *specific feature* (e.g., if the age of the youth is smaller than 18). Each branch represents the outcome of the test, and each leaf node represents the score assigned to all the youth that reach that leaf. Thus, each path from root to leaf represents a classification rule that assigns a unique score to each youth. All youth that reach the same leaf will have the same score. In these policies, the policy-maker selects the depth K of the tree. The vector β collects the set of features to branch on at each node and either the set of feature values that will be assigned to each branch (for categorical features) or the cut-off values (for quantitative features). Thus, these policies partition the space of features into 2^K disjoint subsets. Letting $\mathbb{S}_\ell$ denote the set of all feature values that belong to the ℓth subset and z_ℓ the score assigned to that subset, we have $\pi_\beta(\mathscr{G}, \mathscr{F}) := \sum_{\ell=1}^{2^K} z_\ell \mathcal{I}((\mathscr{G}, \mathscr{F}) \in \mathbb{S}_\ell)$. We note that, while these policies are exponential in K, to maximize interpretability, K should be kept as small as possible. To improve interpretability further, one may require that each feature be branched on at most once.

Example 3 (Decision-Tree-Based Policies enhanced with Linear Branching). A natural variant of the policies from Example 2 is one where policies take again the form of a tree-like structure, but this time, each "test" involves a linear function of several features (e.g., whether a vulnerability measure of the youth is greater than 10). In this setting, the vector β collects the coefficients of the linear function at each node and the cut-off values of the branching.

Example 4 (Decision-Tree-Based Policies enhanced with Linear Leafing). Another variant of the policies from Example 2 is one where rather than having a common score for all youth that reach a leaf, instead, a linear scoring rule is employed on each leaf. Thus, $\pi_\beta(\mathscr{G}, \mathscr{F}) := \sum_{\ell=1}^{2^K} [\beta_{\mathrm{y},\ell}^\top \mathscr{G} + \beta_{\mathrm{h},\ell}^\top \mathscr{F}] \mathcal{I}((\mathscr{G}, \mathscr{F}) \in \mathbb{S}_\ell)$, and the parameters to be optimized are augmented with $\beta_{\mathrm{y},\ell}$ and $\beta_{\mathrm{h},\ell}$ for each ℓ.

In addition to the examples above, one may naturally also consider decision-tree-based policies enhanced with both linear branching and linear leafing.

3   Data-Driven Framework for Policy Calibration

In Sect. 2.1, we proposed a model for the homeless youth housing allocation system and a mathematical formulation (Problem (1)) of the problem of designing fair, efficient, and interpretable policies for allocating these scarce resources. This problem is challenging to solve as it requires knowledge of the distribution of the uncertain parameters. In this section, we propose a data-driven mixed-integer optimization formulation for learning the parameters β of the policy, thus approximating Problem (1).

3.1   A Data-Driven Mixed Integer Optimization Problem

We assume that we have at our disposal a dataset that consists of: *(i)* a (finite) stream $\mathbb{H}$ of housing resources that became available in the past and their associated feature vectors $f_h \in \mathbb{R}^{n_h}$, $h \in \mathbb{H}$; and *(ii)* a (finite) stream $\mathbb{Y}$ of youth waitlisted for a house and their associated feature vectors $g_y \in \mathbb{R}^{n_y}$. We let α_h (resp. τ_y) denote the arrival time of house h (resp. youth y) in the system. For convenience, we define

$$\mathbb{C} := \{(y, h) \in \mathbb{Y} \times \mathbb{H} \; : \; g_y \in \mathbb{M}(f_h)\},$$

and also let $p_{yh} := p(g_y, f_h)$, $\overline{p}_y := \overline{p}(g_y)$, and $\rho_y := \mathscr{R}(y)$. Using this data, the problem of learning (estimating) the parameters β of the policy can be cast as a mixed-integer optimization problem. The main decision variables of the problem are the policy parameters β. Consider the MIP

$$\text{maximize} \sum_{y \in \mathbb{Y}} \left[\sum_{h \in \mathbb{H}} p_{yh} x_{yh} + \overline{p}_y \left(1 - \sum_{h \in \mathbb{H}} x_{yh} \right) \right]$$

$$\text{subject to } \pi_{yh} = \pi_\beta(g_y, f_h), \quad \forall y \in \mathbb{Y}, \; h \in \mathbb{H}$$
$$\forall y \in \mathbb{Y}, \; h \in \mathbb{H},$$

$$x_{yh} = 1 \Leftrightarrow \begin{cases} (y, h) \in \mathbb{C}, \quad \sum_{h' \neq h : \alpha_{h'} \leq \alpha_h} x_{yh'} = 0, \text{ and} \\ \forall y' : (y', h) \in \mathbb{C} \text{ and } \sum_{h' : \alpha_{h'} \leq \alpha_h} x_{y'h'} = 0, \\ (\pi_{yh} > \pi_{y'h}) \text{ or } (\pi_{yh} = \pi_{y'h} \text{ and } \rho_y > \rho_{y'}) \end{cases} \quad (2)$$

$$\beta \in \mathbb{B}, \; x \in \mathbb{F}, \; x_{yh} \in \{0, 1\} \; \forall y \in \mathbb{Y}, \; h \in \mathbb{H}.$$

In addition to β, the decision variables of the problem are the assignment variables x and the scoring variables π. Thus, x_{yh} indicates whether house h is allocated to youth y under the policy with parameters β and π_{yh} corresponds to the score of youth y for house h under the policy. The first (second) part of the objective function corresponds to the probability that youth y will be successful if they do (do not) receive a house under the policy with parameters β. The first constraint in the formulation defines the scoring variables in terms of the parameters β and the features of the youth and the house. The second constraint is used to define the assignment variables in terms of the scores: it stipulates that youth y will receive house h if and only if: *(i)* the two are compatible, *(ii)* youth y is still on the waitlist, and *(iii)* youth y has higher priority over all youth that have not yet been allocated a house in the sense that they score higher using the scoring policy dictated by β (combined with the tie-breaking rule).

Next, we show that if $\mathbb{F}$ is polyhedral, Problem (2) can be solved as a mixed-integer linear optimization problem provided one can define the scores π using linear inequalities. The main decision variables of this problem are the policy parameters β. Consider the MIP

$$\text{maximize} \sum_{y \in \mathbb{Y}} \left[\sum_{h \in \mathbb{H}} p_{yh} x_{yh} + \overline{p}_y \left(1 - \sum_{h \in \mathbb{H}} x_{yh} \right) \right] \quad (3a)$$

$$\text{subject to } \pi_{yh} = \pi_\beta(g_y, f_h) \quad \forall y \in \mathbb{Y}, h \in \mathbb{H} \quad (3b)$$

$$\sum_{h \in \mathbb{H}} x_{yh} \leq 1 \quad \forall y \in \mathbb{Y}, \quad \sum_{y \in \mathbb{Y}} x_{yh} \leq 1 \quad \forall h \in \mathbb{H} \tag{3c}$$

$$z_{yh} = \sum_{h' \in \mathbb{H} \setminus \{h\}} \mathcal{I}\left(\alpha_{h'} \leq \alpha_h\right) x_{yh'} \quad \forall y \in \mathbb{Y}, h \in \mathbb{H} \tag{3d}$$

$$\pi_{yh} - \pi_{y'h} = v^{+}_{yy'h} - v^{-}_{yy'h} \quad \forall y, y' \in \mathbb{Y}, h \in \mathbb{H} \tag{3e}$$

$$v^{+}_{yy'h} \leq M u_{yy'h} \quad \forall y, y' \in \mathbb{Y}, h \in \mathbb{H} \tag{3f}$$

$$v^{-}_{yy'h} \leq M(1 - u_{yy'h}) \quad \forall y, y' \in \mathbb{Y}, h \in \mathbb{H} \tag{3g}$$

$$v^{+}_{yy'h} + v^{-}_{yy'h} \geq \epsilon(1 - u_{yy'h}) \quad \forall y, y' \in \mathbb{Y}, h \in \mathbb{H} : \rho_y > \rho_{y'} \tag{3h}$$

$$v^{+}_{yy'h} + v^{-}_{yy'h} \geq \epsilon u_{yy'h} \quad \forall y, y' \in \mathbb{Y}, h \in \mathbb{H} : \rho_{y'} > \rho_y \tag{3i}$$

$$x_{yh} \leq u_{yy'h} + z_{y'h} \quad \forall y, y' \in \mathbb{Y}, h \in \mathbb{H} \tag{3j}$$

$$1 - z_{yh} \leq \sum_{y':(y',h) \in \mathbb{C}} x_{y'h} \quad \forall (y,h) \in \mathbb{C} \tag{3k}$$

$$x_{yh} = 0 \quad \forall y \in \mathbb{Y}, h \in \mathbb{H} : (y,h) \notin \mathbb{C} \tag{3l}$$

$$x \in \mathbb{F} \tag{3m}$$

$$v^{+}_{yy'h}, \ v^{-}_{yy'h} \geq 0, \ u_{yy'h} \in \{0,1\}, \quad \forall y, y' \in \mathbb{Y}, h \in \mathbb{H} \tag{3n}$$

$$x_{yh}, z_{yh} \in \{0,1\} \quad \forall y \in \mathbb{Y}, h \in \mathbb{H}. \tag{3o}$$

In addition to the policy parameters β, the score variables π and assignment variables x, Problem (3) employs several auxiliary variables (z, v^+, v^-, and u) that are used to uniquely define the assignment variables x based on the scores π. The variables z indicate whether a youth is still waiting at the time a house arrives: $z_{yh} = 1$ if and only if youth y has been allocated a house on or before time α_h. The non-negative variables $v^{+}_{yy'h}$ and $v^{-}_{yy'h}$ denote the positive and negative parts of $\pi_{yh} - \pi_{y'h}$. Finally, the variables u are prioritization variables: $u_{yy'h} = 1$ if and only if either youth y has a higher score than youth y' for house h (i.e., $\pi_{yh} > \pi_{y'h}$) or they have the same score but youth y has priority due to tie-breaking (i.e., $\pi_{yh} = \pi_{y'h}$ and $\rho_y > \rho_{y'}$).

Problems (2) and (3) share the same objective function. An interpretation of the constraints in Problem (3) is as follows. Constraint (3b) is used to define the variables π_{yh}. Constraints (3c) are classical matching constraints. Constraint (3d) is used to define the variables z. Constraints (3e)–(3i) are used to define the prioritization variables u in term of the scores π: constraint (3e) defines v^{+}_{yh} and v^{-}_{yh} as the positive and negative parts of $\pi_{yh} - \pi_{y'h}$, respectively. Constraints (3f) and (3g) stipulate that $u_{yy'h}$ must be 1 if $\pi_{yh} > \pi_{y'h}$ and must be 0 if $\pi_{yh} < \pi_{y'h}$. Constraints (3h) and (3i) ensure that if π_{yh} and $\pi_{y'h}$ are equal then $u_{yy'h} = 1$ if and only if $\rho_y > \rho_{y'}$. Constraint (3j) stipulates that youth y cannot receive house h if there is another youth y' that is still waiting for a house and that has priority for house h over y. Constraint (3k) ensures that if a youth that is compatible with a house has not been served at the time a house arrives, then the house must be assigned to a compatible youth. Finally, constraint (3l) ensures that youth are only assigned houses they are eligible for.

If the scoring variables π can be defined in terms of the policy parameters β (constraint (3b)) using integer linear constraints, then Problem (3) is an MILP.

3.2 Expressing the Policy Values Using Integer Linear Constraints

We now show that for all the interpretable policies from Sect. 2.3, the scoring variables π can be defined using finitely many integer linear constraints, implying that Problem (3) reduces to a mixed-integer linear program if $\mathbb{F}$ is polyhedral.

Example 5 (Linear Scoring Policies). In the case of the linear policies (Example 1), constraint (3b) is equivalent to

$$\pi_{yh} = \beta^{\top}(g_y, f_h) \quad \forall y \in \mathbb{Y},\ h \in \mathbb{H}. \tag{4}$$

To increase interpretability, one may impose a limit K on the number of features employed in the policy by letting

$$\mathbb{B} = \left\{ \beta \in \mathbb{R}^n\ :\ \exists \kappa \in \{0,1\}^n \text{ with } \sum_{i=1}^{n} \kappa_i \leq K,\ |\beta_i| \leq \kappa_i,\ i = 1, \ldots, n \right\},$$

where $\kappa_i = 1$ if and only if the ith feature is employed.

Example 6 (Decision-Tree-Based Scoring Policies). For decision-tree-based scoring policies (Example 2), constraint (3b) is equivalent to

$$\pi_{yh} = \sum_{\ell \in \mathbb{L}} z_{\ell} x_{yh\ell} \quad \forall y \in \mathbb{Y},\ h \in \mathbb{H}, \tag{5}$$

where $\mathbb{L}$ denotes the set of all leafs in the tree, the variables x are leaf assignment variables such that $x_{yh\ell} = 1$ if and only if the feature vectors of youth y and house h belong to leaf ℓ, and z are score variables such that z_{ℓ} corresponds to the score assigned to leaf ℓ. The above constraint is bilinear but can be linearized using standard techniques. Next, we illustrate that the leaf assignment variables can be defined using a system of integer linear inequalities.

Let $\mathbb{I}_c$ and $\mathbb{I}_q$ denote the sets of all categorical and quantitative features (of both the youth and the houses), respectively. Also, let $\mathbb{I} := \mathbb{I}_c \cup \mathbb{I}_q$. Denote with d_{yhi} the value attained by the ith feature of the pair (y, h) and for $i \in \mathbb{I}_c$ let $\mathbb{S}_i$ collect the possible levels attainable by feature i. Finally, let $\mathbb{V}$ denote the set of all branching nodes in the tree and for $\nu \in \mathbb{V}$, let $\mathbb{L}^r(\nu)$ (resp. $\mathbb{L}^l(\nu)$) denote all the leaf nodes that lie to the right (resp. left) of node ν. Consider the system

$$\sum_{i \in \mathbb{I}} p_{\nu i} = 1 \quad \forall \nu \in \mathbb{V} \tag{6a}$$

$$q_{\nu} - \sum_{i \in \mathbb{I}_q} p_{\nu i} d_{yhi} = g^{+}_{yh\nu} - g^{-}_{yh\nu} \quad \forall \nu \in \mathbb{V},\ y \in \mathbb{Y},\ h \in \mathbb{H} \tag{6b}$$

$$g^+_{yh\nu} \le Mw^{\mathrm{q}}_{yh\nu} \quad \forall \nu \in \mathbb{V},\ y \in \mathbb{Y},\ h \in \mathbb{H} \tag{6c}$$

$$g^-_{yh\nu} \le M(1 - w^{\mathrm{q}}_{yh\nu}) \quad \forall \nu \in \mathbb{V},\ y \in \mathbb{Y},\ h \in \mathbb{H} \tag{6d}$$

$$g^+_{yh\nu} + g^-_{yh\nu} \ge \epsilon(1 - w^{\mathrm{q}}_{yh\nu}) \quad \forall \nu \in \mathbb{V},\ y \in \mathbb{Y},\ h \in \mathbb{H} \tag{6e}$$

$$x_{yh\ell} \le 1 - w^{\mathrm{q}}_{yh\nu} + \sum_{i \in \mathbb{I}_{\mathrm{c}}} p_{\nu i} \quad \forall \nu \in \mathbb{V},\ y \in \mathbb{Y},\ h \in \mathbb{H},\ \ell \in \mathbb{L}^{\mathrm{r}}(\nu) \tag{6f}$$

$$x_{yh\ell} \le w^{\mathrm{q}}_{yh\nu} + \sum_{i \in \mathbb{I}_{\mathrm{c}}} p_{\nu i} \quad \forall \nu \in \mathbb{V},\ y \in \mathbb{Y},\ h \in \mathbb{H},\ \ell \in \mathbb{L}^{\mathrm{l}}(\nu) \tag{6g}$$

$$s_{\nu ik} \le p_{\nu i} \quad \forall \nu \in \mathbb{V},\ i \in \mathbb{I}_{\mathrm{c}},\ k \in \mathbb{S}_i \tag{6h}$$

$$w^{\mathrm{c}}_{yh\nu} = \sum_{i \in \mathbb{I}_{\mathrm{c}}} \sum_{k \in \mathbb{S}_i} s_{\nu ik} \mathcal{I}\left(d_{yhi} = k\right) \quad \forall \nu \in \mathbb{V},\ y \in \mathbb{Y},\ h \in \mathbb{H} \tag{6i}$$

$$x_{yh\ell} \le w^{\mathrm{c}}_{yh\nu} + \sum_{i \in \mathbb{I}_{\mathrm{q}}} p_{\nu i} \quad \forall \nu \in \mathbb{V},\ y \in \mathbb{Y},\ h \in \mathbb{H},\ \ell \in \mathbb{L}^{\mathrm{r}}(\nu) \tag{6j}$$

$$x_{yh\ell} \le 1 - w^{\mathrm{c}}_{yh\nu} + \sum_{i \in \mathbb{I}_{\mathrm{q}}} p_{\nu i} \quad \forall \nu \in \mathbb{V},\ y \in \mathbb{Y},\ h \in \mathbb{H},\ \ell \in \mathbb{L}^{\mathrm{l}}(\nu) \tag{6k}$$

$$\sum_{\ell \in \mathbb{L}} x_{yh\ell} = 1 \quad \forall y \in \mathbb{Y},\ h \in \mathbb{H} \tag{6l}$$

in variables $q_\nu \in \mathbb{R}$, $g^+_{yh\nu}, g^-_{yh\nu} \in \mathbb{R}_+$, and $x_{yh\ell}, p_{\nu i}, w^{\mathrm{q}}_{yh\nu}, w^{\mathrm{c}}_{yh\nu}, s_{\nu ik} \in \{0, 1\}$ for all $y \in \mathbb{Y},\ h \in \mathbb{H},\ \ell \in \mathbb{L},\ \nu \in \mathbb{V},\ i \in \mathbb{I},\ k \in \mathbb{S}_i$.

An interpretation of the variables is as follows. The variables p indicate the feature that we branch on at each node. Thus, $p_{\nu i} = 1$ if and only if we branch on feature i at node ν. The variables $q_\nu, g^+_{yh\nu}, g^-_{yh\nu}$, and $w^{\mathrm{q}}_{yh\nu}$ are used to bound $x_{yh\ell}$ based on the branching decisions at each node ν, whenever branching is performed on a *quantitative* feature at that node. The variable q_ν corresponds to the cut-off value at node ν. The variables $g^+_{yh\nu}$ and $g^-_{yh\nu}$ represent the positive and negative parts of $q_\nu - \sum_{i \in \mathbb{I}_{\mathrm{q}}} p_{\nu i} d_{yhi}$, respectively. Whenever branching occurs on a quantitative feature, the variable $w^{\mathrm{q}}_{yh\nu}$ will equal 1 if and only if $q_\nu \ge \sum_{i \in \mathbb{I}_{\mathrm{q}}} p_{\nu i} d_{yhi}$, in which case the data point (y, h) must go left in the branch. The variables $w^{\mathrm{c}}_{yh\nu}$ and $s_{\nu ik}$ are used to bound $x_{yh\ell}$ based on the branching decisions at each node ν, whenever branching is performed on a *categorical* feature at that node. Whenever we branch on categorical feature $i \in \mathbb{I}_{\mathrm{c}}$ at node ν, the variable $s_{\nu ik}$ equals 1 if and only if the points such that $d_{yhi} = k$ must go left in the branch. If we do not branch on feature i, then the variable $s_{\nu ik}$ will equal zero. The variable $w^{\mathrm{c}}_{yh\nu}$ will equal 1 if and only if we branch on a categorical feature at node ν and data point (y, h) must go left at the node.

An interpretation of the constraints is as follows. Constraint (6a) ensures that only one variable is branched on at each node. Constraints (6b)–(6g) are used to bound $x_{yh\ell}$ based on the branching decisions at each node ν, whenever branching is performed on a *quantitative* feature at that node. Constraints (6b)–(6e) are used to define $w^{\mathrm{q}}_{yh\nu}$ to equal 1 if and only if $q_\nu \ge \sum_{i \in \mathbb{I}_{\mathrm{q}}} p_{\nu i} d_{yhi}$. Constraint (6f) stipulates that if we branch on a quantitative feature at node ν and data point (y, h) goes left at the node (i.e., $w^{\mathrm{q}}_{yh\nu} = 1$), then the data point cannot reach any leaf node that lies to the right of the node. Constraint (6g) is symmetric to (6f) for the case when the data point goes right at the node. Constraints (6h)–(6k) are used to bound $x_{yh\ell}$ based on the branching decisions at each node ν, whenever branching is performed on a *categorical* feature at that node. Constraint (6h) stipulates that if we do not branch on feature i at node ν, then $s_{\nu ik} = 0$. Constraint (6i) is used to define $w^c_{yh\nu}$ such that it is equal to 1 if and only if we branch on a particular feature i, the value attained for that feature by data point (y, h) is k and data points with feature value k are assigned to the left branch of the node. Constraints (6j) and (6k) mirror constraints (6f) and (6g), respectively, for the case of categorical features.

Example 7 (Decision-Tree-Based Policies enhanced with Linear Branching). For decision-tree-based policies enhanced with linear branching (Example 3), constraint (3b) can be expressed in terms of linear inequalities using a variant of the formulation from Example 6. Specifically, one can convert the dataset to have only quantitative features using one hot encoding and subsequently enforce constraints (5) and (6b)–(6g) to achieve the desired model.

Example 8 (Decision-Tree-Based Policies enhanced with Linear Leafing). For decision-tree-based policies enhanced with linear leafing (Example 4), constraint (3b) can be expressed in terms of linear inequalities using a variant of the formulation from Example 6 by replacing constraint (5) with

$$\pi_{yh} = \sum_{\ell \in \mathbb{L}} [\beta^\top_{y,\ell} g_y + \beta^\top_{h,\ell} f_h] x_{yh\ell} \quad \forall y \in \mathbb{Y}, \ h \in \mathbb{H}. \tag{7}$$

4   Approximate Solution Approach

Albeit exact, the data-driven MIP (3) scales with the number of youth and houses in the system. In this section, we propose an approximate solution approach that generalizes the one from [2] to decision-tree-based policies, see Examples 1–4. Consider the following relaxation of Problem (3).

$$\begin{aligned}
\text{maximize} \quad & \sum_{y \in \mathbb{Y}} \left[\sum_{h \in \mathbb{H}} p_{yh} x_{yh} + \bar{p}_y \left(1 - \sum_{h \in \mathbb{H}} x_{yh} \right) \right] \\
\text{subject to} \quad & \sum_{h \in \mathbb{H}} x_{yh} \leq 1 \quad \forall y \in \mathbb{Y}, \quad \sum_{y \in \mathbb{Y}} x_{yh} \leq 1 \quad \forall h \in \mathbb{H} \\
& x_{yh} = 0 \quad \forall y \in \mathbb{Y}, \ h \in \mathbb{H} \ : \ (y, h) \notin \mathbb{C} \\
& x \in \mathbb{F}, \ x_{yh} \geq 0 \quad \forall y \in \mathbb{Y}, \ h \in \mathbb{H}
\end{aligned} \tag{8}$$

Contrary to Problem (3) in which the matching is guided by the policy with parameters β, this formulation allows for arbitrary matches and integrality constraints on x are relaxed so that x_{yh} can be interpreted as the probability that house h is offered to youth y. Thus, the optimal policy from (9) is *anticipative* and not implementable in practice. Next, we propose a method that leverages formulation (8) to design an implementable policy. For convenience, we assume that the set of fair matchings is expressible as $\mathbb{F} := \{x : Ax \leq b\}$ for some matrix A and vector b. Moreover, we let λ denote the vector of optimal dual multipliers of the fairness constraints in (8). We define the quantity $C_{yh} := p_{yh} - \bar{p}_y - (\lambda^\top A)_{(y,h)}$ and propose to learn β to approximate C_{yh} using π_{yh}, the score for this match. This problem is expressible as

$$\text{minimize} \left\{ \sum_{y \in \mathbb{Y}} \sum_{h \in \mathbb{H}} |C_{yh} - \pi_{yh}| \; : \; \text{Constraint (3b)} \right\}. \qquad (9)$$

Problem (9) admits an intuitive interpretation. In the absence of fairness constraints, the policy should rank youth according to their probabilities of success. In the presence of fairness constraints, the policy should rank youth to maximize the probability of success while penalizing violations of the fairness constraints. An estimate of the cost at which violating the fairness constraints is not beneficial can be obtained by using the optimal dual multipliers λ. As discussed in Sect. 3.2, for all interpretable policies proposed in Sect. 2.3, constraint (3b) is equivalent to a finite set of linear inequality constraints involving binary variables. Thus, Problem (9) is equivalent to an MILP (an LP for the case of linear policies). Problem (9) is significantly more tractable than (3). While it can still be challenging to solve for large datasets, in the case of tree-based policies (Examples 2–4), it presents an attractive decomposable structure amenable to Bender's decomposition. Thus, x, z, and π are variables of the subproblem and all other variables are decided in the master. Note that integrality constraints in the subproblem may be relaxed to yield an equivalent problem.

5 Numerical Study

We showcase the performance of our approach to design policies for the U.S. homeless youth. We build interpretable policies that maximize efficiency while ensuring fairness across NST scores (see Introduction) and across races, in turn. We use real-world data (10,922 homeless youth and 3474 housing resources) from the HMIS database obtained from Ian De Jong as part of a working group called "Youth Homelessness Data, Policy, Research." The dataset includes 54 features for the youth and each house is of one of two types (rapid rehousing (RRH) or permanent supportive housing (PSH)), see [5]. A youth is considered to have a successful outcome if they are housed one year later. We use 80% (20%) of data for training (testing). We use the training set to learn (using CART) the success probabilities that are fed in our models and to identify the five most significant features. We compare our proposed approach to several baselines: (i) the status-quo policy TAY; (ii) random allocation (Random); (iii) the (interpretable) machine learning approaches without fairness from [5] (Linear and Logistic Regression and CART); (iv) the linear scoring policies with relaxed fairness constraints originally proposed in [2] (Linear RF). To these baselines, we add: (i) Decision-tree-based policies with relaxed fairness constraints (Decision-Tree RF); (ii) Decision-tree-based policies with linear leafing (depth 1 and 2)

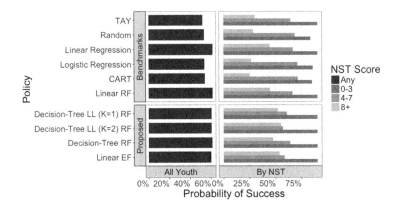

Fig. 1. Success probability across all youth (left) and by vulnerability level (right) when fairness across vulnerability levels is desired.

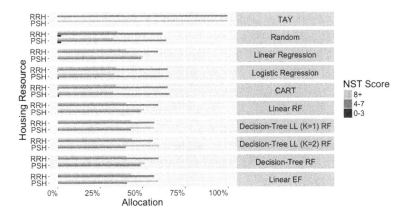

Fig. 2. Housing resources allocated by vulnerability level when fairness across vulnerability levels is desired.

with relaxed fairness constraints (Decision-Tree LL RF); *(iii)* Linear scoring policies with (exact) fairness constraints computed using MILP (3) (Linear EF).

Fairness Across NST Scores. Motivated by TAY which provides the most supportive resources to the most vulnerable youth, we enforce fairness with respect to NST score: independently of their score, youth should be equally likely to transition to a fair and stable exit. We enforce fairness across two groups which were found to have very different chances to remain homeless in the long run: youth with scores 4–7 and 8+, respectively. Youth with scores below 4 are excluded since they have a *higher* estimated success probability when not offered housing. Figure 1 shows the success probability of youth under each policy. The baselines TAY, Random, Logistic Regression, and CART are all very unfair: the probability of success for youth with scores 8+ is uniformly below 30%, while lower

Table 1. Solver times for the proposed approaches for solving to optimality when fairness across vulnerability levels is desired.

Fairness constraints	Type of policy	Decomposition used	Solver time (seconds)
Relaxed	Linear	N/A	932.57
Relaxed	Decision-Tree	Yes (No)	3570.12 (7105.11)
Relaxed	Decision-Tree LL	Yes (No)	9031.32 (14045.45)
Exact	Linear	N/A	36400.98

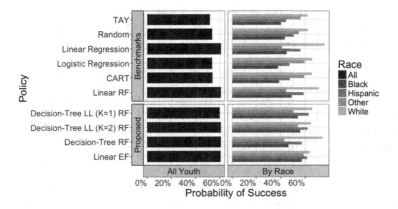

Fig. 3. Success probability across all youth (left) and by race (right) when fairness across races is desired.

risk youth with scores 4–7 have success probability higher than 60%. Linear Regression performs considerably better and introducing relaxed fairness constraints (Linear RF) does not yield any improvement. Our proposed policies outperform all benchmarks in terms of fairness and do so at marginal cost to overall efficiency. Figure 2 shows the percentage of each type of house allocated to each group under each policy. The current policy allocates the most resource-intensive resources (PSH) to the highly vulnerable individuals and the RRH resources to individuals scoring 4–7. Our analysis however shows that some high risk individuals can improve their chances of a successful outcome by receiving an RRH resource. Thus, our policies allocate some RRH (resp. PSH) houses to high (resp. low) risk individuals, resulting in policies that *benefit the most vulnerable* youth, see Figs. 1 and 2. Lastly, Table 1 shows the runtime required to solve each problem.[4] Exact formulations require more runtime than approximations, and more sophisticated policies require greater runtime. Moreover, there are significant benefits in employing our proposed decomposition approach.

---

[4] These experiments were run on a 2.0 GHz Intel Core i7 processor machine with 4 GB RAM and all optimization problems were solved with Gurobi v7.0.

Fairness Across Races. Motivated by the desire to avoid racial discrimination, we seek policies that are fair across races. The results are summarized in Fig. 3 which shows that the current policy and classical machine learning approaches are unfair, with "Whites" having higher success probability than "Blacks" and "Hispanics." In contrast, our proposed policies, in particular Linear EF outperform significantly the state of the art at marginal cost to overall efficiency.

References

1. Adamczyk, M., Sankowski, P., Zhang, Q.: Efficiency of truthful and symmetric mechanisms in one-sided matching. In: Lavi, R. (ed.) SAGT 2014. LNCS, vol. 8768, pp. 13–24. Springer, Heidelberg (2014). https://doi.org/10.1007/978-3-662-44803-8_2

2. Bertsimas, D., Farias, V.F., Trichakis, N.: Fairness, efficiency, and flexibility in organ allocation for kidney transplantation. Oper. Res. **61**(1), 73–87 (2013)

3. Bertsimas, D., King, A., Mazumder, R.: Best subset selection via a modern optimization lens. Ann. Stat. **44**(2), 813–852 (2016)

4. Buchbinder, N., Jain, K., Naor, J.S.: Online primal-dual algorithms for maximizing ad-auctions revenue. In: Arge, L., Hoffmann, M., Welzl, E. (eds.) ESA 2007. LNCS, vol. 4698, pp. 253–264. Springer, Heidelberg (2007). https://doi.org/10.1007/978-3-540-75520-3_24

5. Chan, H., Rice, E., Vayanos, P., Tambe, M., Morton, M.: Evidence from the past: AI decision aids to improve housing systems for homeless youth. In: AAAI 2017. Fall Symposium Series (2017)

6. Che, Z., Purushotham, S., Khemani, R., Liu, Y.: Interpretable deep models for ICU outcome prediction. In: AMIA Annual Symposium Proceedings, vol. 2016, p. 371. American Medical Informatics Association (2016)

7. Devanur, N.R., Jain, K., Sivan, B., Wilkens, C.A.: Near optimal online algorithms and fast approximation algorithms for resource allocation problems. In: Proceedings of the 12th ACM Conference on Electronic Commerce, pp. 29–38. ACM (2011)

8. Dughmi, S., Ghosh, A.: Truthful assignment without money. In: Proceedings of the 11th ACM Conference on Electronic Commerce, pp. 325–334. ACM (2010)

9. Housing and Urban Development (HUD): Coordinated Entry Policy Brief (2015). https://www.hudexchange.info/resources/documents/Coordinated-Entry-Policy-Brief.pdf

10. Huang, L.-T., Gromiha, M.M., Ho, S.-Y.: iPTREE-STAB: interpretable decision tree based method for predicting protein stability changes upon mutations. Bioinformatics **23**(10), 1292–1293 (2007)

11. Kong, N., Schaefer, A.J., Hunsaker, B., Roberts, M.S.: Maximizing the efficiency of the U.S. liver allocation system through region design. Manag. Sci. **56**(12), 2111–2122 (2010)

12. Lakkaraju, H., Bach, S.H., Jure, L.: Interpretable decision sets: a joint framework for description and prediction. In: Proceedings of the 22nd ACM SIGKDD International Conference on Knowledge Discovery and Data Mining, pp. 1675–1684. ACM (2016)

13. Leshno, J.D.: Dynamic matching in overloaded waiting lists (2017)

14. Letham, B., Rudin, C., McCormick, T.H., Madigan, D.: Interpretable classifiers using rules and bayesian analysis: building a better stroke prediction model. Ann. Appl. Stat. **9**(3), 1350–1371 (2015)

15. Lou, Y., Caruana, R., Gehrke, J.: Intelligible models for classification and regression. In: Proceedings of the 18th ACM SIGKDD International Conference on Knowledge Discovery and Data Mining, pp. 150–158. ACM (2012)
16. Lou, Y., Caruana, R., Gehrke, J., Hooker, G.: Accurate intelligible models with pairwise interactions. In: Proceedings of the 19th ACM SIGKDD International Conference on Knowledge Discovery and Data Mining, pp. 623–631. ACM (2013)
17. Mazumder, R., Radchenko, P.: The discrete dantzig selector: estimating sparse linear models via mixed integer linear optimization. IEEE Trans. Inf. Theory **63**(5), 3053–3075 (2017)
18. Mazumder, R., Radchenko, P., Dedieu, A.: Subset selection with shrinkage: sparse linear modeling when the SNR is low. arXiv preprint arXiv:1708.03288 (2017)
19. Mehta, A., Saberi, A., Vazirani, U., Vazirani, V.: Adwords and generalized online matching. J. ACM (JACM) **54**(5), 22 (2007)
20. The U.S. Department of Housing: OFFICE OF COMMUNITY PLANNING Urban Development, and DEVELOPMENT. The 2016 Annual Homeless Assessment Report (AHAR) to Congress (2016)
21. United States Interagency Council on Homelessness. Opening doors: federal strategic plan to prevent and end homelessness (2010)
22. Pearson, C.L.: The applicability of housing first models to homeless persons with serious mental illness: final report. Technical report, U.S. Department of Housing and Urban Development, Office of Policy Development and Research (2007)
23. Rice, E.: Assessment Tools for Prioritizing Housing Resources for Homeless Youth (2017). https://static1.squarespace.com/static/56fb3022d210b891156b3948/t/5887e0bc8419c20e9a7dfa81/1485299903906/Rice-Assessment-Tools-for-Youth-2017.pdf
24. Su, X., Zenios, S.A.: Patient choice in kidney allocation: a sequential stochastic assignment model. Oper. Res. **53**(3), 443–455 (2005)
25. Su, X., Zenios, S.A.: Recipient choice can address the efficiency-equity trade-off in kidney transplantation: a mechanism design model. Manag. Sci. **52**(11), 1647–1660 (2006)
26. Thakral, N.: Matching with stochastic arrival. In: EC, p. 343 (2015)
27. Toro, P.A., Lesperance, T.M., Braciszewski, J.M.: The heterogeneity of homeless youth in America: examining typologies. National Alliance to End Homelessness, Washington, D.C. (2011)
28. Valdes, G., Marcio Luna, J., Eaton, E., Simone, C.B.: MediBoost: a patient stratification tool for interpretable decision making in the era of precision medicine. Sci. Rep. **6** (2016)
29. Wang, T., Rudin, C., Doshi-Velez, F., Liu, Y., Klampfl, E., MacNeille, P.: A bayesian framework for learning rule sets for interpretable classification. J. Mach. Learn. Res. **18**(70), 1–37 (2017)

An Efficient Relaxed Projection Method for Constrained Non-negative Matrix Factorization with Application to the Phase-Mapping Problem in Materials Science

Junwen Bai[1(✉)], Sebastian Ament[1], Guillaume Perez[1], John Gregoire[2], and Carla Gomes[1]

[1] Department of Computer Science, Cornell University, Ithaca, NY 14853, USA
jb2467@cornell.edu
[2] Joint Center for Artificial Photosynthesis, California Institute of Technology, Pasadena, CA 91125, USA

Abstract. In recent years, a number of methods for solving the constrained non-negative matrix factorization problem have been proposed. In this paper, we propose an efficient method for tackling the ever increasing size of real-world problems. To this end, we propose a general relaxation and several algorithms for enforcing constraints in a challenging application: the phase-mapping problem in materials science. Using experimental data we show that the proposed method significantly outperforms previous methods in terms of ℓ_2-norm error and speed.

1 Introduction

Matrix factorization is a well-known method used for data compression and information extraction. Given a matrix A, matrix factorization is the problem of finding two matrices W and H such that $A \approx WH$. As W and H are assumed to be low-rank, the sum of their sizes is usually much smaller than the size of A. Further, the columns of W can be interpreted as basis components, which are linearly combined by columns of H to reconstruct A. The matrix factorization problem occurs in numerous fields, for example topic modeling [1], audio signal processing [2], and crystallography [3]. While successful algorithms for classical matrix factorization have been found, some variants of this problem are challenging. For example, merely restricting W and H to be element-wise non-negative is known to lead to an NP-Hard problem [4], called non-negative matrix factorization (NMF) [1,4]. These variants are important for many practical problems.

Moreover, many real-world problems which can be modeled by NMF also involve domain-specific constraints. A good example is the phase-mapping problem in the field of materials discovery [5]. This problem arises when materials scientists generate potentially novel materials by applying a physical transformation to mixtures of known materials. Individual materials are commonly characterized by a variety of spectrographic techniques, like x-ray diffraction [6] and

© Springer International Publishing AG, part of Springer Nature 2018
W.-J. van Hoeve (Ed.): CPAIOR 2018, LNCS 10848, pp. 52–62, 2018.
https://doi.org/10.1007/978-3-319-93031-2_4

Raman spectroscopy [7]. Each spectrogram produced by these experiments typically corresponds to a mixture of potentially novel materials or phases, whose individual spectrograms are unkown. The phase-mapping problem is then to uncover these unknown phases, from which materials scientists can understand the phase behavior of the associated materials and its relationship to other measured properties. At a high level, the problem can be framed as an NMF problem, where the columns of A are the measured spectrograms Importantly, the matrices W and H of the resulting factorization problem have to respect hard physical constraints for the solution to be meaningful to scientists. This makes the phase-mapping problem particularly challenging from a computational perspective.

In order to incorporate these hard constraints, recent work used mixed-integer programming (MIP) to project the matrices W and H onto the constraint space, after an unconstrained optimization [8] of the NMF problem [9]. Subsequently, W and H were re-optimized without leaving the constraint space. While the results of this work are promising, the method still requires solving time-consuming, hard combinatorial problems for enforcing the constraints. Other previous approaches to this problem were purely based on combinatorial search algorithms applied to the constraint space [5,10,11]. However, these methods are often intractable and deteriorate with the presence of noise in the data. The goal of this paper is to obviate the need for combinatorial optimization techniques for the solution of this problem.

In this paper, we propose a heuristic approach that uses a polynomial-time algorithm **Projected Interleaved Agile Factor Decomposition (PIAFD)** interleaving the enforcement of essential hard-constraints, combined with a relaxation which allows the algorithm to move the optimization variables into the constraint space continuously while optimizing the objective function for the phase-mapping problem. This projection-based algorithm is designed to enforce the three constraints which are most essential in the phase-mapping problem: the Gibbs, Alloying and connectivity constraints. This new algorithm works well in practice, and allows to extract accurate, constraint-satisfying solutions in a very short time. We study the impact of the application of these algorithms on the solution quality, as measured by different objective functions.

This paper is organized as follows. The phase-mapping problem is first described and modeled as an associated matrix factorization problem. After an overview of the state-of-the-art methods for this problem, we present our new method *PIAFD* for enforcing the relevant constraints. Finally, *PIAFD* is applied to several real-world data sets of the phase-mapping problem to demonstrate its performance.

2 Preliminaries

The phase-mapping problem has recently drawn attention because of its great importance to the discovery of new materials [12,13]. In their search for new materials, scientists try to characterize novel materials with spectrographic techniques, like x-ray diffraction (XRD) spectroscopy. For each sample point on an

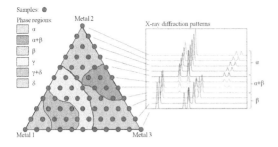

Fig. 1. (Left) the schematic on the left is known as a phase diagram. The sides of the triangle correspond to proportions of the metals which are deposited on the wafer. The corners correspond to regions where only one metal has been deposited, while the center of the triangle corresponds to a location where all metals are deposited in equal proportions. The legend indicates that four phases are present: $\alpha, \beta, \gamma, \delta$. Each colors represents a region in which a unique phase or a unique combination of phases is present. The graph on the right shows how the spectrograms can vary as the proportions of the initial metals are changed. (Right) multiplicative gradient update rules. (Color figure online)

experimental wafer, we denote by $F(q)$ the vector of diffraction intensities as a function of q, the scattering vector. That is, $F(q)$ corresponds to the spectrogram observed at a location on the wafer, and we will refer to it as an XRD pattern. Importantly, each location on the wafer is likely to contain mixtures of new materials. Therefore, $F(q)$ will be a combination of the spectrograms of several unkown materials (i.e. phases), which are generally not observed directly. This makes the phase-mapping problem non-trivial.

This problem is naturally formulated using a matrix A, each of whose columns consists of a XRD pattern $F(q)$ of length Q. If there are N sample points on the wafer, A is of size $Q \times N$. The algorithms of this paper are based on factorizing A using two matrices W and H such that $A \approx WH$. W stores the different phases, while H contains the quantity of each basis pattern at each sample point.

While the non-negativity of the spectrograms puts one constraint on W and H, several other constraints are also present. We briefly describe these additional constraints, and current methods of enforcing them. The *IAFD* method [9] alternates multiplicative gradient updates (Fig. 1) and constraint refinement. The generalized Kullback-Leibler (KL) divergence between A and WH is taken to be the objective function. The update rules above are proven to be non-decreasing, but do not necessarily converge to a stationary point of the objective function. Notably, [14] introduced modifications to these updates which do provably converge to a stationary point.

Shifting. A common phenomenon that occurs in the XRD spectroscopy of certain materials is *shifting*. A basis pattern $F_b(q)$ is shifting with a multiplicative factor λ if the pattern $F_b(\lambda q)$ is present at a sample point, instead of $F_b(q)$. Crucially, a shifted pattern should still be recognized as the original pattern. In

Fig. 1, the signals on the bottom right show the shifting of phase β. In order to model this behavior, one can resample the signal $F(\lambda q)$ on a logarithmic scale, so that the multiplicative shifting becomes additive. Then, a convolutive NMF framework can be readily applied to this problem [9]. This framework allows for multiple shifted copies of the phases. In particular, the columns of W are allowed to shift down with a certain shifting amount m, $W^{\downarrow m}$. Then the factorization problem can be defined as $A \approx \sum_m W^{\downarrow m} H^m$ where H^m is of size $K \times N$ consists of the activation coefficients of corresponding shifted phases. The exact shifting of phase k at sample j is defined by a weighted average of different shifting amounts $\lambda_{kj} = \sum_m m H_{kj}^m / \sum_m H_{kj}^m$.

Gibbs. The Gibbs phase rule puts a limit on the number of phases which can be observed at a sample point in a thermodynamic equilibrium. In particular, the maximum number of different phases at a single location is equal to the number of elements, G, which were deposited on the wafer initially. For example, on the wafer where three different elements were deposited, the maximum number of phases at each sample point is three. This implies that the number of non-zero elements in each column of H should not exceed G. The *IAFD* algorithm projects solutions onto the constraint space by solving a MIP for each column of H containing a bounded maximum number of non-zero entries and minimizing the ℓ_1 distance between the reconstructed sample point and real sample point. However, this method introduced a time-consuming combinatorial problem to solve, and the ℓ_1 distance used in MIP is different from the KL divergence used in the multiplicative gradient updates when factorizing A.

Alloying. The alloying rule states that if shifting is detected at a sample point, the Gibbs phase rule loses a degree of freedom. The number of possible phases for this sample point is then bounded by $G-1$. Furthermore, the shifting parameters λ change continuously across the wafer in the presence of alloying. Similar to the Gibbs constraint, the *IAFD* algorithm repairs solutions by solving a set of MIPs. These MIPs embed the alloying constraint and minimize the absolute distance between the current solution and the measurement data A. Once again, a hard combinatorial problem has to be solved and the objective function is not the same as the one used for optimizing W and H.

Connectivity. The last constraint of the phase-mapping problem is the connectivity constraint. This constraint states that the sample locations where a given phase is present are members of a connected region. For this constraint, the *IAFD* algorithm first defines a graph, using the sample point locations, and a Delaunay triangulation [15] of these points. The triangulation gives a graph in which neighborhood relationships are defined. For a sample j, its neighbors constitute a set ω_j. Then a search is performed to find the connected components for each phase. Only the largest component of every phase is kept, and its complement is zeroed out. This procedure is only applied at the end of the algorithm.

3   Constraint Projection for the Phase-Mapping Problem

Multiplicative updates. PIAFD is based on the multiplicative update rules for convolutive NMF [16]. These update rules are non-decreasing, though they might not converge to a stationary point of the objective function. Future work will be based on the modified update rules introduced in [14] to eliminate this possibility. In addition, a ℓ_1-penalty term on elements of H is included, to suggest a sparse solution. Note that the scaling indeterminacy between W and H, as $(\alpha W)(\frac{1}{\alpha}H)$ leads to the same reconstruction error for all non-zero α. If no further adjustments were made, this property would lead the algorithm to make the elements of H arbitrarily small to minimize the ℓ_1-penalty. In order to avoid this behavior, a normalized version of W is used to derive the multiplicative update rules. See [17] for details. PIAFD method alternates between multiplicative update and projection onto the constraint space. The rest of this section shows how to efficiently project onto the phase-mapping constraints.

Constraint Projection. The projection of a vector y onto a constraint space C is often defined by the following equation:

$$P_C(y) = \arg\min_{x \in C} \|x - y\|_2 \tag{1}$$

Finding the projection of a given point onto a constraint space is often a hard task, but there are some constraints for which efficient algorithms can be defined [18–20]. In this section, we propose to define the projection of each constraint.

Gibbs. First, the projection $P_{Gibbs}(y)$ of the current solution y is the closest point to y which satisfies the Gibbs constraint. This constraint states that no more than G entries in each column of H can have a non-zero value. Each column of H therefore can be projected independently. The closest point satisfying this constraint is the closest point having less or equal to G non-zero values. This point is composed of the G largest elements of the column. Note that the worst-case complexity of finding the G largest component of a vector of size n can be bounded by $O(n + G\log(n))$. Let S_G^j be the vector containing the indexes of the rows, of column j, larger or equal to the G-th largest element of the column. Let $\alpha \in [0, 1]$ be a real value; let the matrix vp be defined by:

$$vp_{ij} = \begin{cases} 1 & \text{if } i \in S_G^j, \\ 1 - \alpha & \text{otherwise} \end{cases} \tag{2}$$

Property 1. *If $\alpha = 1$, then the result of an element-wise multiplication of H and vp respects the Gibbs constraint.*

Property 2. *If $\alpha \in [0, 1]$, then the result of an element-wise multiplication of H and vp is closer to the solution space of the Gibbs constraint than H.*

The parameter α is a relaxation parameter of the constraint. We can set and modify it during the search for a solution. One advantage of having such a parameter is to not drastically modify the current solution, while performing the gradient-based optimization. As shown in the experimental solution, this gives us a great flexibility in practice.

Alloying. The alloying constraint is a conditional constraint. That is, it has to be enforced only when alloying occurs in the data. Alloying occurs if at least one of the phases at a given sample point is shifting. The following equation determines this:

$$Y^j = \sum_{k \in [1,K], n \in \omega_j} H_{kj} \times \max(0, |\lambda_{kj} - \lambda_{kn'}| - \epsilon) \tag{3}$$

If the alloying variable Y^j is bigger than 0, then the jth column loses a degree of freedom regarding the Gibbs constraint. That is, instead of G entries, only $G - 1$ are allowed to be non-zero. This behaviour can be incorporated into the vp matrix, which was previously used for enforcing the Gibbs constraint, by modifying it as follow:

$$vp_{ij} = \begin{cases} 1 & \text{if } Y^j = 0 \wedge i \in S_G^j, \\ 1 & \text{if } Y^j > 0 \wedge i \in S_{G-1}^j, \\ 1 - \alpha & \text{otherwise} \end{cases} \tag{4}$$

The matrix vp can be used to enforce or relax both Gibbs and alloying constraints.

Connectivity. The existing algorithm for enforcing the connectivity constraint set entries of H that do not belong to the largest connected component to zero. As for the two previous constraints, we can relax the connectivity constraint by multiplying these values by $1 - \alpha$ instead of 0. When $\alpha = 1$, the exact constraint is enforced. But when $\alpha \in (0, 1)$, the new point is only closer to, but not equal to the exact projection. Thus, the constraint is not enforced, but the solution is moved closer to the constraint space.

Let C_k be the set of column indices indicating the largest connected component of basis k. As for the two previous constraints, we can use a similar vp matrix:

$$vp_{ij} = \begin{cases} 1 & \text{if } i \in C_i \wedge Y^j = 0 \wedge i \in S_G^j, \\ 1 & \text{if } i \in C_i \wedge Y^j > 0 \wedge i \in S_{G-1}^j, \\ 1 - \alpha & \text{otherwise} \end{cases} \tag{5}$$

The matrix vp is finally used to enforce or relax the Gibbs, alloying, and connectivity constraints. This leads to a simple two-step update for solving the constrained NMF problem. Figure 2 shows the projection method and a high-level schematic of our method. _PIAFD_ starts with unconstrained gradient updates and then interleaves enforcing relaxed constraints and multiplicative updates of the matrices. In each iteration, W and H are first alternatively updated till convergence or for a certain amount of times, whichever happens, and then the relaxed constraints are enforced subsequently. Hard constraints are enforced at the end of the algorithm (Fig. 2).

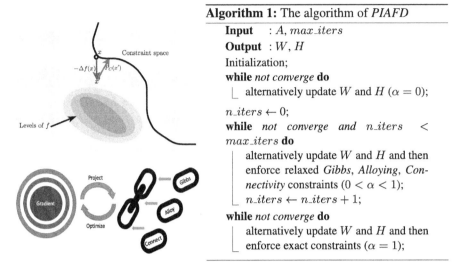

Algorithm 1: The algorithm of *PIAFD*

Input : A, max_iters
Output : W, H
Initialization;
while *not converge* **do**
    alternatively update W and H ($\alpha = 0$);

$n\_iters \leftarrow 0$;
while *not converge and* $n\_iters <$ max_iters **do**
    alternatively update W and H and then enforce relaxed *Gibbs, Alloying, Connectivity* constraints ($0 < \alpha < 1$);
    $n\_iters \leftarrow n\_iters + 1$;
while *not converge* **do**
    alternatively update W and H and then enforce exact constraints ($\alpha = 1$);

Fig. 2. (Upper left) gradient step (green arrow) followed by a projection (red arrow). Solutions updated by the gradient step might leave the constraint space. They are dragged back to the constraint space through a projection. (Lower left) workflow of the *PIAFD* method. It interleaves gradient updates and interleaves projections. (Right) the pseudocode of *PIAFD*. (Color figure online)

4   Experiments

In this section, we compare our method, *PIAFD*, against other methods for solving the phase-mapping problem. Namely, we compare the proposed method against *IAFD* [9], *CombiFD* [5], and *AgileFD* [16]. *CombiFD* is one of the early, purely combinatorial methods for solving the phase-mapping problem. It uses MIP to encode all constraints, and updates solutions in an iterative fashion. However, it is not very efficient, especially compared to more recent methods. *AgileFD* is based on a matrix factorization framework to acquire solutions more efficiently, but it does not encode all the constraints. *IAFD* refines *AgileFD* by alternating multiplicative updates of the matrices and constraint projections using a MIP.

All experiments were run on a server containing 24 nodes, where each node contained an Intel x5650 2.67 GHZ, 48 GB of memory, and was running CentOS 7. The data sets are from real-world experiments from materials science.

Runtime. All methods were tested on 8 different datasets [21] to compare their runtime. In these instances, the inner dimension K of W and H is 6, and $G = 3$. At the beginning of the optimization procedure, α is set to 0 and the optimization is then run to convergence. Subsequently, α is increased to 0.15, a value heuristically found to provide the best reconstruction error, and finally to 1 after a preset number (1000) of iterations is reached. Once α is equal to 1, the algorithm is run to convergence yielding a solution to the constrained optimization

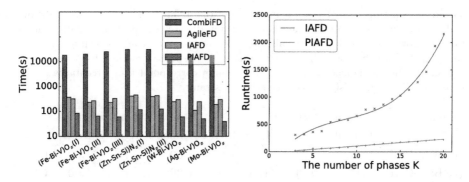

Fig. 3. Runtime comparison of the different methods. The bar chart on the left shows the runtime of *CombiFD*, *AgileFD*, *IAFD*, and *PIAFD* in seconds. The plot on the right compares the runtime of *IAFD*, and *PIAFD* as a function of K, the inner dimension of the matrices W and H.

problem. Figure 3 (left) shows that our new method improves on the runtime of previous methods by at least an order of magnitude.

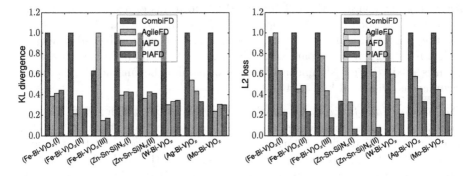

Fig. 4. Accuracy comparison of the different methods (*CombiFD*, *AgileFD*, *IAFD*, and *PIAFD*). The left bar chart shows the minimal error attained for each method using the KL-divergence as the objective function. The right bar chart is similar, just that the ℓ_2-norm was used as the objective function.

Figure 3 (right) shows the runtime of *IAFD* and *PIAFD* as a function of K. Since *IAFD* depends on MIP, it has a worst-case time complexity of $O(\binom{K}{G})$ for each column of H. In other words, if $G = 3$, the worst-case runtime behavior scales proportionally to K^3. Therefore, *IAFD* does not scale well with K. In contrast, *PIAFD* scales linearly in K, as demonstrated by Fig. 3. This improvement in the asymptotic scaling of the constraint projection is crucial for advancing materials science, as the cutting edge of the field deals with datasets of ever increasing size and complexity.

Table 1. Constraint satisfaction comparison of the different methods for the alloying and connectivity constraints. The system column denotes the particular dataset used for each row. The entries in the alloying constraint column are percentages of the columns of H which satisfy the constraint after the respective method has terminated. The entries in the connectivity constraint column correspond the percentages of the phases in W which satisfy the connectivity constraint.

System	Alloying constraint				Connectivity constraint			
	CombiFD	AgileFD	IAFD	PIAFD	CombiFD	AgileFD	IAFD	PIAFD
$(Fe-Bi-V)O_x(I)$	0.44	0.90	**1.00**	**1.00**	**1.00**	0.17	**1.00**	**1.00**
$(Fe-Bi-V)O_x(II)$	0.47	0.76	**1.00**	**1.00**	**1.00**	0.17	**1.00**	**1.00**
$(Fe-Bi-V)O_x(III)$	0.87	0.98	**1.00**	**1.00**	0.83	0.00	**1.00**	**1.00**
$(Zn-Sn-Si)N_x(I)$	0.98	**1.00**	**1.00**	**1.00**	**1.00**	0.17	**1.00**	**1.00**
$(Zn-Sn-Si)N_x(II)$	0.95	0.98	**1.00**	**1.00**	**1.00**	0.00	**1.00**	**1.00**
$(W-Bi-V)O_x$	0.51	0.95	**1.00**	**1.00**	**1.00**	0.00	**1.00**	**1.00**
$(Ag-Bi-V)O_x$	0.19	0.90	**1.00**	**1.00**	0.67	0.00	**1.00**	**1.00**
$(Mo-Bi-V)O_x$	0.55	0.94	**1.00**	**1.00**	0.83	0.00	**1.00**	**1.00**

Accuracy. To compare the different methods in terms of accuracy, we benchmarked all methods using two popular objective functions: the KL-divergence and the ℓ_2-norm. Figure 4 (left) shows that *PIAFD* consistently finds solutions with a KL-divergence comparable to the next best method (*IAFD*). If the ℓ_2-norm is used, *PIAFD* significantly outperforms all other methods in terms of accuracy, as evidenced by Fig. 4 (right).

Constraint Satisfaction. Our goal is to have solutions which satisfy the constraints of the phase-mapping problem. Table 1 demonstrates the ability of the different methods to yield solutions in the constraint space. All the methods respect the Gibbs constraint, so it is not shown in the table. Notably, both *IAFD* and *PIAFD* satisfy all constraints. *CombiFD* encodes the constraints using MIP. As the complexity of the constraints increases, it takes more time to find a satisfactory solution. Within the maximum wall time of the server (4 h) *CombiFD* fails to find a solution satisfying all the constraints but still respects Gibbs and connectivity constraints. The solution generated by *AgileFD* neither satisfy the alloying nor the connectivity constraint, due to the model's limited expressiveness.

5   Conclusion

This paper proposed *PIAFD*, a new method for solving the phase-mapping problem. using a novel algorithm for projecting solutions onto the constraint space. Crucially, the method tends to continuously move the optimization variables towards the constraint space, while minimizing the objective function with a gradient-based optimization procedure. The experimental section shows that this new method is orders of magnitude faster than existing methods and, depending on the choice of objective function, gives comparable or more accurate solutions. Because of this improvement in runtime, problems of previously intractable size become feasible. Consequently, the method has the potential of contributing to accelerating discoveries in materials science.

Acknowledgments. Work supported by an NSF Expedition award for Computational Sustainability (CCF-1522054), NSF Computing Research Infrastructure (CNS-1059284), NSF Inspire (1344201), a MURI/AFOSR grant (FA9550), and a grant from the Toyota Research Institute.

References

1. Lee, D.D., Seung, H.S.: Learning the parts of objects by non-negative matrix factorization. Nature **401**(6755), 788 (1999)
2. Smaragdis, P.: Non-negative matrix factor deconvolution; extraction of multiple sound sources from monophonic inputs. In: Puntonet, C.G., Prieto, A. (eds.) ICA 2004. LNCS, vol. 3195, pp. 494–499. Springer, Heidelberg (2004). https://doi.org/10.1007/978-3-540-30110-3_63
3. Suram, S.K., Newhouse, P.F., Zhou, L., Van Campen, D.G., Mehta, A., Gregoire, J.M.: High throughput light absorber discovery, part 2: establishing structure-band gap energy relationships. ACS Comb. Sci. **18**(11), 682–688 (2016)
4. Vavasis, S.A.: On the complexity of nonnegative matrix factorization. SIAM J. Optim. **20**(3), 1364–1377 (2009)
5. LeBras, R., Damoulas, T., Gregoire, J.M., Sabharwal, A., Gomes, C.P., van Dover, R.B.: Constraint reasoning and kernel clustering for pattern decomposition with scaling. In: Lee, J. (ed.) CP 2011. LNCS, vol. 6876, pp. 508–522. Springer, Heidelberg (2011). https://doi.org/10.1007/978-3-642-23786-7_39
6. Gregoire, J.M., Dale, D., Kazimirov, A., DiSalvo, F.J., van Dover, R.B.: High energy x-ray diffraction/x-ray fluorescence spectroscopy for high-throughput analysis of composition spread thin films. Rev. Sci. Instrum. **80**(12), 123905 (2009)
7. Colthup, N.: Introduction to Infrared and Raman Spectroscopy. Elsevier, Amsterdam (2012)
8. Mørup, M., Schmidt, M.N.: Sparse non-negative matrix factor 2-D deconvolution. Technical report (2006)
9. Bai, J., Bjorck, J., Xue, Y., Suram, S.K., Gregoire, J., Gomes, C.: Relaxation methods for constrained matrix factorization problems: solving the phase mapping problem in materials discovery. In: Salvagnin, D., Lombardi, M. (eds.) CPAIOR 2017. LNCS, vol. 10335, pp. 104–112. Springer, Cham (2017). https://doi.org/10.1007/978-3-319-59776-8_9
10. Ermon, S., Le Bras, R., Suram, S.K., Gregoire, J.M., Gomes, C.P., Selman, B., van Dover, R.B.: Pattern decomposition with complex combinatorial constraints: application to materials discovery. In: AAAI, pp. 636–643 (2015)
11. Ermon, S., Le Bras, R., Gomes, C.P., Selman, B., van Dover, R.B.: SMT-aided combinatorial materials discovery. In: Cimatti, A., Sebastiani, R. (eds.) SAT 2012. LNCS, vol. 7317, pp. 172–185. Springer, Heidelberg (2012). https://doi.org/10.1007/978-3-642-31612-8_14
12. Gregoire, J.M., Van Campen, D.G., Miller, C.E., Jones, R.J.R., Suram, S.K., Mehta, A.: High-throughput synchrotron X-ray diffraction for combinatorial phase mapping. J. Synchrotron Radiat. **21**, 1262–1268 (2014)

13. Hattrick-Simpers, J.R., Gregoire, J.M., Kusne, A.G.: Perspective: composition-structure-property mapping in high-throughput experiments: turning data into knowledge. APL Mater. **4**, 053211 (2016)
14. Lin, C.-J.: On the convergence of multiplicative update algorithms for nonnegative matrix factorization. IEEE Trans. Neural Netw. **18**(6), 1589–1596 (2007)
15. Lee, D.-T., Schachter, B.J.: Two algorithms for constructing a Delaunay triangulation. Int. J. Comput. Inf. Sci. **9**(3), 219–242 (1980)
16. Xue, Y., Bai, J., Le Bras, R., Rappazzo, B., Bernstein, R., Bjorck, J., Longpre, L., Suram, S.K., van Dover, R.B., Gregoire, J.M., et al.: Phase-Mapper: an AI platform to accelerate high throughput materials discovery. In: AAAI, pp. 4635–4643 (2017)
17. Le Roux, J., Weninger, F.J., Hershey, J.R.: Sparse NMF-half-baked or well done? Mitsubishi Electric Research Labs (MERL), Cambridge, MA, USA, Technical report no. TR2015-023 (2015)
18. Duchi, J., Shalev-Shwartz, S., Singer, Y., Chandra, T.: Efficient projections onto the ℓ_1-ball for learning in high dimensions. In Proceedings of the 25th International Conference on Machine Learning, pp. 272–279. ACM (2008)
19. Condat, L.: Fast projection onto the simplex and the l_1 ball. Math. Program. **158**(1–2), 575–585 (2016)
20. Perez, G., Barlaud, M., Fillatre, L., Régin, J.-C.: A filtered bucket-clustering method for projection onto the simplex and the ℓ_1 ball. In: Colloque GRETSI, Juan-les-Pins, France (2017)
21. Le Bras, R., Bernstein, R., Suram, S.K., Gregoire, J.M., Selman, B., Gomes, C.P., van Dover, R.B.: A computational challenge problem in materials discovery: synthetic problem generator and real-world datasets (2014)

Dealing with Demand Uncertainty in Service Network and Load Plan Design

Ahmad Baubaid[1,2(✉)], Natashia Boland[1], and Martin Savelsbergh[1]

[1] Georgia Institute of Technology, Atlanta, USA
baubaid@gatech.edu, {natashia.boland,martin.savelsbergh}@isye.gatech.edu
[2] King Fahd University of Petroleum and Minerals, Dhahran, Saudi Arabia

Abstract. Less-than-Truckload (LTL) transportation carriers plan for their next operating season by deciding: (1) a load plan, which specifies how shipments are routed through the terminal network from origins to destinations, and (2) how many trailers to operate between each pair of terminals in the network. Most carriers also require that the load plan is such that shipments at an intermediate terminal and having the same ultimate destination are loaded onto trailers headed to a unique next terminal regardless of their origins. In practice, daily variations in demand are handled by relaxing this requirement and possibly loading shipments to an alternative next terminal. We introduce the p-alt model, which integrates routing and capacity decisions, and which allows p choices for the next terminal for shipments with a particular ultimate destination. We further introduce and computationally test three solution methods for the stochastic p-alt model, which shows that much can be gained from using the p-alt model and explicitly considering demand uncertainty.

1 Introduction and Motivation

To make their operations economically viable, Less-than-truckload (LTL) carriers consolidate freight from multiple shippers. This freight is routed through a network of terminals before reaching its ultimate destination. At each terminal, freight is sorted and consolidated with other freight before it is loaded onto an outbound trailer to its next destination. An LTL carrier's network typically consists of two types of terminals: *end-of-line* (EOL) terminals which serve only as origin and destination points for freight, and *breakbulk* (BB) terminals which also act as consolidation hubs in the network.

In managing their networks, LTL carriers are faced with numerous tactical decisions. Based on predicted origin-destination (OD) demand for the upcoming operating period, carriers plan for how that demand is to be served. They decide how many trailers to operate between each pair of terminals, and formulate a *load plan*, which determines how freight with a given origin and destination will be routed. This planning process is known in the literature as Service Network Design (for reviews of this topic see [1,2]). To streamline sorting and loading operations at the terminals, many carriers impose a *directed in-tree* structure in

© Springer International Publishing AG, part of Springer Nature 2018
W.-J. van Hoeve (Ed.): CPAIOR 2018, LNCS 10848, pp. 63–71, 2018.
https://doi.org/10.1007/978-3-319-93031-2_5

their load plans for each destination d in the network as in [3–6]. That is, all freight headed to destination d that is at an intermediate terminal – regardless of its origin – is always directed to a *unique* next terminal (as in Fig. 1a).

In practice, carriers often permit an additional option for routing freight at (some) terminals as part of the load plan to handle demand variations. This load plan specifies a *primary* next destination, as well as an *alternative* (or an "alt") next destination, through which some freight might be rerouted if there is not enough planned capacity available on the primary. If there is also not enough capacity available on the alt, then additional capacity, for either the primary or the alt, can be acquired at a higher cost (carriers negotiate deals with independent owner-operators for such eventualities). In practice, these alts are chosen heuristically after the primaries have been determined. Note that throughout the remainder of this paper, for the sake of brevity, we will use the term "alt" to refer to *any* next terminal option – primary or alternative – and no distinction between the two will be made. Figure 1b shows an example of a structure with at most two alts allowed at every terminal for destination d.

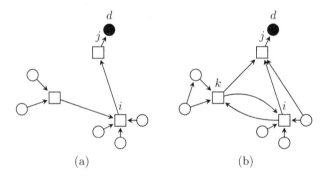

(a)   (b)

Fig. 1. Illustration of 1-alt (in-tree) and 2-alt structures for destination d in a load plan (circles and squares represent EOL and BB terminals, respectively); (a) shows the traditional in-tree (1-alt) structure in which at terminal i, all freight headed for d is sent to j; and (b) shows a 2-alt structure in which at terminal i, all freight headed for d can be sent to *either* j or k.

Some natural questions to explore are: (1) "What are the benefits of choosing alts simultaneously as opposed to sequentially, as is done in practice?" and (2) "What are the benefits of employing a 2-alt design rather than a 1-alt design in an uncertain environment?". In this paper, we address these questions by studying a two-stage stochastic Capacitated Multi-commodity Network Design (CMND) problem [7]. The model we present is different from other stochastic CMND problems studied in the literature (e.g. [8–10]) in two key ways: (1) the recourse, in response to changes in demand, is operating additional trailers acquired at a higher cost, and (2) the load plan can have a pre-specified number of alts (which generalizes the notion of a directed in-tree load plan). The use of more than one alt and its effectiveness in dealing with demand variations has never been

studied; a small number of papers study 1-alt designs in a deterministic setting
(e.g. [3–6]). We are the first to investigate the use of p-alt designs in a rigorous
way, even though such designs are commonly used in practice. We will refer to
the deterministic version of our model as the p-alt model, where the parameter
p represents the number of alts allowed/desired in the load plan.

To solve the stochastic p-alt model, we use Sample Average Approximation
(SAA) [11] as our general framework. However, because the resulting extensive-
form problem in each SAA iteration is difficult to solve in reasonable time, we
consider and compare three approaches: (1) directly solving the extensive-form
with the so-called cut inequalities added a priori; (2) relaxing the integrality of
the second-stage variables, and applying cut inequalities added a priori, and (3)
relaxing the integrality of the second-stage variables and applying slope scaling
[12] on the relaxed second-stage variables.

The rest of the paper is structured as follows. Section 2 presents the stochastic p-alt model formulation. Section 3 outlines the solution approaches we will
compare. This is followed by a computational study in Sect. 4.

2   Problem Description

Let $\mathcal{G} = (\mathcal{N}, \mathcal{A})$ be a digraph representing the transportation network of an
LTL carrier, where $\mathcal{N} = \mathcal{B} \cup \mathcal{E}$ for $\mathcal{B}$ the set of Breakbulk (BB) terminals and $\mathcal{E}$
the set of End-of-Line (EOL) terminals. Let $\delta^+(i)$ denote the set of outbound
arcs and $\delta^-(i)$ denote the set of inbound arcs at node $i \in \mathcal{N}$. Define the set of
commodities $\mathcal{K} \subseteq \{(o, d) : o, d \in \mathcal{E}, o \neq d\}$ to be the set of all EOL OD pairs
for which there may be shipments. Let $o_k, d_k \in \mathcal{N}$ denote commodity k's origin
and destination, respectively. Furthermore, define $\mathcal{D}$ as the set of all destination
EOLs, i.e. $\mathcal{D} = \{d_k : k \in \mathcal{K}\}$, and $\mathcal{K}(d)$ to be the set of commodities in $\mathcal{K}$ with
destination $d \in \mathcal{E}$. Let p be the number of alts allowed for each terminal in the
network and for each possible destination. We define c_{ij} as the cost of operating
a trailer and $\hat{c}_{ij} (>c_{ij})$ as the cost of outsourcing a trailer to operate on the arc
(i, j), respectively, and Q to be the uniform trailer capacity. Let Ω represent the
set of random demands that may be observed, with $\omega \in \Omega$ representing a particular realization of OD demands. Ω may be either discrete or continuous. Define
q_k^ω to be the demand of commodity k in realization ω. Our decision variables are
as follows:

r_{ij} = number of *planned* trailers to operate on arc (i, j),

z_{ij}^ω = number of *outsourced* trailers to operate on arc (i, j) in realization ω,

x_{ijk}^ω = flow on arc (i, j) for commodity $k \in \mathcal{K}$ in realization ω,

$$y_{ijd} = \begin{cases} 1, & \text{if commodities with destination } d \text{ may use arc } (i, j) \text{ as an alt,} \\ 0, & \text{otherwise.} \end{cases}$$

2.1   The Mathematical Formulation

The stochastic p-alt model is

$$\min \sum_{(i,j)\in\mathcal{A}} c_{ij}r_{ij} + \mathbb{E}_\omega\left[Q(r,y,\omega)\right], \tag{1a}$$

$$\text{s.t.} \sum_{(i,j)\in\delta^+(i)} y_{ijd} = \min\{p, |\delta^+(i)|\}, \qquad \forall d \in \mathcal{D},\ i \in \mathcal{N}, \tag{1b}$$

$$y_{ijd} \in \{0,1\}, \qquad \forall d \in \mathcal{D},\ (i,j) \in \mathcal{A}, \tag{1c}$$

$$r_{ij} \in \mathbb{Z}_+, \qquad \forall (i,j) \in \mathcal{A}, \tag{1d}$$

where

$$Q(r,y,\omega) = \min\{\sum_{(i,j)\in\mathcal{A}} \hat{c}_{ij}z_{ij}^\omega\ :\ (z^\omega, x^\omega) \in \mathcal{P}(r,y,\omega)\} \tag{1e}$$

and

$$\mathcal{P}(r,y,\omega) = \{(z^\omega, x^\omega)\ :\ \sum_{k\in\mathcal{K}} x_{ijk}^\omega \le Q(r_{ij} + z_{ij}^\omega), \qquad \forall (i,j) \in \mathcal{A}, \tag{1f}$$

$$\sum_{(i,j)\in\delta^+(i)} x_{ijk}^\omega - \sum_{(j,i)\in\delta^-(i)} x_{jik}^\omega = \begin{cases} q_k^\omega, & \text{if } i = o_k, \\ -q_k^\omega, & \text{if } i = d_k, \\ 0, & \text{otherwise,} \end{cases} \qquad \forall i \in \mathcal{N},\ k \in \mathcal{K}, \tag{1g}$$

$$\sum_{k\in\mathcal{K}(d)} x_{ijk}^\omega \le q_k^\omega y_{ijd}, \qquad \forall d \in \mathcal{D},\ (i,j) \in \mathcal{A}, \tag{1h}$$

$$x_{ijk}^\omega \ge 0, \qquad \forall (i,j) \in \mathcal{A},\ k \in \mathcal{K}, \tag{1i}$$

$$z_{ij}^\omega \in \mathbb{Z}_+, \qquad \forall (i,j) \in \mathcal{A}\}. \tag{1j}$$

The first stage determines the design of the network: installing capacity on arcs while adhering to the desired p-alt freight flow structure. The objective function (1a) minimizes the sum of the cost of operating trailers in the network plus the expected cost of any additional trailers required once demand is realized. Constraints (1b) specify the structure of the freight flow in the network. Note that setting $p = 1$ gives the standard in-tree structure similar to the one depicted in Fig. 1a, while setting $p = 2$ allows up to two next destination options for each terminal and for all possible destinations. The right-hand side of (1b) ensures feasibility if a node's outdegree is less than p.

The second stage routes all realized demand and determines any additional capacity that is required to do so, at minimum cost. Constraints (1g) are the standard flow balance constraints, and also stipulate that all demand is to be met. Constraints (1f) ensure that the capacity on each arc is not exceeded. Constraints (1h) ensure that flows destined for a particular destination are only allowed to use an arc when the corresponding alt is chosen.

3   Model Solution

The stochastic p-alt model is a two-stage mixed-integer stochastic program that may have infinitely many scenarios. Given that even deterministic CMND problems are beyond the reach of modern commercial solvers [10], and can only be solved for small-sized instances, having a two-stage stochastic model with general integer variables in both stages and constraints on the freight flow structure only makes the problem more difficult. Therefore, heuristics are needed, and we thus employ a Sample Average Approximation (SAA) [11] framework.

3.1   Sample Average Approximation

In SAA, a small number of demand scenarios $N << |\Omega|$ is randomly sampled, (with replacement), resulting in a *sample problem*, given by (1) with Ω replaced by the sampled scenarios. This – usually relatively simpler – two-stage stochastic program over N scenarios is solved, and the first-stage design, (r^*, y^*), is evaluated. Evaluation is performed by (independently) sampling a larger number $N' >> N$ of scenarios, and for each such scenario ω, calculating the recourse cost, $Q(r^*, y^*, \omega)$, by solving (1e). The cost of design (r^*, y^*) is approximated as

$$\sum_{(i,j)\in\mathcal{A}} c_{ij} r_{ij}^* + \mathbb{E}_\omega \left[Q(r, y, \omega) \right] \approx \sum_{(i,j)\in\mathcal{A}} c_{ij} r_{ij}^* + \frac{1}{N'} \sum_{n=1}^{N'} Q(r^*, y^*, \omega^n), \quad (2)$$

where $\omega^1, \ldots, \omega^{N'}$ are the N' sampled scenarios. This procedure is repeated M times to get M designs, and the one with the lowest (approximate) cost is chosen as the final solution. We refer the reader to [11] for details and convergence properties of this approach.

3.2   Solving the Sample Problem

In each SAA iteration, the sample problem must be solved. Its *extensive form*, for the sampled scenarios $\mathcal{S} = \{\omega^1, \ldots, \omega^N\}$, is the mixed-integer linear program

$$\min \sum_{(i,j)\in\mathcal{A}} c_{ij} r_{ij} + \frac{1}{N} \sum_{\omega\in\mathcal{S}} \sum_{(i,j)\in\mathcal{A}} \hat{c}_{ij} z_{ij}^\omega,$$
$$\text{s.t.} \quad (1b) - (1d), \ (z^\omega, x^\omega) \in \mathcal{P}(r, y, \omega), \ \forall \omega \in \mathcal{S}.$$

Although simpler than the original stochastic problem, this MIP is still difficult to solve in reasonable time, even for relatively small sample sizes, N. Therefore, we propose two heuristics to solve this problem. Since the primary goal is obtaining a feasible first-stage solution, these heuristics relax the second stage integer variables in an attempt to make the problem easier to solve. Each approach uses a different method to mitigate the effect of this relaxation on the first stage decisions. One uses a slope scaling technique ([12]), which heuristically adjusts the costs of the second stage variables. The other uses cuts that are valid for the problem with integrality in both stages. This gives us three approaches: solving the problem exactly (**Exact**); relaxing the integrality of the second stage variables and using slope scaling (**SS-Heuristic**); and relaxing the integrality of the second stage variables but adding valid inequalities (**Cut-Heuristic**). In all three methods, the evaluation subproblems (1e) in the SAA algorithm are solved exactly, without relaxation.

The valid inequalities we use are the so-called *cut inequalities* [13]. Given a cut in the network, these inequalities stipulate that the capacity crossing the cut should be enough to serve the demand crossing that cut. We adapt these to the stochastic p-alt problem, as follows. Let $d^\omega(V, \bar{V})$ be the total demand that has to traverse the cut defined by $V \subset \mathcal{N}$ in scenario ω. Then the cut inequalities for the stochastic p-alt problem, for a given scenario $\omega \in \Omega$, can be written as

$$\sum_{(i,j)\in\delta^+(V)} (r_{ij} + z_{ij}^\omega) \geq \left\lceil \frac{d^\omega(V, \bar{V})}{Q} \right\rceil, \qquad \forall V \subset \mathcal{N}. \tag{3}$$

Exact. The sample problem is solved using a MIP solver. The cuts (3) are added for every $\omega \in \mathcal{S}$ to the model upfront as lazy constraints. In our experiments, adding these cuts significantly improved the solver performance.

SS-Heuristic. In this approach, we will relax the integrality of the second-stage variables. To compensate for the cost under-approximation introduced by the relaxation, we introduce cost multipliers that will be iteratively adjusted using a slope scaling algorithm, as follows: let ρ_{ij}^t be the cost multiplier for arc (i,j) in iteration t of the slope scaling algorithm and solve

$$\min \sum_{(i,j)\in\mathcal{A}} c_{ij} r_{ij} + \frac{1}{N} \sum_{\omega\in\mathcal{S}} \sum_{(i,j)\in\mathcal{A}} \rho_{ij}^t \hat{c}_{ij} z_{ij}^\omega \tag{4}$$

$$\text{s.t. } (1b) - (1d), \; (z^\omega, x^\omega) \in \mathcal{P}_{LP}(r, y, \omega), \forall \omega \in \mathcal{S}$$

where $\mathcal{P}_{LP}(r, y, \omega)$ denotes the LP relaxation of $\mathcal{P}(r, y, \omega)$.

Slope scaling is terminated when: (1) the true recourse cost estimate (obtained via the evaluation subproblems) and the term $\frac{1}{N} \sum_{\omega\in\mathcal{S}} \sum_{(i,j)\in\mathcal{A}} \rho_{ij}^t \hat{c}_{ij} z_{ij}^\omega$ in the relaxed problem are relatively close, or (2) two successive iterations yield designs with approximately the same costs. If neither of these stopping criteria is met, the multipliers for the next iteration are adjusted as follows:

$$\rho_{ij}^{t+1} = \begin{cases} \dfrac{\sum_{\omega\in\mathcal{S}} \left\lceil \tilde{z}_{ij}^\omega \right\rceil}{\sum_{\omega\in\mathcal{S}} \tilde{z}_{ij}^\omega}, & \text{if } \sum_{\omega\in\mathcal{S}} \tilde{z}_{ij}^\omega > 0, \\ \rho_{ij}^t, & \text{otherwise,} \end{cases} \tag{5}$$

where $\tilde{z}_{ij}^\omega$ is the value of the continuous relaxation of the z variable for (i,j) in scenario ω obtained from the solution of (4) in iteration t.

Cut-Heuristic. We again relax the integrality of the second-stage variables, but add the inequalities (3) for each $\omega \in \mathcal{S}$, to the model, *a priori*. As there is a large number of them, they are added as lazy constraints. These inequalities serve two purposes in the heuristic: they strengthen the relaxed model with respect to the integrality of the first stage variables and, since they are inferred from integrality of the second stage variables, mitigate the effect of relaxing these.

4   Computational Study

In this section, we compare the three solution methods for the stochastic p-alt problem and we assess the benefits of using this model. To be able to solve the extensive-form problems to optimality, we restrict ourselves to small randomly-generated instances (up to 15 terminals). Table 1 shows the characteristics of the instances used. The characteristics are self-explanatory, except for "% of OD Pairs", which indicates the percentage of possible EOL pairs with demand. The SAA parameters used are $M = 10$, $N = 10$, and $N' = 1000$. Algorithms were implemented in Python, and experiments performed on a 2.9 GHz Intel Core i5 computer with 16 GB of RAM using Gurobi 7.5.2 as the IP solver.

Table 1. Instance characteristics

Instance	Number of EOLs	Number of BBs	% of OD Pairs	Demand range	Q	Demand Std. Dev. range
1	8	4	10	$[0, 1]$	1	$(0, 0.3)$
2	7	3	25	$[0, 1]$	1	$(0, 0.3)$
3	7	3	20	$[0, 10]$	3	$(0, 3)$
4	7	3	20	$[0, 10]$	3	$(0, 4.5)$
5	10	5	10	$[0, 1]$	1	$(0, 0.3)$

Table 2 compares the performance of the three methods, and shows the first-stage (design) cost, $\mathbf{F}$, the approximated expected total (design + recourse) cost, $\mathbf{T}$, and the running time (in seconds), $\mathbf{t}$. To reduce bias, each iteration of SAA in the three approaches was performed using the same set of scenarios for the sample problem, and the same set of scenarios for the evaluation subproblems.

We observe that although the methods find different first-stage solutions, the total cost (point) estimates are fairly similar. The main difference is the computational efficiency: Cut-Heuristic consistently outperforms the other two approaches by obtaining solutions that are in a similar range as the other two approaches but in much less time. While that is encouraging, we note that the number of cut inequalities grows exponentially with the number of terminals (and also grows with the number of scenarios), and that separation heuristics will be required when solving larger instances as the separation problem is NP-Hard. We also observe, as expected, that 2-alt designs perform better than their 1-alt counterparts by handling the uncertain demand at a lower cost.

Table 2. Comparison of Exact, SS-Heuristic, and Cut-Heuristic

		Exact			SS-Heuristic			Cut-Heuristic		
		F	T	t	F	T	t	F	T	t
1	1-alt	148.8	163.4	437.2	148.8	163.5	4053.2	148.8	163.4	52.2
	2-alt	148.8	163.2	480.1	148.8	163.2	1257.8	148.8	163.2	225.4
2	1-alt	262.4	266.1	773.7	262.4	266.1	2347.3	262.4	266.3	148.6
	2-alt	248.1	261.8	1085.8	248.1	261.8	2737.0	248.1	261.8	270.6
3	1-alt	595.8	700.0	1422.8	623.3	705.5	1119.1	571.6	703.1	469.4
	2-alt	583.9	674.3	17273.3	580.8	683.4	2567.8	575.4	677.5	419.2
4	1-alt	595.8	732.6	2024.4	635.2	737.4	862.9	576.1	734.3	457.3
	2-alt	588.4	706.6	56627.1	555.4	712.5	2614.6	575.4	703.7	505.3
5	1-alt	185.1	190.6	2505.2	185.1	190.3	30934.2	182.7	190.2	2437.1
	2-alt	175.0	187.5	1013.2	175.0	187.3	3522.0	175.0	187.5	566.2

Table 3. Comparison of sequential vs. integrated optimization

		Sequential		Cut-Heuristic	
		F	T	F	T
1	1-alt	148.8	163.9	148.8	163.4
	2-alt	148.8	163.9	148.8	163.2
2	1-alt	262.4	266.8	262.4	266.3
	2-alt	248.1	262.9	248.1	261.8
3	1-alt	693.7	775.3	571.6	703.1
	2-alt	608.4	697.0	575.4	677.5
4	1-alt	693.6	803.6	576.1	734.3
	2-alt	621.4	726.0	575.4	703.7
5	1-alt	196.1	215.7	182.7	190.2
	2-alt	181.4	191.1	175.0	187.5

In Table 3, we compare the Cut-Heuristic results with the results obtained from sequential optimization using *expected demand*. Sequential optimization involves two steps: (1) solving an optimization problem to decide the number of trailers operating between the terminals, (2) given these numbers, solving an optimization problem to determine the alts – allowing additional trailers to be added. The resulting design is evaluated using a set of 1,000 scenarios. We observe a substantial reduction in the expected total cost when using designs produced by the stochastic p-alt model. This is more pronounced in instances 3–5, which have a larger network and/or higher demand variability.

Overall, these results also demonstrate that there is much to be gained from 2-alt designs, especially in settings with high demand variability. Furthermore, simultaneously choosing the alts and explicitly considering demand uncertainty result in more robust designs, leading to lower expected total costs.

Acknowledgment. Ahmad Baubaid would like to acknowledge financial support from the King Fahd University of Petroleum & Minerals.

References

1. Crainic, T.G.: Service network design in freight transportation. Eur. J. Oper. Res. **122**(2), 272–288 (2000)
2. Wieberneit, N.: Service network design for freight transportation: a review. OR Spectr. **30**(1), 77–112 (2008)
3. Powell, W.B.: A local improvement heuristic for the design of less-than-truckload motor carrier networks. Transp. Sci. **20**(4), 246–257 (1986)
4. Powell, W.B., Koskosidis, I.A.: Shipment routing algorithms with tree constraints. Transp. Sci. **26**(3), 230–245 (1992)
5. Jarrah, A.I., Johnson, E., Neubert, L.C.: Large-scale, less-than-truckload service network design. Oper. Res. **57**(3), 609–625 (2009)
6. Lindsey, K., Erera, A., Savelsbergh, M.: Improved integer programming-based neighborhood search for less-than-truckload load plan design. Transp. Sci. December 2017 (2016). https://doi.org/10.1287/trsc.2016.0700
7. Gendron, B., Crainic, T.G., Frangioni, A.: Multicommodity capacitated network design. In: Sansò, B., Soriano, P. (eds.) Telecommunications Network Planning. CRT, pp. 1–19. Springer, Boston, MA (1999). https://doi.org/10.1007/978-1-4615-5087-7_1
8. Hoff, A., Lium, A.G., Løkketangen, A., Crainic, T.G.: A metaheuristic for stochastic service network design. J. Heuristics **16**(5), 653–679 (2010)
9. Lium, A.G., Crainic, T.G., Wallace, S.W.: A study of demand stochasticity in service network design. Transp. Sci. **43**(2), 144–157 (2009)
10. Crainic, T.G., Fu, X., Gendreau, M., Rei, W., Wallace, S.W.: Progressive hedging-based metaheuristics for stochastic network design. Networks **58**(2), 114–124 (2011)
11. Kleywegt, A.J., Shapiro, A., Homem-de Mello, T.: The sample average approximation method for stochastic discrete optimization. SIAM J. Optim. **12**(2), 479–502 (2002)
12. Kim, D., Pardalos, P.M.: A solution approach to the fixed charge network flow problem using a dynamic slope scaling procedure. Oper. Res. Lett. **24**(4), 195–203 (1999)
13. Bienstock, D., Chopra, S., Günlük, O., Tsai, C.Y.: Minimum cost capacity installation for multicommodity network flows. Math. Program. **81**(2), 177–199 (1998)

Energy-Aware Production Scheduling with Power-Saving Modes

Ondřej Benedikt[1], Přemysl Šůcha[1], István Módos[1(✉)], Marek Vlk[1,2], and Zdeněk Hanzálek[1]

[1] Czech Technical University in Prague, Prague, Czech Republic
benedond@fel.cvut.cz,
{premysl.sucha,istvan.modos,zdenek.hanzalek}@cvut.cz
[2] Charles University, Prague, Czech Republic
vlk@ktiml.mff.cuni.cz

Abstract. This study addresses optimization of production processes where machines have high energy consumption. One efficient way to reduce the energy expenses in production is to turn a machine off when it is not being used or switch it into an energy-saving mode. If the production has several machines and production demand that varies in time, the energy saving can be substantial; the cost reduction can be achieved by an appropriate production schedule that could control the switching between the energy modes with respect to the required production volume. Therefore, inspired by real production processes of glass tempering and steel hardening, this paper addresses the scheduling of jobs with release times and deadlines on parallel machines. The objective is to find a schedule of the jobs and a switching between the power modes of the machines so that the total energy consumption is minimized. Moreover, to further generalize the scheduling problem to other production processes, we assume that the processing time of the jobs is mode-dependent, i.e., the processing time of a job depends on the mode in which a machine is operating. The study provides an efficient Branch-and-Price algorithm and compares two approaches (based on Integer Linear Programming and Constraint Programming) for solving the subproblem.

Keywords: Production scheduling · Energy · Branch-and-Price
Integer Linear Programming · Constraint Programming

1 Introduction

This research is inspired by two existing energy-demanding production processes. The first one is a glass tempering in ERTL Glas company and the second one is steel hardening in ŠKODA AUTO company. In both processes, the material is heated in one of the identical furnaces to a high temperature (hundreds of °C) which consumes a substantial amount of energy. Typically, the furnaces are turned on at the beginning of a scheduling horizon and then continuously

W.-J. van Hoeve (Ed.): CPAIOR 2018, LNCS 10848, pp. 72–81, 2018.
https://doi.org/10.1007/978-3-319-93031-2_6

operate until its end when they are turned off. If the production demand varies within the scheduling horizon, the furnaces remain in an energy-demanding mode even if nothing is being produced and, therefore, wasting the energy.

As identified in [1], a significant energy cost savings could be achieved in manufacturing facilities by switching machines to a power-saving mode when nothing is being produced. Likewise, our preliminary feasibility study for ŠKODA AUTO has shown that about 6% of the production line consumption can be saved using the power-saving modes. However, in the above-mentioned processes, switching to and from the power-saving mode is not immediate since the furnaces in the power-saving mode operate in a lower temperature and re-heating them back to the operational temperature can take dozens of minutes. Thus, the switching has to be planned carefully. Moreover, in some production processes a machine operating in a power-saving mode can still process a material, albeit slower. To take into consideration such processes, we will assume that the processing time is mode-dependent, i.e., the processing time depends on the mode in which a machine is operating.

The problem of scheduling jobs on machines having different energy modes has been already studied in the literature to some extent. The existing works either study a single machine problem [1–3], do not take the transition costs and times into account [4] or propose a time-indexed Integer Linear Programming formulation [5,6] which can optimally solve only small instances. The scheduling with mode-dependent processing times of the jobs is similar to a dynamic voltage scaling [7] in embedded systems, where the processing times of the jobs depend on the operating frequencies of the processors. Since the schedules in the embedded systems are usually event-triggered and the transition times between different operating frequencies is in the order of microseconds, the research results cannot be directly applied to the production processes.

The contribution of this work-in-progress paper is both a formulation and algorithm for solving a general multi-machine scheduling problem with energy modes, where the objective is to minimize the total energy consumption. For this problem, we propose an efficient Branch-and-Price algorithm with a clever representation of the columns to break the symmetries arising due to the identical parallel machines. The algorithm also restricts the structure of transitions between the modes to respect the technological requirements. For the experimental comparison, the subproblem in the Branch-and-Price algorithm is formulated as both Integer Linear Programming (ILP) and Constraint Programming (CP) problems.

2   Problem Statement

Let $M = \{1, \ldots, m\}$ be a set of identical, parallel machines. In every time instant, every machine $i \in M$ is operating in some *mode* $\omega \in \Omega$. While a machine is operating in mode $\omega \in \Omega$, it demands a constant *power* $P_\omega \in \mathcal{R}_{\geq 0}$. The mode of a machine can be switched from one mode to another, however, this transition may take some time during which the machine is not operational, and it incurs a cost in the form of consumed energy. Therefore, for every pair of modes $\omega, \omega' \in \Omega$ we define *transition time* $t_{\omega,\omega'}^{\text{trans}} \in \mathcal{Z}_{\geq 0} \cup \{\infty\}$ and *transition cost* $c_{\omega,\omega'}^{\text{trans}} \in \mathcal{R}_{\geq 0} \cup \{\infty\}$. If $t_{\omega,\omega'}^{\text{trans}} = c_{\omega,\omega'}^{\text{trans}} = \infty$ for some pair of modes $\omega, \omega' \in \Omega$, then a machine cannot be directly switched from ω to ω'. If a machine is operating in some mode $\omega \in \Omega$ for a total time of t, then the *operating cost* is computed as $P_\omega \cdot t$.

Let $J = \{1, \ldots, n\}$ be a set of jobs. The jobs have to be processed on some machine within scheduling *horizon* $H \in \mathcal{Z}_{>0}$, each machine can process at most one job at a time, and the jobs cannot be preempted. A *processing time* of job $j \in J$ depends on the mode $\omega \in \Omega$ in which the assigned machine is operating during processing of the job and is denoted as $p_{j,\omega} \in \mathcal{R}_{\geq 0} \cup \{\infty\}$. If $p_{j,\omega} = \infty$, then job $j \in J$ cannot be processed while the assigned machine is operating in mode $\omega \in \Omega$. A job cannot be processed on a machine during the transition from one mode to another and once a machine starts processing a job, it cannot change its mode until the job completes. Moreover, each job $j \in J$ has *release time* $r_j \in \mathcal{Z}_{\geq 0}$ and *deadline* $d_j \in \mathcal{Z}_{\geq 0}$. The release time and deadline of a job define the time window within which it must be processed.

Due to technological restrictions, such as machine wear, the number of transitions to mode $\omega \in \Omega$ on each machine can be at most $K_\omega \in \mathcal{Z}_{>0}$ within the scheduling horizon.

A *solution* is a tuple $(\boldsymbol{s}, \boldsymbol{\mu}, \boldsymbol{\pi}_1, \ldots, \boldsymbol{\pi}_m, \boldsymbol{t}_1^{\text{mode}}, \ldots, \boldsymbol{t}_m^{\text{mode}})$, where $\boldsymbol{s} \in \mathcal{Z}_{\geq 0}^n$ is a vector of jobs *start times*, $\boldsymbol{\mu} \in M^n$ is a vector of jobs *assignment to the machines*, $\boldsymbol{\pi}_i \in \bigcup_{l \in \mathcal{Z}_{>0}} \Omega^l$ is a *profile* of machine $i \in M$ and $\boldsymbol{t}_i^{\text{mode}} \in \bigcup_{l \in \mathcal{Z}_{>0}} \mathcal{Z}_{\geq 0}^l$ are the *operating times* of machine $i \in M$. Profile $\boldsymbol{\pi}_i$ is a finite sequence of modes which are followed by machine $i \in M$ in the solution. The profiles represent the transition from one mode to another; they do not inform about the time spent operating in a particular mode of a profile. Operating times $\boldsymbol{t}_i^{\text{mode}}$ are a finite sequence of non-negative integers such that $t_{i,k}^{\text{mode}}$ is the operating time of machine $i \in M$ in mode $\pi_{i,k}$. It holds that (i) the length of a profile is the same as the length of the corresponding operating times, i.e., $|\boldsymbol{\pi}_i| = |\boldsymbol{t}_i^{\text{mode}}|, \forall i \in M$ (operator $|\cdot|$ represents the length of a sequence) and (ii) the sum of the total transition times plus the total operating times of a profile equals to the length of the horizon, i.e., $\sum_{k=1}^{|\boldsymbol{\pi}_i|-1} t_{\pi_{i,k},\pi_{i,k+1}}^{\text{trans}} + \sum_{k=1}^{|\boldsymbol{\pi}_i|} t_{i,k}^{\text{mode}} = H, \forall i \in M$. Moreover, the solutions must respect that the number of transitions to every mode is limited, i.e., $|\{k \in \{1, \ldots, |\boldsymbol{\pi}_i|\} \mid \pi_{i,k} = \omega\}| \leq K_\omega, \forall \omega \in \Omega, i \in M$.

The goal of this scheduling problem is to find a solution which minimizes the consumed energy, i.e., the sum of the total transition cost plus the total operating cost

$$\sum_{i \in M} \sum_{k=1}^{|\boldsymbol{\pi}_i|-1} c_{\pi_{i,k},\pi_{i,k+1}}^{\text{trans}} + \sum_{i \in M} \sum_{k=1}^{|\boldsymbol{\pi}_i|} P_{\pi_{i,k}} \cdot t_{i,k}^{\text{mode}}. \tag{1}$$

This scheduling problem is strongly $\mathcal{NP}$-hard due to the underlying strongly $\mathcal{NP}$-complete scheduling problem $1|r_j, d_j|-$ [8].

3   Solution Approach

Although it is possible to model the scheduling problem introduced in Sect. 2 using either ILP or CP, the resulting model would be very large and could solve only small instances. Moreover, due to the parallel identical machines, the scheduling problem has symmetrical solutions, i.e., different solutions may have the same objective value and one solution can be transformed to another by simply re-indexing the machines. Symmetries significantly degrade the efficiency of the models, thus, a specialized approach is necessary. Therefore, we solve this scheduling problem using Branch-and-Price (BaP) methodology [9] which is designed for solving such large-scale optimization problems. Symmetrical solutions are avoided by a clever representation of the columns in the BaP algorithm.

Algorithms based on BaP combine the Branch-and-Bound algorithm with Column Generation (CG) [10]. The CG is a technique for solving Linear Programming (LP) models with many variables. The variables (and the corresponding columns of the constraint matrix) are added lazily until a solution with the restricted set of the variables is optimal for the dual formulation of the model. If the solution is an integer, the whole algorithm terminates. Otherwise, branching on the fractional variables is performed.

The LP model used in the CG is based on a set covering model. Let $A \subseteq \{0, 1\}^n$ be a set of columns, where each *column* $a_l \in A$ represents a particular assignment of the jobs on a machine. For each $a_l \in A$ we can compute its *cost* c_l^{col}, which corresponds to the optimal solution of the single machine scheduling problem with jobs for which $a_{l,j} = 1$. The goal of the *master problem* is to select a subset of columns from A such that every job is assigned to some machine and the total cost of the selected columns is minimized. Since the number of columns is exponential in the number of jobs, we use *restricted column set* $A' \subseteq A$, which is lazily expanded using the CG. The master problem *restricted* to A' then can be modeled as

$$\min \sum_{a_l \in A'} c_l^{col} \cdot x_l \tag{2}$$

$$\text{s.t.} \sum_{a_l \in A'} a_{l,j} \cdot x_l \geq 1, \ j \in J \tag{3}$$

$$\sum_{a_l \in A'} x_l \leq m \tag{4}$$

$$x_l \geq 0, \ a_l \in A'. \tag{5}$$

Variable x_l denotes, whether column a_l is selected in the solution. Notice that the problematic machine symmetries are broken since an assignment of a column to a specific machine is not important and, therefore, not modeled.

The CG operates on the dual formulation of the restricted master problem; consequently, variables x_l diminish. Instead, new variables $\lambda \in \mathcal{R}_{\leq 0} \times \mathcal{R}_{\geq 0}^n$ arise in the dual formulation. If the optimal solution to the dual restricted master problem is feasible for the unrestricted one, the CG terminates, otherwise, column $a_l \in A \backslash A'$ with a negative *reduced cost* has to be found, i.e., $0 > c_l^{\text{col}} - \sum_{j \in J} a_{l,j} \cdot \lambda_j - \lambda_0$. A new column for the restricted master problem is found using another optimization model called a *subproblem* in which the dual variables λ are fixed. Solving such a problem is very similar to solving a single machine variant of the scheduling problem introduced in Sect. 2. The difference is that the subproblem selects a subset of jobs, i.e., new column a_l, such that the reduced cost is minimized.

However, such a subproblem has still a very large solution space, thus selecting the machine profiles is a hard combinatorial problem. If the maximum number of transitions K_ω to the modes is not large (which is true for ŠKODA AUTO and ERTL Glas), it is possible to enumerate all *technologically feasible profiles* $\Pi \subset \bigcup_{l \in \mathcal{Z}_{>0}} \Omega^l$ that do not violate the transition limits, solve the subproblem for every $\pi \in \Pi$ and select the one with the minimum objective value. For each technologically feasible profile $\pi \in \Pi$, we solve the corresponding subproblem using the following ILP model

$$\min \quad \sum_{i \in M} \sum_{k=1}^{|\pi|-1} c_{\pi_k, \pi_{k+1}}^{\text{trans}} + \sum_{i \in M} \sum_{k=1}^{|\pi|} P_{\pi_k} \cdot t_k^{\text{mode}} - \sum_{j \in J} \sum_{k=1}^{|\pi|} y_{j,k} \cdot \lambda_j - \lambda_0 \tag{6}$$

$$\sum_{k=1}^{|\pi|} y_{j,k} \le 1, \ j \in J \tag{7}$$

$$s_1^{\text{mode}} = 0 \tag{8}$$

$$s_k^{\text{mode}} = s_{k-1}^{\text{mode}} + t_{k-1}^{\text{mode}} + t_{\pi_{k-1}, \pi_k}^{\text{trans}}, \ k \in \{2, \dots, |\pi|\} \tag{9}$$

$$s_{|\pi|}^{\text{mode}} + t_{|\pi|}^{\text{mode}} = H \tag{10}$$

$$r_j \le s_j, \ j \in J \tag{11}$$

$$s_j + \sum_{k=1}^{|\pi|} y_{j,k} \cdot p_{j,\pi_k} \le d_j, \ j \in J \tag{12}$$

$$s_k^{\text{mode}} \le s_j + \mathcal{M} \cdot (1 - y_{j,k}), \ j \in J, k \in \{1, \dots, |\pi|\} \tag{13}$$

$$s_j + p_{j,\pi_k} \le s_k^{\text{mode}} + t_k^{\text{mode}} + \mathcal{M} \cdot (1 - y_{j,k}),$$
$$j \in J, k \in \{1, \dots, |\pi|\} \tag{14}$$

$$s_j + p_{j,\pi_k} \le s_{j'} + \mathcal{M} \cdot (3 - y_{j,k} - y_{j',k} - z_{j,j'}),$$
$$j, j' \in J, j < j', k \in \{1, \dots, |\pi|\} \tag{15}$$

$$s_{j'} + p_{j',\pi_k} \le s_j + \mathcal{M} \cdot (2 - y_{j',k} - y_{j,k} + z_{j,j'}),$$
$$j, j' \in J, j < j', k \in \{1, \dots, |\pi|\}. \tag{16}$$

The program uses the following variables: (i) $y_{j,k} \in \{0, 1\}$ denoting whether $j \in J$ is assigned to k-th mode of π, (ii) $s_k^{\text{mode}} \in \mathcal{Z}_{\geq 0}$ is the start time of the time interval in which the machine is operating in k-th mode of profile π, (iii) $t_k^{\text{mode}} \in \mathcal{Z}_{\geq 0}$ is the operating time of k-th mode of profile π, (iv) $s_j \in \mathcal{Z}_{\geq 0}$ is the start time of job $j \in J$ and (v) $z_{j,j'} \in \{0, 1\}$ denotes the relative order between jobs $j, j' \in J$. To see that this subproblem selects a column, notice that $a_{l,j} = \sum_{k=1}^{|\pi|} y_{j,k}$. Constraint (7) allows each job to be assigned to at most one mode, constraints (8)–(10) set the start time and operating time of k-th mode, constraints (11)–(12) assure that the jobs are processed in between its release time and deadline, constraints (13)–(14) ensure that the jobs are fully contained in the time period of k-th mode to which they are assigned, and constraints (15)–(16) ensure that the jobs are not overlapping. To speed-up solving the model, we generate constraints (15)–(16) using *lazy constraints generation*.

The initial set of columns is created using a simple heuristic based on Earliest Deadline First strategy. Since even problem $1|r_j, d_j|-$ is strongly $\mathcal{NP}$-complete, there is no guarantee that any heuristic will find the initial A' that will represent a feasible solution (although the selected Earliest Deadline First strategy is generally better aimed at satisfying the deadlines than cost-based heuristics). Therefore, if the heuristic cannot find a feasible solution, it generates A' such that it contains columns covering all the jobs. In such a situation, it may happen that the master model may not find a feasible solution because of constraint (4). Therefore the master model has to assume this constraint in a slightly different form, i.e., $\sum_{a_l \in A'} x_l \leq m + q$, where $q \geq 0$ is a new decision variable indicating whether the solution of the master problem is feasible or not. Finally, the objective function of the master model is $\sum_{a_l \in A'} c_l^{\text{col}} \cdot x_l + q \cdot \mathcal{C}$, such that $\mathcal{C}$ is much larger than the cost of any feasible column, e.g., $\mathcal{C} = m \cdot H \cdot \max_{\omega \in \Omega} P_\omega + \max_{\pi \in \Pi} |\pi| \cdot \max_{\omega, \omega' \in \Omega} c_{\omega, \omega'}^{\text{trans}}$.

If the optimal solution to the master problem is fractional at the end of the CG, a branching is required. The used branching scheme selects a pair of jobs (j, j') that have not been selected before and have the largest overlap of intervals $[r_j, d_j]$ and $[r_{j'}, d_{j'}]$ (according to preliminary experiments, this branching scheme performed better than random selection). Then the scheme creates two branches, where: (i) jobs j and j' are forbidden to be processed on the same machine and (ii) the same two jobs are required to be processed on the same machine. This scheme generates simple logical constraints that can be included into the subproblem. For each new branch it is necessary to filter out columns $a_l \in A'$ that violate the particular branching decision. However, this step may result in a column set which is not covering all the jobs. Therefore, as in the initialization phase, the algorithm has to add columns such that all jobs are present in A'.

Alternatively, the subproblem can be easily modeled using CP, where efficient filtering techniques for unary resources with optional jobs are employed [11]. Let us introduce two types of interval variables I_k^{mode} and $I_{j,k}$, where $j \in J, k \in \{1, \ldots, |\pi|\}$. Variables $I_{j,k}$ are optional, which means that the presence of $I_{j,k}$ in a schedule is to be decided. The length of $I_{j,k}$ is fixed to p_{j,π_k}, whereas the length of I_k^{mode} is to be determined. Constraints (15)–(16) are substituted by no-overlap constraints, i.e., for each $k \in \{1, \ldots, |\boldsymbol{\pi}|\}$, we add constraint $NoOverlap(\bigcup_{j \in J} I_{j,k})$. The other constraints are straightforward. Note that the state function variable [12] for modeling the modes of a machine cannot be efficiently used as the lengths of the modes are involved in the objective function and the transition times between modes are fixed.

4 Preliminary Experiments

We evaluated the proposed BaP algorithm (with subproblem implemented as both ILP and CP model) on a set of random problem instances that were generated as follows. The scheduling horizon was fixed to $H = 1000$ and the set of assumed machines modes was chosen as $\Omega = \{\mathrm{OFF}, \mathrm{IDLE}, \mathrm{ON}\}$, i.e., the machines have one power-saving mode IDLE and the jobs can be processed only in mode ON. The set of technologically feasible profiles is $\Pi = \{(\mathrm{OFF}, \mathrm{ON}, \mathrm{OFF}), (\mathrm{OFF}, \mathrm{ON}, \mathrm{IDLE}, \mathrm{ON}, \mathrm{OFF})\}$. The processing time of the jobs in mode ON was randomly sampled from discrete uniform distribution $\mathcal{U}\{1, 100\}$. The release times and deadlines were randomly generated in such a way that the generated instances were feasible and the jobs had non-zero overlap.

For each pair $n \in \{15, 20, 25\}$, $m \in \{1, 2, 3, 4\}$, 4 random instances were generated using the scheme described above; each instance had time-limit of 3600 s. The experiments were carried out on an Intel® Core™ i5-3320M CPU @ 2.6 GHz computer with 8 GB RAM. For solving the ILP and CP models, we used Gurobi 7.5 and IBM CP Optimizer 12.7.1 solvers, respectively. The source code of the algorithms with the generated instances are publicly available at https://github.com/CTU-IIG/PSPSM.

The results shown in Table 1 clearly indicate that the BaP algorithm with ILP subproblem (BaP+ILP) outperforms the CP subproblem (BaP+CP). Using continuous variables for the start times of the jobs instead of integer variables led to only a slight deterioration in the computational time, while the number of nodes and the number of columns slightly decreased.

Although it could be possible to compare the proposed algorithm with the time-indexed ILP formulation from the literature [2], the resulting model would be huge for the scheduling granularity usually used in production scheduling (1 min). For example, the time-indexed ILP formulation would require $4 \cdot 25 \cdot 1000 = 100000$ binary variables just to represent the start times of the jobs for the largest problem instance from Table 1.

For the sake of comparison with the global approach, we also evaluated various global CP models. All the global models gave competitive results up to 2 machines, but for 3 or more machines, the global models are strongly suffering from the symmetries caused by parallel identical machines (the best global model time-outed for 9 instances having 4 machines).

Table 1. Experimental results.

Instance	Parameters		BaP+ILP			BaP+CP		
	m	n	Computational time [s]	Nodes	Columns	Computational time [s]	Nodes	Columns
1	1	15	9.50	1	113	58.57	1	115
2	1	15	9.58	1	116	92.80	1	124
3	1	15	9.38	1	102	41.77	1	82
4	1	15	11.37	1	132	50.71	1	118
5	2	15	31.15	13	253	179.53	13	270
6	2	15	14.91	1	93	65.10	1	66
7	2	15	8.66	1	97	45.95	1	90
8	2	15	10.23	1	87	79.79	1	102
9	3	15	40.05	17	173	177.20	17	156
10	3	15	41.59	35	301	210.12	35	341
11	3	15	5.29	1	47	60.41	1	48
12	3	15	24.37	15	200	188.80	15	187
13	4	15	32.43	31	243	278.96	33	213
14	4	15	4.82	1	51	44.23	1	54
15	4	15	7.46	1	77	87.70	1	75
16	4	15	5.43	1	38	45.72	1	42
17	1	20	38.78	1	298	250.04	1	250
18	1	20	79.39	1	323	406.87	1	302
19	1	20	86.69	1	429	453.85	1	437
20	1	20	17.57	1	166	267.30	1	252
21	2	20	372.76	71	1,357	1,719.14	79	1,263
22	2	20	35.91	1	187	169.87	1	118
23	2	20	97.80	3	345	355.56	3	291
24	2	20	195.30	15	413	927.75	15	465
25	3	20	65.91	1	146	288.09	1	144
26	3	20	35.79	1	143	179.21	1	114
27	3	20	38.25	1	156	193.89	1	134
28	3	20	161.99	51	537	1,803.98	385	1,263
29	4	20	30.26	1	94	254.23	1	119
30	4	20	143.86	61	511	965.07	71	623
31	4	20	27.83	1	81	155.68	1	72
32	4	20	81.48	15	213	696.28	15	265
33	1	25	99.17	1	274	383.83	1	258
34	1	25	181.60	1	329	1,008.59	1	288
35	1	25	58.33	1	249	410.03	1	262
36	1	25	87.39	1	287	266.63	1	249
37	2	25	2,362.55	89	4,685	> 3,600.00	1	168
38	2	25	125.78	1	319	647.76	1	314
39	2	25	795.59	101	2,164	2,383.94	53	1,438
40	2	25	111.70	1	273	714.10	1	234
41	3	25	516.39	21	806	1,913.10	21	794
42	3	25	277.58	15	634	1,142.37	15	530
43	3	25	2,657.06	181	4,393	> 3,600.00	61	1,172
44	3	25	1,184.67	39	914	> 3,600.00	39	730
45	4	25	162.21	1	230	755.30	1	258
46	4	25	226.80	49	663	> 3,600.00	221	1,758
47	4	25	99.28	1	170	727.78	1	223
48	4	25	596.11	33	663	3,472.82	79	1,349

5 Conclusion

Reducing energy consumption costs of manufacturing processes can be a significant competitive advantage for producers. This work-in-progress paper provides a Branch-and-Price algorithm for a multi-machine production scheduling problem minimizing energy consumption. The experimental results show that the algorithm can solve instances with four machines and up to 25 jobs in a reasonable time. To be able to solve real production instances, the algorithm can be easily transformed into a heuristic, e.g., by reducing branching. Nevertheless, we work on further improvements of the exact algorithm to make it applicable to larger problem instances. For example, the computational time of the subproblem could be decreased by an online machine learning algorithm [13] that reuses the results of previously solved subproblems.

Acknowledgement. The work in this paper was supported by the Technology Agency of the Czech Republic under the Centre for Applied Cybernetics TE01020197, and partially by the Charles University, project GA UK No. 158216.

References

1. Mouzon, G., Yildirim, M.B., Twomey, J.: Operational methods for minimization of energy consumption of manufacturing equipment. Int. J. Prod. Res. **45**(18–19), 4247–4271 (2007)
2. Shrouf, F., Ordieres-Meré, J., García-Sánchez, A., Ortega-Mier, M.: Optimizing the production scheduling of a single machine to minimize total energy consumption costs. J. Cleaner Prod. **67**(Suppl. C), 197–207 (2014)
3. Gong, X., der Wee, M.V., Pessemier, T.D., Verbrugge, S., Colle, D., Martens, L., Joseph, W.: Integrating labor awareness to energy-efficient production scheduling under real-time electricity pricing: an empirical study. J. Cleaner Prod. **168**(Suppl. C), 239–253 (2017)
4. Ángel González, M., Oddi, A., Rasconi, R.: Multi-objective optimization in a job shop with energy costs through hybrid evolutionary techniques (2017)
5. Selmair, M., Claus, T., Trost, M., Bley, A., Herrmann, F.: Job shop scheduling with flexible energy prices. In: European Conference for Modelling and Simulation (2016)
6. Mitra, S., Sun, L., Grossmann, I.E.: Optimal scheduling of industrial combined heat and power plants under time-sensitive electricity prices. Energy **54**(Suppl. C), 194–211 (2013)
7. Kong, F., Wang, Y., Deng, Q., Yi, W.: Minimizing multi-resource energy for real-time systems with discrete operation modes. In: 2010 22nd Euromicro Conference on Real-Time Systems, pp. 113–122, July 2010
8. Lenstra, J., Kan, A.R., Brucker, P.: Complexity of machine scheduling problems. In: Hammer, P., Johnson, E., Korte, B., Nemhauser, G. (eds.) Studies in Integer Programming. Annals of Discrete Mathematics, vol. 1, pp. 343–362. Elsevier (1977)
9. Feillet, D.: A tutorial on column generation and branch-and-price for vehicle routing problems. 4OR **8**(4), 407–424 (2010)
10. Desrosiers, J., Lübbecke, M.E.: A primer in column generation. In: Desaulniers, G., Desrosiers, J., Solomon, M.M. (eds.) Column Generation, pp. 1–32. Springer, Boston (2005). https://doi.org/10.1007/0-387-25486-2_1

11. Vilím, P., Barták, R., Čepek, O.: Extension of o (n log n) filtering algorithms for the unary resource constraint to optional activities. Constraints **10**(4), 403–425 (2005)
12. Laborie, P., Rogerie, J., Shaw, P., Vilím, P.: Reasoning with conditional time-intervals. Part II: an algebraical model for resources. In: FLAIRS conference, pp. 201–206 (2009)
13. Václavík, R., Novák, A., Sůcha, P., Hanzálek, Z.: Accelerating the branch-and-price algorithm using machine learning. Eur. J. Oper. Res. (2017). under review

EpisodeSupport: A Global Constraint for Mining Frequent Patterns in a Long Sequence of Events

Quentin Cappart[(✉)], John O. R. Aoga, and Pierre Schaus

Université catholique de Louvain, Louvain-La-Neuve, Belgium
{quentin.cappart,john.aoga,pierre.schaus}@uclouvain.be

Abstract. The number of applications generating sequential data is exploding. This work studies the discovering of frequent patterns in a large sequence of events, possibly time-stamped. This problem is known as the Frequent Episode Mining (FEM). Similarly to the mining problems recently tackled by Constraint Programming (CP), FEM would also benefit from the modularity offered by CP to accommodate easily additional constraints on the patterns. These advantages do not offer a guarantee of efficiency. Therefore, we introduce two global constraints for solving FEM problems with or without time consideration. The time-stamped version can accommodate gap and span constraints on the matched sequences. Our experiments on real data sets of different levels of complexity show that the introduced constraints is competitive with the state-of-the-art methods in terms of execution time and memory consumption while offering the flexibility of adding constraints on the patterns.

1 Introduction

The trend in data science is to automate the data-analysis as much as possible. Examples are the *Automating machine learning* project [10], or the commercial products www.automaticstatistician.com and www.datarobot.com. The Auto-Weka [18] and Auto-sklearn [9] modules can automate the selection of a machine learning algorithm and its parameters for solving standard classification or regression tasks. Most of these automated tools target tabular datasets, but not yet sequences and time-series data. Data-mining problems on sequences and time series remain challenging [32] but are nevertheless of particular interest [7,29]. We believe that Constraint Programming (CP), because of the flexibility it offers, may play a role in the portfolio of techniques available for automating data-science on sequential data. As an illustration of this flexibility, Negrevergne and Guns [23] identified some constraints that could be stated on the patterns to discover in a database of sequences: length, exclusion/inclusion on symbols, membership to a regular language [25], etc. The idea of using CP for data-mining is not new. It was already used for item-set mining [11,12,24,28], for Sequential Pattern Mining (SPM) [2,3,16,23] or even for mobility profile mining [17].

J. O. R. Aoga—This author is supported by the FRIA-FNRS, Belgium.

In this paper we address the Frequent Episode Mining (FEM), first introduced with the apriori-like method WINEPI [22] and improved on MINEPI [21], with a CP approach. Contrarily to the traditional SPM, FEM aims at discovering frequent patterns in a single but very long sequence of symbols possibly time-stamped. Assume for instance a non time-stamped sequence $\langle a, b, a, c, b, a, c \rangle$ and we are looking for patterns of length three occurring at least two times. Such a subsequence is $\langle a, b, c \rangle$ that occurs exactly two times. A first occurrence is $\langle \mathbf{a}, \mathbf{b}, a, \mathbf{c}, b, a, c \rangle$ and a second one is $\langle a, \mathbf{b}, a, c, \mathbf{b}, a, \mathbf{c} \rangle$. The attentive reader may wonder why $\langle \mathbf{a}, \mathbf{b}, a, c, b, a, \mathbf{c} \rangle$ is not counted. The reason is that the *head/total frequency* measure [15] avoids duplicate counting by restricting a counting position to the first one. This measure has some interesting properties such as the well known anti-monotonicity which states that if a sequence is frequent all its subsequences are frequent too and reversely. This property makes it possible to design faster data-mining algorithm. Indeed, based on these properties, Huang and Chang [14] proposed two algorithms, MINEPI+ and EMMA. While the first one is only a small adaptation of MINEPI [21], the second uses memory anchors in order to accelerate the mining task with the price of a greater memory consumption. As variants of this problem, episodes can be closed [30,34], and other (interestingness) measures [4,6,19] can be considered. When considering time-stamped sequences such as $\langle (a, 1), (b, 3), (a, 5), (c, 6), (b, 7), (a, 8), (c, 14) \rangle$, one may also want to impose time constraints on the time difference between any two matched symbols or between the first and last matched ones. Such constraints, called *gap* and *span* were also introduced for the SPM [3] with CP.

The problem of discovering frequent pattern in a very long sequence can be reduced do the standard SPM problem [1]. The reduction consists in creating a database of sequences composed of all the suffixes of the long sequence. For our example, the sequence database would be: $\langle a, b, a, c, b, a, c \rangle$, $\langle b, a, c, b, a, c \rangle$, $\langle a, c, b, a, c \rangle$, $\langle c, b, a, c \rangle$, $\langle b, a, c \rangle$, $\langle a, c \rangle$, $\langle c \rangle$. A small adaptation of existing algorithms is required though to match any sequence of the database on its first position in accordance with the *head/total frequency* measure. This reduction has one main drawback. The spatial complexity is $\mathcal{O}(n^2)$ with n the length of the sequence. Such a complexity will quickly exceed the available memory for sequence lengths as small as a few thousands.

The contribution of this paper is a flexible and efficient approach for solving the frequent episode mining problem. WINEPI, MINEPI and EMMA are specialized algorithms not able to accommodate additional constraints. We introduce two global constraints for FEM, which use an implicit decomposition having a $\mathcal{O}(n)$ spatial complexity. Our global constraints are inspired by the state-of-the-art approaches [2,3,16] but keeping the reduction into a suffix database implicit instead of explicit. We propose two versions: with and without considering *gap* and *span* constraints. We are also able to take some algorithmic advantages in the filtering algorithms using the property that the (implicit) database is composed of sorted suffixes from a same sequence. To the best of our knowledge, this work is the first CP-based approach proposed for solving efficiently this family of problems with the benefit that several other constraints can be added.

This paper is organized as follows. Section 2 introduces the technical background related to the FEM problem. It explains how the problem can be modeled using CP and presents our first global constraint (`episodeSupport`). Section 3 shows how time can be integrated into the problem and describes the second global constraint (`episodeSupport` with time). Finally, experiments are carried out on synthetic and real-life datasets in Sect. 4.

2 Mining Episodes in a Non Timed Sequence

2.1 Technical Background

Let $\Sigma = \{1, \ldots, L\}$ be an alphabet representing a set of possible symbols. We define a non timed sequence $s = \langle s_1, \ldots, s_n \rangle$ over Σ as an ordered list of symbols such that $\forall i \in [1, n], s_i \in \Sigma$. Let us consider the following definitions based on the formalization of Aoga et al. [2] and Huang and Chang [14].

Definition 1 (Subsequence relation, Embedding). $\alpha = \langle \alpha_1, \ldots, \alpha_m \rangle$ is a subsequence of $s = \langle s_1, \ldots, s_n \rangle$, denoted by $a \preceq s$, if $m \leq n$ and if there exists a list of indexes $(e_1, \ldots, e_m)$ with $1 \leq e_1 \leq \cdots \leq e_m \leq n$ such that $s_{e_i} = \alpha_i$. Such a list is referred as an embedding of s. Sequence s is also referred as a super-sequence of α.

Example 1. $\langle a, b, c \rangle$ is a subsequence of the sequence $s = \langle a, b, a, c, b, a, c \rangle$ with embeddings $(1, 2, 4)$ or $(1, 2, 7)$ or $(3, 5, 7)$.

Definition 2 (Episode-embedding). *Let us consider* $\alpha = \langle \alpha_1, \ldots, \alpha_m \rangle \preceq s$. *Embedding* $(e_1, \ldots, e_m)$ *is an episode-embedding if it is an embedding of s and if all the other embeddings* $(e_1, e'_2, \ldots, e'_m)$ *are such that* $(e_2, \ldots, e_m) \preceq_L (e'_2, \ldots, e'_m)$ *where* $\preceq_L$ *represents a lexicographic ordering.*

Example 2. $\langle a, b, c \rangle$ is a subsequence of s with $(1, 2, 4)$ and $(3, 5, 7)$ as episode-embeddings. Besides, $(1, 2, 7)$ is not an episode-embedding because $(2, 7)$ is lexicographically greater than $(2, 4)$.

Definition 3 (Support). *The support* $\sigma_s(\alpha)$ *of a subsequence α in a sequence s is the number of episode-embeddings of α in s.*

Example 3. For s of Example 1, we have $\sigma_s(\langle a, b, c \rangle) = 2$.

Frequent Episode Mining (FEM) problem can then be formalised. Let us underline that this definition is related to the *total frequency* measure introduced by Iwanuma et al. [15]. The goal is to count up occurrences without duplication. To do so, we use the concept of prefix-projection introduced in PrefixSpan [13] and used thereafter by Kemmar et al. [16] and Aoga et al. [2] for SPM.

Definition 4 (Frequent Episode Mining (FEM)). *Given a set of symbols Σ, a sequence s over Σ and a threshold θ, the goal is to find all the subsequences α in s such that $\sigma_s(\alpha) \geq \theta$. These subsequences are called episodes.*

Definition 5 (Prefix, Projection, Suffix). *Let* $\alpha = \langle \alpha_1, \ldots, \alpha_k \rangle$ *and* $s = \langle s_1, \ldots, s_n \rangle$ *be two sequences. If* $\alpha \preceq s$, *then the prefix of* s *w.r.t.* α *is the smallest prefix of* s *that remains a super-sequence of* α. *Formally, it is the sequence* $\langle s_1, \ldots, s_j \rangle$ *such that* $\alpha \preceq \langle s_1, \ldots, s_j \rangle$ *and such that there exists no* $j' < j$ *where* $\alpha \preceq \langle s_1, \ldots, s_{j'} \rangle$. *The sequence* $\langle s_{j+1}, \ldots, s_n \rangle$ *is then called the suffix of* s *w.r.t.* α, *or the* α-*projection, and is denoted by* $s|_\alpha$. *If* α *is not a subsequence of* s, *the* α-*projection is empty.*

Example 4. Given sequence s of Example 1 and $\alpha = \langle b \rangle$, sequence $\langle a, b \rangle$ is a prefix of s w.r.t. α and $\langle a, c, b, a, c \rangle$ is a suffix ($s|_\alpha = \langle a, c, b, a, c \rangle$).

Definition 6 (Initial Projection). *An initial projection of a sequence* $s = \langle s_1, \ldots, s_n \rangle$ *w.r.t. a symbol* x, *denoted by* $s|_x^\mathcal{I}$, *is the list of all the suffixes* $s' = \langle s_i, \ldots, s_n \rangle$ *such that* $s_{i-1} = x$ *for all* $i \in (1, n]$.

Example 5. For s and a symbol a, we have $s|_a^\mathcal{I} = \big[\langle b, a, c, b, a, c \rangle, \langle c, b, a, c \rangle, \langle c \rangle\big]$.

Definition 7 (Internal Projection). *Given a list of sequences* Ω, *an internal projection of* Ω *w.r.t. pattern* α, *denoted by* $\Omega|_\alpha$, *is the list of the* α-*projection of all sequences in* Ω. *All the empty sequences are removed from* $\Omega|_\alpha$.

Example 6. For $\alpha = \langle b \rangle$ and $\Omega = \big[\langle b, a, c, b, a, c \rangle, \langle c, b, a, c \rangle, \langle c \rangle\big]$, we obtain $\Omega|_\alpha = \big[\langle a, c, b, a, c \rangle, \langle a, c \rangle\big]$.

Definition 8 (Projected Frequency). *Given the list of sequences* Ω, *and a projection* $s|_\alpha$ *for each sequence* $s \in \Omega$, *the projected frequency of a symbol is the number of* α-*projected sequences where the symbol appears.*

Example 7. Given the internal projection $\Omega|_\alpha$ of Example 6, the projected frequencies are a: 2, b: 1 and c: 2.

In practice, the initial projections and internal projections can be efficiently stored as a list of pointers in the original sequence s. In our example ($s = \langle a, b, a, c, b, a, c \rangle$), we have $s|_a^\mathcal{I} = [2, 4, 7]$ and starting from $\Omega = s|_a^\mathcal{I}$ we can represent $\Omega|_{\langle b \rangle} = [3, 6]$. This representation introduced in PrefixSpan [13] is called the *pseudo projection* representation. The algorithm works as follows. It starts from the empty pattern and successively extends it in a *depth-first search*. At each step, a symbol is added to the pattern, and all the sequences of the database are projected accordingly. A backtrack occurs when all the *projected frequencies* are below the support threshold. When a backtracking is performed during the search, the last appended symbol is removed. This procedure is known as the *pattern growth method* [13]. A new projection is thus built and stored at each step. An important consideration for the efficiency of this method is that the projected sequences do not need to be computed from scratch at each iteration. Instead, the pseudo-projection representation is used and maintained incrementally at each symbol extension of the pattern. Starting from the previous pseudo-projection, when the next symbol is appended, one can start from each position in the pseudo-projection representation and look, for each one, the next matching positions in s equal to this symbol. The new matching positions constitute the new pseudo-projection representation. Since the search follows a depth-first-search strategy, the pseudo projections can be stacked on a same vector allowing to reuse allocated entries on backtrack. This memory management is known as a trailing in CP and was introduced for SPM by Aoga et al. [2,3].

2.2   Problem Modelling

Our first contribution is a global constraint, episodeSupport, dedicated to find frequent patterns (or *episodes* [22]) in a sequence without considering time. Let $s = \langle s_1, \ldots, s_n \rangle$ be a sequence of n symbols over Σ, the set of distinct symbols appearing in s, and θ, the minimum support threshold desired.

Decision Variables. Let $P = \langle P_1, \ldots, P_n \rangle$ be a sequence of variables representing a pattern. The domain of each variable is defined as $P_i = \Sigma \cup \{\varepsilon\}$ for all $i \in [1, n]$. It indicates that each variable can take any symbol appearing in s as value in addition to ε, which is defined as the empty symbol. An assignment of P_i to ε means that P_i has matched no symbol. It is used to model patterns having a length lower than n. A solution is an assignation of each variable in P.

EpisodeSupport Constraint. The episodeSupport(P, s, θ) constraint enforces the three following constraints: (1) $P_1 \neq \varepsilon$, (2) $P_i = \varepsilon \rightarrow P_{i+1} = \varepsilon$, $\forall i \in [2, n)$ and (3) $\sigma_s(P) \geq \theta$. The first constraint states that a pattern cannot begin with the empty symbol. It indicates that a valid pattern must contain at least one symbol. The second constraint ensures that ε can only appear at the end of the pattern. It is used in order to prevent same patterns with ε in different positions to be part of the same solution (such as $\langle a, b, \varepsilon \rangle$ and $\langle a, \varepsilon, b \rangle$). Finally, the last constraint states that a pattern must occur at least θ times in the sequence. The goal is then to find an assignment of each P_i satisfying the three constraints. The episodeSupport constraint filters from the domains of variables P the infrequent symbols in s at each step in order to find an assignment representing a frequent pattern according to θ. All the inconsistent values of the next uninstantiated variables in the pattern are then removed. Assuming the pattern variables are labeled in static order from left to right, the search is failure free when only this constraint must hold (i.e. all the leaf nodes are solution). Besides, episodeSupport is domain consistent: the remaining values in the domain of each variable are part of a solution because all of them have, at least, one support. Additional constraints can also be integrated to the model in order to define properties that the patterns must satisfy. For instance, we can enforce patterns to have at most k symbols or to follow a regular expression.

2.3   Filtering Algorithm

Preprocessing. The index of the last position of each symbol in s is stored into a map (*lastPos*). For instance, $s = \langle a, b, a, c, \mathbf{b}, \mathbf{a}, \mathbf{c} \rangle$ gives $\{(c \rightarrow 7), (a \rightarrow 6), (b \rightarrow 5)\}$. The map can be iterated in a decreasing order by the last positions.

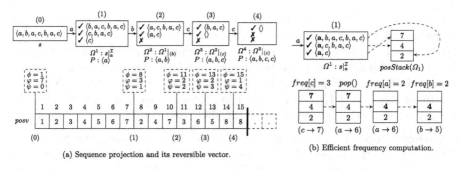

(a) Sequence projection and its reversible vector.

(b) Efficient frequency computation.

Fig. 1. Sequence projection (✓ indicates a match, ✗ otherwise), its reversible vector and frequency computation mechanism.

Sequence Projection and Pseudo Projection.

The key idea is to successively compute a projection from the previous one each time a variable has been assigned. The assignment of the first variable of the pattern (P_1) involves an *initial projection*. It splits s into a list of subsequences such that each one begins with the projected symbol. The assignment of the other variables (P_2 to P_n) implies an *internal projection*. This behavior is illustrated in the upper part of Fig. 1a for an arbitrary example. The steps leading to pattern $\langle a, b, c, c \rangle$ are detailed. Three subsequences are obtained after an *initial projection* of symbol a $((0) \rightarrow (1))$. While there are non empty sequences, *internal projections* are successively performed $((1) \rightarrow (4))$ and the pattern (P) is incrementally built.

In practice, only pointers to the position in each sequence where the prefix has matched are stored. It is the mechanism of *pseudo projection*. As Aoga et al. [2], we implement it with a reversible vector (*posv*) and a trail-based structure (lower part of Fig. 1a). The idea is to use the same vector during all the search inside the propagator, and to only maintain relevant start and stop positions. At each propagator call, three steps are performed. First, the last recorded start and stop positions are taken. Secondly, the propagator records the new information in the vector after the previous stop position. Finally, the new positions are updated in order to retrieve the information added. The reversible vector is then built incrementally. For each projection, the corresponding start index (ϕ) in the vector as well as the number of sequences inside the projection (φ) are stored. In other words, information related to a projection are located between indexes $i \in [\phi, \phi + \varphi)$. Besides, the index of the variable P_i that has been assigned (ψ) is also recorded after each projection step. Before the first assignation ψ is equal to zero. The three variables are implemented as reversible integers. Initially, all the indexes are present in the vector, but all along the pseudo-projections, only the non empty sequences are considered.

Propagation.

The goal is to compute a projection each time a variable has been assigned to a symbol a. Assignments of variables are done successively from the first variable to the last one. The propagator is then called after each assignment. It is shown in Algorithm 1. Initialization of reversible structures is done when

$\psi = 0$ (lines 8–9). If the last assigned variable (P_ψ) has been bound to ε, the algorithm enforces all the next variables to be also bound to ε (lines 10–12). The pattern is then completed and the propagation is finished. Otherwise, after each variable binding, the projected sequence and the *projected frequencies* are computed (line 14). Finally, all the infrequent symbols are removed from the domain of $P_{\psi+1}$ (lines 15–17). *Projected frequency* of each symbol in the domain of $P_{\psi+1}$ (except ε) is compared to the threshold and removed if it is infrequent.

Sequence Projection. Let us now present how sequences are projected (Algorithm 2). First, the *projected frequency* of each symbol for the current pseudo projection is set to 0 (*freq* on line 8). The main loop (lines 10–23) iterates over the previous projection thanks to the reversible integers ϕ and φ. At each iteration a value in *posv* is considered. The next condition (line 12) is used to distinguish the *initial* from an *internal projection*. If a is the first projected symbol and if it does not match with the first symbol of the sequence, then the sequence is not included in the projection. Otherwise, an *internal projection* is applied.

The next expression (lines 13–14) is an optimization we introduced, called *early projection stopping*. This optimization is based on one invariant of our structure: it stores suffixes of s with a decreasing order by their size. Each suffix in a projection is then strictly included in all the previous ones. When a no-match has been detected in a sequence, all the next ones can be directly discarded without being checked. It stops the internal projection as soon as possible. The *early projection stopping* gains importance when the number of sequences is large. Then, if a is not present in the current considered sequence, the loop can be stopped and unnecessary computation is avoided.

At this step, we are sure that a appears at least once in the current sequence in the projection. Lines 16 to 21 make the search for the match, either by *position caching*, or by iteration on the positions. *Position caching* is a second optimization we introduced. Once a match has been detected in a sequence, the position of the match is recorded. Thanks to the aforementioned invariant, we are sure that the match in the next sequence cannot occur before this position. If this position is greater than the start position of the sequence (in *posv*), a match is directly detected. The reversible vectors are then updated (line 23). Variable *sup* is used to store the size of the new projection.

The last loop (lines 24–28) updates the *projected frequency* of each symbol. The *projected frequency* of a symbol in a projection corresponds to the number of sequences of the projection beginning at an index lower than the index of the last position of the symbol. This idea was introduced in LAPIN [33] and exploited by Aoga et al. [2]. It can be implemented efficiently thanks to the invariant and *lastPos* map. It is illustrated in Fig. 1b (upper part) with $lastPos = \{(c \rightarrow 7), (a \rightarrow 6), (b \rightarrow 5)\}$. The position just after each match is pushed in a LIFO structure, *posStack*. The last matched position is located on the top of the stack.

Once the stack is obtained, the idea is to successively compare in a decreasing order the last position of each symbol with the top of the stack. Illustration of this behavior is presented in Fig. 1b (lower part). If the last position of a symbol is greater than the top of the stack, it indicates that the symbol occurs at least

---

Algorithm 1. $propagate(s, \Sigma, a, P)$

---

1 ▷ **Internal State:** $posv$, ϕ, φ and ψ.
2 ▷ **Pre:** s is the initial long sequence of size n.
3 ▷ Σ is the set of symbols and a is a symbol.
4 ▷   If $\psi > 0$ then $\langle P_1, \ldots, P_\psi \rangle \in P$ are bound and P_ψ is assigned to a.
5 ▷ θ is the support threshold.
6 ▷ $posv$, ϕ, φ and ψ are the reversible structures as defined before.
7
8   **if** $\psi = 0$ **then**
9      $\phi := 1$ $\varphi := n$ $\psi := 1$ $posv[i] := i$ $i \in [1, n]$
10   **if** $P_\psi = \varepsilon$ **then**
11   **for** $j \in [\psi + 1, n]$ **do**
12         $P_j.assign(\varepsilon)$
13   **else**
14      $freq := sequenceProjection(s, \Sigma, a)$ ▷ *Detailed in Alg. 2.*
15   **foreach** $b \in Domain(P_{\psi+1})$ **do**
16   **if** $b \neq \varepsilon \wedge freq[b] < \theta$ **then**
17            $P_{\psi+1}.removeValue(b)$

Algorithm 2. $sequenceProjection(s, \Sigma, a)$

---

1 ▷ **Internal State:** $posv$, ϕ, φ and ψ.
2 ▷ **Pre:** s is the initial long sequence.
3 ▷ Σ is the set of symbols.
4 ▷ a is the current projected symbol $(a = P_\psi)$.
5 ▷ $posv$, ϕ, φ and ψ are reversible structures as defined before.
6 ▷ $posv[i]$ with $i \in [\phi, \phi + \varphi)$ is initialized.
7
8   $j := \varphi$ $sup := 0$ $prevPos := -1$ $freq[b] := 0 \ \forall b \in \Sigma$
9   $posStack := Stack()$
10   **for** $i \in [\phi, \phi + \varphi - 1]$ **do**
11      $pos := posv[i]$
12   **if** $\psi > 1 \vee a = s[pos]$ **then**
13   **if** $pos > lastPos[a]$ **then**
14   **break**   ▷ *Early projection stopping*
15   **else**
16   **if** $prevPos < pos$ **then**
17   **while** $a \neq s[pos]$ **do**
18                  $pos := pos + 1$
19               $prevPos := pos$ ▷ *Position caching*
20   **else**
21               $pos := prevPos$
22            $posStack.push(pos + 1)$
23            $posv[j] := pos + 1$ $j := j + 1$ $sup := sup + 1$
24   **foreach** (x, pos_x) **in** $lastPos$ **do**
25   **while** $posStack.notEmpty \wedge posStack.top > pos_x$ **do**
26         $posStack.pop$
27      $freq[x] := posStack.size$ ▷ *Projected frequency*
28   **if** $posStack.isEmpty$ **then break**
29   $\phi := \phi + \varphi$ $\varphi := sup$ $\psi := \psi + 1$
30   **return** $freq$

---

once in the current sequence and consequently in all the previous ones in the stack. The *projected frequency* of this symbol corresponds then to the remaining size of the stack and the next symbol in *lastPos* can be processed. Otherwise, we are sure that the symbol has no occurrence in the current sequence. Its position is popped and the comparison is done with the new top. The resulting *projected frequencies* are $c = 3$, $a = 2$ and $b = 2$. This mechanism has a time complexity of $\mathcal{O}(n + |\Sigma|)$. For comparison, *projected frequencies* are computed in $\mathcal{O}(n \times |\Sigma|)$ by Aoga et al. [2] (each subsequence is scanned at each projection). Finally, the reversible integers are updated and the *projected frequency* map is returned.

Time and Spatial Complexity. Main loop of Algorithm 2 (lines 10–23) is computed in $\mathcal{O}(n^2)$ and the *projected frequencies* (lines 24–28) in $\mathcal{O}(n + |\Sigma|) = \mathcal{O}(n)$ (because the number of different symbols is bounded by the sequence size). In Algorithm 1, lines 8–9 cost $\mathcal{O}(n)$ and the domain pruning (lines 15–17) is performed in $\mathcal{O}(|\Sigma|)$. It gives $\mathcal{O}(n + (n^2 + n) + |\Sigma|) = \mathcal{O}(n^2)$. For the spatial complexity, we have $\mathcal{O}(n + n \times d) = \mathcal{O}(n \times d)$ with d the maximum depth of the search tree, which is the maximum size of the reversible vector. For comparison, an explicit decomposition of the problem gives $\mathcal{O}(n^2 + n \times d) = \mathcal{O}(n^2)$.

3   Mining Episodes in a Timed Sequence

3.1   Technical Background

So far, `episodeSupport` can only deal with *sequences of symbols* where time is not considered. In practice, sequences can also be time-stamped. Such sequences are most often referred as *sequences of events* instead of *sequences of symbols* and new constraints can then be expressed. For instance, we can be interested in finding episodes such that the elapsed time between two events does not exceed one hour. We define a sequence of events $s = \langle (s_1, t_1), \ldots, (s_n, t_n) \rangle$ over Σ as an ordered list of events (s_i) occurred at time t_i such that for all $i \in [1, n]$ we have $s_i \in \Sigma$ and $t_1 \leq t_2 \leq \ldots \leq t_n$. The list containing only the events is denoted by s^s and the list of timestamps by s^t. All the principles defined in the previous sections are reused. Besides, we are now able to enforce time restrictions. Two of them are used in practice: *gap* and *span*. The former (*gap*) restricts the time between two consecutive events while the latter (*span*) restricts the time between the first and the last event. Considering such restrictions cannot be done only by imposing additional constraints in the model [3]. It requires to adapt the subsequence relations (Definition 9) and to design a dedicated propagator. The concept of *extension window* is also defined. The extension window of an embedding contains only events whose timing satisfies *gap* constraint.

Definition 9 (Subsequence under gap/span). *$\alpha = \langle \alpha_1, \ldots, \alpha_m \rangle$ is a subsequence of $s = \langle (s_1, t_1), \ldots, (s_n, t_n) \rangle$ under gap$[M, N]$, denoted by $\alpha \preceq^{gap[M,N]} s$, if and only if s^s is a subsequence of embedding $(e_1, \ldots, e_k)$ according to Definition 1, and if $\forall i \in [2, k]$ we have $M \leq t_{e_i} - t_{e_{i-1}} \leq N$. The embedding $(e_1, \ldots, e_k)$ under $\preceq^{gap[M,N]}$ relation is called a gap$[M, N]$-embedding. $(e_1, \ldots, e_k)$ is an* episode-embedding of α according to Definition 2 where $\preceq^{gap[M,N]}$ is considered for the subsequence relation. The support of α, denoted by $\sigma_s^{gap[M,N]}(e)$, is the number of $gap[M, N]$-embeddings of α in s. Similarly, α is a subsequence of s under $span[W, Y]$, denoted by $\alpha \preceq^{span[W,Y]} s$, if and only if s^s is a subsequence of embedding $(e_1, \ldots, e_k)$ according to Definition 1, and if $W \leq t_{e_k} - t_{e_1} \leq Y$. Relation $\preceq^{span[W,Y]}$ and $\sigma_s^{span[M,N]}(e)$ are also defined similarly.

Example 8. Let us consider $s = \langle (a, 2), (b, 4), (a, 5), (c, 7), (b, 8), (a, 9), (c, 12) \rangle$. $\langle a, b, c \rangle$ is a subsequence of s under $gap[1, 3]$ with embedding $(1, 2, 4)$. $(3, 5, 7)$ is not a $gap[1, 3]$-embedding because $t_{e_3} - t_{e_2} = 12 - 8 > 3$. Besides, $\langle a, b, c \rangle$ is a subsequence of s under $span[6, 10]$ with embedding $(3, 5, 7)$. $(1, 2, 4)$ is not valid because $t_{e_3} - t_{e_1} = 7 - 2 < 6$.

Definition 10 (Extension window). Let $e = (e_1, e_2, \ldots, e_k)$ be any $gap[M, N]$-embedding of a subsequence α in a sequence s. The extension window of this embedding, denoted $ew_e^{gap[M,N]}(s)$, is the subsequence $\langle (s_u, t_u), \ldots, (s_v, t_v) \rangle$ such that $(t_{e_k} + M \leq t_u) \wedge (t_v \leq t_{e_k} + N) \wedge (t_{u-1} < t_{e_k} + M) \wedge (t_{v+1} > t_{e_k} + N)$. Each embedding has a unique extension window, which can be empty.

Example 9. Let $(3, 4)$ be a $gap[2, 6]$-embedding of $\langle a, c \rangle$ in sequence s (Example 8). We have $ew_e^{gap[2,6]}(s) = \langle (a, 9), (c, 12) \rangle$.

The goal is to find the all patterns having a support, possibly under gap and $span$, greater than the threshold. Let $P = \langle P_1, \ldots, P_n \rangle$ be a sequence of variables representing a pattern. the timed version of episodeSupport $(P, s, \theta, M, N, W, Y)$ enforces the four following constraints: (1) $P_1 \neq \varepsilon$, (2) $P_i = \varepsilon \rightarrow P_{i+1} = \varepsilon, \forall i \in [1, n)$, (3) $\sigma_s^{gap[M,N]}(P) \geq \theta$ and (4) $\sigma_s^{span[M,N]}(P) \geq \theta$.

3.2   Filtering Algorithm

Precomputed Structures. The three structures are shown in Fig. 2a. First, the *lastPos* map is adapted from the previous section in order to store the last position of each event that can be matched while satisfying the maximum span (Y). The last position of each event inside each range $[t, t+Y]$ is recorded, where t is the timestamp of the event. Maximum span is then implicitly handled by this structure, which is not done by Aoga et al. [3]. Besides, for each position i in s, the index of the first (u) and the last (v) positions after i such that $t_u \geq t_i + M$ and $t_v \geq t_i + N$ are stored into a map $(nextPosGap)$, where M and N are the minimum and maximum gap. The *nextPosGap* is used after each projection in order to directly access the next extension window. Finally, for each position i in s, the number of times that each event has occurred inside the range $[1, i]$ in s is stored $(freqMap)$. It is used in order to efficiently compute the *projected frequency* of each symbol during a projection. We can be sure that an event a appears at least once in a window of range $[u, v]$ if the occurrence of a at the end of the window is strictly greater than the occurence of a just before the window $(freqMap[v][a] > freqMap[u - 1][a])$. It has not been used by Aoga et al. [3].

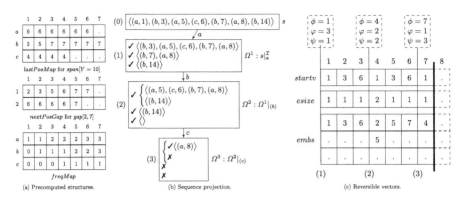

Fig. 2. Data structures used for timed sequences with $gap[2,7]$ and $span[1,10]$.

Storing Several Embeddings. When a *gap* constraint is considered, the anti-monotonicity property does not hold anymore [3]. The main consequence is that all the possible embeddings must be considered, and not only the first one. The projection mechanism (described in Fig. 1a) has then to be adapted. It is illustrated in Fig. 2b. For instance, two embeddings are considered for the projection from (1) to (2) of the first sequence. It is required to record all of the corresponding extension windows in order to miss none supporting event. To do so, a reversible vector (*startv*) recording the start index in s for each sequence is used (Fig. 2c). Besides, other reversible vectors are added: *esize*, which represents the number of embeddings related at each projected sequence and *embs*, which records the start index of the different embeddings. It is a simplified adaptation of the structure proposed by Aoga et al. [3].

Minimum Span. The minimum span is not anti-monotonic. Therefore, we do not consider it during the projection but *a posteriori*: it is only checked when a complete pattern is obtained and not before. It requires slight modifications in the *propagate* method (Algorithm 1). A variable $\gamma_s(P)$ representing the number of supports satisfying the minimum span constraint for P is recorded and computed during the projection. Once the projections are completely done for this episode (after the line 12), $\gamma_s(P)$ is compared with the support threshold and an *inconsistency* is raised if it is below the threshold.

Sequence Projection. Projection mechanism is presented in Algorithm 3. Initially, the *projected frequencies* of each event is set to 0 (line 9). When $\psi = 1$, an initial projection is performed (lines 10–20) and we are looking for events that match with a (line 13). Once a match is detected, reversible vectors are updated (line 14). *Projected frequencies* are computed using *nextPosGap* and *freqMap* structures (lines 15–20). First, the window where the events must be considered is computed. Secondly, the *projected frequency* of each event appearing in the window is incremented. When $\psi > 1$, we have an internal projection (lines 21–43). Each embedding is successively considered (line 26). For each one, the sequence is iterated from the first next position satisfying the minimum gap to the last one satisfying the maximum gap, or to the last symbol of the sequence (line 28). Once a match has been detected, the number of possible embeddings is incremented and its position is recorded (line 30). If the embedding of the current pattern satisfies the minimum span constraint, γ_s is incremented (lines 31–32). It is used in the *propagate* method as explained before. Then, *projected frequencies* are computed as in the initial projection (lines 33–39).

Time and Spatial Complexity. Let us assume k is the maximum length of time window (often $k \ll n$) and d the maximum depth of the tree search ($d \leq k$). Initial projection (lines 11–21) in Algorithm 3 is computed in $\mathcal{O}(n \times |\Sigma|)$: the sequence is completely processed and frequencies are computed at each match. Internal projection (lines 22–44) is computed in $\mathcal{O}(n \times |\Sigma| \times k^2)$. It gives $\mathcal{O}(n \times |\Sigma| + n \times |\Sigma| \times k^2) = \mathcal{O}(n \times |\Sigma| \times k^2)$. For the spatial complexity, vectors have a maximum length of $k \times d$ and there are at most k embeddings, which gives $\mathcal{O}(d \times k^2)$.

4 Experimental Results

This section evaluates the performance of `episodeSupport` on different datasets with and without time consideration. Experiments have been realised on a computer with a 2.7 GHz Intel Core i5 64 bits processor and with a RAM of 8 Go using a 64-Bit HotSpot(TM) JVM 1.8 running on Linux Mint 17.3. Execution time is limited to 1800 s unless otherwise stated. The algorithms have been implemented in Scala with OscaR solver [31] and memory assessment has been performed with *java Runtime* classes. For the reproducibility of results, the implementation of both constraints is open source and available online.[1] One synthetic and three real-data sets are considered: proteins from Uniprot database [5], UCI Unix dataset [20] and UbiqLog [26,27].

Our approach is compared with the existing methods. We identified two ways to mine frequent patterns in a sequence. On the one hand, we can resort to a specialized algorithm. To the best of our knowledge, MINEPI+ and EMMA [14] are the state-of-the-art methods for that. On the other hand, we can explicitly split the sequence into a database and then reduce the problem into an SPM problem. Once done, CP-based methods can be used [2,3,16,23]. Our comparisons are based on the approach of Aoga et al. [2,3] that turns out to be the most efficient. We refer to it as the Decomposed Frequent Episode Mining (DFEM) approach, or DFEMt when time is considered.

Memory Bound Analysis. We applied DFEM and `episodeSupport` on synthetic sequences of different sizes with 100 distinct symbols uniformly distributed in order to define what are the largest sequences that can be processed. We observed that with decomposed approaches, sequences greater than 30000 symbols cannot be processed when memory is limited to 8 GB. With `episodeSupport` memory is not a bottleneck.

---

[1] https://bitbucket.org/projetsJOHN/episodesupport (also available in [31]).

Algorithm 3. $sequenceProjectionTimed(s^s, s^t, \Sigma, a, N, W)$

---

1 ▷ **Internal State:** $startv$, $esize$, $embs$, ϕ, φ, ψ, $\gamma_s(P_{:\psi})$.
2 ▷ **Pre:** s^s and s^t are the event/timestamp list of the initial long sequence.
3 ▷ Σ is the set of symbols.
4 ▷ a is the current projected symbol ($a = P_\psi$).
5 ▷ $startv[i]$, $esize[i]$ and $embs[i]$ with $i \in [\phi, \phi + \varphi)$ are initialized.
6 ▷ $\gamma_s(P_{:\psi}) = 0$ with $P_{:\psi}$ the episode represented by $\langle P_1, \ldots, P_\psi \rangle$.
7 ▷ N and W are the gap max bound and of the span min bound.
8
9 $freq[b] := 0 \quad \forall b \in \Sigma$
10 **if** $\psi = 1$ **then**
11 | $j := 1$
12 | **for** $pos \in [1, |s^s|]$ **do**
13 | | **if** $s^s[pos] = a$ **then**
14 | | | $startv[j] := pos \quad esize[j] := 1 \quad embs[j][1] := pos \quad j := j + 1$
15 | | | $(u, v) := nextPosGap[pos]$ ▷ *Precomputed structure*
16 | | | **if** $u \leq |s^s|$ **then**
17 | | | | **for** $b \in Domain(P_{\psi+1})$ **do**
18 | | | | | $l := \min(v, |s^s|)$
19 | | | | | **if** $freqMap[l][b] > freqMap[u - 1][b]$ **then**
20 | | | | | | $freq[b] := freq[b] + 1$ ▷ *Projected frequency*

21 **else**
22 | $j := \phi + \varphi$
23 | **for** $i \in [\phi, \phi + \varphi - 1]$ **do**
24 | | $id := startv[i] \quad nEmb := 0 \quad k := 1 \quad v := -1 \quad isIncremented := \textbf{false}$
25 | | $isVisited[b] := \textbf{false} \quad \forall b \in \Sigma$
26 | | **while** $v < |s^s| \wedge k \leq esize[i]$ **do**
27 | | | $e := embs[i][k] \quad (pos, \_) := nextPosGap[e]$ ▷ *2nd element unused*
28 | | | **while** $v < |s^s| \wedge pos \leq lastPosMap[id][a] \wedge s^t[pos] \leq s^t[e] + N$ **do**
29 | | | | **if** $s^s[pos] = a$ **then**
30 | | | | | $nEmb := nEmb + 1 \quad embs[j][nEmb] := pos$
31 | | | | | **if not** $isIncremented \wedge s^t[pos] - s^t[id] \geq W$ **then**
32 | | | | | | $isIncremented := \textbf{true} \quad \gamma_s(P_{:\psi}) := \gamma_s(P_{:\psi}) + 1$
33 | | | | | $(u, v) := nextPosGap[pos]$ ▷ *Precomputed structure*
34 | | | | | **if** $u \leq |s^s|$ **then**
35 | | | | | | **for** $b \in Domain(P_{\psi+1})$ **do**
36 | | | | | | | $l := \min(v, |s^s|)$
37 | | | | | | | **if** $\big(freqMap[l][b] > freqMap[u - 1][b]\big) \wedge \textbf{not}$ $isVisited[b]$ **then**
38 | | | | | | | | $isVisited[b] := \textbf{true}$
39 | | | | | | | | $freq[b] := freq[b] + 1$ ▷ *Projected frequency*
40 | | | | $pos := pos + 1$
41 | | | $k := k + 1$
42 | | **if** $nEmb > 0$ **then**
43 | | | $startv[j] := id \quad esize[j] := nEmb \quad j := j + 1$

44 $\phi := \phi + \varphi \quad \varphi := j - \phi$
45 **return** $freq$

Comparison with Decomposed Approaches. Experiments and results with Uniprot and UbiqLog datasets are shown in Fig. 3 and Table 1. The latter presents results for different settings while the former shows the performance profiles [8] for both the memory consumption and the computation time.

We can observe that `episodeSupport` outperforms both decomposed approaches in terms of execution time and memory consumption for most of the instances. Both gains become more important when the sequence is large.

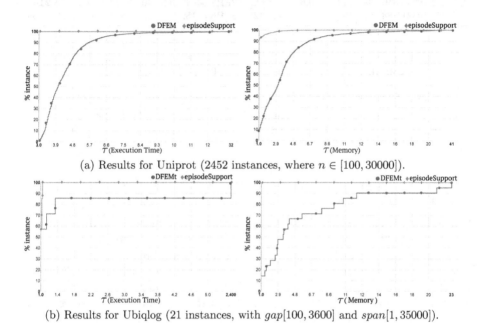

(a) Results for Uniprot (2452 instances, where $n \in [100, 30000]$).

(b) Results for Ubiqlog (21 instances, with $gap[100, 3600]$ and $span[1, 35000]$).

Fig. 3. Performance profiles ($\theta = 5\%$, maximum size of 5, timeout of 600 s).

Table 1. Execution time and memory usage for several datasets and thresholds.

Name $	s	\times	\Sigma	$			Patterns having a maximum size of 5				Name $	s	\times	\Sigma	$			$gap[100, 3600]$ and $span[1, 35000]$			
			Memory (Mb)		Execution time (s)					Memory (Mb)		Execution time (s)	
	θ	nSol	DFEM	episodeSupport	DFEM	episodeSupport		θ	nSol	DFEMt	episodeSupport	DFEMt	episodeSupport
Q08379 1002×20	100	437048	45	**34**	9.08	**2.45**	10Mcomplete 1072×30	300	10	107	**35**	0.24	**0.20**
	90	533395	46	**38**	8.68	**1.97**		100	3113	999	**469**	67.83	**29.53**
	70	645834	35	**16**	9.16	**1.67**		50	10481	1226	**505**	118.56	**57.48**
	50	1128537	67	**43**	13.29	**2.59**		20	51108	1110	**599**	204.98	**80.93**
Q54CU4 11103×20	1110	0	1969	**31**	0.11	**0.02**	9Mcomplete 1128×45	10	6724	244	**139**	1.00	1.35
	999	157003	2057	**33**	113.32	**3.49**		8	11626	318	**173**	1.35	**0.71**
	777	1178939	1980	**33**	657.68	**14.90**		6	19340	339	**142**	1.42	**0.86**
	555	1515789	1849	**31**	1414.83	**18.41**		4	23225	349	**138**	1.48	**1.05**
Q91TU4 18141×20	1814	336842	6898	**38**	946.54	**20.90**	8Mcomplete 3305×44	300	3734	1888	**506**	197.16	**143.94**
	1632	505263	6426	**40**	1146.22	**24.20**		100	38061	3301	**1179**	745.41	**488.32**
	1269	705640	6819	**21**	1674.80	**22.63**		50	123133	3594	**1496**	1489.18	**740.97**
	907	1515791	6819	**21**	timeout	**52.79**		20	516478	3859	**1309**	timeout	**1163.34**

Table 2. Comparison with MINEPI+ and EMMA ($\theta = 5\%$ and $W = 10$).

Databases Features			Execution Time (s)								
name	$\|s\|$	$\|\Sigma\|$	nSol	MINEPI+	EMMA	episodeSupport	name	$\|s\|$	$\|\Sigma\|$	nSol	MINEPI+	EMMA	episodeSupport
USER3	16866	273	46	13.2	0.4	**0.173**	USER5	34821	563	37	4.8	**0.3**	0.724
USER7	17329	449	25	0.6	**0.2**	0.563	USER4	37817	479	48	165.3	1.3	**0.636**
USER2	18738	310	38	43.3	0.6	**0.259**	USER8	54042	706	40	1362.3	9.8	**2.214**
USER1	19881	288	57	93.7	1.2	**0.232**	USER6	64152	609	68	2853.3	14.6	**2.178**

Table 3. Additional constraints on Q08379 Protein (Uniprot).

Only episodeSupport		+ exclusion		+ atLeast		+ regular	
nSol:	Time (s):	nSol:	Time (s):	nSol:	Time (s):	nSol: 2	Time (s):
46221933	83.2	33388768	62.6	104536	0.642		0.002

Besides, decomposed approaches cannot process the largest sequences regarding the time limitation imposed.

Comparison with Specialized Approaches. Experiments on Unix dataset with a threshold of 5% and a maximum span of 10 are provided in [14][2]. Comparisons of these specialized approaches with ours are presented in Table 2. It shows that episodeSupport seems competitive with MINEPI+ and EMMA. For the largest sequences (USER8 and USER6), episodeSupport is the most efficient. For some instances (USER5 and USER7) that are quickly solved, the cost of initializing the data structures with our approach is higher than the gain obtained. In general, the gain becomes more important when sequences are larger or harder to solve. Finally, given that the implementation of MINEPI+ and EMMA is missing, it is difficult to perform a fair comparison of the approaches.

Handling Additional Constraints. Additional constraints can be considered in order to define properties that the patterns must satisfy. No modification of episodeSupport is required. Results of experiments are presented in Table 3. The goal was to find frequent episodes ($\theta \geq 20$) having a maximum length of 6, containing at least three Q (atLeast constraint) but no D (exclusion), and satisfying the regex M(A|T).*F (regular). Two episodes (MTQQQF and MAQQQF) have been discovered. As observed, the additional constraints reduce the execution time as CP takes advantage of the stronger filtering to reduce the search space. This reduction would not be observed with a *generate and filter* approach.

---

[2] Results provided in [14] are directly used since the implementation is not available.

5   Conclusion and Perspective

There is a growing interest for solving data-mining challenges with CP. In addition to the flexibility it brings, recent works have shown that it can provide similar performances, or even better, than specialized algorithms [2,3]. So far CP has not been considered yet for mining frequent episodes. We introduced two global constraints (episodeSupport) for solving this problem with or without time-stamps. It relies on techniques used for SPM such as *pattern growth, pseudo projections* and *reversible vectors* but also on new ideas specific to this problem for improving the efficiency of the filtering algorithms (*early projection stopping, position caching* and *efficient frequency computation*). Experimental results have shown that our approach provides better performances in terms of execution time and memory consumption than state-of-the-art methods, with the additional benefits that it can accommodate additional constraints.

References

1. Agrawal, R., Mannila, H., Srikant, R., Toivonen, H., Verkamo, A.I., et al.: Fast discovery of association rules. Adv. Knowl. Discov. Data Min. **12**(1), 307–328 (1996)
2. Aoga, J.O.R., Guns, T., Schaus, P.: An efficient algorithm for mining frequent sequence with constraint programming. In: Frasconi, P., Landwehr, N., Manco, G., Vreeken, J. (eds.) ECML PKDD 2016. LNCS (LNAI), vol. 9852, pp. 315–330. Springer, Cham (2016). https://doi.org/10.1007/978-3-319-46227-1_20
3. Aoga, J.O.R., Guns, T., Schaus, P.: Mining time-constrained sequential patterns with constraint programming. Constraints **22**(4), 548–570 (2017)
4. Calders, T., Dexters, N., Goethals, B.: Mining frequent itemsets in a stream. In: 2007 Seventh IEEE International Conference on Data Mining, ICDM 2007, pp. 83–92. IEEE (2007)
5. UniProt Consortium: The universal protein resource (UniProt). Nucleic Acids Res. **36**(Suppl. 1), D190–D195 (2008)
6. Cule, B., Goethals, B., Robardet, C.: A new constraint for mining sets in sequences. In: Proceedings of the 2009 SIAM International Conference on Data Mining, pp. 317–328. SIAM (2009)
7. Das, G., Lin, K.I., Mannila, H., Renganathan, G., Smyth, P.: Rule discovery from time series. In: KDD, vol. 98, pp. 16–22 (1998)
8. Dolan, E.D., Moré, J.J.: Benchmarking optimization software with performance profiles. Math. Program. **91**(2), 201–213 (2002). https://doi.org/10.1007/s101070100263
9. Feurer, M., Klein, A., Eggensperger, K., Springenberg, J., Blum, M., Hutter, F.: Efficient and robust automated machine learning. In: Cortes, C., Lawrence, N.D., Lee, D.D., Sugiyama, M., Garnett, R. (eds.) Advances in Neural Information Processing Systems 28, pp. 2962–2970. Curran Associates, Inc. (2015). http://papers.nips.cc/paper/5872-efficient-and-robust-automated-machine-learning.pdf
10. Ghahramani, Z.: Automating machine learning. In: Lecture Notes in Computer Science, vol. 9852 (2016)
11. Guns, T., Dries, A., Tack, G., Nijssen, S., De Raedt, L.: MiningZinc: a modeling language for constraint-based mining. In: Proceedings of the Twenty-Third International Joint Conference on Artificial Intelligence, pp. 1365–1372. AAAI Press (2013)

12. Guns, T., Nijssen, S., De Raedt, L.: Itemset mining: a constraint programming perspective. Artif. Intell. **175**(12–13), 1951–1983 (2011)
13. Han, J., Pei, J., Mortazavi-Asl, B., Pinto, H., Chen, Q., Dayal, U., Hsu, M.: Prefixspan: mining sequential patterns efficiently by prefix-projected pattern growth. In: Proceedings of the 17th International Conference on Data Engineering, pp. 215–224 (2001)
14. Huang, K.Y., Chang, C.H.: Efficient mining of frequent episodes from complex sequences. Inf. Syst. **33**(1), 96–114 (2008)
15. Iwanuma, K., Takano, Y., Nabeshima, H.: On anti-monotone frequency measures for extracting sequential patterns from a single very-long data sequence. In: 2004 IEEE Conference on Cybernetics and Intelligent Systems, vol. 1, pp. 213–217. IEEE (2004)
16. Kemmar, A., Loudni, S., Lebbah, Y., Boizumault, P., Charnois, T.: A global constraint for mining sequential patterns with GAP constraint. In: Quimper, C.-G. (ed.) CPAIOR 2016. LNCS, vol. 9676, pp. 198–215. Springer, Cham (2016). https://doi.org/10.1007/978-3-319-33954-2_15
17. Kotthoff, L., Nanni, M., Guidotti, R., O'Sullivan, B.: Find your way back: mobility profile mining with constraints. In: Pesant, G. (ed.) CP 2015. LNCS, vol. 9255, pp. 638–653. Springer, Cham (2015). https://doi.org/10.1007/978-3-319-23219-5_44
18. Kotthoff, L., Thornton, C., Hoos, H.H., Hutter, F., Leyton-Brown, K.: Auto-WEKA 2.0: automatic model selection and hyperparameter optimization in WEKA. J. Mach. Learn. Res. **17**, 1–5 (2017)
19. Laxman, S., Sastry, P., Unnikrishnan, K.: A fast algorithm for finding frequent episodes in event streams. In: Proceedings of the 13th ACM SIGKDD International Conference on Knowledge Discovery and Data Mining, pp. 410–419. ACM (2007)
20. Lichman, M.: UCI machine learning repository (2013). https://archive.ics.uci.edu/ml/datasets/UNIX+User+Data
21. Mannila, H., Toivonen, H.: Discovering generalized episodes using minimal occurrences. In: KDD, vol. 96, pp. 146–151 (1996)
22. Mannila, H., Toivonen, H., Verkamo, A.I.: Discovering frequent episodes in sequences extended abstract. In: 1st Conference on Knowledge Discovery and Data Mining (1995)
23. Negrevergne, B., Guns, T.: Constraint-based sequence mining using constraint programming. In: Michel, L. (ed.) CPAIOR 2015. LNCS, vol. 9075, pp. 288–305. Springer, Cham (2015). https://doi.org/10.1007/978-3-319-18008-3_20
24. Nijssen, S., Guns, T.: Integrating constraint programming and itemset mining. In: Balcázar, J.L., Bonchi, F., Gionis, A., Sebag, M. (eds.) ECML PKDD 2010. LNCS (LNAI), vol. 6322, pp. 467–482. Springer, Heidelberg (2010). https://doi.org/10.1007/978-3-642-15883-4_30
25. Pesant, G.: A regular language membership constraint for finite sequences of variables. In: Wallace, M. (ed.) CP 2004. LNCS, vol. 3258, pp. 482–495. Springer, Heidelberg (2004). https://doi.org/10.1007/978-3-540-30201-8_36
26. Rawassizadeh, R., Momeni, E., Dobbins, C., Mirza-Babaei, P., Rahnamoun, R.: Lesson learned from collecting quantified self information via mobile and wearable devices. J. Sens. Actuator Netw. **4**(4), 315–335 (2015)
27. Rawassizadeh, R., Tomitsch, M., Wac, K., Tjoa, A.M.: UbiqLog: a generic mobile phone-based life-log framework. Pers. Ubiquit. Comput. **17**(4), 621–637 (2013)
28. Schaus, P., Aoga, J.O.R., Guns, T.: CoverSize: a global constraint for frequency-based itemset mining. In: Beck, J.C. (ed.) CP 2017. LNCS, vol. 10416, pp. 529–546. Springer, Cham (2017). https://doi.org/10.1007/978-3-319-66158-2_34

29. Shokoohi-Yekta, M., Chen, Y., Campana, B., Hu, B., Zakaria, J., Keogh, E.: Discovery of meaningful rules in time series. In: Proceedings of the 21th ACM SIGKDD International Conference on Knowledge Discovery and Data Mining, pp. 1085–1094. ACM (2015)
30. Tatti, N., Cule, B.: Mining closed strict episodes. In: 2010 IEEE 10th International Conference on Data Mining (ICDM), pp. 501–510. IEEE (2010)
31. Team, O.: OscaR: Scala in OR (2012)
32. Yang, Q., Wu, X.: 10 challenging problems in data mining research. Int. J. Inf. Technol. Decis. Making **5**(04), 597–604 (2006)
33. Yang, Z., Wang, Y., Kitsuregawa, M.: LAPIN: effective sequential pattern mining algorithms by last position induction for dense databases. In: Kotagiri, R., Krishna, P.R., Mohania, M., Nantajeewarawat, E. (eds.) DASFAA 2007. LNCS, vol. 4443, pp. 1020–1023. Springer, Heidelberg (2007). https://doi.org/10.1007/978-3-540-71703-4_95
34. Zhou, W., Liu, H., Cheng, H.: Mining closed episodes from event sequences efficiently. In: Zaki, M.J., Yu, J.X., Ravindran, B., Pudi, V. (eds.) PAKDD 2010. LNCS (LNAI), vol. 6118, pp. 310–318. Springer, Heidelberg (2010). https://doi.org/10.1007/978-3-642-13657-3_34

Off-Line and On-Line Optimization Under Uncertainty: A Case Study on Energy Management

Allegra De Filippo[✉], Michele Lombardi, and Michela Milano

DISI, University of Bologna, Bologna, Italy
{allegra.defilippo,michele.lombardi2,michela.milano}@unibo.it

Abstract. Optimization problems under uncertainty arise in many application areas and their solution is very challenging. We propose here methods that merge off-line and on-line decision stages: we start with a two stage off-line approach coupled with an on-line heuristic. We improve this baseline in two directions: (1) by replacing the on-line heuristics with a simple anticipatory method; (2) by making the off-line component aware of the on-line heuristic. Our approach is grounded on a virtual power plant management system, where the load shifts can be planned off-line and the energy balance should be maintained on-line. The overall goal is to find the minimum cost energy flows at each point in time considering (partially shiftable) electric loads, renewable and non-renewable energy generators, and electric storages. We compare our models with an oracle operating under perfect information and we show that both our improved models achieve a high solution quality, while striking different trade-offs in terms of computation time and complexity of the off-line and on-line optimization techniques.

Keywords: Optimization · Uncertainty · Energy management

1 Introduction

Optimization problems under uncertainty arise in many application areas, such as project scheduling, transportation systems, and energy system management. They are challenging to solve, in particular if high quality and robust solutions are desirable. For this reason, they are traditionally solved via off-line methods. There is however a growing interest in on-line algorithms to make decisions over time (without complete knowledge of the future), so as to take advantage of the information that is slowly revealed. In many practical cases, both approaches make sense: "strategic" decisions can be taken off-line, while "tactical" decisions are better left to an on-line approach.

We present *methods to merge off-line and on-line decision stages*: we start with an off-line, two-stage, stochastic optimization approach, coupled with an on-line heuristic. We improve this baseline in two directions: (1) by replacing the

© Springer International Publishing AG, part of Springer Nature 2018
W.-J. van Hoeve (Ed.): CPAIOR 2018, LNCS 10848, pp. 100–116, 2018.
https://doi.org/10.1007/978-3-319-93031-2_8

on-line heuristic with a simple anticipatory method; (2) by making the off-line component aware of the on-line heuristic.

As a case study, we consider a Virtual Power Plant management system, where load shifts can be planned off-line (e.g. the day ahead) and the energy balance should be maintained on-line (i.e. at each time point). The elements of uncertainty stem from uncontrollable deviations from the planned shifts and from the presence of renewable energy sources. The overall goal is to find the minimum cost energy flows at each point in time considering the electric loads, renewable and non-renewable energy generators and electric storages.

We compare our approaches with an oracle operating under perfect information and we show that *both our improved models achieve a high solution quality while striking different trade-offs* in terms of computation time and complexity of the off-line and on-line optimization techniques.

The rest of the paper is organized as follows. Section 2 provides a brief overview of optimization under uncertainty. Section 3 introduces our methods. Section 4 presents the case study. Section 5 discusses our methods in deeper detail, grounded on the use case. Section 6 provides an analysis of experimental results. Concluding remarks are in Sect. 7.

2   Optimization Under Uncertainty

Optimization under uncertainty is characterized by the need to make decisions without complete knowledge about the problem data. This situation is extremely common, but also very challenging: ideally, one should optimize for every possible contingency, which is often impossible or impractical [17].

One extreme (and frequent) method to deal with such issues is to disregard the uncertainty and assume that all parameters are deterministic [21]. When the potential impact of uncertainty is not negligible, however, using stochastic optimization becomes necessary (see [23] for an introduction or [3,12] for an extensive discussion). In this case, *a suitable representation for the uncertainty must be found* and (except in rare cases) some technique must be used to *trade estimation accuracy for a reduction of the computation time.*

Data subject to uncertainty can be often represented via *random variables in a multi-stage decision system.* After taking the decisions for a stage a random event occurs, i.e. some of the random variables are instantiated, and the decisions for the next stage must be taken, and so on.

It is common to use *sampling* to approximate the probability distribution of the random variables [22]. Sampling yields a number of *scenarios*: then, a single set of decisions is associated to the current stage, while separate sets of decisions are associated to each scenario in the next stage. More scenarios result in a better approximation, but a larger computation time. Looking more than one stage ahead also improves the estimation quality, but it requires to repeat the procedure recursively, with major impacts on the solution time.

There is a delicate trade-off between speculating vs. waiting for the uncertainty to be resolved [13]. This leads to an informal (but practical) distinction between *off-line and on-line problems*. On-line algorithms require to make decisions over time as the input is slowly revealed: delaying decisions can either increase the costs or be impossible due to constrained resources.

Off-line problems are often solved via exact solution methods on approximate models with limited look-ahead, e.g. via two-stage scenario-based approaches where both the first-stage and second stage variables are instantiated, or via decomposition based methods [14].

On-line problems are often tackled in practice via greedy heuristics, but more rigorous and effective anticipatory algorithms are also available as long as the temporal constraints are not too tight, e.g. the AMSAA algorithm from [9,15]. Similarly to off-line approaches, on-line anticipatory algorithms take decisions by solving deterministic optimization problems that represent possible realizations of the future. They address the time-critical nature of decisions by making efforts to yield solutions of reasonable quality early on in the search process.

3   Improving Off-Line/On-Line Optimization

In this work, we are interested in optimization problems with both an off-line and an on-line component. Formally, we focus on n-stage problems where the first-stage decisions are "strategic" (and can be taken with relative leisure), while the remaining $n-1$ stages involve "tactical" decisions (with tighter temporal constraints). We will also make the assumption that a greedy heuristic, based on a convex optimization model, is available for the on-line part.

As a baseline, we deal with the off-line decisions by collapsing the $n-1$ on-line stages into a single stage, and then obtaining via sampling a classical two-stage model. The on-line part is tackled with the original heuristic. This results into a relatively efficient approach, but yields solutions of limited quality.

We then investigate two complementary improvement directions: first, *we replace the greedy heuristic with an anticipative method*, once again based on collapsing at stage i the remaining $n-i$ stages. We refer to this approach as Boosted On-line OptimizatioN (BOON). Second, *we make the off-line approach aware of the limitations of the original on-line method*, by: (1) injecting the optimality conditions of the greedy heuristic in the off-line model, and then (2) allowing the off-line model to change certain parameters of the greedy heuristic so as to influence its behavior. We call this approach Master Off-line OptimizatioN (MOON).

The rationale is that: simply collapsing the on-line stages is equivalent to *feed future scenario-based information to the greedy heuristic* when solving the off-line problem. The first improvement direction aims at making the on-line method more similar to its off-line counterpart. The second direction preserves the limitations of the greedy heuristic in the off-line model. The price to pay is an increase on-line solution time in the former case, and an increased off-line time in the latter.

4 Case Study on Virtual Power Plants

The progressive shift towards decentralized generation in power distribution networks has made the problem of optimal Distributed Energy Resources (DERs) operation increasingly constrained. This is due to the integration of flexible (deterministic) energy systems with the strong penetration of (uncontrollable and stochastic) Renewable Energy Sources (RES). The integration of these resources into power system operation requires a major change in the current network control structure. This challenge can be met by using the Virtual Power Plant (VPP) concept, which is based on the idea of aggregating the capacity of many DERs, (i.e. generation, storage, or demand) to create a single operating profile and manage the uncertainty. A VPP is one of the main components of future smart electrical grids, connecting and integrating several types of energy sources, loads and storage devices. A typical VPP is a large plant with high (partially shiftable) electric and thermal loads, renewable energy generators and electric and thermal storages (see Fig. 1).

In a virtual power plant Energy Management System (EMS), the load shifts can be planned off-line, while the energy balance should be maintained on-line by managing energy flows among the grid, the renewable and traditional generators, and the storage systems. This makes VPP EMS a good candidate for grounding our approach. Based on actual energy prices and on the availability of DERs, the EMS decides: (1) how much energy should be produced; (2) which generators should be used for the required energy; (3) whether the surplus energy should be stored or sold to the energy market; (4) the load shifts planned off-line. Optimizing the use of energy can lead to significant economic benefits, and improve the efficiency and stability of the electric system (see e.g. [18]).

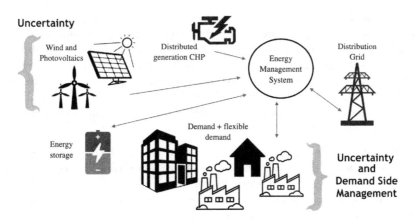

Fig. 1. A typical virtual power plant

Making decisions under uncertainty pervades the planning and operation of energy systems [24]. Optimization techniques have a long tradition in supporting planning and operational decisions in the energy sector, but the recent literature highlights the need for increasing both the scope and the granularity of the decisions, including new factors like distributed generation by renewable sources and smart grids [16].

Both the most popular methods to deal with uncertainty in mathematical programming (i.e. robust optimization and stochastic programming) have been widely applied in energy systems [11,19,26]. One of the most used assumption is that the distribution of future uncertainty is available for sampling, e.g. thanks to historical data and/or predictive models. In particular, the assumption that the distribution of future uncertainty is independent of current decisions is present in a variety of applications [9].

5   Grounding Our Method

In this section we present our methods to deal with mixed off-line/on-line decision making under uncertainty, grounded on the VPP use case. We will assume that the distribution of the random variable is available for sampling, as it is often the case when historical data or a predictive model is available. We also assume that the distribution of the random variables is independent of current decisions, i.e. to be dealing with exogenous uncertainty. This assumption holds in a great variety of applications and has significant computational advantages.

We will first show our baseline solution, which couples a two-stage, off-line, MILP model with a greedy on-line heuristic. We will then discuss the two improvement directions.

5.1   The Baseline Approach

The overall management system is composed by two macro steps: the first (day-ahead) step is a two-stage model designed to plan the load shifts and to minimize the (estimated) cost, and models the prediction uncertainty using sampling and scenarios. The second step is an on-line algorithm, implemented within a simulator, that uses the optimized shifts from the previous step to minimize the (real) operational cost, while fully covering the optimally shifted energy demand and avoiding the loss of energy produced by RES generators. This overall approach has been first presented in [6].

The Off-Line Model: The sources of uncertainty (at each time point, e.g. hour in the day ahead) that we take into account are: (1) errors in predicted power profile or renewable energy sources; (2) uncontrollable deviations from the load demand. We deal with uncertainty via sampling, but we focus on a very small number of "edge" scenarios, thus making our model somewhat close to robust optimization.

Formally, we assume that the error for our generation forecast (resp. load demand) can be considered an independent random variable: this is a reasonable hypothesis, provided that the predictor is good enough. In particular, we assume that the errors follow roughly a Normal distribution $N(0, \sigma^2)$, and that

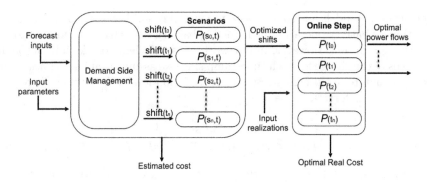

Fig. 2. The two macro steps of our model

the variance σ^2 for each time point is such that the 95% confidence interval corresponds to 10% of the predicted power output (resp. 20% of the load demand value), similarly to [8]. We then sample four scenarios, corresponding to the edge cases where the RES generation and the load demand are both high (at the top of the range), both low, or one high and the other low.

The off-line problem is modeled via Mixed Integer Programming (MILP) (see [6]). In this first step, our goal is to minimize the expected cost of energy for the VPP, estimated via the sample average over the scenarios:

$$min(z) = \frac{1}{|S|} \sum_{s \in S} \sum_{t \in T} c^s(t) \qquad (1)$$

where t is a time point, s is a scenario, and $c^s(t)$ is the (decision-dependent) cost for a scenario/time point pair. T is a set representing the whole horizon, and S is the set of all considered and equally probable scenarios.

In our model, we consider a Combined Heat and Power (CHP) dispatchable generator, with an associated fuel cost. Our approach should decide the amount $P^s_{CHP}(t)$ of generated CHP power for each scenario $s \in S$ and for each time point $t \in T$. We assume physical bounds on $P^s_{CHP}(t)$ due to its Electrical Capability. For simplicity, we assume that the time points represent periods long enough to treat the corresponding CHP decisions as *independent*. Therefore, we can model the generated CHP power with:

$$P^{min}_{CHP} \leq P^s_{CHP}(t) \leq P^{max}_{CHP} \quad \forall t \in T \qquad (2)$$

We assume that the VPP features a battery system, which is modeled by keeping track of the level of energy stored at each timestamp t as a function of the amount of power charged or discharged from the unit. $P^s_{St}(t)$ is the power exchanged between the storage system and the VPP. We actually use two decision variables: $P^s_{St_{In}}(t)$ if the batteries inject power into the VPP (with efficiency η_d) and $P^s_{St_{Out}}(t)$ for the batteries in charging mode (with efficiency η_c). We use a variable $charge^s(t)$ to define for each timestamp the current state of the battery system:

$$charge^s(t) = charge^s(t-1) - \eta_d P^s_{St_{In}}(t) + \eta_c P^s_{St_{Out}}(t) \quad \forall t \in T \qquad (3)$$

More accurate models for storage systems exist in recent literature. For example, [4] optimizes battery operation by modeling battery stress factors and analyzing battery degradation. However, in our work, it is sufficient to take into account the status of the charge for each timestamp since we assume that each timestamp is long enough to avoid high stress and degradation levels. We assume there are physical bounds for $P^s_{St_{In}}(t)$ and $P^s_{St_{Out}}(t)$:

$$P^{min}_{St} \leq P^s_{St_{In}}(t) \leq P^{max}_{St} \quad \forall t \in T \tag{4}$$

$$P^{min}_{St} \leq P^s_{St_{Out}}(t) \leq P^{max}_{St} \quad \forall t \in T \tag{5}$$

The variable $P^s_{Grid}(t)$ represents the current power exchanged with the grid for each scenario and for each timestamp. Similarly, the total power is defined as the sum of two additional variables, namely $P^s_{Grid_{In}}(t)$ if energy is bought from the Electricity Market and $P^s_{Grid_{Out}}(t)$ if energy is sold to the Market.

The Demand Side Management (DSM) of our VPP model (see Fig. 2) aims to modify the temporal consumption patterns, leaving the total amount of required energy constant. Moreover, we assume that the consumption stays unchanged also *over multiple sub-periods of the horizon*: this a possible way to state that demand shifts can make only local alterations of the demand load. The degree of modification is modeled by shifts and the shifted load is given by:

$$\tilde{P}_{Load}(t) = S_{Load}(t) + P_{Load}(t) \quad \forall t \in T \tag{6}$$

where $S_{Load}(t)$ represents the amount of shifted demand, and $P_{Load}(t)$ is the originally planned load for timestamp t (part of the model input). The amount of shifted demand is bounded by two quantities $S^{min}_{Load}(t)$ and $S^{max}_{Load}(t)$. By properly adjusting the two bounds, we can ensure that the consumption can reduce/increase in each time step by (e.g.) a maximum of 10% of the original expected load. We assume that the total energy consumption on the whole optimization horizon is constant. More specifically, we assume that the consumption stays unchanged also *over multiple sub-periods of the horizon*: this a possible way to state that demand shifts can make only local alterations of the demand load. Formally, let T_n be the set of timestamps for the n-th sub-period, then we can formulate the constraint:

$$\sum_{t \in T_n} S_{Load}(t) = 0 \tag{7}$$

Deciding the value of the $\tilde{P}^s_{Load}(t)$ (i.e. the *optimally shifted* demand for each timestamp) variables is the main goal of our off-line optimization step.

At any point in time, the overall shifted load must be covered by an energy mix that includes the generation from the internal sources (we refer to RES production as $P^s_{Ren}(t)$), the storage system, and the power bought from the energy market. Energy sold to the grid and routed to the battery system should be subtracted from the power balance:

$$\tilde{P}^s_{Load}(t) = P^s_{CHP}(t) + P^s_{Ren}(t) + P^s_{Grid_{In}}(t) - P^s_{Grid_{Out}}(t) + P^s_{St_{In}}(t) - P^s_{St_{Out}}(t) \tag{8}$$

The objective of our EMS is to minimize a sum (over all scenarios and over the horizon) of the operational costs $c^s(t)$ (see Eq. 1), which is given by:

$$c^s(t) = c_{Grid}(t)P^s_{Grid_{In}}(t) + c_{CHP}P^s_{CHP}(t) - c_{Grid}(t)P^s_{Grid_{Out}}(t) \qquad (9)$$

where $c_{Grid}(t)$ is the hourly price of electricity on the Market. c_{CHP} is the fuel price for the CHP system, assumed to be constant for each timestamp.

The On-Line Model: Our on-line algorithm runs within a simple simulator that repeatedly: (1) obtains a realization of the random variables (RES generation error and deviation from the planned shift) (2) obtains the power flows via the on-line algorithm; (3) Updates the cumulated costs and the charge level of the storage system.

The algorithm itself is a greedy heuristic based on a simple LP model. At each time point, based on the shifts produced by the off-line step of the on-line method (and adjusted by the simulator to take into account the effect of uncertainty), the algorithms minimizes the real operational cost while fully covering the energy demand. The LP model is the following:

$$P^{min}_{CHP} \leq P_{CHP}(t) \leq P^{max}_{CHP} \quad \forall t \in T \qquad (10)$$

$$charge(t) = charge(t-1) - \eta_d P_{St_{In}}(t) + \eta_c P_{St_{Out}}(t) \quad \forall t \in T \qquad (11)$$

$$P^{min}_{St} \leq P_{St_{In}}(t) \leq P^{max}_{St} \quad \forall t \in T \qquad (12)$$

$$P^{min}_{St} \leq P_{St_{Out}}(t) \leq P^{max}_{St} \quad \forall t \in T \qquad (13)$$

$$\tilde{P}_{Load}(t) = P_{CHP}(t) + P_{Ren}(t) + P_{Grid_{In}}(t) - P_{Grid_{Out}}(t) \qquad (14)$$
$$+ P_{St_{In}}(t) - P_{St_{Out}}(t) \quad \forall t \in T$$

$$z = \sum_{t \in T} c_{Grid}(t)P_{Grid_{In}}(t) + c_{CHP}P_{CHP}(t) - c_{Grid}(t)P_{Grid_{Out}}(t) \qquad (15)$$

In practice, this is the sub-part of the off-line model that takes care of a single time point, in a single scenario. As a consequence, the off-line model makes implicitly the assumption that the on-line stages will be managed under future information, as we have mentioned above. In practice, this will be far from the case, and we will show how the result is a considerable loss of solution quality.

5.2   The BOON Method

Due to this lack of decisions taken in anticipation of future scenarios, we develop a second and more computationally complex version of our on-line simulator.

In this section we investigate a first direction to improve the performance of our off-line/on-line optimization. The main idea is that of making the on-line algorithm behave more similarly to how the off-line model expects it to behave. We achieve this result by making the on-line algorithm anticipatory, and in particular by replacing the greedy heuristic with a two-stage LP model. The new model has exact knowledge of the uncertain quantities for the current time point (say t), and deals with future time points by: (1) using scenarios and (2) collapsing all the remaining $n - t$ decision stages into a single one. The same approach is repeated, yielding to the process described in Fig. 3:

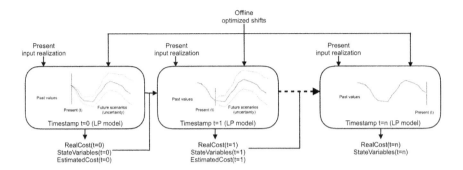

Fig. 3. Online anticipatory simulator

The new two-stage approach is composed of multiple instantiation of the greedy LP model. In detail: (1) for the current time point (say t), corresponding to the fist stage decisions, all the input if fully realized (see Eqs. 10–15); and (2) we deal with the remaining $n - t$ time points (the second-stage decisions) with the scenario-based approach (see Eqs. 1–8). The objective function for each time point is:

$$z = c_{Grid}(t)P_{Grid_{In}}(t) + c_{CHP}P_{CHP}(t) - c_{Grid}(t)P_{Grid_{Out}}(t)$$

$$+ \sum_{s \in S} \sum_{i=t+1}^{n} c_{Grid}(i)P_{Grid_{In}}^{s}(i) + c_{CHP}P_{CHP}^{s}(i) - c_{Grid}(i)P_{Grid_{Out}}^{s}(i) \quad (16)$$

This second version of the on-line algorithm is placed in cascade with the baseline off-line method and it also runs within the simulator.

The idea of using a repeated two-stage approach for on-line optimization is in fact equivalent to the first pass of the AMSAA algorithm [15], assuming that the underlying Markov Decision Process solver starts by making greedy decisions. However, AMSAA itself is not immediately applicable to our case study, since most of our decision variables are not discrete. Indeed, when the decision space is not enumerable (e.g. for continuous variables, as in our case) and each on-line stage requires to take *multiple decisions*, CONSENSUS [2], AMSAA and similar algorithms cannot be applied directly.

This simple look-ahead approach makes it possible to take better decisions based on knowledge of future uncertainty. The price to pay is a higher on-line computation time, even if the improved model remains an LP program and therefore solvable in polynomial time. Since on-line computations may be time constrained, the whole approach may not be viable in some practical cases, thus motivating our second improvement direction.

5.3 The MOON Method

Our second improvement direction is complementary to the first one: rather than trying to mitigate the limitations of the on-line approach, we make the off-line model aware of them, and control them to some degree.

Formally, this is done by taking advantage of the Karush Kuhn Tucker optimality conditions (see e.g. [25]). The KKT conditions are a generalized form of the method of Lagrangian multipliers and, under a few assumptions, they characterize necessary conditions for a solution to be a local optimum. Therefore, *if the problem is convex* and one can find a solution and a set of Lagrangian multipliers that satisfy the conditions, then solution is also a global optimum.

In our case, we add the KKT conditions for the greedy heuristic (which is based on a *Linear* Program) as constraints in the off-line model. In fact, as a consequence of collapsing the on-line stages into a single one, our baseline off-line approach is free to exploit knowledge about the future to optimize the power flow variables. Once the KKT conditions have been added, the power flows are forced to take the values that the greedy heuristic would actually assign to them.

The optimality conditions are specified through the use of a Lagrangian Function by introducing a multiplier $\mu_i \geq 0$ for each inequality constraint and a multiplier ν_j for each equality constraint. Then, if x^* is an optimal solution, there are corresponding values of the multipliers $\mu^* = (\mu_1^*, \ldots, \mu_m^*) \geq 0$ and $\nu^* = (\nu_1^*, \ldots, \nu_p^*)$ that: (1) cancel out the gradient of the Lagrangian Function, and additionally (2) satisfy the so-called complementary slackness conditions. For the LP model corresponding to our greedy heuristic, the conditions boild down to:

Equations (1)−(8)

$$c_{CHP} - \mu_1^s(t) + \mu_2^s(t) - c_{Grid}(t) = 0 \quad \forall t \in T, \forall s \in S \tag{17}$$

$$\alpha_{St}(t) + \mu_3^s(t) - \mu_4^s(t) - \mu_5^s(t) + \mu_6^s(t) - c_{Grid}(t) = 0 \quad \forall t \in T, \forall s \in S \tag{18}$$

$$\mu_1^s(t)(P_{CHP}^{min} - P_{CHP}^s(t)) = 0 \quad \forall t \in T, \forall s \in S \tag{19}$$

$$\mu_2^s(t)(P_{CHP}^s(t) + P_{CHP}^{max}) = 0 \quad \forall t \in T, \forall s \in S \tag{20}$$

$$\mu_3^s(t)(P_{St_{Out}}^s(t) - P_{St_{In}}^s(t) + charge^s(t-1)) = 0 \quad \forall t \in T, \forall s \in S \tag{21}$$

$$\mu_4^s(t)(-P_{St_{Out}}^s(t) + P_{St_{In}}^s(t) - charge^s(t-1)) = 0 \quad \forall t \in T, \forall s \in S \tag{22}$$

$$\mu_5^s(t)(P_{St}^{min} - P_{St_{Out}}^s(t) + P_{St_{In}}^s(t)) = 0 \quad \forall t \in T, \forall s \in S \tag{23}$$

$$\mu_6^s(t)(P_{St_{Out}}^s(t) - P_{St_{In}}^s(t) - P_{St}^{max}) = 0 \quad \forall t \in T, \forall s \in S \tag{24}$$

$$\mu_i^s(t) \geq 0 \quad \forall t \in T, \forall s \in S, for \ i = 1, \ldots, 6 \tag{25}$$

where Eq. (17) requires that the gradient is cancelled out, and Eqs. (19)−(24) correspond to the complementary slackness conditions.

As an additional step, *we treat some of the parameters of the greedy model as decision variables for the off-line problem*. In particular, we introduce a virtual cost $\alpha_{St}(t)$ for the storage system. We then introduce it by modifying the LP greedy model:

$$c(t) = c_{Grid}(t)P_{Grid_{In}}(t) + c_{CHP}P_{CHP}(t) \tag{26}$$
$$- c_{Grid}(t)P_{Grid_{Out}}(t) - \alpha_{St}(t)P_{St_{Out}}(t)$$

With this approach, the off-line solver can avoid producing solutions (i.e. the optimized shifts) that the on-line optimizer may not handle efficiently. Even more, by assigning values to $\alpha_{St}(t)$, the off-line solver can partially *control* the on-line algorithm by handling energy flows in more efficient way for the on-line heuristics, with no increase of computation time in the on-line stages. As a main drawback, injecting the complementary slackness conditions makes the off-line model non linear, potentially with severe adverse effects on the solution time.

6   Results and Discussion

For performing the experiments, we need to obtain realizations for the uncertainties related to loads, PV and wind generation (if it is present). Since we have assumed normally distributed prediction errors, we do this by randomly sampling error values according to the distribution parameters. Specifically, we consider a sample of 100 realizations (enough to ensure that sample average values will follow approximately a Normal distribution [5]) for six different use cases.

For each realization, we obtain a cost value by solving our different two-step approaches (stochastic off-line optimization + on-line algorithm) using Gurobi[1] as a MILP solver. For solving our off-line (non linear) model with KKT optimality conditions we used BARON via the GAMS modeling system[2] on the Neos server for optimization[3], with a time limit of 100 seconds. We then used again Gurobi solver for the second on-line part of our model.

We use data from two public datasets to test our models on a residential plant [7] with only PV energy production for renewable sources and an industrial plant[4] with eolic and PV production. We modify these datasets to obtain use cases for testing our models (see Table 1): (1) Use Case 1 (UC1) is the baseline residential dataset; (2) UC2 is the residential dataset with an increase of renewable (i.e. PV) production; (3) UC3 is dataset UC1 where the market prices are different for the sale/purchase of energy from/to the grid; (4) UC4 is the industrial dataset with also eolic renewable production; (5) in UC5 we increase the renewable production as in UC2; (6) in UC6 we consider UC4 with different market prices as in UC3.

Methodologies for the estimation of hourly global solar radiation have been proposed by many researchers and in this work, we consider as a prediction the average hourly global solar radiation from [20] and we use assumption for wind prediction from [10]. We then assume that the prediction errors in each

---

[1] Available at http://www.gurobi.com.

[2] Available at https://www.gams.com/latest/docs/S_BARON.html.

[3] Available at http://www.neos-server.org/neos/.

[4] Available at https://data.lab.fiware.org/dataset/.

Table 1. Different use cases

Load demand	Baseline dataset	Renewable peak	Different market prices
Residential	UC1	UC2	UC3
Industrial	UC4	UC5	UC6

Table 2. Optimal costs value for different models in different use cases

Model	UC1 μ (K€)	UC2 μ (K€)	UC3 μ (K€)	UC4 μ (K€)	UC5 μ (K€)	UC6 μ (K€)
Oracle	331.362	247.213	393.818	798.388	565.607	856.955
Baseline	404.622	311.145	462.577	923.243	684.197	984.904
Diff. Oracle (%)	**22.114**	**25.914**	**17.560**	**15.568**	**21.933**	**14.998**
BOON	342.061	265.326	404.322	819.249	580.174	874.585
Diff. Oracle (%)	**3.259**	**7.331**	**2.781**	**2.786**	**2.643**	**2.117**
MOON	344.604	263.808	408.721	811.119	573.934	868.764
Diff. Oracle (%)	**4.046**	**6.309**	**3.887**	**1.699**	**1.577**	**1.498**

timestamp can be modeled again as random variables. Specifically, we assume normally distributed variables with a variance such that the 95% confidence interval corresponds to $\pm10\%$ of the prediction value. We assume physical bounds on $P^s_{CHP}(t)$ due to its Electrical Capability based on real generation data [1,7]. The initial battery states and the efficiency values are based on real generation data [1,7] and we assume there are physical bounds for $P^s_{St_{In}}(t)$ and $P^s_{St_{Out}}(t)$ based on real data [1,7].

We compare our model results for each use case with an oracle operating under perfect information. Figure 4 shows the average values of each hourly optimized flow for the oracle and the baseline models over the 100 realizations. We show the optimal flows for both models in UC2 and we can see, in Fig. 4 (baseline), the limits of using a non anticipatory algorithm since it is not possible to acquire energy from the grid in advance (i.e. when the cost is lower) and/or to sell energy to the grid in period of highest price on the market or when more energy is available from renewable sources. Moreover the exchange of energy with the storage system is almost never used, i.e. to store RES energy.

In Table 2 we show the average costs for each use case over the 100 input realizations and for each developed model. From these results and from results in Table 3, we can notice that the baseline model is a relatively efficient approach in term of computation time, but yields solutions of limited quality in terms of cost (i.e. increase of solution quality from 15% to 26%). We improve these results, as explained in Sect. 2, first by replacing the greedy heuristic with the BOON method; second, by making the off-line approach aware of the limitations of the original on-line method by developing the MOON method.

As shown in Fig. 5, it is possible to see that, near the peak of renewable energy production, the EMS of MOON model accumulates energy in the storage and uses in a more balanced way the energy present in the storage system compared

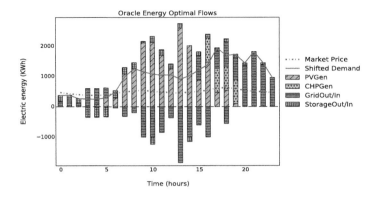

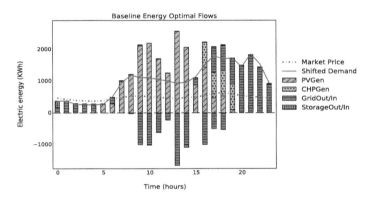

Fig. 4. Oracle optimized flows (at the top) and Baseline optimized flows (at the bottom)

to the baseline model represented in Fig. 4. Furthermore, still looking at Fig. 5, it can be seen that BOON has peaks of energy sold on the network near the increase in electricity prices on the market. With this model we improve quality solutions (in term of optimal costs) but we increase the computational complexity of the on-line stage.

In Table 3 we show the computation time (seconds) for the different model stages. The price to pay is a higher on-line computation time for BOON approach, even if the improved model remains an LP program and therefore solvable in polynomial time. Since on-line computations may be time constrained, the whole approach may not be viable in some practical cases, thus motivating our second improvement direction: the MOON method. In Fig. 5 (at the bottom) it is possible to notice the more homogeneous use of the storage system. We can see that, by optimizing the virtual storage cost in the off-line stage, we can improve solution quality in term of cost (see Table 2) by using the storage system. Since

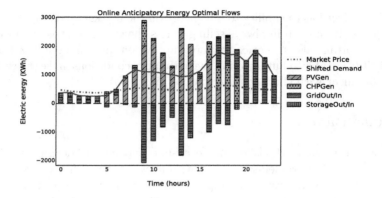

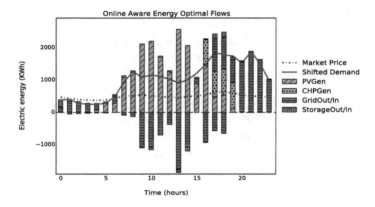

Fig. 5. BOON optimized energy flows (at the top) and MOON optimized flows (at the bottom)

Table 3. Computation time (seconds) for the different models

Model	Residential		Industrial	
	Off-line (day-ahead)	On-line (daily optimiz.)	Off-line (day-ahead)	On-line (daily optimiz.)
Baseline	0.184	0.777	0.346	0.839
BOON	0.184	5.011	0.346	5.429
Diff. Baseline	**0.000**	**+4.233** (545%)	**0.000**	**+4.590** (547%)
MOON	27.885	0.902	58.913	0.983
Diff. Baseline	**+27.701** (15k%)	**+0.124** (16%)	**+58.567** (16k%)	**+0.143** (17%)

the on-line solver has the ability to sell energy on the market, and storing energy has no profit, it ends up in always selling unless the virtual cost is employed.

It is important also to notice that we improve on-line time optimization by reducing it (see Table 3) but injecting the optimality conditions makes the off-line model non linear with effects on the solution time.

7   Conclusion

This work proposes methods to merge off-line and on-line decision stages in energy optimization problems: we start with a two stage off-line approach coupled with an on-line heuristic. We improve this baseline in two directions: (1) by replacing the on-line heuristic with an anticipative method; (2) by making the off-line component aware of the on-line heuristic used.

Our approach is then grounded on a virtual power plant management system where the load shifts can be planned off-line and the energy balance should be maintained on-line. The overall goal is to find the minimum cost energy flows at each point in time considering high (partially shiftable) electric loads, renewable and non-renewable energy generators and electric storages.

We compare our models with an oracle operating under perfect information and we show that both our improved models achieve a high solution quality while striking different tradeoffs in terms of computation time and complexity of the off-line and on-line optimization techniques.

In particular, we show that the BOON Anticipatory model increases the computational complexity of the on-line stage, greatly improving the quality of the baseline model solution. The increased computation time may force one to solve the on-line problems with a reduced frequency.

Resolving the model with the MOON approach, we have a general improvement in the quality of the solution, but the most important thing is that we reduce (as in the baseline model) the time of on-line computational optimization. MOON works more or less as well as BOON in terms of costs but the speed of the on-line part makes MOON the dominant algorithm for this case study.

References

1. Bai, H., Miao, S., Ran, X., Ye, C.: Optimal dispatch strategy of a virtual power plant containing battery switch stations in a unified electricity market. Energies **8**(3), 2268–2289 (2015)
2. Bent, R.W., Van Hentenryck, P.: Scenario-based planning for partially dynamic vehicle routing with stochastic customers. Oper. Res. **52**(6), 977–987 (2004)
3. Birge, J.R., Louveaux, F.: Introduction to Stochastic Programming. Series in Operations Research and Financial Engineering. Springer, New York (1997). https://doi.org/10.1007/978-1-4614-0237-4
4. Bordin, C., Anuta, H.O., Crossland, A., Gutierrez, I.L., Dent, C.J., Vigo, D.: A linear programming approach for battery degradation analysis and optimization in offgrid power systems with solar energy integration. Renew. Energy **101**, 417–430 (2017)

5. Bracewell, R.N.: The Fourier Transform and its Applications, vol. 31999. McGraw-Hill, New York (1986)
6. De Filippo, A., Lombardi, M., Milano, M., Borghetti, A.: Robust optimization for virtual power plants. In: Esposito, F., Basili, R., Ferilli, S., Lisi, F. (eds.) AI*IA 2017. LNCS, vol. 10640, pp. 17–30. Springer, Cham (2017). https://doi.org/10.1007/978-3-319-70169-1_2
7. Espinosa, A.N., Ochoa, L.N.: Dissemination document "low voltage networks models and low carbon technology profiles". Technical report, University of Manchester, June 2015
8. Gamou, S., Yokoyama, R., Ito, K.: Optimal unit sizing of cogeneration systems in consideration of uncertain energy demands as continuous random variables. Energy Convers. Manag. **43**(9), 1349–1361 (2002)
9. Van Hentenryck, P., Bent, R.: Online Stochastic Combinatorial Optimization. The MIT Press, Cambridge (2009)
10. Hodge, B.-M., Lew, D., Milligan, M., Holttinen, H., Sillanpää, S., Gómez-Lázaro, E., Scharff, R., Söder, L., Larsén, X.G., Giebel, G., et al.: Wind power forecasting error distributions: an international comparison. In: 11th Annual International Workshop on Large-Scale Integration of Wind Power into Power Systems as well as on Transmission Networks for Offshore Wind Power Plants Conference (2012)
11. Jurković, K., Pandšić, H., Kuzle, I.: Review on unit commitment under uncertainty approaches. In: 2015 38th International Convention on Information and Communication Technology, Electronics and Microelectronics (MIPRO), pP. 1093–1097. IEEE (2015)
12. Kall, P., Wallace, S.W.: Stochastic Programming. Springer, Heidelberg (1994). ISBN 9780471951087
13. Kaut, M., Wallace, S.W.: Evaluation of scenario-generation methods for stochastic programming. Humboldt-Universität zu Berlin, Mathematisch-Naturwissenschaftliche Fakultät II, Institut für Mathematik (2003)
14. Laporte, G., Louveaux, F.V.: The integer l-shaped method for stochastic integer programs with complete recourse. Oper. Res. Lett. **13**(3), 133–142 (1993)
15. Mercier, L., Van Hentenryck, P.: *Amsaa*: a multistep anticipatory algorithm for online stochastic combinatorial optimization. In: Perron, L., Trick, M.A. (eds.) CPAIOR 2008. LNCS, vol. 5015, pp. 173–187. Springer, Heidelberg (2008). https://doi.org/10.1007/978-3-540-68155-7_15
16. Morales, J.M., Conejo, A.J., Madsen, H., Pinson, P., Zugno, M.: Integrating Renewables in Electricity Markets: Operational Problems, vol. 205. Springer, Boston (2013). https://doi.org/10.1007/978-1-4614-9411-9
17. Powell, W.B.: A unified framework for optimization under uncertainty. In: Optimization Challenges in Complex, Networked and Risky Systems, pp. 45–83. INFORMS (2016). https://doi.org/10.1287/educ.2016.0149
18. Palma-Behnke, R., Benavides, C., Aranda, E., Llanos, J., Sez, D.: Energy management system for a renewable based microgrid with a demand side management mechanism. In: 2011 IEEE Symposium on Computational Intelligence Applications in Smart Grid (CIASG), pp. 1–8, April 2011
19. Reddy, S.S., Sandeep, V., Jung, C.-M.: Review of stochastic optimization methods for smart grid. Front. Energy **11**(2), 197–209 (2017)
20. Kaplanis, S., Kaplani, E.: A model to predict expected mean and stochastic hourly global solar radiation i(h; nj) values. Renew. Energy **32**(8), 1414–1425 (2007)
21. Sahinidis, N.V.: Optimization under uncertainty: state-of-the-art and opportunities. Comput. Chem. Eng. **28**(6), 971–983 (2004). FOCAPO 2003 Special issue

22. Shapiro, A.: Sample average approximation. In: Gass, S.I., Fu, M.C. (eds.) Encyclopedia of Operations Research and Management Science, pp. 1350–1355. Springer, Boston (2013). https://doi.org/10.1007/978-1-4419-1153-7
23. Shapiro, A., Philpott, A.: A tutorial on stochastic programming. Manuscript (2007). www2.isye.gatech.edu/~ashapiro/publications.html
24. Wallace, S.W., Fleten, S.-E.: Stochastic programming models in energy. In: Stochastic Programming. Handbooks in Operations Research and Management Science, vol. 10, pp. 637–677. Elsevier (2003)
25. Winston, W.L., Goldberg, J.B.: Operations Research: Applications and Algorithms, vol. 3. Thomson Brooks/Cole, Belmont (2004)
26. Zhou, Z., Zhang, J., Liu, P., Li, Z., Georgiadis, M.C., Pistikopoulos, E.N.: A two-stage stochastic programming model for the optimal design of distributed energy systems. Appl. Energy **103**, 135–144 (2013)

Reasoning on Sequences
in Constraint-Based Local Search
Frameworks

Renaud De Landtsheer$^{(\boxtimes)}$, Yoann Guyot, Gustavo Ospina, Fabian Germeau,
and Christophe Ponsard

CETIC Research Centre, Charleroi, Belgium
{rdl,yg,go,fg,cp}@cetic.be

Abstract. This paper explains why global constraints for routing cannot be integrated into Constraint-Based Local Search (CBLS) frameworks. A technical reason for this is identified and defined as the *multi-
variable bottleneck*. We solve this bottleneck by introducing a new type
of variables: "sequence of integers". We identify key requirements and
defines a vocabulary for this variable type, through which it communicates with global constraints. Dedicated data structures are designed for
efficiently representing sequences in this context. Benchmarks are presented to identify how to best parametrise those data structures and
to compare our approach with other state-of-the-art local search frameworks: LocalSolver and GoogleCP. Our contribution is included in the
CBLS engine of the open source OscaR framework.

Keywords: Local search · CBLS · Global constraints · Sequences
OscaR.cbls

1   Introduction

A lot of algorithms have been proposed for checking routing-related global constraints in local search engines, see for instance [1–3]. A classic example is a
global constraint that incrementally computes (*maintains*) the length of the
route in routing optimization [1]. It inputs a distance matrix specifying the
distance between each pair of nodes of the routing problem and the current
route. When flipping a portion of route (that is a, b, c, d becomes d, c, b, a or
$a, b, c, d, e, f, g, h, i$ becomes $a, b, c, \mathbf{g}, \mathbf{f}, \mathbf{e}, \mathbf{d}, h, i$), and if the distance matrix is
symmetric, smart algorithms are able to update the route length in $O(1)$-time
because the length of the flipped segment is the same in both directions.

Very often, these algorithms are implemented into custom-made solvers,
notably for benchmarking purposes, and using them on an industrial application
requires re-implementing them within a dedicated custom-made solver.

Constraint-Based Local Search (CBLS) is a modular approach that proposes
to embed constraints into a modular software framework, so that they can be

© Springer International Publishing AG, part of Springer Nature 2018
W.-J. van Hoeve (Ed.): CPAIOR 2018, LNCS 10848, pp. 117–134, 2018.
https://doi.org/10.1007/978-3-319-93031-2_9

easily assembled to build a dedicated solver. Such models are made of *variables* and *invariants* [4–8]. *Invariants* are directed mathematical operators that have input and output variables and that maintain the value of their output variables according to their specification and to the value of their input variables. Decision variables cannot be the output of any invariant. Based on this model, a local search procedure explores neighbourhoods by modifying the value of decision variables and querying the value of a variable representing the objective function. The CBLS framework efficiently updates the value of the variables in model to reflect value changes on the decision variables.

Unfortunately, global constraints algorithms cannot be easily embedded into CBLS frameworks. This is a pity because the CBLS approach can cut down the cost of developing solvers for new optimization problems. This paper identifies a major reason for this problem and proposes a solution to overcome it. We focus on routing optimization since it is a major class of applications of local search.

The bottleneck that prevents global constraints from being embedded into CBLS frameworks arises when a neighbourhood exploration requires repeatedly modifying a large number of variables in a structured way. For instance, flipping a portion of the route in routing optimization problems, which is known as the 2-opt neighbourhood. In a CBLS framework, the route is often modelled using an array of integer variables, each of them is associated to a node, and specifies the next node to be reached by the vehicle. For such models, flipping a portion of the route requires modifying the value of the 'next' variable associated to each node in the flipped portion, as well as the 'next' of the node before this portion. Such move is therefore rasterized into a set of atomic assignment to separated variables, so that the global, symbolic, structure of the move is lost. Any invariant is then only aware of the many changes performed on its input variables, and not about the symbolic nature of the move.

A simple approach to mitigate this bottleneck is to choose the order of exploration of the neighbourhood in such a way that one never has to change a large set of variables [1]. In the case of 2-opt neighbourhood, one can gradually widen the flipped route segment. Each neighbour is then explored in turn, and going from one neighbour to another one requires modifying a constant number of variables. We call this the *circle mode*, as opposed to the *star mode* where the value of each decision variable is rolled back to the current value after each explored neighbour. Circle mode is hard to combine with heuristics such as selecting nodes among the k-nearest ones in vehicle routing, which is often used to reduce the size of a neighbourhood.

This paper introduces a variable of type "sequence of integers" for CBLS frameworks, with a tailored data-structure. Thanks to this type of variable, complex moves such as flips are performed on the value of a single variable and can be described in a symbolic way as an update performed on the previous value of the variable. Global constraints are then aware of changes performed to these variable in a more symbolic form than what was possible using the array of integer approach. The goal is to embed the best global constraints, potentially having $O(1)$-time complexity for evaluating neighbour solutions. The data structure used for representing sequences must exhibit similar complexity. Therefore, adapted data structures for sequence values are also proposed.

The paper is structured as follows: Sect. 2 reviews various CBLS frameworks and the way they support sequences. Section 3 lists the requirements on our "sequence" variables. Section 4 describes our data-structure for representing sequences. Section 5 describes our sequence variable, how it uses the sequence value and how it communicates its updates to global constraints. Section 6 presents benchmarks of our data-structure. Section 7 presents comparative benchmarks. Section 8 sketches further work and concludes.

2   Background and Related Work

This section presents CBLS with more details, introduces a few CBLS frameworks and some global constraints on sequences designed for local search. The section focuses on the OscaR.cbls framework because our contribution was performed in this framework.

2.1   Local Search Frameworks

There are a few local search frameworks including: Comet [4], Kangaroo [5], OscaR.cbls [6,9], LocalSolver [7], InCELL [8], GoogleCP [10], and EasyLocal++ [11].

EasyLocal++ requires a dedicated model to be developed from scratch using ad-hoc algorithms. It provides support for declaring the search procedure [11]. It is not a CBLS framework and hence does not suffer from the multi-variable bottleneck.

Comet, Kangaroo, and OscaR.cbls are CBLS frameworks which support Integers and Sets of Integers. InCELL supports a notion of variable that is a sequence of other variables but does not offer a unified data-structure for modifying the sequence [8].

GoogleCP is a hybrid engine that can be used both as a constraint solver and as a local search engine [10]. It supports a *SequenceVar* type representing sequences of integer variables. Although it is called a variable, it is a view on the underlying integer variables, that are automatically sorted by value.

LocalSolver supports variables of type List of Integers, where each value can occur at most once [7]. Unfortunately, no detail is given about the underlying data structure, nor about the moves that can be efficiently performed on them. LocalSolver supports very few invariants and constraints related to this list variable.

A sequence variable has already been developed for the OscaR.cbls framework in the context of string solving where the length of the string is not known in advance and is potentially very large [12]. The supported incremental update operations are "inserts", "delete" and "set a value to a given position". Insert and delete updates cannot be efficiently performed if the sequence is represented as an array of integers, so that the proposed sequence provides a speed improvement for these updates and solves the multi-variable bottleneck in the context of string solving. However, it does not provide support for large operations such as flipping subsequence's.

2.2 Three Representative Global Constraints on Sequences

A lot of research has been performed on the incremental evaluation of multiple constraints and metrics in routing optimization. We therefore just present three examples, with various properties regarding the information they need about the route and how the route is modified by a search procedure.

The most classical example is the route length constraint that computes the total length of a route, given a distance matrix that is symmetric, and that can quickly update this length when a portion of this route is flipped [1]. The length of the flipped route does not change, since the distance matrix is symmetric, so that the total length of the route is only impacted by the hops that are located at the extremities of the flipped segment. The total distance can therefore be updated in $O(1)$-time.

Another example is the travelling delivery person problem that defines some complex metrics over nodes and driven distance. Pre-computation can be exploited to update this metric in $O(1)$ time for a large proportion of classical routing moves [2].

A third example is routing with time windows, where various algorithms reasoning on the time slack have been proposed [3]. Given a route, the time slack of a node is the amount of time that a vehicle can wait before entering the node such that no deadline will be violated in the subsequent nodes. Before any neighbour exploration, it is computed for each node. It is then exploited to quickly evaluate if some modification of the route leads to a route violating the constraint.

All these algorithms require a symbolic description of the move to perform their update efficiently. Furthermore, they often rely on some pre-computation, that is: there is a first algorithm that generates some intermediary values, and that is executed before a neighbourhood exploration starts. Using these intermediary values, a second algorithm can quickly update the output of the global constraint for each explored neighbour. This second algorithm is also known as a *differentiation*. An invariant might only provide efficient differentiation for a subset of the possible moves; the time slack approach for the time window constraint might be hard to generalize to flip moves for instance.

2.3 Constraint-Based Local Search

A local search algorithm relies on a *model* and a *search procedure*. CBLS frameworks may offer support for both of these two aspects. This paper exclusively focuses on the model and will not cover the search procedure.

In CBLS frameworks, the *model* is composed of *variables* (integers and set of integers at this point), and *invariants*, which are directed constraints maintaining one or more output variables according to the specification they implement and according to the value of one or more input variables. A classical invariant is *Sum*. It has an array of integer variables as input, and a single integer variable as output. The value of the output variable is maintained by the invariant to be the sum of the values of all the input variables. "*Maintained*" means that the invariant sets the value of the output variable, and adjust this value according to the value changes of the input variables. This is generally implemented through incremental algorithms. For instance, when the *Sum* invariant defined here is notified that one of its input variables has changed its value, it computes the delta on this variable between the new value and the old value, and updates the value of its output variable by incrementing it by this delta. OscaR.cbls for instance has a library of roughly 80 invariants.

The model is declared at start-up by some application-specific code that instantiates the variables and invariants. The input and output variables of each invariant are also specified at this stage. When declaring a model, the CBLS framework builds a directed graph, called the *propagation graph*, which we can assume here to be acyclic (DAG). The nodes are the variables and invariants. The edges in the graph represent data flows from variables to listening invariants and from invariants to output variables. The DAG starts with decision variables, that are the output of no invariant, and typically ends at a variable representing the objective function.

During the search, the model is active: if the value of some variable changes, this change is propagated to the other variables through the invariants, and following the propagation graph in some coordinated way called *propagation*. Propagation algorithms require the DAG to be somehow sorted. This sort is performed after the model is completed, by a *close* operation that is triggered by the user code. From this point on, no variable or invariant can be added, modified or removed. The value of all variables can of course change. The search procedure modifies the input variables to explore the search space and query the objective function, which is updated by the propagation.

Figure 1 illustrates a propagation graph for a Travelling Salesman Problem (TSP) with four nodes using the naïve model based on an array of integer variables. Variables (resp. invariants) are represented by grey (resp. white) rounded rectangles, and dependencies using arrows. The $next_i$ variables represent, for each node i, the node that is reached by the vehicle after leaving the node i. They are grouped into an array, represented by the rectangle that encloses them, on the left of the figure. The distance matrix is not represented in the picture. Each *element* invariant accesses the line of the matrix related to the node it is monitoring and maintains its output variable to the value at that line, and on the column designated by the value of $next_i$. Each d_i variable represents the distance that the vehicle drives when leaving the associated $node_i$.

Propagation is performed in such a way that a variable is updated at most once, and only if it needs to be updated. The propagation is driven by the CBLS framework and uses some sorting that is performed when the model is closed: the nodes that needs to be propagated and that are the closest to the decision

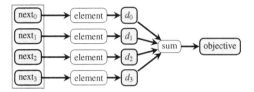

Fig. 1. Propagation graph of a TSP with 4 nodes modelled using an array of integer variables

variables are propagated first. Upon propagation of a node, additional nodes might require to be propagated, and they are queued and prioritized through a heap based again on some distance to the decision variables. The priority can be either based on a topological sort, as implemented e.g. in Comet [4], or based on the distance to the decision variables, as implemented in OscaR.cbls. OscaR.cbls uses a specialized heap that aggregates nodes at the same distance into a list, so that insert and pop operations can often be performed in $O(1)$-time.

CBLS variable notify their value change to its listening invariants, and they cannot change value without notifying it. To implement this behaviour, they internally store two values: the *old value* (v_{old}), and the *new value* (v_{new}). When an invariant queries a variable for its value, it gets v_{old}. When a variable is updated by a neighbourhood or controlling invariant, the updated value is stored into v_{new}. When a variable is propagated, it *notifies* its value change to each invariant that listen to it and then performs $v_{old} := v_{new}$. Such a notification carries a description of how the value has changed. For integer variables, it carries a reference to the variable, v_{old} and v_{new}. For set variables, it also carries the set of integer values that have been added to, and removed from v_{old}.

3   Requirements for a Sequence Variable

We consider standard neighbourhoods used in routing optimization: insert, remove, 1-opt, 2-opt, 3-opt, and composites of these moves such as the Lin-Kernighan, which is a composition of 2-opts, or two-point move neighbourhood used in pick-up and delivery optimization [13,14]. We identify the following requirements for our sequence variable:

immutable-value: The value representing a sequence of integers should be non-mutable: once transmitted to an invariant, it should not be modified. All methods that modify a sequence data-structure should yield a new sequence data-structure, as done in functional programming. A variable can of course change its value.

symbolic-delta: Sequence variables should communicate its incremental value change in a symbolic way to global constraints for the moves specified above. We define three atomic incremental updates of sequence values, that can be combined:

> **insert** an int value at a given position and shift the tail of the sequence accordingly
>
> **delete** the int value at a given position and shift the tail of the sequence accordingly
>
> **seq-move** that moves a sub-sequence to another position in the sequence and optionally flips this subsequence.

fast-exploration: Sequence variables should be updated very quickly in the context of neighbourhood exploration to reflect the moves defined above.

pre-computation: Invariants should get to know around which sequence value a neighbourhood exploration will take place so that they can perform pre-computation based on this value and use differentiation algorithms.

fast-commit: Sequence variables should be updated quickly to reflect moves that are committed, considering the same moves as in *fast-exploration*. This requirement has a lower priority than *fast-exploration* since there are more neighbours explored than moves committed.

4 Implementation of Our Sequence Value

This section presents our data-structure for representing sequence values. The general object model of our sequence is shown in Fig. 2. There is a concrete sequence and there are some updated sequences that can be instantiated in $O(1)$ time to reflect modified sequences in the context of neighbourhood exploration, and comply with the requirement fast-exploration. The concrete sequence can also be updated quickly to comply with the fast-commit requirement.

All sequences, concrete, updated, etc. are non-mutable, based on non-mutable data structures, to comply with the requirement immutable-value. All methods that modify sequence values generate a new sequence value, which usually share a lot of its internal data-structure with the sequence it has been derived from. We rely on a garbage collector for memory management, as usually done in functional programming.

There are three methods declared in the sequence class for updating a sequence: *insert,remove* and *seq-move*. These methods match the delta defined in Sect. 3. They have an additional parameter, called *fast* that specifies if the update should return a concrete sequence or if the update should be stacked by instantiating an updated sequence.

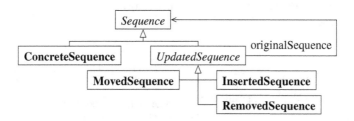

Fig. 2. Class diagram of the sequence value. Classes are in bold, abstract classes in italic.

4.1   Updated Sequences

Updated sequences behave according to the value they represent by translating and forwarding queries they receive to the original non-modified sequence. For instance, considering a *RemovedSequence* obtained by removing the value at position r from sequence s, the query that gets the value at a given position p returns the value at position p in sequence s if $p < r$ and at position $p + 1$ otherwise. Performing queries on Updated Sequences is therefore slightly more time-consuming than on Concrete Sequences. They should therefore not be stacked too much.

4.2   Concrete Sequences

Concrete sequences are represented by a double mapping. The first one maps positions to *internal* positions and the second one maps these internal positions to the actual value. This double mapping is illustrated in Fig. 3.

The *first mapping* is a *piecewise affine bijection with slope +1 or −1*. Its domain and co-domain are $[0..n-1]$ where n is the size of the sequence. Basically, its domain is segmented into a series of consecutive intervals. Within each of these interval, the value of the bijection is defined by an affine function $(ax + b)$ where a is either 1 or −1. Graphical examples of such bijection are given in Figs. 4, 5 and 6.

Concretely, the bijection is represented by the forward function and its reciprocal function. Each of them is defined by a set of *pivots*. Each pivot is defined by a starting value, an offset, and a Boolean slope (± 1). Pivots have no explicit end value. The end value of a pivot is implicitly defined as the last value before the start value of the next pivot. Pivots are stored in balanced binary search trees, indexed by their starting values.

The *second mapping* is implemented by two red-black trees, one that maps internal positions to values, and the other one is the reverse: it maps values to the set of internal positions where they occur.

Let's consider a concrete sequence, of length n, with k pivots in its bijection. Finding the value at some position p requires translating p into the corresponding internal position through the bijection, and finding the value at this internal position, through the second mapping. The first step is performed in $O(\log k)$ time, by searching the pivot with the biggest starting value that is $\leq p$ and applying the affine function associated to this pivot. The second step is performed in $O(\log n)$ time.

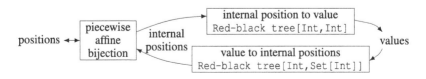

Fig. 3. Data-flow diagram showing the mappings within concrete sequences

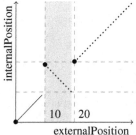
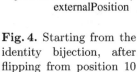
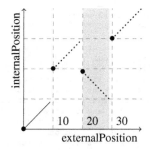
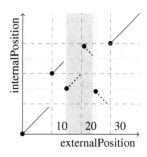

Fig. 4. Starting from the identity bijection, after flipping from position 10 to 19 (both included)

Fig. 5. Starting from the identity bijection, after seq-move (from = 10, to = 19, after = 29, flip = true)

Fig. 6. Flipping subsequence from position 15 to 24 both included from the bijection in Fig. 5

The first mapping can be efficiently updated to capture flipping of shifting of subsequence's, so it contributes to the requirement fast-commit. This is because the number of segments in this bijection is much smaller than the length of the sequence.

Let's consider an initial sequence of 40 values, where the bijection is the identity function. Flipping the subsequence from position 10 to position 19 (both included) can be performed by updating the bijection as shown in Fig. 4. The range of external positions corresponding to the specified subsequence is highlighted in grey. Within the flipped zone, a pivot of slope −1 is introduced, and a new pivot starts just after the flipped zone. The added pivots are drawn with dotted lines. Considering the same initial sequence, Fig. 5 shows the bijection after moving the subsequence located between position 10 and position 19 (both included), inserting it after position 29, and flipping it. The range corresponding to the moved subsequence has a pivot with a negative slope because it was flipped.

To insert a value into a sequence, it is assigned the next available internal position. The bijection is reworked to map the position of the insert to this internal position and shift the tail of the sequence. To delete a value from a sequence, we first swap its internal position with the highest internal position in the sequence and remove it from the sequence. Deletions do not create holes in the range of internal positions.

The algorithm for flipping a subsequence by modifying the forward function is quite representative of all algorithms modifying the bijection and is sketched in the function `flipSubsequence` of Fig. 7. This function only updates the forward part of the bijection. The reverse function of the bijection is lazily recomputed from scratch from the updated forward function. Providing incremental updates for this reverse function is future work. Given the forward function forwardFn of a sequence of length l, and an interval [fromPos; toPos] to flip, it returns an updated function, updatedFn, such that:

```
1    def flipSubsequence(forwardFn:RedBlackTree[Int,Pivot],
2                        fromPos:Int,toPos:Int):RedBlackTree[Int,Pivot]={
3      val cleanCutFunction = Add pivots to forwardFn
4                        at fromPos and toPos+1 if no pivots there
5      val pivotsToFlip = copy values from cleanCutFunction
6                        with keys in [fromPos;toPos] into a list
7      val flippedPivots = flipPivots(pivotsToFlip,toPos,toPos,List.empty)
8      val updatedForwardFunction = update values and keys
9                        between [fromPos;toPos] in cleanCutFunction
10                       to pivots and their startValue from flippedPivots
11     val cleanedUpdatedFunction = remove from updatedForwardFunction
12                       the pivot at fromPos
13                       and the first pivot strictly after toPos
14                       if they are aligned with their predecessors
15     return cleanedUpdatedFunction
16   }
17   def flipSubfunction(pivotList:List[Pivot], movingEnd:Int,
18                       toPos:Int, acc:List[Pivot]):List[Pivot] = {
19     pivotList match {
20       case p1 :: p2 :: tail =>
21         val width = p2.startValue - p1.startValue
22         flipSubfunction(p2 :: tail,movingEnd - width,toPos,
23                         mirrorPivot(p1,width,movingEnd) :: acc)
24       case p1 :: Nil =>
25         val width = toPos - p1.startValue + 1
26         mirrorPivot(p1,width,movingEnd) :: acc
27       case Nil => acc
28     }
29   }
30   def mirrorPivot(p:Pivot,width:Int,newEnd:Int):Pivot = {
31     val newStartValue = newEnd - width + 1
32     val newSlope = !p.f.slope
33     val newOffset = p.f(p.fromValue + newEnd)
34     new Pivot(newStartValue,new LinearTransform(newOffset,newSlope))
35   }
```

Fig. 7. Flipping a portion [from, to] of a concrete sequence by updating its bijection

$$\text{updatedFn}(\text{toPos} - x) = \text{forwardFn}(\text{fromPos} + x) \text{ for } x \in [0; \text{toPos} - \text{fromPos}]$$

$$\text{updatedFn}(x) = \text{forwardFn}(x) \text{ for } x \in [0; \text{fromPos}[\cup]\text{toPos}; l[$$

On line 3, additional pivots are inserted at the limit of the flipped interval defined by [fromPos, toPos], so the pivots within the interval can be reworked without modifying the function out of the interval. This is where the number of pivots in the function increases. On line 5, the pivots with starting values in the updated interval are copied into a list `pivotsToFlip`. On line 7, the selected pivots are flipped, and the result is `flippedPivots`. This is performed by a greedy algorithm iterating over the pivots of `pivotsToFlip` shown in function `flipSubFunction`. On line 8, the red black tree storing the forward function is updated: all pivots in [fromPos, topos] are replaced by pivots from `flippedPivots`. On line 11, it removes the pivots at fromPos and the first pivots after toPos if they are redundant, that is: if their function is aligned with the function of the pivot that precedes them, respectively. The function `flipSubfunction` is a recursion on the list of pivots to flip and uses an accumulator where the new pivots are stored, so that the generated pivots are in reverse order compared to the list of pivots they originate from. The function `mirrorPivot` creates a new pivot mirroring the given one within the flipped interval. We can see for instance that the slope of the function associated with a pivot is a Boolean and that it is negated when a pivot is mirrored.

The update operations performed on a concrete sequence involve updating its bijection and can be chained. Starting from Fig. 5, a flip of the subsequence from position 15 to position 24 (both included) leads to the bijection in Fig. 6. This figure also shows that performing updates can introduce additional pivots. For instance, seq-moves add at most three pivots: a pivot at the start of the moved zone, a pivot after the end of the moved zone, and a pivot after the position where the zone is moved.

When the number of pivots gets too large, the efficiency of queries and update operations is reduced. To avoid a significant increase in the number of pivots, a *regularization* operation is performed to simplify the bijection to the identity function, and re-factor the two balanced binary search trees mapping internal positions and values. This operation has a complexity of $O(n + k \cdot \log n)$ time, where n is the size of the sequence, assuming that each value occurs at most once in the sequence, like in routing optimization. The balanced binary search tree that maps internal positions to value is elaborated as follows: first, we iterate over the sequence and store its content into an array, then a new red-black tree is built from this array in $O(n)$ time. This is possible because the red-black tree is built at once on data that is already sorted. Iterating from one element to the next one is done in $O(\frac{k \cdot \log n}{n})$ time, amortized, as will be explained below.

The regularization is triggered when the ratio k/n gets larger than a maximal authorized value set by default to 3%. It ensures that k is kept significantly smaller than n to benefit from the symbolic nature of the operations performed on the bijection, and avoids triggering regularization too often. The choice of this maximal k/n value is investigated in more detail through some benchmarks in Sect. 6.

4.3   Iterating on Sequences

Iterating through a sequence through a succession of position-to-value queries would be quite slow, given our data structure; each step would require $O(\log n + \log k)$ time. To avoid this problem, our sequence supports a mechanism to speed-up sequence exploration called *explorers*. Considering the balanced binary search tree mapping internal positions to values, an explorer on a concrete sequence stores the path from the root to its current position, so that the next/previous position can be accessed in constant time. When the explorer reaches the end of a segment in the linear bijection, which happens k times during the whole sequence exploration, it must perform a $O(\log n)$ time query on the balanced binary search tree mapping internal positions to values to rebuild the path to its current position in the sequence.

Concrete and updated sequences can generate an explorer pointing to any of their position. Each class of sequence has an explorer class that can access internals of its originating sequence. Explorers can be instantiated from concrete sequences for any position in $O(\log n + \log k)$ time. It can be queried for its

Table 1. Time complexity of queries on a sequence value

	Value at position	Positions of value	Explorer	Explorer.next
Concrete sequence	$O(\log n + \log k)$	$O(\#occ \cdot \log k + \log n)$	$O(\log n + \log k)$	$O(\frac{k \cdot \log n}{n})$ amortized
Added cost for each stacked update	$O(1)$	$O(\#occ)$	$O(1)$	$O(1)$ amortized

Table 2. Time complexity of updates on a sequence value

	Instantiating a stacked update	Concrete update without regularization	Concrete update with amortized regularization
Insert	$O(1)$	$O(\log n + k \cdot \log k)$	$O(\log n + k \cdot \log k + n/k)$
Delete	$O(1)$	$O(\log n + k \cdot \log k)$	$O(\log n + k \cdot \log k + n/k)$
Seq-move	$O(1)$	$O(k \cdot \log k)$	$O(k \cdot \log k + n/k + \log n)$

position in the sequence and the value at this position in $O(1)$ time. It can also be queried for an explorer at the next or previous position in the sequence in $O(\frac{k \cdot \log n}{n})$ time, amortized.

4.4   Time Complexity of Sequence Values

The time complexity of the main queries on our sequence values is summarized in Table 1, where k is the maximal number of pivots in the bijection of the concrete sequence before regularization must occur, n is the size of the sequence, and #occ is the number of occurrences of the considered value in the sequence.

The complexity of the main update operations is summarized in Table 2. The first column shows the complexity of stacked updates. The second column represents the complexities of the concrete updates if no regularization occurs. The third column is the complexity of the concrete updates, considering the amortized complexity of regularization. The regularization operation has a complexity of $O(n + k \cdot \log n)$ time, takes place at most every $\Omega(k)$ updates, so it adds $O(n/k + \log n)$ time complexity, amortized.

5   Implementation of Our Sequence Variable

This section presents the behaviour of our sequence variable. A sequence variable replicates the three operations supported by the sequence value (insert, remove,

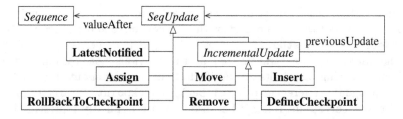

Fig. 8. Class diagram of the notifications used by sequence variables

```
val routes =
    SeqVar(m, 0 to v-1, n-1, knPerc)

val totalDistance =
    ConstantRoutingDistance(
        routes, v, distanceMatrix)

val obj = totalDistance +
    10000*(n - Size(routes))
```

```
val search =
    BestSlopeFirst(
        InsertPtUnroutedFirst(w=10),
        InsertPtRoutedFirst(w=10),
        onePointMove(w=15),
        twoOpt(w=20),
        threeOpt(w=10))
    exhaust threeOpt(w=25)
```

Fig. 9. An example of optimization script using our sequence variable

move) in a mutable form. A sequence variable can also be assigned a sequence value. Incremental updates lead to stacked updates of sequence value, assignment lead to concrete sequences, checkpoint definition lead to concrete sequence by performing all the stacked updates on the underlying concrete sequence.

Sequence variables record the updates performed on them, and include a description of these updates in their notification to listening invariants. The recording and notification are both represented by **SeqUpdate** classes whose inheritance is shown in Fig. 8. All **SeqUpdate** carry a reference to a sequence value (stacked or concrete) that represents the value of the variable after the **SeqUpdate**. A variable can also be assigned a sequence value. Assigns are notified to the listening invariants as an **Assign** update.

Neighbourhoods are expected to perform their exploration in a star mode around an initial sequence value, which is a concrete sequence. The neighbourhood first declares a checkpoint to the variable through a **defineCheckpoint** method. It then performs its exploration by updating the value of the variable through incremental insert, delete and move operations. Between each neighbour, the neighbourhood is expected to perform a roll back to the checkpoint it has defined, by calling a **rollBackToCheckpoint** method. This restores the value of the sequence to its checkpoint by removing stacked updates, and the roll back will be communicated to the listening invariant. When the exploration is completed, the neighbourhood must call a **releasecheckpoint** method.

Upon propagation, sequence variables notify their listening invariants about their value change by calling a notification method of the invariants with a **SeqUpdate** parameter as defined in Fig. 8. At the start of a neighbourhood exploration, listening invariants are notified about a **DefineCheckpoint** and an incremental update representing the first neighbour. The data structure transmitted to the notification procedure is an `IncrementalUpdate`, with a `previousUpdate` reference to a `DefineCheckpoint` with a `previousUpdate` reference to a `LatestNotified`. `LatestNotified` mean that the incremental update starts from the value that was transmitted at the previous notification. For each neighbour, listening invariants are notified about an `IncrementalUpdate` with a `previousUpdate` reference to a `RollBackToCheckpoint`. Invariants are not notified about checkpoint releases because they are implicitly represented by a new checkpoint definition that overrides the previous one.

6   Benchmarking the k/n Factor

This section presents a benchmark to illustrate the efficiency of sequence data structure in this setting and the impact of the k/n factor presented in Sect. 4.2. The benchmark is a symmetric vehicle routing problem with 100 vehicles no other constraint. The total distance driven by all vehicles must be minimized. The problem roughly declares as show in Fig. 9, using various bricks of OscaR.cbls.

The search procedure is build using neighbourhood combinators [15]. It starts with no routed node, but the start points of the vehicles, and uses a mix of insert point, one-point-move, 2-opt and 3-opt. There are two insertion neighbourhoods, one iterates on insertion position first, and then on point to insert; the other iterates on point to insert first. These neighbourhoods have w parameter: when several nodes are considered simultaneously by a neighbourhood, it ensures that the considered nodes are among the w closest nodes of one another. It ends with a 3-opt neighbourhood with a larger w factor. The search strategy does not have any mechanism to escape from local minima; it is exclusively designed to benchmark the k/n factor.

We benchmark with the following values for the k/n factor: 0%, 1%, 2%, 3%, 4%, 5%, and 20% and for the size (n) of the problem: 1k, 3k, 5k, 7k, 9k, and 11k. For each pair (k/n factor,n) we solve 100 randomly generated instances. The benchmarks were executed single-threaded on an isolated and dedicated machine featuring 8 hyper-threaded cores of Intel Xeon ®E5620 @ 2.40 GHz and 32 Gb of RAM. This setting was designed to provide stable computing power.

Figure 10 reports the average run time among the 100 instances for each considered pair (k/n factor,n). We can see that a value of 0% for the k/n factor is suboptimal; it disables the system of piecewise affine bijection, and regularization is triggered every time a move is committed. Above 1%, the impact of this factor on the run time reaches some plateau. The efficiency decreases again if the k/n factor gets too large. A sample value of 20% is illustrated. Another phenomenon is that the efficiency of this mechanism improves with the size of the considered problem; this is probably due to the non-linear nature of the complexities, presented in Sect. 4.4. A last phenomenon that is visible from the table is that the curves seem to reach their minimum around 2 to 3% a bit larger for smaller problems. The default value for the k/n factor is set to 3% and can be changed by the user.

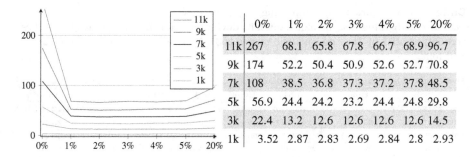

	0%	1%	2%	3%	4%	5%	20%
11k	267	68.1	65.8	67.8	66.7	68.9	96.7
9k	174	52.2	50.4	50.9	52.6	52.7	70.8
7k	108	38.5	36.8	37.3	37.2	37.8	48.5
5k	56.9	24.4	24.2	23.2	24.4	24.8	29.8
3k	22.4	13.2	12.6	12.6	12.6	12.6	14.5
1k	3.52	2.87	2.83	2.69	2.84	2.8	2.93

Fig. 10. Average run time (in seconds) on 100 random instances for various values of the k/n factor $(0\%, 1\%, \dots)$ and various values of n $(1k, 3k, \dots)$

7 Comparing with GoogleCP and LocalSolver

The goal of this section is to illustrate the effectiveness for a CBLS framework of having a sequence variable and its associated invariants. To do this, we compare the efficiency of our implementation with a sequence variable with GoogleCP 6.4, LocalSolver 7.0, and an OscaR.cbls implementation that relies on an array of integer variables as in Fig. 1. We use a symmetric TSP problem on a square map with Euclidean distance and random instances of sizes 500 to 5000 by steps of 500. This problem uses a limited set of constraints, so it only lightly shows the effect of resolving the multi-variable bottleneck. Problems with more constraints would feature more global constraints. Here we deliberately amplify the difference of efficiency by exploring large instances.

GoogleCP was configured to use more or less the same neighbourhoods as OscaR.cbls. We use the TSP script distributed with GoogleCP, and configured it to use only 2-opt, one-point-move, and or-opt. or-opt is a variant of the 3-opt that does not try to flip segments. The script uses an array of integer variables representing the next nodes, and uses a global invariant to maintain the route length. This script encounters a multi-variable bottleneck for the 2-opt. LocalSolver requires virtually no parameters, so that it can be considered as a stable reference. We used the TSP script that is distributed with LocalSolver, which uses the "list" variable discussed in Sect. 2. LocalSolver natively supports multi-threading and some form of simulated annealing. These two features were disabled for the benchmark, to be on the same level as other tools. We have no information on the search procedure used by LocalSolver. The two implementations of OscaR.cbls use the search procedure presented in Sect. 6. The tree engines were assigned a maximal search time of 200 seconds. The benchmarks were executed on the same hardware as the one presented in Sect. 6. One hundred runs were performed for each instance and each solver.

The benchmark must be considered with care because it compares complex tools, whose efficiency are affected by their search procedure and their model, while we only want to illustrate the gain on the efficiency of our model compared to rasterized models. Notable differences are: GoogleCP uses a good heuristic to

Table 3. Benchmark results reporting objective function (obj) and search time in seconds (time) values as: median (standard deviation). A † indicates that all searches were interrupted by the timeout (the min is 200). Min values for obj and time are in bold font.

Size	OscaR.cbls Array[IntVar]		LocalSolver		GoogleCP		OscaR.cbls SeqVar	
	obj	time	obj	time	obj	time	obj	time
500	**17134** (128)	23 (7)	17602 (0)	200 †(0)	17429 (0)	4.4 (0.05)	17211 (72)	**1.42** (0.38)
1000	24173 (127)	65 (14)	24662 (7)	200 †(0)	24230 (0)	34.8 (0.37)	**24171** (137)	**3.75** (0.77)
1500	29279 (143)	139 (23)	30575 (13)	200 †(0)	**29126** (0)	80.1 (0.57)	29140 (111)	**6.03** (1.07)
2000	33712 (151)	198 (27)	42498 (242)	200 †(0)	33785 (0)	200.2 †(0.01)	**33593** (193)	**9.19** (1.07)
2500	37882 (148)	211 †(21)	69276 (570)	200 †(0)	39006 (0)	200.1 †(0.01)	**37772** (143)	**11.4** (1.56)
3000	40925 (147)	204 †(10)	141523 (1925)	200 †(0)	42007 (2)	200.2 †(0.01)	**40737** (131)	**14.8** (1.94)
3500	43563 (147)	202 †(3)	229540 (1378)	200 †(0)	46287 (27)	200.3 †(0.02)	**43317** (158)	**17.8** (2.03)
4000	47122 (155)	202 †(5)	335925 (3885)	200 †(0)	49864 (29)	200.4 †(0.02)	**46723** (137)	**20.5** (3.39)
4500	50184 (147)	202 †(5)	457996 (2680)	200 †(0)	55936 (25)	200.6 †(0.03)	**49647** (148)	**23.9** (3.08)
5000	52940 (147)	204 †(7)	638130 (2467)	200 †(0)	58779 (10)	200.7 †(0.03)	**52545** (153)	**28** (3.02)

build a good initial solution. OscaR.cbls inserts node into the route throughout the search, the w-nearest pruning used in its neighbourhoods might be different from the one used by GoogleCP. GoogleCP and LocalSolver are implemented in C++ and OscaR in Scala 2.11.7. We have no idea of the neighbourhoods used by Localsolver, but they are probably not suited for routing.

The results are presented in Table 3. Our approach based on sequences is faster than the approaches based on arrays of integers. Because of the time-out, it can produce better result for larger problem. The quality on smaller problems is like the one of the other tools. The variation in quality of the two OscaR implementations appear because the search strategy is adaptive and favours faster neighbourhoods, so the succession of explored neighbourhoods varies in the two implementations and find different local minima. This could be improved by using some meta-heuristics.

8   Conclusion

This paper presents an implementation of variables of type "sequence of integers" that is suited for CBLS frameworks. The goal is to embed state-of-the-art global constraints into declarative CBLS frameworks. We also proposed some specific data-structure, so that neighbour solutions can be represented quickly, and queried efficiently.

Our approach was based on a set of requirements that were formulated in the context of a generic CBLS framework. These notably specify that all data-structures must be immutable. We did not consider mutable data-structure. An approach based on mutable data-structure might be considered.

There are a few opportunities of improvement concerning the data structure itself. For instance, some update operations performed on the bijection are not entirely incremental (cfr. Sect. 4), and other data structures might be considered for the concrete sequences. We might consider replacing the large red-black trees of the second mapping with non-mutable arrays.

Our next step is to extend our library of global constraints (including the ones presented in Sect. 2.2) and generic neighbourhoods on sequences. We already proposed a generic framework for defining global routing constraint based on a mathematical structure called a *group* [16]. We also consider integrating the scheduling constraints and neighbourhoods presented in [17], which are based on sequences of tasks and might therefore fit into our sequence framework.

Our sequence variable features a checkpoint mechanism that makes it possible for global constraint to perform pre-computations and implement differentiation. This mechanism is restricted to sequence variables and different variables might declare checkpoints at different points in time in an uncoordinated fashion. Checkpoints should be made pervasive in the model, so that invariants with other types of variables or with multiple input sequence variables could also have differentiation.

With this additional variable, OscaR.cbls will be even more appealing both to users that benefit from highly efficient global constraints in a declarative framework, and to researchers who develop new global constraints and will benefit from the whole framework of OscaR.cbls, so that they can focus on their own contribution. String solving is an active topic of research that might benefit from our framework [18–21], say by integrating our results with those of [12]. Our framework can also offer a common benchmarking environment to compare the efficiency of different algorithms of global constraints on sequences.

Acknowledgements. This research was conducted under the SAMOBI CWALITY (grant nr. 1610019) and the PRIMa-Q CORNET (grant nr. 1610088) research projects from the Walloon Region of Belgium. We thank YourKit profiler and LocalSolver for making their software freely available to us for this research. We also warmly thank Pierre Flener and the anonymous referees for their feedback on earlier versions of this work.

References

1. Glover, F.W., Kochenberger, G.A.: Handbook of Metaheuristics. International Series in Operations Research & Management Science. Springer, New York (2003). https://doi.org/10.1007/b101874
2. Mladenović, N., Urošević, D., Hanafi, S.: Variable neighborhood search for the travelling deliveryman problem. 4OR **11**(1), 57–73 (2013)
3. Savelsbergh, M.W.P.: The vehicle routing problem with time windows: minimizing route duration. ORSA J. Comput. **4**(2), 146–154 (1992)
4. Van Hentenryck, P., Michel, L.: Constraint-Based Local Search. MIT Press, Cambridge (2009)
5. Newton, M.A.H., Pham, D.N., Sattar, A., Maher, M.: Kangaroo: an efficient constraint-based local search system using lazy propagation. In: Lee, J. (ed.) CP 2011. LNCS, vol. 6876, pp. 645–659. Springer, Heidelberg (2011). https://doi.org/10.1007/978-3-642-23786-7_49

6. De Landtsheer, R., Ponsard, C.: OscaR.cbls: an open source framework for constraint-based local search. In: Proceedings of ORBEL'27, 2013
7. Benoist, T., Estellon, B., Gardi, F., Megel, R., Nouioua, K.: LocalSolver 1.x: a black-box local-search solver for 0-1 programming. 4OR **9**(3), 299–316 (2011)
8. Pralet, C., Verfaillie, G.: Dynamic online planning and scheduling using a static invariant-based evaluation model. In: Proceedings of the 23rd International Conference on Automated Planning and Scheduling, ICAPS 2013, Rome, Italy, 10-14 June 2013. AAAI (2013)
9. OscaR Team. OscaR: Operational research in Scala (2012). Available under the LGPL licence from https://bitbucket.org/oscarlib/oscar
10. OR-Tools Team. OR-Tools: Google Optimization Tools. https://code.google.com/p/or-tools/
11. Di Gaspero, L., Schaerf, A.: EASYLOCAL++: an object-oriented framework for the flexible design of local-search algorithms. Softw.: Pract. Exp. **33**(8), 733–765 (2003)
12. Björdal, G.: String variables for constraint-based local search. Master's thesis, Uppsala University (2016). https://uu.diva-portal.org/smash/get/diva2:954752/FULLTEXT01.pdf
13. Lin, S., Kernighan, B.W.: An effective heuristic algorithm for the traveling-salesman problem. Oper. Res. **21**(2), 498–516 (1973)
14. Savelsbergh, M.W.P., Sol, M.: The general pickup and delivery problem. Transp. Sci. **29**, 17–29 (1995)
15. De Landtsheer, R., Guyot, Y., Ospina, G., Ponsard, C.: Combining neighborhoods into local search strategies. In: Amodeo, L., Talbi, E.-G., Yalaoui, F. (eds.) Recent Developments in Metaheuristics. ORSIS, vol. 62, pp. 43–57. Springer, Cham (2018). https://doi.org/10.1007/978-3-319-58253-5_3
16. Meurisse, Q., De Landtsheer, R.: Generic support for global routing constraint in constraint-based local search frameworks. In: Proceedings of the 32th ORBEL Annual Meeting, pp. 129–130, 1–2 February 2018
17. Pralet, C.: An incomplete constraint-based system for scheduling with renewable resources. In: Beck, J.C. (ed.) CP 2017. LNCS, vol. 10416, pp. 243–261. Springer, Cham (2017). https://doi.org/10.1007/978-3-319-66158-2_16
18. Abdulla, P.A., Atig, M.F., Chen, Y.-F., Holík, L., Rezine, A., Rümmer, P., Stenman, J.: Norn: an SMT solver for string constraints. In: Kroening, D., Pǎsǎreanu, C.S. (eds.) CAV 2015. LNCS, vol. 9206, pp. 462–469. Springer, Cham (2015). https://doi.org/10.1007/978-3-319-21690-4_29
19. Fu, X., Powell, M.C., Bantegui, M., Li, C.-C.: Simple linear string constraints. Formal Aspects Comput. **25**(6), 847–891 (2013)
20. Ganesh, V., Kieżun, A., Artzi, S., Guo, P.J., Hooimeijer, P., Ernst, M.: HAMPI: a string solver for testing, analysis and vulnerability detection. In: Gopalakrishnan, G., Qadeer, S. (eds.) CAV 2011. LNCS, vol. 6806, pp. 1–19. Springer, Heidelberg (2011). https://doi.org/10.1007/978-3-642-22110-1_1
21. Scott, J.D., Flener, P., Pearson, J., Schulte, C.: Design and implementation of bounded-length sequence variables. In: Salvagnin, D., Lombardi, M. (eds.) CPAIOR 2017. LNCS, vol. 10335, pp. 51–67. Springer, Cham (2017). https://doi.org/10.1007/978-3-319-59776-8_5

Constraint Programming for High School Timetabling: A Scheduling-Based Model with Hot Starts

Emir Demirović[✉] and Peter J. Stuckey[✉]

School of Computing and Information Systems,
University of Melbourne, Melbourne, Australia
{edemirovic,pstuckey}@unimelb.edu.au

Abstract. High School Timetabling (HSTT) is a well-known and wide-spread problem. It consists of coordinating resources (e.g. teachers, rooms), times, and events (e.g. classes) with respect to a variety of constraints. In this paper, we study the applicability of constraint programming (CP) for high school timetabling. We formulate a novel CP model for HSTT using a scheduling-based point of view. We show that a drastic improvement in performance over the baseline CP model can be achieved by including solution-based phase saving, which directs the CP solver to first search in close proximity to the best solution found, and our hot start approach, where we use existing heuristic methods to produce a starting point for the CP solver. The experiments demonstrate that our approach outperforms the IP and maxSAT complete methods and provides competitive results when compared to dedicated heuristic solvers.

Keywords: Constraint programming · Timetabling · Scheduling
Modeling · Hot start · Warm start · Local search · Phase saving

1 Introduction

The problem of high school timetabling (HSTT) is to coordinate resources (e.g. rooms, teachers, students) with times to fulfill certain goals (e.g. scheduling classes). Every school requires some form of HSTT, making it a wide-spread problem. In a more general sense, timetabling can be found in airports, transportation, and the like. The quality of the timetables is an important issue, since they have a direct impact on the educational system, satisfaction of students and staff, and other matters. Every timetable affects hundreds of students and teachers for prolonged periods of time, as each timetable is used for at least a semester, making HSTT an extremely important and responsible task. However, constructing timetables by hand can be time consuming, difficult, error prone, and in some cases practically impossible. Thus, developing algorithms to produce the best timetables automatically is of utmost importance.

There has been significant research tackling HSTT. However, given that there are many educational systems, each differing in their own ways, much of this

© Springer International Publishing AG, part of Springer Nature 2018
W.-J. van Hoeve (Ed.): CPAIOR 2018, LNCS 10848, pp. 135–152, 2018.
https://doi.org/10.1007/978-3-319-93031-2_10

research was done in isolation targeting only a particular set of rules. It was difficult to compare developed algorithms due to the differences, even though the problems shared similarities. This motivated researchers to develop a general high school timetabling formulation, named XHSTT [24, 25], that allows a variety of education system requirements to be expressed. With the new formulation, researchers now have common ground for fair algorithmic comparison. In 2011, the International Timetabling Competition was dedicated to HSTT and endorsed the said formulation, encouraging further research in the direction of XHSTT. We consider this formulation in our work.

Historically, incomplete algorithms were the most prominent for XHSTT (e.g. [11,12,15]. Recently, complete methods based on integer programming [16] and maximum Boolean satisfiability (maxSAT) [8] have proven to be effective. Their development was essential for the emergence of large neighborhood search algorithms, which combine domain-specific heuristics with complete solving [7,27]. These methods are currently among the most successful ones for XHSTT.

As complete methods play a vital role, it is natural to ask if unexplored complete paradigms could bring more to XHSTT than those currently in use. This is precisely what we study in this work: the applicability of constraint programming for XHSTT. We provide a novel CP model that views XHSTT as a scheduling problem, in contrast to the conventional Boolean formulations. Such a model allows us to exploit high-level global constraints, providing an elegant and possibly more efficient solution. However, to match the state-of-the-art, that is not enough. Indeed, our experimentation demonstrates that a standard CP approach is not competitive. Therefore, further development was required. The first improvement came from employing *solution-based phase saving* [4], an existing technique in maxSAT [1] but not well-known in CP, that directs the CP solver to search in close proximity to the best solution found so far before expanding further. The second was born out of the realization that the community has built sophisticated heuristic algorithms, which can be communicated to the CP solver (so called *hot start*). While usual hot starts in CP would only provide an initial bound to the problem, when combined with the solution phase saving, such an approach offers more: in addition to the bound, it suggests to the solver a region in the search space that is likely to contain high quality solutions. The end result is a complete algorithm that outperforms integer programming and maxSAT methods, while being competitive with dedicated heuristic solvers.

The paper is organized as follows. In the next section we briefly describe the general high school timetabling (XHSTT) problem. In Sect. 3, we provide an overview of the state-of-the-art for XHSTT. The scheduling-based model is given in Sect. 4. Our hot start approach is discussed in Sect. 5, along with solution-based phase saving. Experiments are given in Sect. 6. We conclude in the last section.

2   Problem Description

In our research we consider the general formulation of the High School Timetabling problem (called XHSTT), as described in [25]. This formulation is general enough to be able to model education systems from different countries and was endorsed by the International Timetabling Competition 2011.

The general High School Timetabling formulation specifies three main entities: times, resources and events. Times refer to discrete time units which are available, such as Monday 9:00–10:00, Monday 10:00–11:00, for example. Resources correspond to available rooms, teachers, students, and others. The main entities are the events, which in order to take place require certain times and resources. An event could be a Mathematics class, which requires a math teacher (which needs to be determined) and a specific student group (both the teacher and the student group are considered resources) and two times. Events are to be scheduled into one or more *solution events* or *subevents*. For example, a Mathematics class with total duration of four hours can be split into two subevents with duration two, but can be scheduled as one subevent with duration four (constraints may impose further constraints on the durations of subevents).

The aim of XHSTT is to find a schedule by assigning times and resources to events such that that all hard constraints are satisfied and the sum of soft constraint violations is minimized.

Constraints impose limits on what kind of assignments are legal. These may constrain that a teacher can teach no more than five classes per day, that younger students should attend more demanding subjects (e.g. Mathematics) in the morning, to name a few examples. It is important to differentiate between hard constraints and soft constraints. The former are important constraints which are given precedence over the latter, in the sense that any single violation of any hard constraint is more important than all of the soft constraints combined. Thus, one aims to satisfy as many hard constraints as possible, and then optimize for the soft constraints. In this sense, "hard constraints" are not, in fact, hard as in the usual meaning used in combinatorial optimisation. Each constraint has a nonnegative cost function associated with it, which penalizes assignments that violate it. The goal is to first minimize the hard constraint costs and then minimize the soft constraint costs. In the general formulation, any constraint may be declared hard or soft and no constraint is predefined as such, but rather left as a modeling option based on the concrete timetabling needs. Each constraint has several parameters, such as to which events or resources it applies and to what extent (e.g. how many idles times are acceptable during the week), and other features, allowing great flexibility.

We now give an informal overview of all the constraints in XHSTT (as given in [25]). For more details regarding the problem formulation, see [24,25]. There is a total of 16 constraints (plus preassignments of times or resources to events, which are not listed).

Constraints related to events:

1. *Assigned Time* - assign the specified amount of times to specified events.
2. *Preferred Times* - when assigning times to events, specified times are preferred over others.
3. *Link Events* - specified events must take place at the same time.
4. *Spread Events* - specified events must be spread out during the week.

5. *Distribute Split Events* - limits the number of subevents that may take a particular duration for a given event.
6. *Split Events* - limits the minimum and maximum durations and number of subevents for a given event. Together with *Distribute Split Events* this gives fine control on the subevents.
7. *Order Events* - specified events must be scheduled one after the other with nonnegative time-lag in between them.
8. *Avoid Split Assignments* - for all subevents derived from an event, assign the same resources.

Constraints related to resources:

1. *Assigned Resource* - assign specified resources to specified events.
2. *Avoid Clashes* - specified resources cannot be used by two or more subevents at the same time.
3. *Preferred Resources* - when assigning resources to events, specified resources are preferred over others.
4. *Avoid Unavailable Times* - specified resources cannot be used at specified times.
5. *Limit Workload* - specified resources must have their workload lie between given values.
6. *Limit Busy Times* - the amount of times when a resource is being used within specified time groups should lie between given values.
7. *Cluster Busy Times* - specified resources' activities must all take place within a minimum and maximum amount of time groups.
8. *Limit Idle Times* - specified resources must have their number of idle times lie between given values within specified time groups.

3   Related Work

For HSTT, both heuristic and complete methods have been proposed. Heuristic methods were historically the dominating approach, as they are able to provide good solutions in a reasonable amount of time even when dealing with large instances, albeit not being able to obtain or prove optimality. Recently developed complete methods [5, 26, 30, 31] are successful in obtaining good results and proving bounds but require significantly more time (days or weeks).

The best algorithms from the International Timetabling Competition 2011 (ITC 2011) were incomplete algorithms. The winner was the group GOAL, followed by Lectio and HySST. In GOAL, an initial solution is generated, which is further improved by using Simulated Annealing and Iterated Local Search, using seven different neighborhoods [12]. Lectio uses an Adaptive Large Neighborhood Search [29] with nine insertion methods based on the greedy regret heuristics [32] and fourteen removal methods. HySST uses a Hyper-Heuristic Search [14].

After the competition, the winning team of ITC 2011 developed several new Variable Neighborhood Search (VNS) approaches [11]. All of the VNS approaches have a common search pattern: from one of the available neighborhoods, a random solution is chosen, after which a descent method is applied and the solution is accepted if it is better than the previous best one. Each iteration starts from the best solution. The Skewed Variable Neighborhood was the most successful VSN approach, in which a relaxed rule is used to accept the new solution based on its cost and its distance from the best solution.

Kingston [15] introduced an efficient heuristic algorithm which directly focuses on repairing *defects* (violations of constraints). Defects are examined individually and specialized procedures are developed for most constraints to repair them. KHE14 provides high quality solutions in a low amount of time, but does not necessarily outperform other methods with respect to quality of solution.

Two complete methods have been studied: IP- [16] and maxSAT-based [8] approaches. Neither method strictly dominates the other, as their relative performance depends on the instance. Both models use Boolean variables to explicitly encode if an event is taking place at a particular time, with the main differences being in the expressiveness of the formalism to define the constraints. Overall, the maxSAT approach provides better results but is limited to problems where resources are preassigned to events. A Satisfiability Modulo Theories (SMT) approach has also been investigated in [6] but cannot handle XHSTT instances efficiently. The bitvector-SMT model rather serves as an efficient way of representing XHSTT for local search algorithms, as all constraint costs can be computed using simple bitvector operations.

The developed complete methods were used in large neighborhood search algorithms: IP-based [30] and maxSAT-based [7]. These approaches offer improvements over their complete counterparts when given limited time.

Additionally, several IP- [10,26,28,33] and CP-based [13,19,34] techniques have been introduced for related HSTT problems. There are several notable differences in our work compared to the other CP approaches [19,34]: our modelling is more general as it applies to XHSTT and we demonstrate how to use generic tools to solve XHSTT without intertwining other techniques other than an initial solution procedure, which is clearly decoupled from the rest of our method.

4   Modeling

The key elements of XHSTT are a set of events E, a set of resources R and a set of times T which we shall regard as integers $T = \{0, 1, 2, \ldots |T| - 1\}$. We considered the restricted form of the problem where resources used by each event are predefined. The majority of the benchmarks from the repository of the International Timetabling Competition fall into this category. Earlier models for the general high school timetabling problem used explicit representations for each pair of events and time slots, indicating whether the event is taking place at that particular time. In contrast, we use a scheduling-based modeling approach, where each subevent is linked to two variables: the starting time and duration variable. As a result, we are able to exploit the disjunctive and regular global constraints in our model. We now describe the decision variables and then proceed with the modeling of each constraint.

4.1 Decision Variables

Each event $e \in E$ has a maximum total duration $D(e)$. For each event e, we create $D(e)$ subevents, numbered from $0..D(e) - 1$. Every subevent is associated with two variables indicating its starting time and duration. We label these as $start(e, i)$ and $dur(e, i)$. The special start time $UN = |T|$ is use to denote that a subevent is not used, in which case its corresponding duration is zero. Constraints may impose restrictions on the amount and duration of the subevents.

Example 1. Let e be an event of duration 3. A total of six variables are allocated: three pairs of starting time and duration variables. Let the assignments be as shown in Table 1(a):

Table 1. Decision variable assignments for event e of total duration 3: (a) satisfyies the symmetry breaking constraints, while (b) does not.

i	$start(e, i)$	$dur(e, i)$		i	$start(e, i)$	$dur(e, i)$
0	5	2		0	5	2
1	10	1		1	UN	0
2	UN	0		2	10	1
	(a)				(b)	

The assignments state that for event e, two subevents of durations 2 and 1 are scheduled at starting times 5 and 10, respectively.   □

The following constraints formally define the decision variables for each event e and its subevent i:

$$
\begin{aligned}
&start(e, i) \in T \cup \{UN\}, \\
&dur(e, i) \in \{0, 1, \ldots, D(e)\}, \\
&start(e, i) + dur(e, i) \leq |T|, \\
&start(e, i) = UN \Leftrightarrow dur(e, i) = 0
\end{aligned}
\tag{1}
$$

Symmetry breaking constraints forbid equivalent solutions. These order the subevents by decreasing duration as the primary criteria, and then by start time:

$$
\begin{aligned}
&dur(e, i - 1) \geq dur(e, i), \\
&dur(e, i - 1) = dur(e, i) \Rightarrow start(e, i - 1) \leq start(e, i)
\end{aligned}
\tag{2}
$$

Example 2. Consider the same setting as in the previous example, but with the following assignments shown in Table 1(b). The constraints in Eqs. 1 are satisfied, but symmetry breaking constraints (Eqs. 2) require that subevents are ordered by duration. Furthermore, if $D(e) = 5$ and the $dur(e, 0) + dur(e, 1) + dur(e, 2) = 5$, and $dur(e, 1) = 2$ and $start(e, 1) = 0$ then the assignment would again be invalid, as in case of ties in duration, subevents are to be sorted by increasing time assignments.

Symmetry breaking constraints, apart from preventing equivalent solutions, contribute towards simpler constraints (e.g. see *Split Events* constraint). We note that non-overlapping of subevents is left to the *Avoid Clashes* constraints.

4.2   Additional Notation

Let $EV(r) \subseteq E$ denote the set of events that require resource r. We introduce auxiliary Boolean variables $busy(r, t)$ to indicate that one of the events $e \in EV(r)$ that require $r \in R$ are taking place at time t. This dual viewpoint on the decisions is required for the *Limit Busy Times* and *Cluster Busy Times* constraints. The mapping between the two viewpoints is managed by a straightforward decomposition.

$$busy(r, t) \Leftrightarrow \exists_{e \in EV(r), i \in \{0..D(e)-1\}} start(e, i) \leq t \wedge start(e, i) + dur(e, i) > t$$

A time group $TG \subseteq T$ is a fixed set of times. An event group $EG \subseteq R$ is a fixed set of events.

4.3   Objectives and Constraints

In the problem formulation, constraints are not predefined as hard or soft, but this option is left to the modelers as appropriate for the particular school under consideration. For simplicity of the presentation, we present in the following text the constraints as if they are given as hard constraints. Soft versions are created by reifying the hard formulation. Note that this may mean that global constraints in the soft versions make use of a decomposition instead, explicitly encoding the global constraint using a set of smaller and simpler constraints. We define the predicate *within* to ease further notation:

$$within(x, l, u) = (x \geq l \wedge x \leq u) \tag{3}$$

Soft constraints are similar to their hard counterparts, but instead penalize each constraint violation instead of forbidding it entirely. For example, soft constraints commonly state that a certain value has to be within given bounds. In the soft case, the deviation is penalized based on the linear or quadratic distance (specified by the constraint) of the said value from the imposed bounds. In the linear case, the violation is calculated as

$$within\_viol(x, l, u) = \max\{0, x - u, l - x\}. \tag{4}$$

Hard constraints essentially have large weights compared to soft constraints. In our model, the hard XHSTT constraints are posed as hard CP constraints, meaning the infeasibility value is not tracked. This modelling choice allows us to drastically simplify the modelling, which leads to an increase in performance for CP but also allows advanced CP techniques to be exploited through global constraints. We now proceed with the modeling of each constraint.

Assigned Times: Events must be assigned their prescribed number of times.

$$\sum_{i \in \{0,1,\dots,D(e)-1\}} dur(e,i) = D(e) \tag{5}$$

Preferred Times: Subevents of specified event e may start only within the stated time group TG_e. If an optional parameter d is given, the constraint only applies to subevents of that particular duration.

$$\forall i \in \{0,1,\dots,D(e)-1\} \qquad start(e,i) \neq UN \Rightarrow start(e,i) \in TG_e \tag{6}$$

Link Events: Certain events must simultaneously take place. Let EG be an event group of linked events, all of which have the same total duration TD, i.e. $\forall e \in EG, D(e) = TD$. We make use of the global constraint all_equal [2], which enforces that its input variables must be assigned the same values.

$$\begin{aligned} &\forall i \in \{0,1,\dots,TD-1\} \\ &\text{all\_equal}(\big[start(e,i) \mid e \in EG\big]), \\ &\text{all\_equal}(\big[dur(e,i) \mid e \in EG\big]) \end{aligned} \tag{7}$$

Spread Events: Limits the number of starting times events from specified event groups may have in given time groups. Event and time groups are sets of events and contiguous times, respectively. Let EG and TG denote one such pair with the limits $mine..maxe$.

$$\begin{aligned} &z = \sum_{e \in EG, i \in \{0,\dots,D(e)-1\}} within(start(e,i), \min(TG), \max(TG)) \wedge \\ &within(z, mine, maxe) \end{aligned} \tag{8}$$

Distribute Split Events: Limits the number of subevents of e a given duration d to be in the range $minds_e..maxds_e$.

$$a = \sum_{i \in \{0,1,\dots,D(e)-1\}} (dur(e,i) = d) \;\wedge\; within(a, minds_e, maxds_e) \tag{9}$$

Split Events: Regulates the number of subevents of e between $mins_e..maxs_e$ and the duration of subevents of e between $mind_e..maxd_e$

$$\forall i \in \{0,..,mins_e-1\} : \qquad start(e,i) \neq UN \,\wedge\, dur(e,i) \neq 0 \tag{10}$$

$$\forall i \in \{maxs_e,..,D(e)-1\} : \qquad start(e,i) = UN \wedge dur(e,i) = 0 \tag{11}$$

$$dur(e,i) \leq maxd_e \,\wedge\, dur(e,i) \neq 0 \Rightarrow dur(e,i) \geq mind_e \tag{12}$$

Order Events: For a given pair of events (e_1, e_2), the constraint imposes that e_1 must take place before e_2. In addition, there must be a minimum $mino_{ep}$ and maximum $maxo_{ep}$ units of time apart.

$$oe = min\{start(e_2, i) \mid i \in \{0..D(e_2) - 1\}\}$$
$$- max\{start(e_1, i) + dur(e_1, i) \mid i \in \{0..D(e_1) - 1\}\} \wedge \qquad (13)$$
$$within(eo, min_{ep}, max_{ep})$$

Avoid Clashes: A resource can be used by at most one event at any given time. Here we make use of the global constraint `disjunctive` [9,17], which takes two arrays s and d of variables as input, where $s[i]$ and $d[i]$ represent the starting time and duration of task i, and enforces no overlap between the tasks.

$$\texttt{disjunctive(}$$
$$\big[start(e, i) \mid e \in EV(r), i \in \{0..D(e) - 1\} \big], \qquad (14)$$
$$\big[dur(e, i) \mid e \in EV(r), i \in \{0..D(e) - 1\} \big] \texttt{))}$$

Avoid Unavailable Times: Resources cannot be used at specified times. For each resource r and forbidden time t, this is encoded by creating a dummy event that requires r and is fixed at time t with duration 1. The newly created events are added to $events(r)$ and will be considered in the *Avoid Clashes* constraint (above). For the soft version the duration of these dummy events is 0..1 and the constraint is violated if the duration used in 0.

Limit Busy Times: If a resource r is busy in a time group TG, its number of busy times within the time group is restricted to be in $minb_r..maxb_r$.

$$c = \sum_{t \in TG} (busy(r, t)) \qquad \wedge$$
$$c \neq 0 \Rightarrow within(c, minb_r, maxb_r) \qquad (15)$$

Cluster Busy Times: A resource is busy in a time group TG if it is busy at least one time int the time group. This constraint gives a set of time groups **TG** and limits the total number of time groups $TG \in$ **TG** that the resource may be busy to the range $mint..maxt$. For example, a teacher must finish his or her work within three days.

$$b = \sum_{TG \in \mathbf{TG}} (\exists_{t \in TG} busy(r, t)) \qquad \wedge$$
$$within(b, mint, maxt) \qquad (16)$$

Limit Idle Times: Some resources must not have idle times in their schedule. An idle time occurs at time t within contiguous time group TG if the resource is busy at times from TG before and after t and not busy at time t. This is encoded using the `regular` global constraint [22], which takes as input a sequence of variables that must satisfy the automaton constraint, and a deterministic finite automata A defined by a set of states Q, a set of values of the variable sequence S, a transition function which given a state and value defines the next state to reach, an initial state, and a set of final states. It constrains that the transition sequence defined by the sequence variables starting from the start state leads to a valid final state.

The automata has four states $Q = \{q_0, q_1, q_2, f\}$ with the interpretation: q_0 (*initial*) not yet been busy within the time group, q_1 (*busy*), q_2 (*done*) was busy but is no longer busy, and f is the fail state. The transitions T of the automata are defined as $\{(q_0, 0) \rightarrow q_0, (q_0, 1) \rightarrow q_1, (q_1, 0) \rightarrow q_2, (q_1, 1) \rightarrow q_1, (q_2, 0) \rightarrow q_2, (q_2, 1) \rightarrow f\}$, which enforce that once the resource is busy and then once again idle, it cannot become busy otherwise it ends in a fail state. The final states are $F = \{q_0, q_1, q_2\}$.

$$\texttt{regular}\big(\big[busy(r, t) | t \in TG\big], S, 0..1, T, s_0, F\big) \tag{17}$$

The soft constraint version is done by decomposition. The remaining constraints *Avoid Split Assignments*, *Assigned Resources*, *Prefer Resources*, *Limit Workload* are meaningless for the class of problems we examine where all resources are preassigned.

5   Solution-Based Phase Saving

During the search, the CP solver repeatedly makes decisions on which variable and value to branch on. Variables are chosen based on their activity (VSIDS scheme). Phase-saving [23] is almost universally used in SAT solvers, where the choice of value for a decision used is always the value used the last time the variable was seen in search. Solution-based phase saving [1,4] rather chooses the value for the variable that was used in the last found solution. After a branching variable is selected, the solver is instructed to assign the value that the corresponding variable had in the best solution encountered so far. If that is not possible, it resorts to its default strategy. As a result, the search is focused near the space around the best solution, resembling a large neighbourhood search algorithm, while still remaining complete.

5.1   Hot Starts

At the start of the algorithm, the solver is provided with an initial solution. Due to solution-based phase saving technique, it will immediately focus the search around the given solution. Hence, from the beginning, the search is directed to an area where good solutions reside. As shown in the experimental section, generating the initial solution uses only a small fraction of the total time allocate but our hot start offers significant improvements.

Initial Solution Generation. The same procedure for the starting solution is used as in the maxSAT-LNS approach [7]. We briefly outline it.

The main idea is to exploit existing fast heuristic algorithms. To this end, *KHE14*, a publicly available state-of-the-art incomplete algorithm, is first invoked. The method is designed to provide good solutions rapidly. Thus, it is particularly well-suited for our purpose. In the event that KHE14 does not produce a feasible solution, a pure maxSAT approach that treats all split event constraints as hard and ignores soft constraints is called. Afterwards, a simple simulated annealing local search procedure is executed, which attempts to improve the solution by performing two different moves: swaps (exchange the times of two subevents) and block-swap (if a swap move would cause the subevents to overlap, assign to the second one a time such that the two events appear one after the other). During the course of the algorithm only feasible moves between subevents that share at least one resource are considered. The aim is to use inexpensive techniques to remove easy constraint violations, leaving the more challenging ones to the CP solver. We note that initial solution generation takes only a small faction of the total amount of time allocated.

6   Experimental Results

We provide detailed experimentation with the aim of assessing our proposed approach. We have accordingly set the following goals:

- Evaluate the impact of solution-based phase saving and hot starts for high school timetabling (Sect. 6.3).
- Test if restarting more frequently would lead to an increase in performance (Sect. 6.4).
- Compare the developed approach with other complete methods, namely integer programming and maxSAT (Sect. 6.5).
- Position our method among dedicated heuristic algorithms (Sect. 6.6).

6.1   Benchmarks and Computing Environment

We considered XHSTT benchmarks from the repository of the International Timetabling Competition 2011 (ITC 2011), limited to those where resources are predefined for events. The majority of the benchmarks fall into this category. Resource assignments would drastically increase the search space if done in a straight-forward manner and further specialized techniques would need to be developed. The maxSAT line of work [7,8] follows the same restrictions. These datasets include real-world and artificial timetabling problems and an overview is given in Table 2. For more details we refer the interested reader to [24,25].

We performed all the experiments on an i7-3612QM 2.10 GHz processor with eight GB of RAM, with the exception of the Matheuristic solver (see next section). Each run was given 1200 s and a single core, as during the second phase of the competition, with no experiments running in parallel.

6.2   Solvers

Constraint Programming. We used MiniZinc [21] to model XHSTT and Chuffed [3] as the CP solver, for which we implemented solution-based phase

Table 2. Overview of the datasets used, displaying number of events ($|E|$), times ($|T|$), resources ($|R|$), and sum of event durations ($\sum(dur)$).

Name	$|E|$	$|T|$	$|R|$	$\sum(dur)$
BrazilInstance1	21	25	11	75
BrazilInstance2	63	25	20	150
BrazilInstance3	69	25	24	200
BrazilInstance4	127	25	35	300
BrazilInstance5	119	25	44	325
BrazilInstance6	140	25	44	350
BrazilInstance7	205	25	53	500
FinlandCollege	387	40	111	854
FinlandElementarySchool	291	35	103	445
FinlandHighSchool	172	35	41	297
FinlandSecondarySchool	280	35	64	306
FinlandSecondarySchool2	469	40	79	566
GreeceThirdHighSchoolPatras2010	178	35	113	340
GreeceThirdHighSchoolPreveza2008	164	35	97	340
GreeceWesternUniversityInstance3	210	35	25	210
GreeceWesternUniversityInstance4	262	35	31	262
GreeceWesternUniversityInstance5	184	35	24	184
GreeceFirstHighSchoolAigio2010	283	35	245	532
ItalyInstance1	42	36	16	133
SouthAfricaLewitt2009	185	148	37	838

saving and the hot start approach. The Luby restart scheme [18] was used within Chuffed.

Complete Methods. Integer programming [16] linked with Gurobi 6.5. as the IP solver and the XHSTT-maxSAT formulation [8] with Open-WBO (linear algorithm) [20] as the maxSAT solver.

Dedicated Heuristic Solvers. KHE14 [15], an ejection-chain-based solver. Variable neighborhood search [11]. Matheuristic [27], an adaptive large neighborhood search algorithm with integer programming. MaxSAT large neighborhood search algorithm [7] (uses Open-WBO as its maxSAT solver). These results for these methods were averaged over five runs. The Matheuristic solver was not available and therefore the results shown are from the paper [27], which used a benchmarking tool to set fair normalized computation times.

6.3   Phase Saving and Hot Start Impact

We compared three variants of our CP approach: standard (CP), with solution-based phase saving (CP+PS), and with hot starts (CP+HS). Note that hot starts include solution-based phase saving. The results are given in Table 3, showing the soft constraint violations. In addition to the results, we include the value of the hot started solution ("Initial" column in the table).

Table 3. Comparison of the three CP variants: CP - standard, CP+PS - with solution-based phase saving, CP+HS - with hot starts, showing the number of soft constraint violations in the best solution found within the time limit. The column "Initial" displays the hot started solution value. Overall, PS+HS is the dominating approach on most benchmarks. No solution generated within the time limit is indicated by '—'.

Name	CP	CP+PS	CP+HS	Initial
Brazil1	52	**41**	**41**	83
Brazil2	102	7	**5**	56
Brazil3	187	43	**25**	116
Brazil4	223	110	**88**	176
Brazil5	355	**32**	57	301
Brazil6	442	74	**66**	192
Brazil7	657	230	**181**	252
FinlandCollege	—	—	**9**	917
FinlandESchool	—	—	**3**	9
FinlandHSchool	314	13	**4**	23
FinlandSSchool	—	—	**93**	339
FinlandSSchool2	—	—	**1**	7
GreecePatras	—	—	**0**	12
GreecePreveza	—	—	**279**	630
GreeceUni3	—	—	**5**	10
GreeceUni4	—	—	**7**	13
GreeceUni5	—	—	**0**	2
GreeceAigio	—	—	**597**	689
Italy1	—	—	**12**	1229
SAfricaLewitt	—	—	**0**	330

From the results it is clear that solution-based saving offers improvements over the standard version. The hot start approach further increases the performance. It is of interest to note that in a number of cases the CP solver (with or without phase saving) struggled to find a good solution, but when it was provided with one as a hot start, it managed to provide notable improvements. The additional guidance from phase saving and hot start aids the solver by directing it into fruitful parts of the search space. This approach resembles a local search algorithm, where a better solution is sought for in the vicinity of the best solution. However, unlike local search, the solution-based phase saving (with hot starts) approach is complete.

6.4   Rapid Restarts

Given the similarity between the hot start approach and local search, we decided
to experiment with an extreme version of the algorithm by increasing the restart
frequency. In our previous experiments, the solver used a Luby restart scheme
with the base restart value of 10^4. Now, we set the base restart value to only
100. The results are given in Table 4. The results indicate that rapid restarting
does have an impact for a few benchmarks but no conclusive observations can
be made. Therefore, it is not considered in further experimentation.

Table 4. Analysis of CP variants with rapid restarts (RR in columns). Each entry
shows the soft constraint violations. No solution generated is indicated by '—'.

Name	CP+PS	CP+PS+RR	CP+HS	CP+HS+RR
Brazil1	**41**	**41**	**41**	**41**
Brazil2	7	12	**5**	8
Brazil3	43	26	**25**	26
Brazil4	110	100	88	**85**
Brazil5	32	**30**	57	40
Brazil6	74	103	**66**	77
Brazil7	230	—	**181**	183
FinlandCollege	—	—	**9**	14
FinlandESchool	—	**3**	**3**	**3**
FinlandHSchool	314	13	**4**	23
FinlandSSchool	—	**89**	93	125
FinlandSSchool2	—	—	1	**0**
GreecePatras	—	394	**0**	**0**
GreecePreveza	—	—	279	**140**
GreeceUni3	—	—	**5**	**5**
GreeceUni4	—	—	**7**	**7**
GreeceUni5	—	—	**0**	**0**
GreeceAigio	—	—	bf597	684
Italy1	—	—	**12**	**12**
SAfricaLewitt	—	—	**0**	**0**

6.5   Comparison of Complete Methods

We compare the hot start CP version with the integer programming and maxSAT
approaches in Table 5. The results indicate that the proposed method is indeed
effective for high school timetabling. We note that including solution-based phase
saving and/or hot starts for the competing methods was not done as it would
require modifying the XHSTT solvers to support it.

Table 5. Comparison of complete methods. No solution generated within the timeout is indicated by '—'.

Name	IP	maxSAT	CP+HS
Brazil1	41	**39**	41
Brazil2	76	57	**5**
Brazil3	93	75	**25**
Brazil4	234	214	**88**
Brazil5	135	224	**57**
Brazil6	582	352	**66**
Brazil7	1,045	603	**181**
FinlandCollege	1,731	1,309	**9**
FinlandESchool	**3**	**3**	**3**
FinlandHSchool	179	812	**4**
FinlandSSchool	165	504	**93**
FinlandSSchool2	3,505	3,523	**1**
GreecePatras	25	2,329	**0**
GreecePreveza	2,740	5,617	**279**
GreeceUni3	28	7	**5**
GreeceUni4	51	141	**7**
GreeceUni5	3	224	**0**
GreeceAigio	3,738	4,582	**597**
Italy1	15	**12**	**12**
SAfricaLewitt	—	1,039	**0**

6.6 Comparison with Heuristic Solvers

As previously discussed, solution-based phase saving focuses its search around the current best solution, relating to local search algorithms, while remaining a complete method. Therefore, we decided the evaluate how the approach positions against dedicated heuristic solvers. The results are given in Table 6. Solutions for these randomized (incomplete) methods are averaged over five runs.

Our approach provides competitive results, even when compared against heuristic solvers. Our approach outperforms the competition on most benchmarks, with the exception of the maxLNS approach. When compared to maxLNS, our algorithm outperforms it in two cases. We believe the success of maxLNS is attributed to the fact that it is a dedicated algorithm that incorporates domain-specific knowledge in its search strategy, allowing it to target good solutions quickly. Our approach does not exploit problem-specific details, apart from the initial solution, which both approaches have in common. Nevertheless, it still provides reasonably good results, suggesting that the approach is valuable.

Table 6. Comparison with dedicated heuristic methods. No solution generated within the time limit is indicated by '—'. Solver unavailable listed as '—^{u}'. Insufficient memory noted as '—^{m}'. Values given as pairs (a, b) show the number of hard and soft constraint violations, respectively (for cases where hard constraints are satisfied, only soft constraint violations are displayed). Results averaged over five runs

Name	maxLNS [7]	VNS [11]	KHE14 [15]	CP+HS	Math. [27]
Brazil1	**39**	52.2	54	41	—^{u}
Brazil2	5.4	(1, 44.4)	14	**5**	6
Brazil3	**23**	107.8	116	25	—^{u}
Brazil4	61.4	(17.2, 94.8)	—^{c}	88	**58**
Brazil5	**19.4**	(4, 138.4)	(1, 179)	57	—^{u}
Brazil6	**50.6**	(4, 223.6)	124	66	57
Brazil7	**136.2**	(11.6, 234.6)	179	181	—^{u}
FinlandCollege	54.6	(2.8, 25)	20	**9**	—^{u}
FinlandESchool	3	3	4	3	3
FinlandHSchool	9.8	36.6	29	**4**	—^{u}
FinlandSSchool	95.2	(0.4, 93)	**90**	93	—^{u}
FinlandSSchool2	**0.2**	0.2	2	1	6
GreecePatras	**0**	**0**	**0**	**0**	—^{u}
GreecePreveza	38.2	2	**2**	279	—^{u}
GreeceUni3	7	**5**	7	**5**	6
GreeceUni4	**5**	6.2	8	7	12
GreeceUni5	**0**	**0**	**0**	**0**	**0**
GreeceAigio	368	(0.2, 6.2)	**6**	597	180
Italy1	**12**	21.2	31	**12**	—^{u}
SAfricaLewitt	—^{m}	8	—^{c}	**0**	—^{u}

7   Conclusion

We provide a new CP scheduling-based model for the general high school timetabling problem with preassigned resources. We show that significant improvements over the standard CP approach can be obtained by including solution-based phase saving. Further performance increase is achieved by providing the CP solver with a hot start. The resulting approach outperforms other complete approaches and provides competitive results when compared to dedicated heuristic solvers. The techniques used in our approach and maxLNS [7] do not overlap, indicating that a combination of the two approaches might be worth investigating.

References

1. Abío Roig, I.: Solving hard industrial combinatorial problems with SAT (2013)
2. Global Constraint Catalog: all_equal constraint. http://www.emn.fr/x-info/sdemasse/gccat/Call_equal.html
3. Chu, G.: Improving combinatorial optimization. Ph.D. thesis, The University of Melbourne (2011). http://hdl.handle.net/11343/36679
4. Chu, G., Stuckey, P.J.: LNS = restarts + dynamic search + phase saving. Technical draft
5. Demirović, E., Musliu, N.: Modeling high school timetabling as partial weighted maxSAT. In: The 4th Workshop on Logic and Search (LaSh 2014) (2014)
6. Demirović, E., Musliu, N.: Solving high school timetabling with satisfiability modulo theories. In: Proceedings of the International Conference of the Practice and Theory of Automated Timetabling (PATAT 2014), pp. 142–166 (2014)
7. Demirović, E., Musliu, N.: MaxSAT based large neighborhood search for high school timetabling. Comput. Oper. Res. **78**, 172–180 (2017)
8. Demirović, E., Musliu, N.: Modeling high school timetabling as partial weighted maxSAT. In: Technical Draft - Extended LaSh 2014 Workshop Paper (2017)
9. Global Constraint Catalog: disjunctive constraint. http://www.emn.fr/x-info/sdemasse/gccat/Cdisjunctive.html
10. Dorneles, Á.P., de Araujo, O.C.B., Buriol, L.S.: A fix-and-optimize heuristic for the high school timetabling problem. Comput. Oper. Res. **52**, 29–38 (2014)
11. Fonseca, G.H.G., Santos, H.G.: Variable neighborhood search based algorithms for high school timetabling. Comput. Oper. Res. **52**, 203–208 (2014)
12. da Fonseca, G.H.G., Santos, H.G., Toffolo, T.Â.M., Brito, S.S., Souza, M.J.F.: GOAL solver: a hybrid local search based solver for high school timetabling. Ann. Oper. Res. **239**(1), 77–97 (2016). https://doi.org/10.1007/s10479-014-1685-4
13. Jacobsen, F., Bortfeldt, A., Gehring, H.: Timetabling at German secondary schools: tabu search versus constraint programming. In: Proceedings of the International Conference of the Practice and Theory of Automated Timetabling (PATAT 2006) (2006)
14. Kheiri, A., Ozcan, E., Parkes, A.J.: HySST: hyper-heuristic search strategies and timetabling. In: Proceedings of the International Conference of the Practice and Theory of Automated Timetabling (PATAT 2012), pp. 497–499 (2012)
15. Kingston, J.: KHE14: an algorithm for high school timetabling. In: Proceedings of the International Conference of the Practice and Theory of Automated Timetabling (PATAT 2014), pp. 498–501 (2014)
16. Kristiansen, S., Sørensen, M., Stidsen, T.R.: Integer programming for the generalized high school timetabling problem. J. Sched. **18**(4), 377–392 (2015)
17. Lahrichi, A.: Scheduling: the notions of hump, compulsory parts and their use in cumulative problems. C. R. Acad. Sci. Paris **294**, 209–211 (1982)
18. Luby, M., Sinclair, A., Zuckerman, D.: Optimal speedup of Las Vegas algorithms. Inf. Process. Lett. **47**(4), 173–180 (1993)
19. Marte, M.: Towards constraint-based school timetabling. Ann. Oper. Res. (ANOR) **155**(1), 207–225 (2007)
20. Martins, R., Manquinho, V., Lynce, I.: Open-WBO: a modular MaxSAT solver'. In: Sinz, C., Egly, U. (eds.) SAT 2014. LNCS, vol. 8561, pp. 438–445. Springer, Cham (2014). https://doi.org/10.1007/978-3-319-09284-3_33

21. Nethercote, N., Stuckey, P.J., Becket, R., Brand, S., Duck, G.J., Tack, G.: MiniZinc: towards a standard CP modelling language. In: Bessière, C. (ed.) CP 2007. LNCS, vol. 4741, pp. 529–543. Springer, Heidelberg (2007). https://doi.org/10.1007/978-3-540-74970-7_38

22. Pesant, G.: A regular language membership constraint for finite sequences of variables. In: Wallace, M. (ed.) CP 2004. LNCS, vol. 3258, pp. 482–495. Springer, Heidelberg (2004). https://doi.org/10.1007/978-3-540-30201-8_36

23. Pipatsrisawat, K., Darwiche, A.: A lightweight component caching scheme for satisfiability solvers. In: Marques-Silva, J., Sakallah, K.A. (eds.) SAT 2007. LNCS, vol. 4501, pp. 294–299. Springer, Heidelberg (2007). https://doi.org/10.1007/978-3-540-72788-0_28

24. Post, G., Ahmadi, S., Daskalaki, S., Kingston, J., Kyngas, J., Nurmi, C., Ranson, D.: An XML format for benchmarks in high school timetabling. Ann. Oper. Res. **194**(1), 385–397 (2012)

25. Post, G., Kingston, J.H., Ahmadi, S., Daskalaki, S., Gogos, C., Kyngäs, J., Nurmi, C., Musliu, N., Pillay, N., Santos, H., Schaerf, A.: XHSTT: an XML archive for high school timetabling problems in different countries. Ann. Oper. Res. **218**(1), 295–301 (2014)

26. Santos, H.G., Uchoa, E., Ochi, L.S., Maculan, N.: Strong bounds with cut and column generation for class-teacher timetabling. Ann. Oper. Res. **194**(1), 399–412 (2012)

27. Sørensen, M.: A matheuristic for high school timetabling. In: Timetabling at High Schools, Ph.D. thesis, pp. 137–153. Department of Management Engineering, Technical University of Denmark (2013)

28. Sørensen, M., Dahms, F.H.: A two-stage decomposition of high school timetabling applied to cases in Denmark. Comput. Oper. Res. **43**, 36–49 (2014)

29. Sørensen, M., Kristiansen, S., Stidsen, T.R.: International timetabling competition 2011: an adaptive large neighborhood search algorithm. In: Proceedings of the International Conference on the Practice and Theory of Automated Timetabling (PATAT 2012), pp. 489–492 (2012)

30. Sørensen, M., Stidsen, T.R.: Hybridizing integer programming and metaheuristics for solving high school timetabling. In: Proceedings of the International Conference of the Practice and Theory of Automated Timetabling (PATAT 2014), pp. 557–560 (2014)

31. Sørensen, M., Stidsen, T.R., Kristiansen, S.: Integer programming for the generalized (high) school timetabling problem. In: Proceedings of the International Conference of the Practice and Theory of Automated Timetabling (PATAT 2014), pp. 498–501 (2014)

32. Sørensen, M., Stidsen, T.R.: High school timetabling: modeling and solving a large number of cases in Denmark. In: Proceedings of the International Conference of the Practice and Theory of Automated Timetabling (PATAT 2012), pp. 359–364 (2012)

33. Sørensen, M., Stidsen, T.R.: Comparing solution approaches for a complete model of high school timetabling. Technical report, DTU Management Engineering (2013)

34. Valouxis, C., Housos, E.: Constraint programming approach for school timetabling. Comput. Oper. Res. **30**(10), 1555–1572 (2003)

Epiphytic Trees: Relational Consistency Applied to Global Optimization Problems

Guilherme Alex Derenievicz[(✉)] and Fabiano Silva

Federal University of Paraná, Curitiba, PR, Brazil
{gaderenievicz,fabiano}@inf.ufpr.br

Abstract. Much effort has been spent to identify classes of CSPs in terms of the relationship between network structure and the amount of consistency that guarantees a backtrack-free solution. In this paper, we address Numerical Constrained global Optimization Problems (NCOPs) encoded as ternary networks, characterizing a class of such problems for which a combination of Generalized Arc-Consistency (GAC) and Relational Arc-Consistency (RAC) is sufficient to ensure a backtrack-free solution, called Epiphytic Trees. While GAC is a domain filtering technique, enforcing RAC creates new constraints in the network. Alternatively, we propose a branch and bound method to achieve a relaxed form of RAC, thus finding an approximation of the solution of NCOPs. We empirically show that Epiphytic Trees are relevant in practice. In addition, we extend this class to cover all ternary NCOPs, for which Strong Directional Relational k-Consistency ensures a backtrack-free solution.

1 Introduction

A *Constraint Satisfaction Problem* (CSP) consists of finding an assignment of values to a set of variables that satisfy a constraint network. Local consistency techniques play a central role in solving CSP, pruning values that surely do not constitute a solution of the problem. A wide range of local consistencies have been proposed for finite-domain CSPs, e.g., k-consistency [20], generalized arc-consistency (GAC) [32], hyper-consistency [26] and relational consistency [15].

Many efforts have been spent to identify classes of CSPs by linking the network structure to the level of local consistency that guarantees a backtrack-free solution, i.e., a search that solves the problem without encountering any conflict. Freuder [21] stated that binary networks with *width k* are solved in a backtrack-free manner if achieved strong k-consistency. Jégou [26] extended such results to non-binary networks, showing the relation between hyper-k-consistency and *hypergraph width*. Van Beek and Dechter [3] linked the tightness of a network to the level of relational consistency that guarantees a backtrack-free solution. Although CSP is generally NP-hard, such investigations may identify classes of polynomial problems, like tree-structured CSPs and Horn formulas [14].

In the 1980s, Dechter and Pearl [16] proposed *directional consistency* as a simpler way to enforce local consistency. In short, a variable ordering is used

© Springer International Publishing AG, part of Springer Nature 2018
W.-J. van Hoeve (Ed.): CPAIOR 2018, LNCS 10848, pp. 153–169, 2018.
https://doi.org/10.1007/978-3-319-93031-2_11

to enforce consistency in a specific direction. Thus, only relations between a variable and its predecessors must be analyzed. The classes of tractable CSPs previously introduced are naturally extended to their directional versions.

Most methods for solving finite-domain CSPs are not efficiently applicable to problems over continuous variables (Numerical CSPs). The well-known arc-consistency [44], for the sake of example, is generally uncomputable on continuous case [10,13]. Many alternative methods have been proposed over the last three decades by combining classical CSP concepts with interval analysis [34]. In particular, due to the improvement of computer hardware and all the advancement of constraint propagation, interval methods have been applied on a wide range of Numerical Constrained global Optimization Problems (NCOPs) [9,24,28,29,38,43]. Despite the great progress of interval techniques, optimizers from the mathematical programming community are generally more efficient than interval solvers [2,36], although such optimizers are non-rigorous, i.e., they cannot guarantee the global optimality of the solution.

In this paper, we address NCOPs encoded as ternary networks with an objective function (actually a single variable) to be minimized. We characterize an important class of such problems for which *directional relational arc-consistency* is sufficient to ensure a backtrack-free solution (Sect. 3). We call such a class of *Epiphytic Trees*. In the spirit of Freuder [21] and Jégou [26], we define Epiphytic Trees based on structural properties of the constraint network.

Enforcing relational consistency may create new constraints in the network [14,15], as opposed to domain filtering techniques (e.g. GAC) that only prune variables' domain without changing the network structure. Relational consistency has hardly been used in practice due to the high complexity to be achieved [8,14]. For this reason, we implement an interval branch and bound method to enforce a relaxed form of this consistency, thus finding an approximation for the global minimum of NCOPs (Sect. 4). The goal of our method is to evaluate the proposed class and not to be compared with state-of-art optimizers, since most of advanced techniques of interval solvers are not considered in our implementation. Nonetheless, we are able to solve about 60% of a set of 130 instances proposed in the COCONUT suite [40] with a close approximation (Sect. 5). To our surprise, all tested instances have shown to be encoded as Epiphytic Trees.

To complete, in order to generalize Epiphytic Trees we extend this class to cover all ternary encoded NCOPs, proposing the natural extension of the *width* parameter [21] to ternary networks, namely *epiphytic width*, and proving that networks with epiphytic width k can be solved in a backtrack-free manner if achieved strong directional relational k-consistency (Sect. 6).

2   Background

A *constraint network* $\mathcal{R}$ is a triple $(X, D, \mathcal{C})$ where $X = \{x_1, \ldots, x_n\}$ is a set of variables with respective domains $D = \{D_1, \ldots, D_n\}$ and $\mathcal{C} = \{R_{C_1}, \ldots, R_{C_m}\}$ is a set of constraints such that $C_i = \{x_{i_1}, \ldots, x_{i_k}\} \subseteq X$ is the *scope* of the relation $R_{C_i} \subseteq D_{i_1} \times \cdots \times D_{i_k}$, which defines a set of legal assignments to the variables of C_i. The *arity* of a constraint is the cardinality of its scope. A constraint network $\mathcal{R}$ is said to be *k-ary* if all its constraints have arity k or less. If the domains of the variables are continuous sets of real numbers, $\mathcal{R}$ is said a *numerical constraint network*. The problem of finding an assignment of values from domains of variables that satisfy the (numerical) constraint network is abbreviated to (N)CSP. Given a numerical constraint network $\mathcal{R}$ and an objective function $f : \mathbb{R}^n \mapsto \mathbb{R}$, a Numerical Constrained global Optimization Problem (NCOP) consists of finding a solution of $\mathcal{R}$ that minimizes f in $D_1 \times \cdots \times D_n$.

A *hypergraph* is a structure $\mathcal{H} = (V, E)$ where V is a finite set of vertices and $E \subseteq \{S \subseteq V \mid S \neq \emptyset\}$ is a set of edges. A *Berge-cycle* [7] is an alternating sequence of edges and vertices $(C_1, x_1, C_2, x_2, \ldots, C_m, x_m, C_{m+1} = C_1)$ such that $m \geq 2$ and $x_i \in (C_i \cap C_{i+1})$, for all $1 \leq i \leq m$. A hypergraph is said *Berge-acyclic* if it has no Berge-cycles. The *constraint hypergraph* is a structural representation of the network, where vertices represent variables and edges represent scopes of constraints. For abuse of notation, we will often use the terms *variable* and *vertex* as synonyms, as well as the terms *constraint* and *edge*. Furthermore, we represent the constraint network $\mathcal{R} = (X, D, \mathcal{C})$ by the hypergraph $\mathcal{H} = (X, C)$, where $C = \{C_i \mid R_{C_i} \in \mathcal{C}\}$.

Many local consistencies have been defined for non-binary networks [15, 20, 23, 26, 32]. These techniques prune values that surely do not constitute a solution of the problem. However, as opposed to global consistency, in general they cannot guarantee that a consistent instantiation of a subset of variables can be extended to all remaining variables while satisfying the whole network. In this paper, we focus on two main techniques: generalized arc-consistency [32] and relational arc-consistency [15], both local as only one constraint is considered at once.

Definition 1 (Generalized Arc-Consistency (GAC)).

- *A constraint R_{C_i} is GAC relative to $x_k \in C_i$ iff for every value $a_k \in D_k$ there exists an instantiation I of variables in $C_i \setminus \{x_k\}$ such that $I \cup \langle x_k = a_k \rangle$ satisfies R_{C_i}.*
- *A constraint R_{C_i} is (full) GAC iff it is GAC relative to all $x_k \in C_i$.*
- *A network $\mathcal{R}$ is (full) GAC iff every constraint of $\mathcal{R}$ is GAC.*
- *A network $\mathcal{R}$ is directional GAC, according to an ordering $b = (x_1, \ldots, x_n)$, iff every constraint R_{C_i} is GAC relative to its earliest variable in b.*

Definition 2 (Relational Arc-Consistency (RAC)).

- *A constraint R_{C_i} is RAC relative to $x_k \in C_i$ iff any consistent assignment to all variables of $C_i \setminus \{x_k\}$ has an extension to x_k that satisfies R_{C_i}.*
- *A constraint R_{C_i} is (full) RAC iff it is RAC relative to all $x_k \in C_i$.*
- *A network $\mathcal{R}$ is (full) RAC iff every constraint of $\mathcal{R}$ is RAC.*
- *A network $\mathcal{R}$ is directional RAC, according to an ordering $b = (x_1, \ldots, x_n)$, iff every constraint R_{C_i} is RAC relative to its latest variable in b.*

Dechter and Pear proposed directional consistency motivated by the fact that "full consistency is sometimes unnecessary if a solution is going to be generated by search along a fixed variable ordering" [14, p. 91]. In this work, we address NCOPs by encoding the objective function $x_1 = f(\mathbf{x})$ and the constraint network $\mathcal{R}$ as a ternary NCSP, and performing a search along a fixed variable ordering *starting by* x_1. The idea behind this strategy is that if a consistent instantiation of the encoded network with initial (partial) assignment $\langle x_1 = \min D_1 \rangle$ is achieved by the search, then such solution minimizes f in $D_1 \times \cdots \times D_n$.

2.1   Interval Arithmetic

A *closed interval* $X = [\underline{x}, \overline{x}]$ is the set of real numbers $X = \{x \in \mathbb{R} \mid \underline{x} \leq x \leq \overline{x}\}$, where $\underline{x}, \overline{x} \in \mathbb{R} \cup \{-\infty, \infty\}$ are the *endpoints* of X. This definition is naturally extended for *open* and *half-open* intervals. We denote by $\mathbb{I}$ the set of all intervals in $\mathbb{R}$. A *box* $B \in \mathbb{I}^n$ is a tuple of intervals. A box $B = (X_1, \ldots, X_n)$ is a refinement of the box $D = (Y_1, \ldots, Y_n)$, denoted by $B \subseteq D$, iff $X_i \subseteq Y_i$, for all $1 \leq i \leq n$.

Given two intervals $X, Y \in \mathbb{I}$, the *interval extension* of any binary operator $\circ$ well defined in $\mathbb{R}$ is defined by (1).

$$X \circ Y = \{x \circ y \mid x \in X,\, y \in Y \text{ and } x \circ y \text{ is defined in } \mathbb{R}\}. \tag{1}$$

The set (1) can be an interval, the empty set or a disconnected set of real numbers (union of intervals or a *multi-interval*). It is possible to compute $X \circ Y$, for all algebraic and the most common transcendentals functions, only analyzing the endpoints of X and Y [33], e.g., $[\underline{x}, \overline{x}] + [\underline{y}, \overline{y}] = [\underline{x} + \underline{y}, \overline{x} + \overline{y}]$, $[\underline{x}, \overline{x}] \cdot [\underline{y}, \overline{y}] = [\min\{\underline{xy}, \underline{x}\overline{y}, \overline{x}\underline{y}, \overline{xy}\}, \max\{\underline{xy}, \underline{x}\overline{y}, \overline{x}\underline{y}, \overline{xy}\}]$, etc.

General factorable functions $f : \mathbb{R}^n \mapsto \mathbb{R}$ are also extended to intervals. An interval function $\mathcal{F} : \mathbb{I}^n \mapsto \mathbb{I}$ is an interval extension of $f : \mathbb{R}^n \mapsto \mathbb{R}$ if $\forall B \in \mathbb{I}^n : f[B] \subseteq \mathcal{F}(B)$, where $f[B]$ denotes the image of f under B, i.e. $f[B] = \{f(x_1, \ldots, x_n) \mid (x_1, \ldots, x_n) \in B\}$. Generally, if the interval extension $\mathcal{F}$ is defined by the same expression of f, using the respective interval operators, then $\mathcal{F}(B)$ is a good estimate of $f[B]$. More specifically, if each variable occurs only once in the expression of f, then $\mathcal{F}(B)$ is exactly the image of this function [4]. The reason that an interval extension may overestimate the image of a real functional is that interval operators consider each occurrence of a variable as a different variable.

Example 1. The images of $x_1^2 - x_2$ and $x_1 x_1 - x_2$, under box $([-2, 2], [-1, 1])$ are both $[-1, 5]$. However, the interval extensions $X_1^2 - X_2$ and $X_1 X_1 - X_2$ computes, respectively, $[-2, 2]^2 - [-1, 1] = [-1, 5]$ and $[-2, 2] \cdot [-2, 2] - [-1, 1] = [-5, 5]$.

2.2   Interval Consistency

Since the late eighties, interval arithmetic has been used to deal with numerical constrained problems and several techniques have been extended to NCSP [18,24,25,30,35,38,39,43]. For instance, GAC is easily achieved on ternary constraints of the form $x_1 = x_2 \circ_1 x_3$ by computing the intersection of each relevant domain with the respective *projection function*:

$$\texttt{GAC\_contractor}(x_1 = x_2 \circ_1 x_3) := \begin{cases} D_1 \leftarrow D_1 \cap (D_2 \circ_1 D_3) \\ D_2 \leftarrow D_2 \cap (D_1 \circ_2 D_3) \\ D_3 \leftarrow D_3 \cap (D_1 \circ_3 D_2) \end{cases} \tag{2}$$

where $\circ_2$ and $\circ_3$ maintain $(x_1 = x_2 \circ_1 x_3) \Leftrightarrow (x_2 = x_1 \circ_2 x_3) \Leftrightarrow (x_3 = x_1 \circ_3 x_2)$.

Strictly, the box generated by (2) is not complete due to the finite precision of machine numbers. Furthermore, domains may be disconnected sets of real numbers (union of intervals). In general, there is no consensus in literature about the performance of maintaining multi-interval representation [10,18,41]. Many relaxed consistencies have been proposed to deal with these problems, e.g., hull consistency [6] and box consistency [5], however, the strong property of GAC (the existence of local instantiation given any assignment of a single variable) is lost. Despite the efficiency of the GAC contractor, such procedure may never terminate when applied to the whole network[1] [10,13], unless the hypergraph holds specific structural properties [11,18] or a maximum precision is imposed.

While the GAC contractor shrinks the domain of variables, an RAC contractor must tighten the constraint on $C_i \setminus \{x_k\}$ (if such constraint does not exist, it must be added to the network). For example, the constraint $x_1 = x_2 + x_3$ is not RAC relative to x_1 under the box $([1,3],[0,2],[0,2])$, as for $x_2 = x_3 = 0$ or $x_2 = x_3 = 2$ there is no consistent assignment to x_1. We may turn this constraint RAC by adding the constraints $x_2 + x_3 \geq 1$ and $x_2 + x_3 \leq 3$ to the network.

For binary constraints, GAC and RAC are identical [14].

2.3 Decomposition of Constraint Networks

Generally, in order to apply the GAC contractor on k-ary numerical constraints a procedure of decomposition that transforms the constraint into an equivalent set of ternary constraints is used [19]. Such procedure is efficient and solves the *dependence problem* (multiple occurrences of the same variable) and the *isolation problem* (the projection functions of (2) are hard to obtain if the constraint is a complex combination of many variables). However, the decomposition increases the *locality problem*, as shown in Example 2.

Example 2. Let $x_1(x_2 - x_1) = 0$ such that $D_1 = [0,2]$ and $D_2 = [1,2]$. With an auxiliary variable x_3 this constraint is decomposed into a network of two new constraints: $x_2 - x_1 = x_3$ and $x_1 x_3 = 0$. Contractor (2) is then repeatedly applied to both constraints until a fixed box is obtained, which is $([0,2],[1,2],[-1,2])$. Although this box turns each decomposed constraint consistent, the original constraint is not GAC (for $x_1 = 0.5, \nexists x_2 \in D_2$ such that $x_1(x_2 - x_1) = 0$).

Due to the locality problem, interval methods are composed by three main phases that are repeatedly applied until a solution is found: (i) *prune*: where

---

[1] As occur with the network $x_1 = x_2$ and $x_1 = 0.5 \cdot x_2$, where only the box $B = ([0,0],[0,0])$ turns both the constraints GAC but the contractor does an endless bisection on any initial arbitrary box.

contractors of interval consistency are applied; (ii) *local search*: where cheap search methods are used to detect if the current box contains a solution of the entire network (original problem), e.g., selecting the midpoint of the box or applying the Newton method; and (iii) *branch*: where a variable x is chosen and its domain is bisected to continue the search in a recursive fashion.

Likewise, a NCOP $(f, \mathcal{R})$ can be decomposed into a ternary network by adding a *root* variable $x_1 = f(\mathbf{x})$. For example, the instance (3) can be encoded as (4), where x_1 is the root variable that encodes the objective function.

$$\min \ y_1^2 - y_2 \qquad \text{s.t.} \quad \{y_1 \leq 2y_2\} \tag{3}$$

$$\min \ x_1 \qquad \text{s.t.} \quad \{x_1 = x_2 - y_2, x_2 = y_1^2, y_1 \leq 2y_2\} \tag{4}$$

3   Backtrack-Free Solution of Ternary Encoded NCOPs

We consider the problem of finding a solution of decomposed NCOPs by performing a backtrack-free search along a variable ordering starting by the *root* variable. For this, we classify ternary networks in three primary groups, based on the acyclic behavior of the hypergraph representation.

Let $\mathcal{R}_1$ be a network composed by constraints R_{C_1}, R_{C_2} and R_{C_3}, whose constraint hypergraph is represented in Fig. 1 by black edges, with root x_1.

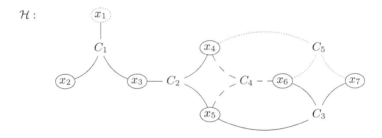

Fig. 1. Constraint hypergraph of networks $\mathcal{R}_1$ (composed by constraints R_{C_1}, R_{C_2} and R_{C_3}), $\mathcal{R}_2$ (composed by constraints $R_{C_1}, \ldots, R_{C_4}$) and $\mathcal{R}_3$ (all constraints).

Considering the initial partial instantiation $\langle x_1 = \min D_1 \rangle$ we want to propagate this value over all the network in a backtrack-free manner, which can be achieved if $\mathcal{R}_1$ is GAC [11,18]. For instance, by constraint R_{C_1} the values of x_2 and x_3 are settled, as GAC ensures that for all $x_1 \in D_1$ there are $x_2 \in D_2$ and $x_3 \in D_3$ that satisfy R_{C_1} (if multiple values of x_2 and x_3 satisfy the constraint an instantiation can be chosen arbitrarily). This process continues for the entire network, processing constraints with one, and only one, already instantiated variable. A possible ordering of processed constraints for $\mathcal{R}_1$ is $(R_{C_1}, R_{C_2}, R_{C_3})$. In general, such an ordering must have as its first constraint the one with the scope containing the root variable and ensure property (5).

$$|C_{p_i} \cap \bigcup_{j=1}^{i-1} C_{p_j}| < 2, \quad \text{for all } 1 < i \le m \text{ in the ordering } (R_{C_{p_1}}, \ldots, R_{C_{p_m}}). \quad (5)$$

Actually, an ordering d with these properties exists iff the constraint hypergraph is Berge-acyclic, because each constraint in d connects with predecessors in only one point (the common variable) [11]. Furthermore, given such a constraint ordering, directional GAC along d is sufficient to ensure this backtrack-free instantiation, instead of full GAC.

Next, let $\mathcal{R}_2$ be the network obtained by adding to $\mathcal{R}_1$ the dashed constraint R_{C_4} in Fig. 1. In this case, there are no guarantees that the previous instantiation is consistent with R_{C_4}. More than that, for this network an ordering ensuring (5) does not exist, which means GAC cannot guarantee the existence of a complete instantiation extending $\langle x_1 = \min D_{x_1} \rangle$.

We may consider another ordering $d' = (R_{C_1}, R_{C_2}, R_{C_4}, R_{C_3})$. Here, variables x_1, x_2, x_3, x_4 and x_5 can be safety instantiate under directional GAC. However, the next constraint R_{C_4} has two variables x_4 and x_5 already valued; in this way there is no guarantee that there exists $a_6 \in D_6$ satisfying the constraint. On the other hand, if R_{C_4} is RAC relative to x_6, instead of GAC, then the existence of $a_6 \in D_6$ satisfying R_{C_4} is guaranteed and the instantiation can be extended. Similarly, this strategy can be applied to R_{C_3}, instantiating the remaining variable x_7. The ordering d' ensures property (6).

$$|C_{p_i} \cap \bigcup_{j=1}^{i-1} C_{p_j}| < 3, \quad \text{for all } 1 < i \le m \text{ in the ordering } (R_{C_{p_1}}, \ldots, R_{C_{p_m}}). \quad (6)$$

In general, a network can be ordered according to (6) if its constraint hypergraph has a "partial acyclic" structure we call *Epiphytic Tree* (Sect. 3.1). Differently from the first example, where a Berge-acyclic hypergraph rarely occurs in practice, a ternary constraint network[2] seems to be usually represented by an Epiphytic Tree. In such an order, for a backtrack-free instantiation, constraints R_{C_i} sharing only one variable with predecessor constraints must satisfy directional GAC while those sharing two variables must satisfy directional RAC.

Finally, all possible orderings of constraints starting by R_{C_1} of the network $\mathcal{R}_3$ obtained by adding to $\mathcal{R}_2$ the constraint R_{C_5} (Fig. 1) do not ensure (5) nor (6), which means neither GAC nor RAC is sufficient to ensure a backtrack-free instantiation of this network. In other words, for all ordering $(R_{C_{p_1}}, \ldots, R_{C_{p_m}})$ such that $x_1 \in C_{p_1}$ there will be some C_{p_i} such that $|C_{p_i} \cap \bigcup_{j=1}^{i-1} C_{p_j}| = 3$.

3.1   Epiphytic Trees

Informally, an Epiphytic Tree (ET) is a hypergraph composed by an ordered set of trees (w.r.t. Berge-cycles), where each tree "sprouts" from its predecessor through a single edge or from the "ground". The root of an ET is the root x_1 of

---

[2] Obtained by encoding NCOPs.

the first tree. In this case, the ET is said to be *according to* x_1. For instance, on Fig. 2 the constraint hypergraph of the network $\mathcal{R}_2$ (Fig. 1) is properly drawn to show that this network is represented by an ET according to x_1 (directed edges indicate the ordering of the trees).

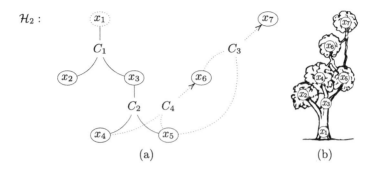

Fig. 2. (a) Epiphytic Tree according to x_1 composed by three trees and (b) an illustration of the botanical inspired nomenclature.

Definition 3 (Epiphytic Tree). *An Epiphytic Tree (ET) according to x_1 is a triple $(\mathcal{A}, \Omega, t)$, where $\mathcal{A} = (\mathcal{A}_1, \ldots, \mathcal{A}_n)$ is an ordered set of disjointed hypergraphs $\mathcal{A}_i = (V_i, E_i)$, Ω is a set of edges not in $\bigcup_{i=1}^{n} E_i$, over the vertices of $\bigcup_{i=1}^{n} V_i$, and $t : \Omega \mapsto V_\Omega$ is a function that associates each edge $C_i \in \Omega$ with one vertex $t(C_i) \in C_i$ such that:*

1. $\mathcal{A}_1$ *is a rooted tree in* x_1 *(i.e., connected and Berge-acyclic).*
2. $\forall \mathcal{A}_{i>1} \in \mathcal{A}$: *there is at most one* $C_i \in \Omega$ *such that* $t(C_i) \in V_i$ *and* $\mathcal{A}_i$ *is a rooted tree in* $t(C_i)$ *(or in any other vertex if such edge C_i does not exist);*
3. *if* $t(C_i) \in V_i$, *then* $C_i \setminus \{t(C_i)\} \subseteq \bigcup_{j=1}^{i-1} V_j$.

The epiphytic height of an ET is the cardinality of its Ω set.

Example 3. The network of Fig. 2 is represented by an ET according to x_1 given by $((\mathcal{A}_1, \mathcal{A}_2, \mathcal{A}_3), \Omega, t)$, where $\mathcal{A}_1 = (\{x_1, \ldots, x_5\}, \{C_1, C_2\})$, $\mathcal{A}_2 = (\{x_6\}, \emptyset)$ and $\mathcal{A}_3 = (\{x_7\}, \emptyset)$ are disjointed, connected and Berge-acyclic hypergraphs, $\Omega = \{C_3, C_4\}$ and t is a function ensuring the conditions of Definition 3 such that $t(C_4) = x_6$ and $t(C_3) = x_7$. The epiphytic height of this ET is 2.

Given a network $\mathcal{R}$ and its ET $(\mathcal{A}, \Omega, t)$, we can efficiently obtain a constraint ordering d of $\mathcal{R}$ that ensures (6), constructing it from first to last element following the order $(\mathcal{A}_1, \ldots, \mathcal{A}_n)$: for each hypergraph $\mathcal{A}_i$, put in d the edges (constraints) following the topological order of $\mathcal{A}_i$ from its root, and then the edge $C_{i+1} \in \Omega$ (if it exists). On the other hand, given a constraint ordering $(R_{C_1}, \ldots, R_{C_m})$ ensuring (6), we can construct an ET by putting in Ω the edges C_i such that $|C_i \cap \bigcup_{j=1}^{i-1} C_j| = 2$ and defining each hypergraph $\mathcal{A}_i$ by the maximal connected component of the hypergraph induced by the remaining edges.

For any ET, the edges of Ω represent the constraints that must be RAC to ensure a backtrack-free solution. In this way, the smaller the cardinality of Ω, the "easier" to solve is the problem. A solution is found in polynomial time when the constraint hypergraph is a single tree, i.e., $\Omega = \emptyset$.

There are problems that cannot be represented by ETs. Given a network $\mathcal{R}$, if there exists a subset of constraints S such that all variables appearing in this subset occur in more than one constraint, then $\mathcal{R}$ does not have such representation. Theorem 1 formalizes this result by the equivalence between ETs and constraint orderings ensuring (6). For this, $\mathcal{H}' = (V', E')$ is said a *partial hypergraph* of $\mathcal{H} = (V, E)$ if $E' \subseteq E$ and $V' = \bigcup_{C_i \in E'} C_i$. We denote by $\mathcal{H} - C_i$ the partial hypergraph with edges set $E \setminus \{C_i\}$.

Theorem 1. *A constraint network $\mathcal{R}$ represented by a hypergraph $\mathcal{H}$ has a constraint ordering starting by C_1 ensuring (6) iff all partial hypergraph $\mathcal{H}'$ of $\mathcal{H}$ with more than one edge have at least one vertex $x' \notin C_1$ with degree 1.*

Proof. ($\Rightarrow$) The proof is by contradiction. Let d be a constraint ordering of $\mathcal{R}$ ensuring (6) and suppose that there is a partial hypergraph $\mathcal{H}'$ of $\mathcal{H}$ with more than one edge such that all vertices $x' \notin C_1$ have degree greater than 1. Let C_k be the latest edge of $\mathcal{H}'$ in d ($k > 1$). Since all vertices $x' \notin C_1$ have degree greater than 1, then all $x' \in C_k$ also belong to another edge of $\mathcal{H}'$. However, all other edges of $\mathcal{H}'$ have index less than k in d and, therefore, $|C_k \cap \bigcup_{j=1}^{k-1} C_j| = 3$. Hence d is not an ordering ensuring (6). ($\Leftarrow$) Let $\mathcal{R}$ be a network such that all partial hypergraphs $\mathcal{H}'$ of $\mathcal{H}$ with more than one edge have at least one vertex $x' \notin C_1$ with degree 1. Let $(\mathcal{H}_1, \ldots, \mathcal{H}_m = \mathcal{H})$ be a sequence of partial hypergraphs such that $\mathcal{H}_1$ has only the edge C_1 and, for all $1 \leq i < m$, the hypergraph $\mathcal{H}_i$ is obtained by removing from $\mathcal{H}_{i+1}$ the edge C_{i+1} that contains a vertex $x \notin C_1$ with degree 1. The sequence of removed edges $(C_2, \ldots, C_m)$ is such that all C_i have at least one vertex x that does not belong to any $C_{j<i}$ (x has degree 1 in $\mathcal{H}_i$). That is, $|C_i \cap \bigcup_{j=1}^{i-1} C_j| < 3$, for all $1 < i \leq m$. Hence the ordering $(R_{C_1}, \ldots, R_{C_m})$ ensures (6). □

The proof of Theorem 1 motivates a method for finding an ordering according to x_1 of a network $\mathcal{R}$, building a sequence d from last to first element, choosing in each step an edge with a vertex of degree 1 in the partial constraint hypergraph obtained by removing processed constraints. Actually, this algorithm is the natural hypergraph extension of the `min-width` method presented in [14] (originally proposed by Freuder [21]), but restricted to always choose a vertex with degree 1. For ternary networks, this method is linear in the number of edges, since we can update in time $O(1)$ the degree of all vertices at each edge removal. Different choices of the picked vertex with degree 1 may result in distinct ETs. Considering the epiphytic height and the "meaning" of the edges in Ω are important tasks to obtain an accurate representation of the original system. In our implementation (Sect. 4), we choose edges in order to minimize the epiphytic height.

4   Achieving Relational Consistency

Enforcing relational consistency generally requires exponential time and space. Indeed, relational consistency can solve NP-Complete problems [14]. One of the first algorithms to achieve such consistency, as observed by Dechter and Rish [17], is the well-known Davis-Putnam procedure for CNF satisfiability. A first practical algorithm for high levels of relational consistency is introduced in [27]. However, "local consistency techniques that only filter domains like GAC tend to be more practical than those that alter the structure of the constraint hypergraph or the constraints' relations" [8, p. 1]. Enforcing directional RAC on ternary constraints adds new binary constraints to the network that changes the structure of the hypergraph[3]. Furthermore, adding new constraints is useful only if these constraints would be processed first, altering the given constraint ordering of the network. Thus, we avoid adding new constraints by applying a domain filtering algorithm that achieves a relaxed form of directional RAC.

4.1   Approximating Directional RAC

A constraint $R_{C_i} \equiv (x_1 = x_2 \circ x_3)$ is RAC relative to x_1 iff $D_1 \supseteq D_2 \circ D_3$, i.e., $\forall a_2 \in D_2, a_3 \in D_3, \exists a_1 \in D_1 \mid a_1 = a_2 \circ a_3$. Without loss of generalization, we can assume that R_{C_i} is GAC relative to x_1 ($D_1 \subseteq D_2 \circ D_3$). Thus, the required RAC condition is reduced to the equality $D_1 = D_2 \circ D_3$. If the constraint is not RAC (i.e., $D_1 \subset D_2 \circ D_3$) an attempt to achieve such consistency, alternatively to adding new constraints to the network, is narrowing the result set of $D_2 \circ D_3$. We can do this by narrowing D_2 or D_3, assured by the *inclusion isotonic*[4] of interval arithmetic. However, any pruning on such domains may exclude feasible instantiations, including the optimal one. We address this problem by using a branch and bound schema with an interval bisection procedure, since there exist infinite sub-boxes of D_2 and D_3 that are RAC under R_{C_i}. However, a tolerance ε must be considered in the equation $D_1 = D_2 \circ D_3$, as there is no guarantee that a bisection procedure can find RAC domains in tractable CPU time. One implication of this relaxation is that an instantiation of these variables may be infeasible in R_{C_i} by a factor of ε, what we call ε-feasible.

When applied to all constraints that must be RAC, this procedure is actually a variant of the usual interval branch and bound for NCOP. By combining this method with the natural evaluation of the network described in Sect. 3 we obtain an approximation of the global minimum of encoded optimization problems. Algorithm 1 describes such procedure. It follows a depth-first search applying the GAC contractor as a local procedure for pruning inconsistent values. If no empty domain is generated, the instantiation of the network starting by $\langle x_1 = \min D_{x_1} \rangle$ is attempt, constructing an ordered set I_Ω of non ε-feasible

---

[3] The same problem occurs with the results of Freuder [21] and Jégou [26]: achieving (hyper-)k-consistency creates new constraints of arity $k - 1$ thus changing the width of the (hyper)graph.

[4] Given $B, B' \in \mathbb{I}^n$ and an interval function $\mathcal{F}$, if $B' \subseteq B$, then $\mathcal{F}(B') \subseteq \mathcal{F}(B)$.

constraints. These constraints must be narrowed to achieve RAC, what is done by the procedure select_and_bisect. Otherwise, the instantiation is a *quasi-solution* of the problem and the search of this branch terminates. Branches that surely have no solution better than the current optimal candidate are not processed by the search, because we added a dynamic constraint $x_1 < z^*$ to the network. A parameter of complexity of this method is the epiphytic height, as it bounds the size of I_Ω and only variables of constraints in this set are bisected.

Algorithm 1. Relaxed Directional RAC   **In:** $(\mathcal{R} = (X, D, \mathcal{C}), x_1, \varepsilon)$ **Out:** z^*

---

push(D, *queue*)
let d be a constraint ordering of $\mathcal{R}$ starting by x_1 ensuring (6)
while *queue* $\neq \emptyset$ **do**
   $D \leftarrow$ pop(*queue*)
   $D \leftarrow$ directional_GAC($X, D, \mathcal{C} \cup \{x_1 < z^*\}, d$)
   if $\forall D_i \in D : D_i \neq \emptyset$ **then**
      $(z, I_\Omega) \leftarrow$ attempt_instantiation($X, D, \mathcal{C}, d, \varepsilon$)
      if $I_\Omega = \emptyset$ **then** $z^* \leftarrow z$
      else
         $(D', D'') \leftarrow$ select_and_bisect($X, D, \mathcal{C}, d, I_\Omega$)
         push(D', *queue*)
         push(D'', *queue*)
      endif
   end if
end while
return z^*

The procedure attempt_instantiation($X, D, \mathcal{C}, d, \varepsilon$) tries to evaluate the network along the ordering d. Constraints R_{C_i} such that $C_i \notin \Omega$ are instantiated using the GAC contractor (2), as the network is directionally GAC along d. Otherwise, let $R'_{C_i} \equiv (x_1 = x_2 \circ x_3)$ be the constraint equivalent to $R_{C_i}, C_i \in \Omega$, such that $t(C_i) = x_1$. Since x_2 and x_3 are already instantiated along d, the procedure evaluates x_1 with the value of D_1 *closest* to $x_2 \circ x_3$. If the absolute error $|x_1 - (x_2 \circ x_3)|$ is greater than ε, then this instantiation is not ε-feasible, and R_{C_i} is put in I_Ω. The search continues for the next constraint of d whenever the partial instantiation is ε-feasible or not.

The procedure select_and_bisect($X, D, \mathcal{C}, d, I_\Omega$) selects a constraint $R_{C_i} \in I_\Omega$ and split the domain of its variables at the midpoint of D, returning two sub-boxes D' and D''. We provide a heuristic for selecting the constraint R_{C_i} by choosing the one with the highest index in I_Ω. Then, variables $x \in C_i \setminus \{t(C_i)\}$ are bisected and the sub-box that *minimizes* the required ε to turn R_{C_i} RAC is defined as the new box D' (and its complement as D'') to continue the search.

Although directional GAC would be enough if the constraints in Ω were RAC, as we are approximating this consistency by a factor of ε then high levels of consistency may be used to improve the branch and bound method. For instance, the procedure `directional_GAC` can be replaced by a full GAC contractor, narrowing the current box in a more effective way than using only the directional approach. In our experiments (Sect. 5), we implemented both the methods.

4.2 Comparison with the Usual Interval Branch and Bound

We compare our method with the usual interval branch and bound for NCOP. For this, we consider the up-to-date survey given on [1]. First, the general form of our procedure `attempt_instantiation` is the called `upper_bounding`, and provides a feasible instantiation of each branched sub-box. Local search methods are generally used to find such instantiations, e.g. applying the Newton method. Our strategy simply tries the natural evaluation of the network, as in [37] and [2], however, it minimizes the objective function by the initial instantiation $\langle x_1 = \min D_1 \rangle$. The main difference is that our approach may not find a solution in this branch, but if it does, then the solution minimizes the encoded objective function in the current box and no further searches in sub-boxes *of this branch* are needed. It is worth noting that others interval solvers [37, 42] also relax constraints on the evaluation phase.

Furthermore, both general interval branch and bound and our approach present the `select_and_bisect` procedure. The choice of which variable and slice of its domain will be the next branch of the search tree is crucial to the effectiveness of backtrack searches. Several heuristic have been proposed to date, e.g., choose the variable with the largest domain or use Smear-based schemas [12]. In this sense, our method provides a non-problem-specific heuristic for choice of variables on branch phase: right-sided variable of non ε-feasible constraints in Ω that maximizes the RAC approximation of the network. Such heuristic is an important contribution of this work, as finding local minima is a great strategy to reduce the search space in branch and bound algorithms.

To complete, on interval branch and bound methods an interval box may or may not contain a real solution of the constraint network. There are some techniques to assure the existence of such solution but, in general, the networks must satisfy some conditions. In our approach, the solution is always an instantiation of real numbers ε-feasible with the problem.

5 Experimental Results

In order to verify the codification of NCOPs as ETs, we executed Algorithm 1 over an experimental set of 130 instances from COCONUT benchmark [40], which consists of academic and real life global optimization problems with industrial relevance. We considered only instances with numerical variables and operators $+, -, *, /$ and *pow*.

First, the experiments showed that Epiphytic Tree is a class of problems relevant in practice, since all tested instances are represented by this structure. Then, we considered a set of parameterized executions with ε varying from 10^{-4}

to 10^1 and a timeout of 7200 s. The method found ε-feasible solutions for 103 instances (about 79%). Using a full GAC contractor[5] instead of the directional one (see Sect. 4.1) this number increased to 108 instances (83%). Of those, 82 instances were solved with a global minimum closest to the best known with an absolute error less than 10 (74 instances using directional GAC). Figure 3 shows these results. Some instances with absolute error above 40 presents a relative error (the ratio of the absolute error to the global minimum) below 60%. The cluster of points in the 100% line are instances for which our method found a solution with value zero, while their global minimum are non-zero values.

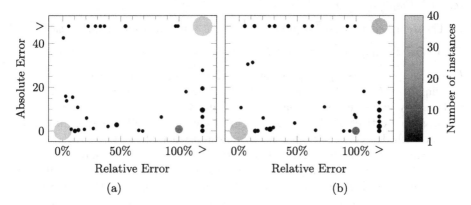

Fig. 3. Absolute error vs. relative error between the solution found and the best known of each instance, using (a) directional GAC and (b) full GAC. The circumference of each point increases according to the number of instances with similar error values.

It is worth remarking that we have not implemented many well-known techniques of current optimization solvers, such as *look-ahead* and *look-back* schemas, high levels of local consistency and more sophisticated local search. Hence, we cannot compare the CPU time of our method to that of state-of-art solvers. Nevertheless, in practice, the proposed method is able to find an approximation of the global minimum of encoded problems; thus, it can be used to improve optimization solvers in means of pre-processing, local search or upper bounding.

6   Extending the Epiphytic Tree Class

We showed that directional RAC is sufficient to solve structures with property (6). Now, we extend this study to hypergraphs that cannot be ordered ensuring such property. For this, we introduce a structural parameter of hypergraphs called *epiphytic width*, considering vertex orderings instead of edge orderings. Many definitions of hypergraph width have been proposed to characterize classes of CSPs. In particular, Gottlob [22] introduced the notion of *hypertree width*

---

[5] Using an AC-3 [31] style algorithm with an iteration counter to avoid looping.

defined over hypertree decompositions; such procedure is generally NP-hard. Alternatively, epiphytic width is a simpler definition being the natural extension of the graph width proposed by Freuder to binary CSPs [20].

Definition 4 (Epiphytic Width). *The epiphytic width of the vertex x_i in the ordering $b = (x_1, \ldots, x_n)$ is the number of edges $C_1, \ldots, C_k$ such that x_i is the latest variable of any $C_1, \ldots, C_k$ (w.r.t. b). The epiphytic width of the ordering b is the maximum epiphytic width over all x_i. The epiphytic width of $\mathcal{H}$ is the minimum epiphytic width over all vertex ordering starting by x_1.*

Definition 5 (k-Epiphytic Trees). *A hypergraph with epiphytic width k is said a k-Epiphytic Tree.*

Theorem 2 shows that k-Epiphytic Tree is a generalization of Epiphytic Tree.

Theorem 2. *Epiphytic Tree $\equiv$ 1-Epiphytic Tree.*

Proof. We proof this equivalence showing that it is possible to convert an edge ordering $d = (C_1, \ldots, C_m)$ ensuring (6) into a vertex ordering $b = (x_1, \ldots, x_n)$ with epiphytic width 1 such that $x_1 \in C_1$, and vice versa. ($\Rightarrow$) for $k = 1$ to m put in b the vertices of C_k not already in b, such that the last vertex placed is the one with degree 1 in the partial hypergraph with edges $\{C_1, \ldots, C_k\}$ (such vertex exists as d satisfies (6)). The first two vertices, if placed, have epiphytic width 0, because they are not the latest vertex of C_k; by the same reason, if they are already in b, then their epiphytic width will not change. On the other hand, the third vertex is being placed in b for the first time, because it has degree 1 in this sub-order. Since it is the latest vertex of C_k in b, then its epiphytic width is 1. ($\Leftarrow$) for $k = 1$ to n, mark x_k. If this turns out that all vertices of an edge C_i not in d are marked, put C_i in d. Each edge C_i placed in d have at least the vertex x_k of degree 1 in this sub-order, otherwise it should be already marked. Hence $|C_i \cap \bigcup_{j=1}^{i-1} C_j| < 3$.   □

Definition 6 (Directional Relational k-Consistency). *Given a variable ordering $b = (x_1, \ldots, x_n)$, a network $\mathcal{R}$ is k-directionally relationally consistent iff for every subset of constraints $\{R_{C_1}, \ldots, R_{C_k}\}$ where the latest variable in any C_i is x_l, and for every $A \subseteq \{x_1, \ldots, x_{l-1}\}$, every consistent assignment to A can be extended to x_l while simultaneously satisfying all the relevant constraints in $\{R_{C_1}, \ldots, R_{C_k}\}$. A network is strongly directional relational $k-$consistent iff it is directional relational j-consistent for every $j \leq k$.*

Theorem 3. *A ternary constraint network $\mathcal{R}$ represented by a k-Epiphytic Tree can be solved in a backtrack-free manner if $\mathcal{R}$ is strongly directional relational k-consistent.*

Proof. Let $b = (x_1, \ldots, x_n)$ be a variable ordering of $\mathcal{R}$ with epiphytic width k in the constraint hypergraph. Thus, every variable x_l is the latest variable of at most k constraints $C_1, \ldots, C_{k'}, k' \leq k$. A consistent instantiation of $(x_1, \ldots, x_{l-1})$ can be extended to x_l iff every consistent instantiation of $\bigcup_{j=1}^{k'} C_j \setminus \{x_l\} \subseteq \{x_1, \ldots, x_{l-1}\}$ can be extended to x_l, which is achieved by strong directional relational k-consistency.   □

Algorithm 1 can be naturally extended to deal with k-Epiphytic Trees. In general, instead of a single constraint in Ω for each tree of the structure there will be k constraints and all of them must be satisfied under ε.

7 Conclusion

We characterized a class of ternary networks called k-Epiphytic Trees, for which strong directional relational k-consistency is sufficient to ensure a backtrack-free solution. We empirically showed that NCOPs seems to be usually represented by 1-Epiphytic Trees. Despite the importance of these theoretical results, enforcing relational consistency is generally impractical. We proposed a branch and bound algorithm to achieving a relaxed form of directional RAC and evaluated such algorithm by executing it on a set of 130 instances from the COCONUT benchmark, of which about 60% were solved with a close approximation.

Many works have linked the network structure to the level of local consistency that guarantees a backtrack-free solution [21,26]. However, to the best of our knowledge this is the first time that the relationship between optimization problems and relational consistency is addressed. Previously, Sam-Haroud and Faltings [39] found that $(3,2)$-relational consistency[6] on *convex* ternary networks is a sufficient condition for global consistency. Besides the convexity restriction (which our result does not impose), the main goal of their work was construct a compact description of the complete solution space, instead of finding the optimum solution. A structure similar to Epiphytic Trees is the *cycle-cutset decomposition* used on finite-domain CSPs (see [14]), in which a consistent instantiation of the variables that constitute a cycle (our Ω set) is first addressed by some enumeration method to then instantiate the remaining variables in a backtrack-free manner. However, as the variable ordering initiates by the cycle-cutset variables, instead of the one representing the objective function, there are no guarantees that a solution found is optimal.

In future works, we want to deepen the relation between Epiphytic Trees and other network structures, such as hypergraph's width and the cycle-cutset schema. Also, we wish to study which Epiphytic Tree representing the network is better for the performance of the proposed algorithm. Furthermore, our method provides a non-problem-specific heuristic for interval branch and bound methods; thus, it can be used to improve optimization solvers in means of pre-processing, local search or upper bounding.

Acknowledgments. This work was supported by CAPES - Brazilian Federal Agency for Support and Evaluation of Graduate Education within the Ministry of Education of Brazil.

---

[6] A variant of 3-relational consistency where the consistency is considered with relation to two variables instead of one (see [15]).

References

1. Araya, I., Reyes, V.: Interval branch-and-bound algorithms for optimization and constraint satisfaction: a survey and prospects. J. Glob. Optim. **65**(4), 837–866 (2016)
2. Araya, I., Trombettoni, G., Neveu, B., Chabert, G.: Upper bounding in inner regions for global optimization under inequality constraints. J. Glob. Optim. **60**(2), 145–164 (2014)
3. van Beek, P., Dechter, R.: Constraint tightness and looseness versus local and global consistency. J. ACM **44**(4), 549–566 (1997)
4. Benhamou, F., Granvilliers, L.: Continuos and interval constraints. In: Rossi, F., van Beek, P., Walsh, T. (eds.) Handbook of Constraint Programming. Foundations of Artificial Intelligence, pp. 569–601. Elsevier, New York (2006)
5. Benhamou, F., McAllester, D., van Hentenryck, P.: CLP (intervals) revisited. In: International Symposium on Logic Programming, pp. 124–138. MIT Press, Cambridge (1994)
6. Benhamou, F., Older, W.J.: Applying interval arithmetic to real, integer, and Boolean constraints. J. Logic Program. **32**(1), 1–24 (1997)
7. Berge, C.: Graphs and Hypergraphs. Elsevier Science, Oxford (1985)
8. Bessiere, C., Stergiou, K., Walsh, T.: Domain filtering consistencies for non-binary constraints. Artif. Intell. **172**(6), 800–822 (2008)
9. Bhurjee, A.K., Panda, G.: Efficient solution of interval optimization problem. Math. Methods Oper. Res. **76**(3), 273–288 (2012)
10. Chabert, G., Trombettoni, G., Neveu, B.: New light on arc consistency over continuous domains. Technical report RR-5365, INRIA (2004)
11. Cohen, D.A., Jeavons, P.G.: The power of propagation: when GAC is enough. Constraints **22**(1), 3–23 (2017)
12. Csendes, T., Ratz, D.: Subdivision direction selection in interval methods for global optimization. SIAM J. Numer. Anal. **34**(3), 922–938 (1997)
13. Davis, E.: Constraint propagation with interval labels. Artif. Intell. **32**(3), 281–331 (1987)
14. Dechter, R.: Constraint Processing. Morgan Kaufmann, San Francisco (2003)
15. Dechter, R., van Beek, P.: Local and global relational consistency. Theoret. Comput. Sci. **173**(1), 283–308 (1997)
16. Dechter, R., Pearl, J.: Network-based heuristics for constraint-satisfaction problems. Artif. Intell. **34**(1), 1–38 (1987)
17. Dechter, R., Rish, I.: Directional resolution: the Davis-Putnam procedure, revisited. In: 4th International Conference on Principles of Knowledge Representation and Reasoning, pp. 134–145. Morgan Kaufmann, San Francisco (1994)
18. Faltings, B.: Arc-consistency for continuous variables. Artif. Intell. **65**(2), 363–376 (1994)
19. Faltings, B., Gelle, E.M.: Local consistency for ternary numeric constraints. In: 15th International Joint Conference on Artificial Intelligence, pp. 392–397. Morgan Kaufmann, San Francisco (1997)
20. Freuder, E.C.: Synthesizing constraint expressions. Commun. ACM **21**(11), 958–966 (1978)
21. Freuder, E.C.: A sufficient condition for backtrack-free search. J. ACM **29**(1), 24–32 (1982)
22. Gottlob, G., Leone, N., Scarcello, F.: Hypertree decompositions and tractable queries. J. Comput. Syst. Sci. **64**(3), 579–627 (2002)

23. Gyssens, M.: On the complexity of join dependencies. ACM Trans. Database Syst. **11**(1), 81–108 (1986)
24. Hansen, E., Walster, G.W.: Global Optimization Using Interval Analysis. Monographs and Textbooks in Pure and Applied Mathematics. Marcel Dekker, New York (2004)
25. van Hentenryck, P., McAllester, D., Kapur, D.: Solving polynomial systems using a branch and prune approach. SIAM J. Numer. Anal. **34**(2), 797–827 (1997)
26. Jégou, P.: On the consistency of general constraint-satisfaction problems. In: 11th National Conference on Artificial Intelligence, pp. 114–119. AAAI Press, Washington, D.C. (1993)
27. Karakashian, S., Woodward, R.J., Reeson, C., Choueiry, B.Y., Bessiere, C.: A first practical algorithm for high levels of relational consistency. In: 24th AAAI Conference on Artificial Intelligence, pp. 101–107. AAAI Press, California (2010)
28. Kearfott, R.B.: An interval branch and bound algorithm for bound constrained optimization problems. J. Glob. Optim. **2**(3), 259–280 (1992)
29. Lebbah, Y., Michel, C., Rueher, M.: An efficient and safe framework for solving optimization problems. J. Comput. Appl. Math. **199**(2), 372–377 (2007)
30. Lhomme, O.: Consistency techniques for numeric CSPS. In: 13th International Joint Conference on Artificial Intelligence, pp. 232–238. Morgan Kaufmann, San Francisco (1993)
31. Mackworth, A.K.: Consistency in networks of relations. Artif. Intell. **8**(1), 99–118 (1977)
32. Mackworth, A.K.: On reading sketch maps. In: 5th International Joint Conference on Artificial Intelligence, pp. 598–606. Morgan Kaufmann, San Francisco (1977)
33. Moore, R.E., Kearfott, R.B., Cloud, M.J.: Introduction to Interval Analysis. Society for Industrial and Applied Mathematics, Philadelphia (2009)
34. Moore, R.E.: Interval Analysis. Prentice-Hall, New Jersey (1966)
35. Neumaier, A.: Interval Methods for Systems of Equations. Encyclopedia of Mathematics and Its Applications. Cambridge University Press, Cambridge (1991)
36. Neumaier, A., Shcherbina, O., Huyer, W., Vinkó, T.: A comparison of complete global optimization solvers. Math. Program. **103**(2), 335–356 (2005)
37. Ninin, J., Messine, F., Hansen, P.: A reliable affine relaxation method for global optimization. 4OR **13**(3), 247–277 (2015)
38. Ratschek, H., Rokne, J.: New Computer Methods for Global Optimization. Halsted Press, New York (1988)
39. Sam-Haroud, D., Faltings, B.: Consistency techniques for continuous constraints. Constraints **1**(1), 85–118 (1996)
40. Shcherbina, O., Neumaier, A., Sam-Haroud, D., Vu, X.-H., Nguyen, T.-V.: Benchmarking global optimization and constraint satisfaction codes. In: Bliek, C., Jermann, C., Neumaier, A. (eds.) COCOS 2002. LNCS, vol. 2861, pp. 211–222. Springer, Heidelberg (2003). https://doi.org/10.1007/978-3-540-39901-8_16
41. Sidebottom, G., Havens, W.S.: Hierarchical arc consistency for disjoint real intervals in constraint logic programming. Comput. Intell. **8**(4), 601–623 (1992)
42. Trombettoni, G., Araya, I., Neveu, B., Chabert, G.: Inner regions and interval linearizations for global optimization. In: 25th AAAI Conference on Artificial Intelligence, pp. 99–104. AAAI Press, California (2011)
43. Van Hentenryck, P.: Numerica: a modeling language for global optimization. In: 15th International Joint Conference on Artificial Intelligence, pp. 1642–1647. Morgan Kaufmann, San Francisco (1997)
44. Waltz, D.: Understanding line drawings of scenes with shadows. In: Winston, P.H. (ed.) The Psychology of Computer Vision, pp. 19–91. McGraw-Hill, New York (1975)

Learning Heuristics for the TSP by Policy Gradient

Michel Deudon[1], Pierre Cournut[1], Alexandre Lacoste[2], Yossiri Adulyasak[3(✉)], and Louis-Martin Rousseau[4]

[1] Polytechnique (France), Palaiseau, France
{michel.deudon,pierre.cournut}@polytechnique.edu
[2] Element AI, Montreal, Canada
allac@elementai.ca
[3] HEC Montréal, Montreal, Canada
yossiri.adulyasak@hec.ca
[4] Polytechnique Montréal, Montreal, Canada
louis-martin.rousseau@cirrelt.ca

Abstract. The aim of the study is to provide interesting insights on how efficient machine learning algorithms could be adapted to solve combinatorial optimization problems in conjunction with existing heuristic procedures. More specifically, we extend the neural combinatorial optimization framework to solve the traveling salesman problem (TSP). In this framework, the city coordinates are used as inputs and the neural network is trained using reinforcement learning to predict a distribution over city permutations. Our proposed framework differs from the one in [1] since we do not make use of the Long Short-Term Memory (LSTM) architecture and we opted to design our own critic to compute a baseline for the tour length which results in more efficient learning. More importantly, we further enhance the solution approach with the well-known 2-opt heuristic. The results show that the performance of the proposed framework alone is generally as good as high performance heuristics (OR-Tools). When the framework is equipped with a simple 2-opt procedure, it could outperform such heuristics and achieve close to optimal results on 2D Euclidean graphs. This demonstrates that our approach based on machine learning techniques could learn good heuristics which, once being enhanced with a simple local search, yield promising results.

Keywords: Combinatorial optimization · Traveling salesman
Policy gradient · Neural networks · Reinforcement learning

1 Introduction

Combinatorial optimization is a topic that consists of finding an optimal object from a finite set of objects. Sequencing problems are those where the best order for performing a set of tasks must be determined, which in many cases leads to a NP-hard problem. Specific variations include single machine scheduling and the

© Springer International Publishing AG, part of Springer Nature 2018
W.-J. van Hoeve (Ed.): CPAIOR 2018, LNCS 10848, pp. 170–181, 2018.
https://doi.org/10.1007/978-3-319-93031-2_12

Traveling Salesman Problem (TSP). Sequencing problems are among the most widely studied problems in Operations Research (OR). They are prevalent in manufacturing and routing applications.

As in [2], our work is motivated by the fact that many real world problems arising from the OR community are solved daily from scratch with hand-crafted features and man-engineered heuristics. We propose a generic framework to learn heuristics for combinatorial tasks where the output is an ordering of the input. Our focus in this paper is a data-driven heuristic that can effectively solve the TSP, the well-known combinatorial problem (due to limited space, the review on optimization algorithms on TSP is provided in Appendix).

We first review some recent Reinforcement Learning (RL) approaches to solve the TSP in Sect. 2. We then present our proposed method in Sects. 3 and 4. Finally, we describe experiments and discuss results in Sect. 5.

2   Reinforcement Learning Perspective for the TSP

Reinforcement learning (RL) is a general-purpose framework for decision making in a scenario where a learner actively interacts with an environment to achieve a certain goal. In response to an action, the learner receives two types of information: his new state in the environment, and a real-valued reward, which is specific to the task and its corresponding goal. Successful examples include playing games at high level (Atari [8], Go [9,10]), navigating 3D worlds or labyrinths, controlling physical systems and interacting with users.

Combinatorial problems such as the TSP are often solved sequentially. Typically, a state is a partial solution (a sequence of visited cities) and an action is the next city to visit (among those not yet visited). In response to an action, the new state is the updated solution and the reward signal could either come when a tour is completed or be incremental. An RL agent builds on its own experience - sequences (state, action; reward, state) - to maximize future rewards. In practice, one could either learn directly a (deterministic or stochastic) mapping from state to action, called a policy $\pi(a|s)$, or learn an auxiliary evaluation function (Value or Q function) measuring the quality of a state and used to discriminate among actions based on their usefulness. In both cases, the combinatorial structure of the state space S is intractable and calls for the use of function approximators such as Deep Neural Networks. Deep Learning (DL) is a general-purpose framework for representation learning. Given an objective, a Neural Network learns the representation that is required to achieve the objective. Neural Networks compute hierarchical, abstract representations of the data (through linear transformations and non-linear activation functions) and learn features (at several levels of abstractions) by back-propagating gradients of the loss w.r.t. the parameters, using the chain rule and Stochastic Gradient Descent.

Recurrent Neural Networks (RNN) with Long Short Term Memory (LSTM) cells [11] have been successfully used for structured inputs and/or outputs with long term dependencies. More recently, attention based Neural Networks have significantly improved models in computer vision [12], image [13] or video

[14] captioning, machine translation [15], speech recognition [16] and question answering [17]. Rather than processing a signal once, attention allows to process step by step some regions or features of the signal at high resolution to acquire information when and where needed. At each step, next location is chosen based on past information and demands for the task. Google Brain's Pointer Network [18] is a neural architecture to learn the conditional probability of an output sequence with elements that are discrete tokens corresponding to positions in an input sequence. The neural network comprises a RNN encoder-decoder connected with hard attention. At each decoding step, a "pointer" is used to sample from the action space (in our case, a probability distribution over cities to visit). It overall parametrizes a stochastic policy over city permutations $p_\theta(\pi|s)$ and can be used for problems such as sorting variable sized sequences, and various combinatorial optimization problems. Google Brain's Pointer Network trained by Policy Gradient [1] could determine good quality solutions for 2D Euclidean TSP with up to 100 nodes.

In [2], the authors use a graph embedding network called structure2vec (S2V) to featurize nodes in the graph in the context of their neighbourhood. The learned greedy algorithm constructs solutions sequentially and is trained by fitted Q-learning to learn the policy together with the graph embedding network. For the TSP task, Google Brain's Pointer Network trained by Policy Gradient performs on par with the S2V network trained by fitted Q-learning.

Based on the recent work [1], we further enhance the approach in several ways. In particular, instead of relying on the LSTM architecture, our model is based solely on attention mechanisms. This result in a more efficient learning. The framework is further enhanced with a simple 2-opt procedure and the approach shows promising results on the TSP. We believe that the outcome of this study sheds light on a data-driven hybrid heuristic that makes use of ML and local search techniques to tackle combinatorial optimization.

3   Neural Architecture for TSP

Given a set of n cities s, the Traveling Salesman Problem (TSP) consists in finding a minimum cost tour visiting all n cities exactly once. The total cost of a tour is the total distance traveled in the tour. Following [1], we aim to learn the parameters θ of a stochastic policy over city permutations $p_\theta(\pi|s)$, using Neural Networks and Policy Gradient. Given an input set of points s, the key idea is to assign higher probability to "good" tours π^+ and lower probability to "undesirable" tours π^-.

We follow the general encoder-decoder perspective. The encoder maps an input set $I = (i_1, ..., i_n)$ to a set of continuous representations $Z = (z_1, ..., z_n)$. Given Z, the decoder then generates an output sequence $O = (o_1, ..., o_n)$ of symbols one element at a time. At each step the model is auto-regressive, using the previously generated symbols as additional input when generating the next.

3.1 TSP Setting and Input Preprocessing

In this paper, we focus on the 2D Euclidean TSP. Each $city_i$ is described by its $2D$ coordinates (x_i, y_i) in a Euclidean space. We use Principal Component Analysis (PCA) on the centered input coordinates to exploit spatial invariance by rotation of all cities. This way, the learned heuristic does not depend on the orientation of the input $s = ((x_i, y_i))_{i \in [1,n]}$.

3.2 Encoder

The purpose of our encoder is to obtain a representation for each action (city) given its context. The output of our encoder is a set of action vectors $A = (a_1, ..., a_n)$, each representing a city interacting with other cities. Our neural encoder takes inspiration from recent advances in Neural Machine Translation. Similarly to [19], our actor and our critic use neural attention mechanisms to encode cities as a set (rather than a sequence as in [1]).

TSP Encoder. We use the encoder proposed in [19], which relies on attention mechanisms in place of the traditional convolutions or recurrences. Our self attentive encoder takes as input an embedded and batch normalized [20] set of n cities $s = (city_i)_{i \in [1,n]}$ (d-dimensional space). It overall consists in a stack of N identical layers as shown in Fig. 1 in Appendix. Each layer has two sublayers. The first sublayer *Multi-head Attention* is detailed in the next paragraph. The second sublayer *Feed-Forward* consists of two position-wise linear transformations with a ReLU activation in between. The output of each sublayer is $LayerNorm(x + Sublayer(x))$, where $Sublayer(x)$ is the function implemented by the sublayer itself and $LayerNorm()$ stands for layer normalization [20].

Multi Head Attention. Neural attention mechanisms allow queries to interact with key-value pairs. For the TSP, queries and key-value pairs $q_i, k_i, v_i \in \mathbb{R}^d$ are obtained by linearly transforming each $city_i \in \mathbb{R}^d$ and applying a ReLu non linearity. Following [19], our attention mechanism is defined as

$$Attention(Q; K; V) = softmax(\frac{QK^T}{\sqrt{d}})V \tag{1}$$

where $Q = [q_1, ..., q_n]$, $K = [k_1, ..., k_n]$, $V = [v_1, ..., v_n]$. Our Multi Head Attention sublayer outputs a new representation for each city, computed as a weighted sum of city values, where the corresponding weights are defined by an affinity function between cities' queries and keys. As suggested in [19], queries, keys and values are linearly projected on h different learned subspaces (hence the name *Multi Head*). We then apply the attention mechanism on each of these new set of representations to obtain h d_h-dimensional output values for each city which are concatenated into the final values.

3.3   Decoder

Following [1], our neural network architecture uses the chain rule to factorize the probability of a tour as

$$p_\theta(\pi|s) = \prod_{t=1}^{n} p_\theta(\pi(t)|\pi(<t), s) \qquad (2)$$

Each term on the right hand side of Eq. (2) is computed sequentially with softmax modules. As opposed to [1] which summarizes all previous actions in a fixed-length vector, our model explicitly forgets after $K = 3$ steps, dispensing with LSTM networks. At each output time t, we map the three last sampled actions (visited cities) to the following query vector:

$$q_t = ReLu(W_1 a_{\pi(t-1)} + W_2 a_{\pi(t-2)} + W_3 a_{\pi(t-3)}) \in \mathbb{R}^{d'} \qquad (3)$$

Similar to [1], our query vector q_t interacts with a set of n vectors to define a pointing distribution over the action space. Once the next city is sampled, the trajectory q_{t+1} is updated with the selected action vector and the process ends when the tour is completed. See Fig. 2 in Appendix.

Pointing Mechanism. We use the same pointing mechanism as in [1] to predict a distribution over cities given encoded actions (cities) and a state representation (query vector). Pointing to a specific position in the input sequence allows to adapt the same framework to variable length tours. As in [1], our pointing mechanism is parameterized by two attention matrices $W_{ref} \in \mathbb{R}^{dd''}, W_q \in \mathbb{R}^{d'd''}$ and an attention vector $v \in \mathbb{R}^{d''}$ as follows:

$$\forall i \leq n, u_i^t = \begin{cases} v^T tanh(W_{ref} a_i + W_q q_t) & if \quad i \notin \{\pi(0), ..., \pi(t-1)\} \\ -\infty & otherwise. \end{cases} \qquad (4)$$

$$p_\theta(\pi(t)|\pi(<t), s) = softmax(C \, tanh(u^t/T)) \qquad (5)$$

$p_\theta(\pi(t)|\pi(<t), s)$ predicts a distribution over the set of n action vectors, given a query q_t. Following [1], we use a mask to set the logits (aka log-probabilities) of cities that already appeared in the tour to $-\infty$, as shown in Eq. (4). This ensures that our model outputs valid permutations of the input. As suggested in [1], clipping the logits in $[-C, +C]$ is a way to control the entropy. T is a temperature hyper-parameter used to control the certainty of the sampling. $T = 1$ during training and $T > 1$ during inference.

4   Training the Model

Supervised learning for NP-hard problems such as the TSP and its variants is undesirable because the performance of the model is tied to the quality of the supervised labels and getting supervised labels is expensive (and may be infeasible). By contrast, RL provides an appropriate and simple paradigm for training a Neural Network. An RL agent explores different tours and observes their corresponding rewards.

Following [1], we train our Neural Network by Policy Gradient using the REINFORCE learning rule [21] with a critic to reduce the variance of the gradients. For the TSP, we use the tour length as reward $r(\pi|s) = L(\pi|s)$ (which we seek to minimize).

Policy Gradient and REINFORCE: Our training objective is the expected reward, which given an input graph s is defined as:

$$J(\theta|s) = \mathbb{E}_{\pi \sim p_\theta(.|s)}[r(\pi|s)] \tag{6}$$

During training, our graphs are drawn from a distribution S and the total training objective is defined as:

$$J(\theta) = \mathbb{E}_{s \sim S}[J(\theta|s)] \tag{7}$$

To circumvent non-differentiability of hard-attention, we resort to the well-known REINFORCE learning rule [21] which provides an unbiased gradient of (6) w.r.t. the model's parameters θ:

$$\nabla_\theta J(\theta|s) = \mathbb{E}_{\pi \sim p_\theta(.|s)}[(r(\pi|s) - b_\phi(s))\nabla_\theta log(p_\theta(\pi|s))] \tag{8}$$

where $b_\phi(s)$ is a parametric baseline implemented by a *critic* network to reduce the variance of the gradients while keeping them unbiased. With Monte-Carlo sampling, the gradient of (7) is approximated by:

$$\nabla_\theta J(\theta) \approx \frac{1}{B} \sum_{k=1}^{B} (r(\pi_k|s_k) - b_\phi(s_k))\nabla_\theta log(p_\theta(\pi_k|s_k)) \tag{9}$$

We learn the actor's parameters θ by starting from a random policy and iteratively optimizing them with the REINFORCE learning rule and Stochastic Gradient Descent (SGD), on instances generated on the fly.

Critic: Our critic uses the same encoder as our actor. It uses once the pointing mechanism with $q = 0_{d'}$. The critic's pointing distribution over cities $p_\phi(s)$ defines a *glimpse* vector gl_s computed as a weighted sum of the action vectors $A = (a_1, ..., a_n)$

$$gl_s = \sum_{i=1}^{n} p_\phi(s)_i a_i \tag{10}$$

The glimpse vector gl_s is fed to a 2 fully connected layers with ReLu activations. The critic is trained by minimizing the Mean Square Error between its predictions and the actor's rewards.

5   Experiments and Results

We conduct experiments to investigate the behavior of the proposed method. We consider a benchmarked test set of 1,000 Euclidean TSP20, TSP50 and TSP100 graphs. Points are drawn uniformly at random in the 2D unit square. Experiments are conducted using Tensorflow 1.3.0. We use mini-batches of 256 sequences of length $n = 20$, $n = 50$ and $n = 100$. The actor and critic embed each city in a 128-dimensional space. Our self attentive encoder consists of 3 stacks with $h = 16$ parallel heads and $d = 128$ hidden dimensions. For each head we use $d_h = d/h = 8$. Our FFN sublayer has input and output dimension d and its inner-layer has dimension $4d = 512$. Queries for the pointing mechanisms are 360-dimensional vectors ($d' = 360$). The pointing mechanism is computed in a 256-dimensional space ($d" = 256$). The critic's feed forward layers consist in 256 and 1 hidden units. Parameters θ are initialized with xavier_initializer [22] to avoid saturating the non-linear activation functions and to keep the scale of the gradients roughly the same in all layers. We clip our tanh logits to $[-10,10]$ for the pointing mechanism. Temperature is set to 2.4 for TSP20 and 1.2 for TSP50 and TSP100 during inference. We use Adam [23] optimizer for SGD with $\beta_1 = 0.9$, $\beta_2 = 0.99$ and $\epsilon = 10^{-9}$. Our initial learning rate of 10^{-3} is decayed every 5000 steps by 0.96. Our model was trained for 20000 steps on two Tesla K80 (approximately 2h).

Results are compared in terms of solution quality to Google OR tools, a Vehicle Routing Problem (VRP) solver that combines local search algorithms (*cheapest insertion*) and meta-heuristics, the heuristic of Christofides [24], the well-known Lin-Kernighan heuristic (LK-H) [25] and Concorde exact TSP solver (which yields an optimal solution). Note that, even though LK-H is a heuristic, the average tour lengths of LK-H are very close to those of Concorde. Thus, the LK-H results are omitted from the Table. We refer to Google Brain's Pointer Network trained with RL [1] as *Ptr*. For the TSP, we run the actor on a batch of a single input graph. The more we sample, the more likely we will visit the optimal tour. Table 1 compares the different sampling methods with a batch of size 128. 2-opt is used to improve the best tour found by our model (model+2opt). For the instances TSP100, experiments were conducted with our model trained on TSP50. In terms of computing time, all the instances were solved within a fraction of second. For TSP50, the average computing time per instance are $0.05s$ for Concorde, $0.14s$ for LK-H and $0.02s$ for OR-Tools on a single CPU, as well as $0.30s$ for the pointer network and $0.06s$ for our model on a single GPU.

The results clearly show that our model is competitive with existing heuristics for the TSP both in terms of optimality and running time. We provide the following insights based on the experimental results.

- **LSTM vs. Explicitly forgetting:** Our results suggest that keeping in memory the last three sampled actions during decoding performs on par with [1] which uses a LSTM network. More generally, this raises the question of what information is useful to take optimal decisions.

Table 1. Average tour length (lower is better)

Task	Model	Model + 2opt	Ptr supervised	Ptr greedy	Ptr sample	Christofides	OR tools	Concorde (optimal)
TSP20	3.84	3.82	3.88	3.89	-	4.30	3.85	3.82
TSP50	5.81	5.77	6.09	5.95	5.80	6.62	5.80	5.68
TSP100	8.85*	8.16*	10.81	8.30	8.05	9.18	7.99	7.77

*N.B.: Results for TSP100 were obtained with our model trained on TSP50

- **Results for TSP100:** For TSP100 solved by our model pre-trained on TSP50 (see also Fig. 3 in Appendix), our approach performs relatively well even though it was not directly trained on the same instance size as in [1]. This suggests that our model can generalize heuristics to unseen instances.
- **AI-OR Hybridization:** As opposed to [1] which builds an end-to-end deep learning pipeline for the TSP, we combine heuristics learned by RL with local-search (2-opt) to quickly improve solutions sampled from our policy without increasing the running time during inference. Our actor together with this hybridization achieves approximately 5x speedup compared to the framework of [1].

6 Conclusion

Solving Combinatorial Optimization is difficult in general. Thanks to decades of research, solvers for the Traveling Salesman Problem (TSP) are highly efficient, able to solve large instances in a few computation time. With little engineering and no labels, Neural Networks trained with Reinforcement Learning are able to learn clever heuristics (or distribution over city permutations) for the TSP. Our code is made available on Github[1].

We plan to investigate how to extend our model to constrained variants of the TSP, such as the TSP with time windows, an important and practical TSP variant which is much more difficult to solve. We believe that Markov Decision Processes (MDP) provide a sound framework to address feasibility in general. One contribution we would like to emphasize here is that simple heuristics can be used in conjunction with Deep Reinforcement Learning, shedding light on interesting hybridization between Artificial Intelligence (AI) and Operations Research (OR). We encourage more contributions of this type.

Acknowledgment. We would like to thank Polytechnique Montreal and CIRRELT for financial and logistic support, Element AI for hosting weekly meetings as well as Compute Canada, Calcul Quebec and Telecom Paris-Tech for computational resources. We are also grateful to all the reviewers for their valuable and detailed feedback.

---

[1] https://github.com/MichelDeudon/encode-attend-navigate.

Appendix: supplementary materials

Literature review on optimization algorithms for the TSP

The best known exact dynamic programming algorithm for the TSP has a complexity of $O(2^n n^2)$, making it infeasible to scale up to large instances (e.g., 40 nodes). Nevertheless, state of the art TSP solvers, thanks to handcrafted heuristics that describe how to navigate the space of feasible solutions in an efficient manner, can provably solve to optimality symmetric TSP instances with thousands of nodes. Concorde [3], known as the best exact TSP solvers, makes use of cutting plane algorithms, iteratively solving linear programming relaxations of the TSP, in conjunction with a branch-and-bound approach that prunes parts of the search space that provably will not contain an optimal solution.

The MIP formulation of the TSP allows for tree search with Branch & Bound which partitions (Branch) and prunes (Bound) the search space by keeping track of upper and lower bounds for the objective of the optimal solution. Search strategies and selection of the variable to branch on influence the efficiency of the tree search and heavily depend on the application and the physical meaning of the variables. Machine Learning (ML) has been successfully used for variable branching in MIP by learning a supervised ranking function that mimics Strong Branching, a time-consuming strategy that produces small search trees [4]. The use of ML in branching decisions in MIP has also been studied in [5].

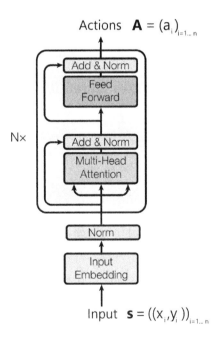

Fig. 1. Our neural encoder. Figure modified from [19].

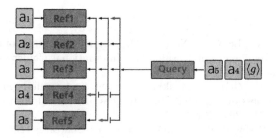

Fig. 2. Our neural decoder. Figure modified from [1].

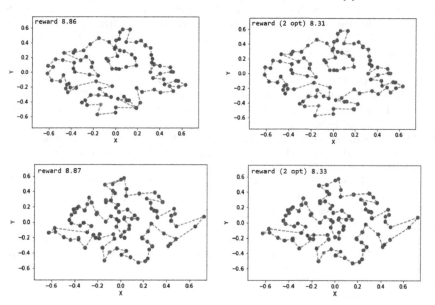

Fig. 3. 2D TSP100 instances sampled with our model trained on TSP50 (left) followed by a 2opt post processing (right)

For constrained based scheduling, filtering techniques from the OR community aim to drastically reduce the search space based on constraints and the objective. For instance, one could identify mandatory and undesirable edges and force edges based on degree constraint as in [6]. Another approach consists in building a relaxed Multivalued Decision Diagrams (MDD) that represents a superset of feasible orderings as an acyclic graph. Through a cycle of filtering and refinement, the relaxed MDD approximates an exact MDD, i.e., one that exactly represents the feasible orderings [7].

References

1. Bello, I., Pham, H., Le, Q.V., Norouzi, M., Bengio, S.: Neural combinatorial optimization with reinforcement learning. In: International Conference on Learning Representations (ICLR 2017) (2017)
2. Khalil, E., Dai, H., Zhang, Y., Dilkina, B., Song, L.: Learning combinatorial optimization algorithms over graphs. In: Advances in Neural Information Processing Systems, pp. 6351–6361 (2017)
3. Applegate, D., Bixby, R., Chvatal, V., Cook, W.: Concorde TSP solver (2006)
4. Khalil, E.B., Le Bodic, P., Song, L., Nemhauser, G.L., Dilkina, B.N.: Learning to branch in mixed integer programming. In: AAAI, pp. 724–731, February 2016
5. Di Liberto, G., Kadioglu, S., Leo, K., Malitsky, Y.: Dash: dynamic approach for switching heuristics. Eur. J. Oper. Res. **248**(3), 943–953 (2016)
6. Benchimol, P., Van Hoeve, W.J., Régin, J.C., Rousseau, L.M., Rueher, M.: Improved filtering for weighted circuit constraints. Constraints **17**(3), 205–233 (2012)
7. Bergman, D., Cire, A.A., van Hoeve, W.J., Hooker, J.: Sequencing and single-machine scheduling. In: Bergman, D., Cire, A.A., van Hoeve, W.J., Hooker, J. (eds.) Decision Diagrams For Optimization, pp. 205–234. Springer, Cham (2016). https://doi.org/10.1007/978-3-319-42849-9_11
8. Mnih, V., Kavukcuoglu, K., Silver, D., Rusu, A.A., Veness, J., Bellemare, M.G., Petersen, S.: Human-level control through deep reinforcement learning. Nature **518**(7540), 529 (2015)
9. Silver, D., Huang, A., Maddison, C.J., Guez, A., Sifre, L., Van Den Driessche, G., Schrittwieser, J., Antonoglou, I., Panneershelvam, V., Lanctot, M., Dieleman, S.: Mastering the game of Go with deep neural networks and tree search. Nature **529**(7587), 484–489 (2016)
10. Silver, D., Schrittwieser, J., Simonyan, K., Antonoglou, I., Huang, A., Guez, A., Chen, Y.: Mastering the game of go without human knowledge. Nature **550**(7676), 354 (2017)
11. Hochreiter, S., Schmidhuber, J.: Long short-term memory. Neural Comput. **9**(8), 1735–1780 (1997)
12. Mnih, V., Heess, N., Graves, A.: Recurrent models of visual attention. In: Advances in Neural Information Processing Systems, pp. 2204–2212 (2014)
13. Xu, K., Ba, J., Kiros, R., Cho, K., Courville, A., Salakhudinov, R., Zemel, R., Bengio, Y.: Show, attend and tell: neural image caption generation with visual attention. In: International Conference on Machine Learning, pp. 2048–2057, June 2015
14. Gao, L., Guo, Z., Zhang, H., Xu, X., Shen, H.T.: Video captioning with attention-based lstm and semantic consistency. IEEE Trans. Multimedia **19**(9), 2045–2055 (2017)
15. Bahdanau, D., Cho, K., Bengio, Y.: Neural machine translation by jointly learning to align and translate. In: ICLR 2015 (2015)
16. Chan, W., Jaitly, N., Le, Q., Vinyals, O.: Listen, attend and spell: a neural network for large vocabulary conversational speech recognition. In: 2016 IEEE International Conference on Acoustics, Speech and Signal Processing (ICASSP), pp. 4960–4964. IEEE, March 2016
17. Xu, H., Saenko, K.: Ask, attend and answer: exploring question-guided spatial attention for visual question answering. In: Leibe, B., Matas, J., Sebe, N., Welling, M. (eds.) European Conference On Computer Vision. LNCS, pp. 451–466. Springer, Cham (2016). https://doi.org/10.1007/978-3-319-46478-7_28

18. Vinyals, O., Fortunato, M., Jaitly, N.: Pointer networks. In: Advances in Neural Information Processing Systems, pp. 2692–2700 (2015)
19. Vaswani, A., Shazeer, N., Parmar, N., Uszkoreit, J., Jones, L., Gomez, A.N., Kaiser, L., Polosukhin, I.: Attention is all you need. In: Advances in Neural Information Processing Systems, pp. 6000–6010 (2017)
20. Ioffe, S., Szegedy, C.: Batch normalization: accelerating deep network training by reducing internal covariate shift. In: International Conference on Machine Learning, pp. 448–456, June 2015
21. Williams, R.J.: Simple statistical gradient-following algorithms for connectionist reinforcement learning. In: Sutton, R.S. (ed.) Reinforcement Learning, pp. 5–32. Springer, Boston (1992). https://doi.org/10.1007/978-1-4615-3618-5_2
22. Glorot, X., Bengio, Y.: Understanding the difficulty of training deep feedforward neural networks. In: Proceedings of the Thirteenth International Conference on Artificial Intelligence and Statistics, pp. 249–256, March 2010
23. Kingma, D.P., Ba, J.: Adam: a method for stochastic optimization. In: ICLR 2015 (2015)
24. Christofides, N.: Worst-case analysis of a new heuristic for the travelling salesman problem (No. RR-388). Carnegie-Mellon Univ Pittsburgh Pa Management Sciences Research Group (1976)
25. Lin, S., Kernighan, B.W.: An effective heuristic algorithm for the traveling-salesman problem. Oper. Res. 21(2), 498–516 (1973)

Three-Dimensional Matching Instances Are Rich in Stable Matchings

Guillaume Escamocher[(✉)] and Barry O'Sullivan

Insight Centre for Data Analytics, Department of Computer Science,
University College Cork, Cork, Ireland
guillaume.escamocher@insight-centre.org,
barry.osullivan@insight-centre.org

Abstract. Extensive studies have been carried out on the Stable Matching problem, but they mostly consider cases where the agents to match belong to either one or two sets. Little work has been done on the three-set extension, despite the many applications in which three-dimensional stable matching (3DSM) can be used. In this paper we study the Cyclic 3DSM problem, a variant of 3DSM where agents in each set only rank the agents from one other set, in a cyclical manner. The question of whether every Cyclic 3DSM instance admits a stable matching has remained open for many years. We give the exact number of stable matchings for the class of Cyclic 3DSM instances where all agents in the same set share the same master preference list. This number is exponential in the size of the instances. We also show through empirical experiments that this particular class contains the most constrained Cyclic 3DSM instances, the ones with the fewest stable matchings. This would suggest that not only do all Cyclic 3DSM instances have at least one stable matching, but they each have an exponential number of them.

1 Introduction

1.1 Different Kinds of Stable Matchings Problems

Stable matching is the problem of establishing groups of agents according to their preferences, such that there is no incentive for the agents to change their groups. It has a plethora of applications; the two most commonly mentioned are the assignment of students to universities and of residents to hospitals [11].

Most of the research done on stable matching focuses on the cases where the agents belong to either one set (the stable roommates problem) or two (the stable marriage problem). Far fewer studies have looked at the three-dimensional version, where every agent in each set has a preference order over *couples* of agents from the two other sets, even though it is naturally present in many situations. It can be used for example to build market strategies that link suppliers, firms and buyers [15], or in computer networking systems to match data sources, servers and end users [4]. Even some applications like kidney exchange, which

© Springer International Publishing AG, part of Springer Nature 2018
W.-J. van Hoeve (Ed.): CPAIOR 2018, LNCS 10848, pp. 182–197, 2018.
https://doi.org/10.1007/978-3-319-93031-2_13

is traditionally associated with the stable roommates problem, can be easily represented in a three-dimensional form [2].

One of the possible reasons for this lack of interest is that, while it is well-known that every two-dimensional matching instance admits at least one stable matching [7], some three-dimensional matching instances do not [1]. In fact, determining whether a given three-dimensional matching instance has a stable matching is NP-Complete [12,16], even when each agent's preference order is required to be consistent, or when ties are allowed in the rankings [8].

Due to the hardness of the general problem, other restrictions on the preferences have been proposed. With lexicographically acyclic preferences, then there is always a stable matching, which can easily be found in quadratic time [5]. If the preferences are lexicographically cyclic, then the complexity of determining whether a given instance admits a stable matching is still open. For the latter kind of preferences, some instances with no stable matching have been found [3].

Most of the work in this paper is about the cyclic Three-Dimensional Stable Matching Problem. In this version, agents from the first set only rank agents from the second set, agents from the second set only rank agents from the third set, and agents from the third set only rank agents from the first set. Hardness results are also known for this variant: imposing a stronger form of stability [9], or allowing incomplete preference lists [2] both make it NP-Complete to determine whether an instance admits a stable matching. However, the standard problem with complete preference lists is still open. It has actually been around for decades and is considered "hard and outstanding" [17]. Few results about it have been found since its formulation. To date, it is only known that there always exists a stable matching for instances with at most 3 agents in each set [3], a result that has been subsequently improved to include instances with sets of size 4 [6].

1.2 Master Preference Lists

Whatever the type of matching problem studied, it is generally assumed that the preferences of each agent are independent from the preferences of the other agents in the same set. However, this is often not the case in real-life settings. Indeed, it is not hard to imagine that in many cases hospitals will have close, if not identical, preferences over which residents they want to accept, or that firms will often compete for the same top suppliers. Shared preference lists have also been used to assign university students to dormitory rooms, where the students were ranked according to a combination of academic record and socio-economic characteristics [13].

Imposing a master preference list on all agents within a same set leads to a much more constrained problem. In most cases, the only stable matching is obtained by grouping the best ranked agent of each set together, the second best ones together, and so on. This is true for the two-dimensional matching problem [10]. As we explain in Sect. 3.4, this is also true for many versions of the three-dimensional Stable Matching (3DSM) problem.

We will show in this paper that the cyclic 3DSM problem is singular with regard to the number of stable matchings for instances with master preference lists. Not only is this number more than one, but it is extremely large, exponential in the size of the instances. We will also demonstrate through experiments that cyclic 3DSM instances with master preference lists are the most constrained cyclic 3DSM instances, the ones with the fewest stable matchings. Combining these two results would indicate that it is the natural behavior of *all* cyclic 3DSM instances to have an exponential number of stable matchings, making cyclic 3DSM an attractive problem when looking for tractable three-dimensional matching classes.

We divide the paper in the following manner. In Sect. 2, we recall the standard definitions of stability for the cyclic 3DSM problem, as well as the notion of master preference list. In Sect. 3, we give the exact number of stable matchings for instances with master preference lists. The bulk of Sect. 3 is about the cyclic 3DSM problem, but we also take a look at instances from other matching conventions to see how they compare. We empirically show in Sect. 4 that in cyclic 3DSM, instances with more balanced preferences have more stable matchings, while instances with the most unanimous preferences (master lists) have the fewest stable matchings. We also observe this behavior in other matching problems. Finally, we reflect on these findings in the conclusion.

2   General Definitions

Definition 1. *A* cyclic three dimensional stable matching instance, *or* cyclic 3DSM instance *comprises:*

- *Three* agent sets $A = \{a_1, a_2, \ldots, a_n\}$, $B = \{b_1, b_2, \ldots, b_n\}$ and $C = \{c_1, c_2, \ldots, c_n\}$ *each containing* n agents.
- *For each agent* $a \in A$, *a strict preference order* $>_a$ *over the agents from the set* B. *For each agent* $b \in B$, *a strict preference order* $>_b$ *over the agents from the set* C. *For each agent* $c \in C$, *a strict preference order* $>_c$ *over the agents from the set* A.

The number n *of agents in each agent set is the* size *of the instance.*

Definition 2. *A* master list cyclic 3DSM instance *is a cyclic 3DSM instance where all agents within a same set have the same preference order.*

All master list instances of a same size are isomorphic, so from now on we will assume that every master list cyclic 3DSM instance of size n satisfies the following condition: for all i and j such that $1 \leq i < j \leq n$, for all agents $a \in A$, $b \in B$ and $c \in C$, we have $b_i >_a b_j$, $c_i >_b c_j$ and $a_i >_c a_j$. In other words, the agents in each agent set are ranked according to the preferences of the previous agent set.

Definition 3. *Let* I *be a cyclic 3DSM instance of size* n. *A* matching *for* I *is a set* $M = \{t_1, t_2, \ldots, t_n\}$ *of* n *triples such that each triple contains exactly one agent from each agent set of* I, *and each agent of* I *is represented exactly once in* M.

Definition 4. *Let I be a cyclic 3DSM instance and let M be a matching for I. Let t be a triple containing the three agents $a \in A$, $b \in B$ and $c \in C$. Let $a_M \in A$, $b_M \in B$ and $c_M \in C$ be three agents such that a and b_M are in the same triple of M, b and c_M are in the same triple of M, and c and a_M are in the same triple of M. Then we say that t is a blocking triple for M if $b >_a b_M$, $c >_b c_M$ and $a >_c a_M$.*

Note that, from the definition, no two agents in a blocking triple t can be in the same triple t_i in the matching M.

Definition 5. *Let I be a cyclic 3DSM instance and let M be a matching for I. We say that M is a stable matching for I if there is no blocking triple for M.*

We present an example of a master list cyclic 3DSM instance and of a matching in Fig. 1. The dots represent the agents and the lines represent the triples in the matching $M = \{\langle a_1, b_3, c_2 \rangle, \langle a_2, b_1, c_1 \rangle, \langle a_3, b_4, c_4 \rangle, \langle a_4, b_2, c_3 \rangle\}$. The triple $\langle a_1, b_2, c_1 \rangle$ is a blocking triple because a_1 prefers b_2 over the agent it got in M, b_2 prefers c_1 over the agent it got in M, and c_1 prefers a_1 over the agent it got in M. Therefore M is not stable.

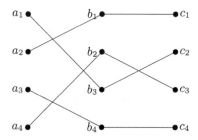

Fig. 1. A matching M for a master list cyclic 3DSM instance of size 4.

3   Stable Matchings for Master List Instances

3.1   Preliminary Notions

In this section we present the main theoretical result of the paper: a function f such that $f(n)$ is the exact number of stable matchings for a master list cyclic 3DSM instance of size n. In the proof, we will consider two kinds of matchings: divisible and indivisible.

Definition 6. *Let I be a master list cyclic 3DSM instance of size n with three agent sets $A = \{a_1, a_2, \ldots, a_n\}$, $B = \{b_1, b_2, \ldots, b_n\}$ and $C = \{c_1, c_2, \ldots, c_n\}$, and let M be a matching for I. We say that M is divisible if there exists some p such that $0 < p < n$ and:*

– *for all i, j such that a_i and b_j are in the same triple of M, we have $i \leq p \Leftrightarrow j \leq p$.*
– *for all i, j such that a_i and c_j are in the same triple of M, we have $i \leq p \Leftrightarrow j \leq p$.*

We also say that p is a divider *of M.*

We say that a matching that is not divisible is *indivisible*. Note that a same divisible matching can have several dividers. To illustrate the notion of divisible matching, we present in Fig. 2 two examples of divisible matchings for a master list cyclic 3DSM instance of size 5. The first matching has two dividers, 1 and 4, while the second matching has one divider, 3. The matching from Fig. 1 was an example of an indivisible matching.

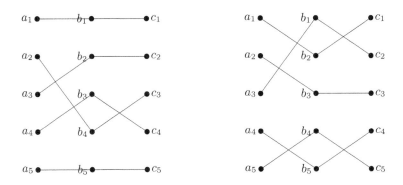

Fig. 2. Two divisible matchings.

3.2 Indivisible Matchings

Before presenting the function f that counts the number of total matchings for a given size n, we look at the function g that counts the number of indivisible matchings. It turns out that this function is very simple: $g(n) = 1$ if $n = 1$ and $g(n) = 3$ otherwise.

Proposition 1. *Let $n \geq 2$ be an integer and let I be a master list cyclic 3DSM instance of size n. Then there are exactly 3 indivisible stable matchings for I.*

To prove the proposition, we are going to define a matching $IndMat_n$ for each size n, then show that $IndMat_n$ is stable, and finally show that any indivisible stable matching for a master list cyclic 3DSM instance of size n is either $IndMat_n$ or one of the two matchings that are isomorphic to $IndMat_n$ by rotation of the agent sets.

Definition 7. *Let I be a master list cyclic 3DSM instance with three agent sets $A = \{a_1, \ldots, a_n\}$, $B = \{b_1, \ldots, b_n\}$ and $C = \{c_1, \ldots, c_n\}$ of size $n > 0$. We call $IndMat_n$ the matching $\{t_1, t_2, \ldots, t_n\}$ for I defined in the following way:*

1. If $n = 1$, $IndMat_n = \{\langle a_1, b_1, c_1\rangle\}$. If $n = 2$, $IndMat_n = \{\langle a_1, b_2, c_1\rangle,$ $\langle a_2, b_1, c_2\rangle\}$.
   If $n = 3$, $IndMat_n = \{\langle a_1, b_2, c_1\rangle, \langle a_2, b_3, c_3\rangle, \langle a_3, b_1, c_2\rangle\}$.
2. If $n > 3$: $t_1 = \langle a_1, b_2, c_1\rangle$, $t_2 = \langle a_2, b_3, c_4\rangle$ and $t_3 = \langle a_3, b_1, c_2\rangle$.
3. If $n > 3$ and $n \equiv 1 \mod 3$: $t_n = \langle a_n, b_n, c_{n-1}\rangle$.
4. If $n > 3$ and $n \equiv 2 \mod 3$: $t_{n-1} = \langle a_{n-1}, b_n, c_{n-2}\rangle$ and $t_n = \langle a_n, b_{n-1}, c_n\rangle$.
5. If $n > 3$ and $n \equiv 0 \mod 3$: $t_{n-2} = \langle a_{n-2}, b_{n-1}, c_{n-3}\rangle$, $t_{n-1} = \langle a_{n-1}, b_n, c_n\rangle$ and $t_n = \langle a_n, b_{n-2}, c_{n-1}\rangle$.
6. If $i \equiv 1 \mod 3$, $i > 3$ and $i \leq n - 3$: $t_i = \langle a_i, b_{i+1}, c_{i-1}\rangle$, $t_{i+1} = \langle a_{i+1}, b_{i+2}, c_{i+3}\rangle$ and $t_{i+2} = \langle a_{i+2}, b_i, c_{i+1}\rangle$.

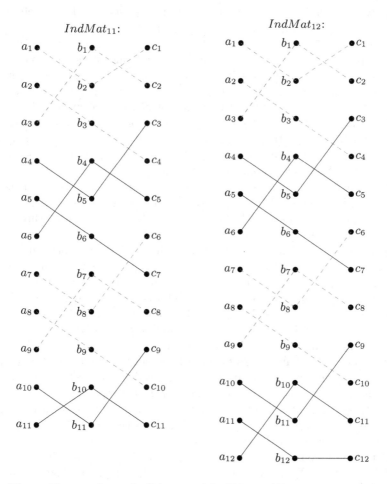

Fig. 3. The matchings $IndMat_{11}$ and $IndMat_{12}$. (Color figure online)

$IndMat_n$ can be seen as a set of $n/3$ gadgets G_i, with each G_i composed of the three triples t_i, t_{i+1} and t_{i+2} for each i such that $i \equiv 1 \mod 3$. All these gadgets are isomorphic by translation, apart from the first one and the last one. Figure 3 shows two examples of $IndMat_n$ matchings, one with $n = 11$ and the other with $n = 12$. Both matchings are almost identical, but they each illustrate a different type of final gadget. To help the reader clearly visualize the structure of the matchings, we alternated the colors and line styles of the gadgets.

Lemma 1. *Let I be a master list cyclic 3DSM instance of size n. Then $IndMat_n$ is a stable matching for I.*

Proof. For each i such that $i \equiv 1 \mod 3$ and $i \leq n$, let G_i be the gadget composed of the three triples t_i, t_{i+1} and t_{i+2}. Let $t = \langle a, b, c \rangle$ be a blocking triple for $IndMat_n$.

Suppose first that a, b and c are from the same gadget G_i. From Definition 7, we have $t_i = \langle a_i, b_{i+1}, c_j \rangle$ with either $j = i - 1$ or $j = i$, $t_{i+1} = \langle a_{i+1}, b_{i+2}, c_k \rangle$ with either $k = i + 2$ or $k = i + 3$, and $t_{i+2} = \langle a_{i+2}, b_i, c_{i+1} \rangle$. Two agents from a same triple in a matching cannot be part of a same blocking triple for this matching, and there are three triples in G_i, therefore at least one agent from each triple is part of t. b_{i+1} cannot be part of t because it got assigned c_i, which is the best ranked agent among the agents of C that are in G_i. Likewise, c_j cannot be part of t because it got assigned a_j, which is the best ranked agent among the agents of A that are in G_i. So $a = a_i$. b_{i+2} cannot be part of t, because it is not as well ranked as b_{i+1}, the agent from B that got assigned to a_i in $IndMat_n$. So $b = b_i$. So c_{i+1} cannot be part of t, because it shares a triple in $IndMat_n$ with b_i. So $c = c_k$. So either $c = c_{i+2}$ or $c = c_{i+3}$. However, neither c_{i+2} nor c_{i+3} is as well ranked as c_{i+1}, the agent from B that got assigned to b_i. So it is not possible to have a blocking triple for $IndMat_n$ with all three agents of the triple in the same gadget G_i.

Suppose now that a, b and c are not all in the same three-triple gadget. Then there must be i and j with $i < j$ such that a is in G_i and b is in G_j, or b is in G_i and c is in G_j, or c is in G_i and a is in G_j. From Definition 7 and by construction of the gadgets, if a is in G_i and b is in G_j then the agent from B that got assigned to a in $IndMat_n$ is better ranked in the preference order of a than b is, so t cannot be a blocking triple. Similarly, if c is in G_i and a is in G_j then the agent from A that got assigned to c in $IndMat_n$ is better ranked in the preference order of c than a is, so t cannot be a blocking triple. So b is in G_i and c is in G_j. The only way for c to be better ranked in the preference order of b than the agent from C that got assigned to b in $IndMat_n$ is if $j = i+3$, $b = b_{i+2}$ and $c = c_{j-1} = c_{i+2}$. Let r be such that $a = a_r$ and let s be such that b_s got assigned to a_r in $IndMat_n$. If $r \leq i + 2$, then from Definition 7 we have $b_s \geq_a b$ and a prefers the agent of B it got assigned in $IndMat_n$ over b. If $r \geq i+3$, then from Definition 7 we have $a_{i+1} \geq_c a$ and c prefers the agent of A it got assigned in $IndMat_n$ over a. Either way, t cannot be a blocking triple and we have the result. $\square$

Lemma 2. *Let I be a master list cyclic 3DSM instance of size n and let M be a matching for I. If there are some a_i (respectively b_i, c_i) and b_j (respectively c_j, a_j) in the same triple of M such that $j \geq i + 2$, then M is not stable.*

Proof. We only do the proof for a_i and b_j, as the other two cases are exactly the same after rotation of the agent sets.

Suppose that we have a matching $M = \{t_1, t_2, \ldots, t_n\}$ for I such that $t_i = \langle a_i, b_j, c_k \rangle$ with $j \geq i + 2$. The triples $t_{i'} = \langle a_{i'}, b_{j'}, c_{k'} \rangle$ and $t_{i''} = \langle a_{i''}, b_{j''}, c_{k''} \rangle$ will be used in the proof. We distinguish the two cases $k \leq i$ and $k > i$.

- $k \leq i$: from the pigeonhole principle we know that at least one of the $i + 1$ agents $\{c_1, c_2, \ldots, c_{i+1}\}$ got assigned an agent $a_{i'} \in A$ such that $i < i'$. Let $k' \leq i + 1$ be such that $c_{k'}$ is one such agent. There cannot be a bijection in M among the sets $\{b_1, b_2, \ldots, b_{i+1}\}$ and $\{c_1, c_2, \ldots, c_{i+1}\}$ because b_j is not in the former but got assigned an agent from the latter. So we know that at least one of the agents $b_1, b_2, \ldots, b_{i+1}$ got assigned an agent $c_{k''} \in C$ such that $k'' > i + 1$. Let $j'' \leq i + 1$ be such that $b_{j''}$ is such an agent. We have $b_{j''} >_{a_i} b_j$ (because $j \geq i + 2$), $c_{k'} >_{b_{j''}} c_{k''}$ (because $k' \leq i + 1 < k''$) and $a_i >_{c_{k'}} a_{i'}$ (because $i < i'$), so $\langle a_i, b_{j''}, c_{k'} \rangle$ is a blocking triple for M and M is not stable.

- $k > i$: none of the i agents $\{c_1, c_2, \ldots, c_i\}$ got assigned a_i, so from the pigeonhole principle we know that at least one of them got assigned an agent $a_{i'} \in A$ such that $i < i'$. Let $k' \leq i$ be such that $c_{k'}$ is one such agent. Also from the pigeonhole principle, we know that among the $i + 1$ agents $\{b_1, b_2, \ldots, b_{i+1}\}$ at least one of them got assigned an agent $c_{k''} \in C$ such that $k'' > i$. Let $j'' \leq i + 1$ be such that $b_{j''}$ is such an agent. We have $b_{j''} >_{a_i} b_j$ (because $j \geq i + 2$), $c_{k'} >_{b_{j''}} c_{k''}$ (because $k' \leq i < k''$) and $a_i >_{c_{k'}} a_{i'}$ (because $i < i'$), so $\langle a_i, b_{j''}, c_{k'} \rangle$ is a blocking triple for M and M is not stable. □

Lemma 3. *Let I be a master list cyclic 3DSM instance of size n and let M be an indivisible stable matching for I. Then either M is $IndMat_n$, or M is one of the two matchings that are isomorphic to $IndMat_n$ by rotation of the agent sets.*

Proof. Let $M = \{t_1, t_2, \ldots, t_n\}$ be an indivisible matching for I. Without loss of generality, assume that a_i is in t_i for every i. We are going to show by induction that every t_i is equal to the ith triple of $IndMat_n$, modulo rotation of the agent sets.

Case i $= 1$: if a_1, b_1 and c_1 are in three different triples of M, then they form a blocking triple for M because they each prefer each other over the agent they got assigned in M. If they are in the same triple of M, then $n = 1$ because M is indivisible; since $IndMat_1 = \langle a_1, b_1, c_1 \rangle$ from Definition 7.1, we have the Lemma. So we can assume from now on that $n > 1$ and that exactly two agents among a_1, b_1 and c_1 are in the same triple in M. We will assume that a_1 and c_1 are the ones in the same triple. All three cases are isomorphic by rotation of the agent sets; the case we chose will lead to $IndMat_n$, while the other two would have led to one of the matchings that are isomorphic to $IndMat_n$ by rotation of the agent sets.

We know that $t_1 = \langle a_1, b_j, c_1 \rangle$ with $j \geq 2$. From Lemma 2, we know that $j \leq 2$. So $t_1 = \langle a_1, b_2, c_1 \rangle$.

Inductive step: suppose now that the triples $t_1, t_2, \ldots, t_p$ are the same as the first p triples of $IndMat_n$ for some p such that $1 \leq p < n$. We are going to prove that t_{p+1} is the same triple as the $(p+1)^{th}$ triple in $IndMat_n$. With p_{last} the highest $p < n$ such that $p \equiv 1 \mod 3$, there are five possibilities to consider.

- $p \equiv 1 \mod 3$ and $p < p_{last}$: since p_{last} is also congruent to 1 modulo 3, we also have $p \leq n - 3$. From Definition 7 we know that the agents that have already been assigned are $a_1, a_2, \ldots, a_p$ from A, $b_1, b_2, \ldots, b_{p-1}$ and b_{p+1} from B, and $c_1, c_2, \ldots, c_p$ from C. So b_p and some c_k are in the same triple, with $k \geq p + 1$. From Lemma 2, we know that $k \leq p + 1$. So b_p and c_{p+1} are in the same triple of M. a_{p+1} cannot be in this triple, because otherwise either n would be equal to $p + 1$ or $p + 1$ would be a divider of M. So some a_i is assigned to c_{p+1} with $i \geq p + 2$. From Lemma 2, we know that $i \leq p + 2$. So a_{p+2} is assigned to c_{p+1} and $t_{p+2} = \langle a_{p+2}, b_p, c_{p+1} \rangle$ (which proves the next bullet point). So $t_{p+1} = \langle a_{p+1}, b_j, c_k \rangle$ for some $j \geq p + 2$ and $k \geq p + 2$. From Lemma 2, we have $j \leq p + 2$ and therefore $b_j = b_{p+2}$. We cannot have $k = p + 2$, because otherwise either n would be equal to $p + 2$ or $p + 2$ would be a divider of M. So $k \geq p + 3$. From Lemma 2, $k \leq p + 3$ and therefore $t_{p+1} = \langle a_{p+1}, b_{p+2}, c_{p+3} \rangle$, which from Definition 7.2 and 7.6 is the same triple as the $(p+1)^{th}$ triple of $IndMat_n$.
- $p \equiv 2 \mod 3$ and $p < p_{last}$: let $p' = p - 1$. So $p' \equiv 1 \mod 3$ and $p' < p_{last}$. So from the proof of the previous bullet point we know that $t_{p'+2} = \langle a_{p'+2}, b_{p'}, c_{p'+1} \rangle$. So $t_{p+1} = \langle a_{p+1}, b_{p-1}, c_p \rangle$, which from Definition 7.2 and 7.6 is the same triple as the $(p+1)^{th}$ triple of $IndMat_n$.
- $p \equiv 0 \mod 3$ and $p < p_{last}$: from Definition 7 we know that the agents that have already been assigned are $a_1, a_2, \ldots, a_p$ from A, $b_1, b_2, \ldots, b_p$ from B, and $c_1, c_2, \ldots, c_{p-1}$ and c_{p+1} from C. So c_p is in the same triple as some a_i with $i \geq p + 1$. From Lemma 2, we know that $i \leq p + 1$. So $i = p + 1$ and $t_{p+1} = \langle a_{p+1}, b_j, c_p \rangle$ for some $j \geq p + 1$. From Lemma 2, we know that $j \leq p + 2$ so either $b_j = b_{p+1}$ or $b_j = b_{p+2}$. If $n = p + 1$, then we have $t_{p+1} = \langle a_{p+1}, b_{p+1}, c_p \rangle$ which from Definition 7.3 is equal to the $(p+1)^{th}$ triple of $IndMat_n$. If $n > p + 1$, then $b_j = b_{p+2}$, because otherwise $p + 1$ would be a divider of M, and $t_{p+1} = \langle a_{p+1}, b_{p+2}, c_p \rangle$. So from Definition 7.5 and 7.6, t_{p+1} is equal to the $(p+1)^{th}$ triple of $IndMat_n$.
- $p = p_{last}$: since $p < n$, either $n = p + 1$ or $n = p + 2$. If $n = p + 1$, then from Definition 7 only the agents $a_{p+1} \in A$, $b_p \in B$ and $c_{p+1} \in C$ have not been assigned. So $t_{p+1} = \langle a_{p+1}, b_p, c_{p+1} \rangle$, which from Definition 7.4 is the same as the $(p+1)^{th}$ tuple of $IndMat_n$. If on the other hand $n = p + 2$, then from Definition 7 only the agents a_{p+1} and a_{p+2} in A, b_p and b_{p+2} in B, and c_{p+1} and c_{p+2} in C remain to be assigned. From Lemma 2, c_{p+2} cannot be assigned to b_p, so c_{p+1} is assigned to b_p. a_{p+1} cannot be in the same triple as these two agents, because otherwise $p + 1$ would be a divider of M. So a_{p+2} is in the same triple as b_p and c_{p+1} and $t_{p+2} = \langle a_{p+2}, b_p, c_{p+1} \rangle$. Consequently, the three other remaining agents are assigned together in the triple $t_{p+1} = \langle a_{p+1}, b_{p+2}, c_{p+2} \rangle$. This is from Definition 7.5 the same triple as the $(p+1)^{th}$ triple of $IndMat_n$.

– $p = p_{last} + 1$: since $p < n$, $n = p + 1$ and only one agent from each agent set has not been assigned. From Definition 7, we know that these agents are $a_{p+1} \in A$, $b_{p-1} \in B$ and $c_p \in C$. So $t_{p+1} = \langle a_{p+1}, b_{p-1}, c_p \rangle$, which is from Definition 7.5 the same triple as the $(p+1)^{th}$ triple of $IndMat_n$.

We did not consider the case where $p = p_{last} + 2$, because it cannot happen if $p < n$.

We have shown that t_1 is equal to the first triple in $IndMat_n$ and that if $n > 1$ and the first p triples of M are equal to the first triples of $IndMat_n$ for $1 \leq p < n$, then t_{p+1} is equal to the $(p+1)^{th}$ triple of $IndMat_n$. By induction, this completes the proof.   □

Lemmas 1 and 3 together prove Proposition 1.

3.3   Main Theorem

Before introducing the Theorem, we need one last Lemma.

Lemma 4. *Let I be a master list cyclic 3DSM instance of size n and let p be an integer such that $1 \leq p < n$. Then the number of stable matchings for I that admit p as their lowest divider is equal to $f(n-p)$ times the number of indivisible stable matchings for a master list cyclic 3DSM instance of size p.*

Proof. Let $M = \{t_1, t_2, \ldots, t_n\}$ be a matching for I such that p is the lowest divider of M and $a_i \in t_i$ for each i. Let $M_1 = \{t_1, t_2, \ldots, t_p\}$ and let $M_2 = \{t_{p+1}, t_{p+2}, \ldots, t_n\}$. Since p is the lowest divider of M, M_1 is indivisible. We show that a triple $\langle a_i, b_j, c_k \rangle$ cannot be a blocking triple for M if it is across the divider p, that is if it fulfills one of the three following conditions: $i \leq p$ and $j > p$, $j \leq p$ and $k > p$, $k \leq p$ and $i > p$. Let t be such a triple. Without loss of generality, assume that $i \leq p$ and $j > p$. Let b_m be the agent of B assigned to a_i in M. Since p is a divider of M, we have $m \leq p < j$. So $b_m >_{a_i} b_j$. So t cannot be a blocking triple for M. So any blocking triple for M is either a blocking triple for M_1 or a blocking triple for M_2. So M is stable if and only if both M_1 and M_2 are stable, and we have the result.   □

We now have all the tools we need to state and prove the Theorem:

Theorem 1. *Let f be the function from $\mathbb{N}$ to $\mathbb{N}$ such that $f(1) = 1$, $f(2) = 4$ and for every n such that $n > 2$ we have $f(n) = 2f(n-2) + 2f(n-1)$. Let $n > 0$ be an integer and let I be a master list cyclic 3DSM instance of size n. Then there are exactly $f(n)$ stable matchings for I.*

Proof. For $n = 1$ and $n = 2$ there are 1 and 4 matchings respectively, and they are all trivially stable. Suppose now that $n > 2$. Let g be the function such that for each integer $q \leq n$, $g(q)$ is the number of indivisible matchings for master list cyclic 3DSM instances of size q. For each p such that $1 \leq p < n$, let f_p be the function such that $f_p(n)$ is the number of stable matchings for a master list cyclic 3DSM instance of size n that have p as their lowest divider. We have:

$$f(n) = (\sum_{p=1}^{n-1} f_p(n)) + g(n)$$

From Lemma 4 we have:

$$f(n) = (\sum_{p=1}^{n-1} g(p)f(n-p)) + g(n)$$

From Proposition 1, we know that $g(1) = 1$ and that $g(p) = 3$ for every $p \geq 2$. Therefore we have:

$$f(n) = f(n-1) + 3f(n-2) + (\sum_{p=3}^{n-1} 3f(n-p)) + 3$$

$$= 2f(n-2) + f(n-1) + f(n-2) + (\sum_{p=2}^{n-2} 3f(n-1-p)) + 3$$

$$= 2f(n-2) + f(n-1) + (\sum_{p=1}^{n-2} g(p)f(n-1-p)) + g(n-1)$$

$$= 2f(n-2) + 2f(n-1)$$

$\square$

Note that $f(n) > 2f(n-1)$ for all n, so master list cyclic 3DSM instances have a number of stable matchings which is exponential in their size.

3.4 Other Matching Problems

An obvious follow-up to our main theorem would be to determine how the number of stable matchings for master list cyclic 3DSM instances compares to the number of stable matchings for master list instances of matching problems with different rules. We first look at what happens when imposing a stronger form of stability, which is based on the notion of weakly blocking triple [2].

Definition 8. *Let I be a cyclic 3DSM instance and let M be a matching for I. Let t be a triple containing the three agents $a \in A$, $b \in B$ and $c \in C$ such that t does not belong to M. Let $a_M \in A$, $b_M \in B$ and $c_M \in C$ be three agents such that a and b_M are in the same triple of M, b and c_M are in the same triple of M, and c and a_M are in the same triple of M. Then we say that t is a weakly blocking triple for M if $b \geq_a b_M$, $c \geq_b c_M$ and $a \geq_c a_M$.*

Informally, a triple t is weakly blocking for some matching M if each agent of t either prefers t over the triple it got assigned to in M, or is indifferent. Note that since we explicitly require t not to belong in M, at least one of the three preferences will be strict.

Definition 9. *Let I be a cyclic 3DSM instance and let M be a matching for I. We say that M is a strongly stable matching for I if there is no weakly blocking triple for M.*

Strong stability is more restrictive than standard stability, therefore we can expect a lower number of stable matchings. Indeed, the number of strongly stable matchings is always equal to 1 for master list instances.

Proposition 2. *Let I be a master list cyclic 3DSM instance of size n. Then the number of strong stable matchings for I is equal to 1.*

Proof. Let M be a strongly stable matching for I. Let p be the largest integer such that $0 \leq p \leq n$ and for each $0 < q \leq p$ the triple $\langle a_q, b_q, c_q \rangle$ belongs to M. Suppose that $p < n$. Therefore the triple $t = \langle a_{p+1}, b_{p+1}, c_{p+1} \rangle$ does not belong to M. Let i, j and k be such that a_i is assigned to c_{p+1} in M, b_j is assigned to a_{p+1} in M and c_k is assigned to b_{p+1} in M. We know that for each q such that $1 \leq q \leq p$, a_q, b_q and c_q have been assigned to each other in M. So $i \geq p + 1$, $j \geq p + 1$ and $k \geq p + 1$. So $b_{p+1} \geq_a b_j$, $c_{p+1} \geq_b c_k$ and $a_{p+1} \geq_c a_i$. So from Definition 8, t is a weakly blocking triple for M. Therefore $p = n$ and the only possible strongly stable matching is the matching M_0 which contains the triple $\langle a_p, b_p, c_p \rangle$ for each $1 \leq p \leq n$.

It only remains to prove that M_0 is strongly stable. Let $t = \langle a_i, b_j, c_k \rangle$ be a triple. If $i > j$, then $b_i <_{a_i} b_j$ and a_i strictly prefers b_i, the agent from B it got assigned to in M_0, over b_j, the agent from B it got assigned to in t, which means that t cannot be a weakly blocking triple for M_0. So if t is weakly blocking for M_0, then $i \leq j$. By the same reasoning, if t is weakly blocking for M_0, then $j \leq k$ and $k \leq i$. So if t is weakly blocking for M_0, then $i = j = k$. But in this case, t is in M_0 by construction and therefore cannot be a weakly blocking triple for M_0. So there is no weakly blocking triple for M_0. Therefore M_0 is strongly stable, which completes the proof.   □

The same very simple proof can be used to show that for at least two more matching problems, namely lexicographically cyclic 3DSM (defined in [3]) and lexicographically acyclic 3DSM (defined in [5]), master list instances of size n have exactly one stable matching, which is also of the form $\{\langle a_1, b_1, c_1 \rangle, \langle a_2, b_2, c_2 \rangle, \ldots, \langle a_n, b_n, c_n \rangle\}$. This result holds for the extensively studied two-dimensional stable matching (2DSM) too [10]. This indicates that master list cyclic 3DSM instances offer many more stable matchings than their master list counterparts in some others of the most widely used matching problems.

4 Stable Matchings for Instances Without Master Preference Lists

If master list cyclic 3DSM instances have fewer stable matchings than other instances from the same problem, then our main Theorem implies that all cyclic 3DSM instances have a number of stable matchings exponential in their size. This is not a trivial assumption, so we need further study to determine what happens to the number of stable matchings when considering other instances.

In this section, we empirically investigate the evolution of the number of stable matchings when going from a master list cyclic 3DSM instance, which can be seen as an instance with unanimous preferences, to its opposite: a cyclic 3DSM instance with evenly split preferences. We will need a few definitions to formally describe our procedure.

Definition 10. *Let I be a cyclic 3DSM instance of size n. Let g and g' be two agents of I, such that strictly more than $n/2$ agents prefer g over g'. Then we call* adding an ML-step *to I the act of switching g and g' in the preference list of an agent that prefers g over g'.*

Definition 11. *We say that a cyclic 3DSM instance is* perfectly split *if it is not possible to add an ML-step to I.*

Our experiments consist in starting from a master list cyclic 3DSM instance and randomly adding ML-steps until we reach a perfectly split instance. We summarize our results in Figs. 4 and 5. In Fig. 4, we added ML-steps to 1000 starting master list cyclic 3DSM instances of size 8, until getting a perfectly split instance. Note that the number of steps required to arrive to a perfectly split instance is not the same in each of the runs, so the last few data points represent fewer than 1000 instances. This explains why the "minimum" and "maximum" plots seem to converge towards the "average" one at the end. The exact number of instances represented by each data point can be found in Table 1. The numbers of stable matchings were obtained using Cachet [14], an exact SAT model counter.

The figure clearly confirms what we suspected: the cyclic 3DSM instances with the fewest stable matchings are the ones with master preference lists, or are at least very similar to these instances. More precisely, the number of stable matchings seems to initially increase steadily when going away from master list instances, before plateauing when a certain number of ML-steps has been added.

Table 1 contains the exact numbers for Cyclic 3DSM instances. The last line, not represented in Fig. 4, describes the number of stable matchings for 1000 completely random cyclic 3DSM instances, whose construction was not related in any way to master list instances or ML-steps. This serves as a control experiment, to make it clear that our results are not dependent on the particular way that we build our instances.

Figure 5 illustrates the results of the same experiments on two other stable matching variants: 2DSM and cyclic 3DSM with strong stability, both of size 8.

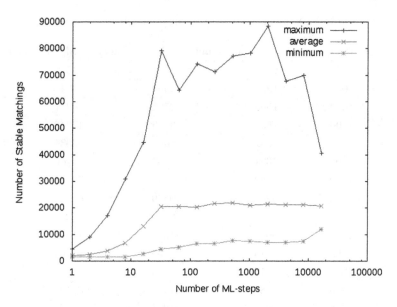

Fig. 4. Number of stable matchings when adding ML-steps to cyclic 3DSM instances.

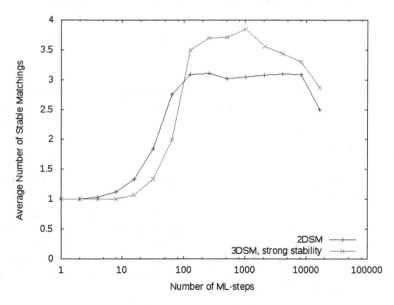

Fig. 5. Number of stable matchings for 2DSM instances and of strongly stable matchings for cyclic 3DSM instances.

Here again, we added ML-steps to 1000 starting master list instances of each problem. The numbers of (strongly) stable matchings are much lower for these two problems, yet we can observe the same behavior of a steady increase followed

Table 1. Number of stable matchings in 3DSM instances.

# ML-steps	# instances	# SM (minimum)	# SM (average)	# SM (maximum)
0	1000	1552	1552	1552
1	1000	1552	2009	4544
2	1000	1508	2561	8917
4	1000	1552	3779	17242
8	1000	1552	6669	30831
16	1000	2681	13095	44766
32	1000	4529	20442	79129
64	1000	5201	20599	64234
128	1000	6615	20216	74233
256	1000	6615	21683	71376
512	1000	7716	21965	77204
1024	993	7515	21084	78257
2048	953	6989	21478	88481
4096	760	7085	21235	67604
8192	301	7515	21201	69996
16384	24	11904	20713	40688
Random	1000	4932	20521	105070

by a plateau. This empirically shows that instances with master preference lists are linked with very high constrainedness in many stable matching problems.

5   Conclusion

We have given the exact number of stable matchings for cyclic 3DSM instances with master preference lists. This number is 1 for many other stable matching problems, but it is exponential in the case of the Cyclic 3DSM problem.

We have also shown through experiments that despite their high number of stable matchings, cyclic 3DSM instances with master preference lists are the most constrained instances of the cyclic 3DSM problem, the ones with the fewest stable matchings, a behavior that mirrors what can be observed in other standard matching problems.

Combining these two results, we propose the following conjecture: each cyclic 3DSM instance has a number of stable matchings exponential in its size. If true, this would make the cyclic 3DSM problem a very interesting object of research when looking for positive and/or tractable three-dimensional matching results.

Acknowledgements. This research has been funded by Science Foundation Ireland (SFI) under Grant Number SFI/12/RC/2289.

References

1. Alkan, A.: Nonexistence of stable threesome matchings. Math. Soc. Sci. **16**(2), 207–209 (1988)
2. Bíró, P., McDermid, E.: Three-sided stable matchings with cyclic preferences. Algorithmica **58**(1), 5–18 (2010)
3. Boros, E., Gurvich, V., Jaslar, S., Krasner, D.: Stable matchings in three-sided systems with cyclic preferences. Discrete Math. **289**(1–3), 1–10 (2004)
4. Cui, L., Jia, W.: Cyclic stable matching for three-sided networking services. Comput. Netw. **57**(1), 351–363 (2013)
5. Danilov, V.: Existence of stable matchings in some three-sided systems. Math. Soc. Sci. **46**(2), 145–148 (2003)
6. Eriksson, K., Sjöstrand, J., Strimling, P.: Three-dimensional stable matching with cyclic preferences. Math. Soc. Sci. **52**(1), 77–87 (2006)
7. Gale, D., Shapley, L.S.: College admissions and the stability of marriage. Am. Math. Mon. **69**(1), 9–15 (1962)
8. Huang, C.-C.: Two's company, three's a crowd: stable family and threesome roommates problems. In: Arge, L., Hoffmann, M., Welzl, E. (eds.) ESA 2007. LNCS, vol. 4698, pp. 558–569. Springer, Heidelberg (2007). https://doi.org/10.1007/978-3-540-75520-3_50
9. Huang, C.: Circular stable matching and 3-way kidney transplant. Algorithmica **58**(1), 137–150 (2010)
10. Irving, R.W., Manlove, D., Scott, S.: The stable marriage problem with master preference lists. Discrete Appl. Math. **156**(15), 2959–2977 (2008)
11. Manlove, D.F.: Algorithmics of Matching Under Preferences, Series on Theoretical Computer Science, vol. 2. WorldScientific, Singapore (2013)
12. Ng, C., Hirschberg, D.S.: Three-dimensional stable matching problems. SIAM J. Discrete Math. **4**(2), 245–252 (1991)
13. Perach, N., Polak, J., Rothblum, U.G.: A stable matching model with an entrance criterion applied to the assignment of students to dormitories at the Technion. Int. J. Game Theory **36**(3–4), 519–535 (2008)
14. Sang, T., Bacchus, F., Beame, P., Kautz, H.A., Pitassi, T.: Combining component caching and clause learning for effective model counting. In: The Seventh International Conference on Theory and Applications of Satisfiability Testing, SAT 2004, Vancouver, BC, Canada, 10–13 May 2004, Online Proceedings (2004)
15. Stuart Jr., H.W.: The supplier—firm—buyer game and its m-sided generalization. Math. Soc. Sci. **34**(1), 21–27 (1997)
16. Subramanian, A.: A new approach to stable matching problems. SIAM J. Comput. **23**(4), 671–700 (1994)
17. Woeginger, G.J.: Core stability in hedonic coalition formation. In: van Emde Boas, P., Groen, F.C.A., Italiano, G.F., Nawrocki, J., Sack, H. (eds.) SOFSEM 2013. LNCS, vol. 7741, pp. 33–50. Springer, Heidelberg (2013). https://doi.org/10.1007/978-3-642-35843-2_4

From Backdoor Key to Backdoor Completability: Improving a Known Measure of Hardness for the Satisfiable CSP

Guillaume Escamocher, Mohamed Siala$^{(\boxtimes)}$, and Barry O'Sullivan

Insight Centre for Data Analytics, Department of Computer Science,
University College Cork, Cork, Ireland
{guillaume.escamocher,mohamed.siala,barry.osullivan}@insight-centre.org

Abstract. Many studies have been conducted on the complexity of Constraint Satisfaction Problem (CSP) classes. However, there exists little theoretical work on the hardness of *individual* CSP instances. In this context, the backdoor key fraction (BKF) [17] was introduced as a quantifier of problem hardness for individual satisfiable instances with regard to backtracking search. In our paper, after highlighting the weaknesses of the BKF, we propose a better characterization of the hardness of an individual satisfiable CSP instance based on the ratio between the size of the solution space and that of the search space. We formally show that our measure is negatively correlated with instance hardness. We also show through experiments that this measure evaluates more accurately the hardness of individual instances than the BKF.

1 Introduction

Finding a solution to a CSP instance is well known to be NP-hard, even when considering satisfiable instances [3]. The complexity of CSP instances has been extensively cataloged in the framework of complexity theory. However, attempts to formally define instance hardness, to find out what makes some CSP instances difficult to solve have been scarcer. A number of studies have been proposed based, in particular, on the notion of constrainedness [9,10], to predict the behavior of large sets of instances. Constrainedness compares the expected number of solutions of constraint instances to their average size. It is straightforward to compute, and is well-suited for large classes of problems, but it does not establish a distinction between individual instances that have the same average tightness but different solution spaces. On the other hand, only considering the solution space is not enough to accurately predict instance complexity [18].

The *Backdoor Key Fraction* was proposed in [17] to characterize the hardness of a given satisfiable CSP instance with respect to backtracking search based on the notion of *backdoor*. A backdoor is a set of variables that, when properly assigned, allows us to decide the remainder of the problem in polynomial time

© Springer International Publishing AG, part of Springer Nature 2018
W.-J. van Hoeve (Ed.): CPAIOR 2018, LNCS 10848, pp. 198–214, 2018.
https://doi.org/10.1007/978-3-319-93031-2_14

using a given sub-solver [19]. The backdoor key fraction is based on the backdoor key set. A variable is in the backdoor key set if its value is logically determined by the settings of other backdoor variables. The key fraction is the ratio of the backdoor key set size to the corresponding backdoor size. Unfortunately, as we explain in Sect. 2, there are many classes of instances for which the backdoor key fraction is ill-fitted. The main motivation behind our paper is to revisit this measure by proposing a better characterization of instance hardness.

In this paper, we propose an improvement over the backdoor key fraction. Intuitively, a solver finds an instance difficult if it contains many paths that do not lead to a solution, where a path is a possible sequence of choices made by the solver. Therefore, we define our *completability* measure as the number of paths that are completable, meaning that they lead to a solution, divided by the number of paths actually explored by the solver. This can be viewed as the ratio between the solution space and the search space. If the ratio is close to 1, the solver mostly branches on completable paths and can easily solve the instance. If, however, the ratio is low, the solver explores a lot of dead-ends and finds the instance hard. Completability can be seen as an improvement over the backdoor key set. Indeed, as we explain in Sect. 2, both our metric and the backdoor key fraction are composed of a ratio between a numerator that takes into account the global interactions between the backdoor and the rest of the instance, and a denominator that only relies on the internal structure of the backdoor.

The notion of completability has previously been used by [7], albeit in a completely different way. The author proposed to add constraints to CSP instances to transform them into equivalent minimal instances, where an instance is minimal if any partial solution of size bounded by some predefined constant can be extended to a solution [16]. On the other hand, we are computing a theoretical measure and are not modifying any part of the observed instances. Nonetheless, it is interesting to note how his intention was to make CSP instances easier by, in essence, increasing the completability ratio of small subsets of variables. We show that when some particular small subsets of variables, namely backdoors, have a high completability ratio, then backtracking solvers have an easier time finding a solution. His paper and ours are, therefore, consistent with each other in their approach of completability. Another work closely related to the idea of completable partial solutions can be found in [4]. The authors also consider the ratio between the solution space and the search space within small sets of variables (although not backdoors), but their measure is restricted to minimal CSP instances, while our metric can measure any satisfiable CSP instance.

In the next section, we give the different notions used throughout the paper and we highlight the limitations of the backdoor key fraction. In Sect. 3 we adopt a theoretical approach to justify our measure. Finally, we present our experimental study in Sect. 4.

2   CSP, Backdoor Key Fraction, and Backdoor Completability

A CSP instance is a triplet $\langle \mathcal{X}, \mathcal{D}, \mathcal{C} \rangle$ where $\mathcal{X}$ is a set of variables $\{v_1, v_2, \ldots, v_n\}$, $\mathcal{D}$ is a set of domains $\{D_{v_1}, D_{v_2}, \ldots, D_{v_n}\}$, and $\mathcal{C}$ is a set of constraints. A domain D_v is a set of integers (values) associated with the variable v. A constraint C of arity $k \geq 1$ is a pair $(\mathcal{X}(C), \mathcal{R}(C))$, where $\mathcal{X}(C)$ is a sequence of k variables, and $\mathcal{R}(C) \subseteq \mathbb{Z}^k$. The set of variables in $\mathcal{X}(C)$ is called the scope of C. The constraint C is universal if every k-tuple of integers is in $\mathcal{R}(C)$. An assignment is a pair $\langle v, a \rangle$ where v is a variable and $a \in D_v$. A value a is said to be assigned to a variable v if $D_v = \{a\}$. An *instantiation* is a set of assignments where each variable appears at most once. The scope of an instantiation S is the set of variables $\{v \mid \exists \langle v, a \rangle \in S\}$. Let S be an instantiation and C be a constraint such that $\mathcal{X}(C) = [v_{i_1}, v_{i_2}, \ldots, v_{i_k}]$. We say that S violates C if $\forall l \in [1, k]$, there exists a_l such that $\langle v_{i_l}, a_l \rangle \in S$ and $\langle a_1, a_2, \ldots a_k \rangle \notin \mathcal{R}(C)$. The instantiation S is said to satisfy C if S does not violate C. An instantiation that does not violate any constraint is called a *partial solution*. A *solution* to a CSP instance $\langle \mathcal{X}, \mathcal{D}, \mathcal{C} \rangle$ is a partial solution with a scope equal to $\mathcal{X}$. Not every partial solution can be extended to a solution. A CSP instance that admits a solution is called *satisfiable*. The arity of a CSP instance is the greatest arity of its constraints. When an instance is binary (i.e., of arity 2), two assignments $\langle v, a \rangle$ and $\langle v', a' \rangle$ are incompatible if there exists a constraint C such that $\mathcal{X}(C) = [v, v']$ such that $\langle a, a' \rangle \notin \mathcal{R}(C)$. Two assignments are compatible if they are not incompatible.

Definition 1. *Let I be a CSP instance with n variables and let p be an integer such that $1 \leq p < n$. We say that I is $(p, 1)$-consistent if for any partial solution S of size p and for any variable v not in the scope of S, there is a value $a \in D_v$ such that $S \cup \{\langle v, a \rangle\}$ is a partial solution. We also say that I is strongly $(p, 1)$-consistent if it is $(q, 1)$-consistent for all q such that $1 \leq q \leq p$.*

A backdoor [19] is defined with regard to a particular sub-solver. We define a sub-solver and the other notions the same way that [17] did.

Definition 2. *An algorithm A that takes a CSP instance as input is a sub-solver if:*

1. *For any CSP instance I, either A rejects I or A correctly recognizes I as satisfiable or unsatisfiable. If I is recognized as satisfiable, then A also returns a solution to I.*
2. *A runs in time polynomial in the size of I.*

Now that we have explained the concept of sub-solvers, we can properly define the notion of a backdoor to a CSP.

Definition 3. *Let A be a sub-solver. Let I be a CSP instance. Let V be a subset of the variables of I. We say that V is a backdoor for A of I if there exists a partial solution S_p of scope V such that the instance I' obtained from I after assigning the value a to the variable v for each assignment $\langle v, a \rangle \in S_p$ is recognized as satisfiable by A.*

Informally, a backdoor is a (small) set of variables that, when properly assigned, makes the rest of the instance easy. When the sub-solver A is clear from the context, we shall use "backdoor" instead of "backdoor for A". Technically, the set of all variables in an instance is always a trivial backdoor. Therefore, we mainly focus on backdoors of minimal size.

Definition 4. *Let A be a sub-solver and let I be a satisfiable CSP instance. Let B be a set of variables of I. We say that a backdoor B for A of I is a* minimal *backdoor if $\forall v \in B$, $B \backslash \{v\}$ is not a backdoor for A of I.*

Before presenting our backdoor completability measure, we describe the existing metric that is closest to our own. This is the backdoor key fraction, introduced in [17]. Backdoor keys are sets of dependent variables, where a dependent variable is defined as follows:

Definition 5. *Let I be a satisfiable CSP instance, let B be a subset of variables of I, and let $v \in B$ be a variable. Let S be a solution to I, and let $S_p \subset S$ be a partial solution of scope $B \backslash \{v\}$. We say that v is a* dependent variable with respect to S_p *if there is exactly one value a in D_v such that $S_p \cup \{\langle v, a \rangle\}$ can be extended to a solution to I.*

Definition 6. *Let I be a satisfiable CSP instance, let B be a backdoor for I, and let $v \in B$ be a variable. Let S be a solution to I, and let $S_B \subset S$ be a partial solution of scope B and let S_v be equal to S_B restricted to $B \backslash \{v\}$. We say that v is in the* backdoor key *set of B with respect to S_B if v is a dependent variable with respect to S_v.*

The backdoor key fraction can now be defined.

Definition 7. *Let B be a backdoor for a satisfiable CSP instance I, and let S_B be a partial solution to B. The* backdoor key fraction *of B with respect to S_B is the ratio between the number of variables in the backdoor key set of B with respect to S_B and the total number of variables in B. If B is empty, we say that the backdoor key fraction of B is 0.*

The last sentence is our own addition to account for the cases when the backdoor is empty. Note that this follows the intuition of their paper. Indeed, empty backdoors are associated with very easy instances, and their intention was for the backdoor key fraction to be positively correlated with the hardness.

There are many cases where the backdoor key fraction is not useful. The authors of [17] mentioned the case where given any backdoor and its corresponding solution, one can always flip the truth assignment of any variable in the backdoor and still extend the backdoor to a solution. In such instances, the backdoor key fraction is equal to 0 for any backdoor. Another issue arises when a given CSP instance I only has one solution, the backdoor key fraction of any non-empty backdoor of I is by Definition 7 always 1. More generally, any backdoor variable which is also part of the instance backbone (the set of variables that are assigned the same values in all solutions [15]) is a dependent variable, and therefore is in the backdoor key.

In general, hard instances with regard to backtracking algorithms are the ones that offer many potentially wrong choices and few potentially right ones to solvers. Therefore, if we want to quantify hardness, it makes sense to build a measure that keeps track of both the size of the search space (the choices that the solver can make) and the size of the solution space (the right choices). What we define as the search space is the set of partial, local solutions while the solution space is simply the set of global solutions.

Definition 8. *Let I be a CSP instance and let B be a non-empty set of variables of I. We say that the* completability ratio *of B is the ratio $\frac{\#completable}{\#partial}$ where:*

- *#partial is the number of partial solutions of scope B.*
- *#completable is the number of partial solutions of scope B that can be extended to a solution to I.*

We also say that the completability ratio of an empty set of variables is 1, and that the completability ratio of a set of variables with no partial solution is 0.

From now on, we only apply the notion of completability ratio on minimal backdoors within satisfiable CSP instances. However, this concept is general and could be applied to any set of variables in any CSP instance (although the ratio is trivially always 0 in unsatisfiable instances). In particular, the completability ratio is not dependent on a particular sub-solver, and is unique for each set of variables within a particular CSP instance. We define now our measure for a whole instance.

Definition 9. *Let A be a sub-solver and let I be a satisfiable CSP instance. The* backdoor completability *for A of I is the average of the completability ratios of all minimal backdoors for A of I.*

Observe that backdoor completability can be used to study satisfiable CSP instances of any arity. This is also the case for backdoor key fraction. Recall that finding a solution to a satisfiable CSP instance is an NP-hard problem [3], so such a restriction does not diminish the usefulness of either measure. It should be noted also that we are not limited to binary instances. Indeed, in Sect. 4, we present an experimental study on both binary and non-binary instances.

3   Theoretical Justification

In order to be a valid measure of hardness, backdoor completability needs to correctly recognize both easy and hard classes. In the former case, this is done by returning a high value for tractable classes. In the latter case, this is done by returning a low value for a subset of decent size in each non-tractable class. However, both tractability and backdoors are defined with regard to a specific algorithm. So ideally backdoor completability should tag a class as tractable if and only if the sub-solver used to define a backdoor solves the class.

In this section, we present an example to further explain what result we are aiming for, then we state our main Theorem. We refer to *(primal) constraint graphs*, *tree decompositions* and *treewidth*. We now recall the definitions of these four concepts.

Definition 10. *Let I be a CSP instance. The* primal constraint graph *of I is the graph G such that:*

- *The vertices of G are the variables of I.*
- *There is a an edge between two vertices v_i and v_j of G if and only if there is a non-universal constraint C of I such that $\mathcal{X}(C)$ contains both v_i and v_j.*

In the case of binary instances, the primal constraint graph is called the constraint graph. Part of the algorithms that we present is to build the *tree decomposition* of some (primal) constraint graph.

Definition 11. *Let G be a graph. Let T be a tree such that each vertex of T is a set of vertices of G. We say that T is a* tree decomposition *of G if:*

1. *Each vertex of G belongs to at least one vertex of T.*
2. *If two vertices v_1 and v_2 are connected in G, then there is a vertex of T containing both v_1 and v_2.*
3. *If two vertices t_1 and t_2 in T both contain some vertex v of G, then all the vertices of T in the path between t_1 and t_2 also contain v.*

Definition 12. *Let G be a graph. The* width *of a given tree decomposition of G is the number of vertices of G in the largest vertex of this tree decomposition, minus one. The* treewidth *of G is the lowest width of all possible tree decompositions of G.*

To illustrate the validity of our measure on one very specific example, consider the class $\mathcal{C}_{\text{tree}}$ composed of the satisfiable binary CSP instances whose constraint graph is a tree, the class $\mathcal{C}_{\text{all}}$ composed of all satisfiable CSP instances of any arity and with any (primal) constraint graph, and the sub-solver described by Algorithm 1.

Algorithm 1 builds a solution to instances with a tree as a primal constraint graph by starting from a random root after establishing $(1,1)$-consistency and then following along the branches of the tree. It correctly solves the class $\mathcal{C}_{\text{tree}}$ [5], but not the class $\mathcal{C}_{\text{all}}$. Therefore, in order to be a valid measure of hardness for these two classes and Algorithm 1, backdoor completability needs to be high for all instances of $\mathcal{C}_{\text{tree}}$ and very low for at least some instances of $\mathcal{C}_{\text{all}}$. As we show in a generalized version of this example in Theorem 1, this is what indeed happens.

Note that we do not require backdoor completability to return a high value for *all* instances of a given hard CSP class. A CSP class does not need to have

Data: A satisfiable instance I with n variables $v_1, v_2, \ldots, v_n$.
Result: Either REJECT or SATISFIABLE.
Build primal constraint graph G of I;
if *G is a tree* **then**
  Sort the n vertices of G to get an ordering $v'_1, v'_2, \ldots, v'_n$ such that for each i
  there is at most one j such that $j < i$ and v'_i is connected to v'_j;
else
  return REJECT;
end
Establish $(1,1)$-consistency on I;
for $i \leftarrow 1$ **to** n **do**
  Assign to the variable v'_i the lowest value a_i left in $D_{v'_i}$ such that
  $\{\langle v'_1, a_1 \rangle, \langle v'_2, a_2 \rangle, \ldots, \langle v'_i, a_i \rangle\}$ is a partial solution;
  if *no such value exists* **then**
    return REJECT;
  end
end
return SATISFIABLE;

Algorithm 1. A simple sub-solver based on $(1,1)$-consistency.

all of its instances hard to be considered hard. The general CSP is hard for all solvers, even though most CSP instances are easy in practice.

Our Theorem covers the set of CSP classes $\{\mathcal{C}_1, \mathcal{C}_2, \ldots\}$, such that each $\mathcal{C}_p$ is the set of satisfiable CSP instances whose primal constraint graph treewidth is upper bounded by p. These classes are hierarchically ordered by inclusion: for each p, $\mathcal{C}_p \subset \mathcal{C}_{p+1}$. Therefore, any given sub-solver finds all classes up to some p easy, meaning that it returns a solution to all instances from these classes, and all larger classes hard, meaning that it rejects at least some instances from each one of these subsequent classes. Note that the union of all the $\mathcal{C}_p$ is equal to the NP-hard satisfiable CSP, so no sub-solver can find all the classes easy, unless $P = NP$.

We are going to prove that for any sub-solver A belonging to a specific set of algorithms based on local consistency, and for any aforementioned class $\mathcal{C}_p$, backdoor completability returns a very low value with regard to A for some of the instances in $\mathcal{C}_p$ if and only if $\mathcal{C}_p$ is a hard class for A. We define "very low" as exponentially inverse to the number of variables.

Theorem 1. *Let p and p' be such that $p, p' > 0$. Let $A_{p,p'}$ be the sub-solver described by Algorithm 2 and let $\mathcal{C}_{p'}$ be the set of satisfiable CSP instances whose primal constraint graph treewidth is upper bounded by p'. Then exactly one of the two following statements is true:*

- *$p \geq p'$ and for every instance $I \in \mathcal{C}_{p'}$, the backdoor completability for $A_{p,p'}$ of I is equal to 1.*
- *$p < p'$ and for every integer N there is an instance I in $\mathcal{C}_{p'}$ with n variables such that $n > N$ and the backdoor completability for $A_{p,p'}$ of I is equal to $O(\frac{1}{2^{n/p'}})$.*

Data: A satisfiable instance I with n variables $v_1, v_2, \ldots, v_n$.
Result: Either REJECT or SATISFIABLE.
Build primal constraint graph G of I;
if *(treewidth of G)$\leq p'$* **then**
   | Build a p'-wide tree decomposition T of G;
else
   | **return** REJECT;
end
Sort the n' vertices of T to get an ordering $v'_1, v'_2, \ldots, v'_{n'}$ such that for each i
there is at most one j such that $j < i$ and v'_i is connected to v'_j;
Establish strong $(p, 1)$-consistency on all sets of variables that are entirely
contained within a single vertex v_i;
for $i \leftarrow 1$ **to** n' **do**
   | **for** *each variable v_j in the vertex v'_i of T* **do**
   | | Assign to v_j the lowest value a_j left in D_{v_j} such that
   | | $\{\langle v_1, a_1 \rangle, \langle v_2, a_2 \rangle, \ldots, \langle v_j, a_j \rangle\}$ is a partial solution;
   | | **if** *no such value exists* **then**
   | | | **return** REJECT;
   | | **end**
   | **end**
end
return SATISFIABLE;

Algorithm 2. A sub-solver based on $(p, 1)$-consistency.

To prove the second point of the Theorem, we shall build instances with a low
enough backdoor completability for $A_{p,p'}$. Using binary constraints is enough to
do so, however we emphasize that instances in the class $\mathcal{C}_{p'}$ can be of any arity,
so the scope of our result is not restricted to binary instances.

Definition 13. *Let $N > 1$ and $p' > 1$ be two integers. Then we call $I_{N,p'}$ the
binary CSP instance defined in the following way:*

1. *Variables:* $\quad 1 \ + \ p'N \quad$ *variables* $\quad v_0, v_{1,1}, v_{1,2}, \ldots, v_{1,p'}, v_{2,1}, \ldots, v_{2,p'},$
   $v_{3,1}, \ldots, v_{N,p'}$.
2. *Domains: For each $1 \leq i \leq N$ and each $1 \leq j \leq p'$, $D_{v_{i,j}} = \{1, 2, \ldots, p'\}$.
   Furthermore, $D_{v_0} = \{1, 2, 3, \ldots, N + 1\}$.*
3. *Constraints: For each $1 \leq i \leq N$, for all $1 \leq i_1, i_2 \leq p'$ such that $i_1 \neq i_2$, for
   each $1 \leq a < p'$, the assignments $\langle v_{i,i_1}, a \rangle$ and $\langle v_{i,i_2}, a \rangle$ are incompatible.*
4. *Constraints: For each $1 \leq a \leq N$, for each $1 \leq i \leq p'$, the assignments $\langle v_0, a \rangle$
   and $\langle v_{a,i}, p' \rangle$ are incompatible.*
5. *Constraints: For each $1 \leq i \leq N$, for each $1 \leq j \leq p'$, for each $1 \leq a < p'$,
   the assignments $\langle v_0, N + 1 \rangle$ and $\langle v_{i,j}, a \rangle$ are incompatible.*
6. *Constraints: All pairs of assignments that have not been mentioned yet are
   compatible.*

*In addition, for each $1 \leq i \leq N$, we call V_i the set of variables $\{v_{i,1}, v_{i,2}, \ldots, v_{i,p'}\}$
and we call t_i the set of variables $\{v_0\} \cup V_i$.*

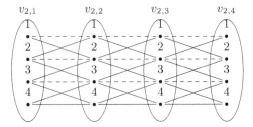

Fig. 1. The variables of V_2 in $I_{7,4}$, their domains and the related constraints.

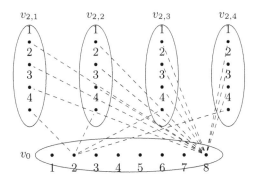

Fig. 2. The constraints between the variable v_0 and the set of variables V_2 in $I_{7,4}$.

Note that any non-universal constraint can only be between two variables of a same set t_i for some i. We give two figures to illustrate the constraints within the variables of a set t_i in an instance $I_{N,p'}$, in this case t_2 in the instance $I_{7,4}$. Figure 1 illustrates points 3 and 6 from Definition 13, while Fig. 2 illustrates points 4, 5 and 6. In both figures, a circle represents the domains of a variable, a dot represents a value in a domain and a dashed line connects two incompatible assignments. In Fig. 1, a continuous line connects two compatible assignments. In order to not clutter the figure, only the most representative pairs of (in)compatible assignments are connected. For the same reason, in Fig. 2 pairs of compatible assignments are not shown, and only pairs of incompatible assignments involving a value from the domain of v_0 are connected.

To simplify the proof of the theorem, we first give some preliminary results concerning the instances $I_{N,p'}$.

Lemma 1. *Let $N > 1$ and $p' > 1$ be two integers. Then the instance $I_{N,p'}$ has exactly one solution, consisting of the assignment $\langle v_0, N+1 \rangle$ and of the assignments $\langle v_{i,j}, p' \rangle$ for all $1 \le i \le N$ and $1 \le j \le p'$.*

Proof. It is easy to check that the set of assignments described in the statement of the Lemma is indeed a solution. We prove that it is the only one. Let S be a solution to I. let s_0 be the value such that $\langle v_0, s \rangle \in S$ and for all i and j such that $1 \leq i \leq N$ and $1 \leq j \leq p'$, let $s_{i,j}$ be the value such that $\langle v_{i,j}, s_{i,j} \rangle \in S$. For each $1 \leq i \leq N$, we know from Definition 13.3. that at least one of the $s_{i,1}, s_{i,2}, \ldots, s_{i,p'}$ is equal to p'. So from Definition 13.4 we know that for each $1 \leq i \leq N$, $s_0 \neq i$. So $s_0 = N + 1$. So from Definition 13.5 we know that $s_{i,j} = p'$ for all i and j such that $1 \leq i \leq N$ and $1 \leq j \leq p'$. So we have shown that the one and only solution to I is the one described in the Lemma. □

Lemma 2. *Let $N > 1$ and $p' > 1$ be two integers. Then the instance $I_{N,p'}$ belongs to $\mathcal{C}_{p'}$.*

Proof. $\mathcal{C}_{p'}$ is the set of satisfiable CSP instances whose primal constraint graph treewidth is upper bounded by p'. From Lemma 1, we know that $I_{N,p'}$ is satisfiable, so it only remains to prove that the treewidth of its constraint graph is upper bounded by p'.

Let T be a graph with N vertices $t_1, t_2, \ldots, t_N$ and $N - 1$ edges, such that:

- Each vertex t_i is as defined in Definition 13: the set of the $p' + 1$ variables $\{v_0, v_{i,1}, v_{i,2}, \ldots, v_{i,p'}\}$.
- For each $1 \leq i < N$, the pair (t_i, t_{i+1}) is an edge of T.

From the second point, T is a tree. Each variable $v_{i,j}$ of $I_{N,p'}$ is in the vertex t_i of T and v_0 is in every vertex of T, so the first condition in the definition of a tree decomposition (Definition 11) is fulfilled. Since each non-universal constraint of $I_{N,p'}$ either involves v_0 or is between two variables of a same set V_i, each edge in the constraint graph of $I_{N,p'}$ is contained in a vertex of T and the second condition of Definition 11 is fulfilled. Furthermore, the only variable of $I_{N,p'}$ that appears in several vertices of T is v_0, which appears in all vertices of T, so the third condition in Definition 11 is fulfilled. So T is a tree decomposition of the constraint graph of $I_{N,p'}$. Since each vertex of T contains $p' + 1$ variables of I, the treewidth of the constraint graph of $I_{N,p'}$ is (at most) p'. □

Lemma 3. *Let N, p and p' be three integers such that $N > 1$ and $0 < p < p'$. Let B be a set of variables of $I_{N,p'}$ such that $v_0 \notin B$. Then B is a backdoor for $A_{p,p'}$ of $I_{N,p'}$ if and only if B contains a variable from each set $V_i = \{v_{i,1}, v_{i,2}, \ldots, v_{i,p'}\}$, with at most one exception.*

Proof. – B is a backdoor for $A_{p,p'}$ of $I_{N,p'} \Rightarrow B$ contains a variable from each set V_i, with at most one exception:

We first show that for each $1 \leq i \leq N$, t_i is strongly $(p, 1)$-consistent. Let i be an integer such that $1 \leq i \leq N$ and let S be a partial solution of scope $W = \{w_1, w_2, \ldots, w_q\}$, with $q \leq p$ and $W \subset t_i$. Let v be a variable from $t_i \backslash W$. We need to show that there is a value $a \in D_v$ such that $S \cup \{\langle v, a \rangle\}$ is a partial solution. There are three cases to consider:

- v is v_0: from Definition 13, $\langle v_0, j \rangle$ is compatible with all assignments on variables from V_i if $i \neq j$. So $S \cup \{\langle v, 2 \rangle\}$ is a partial solution if $i = 1$ and $S \cup \{\langle v, 1 \rangle\}$ is a partial solution otherwise.

- One of the variables from W is v_0: let s_0 be such that $\langle v_0, s_0 \rangle \in S$. There are three possibilities for the value of s_0. First possibility, $s_0 = N + 1$. In this case, we know from Definition 13.5 that all the other assignments in S are of the form $\langle w_j, p' \rangle$, so $S \cup \{\langle v, p' \rangle\}$ is a partial solution. Second possibility, $s_0 = i$. In this case, we know from Definition 13.4 that none of the other assignments in S is of the form $\langle w_j, p' \rangle$, and $S \cup \{\langle v, a \rangle\}$ is a partial solution, with a a value such that $a \neq p'$ and $a \neq b$ for each assignment $\langle w_j, b \rangle \in S$ such that $w_j \neq v_0$. There is always such a value a, because W has at most p variables, so $W \backslash \{v_0\}$ has at most $p - 1 \leq p' - 2$ variables. Third and last possibility, either $s_0 < i$ or $i < s_0 < N + 1$. In this case, $S \cup \{\langle v, p' \rangle\}$ is a partial solution.
- Neither v nor any of the variables from W is v_0: we know from Definition 13 that $S \cup \{\langle v, p' \rangle\}$ is a partial solution.

So for each variable $v \in t_i \backslash W$, there is a value $a \in D_v$ such that $S \cup \{\langle v, a \rangle\}$ is a partial solution. So t_i is strongly $(p, 1)$-consistent.

Let B' be a set of variables of $I_{N,p'}$ such that $v_0 \notin B'$. Suppose that there are some i and j with $i \neq j$ such that no variable from $V_i \cup V_j$ is in B'. We have just shown that t_i (which we recall is $V_i \cup \{v_0\}$) and t_j (which is $V_j \cup \{v_0\}$) are strongly $(p, 1)$-consistent. So establishing strong $(p, 1)$-consistency after assigning the value p' to the variables in B' leaves at least the three values i, j (because $\langle v_0, i \rangle$ and $\langle v_0, j \rangle$ are compatible with all assignments $\langle v, p' \rangle$ for all variables v not in t_i nor t_j) and $N + 1$ (because $\langle v_0, N + 1 \rangle$ is compatible with all assignments $\langle v, p' \rangle$ for all variables v of $I_{N,p'}$) in D_{v_0}. However, the only assignments in the unique solution to I are of the form $\langle v, a \rangle$, with a the highest value in D_v, and $A_{p,p'}$ picks the lowest available value in each domain. So even after assigning the correct values to B' and establishing strong $(p, 1)$-consistency, and whatever the order in which the sub-solver sorts the variables, $A_{p,p'}$ will pick a wrong value when making its first choice within $t_1 \cup t_2$, and will eventually reject I. So B' is not a backdoor for $A_{p,p'}$ of $I_{N,p'}$. So any backdoor B for $A_{p,p'}$ of $I_{N,p'}$ not containing v_0 contains at least one variable in every set V_i, with at most one exception.

- B contains a variable from each set V_i, with at most one exception $\Rightarrow$ B is a backdoor for $A_{p,p'}$ of $I_{N,p'}$:

Let B be a set of $N - 1$ variables, each in a different set V_i, and none being v_0. Let i be the integer such that no variable from V_i is in B. Once we have assigned the value p' to all variables of B' and established strong $(p, 1)$-consistency, we know from Definition 13.4 that all the values $1, 2, 3, \ldots, i - 1, i + 1, \ldots, N - 1, N$ will be removed from D_{v_0}. Let $a \in D_v$ and $b \in D_w$ be two values from the domains of two different variables v and w of $t_i \backslash \{v_0\}$, such that $a < p'$ and $b = p'$. Since the only two remaining values in D_{v_0} are i and $N + 1$, we know from Definition 13.4 and 13.5 respectively that neither $\langle v_0, i \rangle$ nor $\langle v_0, N + 1 \rangle$ is compatible with both $\langle v, a \rangle$ and $\langle w, b \rangle$. So establishing strong $(p, 1)$-consistency will make incompatible all such pairs of assignments $\langle v, a \rangle$ and $\langle w, b \rangle$ with v and w in $t_i \backslash \{v_0\}$. From Definition 13.3, we know that there is no partial solution to V_i that contains an assignment $\langle v, a \rangle$

with $a \neq p'$. So establishing strong $(p, 1)$-consistency will make incompatible all pairs of assignments $\langle v, a \rangle$ and $\langle w, b \rangle$ with v and w from $t_i \backslash \{v_0\}$ and either $a \neq p'$ or $b \neq p'$, and will eventually remove all values other than p' from the domains of $t_i \backslash \{v_0\}$. So from Definition 13.4, the value i will also be removed from the domain of v_0, leaving only the value $N + 1$ in this domain. Lastly, from Definition 13.5, all values other than p' will be removed from all other domains, leaving only one value in each domain after establishing strong $(p, 1)$-consistency. So B is a backdoor for $A_{p,p'}$ of $I_{N,p'}$.

$\square$

Now that we have the results we need, we can prove the main theorem.

Proof of Theorem 1. Through the proof, we assume that "backdoor" and "backdoor completability" are implicitly "backdoor for $A_{p,p'}$" and "backdoor completability for $A_{p,p'}$" respectively.

– Suppose that $p \geq p'$. Since p' is a constant, building a tree decomposition of width p' is polynomial [2]. In general, it is well-known that using strong $(p, 1)$-consistency alongside a p-wide tree decomposition solves the CSP restricted to instances whose primal constraint graph has a treewidth bounded by p [6]. Therefore, the empty set is a minimal backdoor of I, and from Definition 9 the backdoor completability of I is 1.
– Suppose that $p < p'$. It is enough to show that for each integer $N > 1$, there is an instance $I \in C_{p'}$ with $n = p'N + 1$ variables such that the backdoor completability of I is equal to $O(\frac{1}{2^{n/p'}})$. Let $N > 0$ be an integer and let I be the instance $I_{N,p'}$ defined in Definition 13.
  • I is in $C_{p'}$: from Lemma 2.
  • I has n variables: from Definition 13.1.
  • The backdoor completability of I is equal to $O(\frac{1}{2^{n/p'}})$: let B be a minimal backdoor of I. There are two possibilities for B:
    * B contains v_0. If we assign the correct value $N + 1$ to the backdoor variable v_0, then we know from Definition 13.5 that after $A_{p,p'}$ establishes strong $(p, 1)$-consistency, only one value will remain in every other domain. So any set of variables containing v_0 is a backdoor of I, but among them only $\{v_0\}$ is a minimal backdoor of I. In this case, there are $N + 1$ partial solutions of scope B (one for each value in D_{v_0}), and exactly one of them is a subset of a solution ($\{\langle v_0, N+1 \rangle\}$). Therefore the completability ratio of B is $\frac{1}{N+1}$.
    * B does not contain v_0. From Lemma 3, we know that B contains at least one variable from each set $V_i = \{v_{i,1}, v_{i,2}, \ldots, v_{i,p'}\}$, with at most one exception. From the same Lemma we also know that containing one variable from each set V_i except one is a sufficient condition for a backdoor. So, since B is minimal, B contains exactly one variable from each set V_i except one. Note that all constraints between any two variables of B are universal and that every domain of B contains exactly p' values, so there are p'^{N-1} possible partial solutions of scope B. From Lemma 1, we know that there is only one solution to I, so

only one partial solution of scope B can be extended to a solution and therefore the completability ratio of B is $\frac{1}{p'^{N-1}}$.

We have shown that either $B = \{v_0\}$ and has a completability ratio of $\frac{1}{N+1}$ or B is composed of $N-1$ variables from $N-1$ different sets V_i and has a completability ratio of $\frac{1}{p'^{N-1}}$. There are Np'^{N-1} possible backdoors of the latter kind, so the backdoor completability of I is lower than $\frac{2}{p'^{N-1}}$. Recall than $n = p'N + 1$, so $N = \frac{n-1}{p'}$ and therefore $\frac{2}{p'^{N-1}} = O(\frac{2}{p'^{n/p'}})$. Since $p' > p \geq 1$, $O(\frac{2}{p'^{n/p'}}) = O(\frac{1}{2^{n/p'}})$ and therefore the backdoor completability of I is equal to $O(\frac{1}{2^{n/p'}})$.

We have exhibited a satisfiable instance $I \in \mathcal{C}_{p'}$ with n variables and a backdoor completability equal to $O(\frac{1}{2^{n/p'}})$.   □

We have formally proved that backdoor completability correctly measures hardness for some precisely defined classes of instances and sub-solvers, namely satisfiable CSP instances with bounded treewidth and sub-solvers based on local consistency. Tractable classes relying on bounded treewidth are common [5,8], while consistency is a ubiquitous tool in modern solvers [13], so our result shows that at least for some widely known algorithms and instance classes, backdoor completability is a valid measure.

4   Comparison with the Backdoor Key Fraction

We present an experimental comparison between the backdoor key fraction and the backdoor completability. Our experiments cover two different sets of problems: Quasigroup Completion With Holes (QWH) [1,12], and random satisfiable CSP. We used the Mistral [11] solver for both experiments with default settings.

For QWH, we generated 1100 instances of order 22, with a number of holes equally spread over the range 192 to 222. This range was chosen to capture the instances at hand around the observed peak of difficulty at 204 holes. Note that [17] also used QWH instances to test the backdoor key fraction.

The inequality constraints are posted through the AllDifferent constraint with the bound consistency algorithm of [14]. The sub-solver that we chose has exactly the same configuration, however, with 500 failure limit. Note that the sub-solver is guaranteed to run in polynomial time since constraint propagation is polynomial as well. For each instance, we generated 100 minimal backdoors, using the methodology described in [17]. For each backdoor, we randomly sampled 20 partial solutions.

Figure 3a represents the results of our experiments on QWH instances. The x-axis represents the number of decisions required to find a solution, while the left y-axis represents the backdoor key fraction and the right y-axis represents backdoor completability. Each point represents an average of 50 instances, with the 50 easiest instances in one group, the 50 next easiest ones in a second group, and so on until the 50 hardest instances.

The correlation for the backdoor key fraction is good, with a Pearson coefficient of .876. This is consistent with the results from [17]. The correlation for

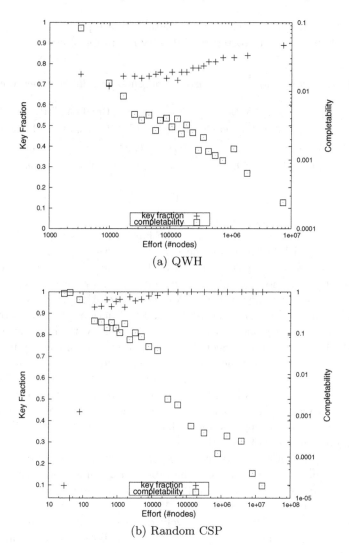

Fig. 3. Experimental results for QWH instances (top) and random CSP (bottom)

backdoor completability is even better, with a coefficient of $-.943$; we recall that the further away from 0 the coefficient is, the more correlated the values are. This demonstrates that backdoor completability can prove a better quantifier of instance difficulty than the backdoor key fraction, even in problems where the latter has a good track record.

The second problem we studied is composed of random satisfiable CSP instances with 60 variables and 1770 constraints. We generated 1200 such instances, with an average tightness in each constraint equally spread over the range 5% to 16%, with an observed peak of difficulty at 8%. The correlations for backdoor key fraction and backdoor completability are presented in Fig. 3b. As in Fig. 3a, the instances are sorted by difficulty in groups of 50.

Table 1. Correlations between instance hardness and different measures.

	QWH			Random CSP		
	Pearson CC	RMSE	MAE	Pearson CC	RMSE	MAE
Backdoor key fraction	.876	.053	.032	.590	.186	.158
Backdoor completability	−.943	.037	.029	−.975	.051	.044

We can see from these results that the backdoor key fraction is not adequate for that type of instance. It grows with the hardness, as is expected, but does not clearly distinguish between instances above a certain threshold of difficulty. Indeed, we observed that most of the hardest instances in this set have only a few variables that are not part of the backbone. In many cases all variables are in the backbone and the instance has exactly one solution; as explained in the remarks following Definition 7, the backdoor key fraction will always output 1 for instances with exactly one solution. This case (along with the other one mentioned in the last paragraph of Sect. 2) shows a limitation of the key fraction to capture hardness in some instances. On the other hand, backdoor completability does not study the variables separately, but examines the properties of the partial solutions to a whole backdoor. It is therefore more refined than the backdoor key fraction, in particular when looking at individual variables is not enough, for example when comparing instances that have a backbone of similar (large) size but different degrees of difficulty.

Table 1 contains the summary of our experiments. In addition to the Pearson correlation coefficients (Pearson CC), the table also includes the normalized values for the root mean square error (RMSE) and mean absolute error (MAE), two error measures for linear regression that [17] also reported. The results confirm that backdoor completability is negatively correlated with instance hardness, and that it measures the difficulty of instances in both sets more accurately than the backdoor key fraction does.

5   Conclusion

We have introduced a new measure, backdoor completability, that characterizes the hardness of an individual satisfiable CSP instance with regard to a given solver. Backdoor completability can be viewed as an index of hardness; the lower the value, the harder the instance. This measure is a crucial step towards the understanding of what makes a particular instance difficult.

We provided a theoretical justification of our measure. We proved that for some widespread classes of instances, namely CSP instances with bounded treewidth, backdoor completability captures exactly the limits of tractability.

We also presented an empirical comparison between our metric and the existing backdoor key fraction, and showed that for some kinds of CSP instances, backdoor completability is reliable even though the backdoor key fraction is not.

The main motivation of our work is to revisit and to improve the backdoor key fraction measure. In the future, it would be interesting to study the practical usefulness of completability as it provides insights for designing search strategies. Moreover, we believe that completability could eventually be useful in generating hard instances since hard instances are the ones with low completability.

Acknowledgements. This research has been funded by Science Foundation Ireland (SFI) under Grant Number SFI/12/RC/2289.

References

1. Achlioptas, D., Gomes, C.P., Kautz, H.A., Selman, B.: Generating satisfiable problem instances. In: Proceedings of AAAI, IAAI, Austin, Texas, USA, 30 July–3 August 2000, pp. 256–261 (2000)
2. Bodlaender, H.L.: A linear-time algorithm for finding tree-decompositions of small treewidth. SIAM J. Comput. **25**(6), 1305–1317 (1996)
3. Dechter, R.: Constraint Processing. Elsevier/Morgan Kaufmann, New York City/Burlington (2003)
4. Escamocher, G., O'Sullivan, B.: On the minimal constraint satisfaction problem: complexity and generation. In: Lu, Z., Kim, D., Wu, W., Li, W., Du, D.-Z. (eds.) COCOA 2015. LNCS, vol. 9486, pp. 731–745. Springer, Cham (2015). https://doi. org/10.1007/978-3-319-26626-8_54
5. Freuder, E.C.: A sufficient condition for backtrack-free search. J. ACM **29**(1), 24–32 (1982)
6. Freuder, E.C.: Complexity of K-tree structured constraint satisfaction problems. In: Proceedings of AAAI, Boston, Massachusetts, 29 July–3 August 1990, vol. 2, pp. 4–9 (1990)
7. Freuder, E.C.: Completable representations of constraint satisfaction problems. In: Proceedings of KR, Cambridge, MA, USA, 22–25 April 1991, pp. 186–195 (1991)
8. Ganian, R., Ramanujan, M.S., Szeider, S.: Combining treewidth and backdoors for CSP. In: 34th Symposium on Theoretical Aspects of Computer Science, STACS 2017, Hannover, Germany, 8–11 March 2017, pp. 36:1–36:17 (2017)
9. Gent, I.P., MacIntyre, E., Prosser, P., Walsh, T.: The constrainedness of search. In: Proceedings of AAAI, IAAI, Portland, Oregon, 4–8 August 1996, vol. 1, pp. 246–252 (1996)
10. Gomes, C.P., Fernández, C., Selman, B., Bessière, C.: Statistical regimes across constrainedness regions. Constraints **10**(4), 317–337 (2005)
11. Hebrard, E.: Mistral, a constraint satisfaction library. In: Proceedings of the Third International CSP Solver Competition, vol. 3, p. 3 (2008)
12. Kautz, H.A., Ruan, Y., Achlioptas, D., Gomes, C.P., Selman, B., Stickel, M.E.: Balance and filtering in structured satisfiable problems. In: Proceedings of IJCAI, Seattle, Washington, USA, 4–10 August 2001, pp. 351–358 (2001)
13. Larrosa, J., Schiex, T.: Solving weighted CSP by maintaining arc consistency. Artif. Intell. **159**(1–2), 1–26 (2004)

14. López-Ortiz, A., Quimper, C., Tromp, J., van Beek, P.: A fast and simple algorithm for bounds consistency of the alldifferent constraint. In: Proceedings of IJCAI, Acapulco, Mexico, 9–15 August 2003, pp. 245–250 (2003)
15. Monasson, R., Zecchina, R., Kirkpatrick, S., Selman, B., Troyansky, L.: Determining computational complexity from characteristic 'phase transitions'. Nature **400**(8), 133–137 (1999)
16. Montanari, U.: Networks of constraints: fundamental properties and applications to picture processing. Inf. Sci. **7**, 95–132 (1974)
17. Ruan, Y., Kautz, H.A., Horvitz, E.: The backdoor key: a path to understanding problem hardness. In: Proceedings of AAAI, IAAI, San Jose, California, USA, 25–29 July 2004, pp. 124–130 (2004)
18. Valiant, L.G., Vazirani, V.V.: NP is as easy as detecting unique solutions. Theoret. Comput. Sci. **47**(3), 85–93 (1986). https://doi.org/10.1016/0304-3975(86)90135-0
19. Williams, R., Gomes, C.P., Selman, B.: Backdoors to typical case complexity. In: Proceedings of IJCAI, Acapulco, Mexico, 9–15 August 2003, pp. 1173–1178 (2003)

Constrained-Based Differential Privacy: Releasing Optimal Power Flow Benchmarks Privately

Ferdinando Fioretto$^{(\boxtimes)}$ and Pascal Van Hentenryck

University of Michigan, Ann Arbor, MI, USA
{fioretto,pvanhent}@umich.edu

Abstract. This paper considers the problem of releasing optimal power flow benchmarks that maintain the privacy of customers (loads) using the notion of *Differential Privacy*. It is motivated by the observation that traditional differential-privacy mechanisms are not accurate enough: The added noise fundamentally changes the nature of the underlying optimization and often leads to test cases with no solution. To remedy this limitation, the paper introduces the framework of *Constraint-Based Differential Privacy (CBDP)* that leverages the post- processing immunity of differential privacy to improve the accuracy of traditional mechanisms. More precisely, CBDP solves an optimization problem to satisfies the problem-specific constraints by redistributing the noise. The paper shows that CBDP enjoys desirable theoretical properties and produces orders of magnitude improvements on the largest set of test cases available.

1 Introduction

In the last decades, scientific advances in artificial intelligence and operations research have been driven by competitions and collections of test cases. The MIPLIB library for mixed-integer programming and the constraint-programming, planning, and SAT competitions have significantly contributed to advancing the theoretical and experimental branches of the field. Recent years have also witnessed the emergence of powerful platforms (such as Kaggle [1]) to organize competitions between third-parties. Finally, the release of data sets may become increasingly significant in procurement where third-parties compete to demonstrate their capabilities. The desire to release data sets for scientific research, competitions, and procurement decisions is likely to accelerate. Indeed, with ubiquitous connectivity, many organizations are now collecting data at an unprecedented scale, often on large socio-technical systems such as energy networks. This data is often used as input to complex optimization problems.

The release of such rich data sets however raises some fundamental privacy concerns. For instance, the electrical load of an industrial customer in the power grid typically reflects its production and may reveal sensitive information on its economic strategy.

© Springer International Publishing AG, part of Springer Nature 2018
W.-J. van Hoeve (Ed.): CPAIOR 2018, LNCS 10848, pp. 215–231, 2018.
https://doi.org/10.1007/978-3-319-93031-2_15

Differential Privacy (DP) [5,6] is a general framework that addresses the sensitivity of such information and can be used to generate privacy-preserving data sets. It introduces carefully calibrated noise to the entries of a data set to prevent the disclosure of information related to those providing it. However, when these private data sets are used as inputs to complex optimization algorithms, they may produce results that are fundamentally different from those obtained on the original data set. For instance, the noise added by differential privacy may make the optimization problem infeasible or much easier to solve. As a result, despite its strong theoretical foundations, adoptions of differential privacy in industry and government have been rare. Large-scale practical deployments of differential privacy have been limited to big-data owners such as Google [7] and Apple [10]. In their applications, however, only internal users can access the private data by evaluating a pre-defined set of *queries*, e.g., the count of individuals satisfying certain criteria. This is because constructing a private version of the database is equivalent to simultaneously answering all possible queries and thus requires a large amount of noise.

This paper is motivated by the desire of releasing *Optimal Power Flow* (OPF) benchmarks that maintain the privacy of customers loads and the observation that traditional differential-privacy mechanisms are not accurate enough: The added noise fundamentally changes the nature of the underlying optimization and often leads to test cases with no solution. The paper proposes the framework of *Constraint-Based Differential Privacy* (CBDP) that leverages the post-processing immunity of DP to redistribute the noise introduced by a standard DP-mechanism, so that the private data set preserves the salient features of the original data set. More precisely, CBDP solves an optimization problem that minimizes the distance between the post-processed and original data, while satisfying constraints that capture the essence of the optimization application.

The paper shows that the CBDP has strong theoretical properties: It achieves ϵ-differential privacy, ensures that the released data set can produce feasible solutions for the optimization problem of interest, and is a constant factor away from optimality. Finally, experimental results show that the CBDP mechanism can be adopted to generate private OPF test cases: On the largest collection of OPF test cases available, it improves the accuracy of existing approaches of at least one order of magnitude and results in solutions with similar optimality gaps to those obtained on the original problems.

2   Differential Privacy

A *data set* D is a multi-set of elements in the *data universe* $\mathscr{U}$. The set of every possible data set is denoted $\mathscr{D}$. Unless stated otherwise, $\mathscr{U}$ is a cross product of multiple *attributes* $U_1, \ldots, U_n$ and has *dimension* n. For example, $\mathscr{U} = \mathbb{R}^n$ is the numeric data universe consisting of n-dimensional real-vectors. A *numeric query* is a function that maps a data set to a result in $\mathscr{R} \subseteq \mathbb{R}^r$.

Two data sets $D_1, D_2 \in \mathscr{D}$ are called *neighbors* (written $D_1 \sim D_2$) if D_1 and D_2 differ by at most one element, i.e., $|(D_1 - D_2) \cup (D_2 - D_1)| = 1$.

Definition 1 (Differential Privacy [5]**).** *A randomized mechanism* $\mathcal{M} : \mathscr{D} \to \mathscr{R}$ *with domain $\mathscr{D}$ and range $\mathscr{R}$ is ϵ-differentially private if, for any event $\mathcal{S} \subseteq \mathscr{R}$ and any pair $D_1, D_2 \in \mathscr{D}$ of neighboring data sets:*

$$Pr[\mathcal{M}(D_1) \in \mathcal{S}] \leq \exp(\epsilon) Pr[\mathcal{M}(D_2) \in \mathcal{S}], \tag{1}$$

where the probability is calculated over the coin tosses of $\mathcal{M}$.

A Differential Privacy (DP) mechanism maps a data set to distributions over the output set. The released DP output is a single random sample drawn from these distributions. The level of privacy is controlled by the parameter $\epsilon \geq 0$, called the *privacy budget*, with values close to 0 denoting strong privacy.

DP satisfies several important properties. Composability ensures that a combination of differentially private mechanisms preserve DP [6].

Theorem 1 (Composition). *Let $M_i : \mathscr{D} \to \mathscr{R}_i$ be an ϵ_i-differentially private mechanism for $i = 1, \ldots, k$. Then, their composition, defined as $\mathcal{M}(D) = (\mathcal{M}_i(D), \ldots, \mathcal{M}_k(D))$, is $(\sum_{i=1}^{k} \epsilon_i)$-differentially private.*

Post-processing immunity ensures that privacy guarantees are preserved by arbitrary post-processing steps [6].

Theorem 2 (Post-processing Immunity). *Let $\mathcal{M} : \mathscr{D} \to \mathscr{R}$ be a mechanism that is ϵ-differentially private and $g : \mathscr{R} \to \mathscr{R}'$ be an (arbitrary) mapping. The mechanism $g \circ \mathcal{M}$ is ϵ-differentially private.*

The Laplace Distribution with 0 mean and scale b has a probability density function $\text{Lap}(x|b) = \frac{1}{2b} e^{-\frac{|x|}{b}}$. The *sensitivity* of a query Q, denoted by Δ_Q, is defined as $\Delta_Q = \max_{D_1 \sim D_2} \|Q(D_1) - Q(D_2)\|_1$. The following theorem gives a differentially private mechanism for answering numeric queries [5].

Theorem 3 (Laplace Mechanism). *Let $Q : \mathscr{D} \to \mathscr{R}$ be a numerical query. The Laplace mechanism, defined as $\mathcal{M}_{Lap}(D; Q, \epsilon) = Q(D) + z$ where $z \in \mathscr{R}$ is a vector of i.i.d. samples drawn from $\text{Lap}(\frac{\Delta_Q}{\epsilon})$ achieves ϵ-differential privacy.*

The Laplace mechanism is a particularly useful building block for DP [6]. Koufogiannis et al. [16] proved its optimality by showing that it minimizes the mean-squared error for both the L_1 and L_2 norms among all private mechanisms that use additive and input-independent noise. In the following, $\text{Lap}(\lambda)^n$ denotes the i.i.d. Laplace distribution over n dimensions with parameter λ.

Lipschitz Privacy. The concept of Lipschitz privacy is appropriate when a data owner desire to protect individual quantities rather than individual participation in a data set. Application domains where Lipschitz privacy has been successful include location-based systems [3,20] and epigenetics [4].

Definition 2 (Lipschitz privacy [16]**).** *Let $(\mathscr{D}, d)$ be a metric space and $\mathcal{S}$ be the set of all possible responses. For $\epsilon > 0$, a randomized mechanism $\mathcal{M} : \mathscr{D} \to \mathscr{R}$ is ϵ-Lipschitz differentially private if:*

$$Pr[\mathcal{M}(D_1) \in \mathcal{S}] \leq \exp(\epsilon d(D_1, D_2) Pr[\mathcal{M}(D_2) \in \mathcal{S}],$$

for any $\mathcal{S} \subseteq \mathscr{R}$ and any two inputs $D_1, D_2 \in \mathscr{D}$.

Table 1. Power network nomenclature.

N	The set of nodes in the network	θ^{Δ}	Phase angle difference limits
E	The set of *from* edges in the network	$S^d = p^d + iq^d$	AC power demand
E^R	The set of *to* edges in the network	$S^g = p^g + iq^g$	AC power generation
i	imaginary number constant	c_0, c_1, c_2	Generation cost coefficients
I	AC current	$\Re(\cdot)$	Real component of a complex number
$S = p + iq$	AC power	$\Im(\cdot)$	Imaginary component of a complex number
$V = v\angle\theta$	AC voltage	$(\cdot)^*$	Conjugate of a complex number
$Y = g + ib$	Line admittance	$\lvert \cdot \rvert$	Magnitude of a complex number
$W = w^R + iw_I$	Product of two AC voltages	$\angle$	Angle of a complex number
s^u	Line apparent power thermal limit	x^l, x^u	Upper and lower bounds of x
θ_{ij}	Phase angle difference (i.e., $\theta_i - \theta_j$)	$\boldsymbol{x}$	A constant value

The Laplace mechanism with parameter ϵ achieve ϵ-Lipschitz DP.

3   Optimal Power Flow

Optimal Power Flow (OPF) is the problem of determining the best generator dispatch to meet the demands in a power network. A power network is composed of a variety of components such as buses, lines, generators, and loads. The network can be viewed as a graph (N, E) where the set of buses N represent the nodes and the set of lines E represent the edges. Note that E is a set of directed arcs and E^R is used to denote those arcs in E but in reverse direction. Table 1 reviews the symbols and notation adopted in this paper. Bold-faced symbols are used to denote constant values.

The AC Model. The AC power flow equations are based on complex quantities for current I, voltage V, admittance Y, and power S. The quantities are linked by constraints expressing Kirchhoff's Current Law (KCL), i.e., $I_i^g - I_i^d = \sum_{(i,j) \in E \cup E^R} I_{ij}$, Ohm's Law, i.e., $I_{ij} = \boldsymbol{Y}_{ij}(V_i - V_j)$,, and the definition of AC power, i.e., $S_{ij} = V_i I_{ij}^*$. Combining these three properties yields the AC Power Flow equations, i.e.,

---

Model 1. The AC Optimal Power Flow Problem (AC-OPF)

$$\textbf{variables:} \quad S_i^g, V_i \;\; \forall i \in N, \;\; S_{ij} \;\; \forall (i,j) \in E \cup E^R$$

$$\textbf{minimize:} \quad \sum_{i \in N} c_{2i}(\Re(S_i^g))^2 + c_{1i}\Re(S_i^g) + c_{0i} \tag{2}$$

$$\textbf{subject to:} \quad \angle V_r = 0, \;\; r \in N \tag{3}$$

$$v_i^l \le |V_i| \le v_i^u \;\; \forall i \in N \tag{4}$$

$$-\theta_{ij}^{\Delta} \le \angle(V_i V_j^*) \le \theta_{ij}^{\Delta} \;\; \forall (i,j) \in E \tag{5}$$

$$S_i^{gl} \le S_i^g \le S_i^{gu} \;\; \forall i \in N \tag{6}$$

$$|S_{ij}| \le s_{ij}^u \;\; \forall (i,j) \in E \cup E^R \tag{7}$$

$$S_i^g - S_i^d = \sum_{(i,j) \in E \cup E^R} S_{ij} \;\; \forall i \in N \tag{8}$$

$$S_{ij} = Y_{ij}^*|V_i|^2 - Y_{ij}^*V_iV_j^* \;\; \forall (i,j) \in E \cup E^R \tag{9}$$

$$S_i^g - S_i^d = \sum_{(i,j) \in E \cup E^R} S_{ij} \;\; \forall i \in N$$

$$S_{ij} = Y_{ij}^*|V_i|^2 - Y_{ij}^*V_iV_j^* \;\; (i,j) \in E \cup E^R$$

These non-convex nonlinear equations are a core building block in many power system applications. Practical applications typically include various operational constraints on the flow of power, which are captured in the AC OPF formulation in Model 1. The objective function (2) captures the cost of the generator dispatch. Constraint (3) sets the reference angle for some arbitrary $r \in N$, to eliminate numerical symmetries. Constraints (4) and (5) capture the voltage and phase angle difference operational constraints. Constraints (6) and (7) enforce the generator output and line flow limits. Finally, Constraints (8) capture KCL and constraints (9) capture Ohm's Law.

Notice that this is a non-convex nonlinear optimization problem and is NP-Hard [17,23]. Therefore, significant attention has been devoted to finding convex relaxations of Model 1.

The SOC Relaxation. The SOC relaxation [14] lifts the product of voltage variables $V_iV_j^*$ into a higher dimensional space (i.e., the W-space):

$$W_i = |V_i|^2 \;\; i \in N \tag{10a}$$

$$W_{ij} = V_iV_j^* \;\; \forall (i,j) \in E \tag{10b}$$

It takes the absolute square of each constraint (10b), refactors it, and relaxes the equality into an inequality:

$$|W_{ij}|^2 \le W_iW_j \;\; \forall (i,j) \in E \tag{11}$$

Constraint (11) is a second-order cone constraint, which is widely supported by industrial strength convex optimization tools (e.g., Gurobi [11], CPlex [13],

Model 2. The SOC Relaxation of AC-OPF (SOC-OPF)

variables: $S_i^g, W_i \ \forall i \in N, \ W_{ij} \ \forall (i,j) \in E, S_{ij} \ \forall (i,j) \in E \cup E^R$

minimize: (2)

subject to: (6), (7), (8)

$$(v_i^l)^2 \le W_i \le (v_i^u)^2 \ \forall i \in N \tag{12}$$

$$\tan(-\theta_{ij}^\Delta) \Re(W_{ij}) \le \Im(W_{ij}) \le \tan(\theta_{ij}^\Delta) \Re(W_{ij}) \ \forall (i,j) \in E \tag{13}$$

$$S_{ij} = \mathbf{Y}_{ij}^* W_i - \mathbf{Y}_{ij}^* W_{ij} \ (i,j) \in E \tag{14}$$

$$S_{ji} = \mathbf{Y}_{ij}^* W_j - \mathbf{Y}_{ij}^* W_{ij}^* \ (i,j) \in E \tag{15}$$

$$|W_{ij}|^2 \le W_i W_j \ \forall (i,j) \in E \tag{16}$$

Mosek [21]). The SOC relaxation of AC-OPF is presented in Model 2 SOC-OPF. The constraints for the generator output limits (6), line flow limits (7), and KCL (8), are identical to those in the AC-OPF model. Constraints (12) and (13) capture the voltage and phase angle difference operational constraints. Constraints (14) and (15) capture the line power flow in the W-space. Finally, constraints (16) strengthen the relaxation with second-order cone constraints for voltage products.

The Quadratic Convex (QC) Relaxation. The QC relaxation was introduced to preserve stronger links between the voltage variables [12]. It represents the voltages in polar form (i.e., $V = v \angle \theta$) and links these real variables to the W variables using the following equations:

$$W_{ii} = v_i^2 \ i \in N \tag{17a}$$

$$\Re(W_{ij}) = v_i v_j \cos(\theta_i - \theta_j) \ \forall (i,j) \in E \tag{17b}$$

$$\Im(W_{ij}) = v_i v_j \sin(\theta_i - \theta_j) \ \forall (i,j) \in E \tag{17c}$$

The QC relaxation relaxes these equations by taking tight convex envelopes of their nonlinear terms, exploiting the operational limits for $v_i, v_j, \theta_i - \theta_j$. In particular, it uses the convex envelopes for the square $\langle x^2 \rangle^T$ and product $\langle xy \rangle^M$ of variables, as defined in [19]. Under the assumption that the phase angle difference bound is within $-\pi/2 \le \theta_{ij}^l \le \theta_{ij}^u \le \pi/2$, relaxations for sine $\langle \sin(x) \rangle^S$ and cosine $\langle \cos(x) \rangle^C$ are given in reference [12]. Convex envelopes for Eqs. (17a)–(17c) can be obtained by composing the convex envelopes of the functions for square, sine, cosine, and the product of two variables, i.e.,

$$W_{ii} = \langle v_i^2 \rangle^T \ i \in N \tag{21a}$$

$$\Re(W_{ij}) = \langle \langle v_i v_j \rangle^M \langle \cos(\theta_i - \theta_j) \rangle^C \rangle^M \ \forall (i,j) \in E \tag{21b}$$

$$\Im(W_{ij}) = \langle \langle v_i v_j \rangle^M \langle \sin(\theta_i - \theta_j) \rangle^S \rangle^M \ \forall (i,j) \in E \tag{21c}$$

The QC relaxation also proposes to strengthen these convex envelopes with a second-order cone constraint from the SOC relaxation ((10a), (10b), (11)). The complete QC relaxation is presented in Model 3.

Model 3. The QC Relaxation of AC-OPF (QC-OPF)

variables: $S_i^g, V_i = v_i \angle \theta_i, \ \forall i \in N, \ W_{ij} \ \forall (i,j) \in E, \ S_{ij} \ \forall (i,j) \in E \cup E^R$

minimize: (2)

subject to: $(3)-(8),(14)-(16)$

$$W_{ii} = \langle v_i^2 \rangle^T \quad i \in N \tag{18}$$

$$\Re(W_{ij}) = \langle\langle v_i v_j \rangle^M \langle\cos(\theta_i - \theta_j)\rangle^C\rangle^M \quad \forall (i,j) \in E \tag{19}$$

$$\Im(W_{ij}) = \langle\langle v_i v_j \rangle^M \langle\sin(\theta_i - \theta_j)\rangle^S\rangle^M \quad \forall (i,j) \in E \tag{20}$$

Model 4. The DC Relaxation of the AC OPF (DC-OPF)

variables: $\Re(S_i^g), \theta_i \ \forall i \in N, \ S_{ij} \ \forall (i,j) \in E \cup E^R$

minimize: (2)

subject to: (3)

$$|\theta_i| \leq \boldsymbol{\theta}_i^u \quad \forall i \in N \tag{22}$$

$$\Re(\boldsymbol{S}_i^{g\,l}) \leq \Re(S_i^g) \leq \Re(\boldsymbol{S}_i^{g\,u}) \quad \forall i \in N \tag{23}$$

$$\Re(|S_{ij}|) \leq \Re(\boldsymbol{s}_{ij}^u) \quad \forall (i,j) \in E \cup E^R \tag{24}$$

$$\Re(S_{ij}) = -\boldsymbol{b}_{ij}(\theta_i - \theta_j) \quad \forall (i,j) \in E \cup E^R \tag{25}$$

$$\Re(S_i^g) - \Re(\boldsymbol{S}_i^d) = \sum_{(i,j)\in E \cup E^R} S_{ij} \quad \forall i \in N \tag{26}$$

The DC Model. The DC model is an extensively studied linear approximation to the AC power flow [24]. The DC load flow relates real power to voltage phase angle, ignores reactive power, and assumes voltages are close to their nominal values (1.0 in per unit notation). The DC OPF is presented in Model 4. Constraints (22) capture the phase angles operational constraints. Constraints (23) and (24) enforce the generator output and line flow limits. Constraints (25) captures the KCL and constraints (26) the Ohm's Law.

4 The Differential Privacy Challenge for OPF

When releasing private OPF test cases, it is not critical to hide user participation: The location of a load is public knowledge. However, the magnitude of a load is sensitive: It is associated with the activity of a particular customer (or group of customers) and may indirectly reveal production levels and hence strategic investments, decreases in sales, and other similar information. Indirectly, it may also reveal how transmission operators operate their networks, which should not be public information. As a result, the concept of *Lipschitz differential privacy* is particularly suited to the task.

As mentioned in Sect. 2, the Laplace mechanism can be used to achieve Lipschitz DP. However, its application on load profile queries results in a new output vector of loads which produces undesirable outcomes when used as input to an OPF problem. Indeed, Fig. 1 illustrates the average error (measured as the L_1 distance) between the original load and the private load for a set of 44 networks.[1] With a privacy budget of $\epsilon = 0.1$, the average error is about 10–implying a significantly higher load than the actual demand. The numbers reported on each bar represent the percentage of feasible private instances for the AC OPF problem: It reveals severe feasibility issues with the private instances.

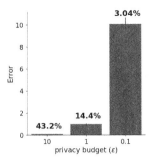

Fig. 1. Average L_1 error reported by the Laplace Mechanism. The percentages express the AC-OPF instances with satisfiable solution.

These results highlight the challenges that arise when traditional differential privacy is applied to inputs of complex optimization tasks. For instance, the Laplacian mechanism is oblivious to the structure of the data set (e.g., the generation capabilities should be large enough to serve the load) and the constraints and objectives of the optimization application (e.g., the transmission network should have the ability to transport electricity from generators to loads). As a result, it produces private data sets that are typically not useful and not representative of actual OPFs. *What is needed is a differential-privacy mechanism that preserves the structure of the optimization model and its computational properties such as the optimality gap between the solutions produced by MINLP solvers and convex relaxations.*

5 Constrained-Based Differential Privacy

This section introduces Constraint-Based Differential Privacy (CBDP) to remedy the limitations identified in the previous section. It considers an optimization problem $\mathcal{O}(D)$:

$$\text{minimize}_{\mathbf{x} \in \mathbb{R}^n} \quad f(D, \mathbf{x})$$
$$\text{subject to} \quad g_i(D, \mathbf{x}) \leq 0, \quad i = 1, \ldots, p$$

where $f : \mathscr{D} \times \mathbb{R}^n \to \mathbb{R}$ is the *objective function* to minimize over variables $\mathbf{x}$ and $g_i(D, x) \leq 0$ $(i = 1, \ldots, p)$ are the problem constraints.

This paper studies the following setting. The data owner desires to release a private data set $\hat{D}$ such that the optimization problems $\mathcal{O}(D)$ and $\mathcal{O}(\hat{D})$ are closely related. In particular, the optimal objective value of $\mathcal{O}(\hat{D})$ must be *close* to the optimal value of the original problem $f(D, \mathbf{x}^*)$ (which is a public information), where $\mathbf{x}^* \in \mathbb{R}^n$ is the optimal solution of the original optimization problem.

---

[1] The experimental settings are reported in all details in Sect. 7.

Hence the private data set must satisfy the following desiderata: (1) *data privacy*: The data set to be released must be private; (2) *faithfulness*: The private data must be faithful to the objective function; (3) *consistency*: The private data must satisfy the constraints arising from the data and/or from the problem of interest. To address such challenges the following definition is introduced.

Definition 3 ((ϵ, β)-CBDP). *Given $\epsilon > 0, \beta \geq 0$, a DP-data-release mechanism $\mathcal{M} : \mathscr{D} \rightarrow \mathscr{D}$ is (ϵ, β)-CBDP iff, for each private database $\hat{D} = \mathcal{M}(D)$, there exists a solution $\mathbf{x}$ such that*

1. *ϵ-privacy: $\mathcal{M}$ satisfies ϵ-DP;*
2. *β-faithfulness: $|f(\hat{D}, \mathbf{x}) - f(D, \mathbf{x}^*)| \leq \beta$;*
3. *Consistency: Constraints $g_i(\hat{D}, \mathbf{x}) \leq 0$ ($i = 1, \dots, p$) are satisfied.*

The Input. To balance between utility and privacy, the mechanism takes as input the data set D, as well as two non-negative real numbers: ϵ which determines the *privacy value* of the private data and β which determines the required *faithfulness* of the optimization problem over the private data. Additionally, the data owner provides the optimization problem and the optimal objective value $f^* = f(D, \mathbf{x}^*)$, which are typically considered public information in competitions. For simplicity, this section assumes that $\mathscr{D} = \mathbb{R}^n$.

$$\text{minimize}_{\hat{D}, \mathbf{x} \in \mathbb{R}^n} \|\hat{D} - \tilde{D}\|_2^2 \tag{O1}$$

$$\text{subject to} \quad |f(\hat{D}, \mathbf{x}) - f^*| \leqslant \beta \tag{O2}$$

$$g_i(\hat{D}, \mathbf{x}) \leqslant 0, i = 1, \dots, p \tag{O3}$$

Fig. 2. The CBDP post-processing step.

The Mechanism. The CBDP mechanism first injects Laplace noise with privacy parameter ϵ to each query on each dimension of the data set:

$$\mathcal{M}_{\text{Lap}}(D, Q, \epsilon) = \tilde{D} = D + \text{Lap}(1/\epsilon)^n,$$

where $\tilde{D} = (\tilde{c}_1, \dots, \tilde{c}_n)$ is the vector of noisy values. These values are then post-processed by the optimization algorithm specified in Fig. 2 to obtain a value vector $\hat{D} = (\hat{x}_1, \dots, \hat{x}_n) \in \mathbb{R}^n$. Finally, the CBDP mechanism outputs $\hat{D}$.

The CBDP mechanism thus solves a constrained optimization problem whose decision variables include vectors of the form $\hat{D} = (\hat{x}_1, \dots, \hat{x}_n)$ that correspond to the post-processed result of the private query on each dimension of the universe. In other words, each original data (that must remain private) is replaced by a decision variable representing its post-processed private counterpart. The objective (O1) minimizes the L_2-norm between the private query result $\tilde{D}$ and its post-processed version $\hat{D}$. Constraint (O2) forces the post-processed values to be β-faithful with the respect to the objective value, and Constraints (O3) enforce the optimization constraints of the original model.

Additional constraints capturing auxiliary public information about the data can be integrated in this model. For instance, in the OPF problem, the total power load is public. Thus, an additional constraint on the sum of values of the variables $(\hat{x}_1, \ldots \hat{x}_n)$ can be enforced to be equal to this public information.

The CBDP post-processing can be thought as redistributing the noise of the Laplace mechanism to obtain a data set which is consistent with the problem constraints and objective. It searches for a feasible solution that satisfies the problem constraints $g_i(\hat{D}, \mathbf{x}) \leq 0$ and the β-faithfulness constraint. *A feasible solution always exists, since the original values D trivially satisfy all constraints.*

It is important to notice that the post-processing step of CBDP uses exclusively the private data set $\tilde{D}$ and additional public information (i.e., the optimization problem and its optimal solution value). Its privacy guarantees are discussed below.

5.1   Theoretical Properties

Theorem 4. *The mechanism above is (ϵ, β)-CBDP.*

Proof. Each $\tilde{c}_i$ obtained from the Laplace mechanism is ϵ-differentially-private by Theorem 3. The combination of these results $(\tilde{c}_1, \ldots, \tilde{c}_n)$ is ϵ-differentially-private by Theorem 1. The β-faithfulness and the consistency properties is satisfied by constraint (O2) and (O3) respectively. The result follows from post-processing immunity (Theorem 2).

As mentioned earlier, additional constraints can be enforced, e.g., to ensure the consistency that the sums of individual quantities equals their associated aggregated quantity. In this case, the aggregated quantities must return private counts and a portion of the privacy budget must be used to answer such queries.

Theorem 5. *The optimal solution $\langle \hat{D}^+, \mathbf{x}^+ \rangle$ to the optimization model (O1–O3) satisfies $\|\hat{D}^+ - D\|_2 \leq 2\|\tilde{D} - D\|_2$.*

Proof. We have

$$\|\hat{D}^+ - D\|_2 \leq \|\hat{D}^+ - \tilde{D}\|_2 + \|\tilde{D} - D\|_2 \tag{27}$$

$$\leq 2\|\tilde{D} - D\|_2. \tag{28}$$

where the first inequality follows from the triangle inequality on norms and the second inequality follows from

$$\|\hat{D}^+ - \tilde{D}\|_2 \leq \|\tilde{D} - D\|_2$$

by optimality of $\langle \hat{D}^+, \mathbf{x}^+ \rangle$ and the fact that $\langle D, \mathbf{x}^* \rangle$ is a feasible solution to constraints (O2) and (O3).

The following result follows from the optimality of the Laplace mechanism [16].

Corollary 1. *The CBDP mechanism is at most a factor 2 away from optimality.*

Model 5. The CBDP mechanism for the AC-OPF

$$\text{variables: } S_i^g, V_i, \dot{S}_i^l \ \forall i \in N, \ S_{ij} \ \forall (i,j) \in E \cup E^R$$

$$\text{minimize: } \|\dot{S}^l - \tilde{S}^l\|_2^2 \tag{s_1^l}$$

$$\text{subject to: } (3)-(9)$$

$$\left| \sum_{i \in N} c_{2i}(\Re(S_i^g))^2 + c_{1i}\Re(S_i^g) + c_{0i} - f^* \right| \leq \beta \tag{s_2}$$

$$\sum_{i \in N} \dot{S}_i^l = L \tag{s_3}$$

6 Application to the Optimal Power Flow

The CBDP optimization model for the AC-OPF is p resented in Model 5. In addition to the variables of Model 5, it takes as inputs the variables $\dot{S}_i^l$ representing the post-processed values of the loads for each bus in $i \in N$. The optimization model minimizes the L_2-norm between the variables $\dot{S}^l \in \mathbb{R}^n$ and the noisy loads $\tilde{S}^l \in \mathbb{R}^n$ resulting from the application of the Laplace mechanism to the original load values. Model 5 is subject to the same constraint of Model 1, with the addition of the β-faithfulness constraint (s_2) and the constraint enforcing consistency of the aggregated load values $L \in \mathbb{R}$ (s_3), which is typically public knowledge.

7 Experimental Results

This section presents an evaluation of the CBDP mechanism on the case study. It first presents the experimental setup and then compares the CBDP mechanism with the Laplace mechanism.

Data Sets and Experimental Setup. The experimental results concern the *NESTA* power network test cases (https://gdg.engin.umich.edu). The test cases comprise 44 networks whose number of buses ranges from 3 to 9241. This section categorizes them in *small* (networks with up to 100 buses), *medium* (networks with more than 100 buses and up to 2000 buses) and *large* (networks with more than 2000 buses).

In the following, D denotes the original data set and $\tilde{D}$ its private version (i.e., the data set resulting through the application of a DP mechanism). Moreover, $\mathbf{c}_m(D)$ and $\mathbf{c}_m(\tilde{D})$ denote the cost of the dispatch obtained by an OPF given model m (i.e., AC, QC, SOC, or DC) on the original data set D and on its private version $\tilde{D}$, respectively.

The results obtained by the AC-OPF and its relaxations/approximations (QC, SOC, and DC) are evaluated using both the original and the private data sets, analyzing the dispatch cost (c) and the optimality gap, i.e., the ratio $G_R(D) = \frac{|c_R(D) - c_{AC}(D)|}{c_{AC}(D)}$, where $c_{AC}(D)$ and $c_R(D)$ denote the best-known solution cost of the problem instance and of the relaxation R over data set D.

The baseline $\mathcal{M}_{\mathrm{Lap}}$ is the Laplace mechanism applied to each load of the network. To obtain a private version of the loads, $\mathcal{M}_{\mathrm{Lap}}$ is first used to construct a private value for the active loads $p_i^l = \Re(S_i^l)$ as $\tilde{p}_i^l = p_i^l + \mathrm{Lap}(100/\epsilon)$, ($\forall i \in N$) where 100 is the change in MWs protected by Lipschitz DP. The reactive load $q_i^l = \Im(S_i^l)$ is set as $\tilde{q}_i^l = \tilde{p}_i^l \, r_i$, with $r_i = q_i^l/p_i^l$ is the power load factor and is considered to be public knowledge, as is natural in power systems. The CBDP mechanism $\mathcal{M}_C$ uses the output of the Laplace mechanism and the post-processing step to obtain the private loads $\dot{S}_i^l$ for all $i \in N$.

The mechanisms are evaluated for privacy budgets $\epsilon \in \{0.1, 1.0, 10.0\}$ and faithfulness parameter $\beta \in \{0.01, 1.0, 100.0\}$. Smaller values for ϵ increase privacy guarantees at the expense of more noise introduced by the Laplace mechanism. All experimental results are reported as the average of 30 runs. This gives a total of 47,520 experiments, which are analyzed below.

Error Analysis on the OPF Cost. The first experimental result measure the error introduced by a mechanism as the average distance, in percentage, between the exact and the private costs of the OPF dispatches. The reported error is expressed as $\frac{|\mathbf{c}_m(D) - \mathbf{c}_m(\tilde{D})|}{\mathbf{c}_m(D)} \cdot 100$, for $m = \{AC, QC, SOC, DC\}$.

Figure 3 illustrates the error of the private mechanisms for varying privacy budgets. Rows show the results for the different network sizes (small, medium, and large), while columns show the results for the different faithfulness level values ($\beta = 0.01, 1.0, 100.0$). The results are shown in log scale. Each sub-figure also presents the results for three privacy budgets $(10, 1, 0.1)$.

For all privacy budgets and all faithfulness-levels, the CBDC mechanism outperforms the Laplace mechanism by one to two orders of magnitude. For every OPF model, the Laplace mechanism produces OPF values which are, in general, more than 10% away (and exceeding 100% in many instances) from the values reported by the model ran on the original data, with the exception for the largest privacy budget, whose results produces differences slightly below 10%. In contrast, CBDP produces OPF values close (within 10%) to those produced on the original data, with all the OPF models adopted.

The errors reported by the Laplace mechanism on small networks are larger than those reported on medium and large networks. This is due to the fact that the dispatch costs of larger networks are typically much higher than those of small networks, and thus the relative distance of the error accumulated by the DP mechanism is more pronounced for the smaller test cases. Despite this, the CBDP mechanism produces solutions with small error costs, even for the small network instances.

The CBDP mechanism preserves the objectives of the AC OPF problems accurately (within 1%), demonstrating its benefits for small beta values. The OPF value differences increase as the faithfulness parameter β increases.

Analysis of the Optimality Gap. In a competition setting, it is also critical to preserve the computational difficulty of the original test case. The next set of results show how well CBDP preserves the optimality gap of the instance and its relationship to well-known approximations such as the DC model.

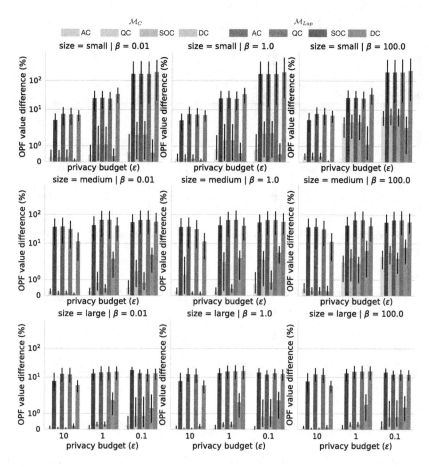

Fig. 3. Average OPF objective differences for the CBDP mechanism (light colors) and the Laplace mechanism (dark colors). (Color figure online)

Figure 4 shows the differences $G_R(\tilde{D}) - G_R(D)$, for the AC OPF relaxation/approximation models $R = \{QC, SOC, DC\}$, where the private data set $\tilde{D}$ is produced by the CBDP mechanism. It compares four CBDP post-processing models: $M5$, which solves the CBDP of Model 5; $M5_{+g}$, which extends the Model 5 by modifying the objective (s_1^l) by adding the terms $\|S^g - \boldsymbol{S}^g\|_2^2$ to minimize the distance from the generator setpoints; $M5_{-\beta}$, which solves the Model 5 without the beta faithfulness constraint (s_2); and $M5_{+g,-\beta}$, which excludes constraint (s_2) but includes the terms $\|S^g - \boldsymbol{S}^g\|_2^2$ into its objective. Rows show the results for the different privacy faithfulness levels $(\beta = 0.01, 1.0, 100.0)$, while columns show the results for the different privacy budgets $(10, 1, 0.1)$. The results are shown in log scale.

For all settings, $M5$ produces instances whose optimality gaps are close to those of the original ones (their distance is <1 for $\epsilon = 10.0$, and 1.0, and <3 for

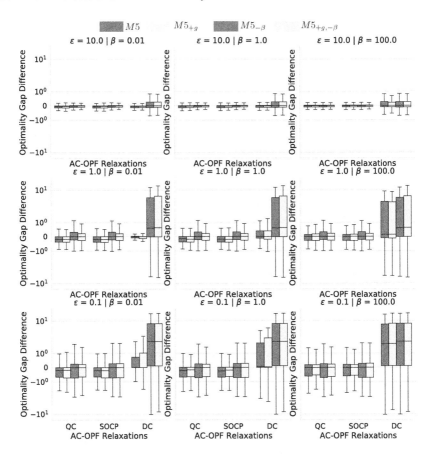

Fig. 4. Optimality GAP error for the QC, SOC, and DC, relaxations of the AC-OPF, on different CBDP post-processing models.

$\epsilon = 0.1$). $M5_{+g}$ produces very similar optimality gaps, showing that the CBDP mechanism does not need to take into account the generator setpoint values. For the relaxations, both $M5_{-\beta}$ and $M5_{+g,-\beta}$ produce quite similar results to $M5$. It is only for the DC-approximation that the faithfulness constraint is important. Moreover, note that the faithfulness constraint must be relatively tight even for $M5$ to preserve the result of the DC model. The DC model ignores many aspects of the power systems and hence it is not a surprise that it is more brittle. *These results indicate that the CBDP mechanism is capable of preserving the optimality gap of relaxations (and the quality of the DC approximation) with high fidelity.*

Analysis of the Private Network Loads. The last set of results reports the effect of the CBDP mechanism on the network load profiles. Figure 5 depicts the percentage of load increase when applying the CBDP mechanism on three example networks with 4-buses (left), 73 buses (center), and 300 buses (right) for various privacy budgets. The results illustrate that load variation is often

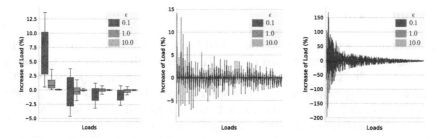

Fig. 5. Percentage of load increase in the 4-bus (left) 73-bus (center), and 300-bus (right) NESTA test cases.

significant for a portion of the loads, although the CBDP mechanism preserves the problem structure accurately. Moreover, some of the loads exhibit a positive or negative bias. A detailed examination of these test cases reveals that this is due to the underlying network characteristics. For example, in the 4-bus test case (Fig. 5(left)), the first load (load 1) tends to be higher than its original value. This is explained by the fact that such load resides on the same bus as the cheaper generator which also has a very high generation capacity [9] (pp. 337–338). As a result, the CBDP mechanism has significant flexibility to increase this load to redistribute the noise appropriately.

8   Related Work

There is rich literature on theoretical results of DP (see e.g., [6,22]). The literature on DP applied to energy systems includes considerably fewer efforts. Ács and Castelluccia [2] exploited a direct application of the Laplace mechanism to hide user participation in smart meter data sets, achieving ϵ-DP. Zhao et al. [25] studied a DP schema that exploits the ability of households to charge/discharge a battery to hide the real energy consumption of their appliances. Liao et al. [18] introduce Di-PriDA, a privacy-preserving mechanism for appliance-level peak-time load balancing control in the smart grid, aimed at masking the consumption of the top-k appliances of a household.

Karapetyan et al. [15] conduct an empirical study on quantifying the trade-off between privacy and utility in demand response systems. The authors analyze the effects of a simple Laplace mechanism on the objective value of the demand response optimization problem. Their experiments on a 4-bus micro-grid show drastic results: the optimality gap approaches nearly 90% in some cases.

A DP schema that uses constrained post-processing was recently introduced by Fioretto et al. [8] and adopted to release private mobility data. In contrast, the proposed CBDP schema proposed in this work releases the private data set through a mechanism that imposes constraints to ensure the problem solution cost is close to the solution cost of the original problem, and that the underlying optimal power flow constraints are satisfiable.

9   Conclusions

This paper introduced the Constraint-Based Differential Privacy (CBDP) mechanism, an approach to Differential Privacy (DP) which aims at releasing optimal power flow benchmarks that retain the privacy of the customers (loads). CBDP leverages the post-processing immunity of DP to cast the production of a private data set as an optimization problem that redistributes the noise introduced by a randomized mechanism to satisfy problem-specific constraints.

The proposed mechanism enjoys desirable theoretical properties: It achieves ϵ-DP, ensures that the released data set can produce feasible solutions for the optimization problem of interest, and is a constant factor away from optimality. CBDP has been evaluated on the largest collection of OPF test cases available. Experimental results show that CBDP improves the accuracy of traditional approaches (e.g., the Laplace mechanism) by orders of magnitude and preserves some salient computational features of the test cases, such as the optimality gap. These results are significant and indicate that CBDP has the potential to become an important tool to release data sets for competition settings.

Although the paper focused on the applicability of CBDP to OPF problems, the proposed mechanism is general and can be used for other applications where a private data set is the input to a complex optimization problem.

Acknowledgments. The authors would like to thank the anonymous reviewers for their valuable comments. This research is partly funded by the ARPA-E Grid Data Program under Grant 1357-1530. The views and conclusions contained in this document are those of the authors only.

References

1. Kaggle: Your home for data science. https://www.kaggle.com
2. Ács, G., Castelluccia, C.: I have a DREAM! (DiffeRentially privatE smArt Metering). In: Filler, T., Pevný, T., Craver, S., Ker, A. (eds.) IH 2011. LNCS, vol. 6958, pp. 118–132. Springer, Heidelberg (2011). https://doi.org/10.1007/978-3-642-24178-9_9
3. Andrés, M.E., Bordenabe, N.E., Chatzikokolakis, K., Palamidessi, C.: Geo-indistinguishability: differential privacy for location-based systems. In: Proceedings of the 2013 ACM SIGSAC Conference on Computer & Communications Security, pp. 901–914. ACM (2013)
4. Backes, M., Berrang, P., Hecksteden, A., Humbert, M., Keller, A., Meyer, T.: Privacy in epigenetics: temporal linkability of MicroRNA expression profiles. In: USENIX Security Symposium, pp. 1223–1240 (2016)
5. Dwork, C., McSherry, F., Nissim, K., Smith, A.: Calibrating noise to sensitivity in private data analysis. In: Halevi, S., Rabin, T. (eds.) TCC 2006. LNCS, vol. 3876, pp. 265–284. Springer, Heidelberg (2006). https://doi.org/10.1007/11681878_14
6. Dwork, C., Roth, A.: The algorithmic foundations of differential privacy. Theor. Comput. Sci. **9**(3–4), 211–407 (2013)
7. Fanti, G., Pihur, V., Erlingsson, Ú.: Building a rappor with the unknown: privacy-preserving learning of associations and data dictionaries. Proc. Priv. Enhancing Technol. **2016**(3), 41–61 (2016)

8. Fioretto, F., Lee, C., Van Hentenryck, P.: Constrained-based differential privacy for private mobility. In: Proceedings of the International Joint Conference on Autonomous Agents and Multiagent Systems (AAMAS) (2018)
9. Grainger, J.J.S., Grainger, W.D.J.J., Stevenson, W.D.: Power System Analysis. McGraw-Hill Education, New York City (1994)
10. Greenberg, A.: Apple's 'differential privacy' is about collecting your data—but not your data, 13 June 2016. https://www.wired.com/2016/06/apples-differential-privacy-collecting-data/. Accessed 21 Sept 2016
11. Gurobi. Gurobi software. http://www.gurobi.com/
12. Hijazi, H., Coffrin, C., Van Hentenryck, P.: Convex quadratic relaxations of nonlinear programs in power systems. Math. Program. Comput. $32(5)$, 3549–3558 (2017)
13. IBM. ILOG CPLEX software. http://www.ibm.com/
14. Jabr, R.: Radial distribution load flow using conic programming. IEEE Trans. Power Syst. $21(3)$, 1458–1459 (2006)
15. Karapetyan, A., Azman, S.K., Aung, Z.: Assessing the privacy cost in centralized event-based demand response for microgrids. CoRR, abs/1703.02382 (2017)
16. Koufogiannis, F., Han, S., Pappas, G.J.: Optimality of the Laplace mechanism in differential privacy. arXiv preprint arXiv:1504.00065 (2015)
17. Lehmann, K., Grastien, A., Van Hentenryck, P.: AC-feasibility on tree networks is NP-hard. IEEE Trans. Power Syst. 99, 1–4 (2015)
18. Liao, X., Srinivasan, P., Formby, D., Beyah, A.R.: Di-PriDA: differentially private distributed load balancing control for the smart grid. IEEE Trans. Dependable Secure Comput. (2017). https://doi.org/10.1109/TDSC.2017.2717826
19. McCormick, G.: Computability of global solutions to factorable nonconvex programs: part i - convex underestimating problems. Math. Program. 10, 146–175 (1976)
20. Mir, D.J., Isaacman, S., Cáceres, R., Martonosi, M., Wright, R.N.: DP-WHERE: differentially private modeling of human mobility. In: 2013 IEEE International Conference on Big Data, pp. 580–588. IEEE (2013)
21. MOSEK ApS. The MOSEK optimization toolbox (2015)
22. Vadhan, S.: The complexity of differential privacy. Tutorials on the Foundations of Cryptography. ISC, pp. 347–450. Springer, Cham (2017). https://doi.org/10.1007/978-3-319-57048-8_7
23. Verma, A.: Power grid security analysis: an optimization approach. Ph.D. thesis, Columbia University (2009)
24. Wood, A.J., Wollenberg, B.F.: Power Generation, Operation, and Control. Wiley, Hoboken (1996)
25. Zhao, J., Jung, T., Wang, Y., Li, X.: Achieving differential privacy of data disclosure in the smart grid. In: INFOCOM, 2014 Proceedings, pp. 504–512. IEEE (2014)

Chasing First Queens by Integer Programming

Matteo Fischetti and Domenico Salvagnin$^{(\boxtimes)}$

Department of Information Engineering (DEI), University of Padova, Padua, Italy
{matteo.fischetti,domenico.salvagnin}@unipd.it

Abstract. The n-queens puzzle is a well-known combinatorial problem that requires to place n queens on an $n \times n$ chessboard so that no two queens can attack each other. Since the 19th century, this problem was studied by many mathematicians and computer scientists. While finding any solution to the n-queens puzzle is rather straightforward, it is very challenging to find the lexicographically first (or smallest) feasible solution. Solutions for this type are known in the literature for $n \leq 55$, while for some larger chessboards only partial solutions are known. The present paper was motivated by the question of whether Integer Linear Programming (ILP) can be used to compute solutions for some open instances. We describe alternative ILP-based solution approaches, and show that they are indeed able to compute (sometimes in unexpectedly-short computing times) many new lexicographically optimal solutions for n ranging from 56 to 115.

Keywords: n-Queens problem · Mixed-integer programming
Lexicographic simplex

1 Introduction

The n-queens puzzle is a well-known combinatorial problem that requires to place n queens on an $n \times n$ chessboard so that no two queens can attack each other, i.e., no two queens are on the same row, column or diagonal of the chessboard. Initially stated for the regular 8×8 chessboard in 1848 [5], it was soon generalized to the $n \times n$ case [17], and has attracted the interest of many mathematicians (including Carl Friedrich Gauss) and, more recently, by Edsger Dijkstra who used it to illustrate a depth-first backtracking algorithm. As a decision problem, the n-queens puzzle is rather trivial, as a solution exists for all $n > 3$, and there are closed formulas to compute such solutions; see, e.g., the survey in [4]. On the other hand, the counting version of the problem, i.e., to determine the number of different ways to put n queens on a $n \times n$ chessboard turns out to be extremely challenging. The sequence, labelled A000170 on the Online Encyclopedia of Integer Sequences (OEIS) [20], is currently known only up to $n = 27$. The related problem of finding all solutions to the problem was shown in [14] to be beyond the #P-class.

© Springer International Publishing AG, part of Springer Nature 2018
W.-J. van Hoeve (Ed.): CPAIOR 2018, LNCS 10848, pp. 232–244, 2018.
https://doi.org/10.1007/978-3-319-93031-2_16

Another variant of the problem, which is somewhat related to the one addressed in this paper, is the n-queens completion problem, in which some queens are already placed on the chessboard and the solver is required to place the remaining ones, or show that it is not possible. The n-queens completion problem is both NP-complete and #P-complete, as proved in [10].

Following a suggestion of Donald Knuth [16], in this paper we study another very challenging version of the n-queens problem, namely, finding the lexicographically-first (or smallest) feasible solution. This is sequence A141843 on OEIS. Solutions for this variant are known only for $n \leq 55$ [19], while for some larger chessboards only partial solutions are known.

It is worth noting that the lexicographically optimal solution is known for the case of a chessboard of infinite size. Indeed, such a sequence can be easily computed by a simple greedy algorithm that iterates over the anti-diagonals of the chessboard and places a queen in each anti-diagonal in the first available position (this is sequence A065188 on OEIS). Interestingly, as the size of the chessboard increases, its lexicographically optimal solution overlaps more and more with this greedy sequence.

The outline of the paper is as follows. In Sect. 2 we describe the basic Integer Linear Programming (ILP) formulation for the n-queens model, as well as potential families of valid inequalities. In Sect. 3 we describe the different methods developed to solve the instances to lexicographic optimality, while computational results are given in Sect. 4. Conclusions and future directions of research are drawn in Sect. 5. Finally, we list in Appendix all the new optimal solutions we found for n ranging from 56 to 115.

2   An ILP Model

A basic ILP model for the n-queens problem can be obtained by introducing the binary variables $x_{ij} = 1$ iff a queen is placed in row i and column j of the chessboard, for each $i, j = 1, \ldots, n$. Constraints in the basic model stipulate that (i) there is exactly one $x_{ij} = 1$ in each row i; (ii) there is exactly one $x_{ij} = 1$ in each column j; and (iii) there is at most one $x_{ij} = 1$ in each diagonal of the chessboard. Note that all such constraints are *clique* constraints.

In principle, it would be possible to encode the (row-wise) lexicographically minimum requirement by just adding the objective function:

$$\sum_{i=1}^{n} \sum_{j=1}^{n} 2^{ni+j} x_{ij} \tag{1}$$

and solve the problem with a black-box ILP solver. However, the size of the coefficients makes such a method practical only for the smallest chessboards. Still, this simple model, without the objective (1), is the basis of all the methods that will be discussed in Sect. 3.

A compact way to represent a feasible solution is to use a permutation $\pi = (\pi_1, \ldots, \pi_n)$ of the integers $1, \ldots, n$ defined as follows:

$$\pi_i := \sum_{j=1}^{n} j\, x_{ij}, \quad i = 1, \ldots, n. \tag{2}$$

Among all permutations π that correspond to a feasible x, we then look for the lexicographically smallest one. For example, the lex-optimal solution for $n = 10$, depicted in Fig. 1, can be described as

$$(1, 3, 6, 8, 10, 5, 9, 2, 4, 7).$$

Fig. 1. Lexicographically optimal solution for $n = 10$.

The n-queens problem can also be easily reformulated as a maximum independent set problem, as noted for example in [8]. Indeed, one just needs to construct a graph in which there is a node for each square of the chessboard and an edge for each pair of conflicting squares, i.e., for any two squares in the same row, column or diagonal. Then any independent set of cardinality n is a solution to the puzzle. The independent set reformulation immediately suggests classes of valid inequalities for the n-queens problem, namely all that are valid for the stable set polytope, such as *clique* and *odd-cycle* [13] inequalities.

Among clique inequalities, the following (polynomial in n) family is particularly relevant for our problem:

$$x_{ij} + x_{i,j+h} + x_{i+h,j} + x_{i-h,j} + x_{i,j-h} \leq 1 \tag{3}$$
$$x_{ij} + x_{i+h,j+h} + x_{i-h,j+h} + x_{i-h,j-h} + x_{i+h,j-h} \leq 1 \tag{4}$$
$$x_{ij} + x_{i+h,j} + x_{i+h,j+h} + x_{i,j+h} \leq 1 \tag{5}$$

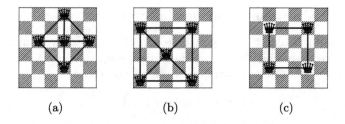

(a)   (b)   (c)

Fig. 2. Three different families of clique cuts for n-queens.

where $i, j, h \in \{1, \ldots, n\}$; of course, variables x_{uv} corresponding to a position (u, v) outside the $n \times n$ chessboard are removed from the summations. The three different types of cliques in this family are depicted in Fig. 2.

Clique inequalities (3)–(5) can be trivially separated in time that is polynomial in n. In addition, in preliminary experiments we implemented a general-purpose exact clique separator based on the solution of an auxiliary ILP model, and it never produced any additional violated clique inequality for the instances in our testbed.

A second class of inequalities contains the so-called *odd-cycle* inequalities. Given any odd cycle O in the graph, the following inequality:

$$\sum_{k \in O} x_k \leq \frac{|O| - 1}{2} \tag{6}$$

is valid for the stable set polytope. Odd-cycle inequalities can be easily separated as $\{0, 1/2\}$-cuts with the combinatorial procedures described in [2,6,7]. An example of odd-cycle inequality occurring in the n-queens problem is illustrated in Fig. 3.

Fig. 3. Example of odd-cycle inequality for n-queens: no more than two of the five positions can be occupied by a queen.

3   Solution Methods

We next describe the solution algorithms that we implemented.

3.1   Using a Constraint Programming Solver

The n-queens puzzle can be easily modeled as a Constraint Programming (CP) problem. Indeed, working directly on the variables π_i, the puzzle can be formulated by just three alldifferent [18,21] global constraints:

$$\text{alldifferent}(\pi_i, i = 1, \ldots, n) \tag{7}$$

$$\text{alldifferent}(\pi_i + i, i = 1, \ldots, n) \tag{8}$$

$$\text{alldifferent}(\pi_i - i, i = 1, \ldots, n). \tag{9}$$

We implemented the model above with Gecode [9]. In order to enforce the model to find the lexicographically-smallest solution, we use Depth-First Search (DFS) as search strategy, always branching on the first unfixed variable π_i and picking values in increasing order—in Gecode terminology, that amounts to using a *brancher* specified by INT_VAR_NONE() and INT_VALUES_MIN(). In the following, we will refer to this solution method as CP.

3.2   Using an Exact ILP Solver

A simple algorithm to compute the lex-optimal solution by iteratively using a black-box ILP solver is as follows: We scan all the chessboard positions (i, j) in lexicographical order, i.e., row by row. For each (i, j), we are given the queens already positioned in the previous iterations (i.e., we have a number of fixed x variables), and our order of business is to decide whether a queen can be placed in (i, j) or not. This in turn requires solving the basic ILP model with some variables fixed in the previous iterations, by *maximizing* x_{ij}: if the final optimal solution has value 1, we place a new queen in position (i, j) by fixing $x_{ij} = 1$, otherwise we fix $x_{ij} = 0$ and proceed with the next chessboard position[1]. This approach requires solving n^2 ILPs.

In our actual implementation, a more effective scheme is used that exploits representation (2). To be specific, we scan the rows $i = 1, \ldots, n$, in sequence. For each i, we have already fixed in the previous iterations the lex-optimal sequence $\pi_1, \cdots, \pi_{i-1}$ and the corresponding x variables, and we want to compute the *smallest* feasible integer π_i. To this end we solve the basic ILP model, with some variables fixed in the previous iterations, by *minimizing* the objective function (2), fix all the x_{ij} variables in row i accordingly, and proceed with the next row. In this way, only n ILPs need to be solved. In the following, we will refer to this solution method as ILP-ITER.

3.3   Using a Truncated ILP Solver

We also implemented an explicit depth-first backtracking algorithm to build the lex-optimal permutation π, very much in the spirit of the CP approach described

---

[1] Alternatively, one could fix $x_{ij} = 1$, check the resulting model for feasibility, and then move to the next position.

in Subsect. 3.1. At each iteration (i.e., at each node of the branching tree) we have tentatively fixed a lex-minimal, but possibly infeasible sequence, $(\overline{\pi}_1, \ldots, \overline{\pi}_{i-1})$, and the corresponding x variables, and we have to decide the next value in position i. This is in turn obtained by solving a *relaxation* of the current ILP with objective function (2), to be minimized, i.e., by applying the following three steps:

(i) invoke the ILP solver (with its default cutting-plane generation and prepro-cessing) for a limited number of nodes, say NN;
(ii) define $\overline{\pi}_i$ as the best *lower bound* available at the node limit (rounded up);
(iii) tentatively fix $\overline{\pi}_i$, along with the corresponding x variables, as the i-th value in the sequence.

As a lower bound (instead of the true value) is used, it may happen that, at a later iteration, the current ILP becomes infeasible, proving that the current tentative subsequence $(\overline{\pi}_1, \cdots, \overline{\pi}_k)$ till position k (say) is infeasible as well. In this case, a backtracking operation takes place, that consists in imposing that the k-th position must hold a value strictly larger than $\overline{\pi}_k$. The latter requirement can easily be enforced in the ILP model by setting $x_{kj} = 0$ for $j = 1, \ldots, \overline{\pi}_k$. The algorithm ends as soon as the first feasible complete permutation $(\overline{\pi}_1, \ldots, \overline{\pi}_n)$ is found.

After some preliminary tests, we decided to set $NN = 0$, i.e., to only solve the root node of the ILP at hand. Note that this is *not* equivalent to solving the LP relaxation of the ILP, as cutting planes and (most importantly) preprocessing play a crucial role here. According to our computational experience, solving just the LP relaxation is indeed mathematically correct and very fast, as the dual simplex can be used to reoptimize each LP, but the number of backtrackings becomes too large to have a competitive implementation. In the following, we will refer to this solution method as ILP-TRUNC.

3.4   An Enumerative Method Based on Lexicographic Simplex

Finally, given the strong lexicographic nature of the problem at hand, we decided to implement a custom enumerative algorithm based on the lexicographic sim-plex method [11, 12]. The lexicographic simplex method not only finds an optimal solution to a given LP, but it guarantees to return the lexicographically smallest (or greatest) one among all optimal solutions. The lexicographic variant of the simplex method can be implemented quite easily on top of a black-box regular simplex solver, as described for example in [3, 22]. The idea is as follows. Given an ordered sequence of objective functions f_k to optimize lexicographically, at each step we impose to stay on the optimal face of the current objective by fixing all variables (including the artificial variables associated to inequality constraints) with nonzero reduced cost, move to the next objective and reoptimize. Once all objectives have been optimized, in sequence, the original bounds for all variables are restored, which does not change the optimality status of the final basis, which is the lex-optimal one.

In our n-queens case, given our encoding of the permutation variables π as x_{ij}, we are interested in the lexicographically maximal solution in the x space or, equivalently, the sequence of objective functions to be minimized is $-x_{ij}$, for all $i, j = 1, \ldots, n$.

Using a lexicographic simplex method within an enumerative DFS scheme, in which again we always branch on the first unfixed variable and explore the 1-branch first, provides the following advantages over using a "regular" simplex method:

- Whenever the LP relaxation turns out to be integer, i.e., there are no fractional variables, we are guaranteed that this is the lex-optimal integer solution within the current subtree, hence we can prune the node. Given our branching and exploration strategy, this also implies that we are done.
- If the first unfixed variable at the current node gets a value strictly less than one, then we can fix the variable to zero. This is easily proved using the lex-optimality of the LP solution as an argument. Being the first unfixed variable, this is the first objective to be considered by the lexicographic simplex at the current node, so a lex-optimal value <1 means that there is no feasible solution (in the current subtree) in which this variable takes value 1. Note that this reduction can be applied iteratively until the first unfixed variable gets a value of 1. We call this process *mini-cutloop*.

The basic scheme above can be improved with some additional modifications. First of all, we do not need to branch on single variables but we can branch directly on rows, again always picking the first row that contains an unfixed variable. For example, let the first unfixed variable be x_{ij}: instead of branching on the binary dichotomy $x_{ij} = 1 \vee x_{ij} = 0$, we use the n-way branching $x_{i1} = 1 \vee x_{i2} = 1 \vee \ldots \vee x_{in} = 1$. Of course, variables that are already fixed are removed from the list. This basically mimics the branching that would have been done by working directly with the π variables, as done by the CP solver.

Note that, because of our rigid branching strategy, there is no need for a full lexicographic optimization at each node. Indeed, for the purpose of branching, we can stop the lexicographic optimization at the first fractional variable, as we will be forced to branch on its row, or on a previous one. For this very reason, and because of the n-queens structure, we implemented a specialized lexicographic simplex method, where instead of optimizing one variable at the time, we optimize row by row, also integrating the mini-cutloop in the process. In particular, we do the following:

1. Let i^* be the first row with an unfixed variable. Set the objective function to $\sum_{j=1}^{n} j x_{i^*j}$ and minimize it.
2. Apply the mini-cutloop, by iteratively fixing the first unfixed variable in the row if its fractional value is <1 and by reoptimizing with the dual simplex.
3. If all variables in the current row are fixed this way, then we can move to the next row and go to step (1). Otherwise stop.

Note that the method above does not need to temporarily fix variables as the regular lexicographic simplex would. It is also important to note that, in the loop above, if the current fractional solution is integer, we are no longer guaranteed that this is the lexicographically optimal solution. In this (rare) case, we resort to a full-blown lexicographic simplex method to tell whether we can prune the node or need to branch.

The effectiveness of the node processing above greatly depends on the mini-cutloop, which in turn relies on being able to recognize fixed variables, i.e., to distinguish between a variable that happens to be zero or one in the current fractional solution, and a variable that is actually fixed at that value in the current node. For this purpose, we implemented a specialized propagator for the clique constraints of the basic model—while there is no need to propagate the clique constraints (3)–(5) as those can never lead to additional fixings.

Finally, separation of the clique inequalities (3)–(5) and odd-cycle inequalities has also been implemented and added to the node processing code. In the following, we will refer to this solution method as LEX-DFS.

4   Computational Comparisons

We implemented our ILP models with the MIP solver IBM ILOG CPLEX 12.7.1 [15], while we used Gecode 5.1.0 [9] as the CP solver for model (7)–(9). All experiments were done on a cluster of 24 identical machines, each equipped with an Intel Xeon E3-1220 V2 quad-core PC and 16 GB of RAM.

The testbed is made of all instances with n ranging from 21 to 60. A time limit of 2 days was given for each instance to each method. Detailed results are given in Table 1, where we report the running time, in seconds, for all of our methods. The last two rows of the table report the shifted geometric mean [1] of the computing time (with a shift of 10 s) and the number of solved instances. According to the table, the CP model is able to solve models up to size 40 in a reasonable amount of time, after which it can no longer solve any model. Comparing with the numbers reported in [19], this can be already considered a good achievement, and a testament to how efficient Gecode's implementation is. On the other hand, all methods based on ILP, while initially slower, turn out to be able to solve almost all models in the testbed. Among the ILP methods, ILP-ITER, while being the easiest to implement, is also the slowest method, while ILP-TRUNC and LEX-DFS are the fastest methods, with very similar average running times.

As already noted in [19], the size of the chessboard is not a direct indicator of instance difficulty, as some bigger chessboards can be solved significantly faster than smaller ones. This is true in particular for ILP-based methods, where for example $n = 48$ is unsolved while $n = 49$ can be cracked in a few seconds. Interestingly, chessboards with even n seem to be consistently harder than the ones with odd n.

As for the advanced techniques implemented in LEX-DFS, we have to admit that for some of them the overall effect was rather disappointing. In particular, the separation of clique and odd-cycle inequalities, while able to reduce the number of enumerated nodes by more than a factor of 2, does not lead to a

Table 1. Comparison of different methods for $n = 21, \ldots, 60$, with a time limit of 172800 s (2 days).

n	Methods			
---	CP	ILP-ITER	ILP-TRUNC	LEX-DFS
21	0.01	0.30	0.45	0.08
22	0.95	1.63	16.67	9.20
23	0.02	0.40	0.60	0.11
24	0.20	0.60	2.95	0.82
25	0.03	0.49	0.79	0.12
26	0.16	0.84	1.59	0.42
27	0.17	0.59	0.90	0.09
28	0.84	1.13	2.08	1.06
29	0.39	1.05	1.36	0.32
30	15.80	13.16	77.05	16.35
31	2.86	1.50	3.31	0.87
32	19.45	4.97	42.73	5.23
33	29.82	28.47	56.76	13.83
34	593.60	342.32	4558.02	228.07
35	33.70	11.21	30.46	4.67
36	5199.27	1882.10	20901.43	1196.59
37	185.37	2.06	7.49	0.54
38	2485.20	101.30	151.86	130.16
39	1642.30	143.02	184.50	44.79
40	$t.l.$	9604.18	117591.20	7068.84
41	1543.84	20.91	105.47	5.69
42	$t.l.$	$t.l.$	$t.l.$	$t.l.$
43	23528.50	21.65	162.13	6.08
44	$t.l.$	1013.43	14838.95	2220.52
45	$t.l.$	3604.37	4560.69	1388.93
46	$t.l.$	$t.l.$	$t.l.$	$t.l.$
47	$t.l.$	1602.10	5057.63	601.54
48	$t.l.$	$t.l.$	$t.l.$	$t.l.$
49	$t.l.$	23.26	460.07	10.35
50	$t.l.$	28011.44	$t.l.$	61679.70
51	$t.l.$	4.63	874.16	0.88
52	$t.l.$	30306.67	27701.60	75659.40
53	$t.l.$	5.05	285.65	0.96
54	$t.l.$	64.09	19784.91	67031.40
55	$t.l.$	44.50	569.58	18.42
56	$t.l.$	28026.57	13386.85	101936.00
57	$t.l.$	10.03	3961.54	5.87
58	$t.l.$	$t.l.$	129596.69	$t.l.$
59	$t.l.$	49647.30	$t.l.$	18795.70
60	$t.l.$	$t.l.$	39143.49	$t.l.$
shmean	2945.76	780.20	251.85	263.79
#solved	21	35	35	35

faster algorithm overall. To the contrary, disabling cut separation leads to a slightly faster method with an average runtime of 246 s. Note that this is not due to the complexity of separating cuts, separation being extremely fast for both classes of inequalities, but rather for the reduced node throughput.

5 Conclusions and Future Directions of Work

Finding a lexicographically minimal (also called "first") solution of the n-queens puzzle is a very difficult problem that attracted some research interest in recent years. Following a suggestion by Donald E. Knuth, we have developed new solution methods based on Integer Linear Programming, and have been able to provide the optimal solution for several open problems.

The two main outcomes of our research are as follows: (1) ILP has been able to solve many previously unsolved models for this problem, sometimes in unexpectedly-short computing times; (2) the yet-unsolved cases provide excellent benchmark examples on which to base the next advances in ILP technology. In addition, we think that improving our understanding on how to solve lexicographic variants of combinatorial problems is an interesting topic on its own.

Future research should address the unsolved cases, and in particular should try to better understand the reason why, in the ILP setting, the instances with even n seem to be much more difficult to solve than those with n odd.

Acknowledgements. This research was partially supported by MiUR, Italy, through project PRIN2015 "Nonlinear and Combinatorial Aspects of Complex Networks". We thank Donald E. Knuth for having pointed out the problem to us, and for inspiring discussions on the role of Integer Linear Programming in solving combinatorial problems arising in digital tomography.

A New Solutions

Here are the solutions we found for some open problems from the literature:

n	Solution
56	1 3 5 2 4 9 11 13 15 6 8 19 7 22 10 25 27 29 31 33 42 44 46 43 51 53 55 45 54 50 47 56 48 52 49 12 14 23 21 32 34 26 16 30 17 24 18 37 28 40 20 39 41 35 38 36
57	1 3 5 2 4 9 11 13 15 6 8 19 7 22 10 25 27 29 31 12 34 43 45 47 50 52 54 44 57 49 46 56 51 48 55 53 14 28 17 33 23 16 18 30 24 37 20 32 21 26 40 35 41 39 42 36 38
58	1 3 5 2 4 9 11 13 15 6 8 19 7 22 10 25 27 29 31 12 42 45 48 52 54 43 53 55 49 44 46 50 57 47 51 58 56 28 26 20 34 30 18 14 17 24 21 16 35 23 40 33 36 38 32 41 39 37
59	1 3 5 2 4 9 11 13 15 6 8 19 7 22 10 25 27 29 31 12 34 36 45 47 49 52 56 53 46 57 59 48 51 54 50 55 58 16 14 17 32 23 26 20 18 33 35 28 21 43 41 37 24 40 44 30 39 42 38

n	Solution
60	1 3 5 2 4 9 11 13 15 6 8 19 7 22 10 25 27 29 31 12 34 44 46 48 45 51 54 58 50 59 57 60 47 49 52 55 53 56 18 33 23 32 28 16 20 17 21 37 35 26 24 30 14 42 38 43 41 39 36 40
61	1 3 5 2 4 9 11 13 15 6 8 19 7 22 10 25 27 29 31 12 14 35 45 47 49 52 54 56 50 60 46 61 58 48 51 53 55 57 59 23 32 16 33 21 17 26 36 18 20 38 24 28 34 40 30 41 44 42 37 39 43
63	1 3 5 2 4 9 11 13 15 6 8 19 7 22 10 25 27 29 31 12 14 35 37 47 49 51 53 59 57 52 60 62 48 50 54 63 55 58 56 61 32 16 33 17 21 26 36 20 18 38 28 23 40 24 30 34 41 39 44 46 43 45 42
65	1 3 5 2 4 9 11 13 15 6 8 19 7 22 10 25 27 29 31 12 14 35 37 39 49 51 53 50 56 59 63 55 64 62 65 52 54 57 60 58 61 16 30 17 21 26 36 33 20 18 41 38 23 32 24 28 48 46 34 43 40 44 47 45 42
67	1 3 5 2 4 9 11 13 15 6 8 19 7 22 10 25 27 29 31 12 14 35 37 39 41 51 53 55 52 58 61 65 57 66 64 67 54 56 59 62 60 63 16 18 34 30 38 20 24 17 21 23 43 32 40 33 36 26 28 46 48 50 44 47 45 42 49
69	1 3 5 2 4 9 11 13 15 6 8 19 7 22 10 25 27 29 31 12 14 35 37 39 41 43 53 55 57 54 60 63 67 59 68 66 69 56 58 61 64 62 65 17 20 16 30 24 33 40 38 18 21 34 26 23 42 49 28 32 50 36 51 46 44 52 48 45 47
71	1 3 5 2 4 9 11 13 15 6 8 19 7 22 10 25 27 29 31 12 14 35 37 39 41 16 53 55 57 54 56 62 68 66 69 59 70 67 58 71 61 64 60 65 63 21 30 17 40 18 24 36 20 42 44 26 34 23 33 38 32 28 49 51 45 47 52 50 48 46 43
73	1 3 5 2 4 9 11 13 15 6 8 19 7 22 10 25 27 29 31 12 14 35 37 39 41 16 44 55 57 59 56 58 63 67 69 71 73 61 70 72 65 60 62 64 66 68 20 34 21 18 42 17 38 24 43 23 28 45 33 40 36 26 32 30 54 47 50 52 46 48 53 51 49
77	1 3 5 2 4 9 11 13 15 6 8 19 7 22 10 25 27 29 31 12 14 35 37 39 41 16 18 45 57 59 61 58 60 65 68 72 74 76 73 75 63 67 64 62 77 70 66 71 69 38 40 28 17 21 24 26 20 43 46 42 23 36 34 32 30 44 33 52 55 47 50 53 56 54 48 51 49
79	1 3 5 2 4 9 11 13 15 6 8 19 7 22 10 25 27 29 31 12 14 35 37 39 41 16 18 45 47 59 61 63 60 62 67 70 74 71 77 79 76 78 64 68 65 69 66 73 75 72 20 38 17 21 44 24 30 23 46 48 36 42 40 34 26 28 33 50 32 53 43 57 52 58 56 54 51 49 55
85	1 3 5 2 4 9 11 13 15 6 8 19 7 22 10 25 27 29 31 12 14 35 37 39 41 16 18 45 17 48 50 63 65 67 64 66 71 73 75 80 82 84 81 83 72 70 68 85 69 78 74 77 79 76 20 23 43 24 21 49 44 42 34 46 28 30 52 26 38 51 32 40 33 61 47 60 36 53 58 54 57 59 56 62 55
91	1 3 5 2 4 9 11 13 15 6 8 19 7 22 10 25 27 29 31 12 14 35 37 39 41 16 18 45 17 48 20 51 53 67 69 71 68 70 75 77 79 81 85 87 90 86 91 89 72 74 76 73 80 82 84 78 83 88 21 34 26 49 46 24 47 52 43 23 30 33 55 28 42 32 54 40 36 44 64 50 38 59 61 65 57 66 60 63 56 58 62
93	1 3 5 2 4 9 11 13 15 6 8 19 7 22 10 25 27 29 31 12 14 35 37 39 41 16 18 45 17 48 20 51 53 55 69 71 73 70 72 77 79 81 83 87 89 92 88 93 91 74 76 78 75 82 84 86 80 85 90 24 21 23 46 49 47 52 38 30 56 33 26 28 43 32 54 57 42 44 36 34 40 50 61 68 65 62 59 63 58 67 64 66 60

n	Solution
97	1 3 5 2 4 9 11 13 15 6 8 19 7 22 10 25 27 29 31 12 14 35 37 39 41 16 18 45 17 48 20 51 53 21 56 71 73 75 72 74 79 81 83 85 87 89 93 95 97 94 96 76 80 77 86 78 82 84 91 88 90 92 46 24 28 52 23 49 47 34 30 26 57 50 33 61 42 44 36 32 55 43 38 54 60 66 40 70 68 63 58 69 62 65 67 64 59
101	1 3 5 2 4 9 11 13 15 6 8 19 7 22 10 25 27 29 31 12 14 35 37 39 41 16 18 45 17 48 20 51 53 21 56 58 60 75 77 79 76 78 83 85 87 89 91 93 97 99 101 98 100 80 84 81 90 82 86 88 95 92 94 96 23 26 28 40 43 54 57 24 32 47 50 42 59 33 30 34 52 62 68 46 38 36 44 55 66 71 74 70 49 73 63 72 67 61 64 69 65
103	1 3 5 2 4 9 11 13 15 6 8 19 7 22 10 25 27 29 31 12 14 35 37 39 41 16 18 45 17 48 20 51 53 21 56 58 60 62 77 79 81 78 80 85 87 89 91 93 95 99 101 103 100 102 82 86 83 92 84 88 90 97 94 96 98 23 26 24 30 28 36 46 55 59 52 54 44 61 34 66 33 42 32 47 49 40 38 57 73 71 63 72 43 64 70 75 50 69 67 76 74 68 65
109	1 3 5 2 4 9 11 13 15 6 8 19 7 22 10 25 27 29 31 12 14 35 37 39 41 16 18 45 17 48 20 51 53 21 56 58 60 23 63 65 81 83 85 82 84 89 91 93 95 86 100 104 106 101 109 107 105 108 88 92 87 96 90 97 102 94 98 103 99 26 24 32 28 36 55 57 40 64 61 54 50 30 66 34 42 38 33 49 43 67 59 62 77 52 44 47 75 71 46 76 80 73 70 79 69 78 72 74 68
115	1 3 5 2 4 9 11 13 15 6 8 19 7 22 10 25 27 29 31 12 14 35 37 39 41 16 18 45 17 48 20 51 53 21 56 58 60 23 63 24 66 68 85 87 89 86 88 93 95 97 99 90 102 108 111 113 107 109 112 115 91 114 98 101 92 94 96 100 105 103 110 106 104 26 28 30 32 36 50 59 62 64 55 43 34 72 67 52 33 40 65 57 44 42 38 74 54 61 46 83 47 77 69 49 82 79 75 84 71 80 78 81 73 70 76

References

1. Achterberg, T.: Constraint integer programming. Ph.D. thesis, Technische Universität Berlin (2007)
2. Andreello, G., Caprara, A., Fischetti, M.: Embedding {0, 1/2}-cuts in a branch-and-cut framework: a computational study. INFORMS J. Comput. **19**(2), 229–238 (2007)
3. Balas, E., Fischetti, M., Zanette, A.: On the enumerative nature of Gomory's dual cutting plane method. Math. Program. **125**, 325–351 (2010)
4. Bell, J., Stevens, B.: A survey of known results and research areas for n-Queens. Discret. Math. **309**(1), 1–31 (2009)
5. Bezzel, M.: Proposal of 8-Queens problem. Berl. Schachzeitung **3**, 363 (1848)
6. Caprara, A., Fischetti, M.: {0, $\frac{1}{2}$}-Chvátal-Gomory cuts. Math. Program. **74**, 221–235 (1996)
7. Caprara, A., Fischetti, M.: Odd cut-sets, odd cycles, and 0–1/2 Chvatal-Gomory cuts. Ricerca Operativa **26**, 51–80 (1996)
8. Foulds, L.R., Johnston, D.G.: An application of graph theory and integer programming: chessboard non-attacking puzzles. Math. Mag. **57**, 95–104 (1984)
9. Gecode Team. Gecode: Generic constraint development environment (2017). http://www.gecode.org
10. Gent, I.P., Jefferson, C., Nightingale, P.: Complexity of n-Queens completion. J. Artif. Intell. Res. **59**, 815–848 (2017)
11. Gomory, R.E.: Outline of an algorithm for integer solutions to linear programs. Bull. Am. Math. Soc. **64**, 275–278 (1958)

12. Gomory, R.E.: An algorithm for the mixed integer problem. Technical report RM-2597, The RAND Cooperation (1960)
13. Grötschel, M., Lovász, L., Schrijver, A.: Geometric Algorithms and Combinatorial Optimization. Springer, Heidelberg (1988). https://doi.org/10.1007/978-3-642-97881-4
14. Hsiang, J., Frank Hsu, D., Shieh, Y.-P.: On the hardness of counting problems of complete mappings. Discret. Math. **277**(1–3), 87–100 (2004)
15. IBM. ILOG CPLEX 12.7 User's Manual (2017)
16. Knuth, D.E.: Private communication, November 2017
17. Lionnet, F.J.E.: Question 963. Nouvelles Annales de Mathématiques **8**, 560 (1869)
18. Régin, J.-C.: A filtering algorithm for constraints of difference in CSPs. In: Artificial Intelligence, vol. 1, pp. 362–367 (1994)
19. Schubert, W.: Wolfram Schubert's N-Queens page. http://m29s20.vlinux.de/~wschub/nqueen.html. Accessed Dec 2017
20. Sloane, N.J.A.: The on-line encyclopedia of integer sequences (2017)
21. van Hoeve, W.F.: The alldifferent constraint: a survey. CoRR (2001)
22. Zanette, A., Fischetti, M., Balas, E.: Lexicography and degeneracy: can a pure cutting plane algorithm work? Math. Program. **130**, 153–176 (2011)

Accelerating Counting-Based Search

Samuel Gagnon[(⊠)] and Gilles Pesant[(⊠)]

Polytechnique Montréal, Montreal, Canada
{samuel-2.gagnon,gilles.pesant}@polymtl.ca

Abstract. Counting-based search, a branching heuristic used in constraint programming, relies on computing the proportion of solutions to a constraint in which a given variable-value assignment appears in order to build an integrated variable- and value-selection heuristic to solve constraint satisfaction problems. The information it collects has led to very effective search guidance in many contexts. However, depending on the constraint, computing such information can carry a high computational cost. This paper presents several contributions to accelerate counting-based search, with supporting empirical evidence that solutions can thus be obtained orders of magnitude faster.

1 Introduction

Constraint programming builds concise models from high-level constraints that reveal much of the combinatorial structure of a problem. That structure is used to prune the search space through domain filtering algorithms, to guide its exploration through branching heuristics, and to learn from previous attempts at finding a solution. *Counting-based search* [12] represents a family of branching heuristics that guide the search for solutions by identifying likely variable-value assignments in each constraint. Given a constraint $c(x_1, \ldots, x_n)$, its number of solutions $\#c(x_1, \ldots, x_n)$, respective finite domains D_i $1 \leq i \leq n$, a variable x_i in the scope of c, and a value $v \in D_i$, we call

$$\sigma(x_i, v, c) = \frac{\#c(x_1, \ldots, x_{i-1}, v, x_{i+1}, \ldots, x_n)}{\#c(x_1, \ldots, x_n)} \qquad (1)$$

the *solution density* of pair (x_i, v) in c, i.e. how often a certain assignment is part of a solution to c. Though that concept was originally introduced for satisfaction problems, it has been extended to optimization problems as well [9].

Solution densities from every constraint $c \in C$ in a model can be combined in many ways to produce a branching heuristic—one simple combination that works well in practice, called *maxSD* [10], branches on $x_i^\star = v^\star$ where

$$(x_i^\star, v^\star, c^\star) = \operatorname*{argmax}_{c(x_1, \ldots, x_n) \in C,\ i \in \{1, \ldots, n\},\ v \in D_i} \sigma(x_i, v, c) \qquad (2)$$

and on $x_i^\star \neq v^\star$ upon backtracking. The computational cost of solution densities depends on the constraint: for some it is only marginally more expensive than its

© Springer International Publishing AG, part of Springer Nature 2018
W.-J. van Hoeve (Ed.): CPAIOR 2018, LNCS 10848, pp. 245–253, 2018.
https://doi.org/10.1007/978-3-319-93031-2_17

existing filtering algorithm (e.g. `regular`) while for others exact computation is intractable (e.g. `alldifferent`). Given its effectiveness at guiding search, finding more efficient ways to compute solution densities is desirable.

This paper presents several contributions to accelerate counting-based search. We first discuss specific improvements for the `alldifferent` and `spanningtree` counting algorithms in Sects. 2 and 3. Then a generic method for accelerating search is presented in Sect. 4. All discussed algorithms are implemented using Gecode [6] and available in [5].

2   Alldifferent Constraints

An instance of an `alldifferent`$(x_1, \ldots, x_n)$ constraint is equivalently represented by an incidence matrix $A = (a_{iv})$ with $a_{iv} = 1$ whenever $v \in D_i$ and $a_{iv} = 0$ otherwise. For notational convenience and without loss of generality, we identify domain values with consecutive natural numbers. Because we will want A to be square (with $m = |\bigcup_{x_i \in X} D_i|$ rows and columns), if there are fewer variables than values we add enough rows, say p, filled with 1s. It is known that counting the number of solutions to the `alldifferent` constraint is equivalent to computing the *permanent* of that square matrix (dividing the result by $p!$ to account for the extra rows) [13]:

$$perm(A) = \sum_{v=1}^{m} a_{1v} \cdot perm(A^{1v}) \tag{3}$$

where A^{ij} denotes the sub-matrix obtained from A by removing row i and column j.

Since computing the permanent is #P-complete [11], Zanarini and Pesant proposed approximate counting algorithms for the `alldifferent` constraint based on sampling [12] and upper bounds [10]. Algorithm 1 reproduces the latter using notation adapted for this article. As each assignment $x_i = v$ in the `alldifferent` constraint induces a different incidence matrix, a naive approach to compute solution densities is to recompute the permanent upper bound for each assignment. However, our upper bounds are a product of factors F for each variable x_i which depend only on the size of its domain $d_i = |D_i|$ (line 1). Hence if we account for the domain reduction of the assigned variable (line 5) and of each variable which could have taken that value (line 6)—simulating forward checking—we can compute the solution density of each assignment (line 10) by updating the upper bound UB$_A$ calculated for the whole constraint (line 1). Reusing UB$_A$ avoids recomputing upper bounds from scratch. Let c_v denote the number of 1s in column v of A. Given that we can precompute the factors, the total computational effort is dominated by line 6 where we do a total of $\Theta(\sum_{v=1}^{m} c_v^2)$ operations: for a given value v, u_v is computed c_v times by multiplying $c_v - 1$ terms.

```
1  UB_A = ∏_{x_i} F[d_i]                                    ▷ Constraint upper bound
2  foreach x_i ∈ X do
3  |   total = 0                                             ▷ Normalization factor
4  |   foreach v ∈ D_i do
5  |   |   u_{x_i} = F[1]/F[d_i]                             ▷ Variable assignment update
6  |   |   u_v = ∏_{k≠i : v∈D_k} F[d_k−1]/F[d_k]            ▷ Value assignment update
7  |   |   UB_{x_i=v} = UB_A · u_{x_i} · u_v                ▷ Assignment upper bound
8  |   |   total += UB_{x_i=v}
9  |   foreach v ∈ D_i do
10 |   |   SD[i][v] = UB_{x_i=v} / total
11 return SD
```

Algorithm 1. Solution densities for `alldifferent`, adapted from [10]

```
1  UB_v = 1, ∀v ∈ {1, 2, ..., m}
2  foreach x_i ∈ X do
3  |   foreach v ∈ D_i do
4  |   |   UB_v *= F[d_i−1]/F[d_i]
5  foreach x_i ∈ X do
6  |   total = 0
7  |   foreach v ∈ D_i do
8  |   |   UB_{x_i=v} = F[1]/F[d_i−1] · UB_v
9  |   |   total += UB_{x_i=v}
10 |   foreach v ∈ D_i do
11 |   |   SD[i][v] = UB_{x_i=v} / total
12 return SD
```

Algorithm 2. Improved version of Algorithm 1

2.1 Improved Algorithm

The product at line 6 of Algorithm 1 can be rewritten to depend only on v:

$$u_v = \frac{F[d_i]}{F[d_i-1]} \prod_{k \,:\, v \in D_k} \frac{F[d_k-1]}{F[d_k]} \,. \tag{4}$$

This allows us, as shown in Algorithm 2, to precompute this product for every value (line 1–4) as it does not depend on i anymore, leading to each $UB_{x_i=v}$ being computed in constant time (line 8). We also avoid computing UB_A since that factor cancels out during normalization. Algorithm 2 runs in $\Theta(\sum_{v=1}^{m} c_v)$ time, which is asymptotically optimal if we need to compute every solution density (since $\sum_{v=1}^{m} c_v = \sum_{i=1}^{n} d_i$).

2.2 Computing Maximum Solution Densities Only

Some search heuristics, such as *maxSD*, only really need the highest solution density from each constraint in order to make a branching decision. In such a case it may be possible to accelerate the counting algorithm further. We present such an acceleration for the `alldifferent` constraint.

The factors F in our upper bounds are strictly increasing functions, meaning that for a given value v, the highest solution density will occur for the variable with the smallest domain. Algorithm 3 identifies that peak for each value, knowing that the highest one will be included in this subset. Note however that because we don't compute a solution density for each value in the domain of a given variable, we cannot normalize them as before (though we at least adjust for the p extra rows). So we may loose some accuracy but what we were computing was already an estimate, not the exact density. The asymptotic complexity of this algorithm remains the same as the previous one, but makes fewer computations: we iterate on each variable and value once instead of three times.

---

1 $\text{UB}_v = (\frac{F[n-1]}{F[n]})^p,\ \min_v = 1,\ \forall v \in \{1, 2, \ldots, m\}$
2 **foreach** $x_i \in X$ **do**
3 $\quad$ **foreach** $v \in D_i$ **do**
4 $\quad\quad$ $\text{UB}_v\ \mathbin{*}= \frac{F[d_i-1]}{F[d_i]}$
5 $\quad\quad$ **if** $d_i < d_{\min_v}$ **then**
6 $\quad\quad\quad$ $\min_v = i$
7 $\text{maxSD} = \{var = 0,\ val = 0,\ dens = 0\}$
8 **foreach** $v \in \{1, 2, \ldots, m\}$ **do**
9 $\quad$ $\text{SD}[\min_v][v] = \frac{F[1]}{F[d_{\min_v}-1]} \cdot \text{UB}_v$
10 $\quad$ **if** $\text{SD}[\min_v][v] > \text{maxSD}.dens$ **then**
11 $\quad\quad$ $\text{maxSD} = \{\min_v,\ v,\ \text{SD}[\min_v][v]\}$
12 **return** maxSD

Algorithm 3. Maximum solution density for `alldifferent`

2.3 Experiments on the Quasigroup Completion Problem

The Quasigroup Completion Problem (#67 in CSPLib) can be described using an `alldifferent` constraint on each row and column, making it ideal for testing the above counting algorithms. Gecode's distribution already includes a model with branching heuristics *afc* (weighted degree) and *size* (smallest domain), both with lexicographic value selection. We consider heuristic *maxSD* using Algorithms 1, 2 and 3. We use 20 instances of size 90 to 110 with 25% of entries filled, generated as in [7]. That ratio of filled entries may not yield the hardest instances for that size but our goal here is to have a lot of shared values between variables in order to emphasize the improvement of Algorithm 2 over 1.

First we observe that *maxSD* guides search more effectively by solving all instances in several orders of magnitude fewer failures than *afc* and *size*. As expected Algorithm 3 is less accurate than the other two (which share the same number of failures), but still about one order of magnitude faster (Fig. 1).

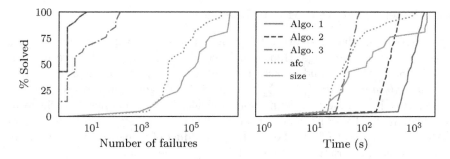

Fig. 1. Percentage of Quasigroup Completion instances solved w.r.t. time and number of failures.

3 Spanning Tree Constraints

Brockbank et al. introduced an algorithm to compute solution densities for the `spanningTree` constraint in [1]. The graph is represented as a *Laplacian matrix* L (vertex degrees on the diagonal and edges indicated by -1 entries) and Kirchhoff's Matrix-Tree Theorem [2] is used to compute solution densities for every edge (u, v) using the following formula:

$$\sigma((u, v), 1, \texttt{spanningTree}(G, T)) = m^u_{v'v'} \tag{5}$$

with $M^u = (m^u_{ij})$ defined as the inverse of the sub-matrix L^u obtained by removing row and column u from L and v' equal to v if $v < u$ and to $v - 1$ otherwise. Given a vertex cover of size γ on a graph over n nodes, computing all solution densities takes $\mathcal{O}(\gamma n^3)$ time. Figure 2 shows an example graph and its Laplacian matrix.

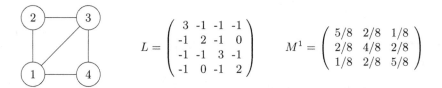

Fig. 2. Graph and its Laplacian matrix with $\gamma = 2$, meaning we can get every edge density by inverting two submatrices, e.g. L^1 (with inverse M^1 shown) and L^3.

With this formula counting-based search heuristics can be used on problems such as degree constrained spanning trees (and Hamiltonian paths in particular) with very good results [1]. However, they become impractical for large instances because of repeated matrix inversion. The following sections address this problem by proposing two improvements. Note that these improvements remain valid for the recent generalization to weighted spanning trees [4].

3.1   Faster Specialized Matrix Inversion

By construction, the sub-matrix L^u we invert has a special form that enables us to use a specialized algorithm. It is Hermitian (more precisely, integer symmetric). Since the row and column removed from L have the same index u, it is *diagonally dominant*: $|\ell_{ii}^u| \geq \sum_{j \neq i} |\ell_{ij}|, \forall i$. Its diagonal entries are positive. Therefore it is positive semidefinite or, equivalently, has non-negative eigenvalues. The Matrix-Tree Theorem states that the number of spanning trees is equal to the determinant of L^u, itself equal to the product of its eigenvalues. Therefore each eigenvalue is strictly positive and L^u is positive definite.

A Hermitian positive definite matrix can be inverted via Cholesky factorization instead of the standard LU factorization. Inverting a positive definite matrix requires approximately $\frac{1}{3}n^3$ (Cholesky factorization) $+ \frac{2}{3}n^3$ floating-point operations whereas inverting a general matrix requires approximately $\frac{2}{3}n^3$ (LU factorization) $+ \frac{4}{3}n^3$ floating-point operations [3]. We therefore expect a two-fold improvement in runtime.

3.2   Inverting Smaller Matrices Through Graph Contraction

When branching, if an edge (u, v) is fixed, the Laplacian matrix must be updated to reflect this change. The technique described by Brockbank, Pesant and Rousseau is the following: if it is forbidden, we set $l_{uv} = 0$; if it is required, we must contract it in the graph, meaning we transfer the edges of vertices u and v to a representative vertex in the same connected component while keeping a 1 on the diagonal of these vertices to keep the matrix invertible. Figure 3 shows an example.

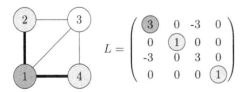

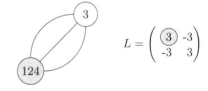

Fig. 3. Contraction following assignments $e(1, 2) = 1$ and $e(1, 4) = 1$, with 1 as the representative vertex in connected component $\{1, 2, 4\}$.

Fig. 4. Connected component $\{1, 2, 4\}$ as a single vertex.

This way of updating L works but still requires that we invert $(n-1) \times (n-1)$ matrices to compute solution densities throughout the search. However, as shown in Fig. 4, we can view each connected component as a single vertex, leading to smaller Laplacian matrices, and thus smaller matrices to invert as we fix edges.

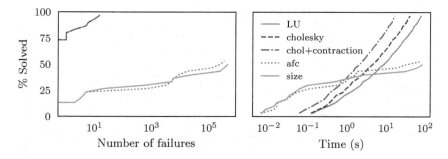

Fig. 5. Percentage of Hamiltonian Path instances solved w.r.t. time and number of failures with a 5 min cutoff.

3.3 Experiments on the Hamiltonian Path Problem

To test our two improvements we designed a simple model to find Hamiltonian paths with Gecode's `path` constraint and a redundant `spanningTree` constraint that computes solution densities according to our `spanningTree` counting algorithm. The `spanningTree` constraint is expressed on binary edge variables as opposed to vertex successor variables for the `path` constraint—the two variable representations are channelled.

In Fig. 5 improvements to the `spanningTree` counting algorithm are tested on this model with *maxSD* branching on binary edge variables—for comparison we also tried heuristics *afc* and *size* branching on vertex successor variables (and trying values in lexicographic order). We use 30 graphs over 60 to 234 vertices taken from the FHCP Challenge Set [8]. Again we observe that *maxSD* (with the curves of its three algorithms coinciding in the first graph) guides search much more effectively by solving the instances in several orders of magnitude fewer failures than *afc* and *size*. At one second of computation time, all three heuristics solve about the same number; at ten seconds, *maxSD* solves almost all of them whereas *afc* and *size* solve about half.

4   Avoiding Systematic Recomputation

The improvements we presented so far are specific to the `alldifferent` and `spanningTree` constraints. In this section we present an additional technique applicable to any constraint in order to avoid recomputation but at the expense of accuracy. Usually at every node of the search tree, before branching, we systematically call the counting algorithm for each constraint. Suppose we have a `spanningTree` constraint on a graph with hundreds of vertices and thousands of edges: we may have fixed a single edge with very few changes propagated since the last call to its counting algorithm but the whole computation, involving the expensive inversion of large matrices, will be undertaken again even though the resulting solution densities are likely to be very similar. To avoid this we propose a simple dynamic technique: while the variable domains involved remain about

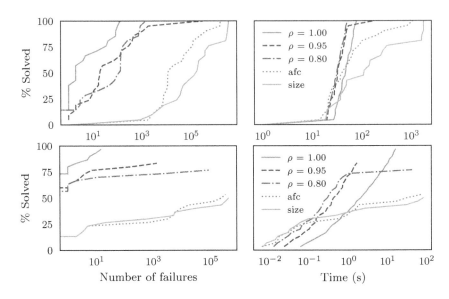

Fig. 6. The effect of different recomputation ratios for the Quasigroup Completion (first row) and Hamiltonian Path Problem (second row), using same instances as before.

the same, we do not recompute solution densities for a constraint but use the latest ones as an estimate instead. For any given node k in the search tree and constraint c, let $S_c^k = \sum_{x_i \in c} d_i$. We recompute only if $S_c^k \leq \rho S_c^j$, with $0 < \rho \leq 1$ some appropriate ratio and j the last node above k but on the same path from the root for which we computed solution densities for c. Note that as opposed to a static criterion such as calling the counting algorithm of every constraint at fixed intervals of depth in the search tree, our approach adapts dynamically to individual constraints and to how quickly the domains of the variables in their scope shrink.

Figure 6 show the performance of different recomputation ratios with $maxSD$ for the previous experiments ($\rho = 1$ means always recompute). As expected, with a lower ρ we loose accuracy as seen in the number of fails but overall we gain speed.

5   Conclusion

To accelerate counting-based search we proposed various improvements to the `alldifferent` and `spanningtree` counting algorithms and a simple generic method to avoid systematic recomputation during search. In our experiments these improvements brought 30- to 100-fold speed ups on Quasigroup Completion and Hamiltonian Path problems.

Financial support for this research was provided by an NSERC postgraduate scholarship and NSERC Discovery Grant 218028/2017.

References

1. Brockbank, S., Pesant, G., Rousseau, L.-M.: Counting spanning trees to guide search in constrained spanning tree problems. In: Schulte, C. (ed.) CP 2013. LNCS, vol. 8124, pp. 175–183. Springer, Heidelberg (2013). https://doi.org/10.1007/978-3-642-40627-0_16

2. Chaiken, S., Kleitman, D.J.: Matrix tree theorems. J. Comb. Theory Ser. A **24**(3), 377–381 (1978)

3. Choi, J., Dongarra, J.J., Ostrouchov, L.S., Petitet, A.P., Walker, D.W., Whaley, R.C.: Design and implementation of the ScaLAPACK LU, QR, and Cholesky factorization routines. Sci. Program. **5**(3), 173–184 (1996)

4. Delaite, A., Pesant, G.: Counting weighted spanning trees to solve constrained minimum spanning tree problems. In: Salvagnin, D., Lombardi, M. (eds.) CPAIOR 2017. LNCS, vol. 10335, pp. 176–184. Springer, Cham (2017). https://doi.org/10.1007/978-3-319-59776-8_14

5. Gagnon, S.: Gecode extension for counting-based search (2017). https://github.com/SaGagnon/gecode-5-extension

6. Gecode Team: Gecode: generic constraint development environment (2017). http://www.gecode.org

7. Gomes, C., Shmoys, D.: Completing quasigroups or Latin squares: a structured graph coloring problem. In: Computational Symposium on Graph Coloring and Generalizations, January 2002

8. Haythorpe, M.: FHCP challenge set (2015). http://fhcp.edu.au/fhcpcs

9. Pesant, G.: Counting-based search for constraint optimization problems. In: Schuurmans, D., Wellman, M.P. (eds.) AAAI, pp. 3441–3448. AAAI Press (2016)

10. Pesant, G., Quimper, C.G., Zanarini, A.: Counting-based search: branching heuristics for constraint satisfaction problems. J. Artif. Int. Res. **43**(1), 173–210 (2012)

11. Valiant, L.G.: The complexity of computing the permanent. Theoret. Comput. Sci. **8**(2), 189–201 (1979)

12. Zanarini, A., Pesant, G.: Solution counting algorithms for constraint-centered search heuristics. In: Bessière, C. (ed.) CP 2007. LNCS, vol. 4741, pp. 743–757. Springer, Heidelberg (2007). https://doi.org/10.1007/978-3-540-74970-7_52

13. Zanarini, A., Pesant, G.: More robust counting-based search heuristics with alldifferent constraints. In: Lodi, A., Milano, M., Toth, P. (eds.) CPAIOR 2010. LNCS, vol. 6140, pp. 354–368. Springer, Heidelberg (2010). https://doi.org/10.1007/978-3-642-13520-0_38

Model Agnostic Solution of CSPs via Deep Learning: A Preliminary Study

Andrea Galassi[(✉)] [ID], Michele Lombardi, Paola Mello, and Michela Milano

Department of Computer Science and Engineering (DISI),
University of Bologna, Bologna, Italy
{a.galassi,michele.lombardi2,paola.mello,michela.milano}@unibo.it

Abstract. Deep Neural Networks (DNNs) have been shaking the AI scene, for their ability to excel at Machine Learning tasks without relying on complex, hand-crafted, features. Here, we probe whether a DNN can learn how to construct solutions of a CSP, without any explicit symbolic information about the problem constraints. We train a DNN to extend a feasible solution by making a single, globally consistent, variable assignment. The training is done over intermediate steps of the construction of feasible solutions. From a scientific standpoint, we are interested in whether a DNN can learn the structure of a combinatorial problem, even when trained on (arbitrarily chosen) construction sequences of feasible solutions. In practice, the network could also be used to guide a search process, e.g. to take into account (soft) constraints that are implicit in past solutions or hard to capture in a traditional declarative model. This research line is still at an early stage, and a number of complex issues remain open. Nevertheless, we already have intriguing results on the classical Partial Latin Square and N-Queen completion problems.

1   Introduction

Deep Neural Networks (DNNs) [12], are characterized by the ability to learn high-level concepts without the need of symbolic features. In this paper, we investigate the idea that DNNs could be capable of learning how to solve combinatorial problems, *with no explicit information about the problem constraints*. This is partially motivated by the results achieved in a previous work regarding the application of DNNs to a board game [3]. In particular, we train a DNN to extend a feasible partial solution by making a single, globally consistent, variable assignment.

   In principle, *such a network could be used to guide a search process*: this may be used to take into account constraints that are either implicit in the training solutions, or too difficult to capture in a declarative model. In this sense, the approach is complementary to Empirical Model Learning (EML) [14], where the goal is instead to learn a constraint. The method presented here is applicable even when only positive examples (i.e. feasible solutions) are available. Moreover, using the DNN to guide search may also provide a speed-up when solving multiple instances of the same problem. *Practical applications are not our only*

© Springer International Publishing AG, part of Springer Nature 2018
W.-J. van Hoeve (Ed.): CPAIOR 2018, LNCS 10848, pp. 254–262, 2018.
https://doi.org/10.1007/978-3-319-93031-2_18

driver, however: there is a strong scientific interest in assessing to what extent a sub-symbolic technique, trained on arbitrarily chosen solution construction sequences, can learn something of the problem structure.

This line of research is at an early stage, and there are many complex issues to be solved before reaching practical viability. *So far, we have focused on two classical Constraint Satisfaction Problems* (CSPs), namely N-queen completion and Partial Latin Square. For these benchmarks we have intriguing results, the most striking being an impressive discrepancy between the (low) DNN accuracy and its (very high) ability to generate feasible assignments: this suggest that the network is indeed learning something about the problem structure, even if it has been trained to "mimic" specific solution construction sequences.

This is not the first time that Neural Networks have been employed to solve CSPs. For example, Guarded Discrete Stochastic networks [1] can solve generic CSPs in an unsupervised way. They rely on a Hopfield network to find consistent variable assignment, and on a "guard" network to force the assignment of all the variables. The GENET [16] method can construct neural networks capable of solving binary CSPs, and was later extended in EGENET [13] to support non-binary CSPs. In [2], a CSP is first reformulated as a quadratic optimization problem. Then, a heuristic is used to guide the evolution of a Hopfield network from an initial state representing an infeasible solution to a final feasible state. Crucially, *all these methods rely on full knowledge of the problem constraints to craft both the structure and the weights of the networks*. What we are trying to do is in fact radically different.

2   General Method and Grounding

General Approach. We train a DNN to extend a partial solution of a combinatorial problem, *by making a single additional assignment that is globally consistent*, i.e. that can be extended to a full solution.

We use simple bit vectors for both the network input and output. We represent assignments using a one-hot encoding, i.e. for a variable with n values we reserve n bits; raising the i-th bit corresponds to assigning the i-th domain value. If no bit is raised, the variable is unassigned. Using such a simple format makes our input encoding *general* (any set of finite domain variables can be encoded), and truly *agnostic to the problem constraints*. As a major drawback, the method is currently restricted to problems of a pre-determined size.

Our training examples are obtained by *deconstructing a comparatively small set of solutions*. We considered two different strategies, referred to as *random* and *systematic deconstruction*, as described in Algorithms 1 and 2. Both methods operate by processing a partial solution s and populate a dataset T with pairs of partial solutions and assignments. In the pseudo code, s_i refers to the value of the i-th variable in s, and $s_i = \bot$ if the variable is unassigned. The random strategy generates in a backward fashion one arbitrary construction sequence for the solution. The systematic strategy generates all possible construction sequences. *When all the original solutions have been deconstructed, we prune the dataset* by considering all groups of examples sharing the same partial solution, and selecting a single representative at random.

Alg. 1 RandomDeconstruction(s)	**Alg. 2** SystematicDeconstruction(s)
Randomly choose a variable index i	**for all** variable indices i **do**
$s' = s$ (copy the partial solution)	$\quad s' = s$ (copy the partial solution)
$s'_i = \perp$ (undo one assignment)	$\quad s'_i = \perp$ (undo one assignment)
Insert (s', s_i) in T	$\quad$ Insert (s', s_i) in T
RandomDeconstruction(s')	$\quad$ SystematicDeconstruction(s')

The DNN is trained for a classification task: for each example, the target vector (i.e. the class label) contains a single raised bit, corresponding to the assignment s_i in the dataset. The network yields a normalized score for each bit in the output vector, which can be interpreted as a probability distribution. The bit with the highest score corresponds to the suggested next assignment. We take no special care to prevent the network from trying to re-assign an already assigned variable. These choices have three important consequences: (1) the network is *agnostic to the problem structure*; (2) the network is technically *trained to mimic specific construction sequences* of arbitrarily chosen solutions; (3) assuming that the DNN is used to guide a search process, it is *easy to take into account propagation* by disregarding the scores for variable-value pairs that have been pruned. As an adverse effect, we are forsaking possible performance advantages that could come by including information about the problem structure.

Grounding (Benchmark Problems). So far, we have grounded our approach on two classical CSPs, namely the N-queen completion and Partial Latin Square (PLS, see [4]) problems. Classical problems let us work in a controlled setting with well known properties [6–8], and simplifies drawing scientific conclusions.

The N-queen completion problem consist in placing n queens pieces on a $n \times n$ chessboard, so that no queen threats another. The PLS problem consist in filling an $n \times n$ square with numbers from 1 to n so that the same number appears only once per row and column. In both cases, some variables may be pre-assigned. We focus on N-queen problems of size 8 and PLSs of size 10. In both cases, we model assignments using a one-hot encoding, leading to vector of size $8 \times 8 = 64$ for the n-queens and $10 \times 10 \times 10 = 1,000$ for the PLS.

For the 8-queen problem, we have used 1/4 of the 12 non-symmetric solution to seed the training set, and the remaining ones for the test set. Both the training and the test set are then obtained by generating all the symmetric equivalents, and then by applying systematic deconstruction to the resulting solutions.

For the PLS, we have used an unbiased random generation method to obtain two "raw" datasets, respectively containing 10,000 and 20,000 solutions. The numbers are considerably large in this case, but they are very small compared to the number of size 10 PLS ($\sim 10^{31}$). As comparison, it is a bit like making sense of the layout of Manhattan from ~0.75 square nanometers of surface scattered all over the place. Each of the raw datasets is split into a training and test set, containing respectively 1/4 and 3/4 of the solutions. The actual examples have then been obtained by random deconstruction.

Grounding (Networks and Training). Due to the impressive results obtained in computer vision tasks, we have chosen to use pre-activated Residual Networks [5,9,10]. We have adapted the architecture to use fully-connected layers rather than convolutional ones, as the latter are not well suited to deal with generic combinatorial problems.

We have trained the networks in a supervised fashion, using 10% of the examples (chosen at random) as a validation set. The loss function is the negative log-likelihood of the target class, with a 10^{-4} L1 regularization coefficient. The choice of the network and training hyper-parameters has been made after an informal tuning. We have eventually settled for using the Adam [11] optimizer, with parameters $\beta_1 = 0.9$ and $\beta_2 = 0.99$. The initial learning rate α_0, was progressively annealed through epochs with decay proportional to training epoch t, resulting in a learning rate $\alpha = \frac{\alpha_0}{1+k \times t}$ with $k = 10^{-3}$. Training was stopped after there was no improvement on the validation accuracy for e epochs.

For the 8-queens problem we have used an initial layer of 200 neurons, than 100 residual blocks, each one composed by two layers of 500 and 200 neurons, and finally an output layer of 64 neurons, for a total of more than 200 layers. Batch optimization has been employed, using a initial learning rate of $\alpha_0 = 0.1$ and a patience of $e = 200$ epochs. Dropout [15] has been applied to each input and hidden neuron with probability $p = 0.1$.

For the PLS problem, we have used a smaller network because of the bigger input/output vectors and the larger datasets would have required too much training time. Therefore we have used an initial layer of 200 neurons, then 10 residual blocks, each one composed by two layers of 300 and 200 neurons, and a final output layer of 1000 neurons, for a total of 22 layers. Mini-batch optimization has been employed, using shuffling in each epoch, using a initial learning rate of $\alpha_0 = 0.03$ and a patience of $e = 50$ epochs. Dropout has been applied to hidden neuron with probability $p = 0.1$. The size of the mini batch has been setted to 50,000 for the training on the 10k dataset and to 100,000 for the 20k dataset.

3   Experimentation

We designed our experiments to address four main questions. First, we want to assess how well the DDNs are actually learning their designated task, i.e. to guess the "correct" assignment according to the employed deconstruction method. Second, we are interested in whether the DNNs learn to generate feasible assignments, no matter whether those are "correct" according to datasets. Third, assuming that the networks are actually learning something about the problem constraints, it makes sense to check whether some constraint types are learned

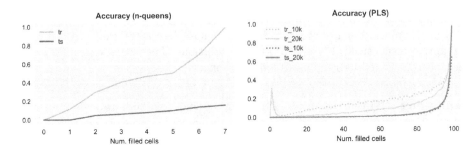

Fig. 1. Accuracy on the training and test sets

better than others. Finally, we want to investigate whether using the DNNs to guide an actual tree search process leads to a reduction in the number of fails.

Network Accuracy. Here we are interested in assessing the performance of our DNNs in their natural task, i.e. learning "correct" variable-value assignment, as defined by our deconstruction procedure. Figure 1 shows the accuracy reached by our DNNs on both the training and the test sets, grouped by the number of pre-assigned variables in the example input. For comparison, random guessing would reach an accuracy of $1/64 \simeq 0.015$ for the 8-queens and $1/1,000$ for the PLS. There are three notable facts:

1. The accuracy is at least one order or magnitude larger than random guessing, but still rather low, in particular for the PLS; this suggest that *the networks are not doing particularly well at the task they are being trained for.*
2. Second, *the accuracy on the test set if considerably lower than on the training set*; normally this is symptomatic of overfitting, but in this case there is also *a structural reason.* The pruning in the last phase of dataset generation introduces a degree of ambiguity in our training: as an extreme case, for the same partial assignment, the training and the test set may report different "correct" assignments that cannot be both predicted correctly.
3. Third, the accuracy tends to increase with the number of filled cells. Having many filled cells means having very few feasible completions, and therefore it is more unlikely for the same instance to appear both in the test and train set with a different target. In this situation it is intuitively easier for the network to label a specific assignment as the "correct" one.

The third observation leaves an open question: while the small number of feasible completions can explain why the accuracy raises, it fails to explain the magnitude of the increase. *The result would be much easier to explain by assuming that the DNN has somehow learned something about the problem constraints.*

Feasibility Ratio. It makes sense to evaluate the ability of the DNNs to yield globally consistent assignments, even if those are not chosen as "correct", since

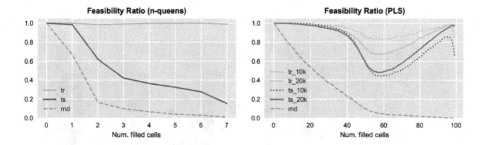

Fig. 2. Feasibility ratios on the training and test sets

this is our primary goal. Figure 2 show the ratio of predictions of the DNNs (both on the training and test set) that could be expanded to full solutions. For comparison, the figures report also the results that can be obtained by guessing at random on the test sets. There are three very relevant observations to make:

1. There is *a striking difference between the accuracy values from Fig. 1 and the feasibility ratios.*
2. Such discrepancy may be due to the fact that the more a partial solution is empty, the more are the feasible assignments that can be found even by guessing. However, *the reported feasibility ratio are also significantly higher than the random baseline.* This is hard to explain, unless we assume that the DNNs have somehow learned the semantic of the problem constraints.
3. The feasibility ratios for the PLS networks have a dip between 50 and 60 pre-assigned variables, and then tend to raise again. *This is exactly the behavior that one would expect thank to constraint propagation:* when many variables are bound many values are pruned and the number of available assignments is reduced. However, *the DNNs at this stage do not rely on propagation at all.* Even the higher accuracy from Fig. 1 is not enough to justify how much the feasibility rations tend to increase for almost full solutions. Assuming that the DNN has learned the problem constraints can explain the increase, but not so easily the dip.

Constraints Preference. Next, we have designed an experiment to investigate whether some constraints are handled better than others. We start by generating a pool of (partial) solutions by using the DNN to guide a randomized constructive heuristic. Given a partial solution, we use the DNN to obtain a probability distribution over all possible assignment, one of which is chosen randomly and performed. Starting from an empty solution, the process is repeated as many times as there are variables, and relies on our low-level, bit vector, representation of the partial solution. As a consequence, at the end of the process there may be variables that have been "assigned multiple times", and therefore also unassigned variables. We have used this approach to generate 10,000 solutions for each DNN, and for comparison we have done the same using a uniform distribution.

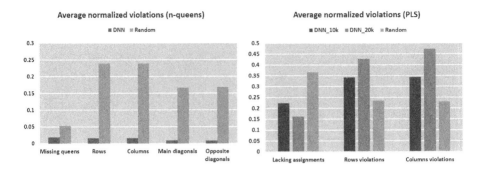

Fig. 3. Average violations on the two problems

Once we have such a pool of partial solution, we count the average degree of violation of each abstract problem constraints, e.g. the number of rows with multiple queens. Each quantity is then normalized over the corresponding maximum (i.e. the number of row/columns, or the number of variables). Looking at the average violations for the random baseline intuitively tells the natural difficulty of satisfying a constraint type. Comparing such values with those of the DNN allows to evaluate how well the DNN is faring.

As reported in Fig. 3, for the N-queens problem the network gets much closer to feasibility than the random baseline and all problem constraints are handled equally well. For the PLS problem, the DNNs violates the row and column constraints significantly more than the random baseline, but they also tend to leave fewer variable unassigned. There is a logic correlation between these two values, since assigning more variables increases the probability to violate a row or a column constraint.

Guiding Tree Search. Finally, we have tried using our DNNs to guide a Depth First Search process for the Partial Latin Square[1]. In particular, we always make the assignment with the largest score, excluding bounded variables and values pruned by propagation.

We employ a classic CP model for the PLS (one finite domain variable per cell), and use the GAC ALLDIFFERENT propagator for the row and column constraint. We compare the results of the two DNNs with those of heuristic that pick uniformly at random both the variable and the value to be assigned. Given that our research is at a early stage, we have opted for a simple (but inefficient) implementation relying on the Google or-tools python API: for this reason we focus our evaluation on the number of fails. As a benchmarks, we have sampled 4,000 partial solutions from the 20k training and test set, at the complexity peak. All instances have been solved with a cap at 10,000 fails.

The results of this experimentation are reported in Fig. 4, using box plots. Apparently, the DNN trained on the 10k dataset is more efficient than the

---

[1] The 8-queens problem is too easy to provide meaningful measurements.

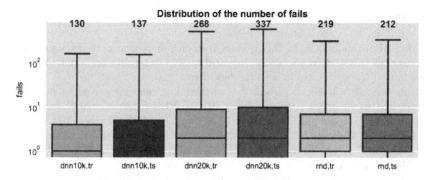

Fig. 4. Distribution of the number of fails on the train and test sets. At the top of each box we report the number of times the fail cap was reached.

random baseline, but the opposite holds for the one trained in the 20k dataset. This matches the results obtained in our analysis of violated constraints, but not those obtained for the feasibility ratio. We suspect however that explaining the performance (and obtaining practical speed-up) will require to take into account the complex trade-off making choice that are likely feasible, and recovering quickly from the inevitable mistakes. This is a well know open problem in Constraint Programming, that we plan to tackle as part of future work.

4   Conclusions

We have performed a preliminary investigation to understand whether DNNs can learn how to solve combinatorial problems. We have adopted a general setup, totally agnostic to the problem structure, and we have trained the networks on arbitrarily chosen solution construction sequences.

Our experimentation has provided evidence that, despite having low accuracy at training time, a DNN can become capable of generating globally consistent variable-value assignments. This cannot be explained assuming that the networks only mimic the assignment sequences in the training set, but it compatible with the hypothesis that the DNNs have learned something about the problem structure. The networks do not seem to favor any abstract constraint in particular, suggesting that what they are learning does not match our usual understanding of CSPs. When used for guiding a search process, our DNNs have provided mixed results, highlighting that achieving performance improvements may require to deal more explicitly with the peculiarities of a specific solution technique (e.g. constraint propagation).

This research line is still at an early stage: there are considerable overheads that make practical applications still far, and the method is currently limited to problems of fixed size, a problem that maybe could be solved using only convolutional layers. However, we believe the approach to have enough potential to deserve further investigation.

References

1. Adorf, H.M., Johnston, M.D.: A discrete stochastic neural network algorithm for constraint satisfaction problems. In: 1990 IJCNN International Joint Conference on Neural Networks, vol. 3, pp. 917–924, June 1990
2. Bouhouch, A., Chakir, L., Qadi, A.E.: Scheduling meeting solved by neural network and min-conflict heuristic. In: 2016 4th IEEE International Colloquium on Information Science and Technology (CiSt), pp. 773–778, October 2016
3. Chesani, F., Galassi, A., Lippi, M., Mello, P.: Can deep networks learn to play by the rules? A case study on nine men's morris. IEEE Trans. Games **PP**(99), 1 (2018). https://doi.org/10.1109/TG.2018.2804039
4. Colbourn, C.J.: The complexity of completing partial latin squares. Discret. Appl. Math. **8**(1), 25–30 (1984)
5. Ebrahimi, M.S., Abadi, H.K.: Study of residual networks for image recognition. arXiv preprint arXiv:1805.00325 (2018)
6. Gent, I.P., Jefferson, C., Nightingale, P.: Complexity of n-Queens completion. J. Artif. Intell. Res. **59**, 815–848 (2017)
7. Gomes, C.P., Selman, B., Crato, N.: Heavy-tailed distributions in combinatorial search. In: Smolka, G. (ed.) CP 1997. LNCS, vol. 1330, pp. 121–135. Springer, Heidelberg (1997). https://doi.org/10.1007/BFb0017434
8. Gomes, C.P., Selman, B., Kautz, H.A.: Boosting combinatorial search through randomization. In: Proceedings of the Fifteenth National Conference on Artificial Intelligence and Tenth Innovative Applications of Artificial Intelligence Conference, AAAI 1998, IAAI 1998, 26–30 July 1998, Madison, Wisconsin, USA, pp. 431–437 (1998). http://www.aaai.org/Library/AAAI/1998/aaai98-061.php
9. He, K., Zhang, X., Ren, S., Sun, J.: Deep residual learning for image recognition. In: Proceedings of the IEEE Conference on Computer Vision and Pattern Recognition, pp. 770–778 (2016)
10. He, K., Zhang, X., Ren, S., Sun, J.: Identity mappings in deep residual networks. In: Leibe, B., Matas, J., Sebe, N., Welling, M. (eds.) ECCV 2016. LNCS, vol. 9908, pp. 630–645. Springer, Cham (2016). https://doi.org/10.1007/978-3-319-46493-0_38
11. Kingma, D.P., Ba, J.: Adam: a method for stochastic optimization. CoRR abs/1412.6980 (2014). http://arxiv.org/abs/1412.6980
12. LeCun, Y., Bengio, Y., Hinton, G.: Deep learning. Nature **521**(7553), 436–444 (2015)
13. Lee, J.H.M., Leung, H.F., Won, H.W.: Extending GENET for non-binary CSP's. In: Proceedings of 7th IEEE International Conference on Tools with Artificial Intelligence, pp. 338–343, November 1995
14. Lombardi, M., Milano, M., Bartolini, A.: Empirical decision model learning. Artif. Intell. **244**, 343–367 (2017). https://doi.org/10.1016/j.artint.2016.01.005
15. Srivastava, N., Hinton, G.E., Krizhevsky, A., Sutskever, I., Salakhutdinov, R.: Dropout: a simple way to prevent neural networks from overfitting. J. Mach. Learn. Res. **15**(1), 1929–1958 (2014)
16. Wang, C.J., Tsang, E.P.K.: Solving constraint satisfaction problems using neural networks. In: 1991 Second International Conference on Artificial Neural Networks, pp. 295–299, November 1991

Boosting Efficiency for Computing the Pareto Frontier on Tree Structured Networks

Jonathan M. Gomes-Selman[1], Qinru Shi[2], Yexiang Xue[3],
Roosevelt García-Villacorta[4], Alexander S. Flecker[4], and Carla P. Gomes[3(✉)]

[1] Department of Computer Science, Stanford University, Stanford, USA
jgs8@stanford.edu
[2] Center for Applied Mathematics, Cornell University, Ithaca, USA
qs63@cornell.edu
[3] Department of Computer Science, Cornell University, Ithaca, USA
yx247@cornell.edu, gomes@cs.cornell.edu
[4] Department of Ecology and Evolutionary Biology, Cornell University, Ithaca, USA
rg676@cornell.edu, asf3@cornell.edu

Abstract. Multi-objective optimization plays a key role in the study of real-world problems, as they often involve multiple criteria. In multi-objective optimization it is important to identify the so-called Pareto frontier, which characterizes the trade-offs between the objectives of different solutions. We show how a divide-and-conquer approach, combined with batched processing and pruning, significantly boosts the performance of an exact and approximation dynamic programming (DP) algorithm for computing the Pareto frontier on tree-structured networks, proposed in [18]. We also show how exploiting restarts and a new instance selection strategy boosts the performance and accuracy of a mixed integer programming (MIP) approach for approximating the Pareto frontier. We provide empirical results demonstrating that our DP and MIP approaches have complementary strengths and outperform previous algorithms in efficiency and accuracy. Our work is motivated by a problem in computational sustainability concerning the evaluation of trade-offs in ecosystem services due to the proliferation of hydropower dams throughout the Amazon basin. Our approaches are general and can be applied to computing the Pareto frontier of a variety of multi-objective problems on tree-structured networks.

Keywords: Multi-objective optimization · Pareto frontier
Approximation algorithms · Dynamic programming
Mixed-integer programming

J.M. Gomes-Selman and Q. Shi—These authors are contributed Equally.

W.-J. van Hoeve (Ed.): CPAIOR 2018, LNCS 10848, pp. 263–279, 2018.
https://doi.org/10.1007/978-3-319-93031-2_19

1   Introduction

In recent years there has been a rapid proliferation of hydropower dams through-out the Amazon basin, which dramatically affects a variety of ecosystem services provided by the river network such as biodiversity, nutrient and sediment transport, freshwater fisheries, navigation, and energy production [4,17,21,22] (see Fig. 1). Hydropower dam placement is a good example of a challenging real-world problem in computational sustainability [6], which often involves multiple objective problems concerning the balancing of environmental, economic, and societal needs. More concretely, hydropower dam placement is a multi-objective optimization problem concerning the placement of dams throughout a river network, which is naturally a tree-structured network, trading-off various ecological, social, and economic goals. In multi-objective optimization, the so-called *Pareto frontier* captures the trade-offs among multiple objectives. The Pareto frontier is the set of all *Pareto optimal solutions*; a solution is considered Pareto optimal if its vector of objective values is not *dominated* by any other feasible solution. See Fig. 1 for an example of a 2-dimensional Pareto frontier.

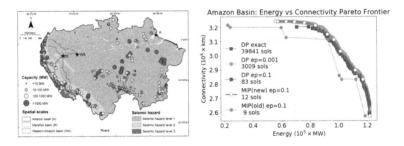

Fig. 1. Left panel: Amazon basin: around 300 hydropower dams are proposed or planned. Three objectives are depicted: (1) Energy (dot sizes denote dam capacity in MW); (2) Longitudinal connectivity (continuous unobstructed river segments from the root of the basins, which are marked with stars); and (3) Seismic risk (map background colors). **Right panel:** The approximate DP ($\epsilon = 0.001$) overlaps the exact DP. The new MIP approach is substantially better than the previous MIP approach ($\epsilon = 0.1$) and its solutions are on the exact Pareto curve, slightly better than the DP approximation for the same $\epsilon = 0.1$. For a given ϵ, DP produces substantially more Pareto solutions than MIP.

Recently there has been considerable interest in the study of multi-objective optimization problems (see e.g., [2,3,8,10,11,13,14,16,20]). Existing approaches are primarily heuristic, based on local search or evolutionary algorithms, without theoretical guarantees, and do not exploit the tree structure. [5] provides a constraint programming exact algorithm which is extended by [12] with large neighborhood search. Both algorithms are designed for general problems. [1] provides data structures to store Pareto optimal policies in an exact algorithm.

This paper focuses on *computing the Pareto frontier, both exact and with approximation guarantees, on tree-structured networks*. In [18] we proposed a dynamic programming algorithm for trees which computes the exact Pareto frontier, as well as a rounding technique applied to the exact dynamic programming algorithm that provides a fully polynomial-time approximation scheme (FPTAS). The FPTAS finds a solution set of polynomial size, which approximates the Pareto frontier within an arbitrary small ϵ factor and runs in time that is polynomial in the size of the instance and $1/\epsilon$. We also formulated the problem of optimizing the placement of dams as a mixed integer programming problem (MIP) and used it to approximate the Pareto frontier. While the results in [18] are encouraging, there is room for improvement.

Our Contributions: (1) A key component of our DP algorithm is the pruning of dominated solutions. We provide a **divide-and-conquer approach** that **significantly improves the efficiency of the pruning of dominated solutions** and outperforms the previous approach, leading to **speed-ups of two to three orders of magnitude**, in practice; (2) To cope with the large memory requirements of a multi-objective Pareto frontier, we propose **batching to identify and prune dominated solutions incrementally**, scaling up to much larger problems; (3) We also propose a **new MIP based approximation scheme** that exploits restarts and a new instance selection strategy, which boosts the performance and accuracy of the previous MIP approach for approximating the Pareto frontier. (4) We design a visualization tool of our results intended for decision makers. (5) We provide empirical results showing that **our proposed algorithms significantly outperform previous approaches.**

Preview of Results: Our DP and MIP Pareto frontier algorithms are complementary and scale up to much larger real-world instances than previous algorithms: **the DP can now approximate the Pareto frontier for the entire Amazon basin**, when optimizing for energy, connectivity (a proxy for e.g., unimpeded fish migrations and transportation), seismic risk, and sediment, in around 5 days, with a coverage of $2,193,314$ non-dominated solutions, with the guarantee that the solutions are within at most 5% of the true optimum ($\epsilon = 0.05$); in less than 6 h, the DP provides a coverage of $491,578$ non-dominated solutions ($\epsilon = 0.1$); in around 6.5 min, the DP provides a coverage of $23,019$ non-dominated solutions ($\epsilon = 0.25$); for the same $\epsilon = 0.25$, the MIP approach approximates the Pareto frontier in around 25 min, with a smaller coverage of 95 non-dominated solutions, but the MIP approach provides more flexibility when considering additional constraints and, in practice, its solutions tend to be closer to the exact Pareto frontier for a given ϵ. Our overall goal is to enable more informed decisions concerning the trade-offs of multiple objectives of optimization problems.

2   Problem Formulation

In this section, we first introduce the hydropower dam placement problem as an example of a multi-objective optimization problem on a tree structured network. Then, we show the general formulation of such problems.

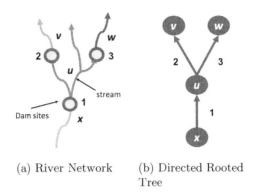

(a) River Network   (b) Directed Rooted
                               Tree

Fig. 2. Converting a river network (left) into a more compact directed rooted tree (right: x is the root). Each contiguous region of the river network (represented by different colors, and labeled x, u, v, w) is converted into a node, also referred to as a hypernode (labeled with the corresponding letter, x, u, v, w) in the tree network. Each potential dam site (represented by a red-yellow circle) is represented by an edge in the directed rooted tree. (Color figure online)

2.1   Hydropower Dam Placement Problem

We are given a set of planned dams and need to decide the optimal subset of dams to build. We refer to this problem as the hydropower dam placement problem. We first point out that a river network is a directed tree-structure network and that, for the purposes of our hydropower dam placement problem, we don't need to explicitly consider every river segment. So, we first abstract the river network and potential dam locations into a more compact directed rooted tree that captures the key problem information. Each contiguous section of the river network uninterrupted by existing or potential dam locations is represented by a node (we also call it hypernode to emphasize that it encapsulates a river sub network). Each existing or potential dam location is represented by a directed edge pointing from downstream to upstream. See Fig. 2 for an example of our conversion of a river network into a more compact directed rooted tree.

A policy (or solution) π is a subset of potential dam sites to be built. We can encode many environmental and economical objectives as a function of π. In this paper, we focus on the following four objectives:

Energy (E)**:** Given a solution π, the total hydropower produced by the selected dams is $E(\pi) = \sum_{e \in \pi} h_e$, where h_e is the hydropower of the dam represented by edge e. We want to maximize this objective.

Longitudinal Connectivity (C)**:** For a given solution π, the connectivity of a river network is measured by the total length of the unobstructed stream segments that one can travel starting from the root (river mouth) without passing any dam site in π. We want to maximize this objective.

Sediment (S_d)**:** For a given solution π, this objective represents the total amount of sediment transported to the river mouth (the ocean in the case of the entire Amazon basin). We assume that each node produces a fixed amount of sediment and each dam traps a certain percentage of the sediment from upstreams. We want to maximize this objective.

Seismic Risk (S_e)**:** Each dam is associated with a seismic risk factor computed based on its location and its capacity. Given a solution π, this objective is just the sum of seismic risk factors of all dams in π. We want to minimize this objective.

The goal of the hydropower dam placement problem is then to optimize the objective function: $(E(\pi), C(\pi), S_d(\pi), S_e(\pi))$. Here we are not just looking for a single solution. For every possible solution, we say that the solution is optimal as long as there does not exist another solution that is superior in every aspect.

2.2   General Formulation

In general, a multi-objective optimization problem is to optimize a given multi-objective function: $(z^1(\pi), z^2(\pi), \ldots, z^d(\pi))$, where the value of each function z^i depends on a common *solution* π, also referred to as a *policy*. Without loss of generality, we only consider the problem of *maximizing* objective functions. Minimizing objective functions can be treated in a similar fashion.

Pareto Dominance: Given two policies π and π', we say that π dominates π' if the following two conditions hold: (1) for all i, $z^i(\pi) \geq z^i(\pi')$; (2) at least one strict inequality holds for some i.

Pareto Frontier: A Pareto optimal policy is the one that is not dominated by any other policies. The Pareto frontier is the set containing all Pareto optimal policies.

An Example: Consider a multi-objective function (z^1, z^2, z^3) and policies π_1, π_2, and π_3, leading to: $(z^1(\pi_1), z^2(\pi_1), z^3(\pi_1)) = (5, 7, 10)$, $(z^1(\pi_2), z^2(\pi_2)$, $z^3(\pi_2)) = (4, 7, 9)$, and $(z^1(\pi_3), z^2(\pi_3), z^3(\pi_3)) = (6, 6, 9)$. π_1 dominates π_2 because it has higher or equal values in all objectives, with some objectives with higher values. π_1 does not dominate π_3 because of the first objective. π_3 does not dominate π_1 because of the second and third objectives. π_1 and π_3 are Pareto optimal and form the Pareto frontier.

Multi-objective Function on a Tree: We now give a formal definition of the multi-objective optimization problem on a tree structured network: the hydropower dam placement problem is a particular example. Given a *tree structured network* (such as a river network), the objective function z^i is defined recursively. Node rewards $r_v^1, \ldots, r_v^d$ are associated with each node v in the tree. The objective function defined on a leaf node v is its corresponding reward, i.e.,

$z_v^i(\pi) = r_v^i$. Each edge is associated with a transfer coefficient that is affected by whether the corresponding dam is built or not. If the dam represented by (u, v) is built, then (u, v) has a transfer coefficient of p_{uv}^i; otherwise, q_{uv}^i. Also associated with each edge (u, v) is a reward s_{uv}^i and an indicator variable denoting whether the corresponding edge is in π or not. The objective function on a non-leaf node u is defined recursively:

$$z_u^i(\pi) = r_u^i + \sum_{v \in ch(u)} I(uv \in \pi)s_{uv}^i + \sum_{v \in ch(u)} \left(I(uv \in \pi)p_{uv}^i + I(uv \notin \pi)q_{uv}^i \right) z_v^i(\pi). \quad (1)$$

Here, $I(\cdot)$ is an indicator function. $ch(u)$ is the child set of u. The objective function for the entire tree network $\mathcal{T}$ is the function at the root node s, i.e., $z^i(\pi) = z_s^i(\pi)$. Given a multi-objective function defined on a tree network $\mathcal{T}$, our **multi-objective optimization problem on a tree structured network** is to find the Pareto frontier consisting of all non-dominated policies, which is NP-hard even though it is defined on a tree. See [18] for further details.

Application to the Hydropower Dam Placement Problem: In the hydropower dam placement problem, when modeling **connectivity** (i.e., $i =$ connectivity), we set r_u^i to be the total lengths of all stream segments in the region represented by node u. We set $p_{uv}^i = 0$ and $q_{uv}^i = 1$; that is, we either acquire all upstream segments (when the dam corresponding to edge (u, v) is not built) or lose all of them (when the dam is built). We set $s_{uv}^i = 0$. When modeling energy (i.e., $i =$ energy), we set $s_{uv}^i = h_{uv}$, in which h_{uv} is the hydropower produced by the dam site (u, v). r_u^i is set to 0, p_{uv}^i and q_{uv}^i are both set to 1.

3   DP-Based Pareto Frontier

In [18] we proposed a dynamic programming (DP) algorithm (shown in Algorithm 1) that recursively computes the Pareto optimal partial solutions from leaf nodes up to the root. The key insight is that at a given node u we only need to keep the Pareto optimal partial solutions [18]. To increase incremental pruning, we convert the original tree into an equivalent binary tree. Given a binary tree, we first compute Pareto optimal solutions for the two children of u (line 6 and 7), enumerate the partial policies from the children and consider four different combinations of whether to include each of the edges from the children, computing the objective values based on Eq. 1, and adding them to the policy set P (line 8). We then remove all dominated policies (line 9). So, the remaining policies are Pareto optimal for the parent node.

In [18] we also proposed a rounding scheme applied to the exact DP algorithm, which provides a fully polynomial-time approximation scheme (FPTAS) that finds a polynomially succinct policy set, which approximates the Pareto frontier within an arbitrary small ϵ factor and runs in time polynomial in the size of the instance and $1/\epsilon$. The key idea is to project objective values that are ϵ-close into one, which decreases the number of Pareto solutions.

Algorithm 1. $\text{Pareto}_T(u)$: compute the Pareto frontier for the value function defined on the subtree of T rooted at node u.

---

1 **if** *is_leaf(u)* **then**
2   return $\{(r_u^1, \ldots, r_u^m)\}$;
3 **else**
4     $l \leftarrow u.left\_child$;
5     $r \leftarrow u.right\_child$;
6     $Pleft \leftarrow \text{Pareto}_T(l)$;
7     $Pright \leftarrow \text{Pareto}_T(r)$;
8     $S_u \leftarrow$ the set of all possible partial solutions at u obtained by combining solutions from $Pleft$ and $Pright$ and possible policies on (u, l) and (u, r);
9   return $\text{Non\_Dominated}(S_u)$;
10 **end**

3.1 Divide-and-Conquer for Identifying Dominated Solutions

The major runtime bottleneck of Algorithm 1 is pruning the dominated solutions (line 9). Let d be the number of objectives. Assume we generate n partial solutions and get m non-dominated partial solutions, then the naive pruning step takes $O(mnd)$ time. Here we describe a strategy that *significantly boost the efficiency of the overall DP algorithms for computing the Pareto frontier*, which leverages the dimensionality of solutions to efficiently identify the subset of non-dominated solutions from a set of candidate solutions. The new divide-and-conquer based algorithm for finding the non-dominated solutions runs in $O(n(\log n)^{d-1})$ if we use comparison-based sort or $O(n(\log n)^{d-2})$ if the data is stored in the Lattice Latin Hypercube (LLH) form [19]. Here d is the number of criteria we are considering. This algorithm is inspired by an approach proposed in [19].

To simplify the description of our algorithms, we assume that the values of each criterion never repeat. In practice, it is fairly trivial to consider the corner case. Specifically, when splitting the set S based on the dth criterion, we implemented a modified sorting routine that sorts the solutions in lexicographic order, based on the dth criterion. If two solutions have the same d^{th} criterion, we sort them based on the $(d-1)^{th}$ criterion, etc. Note that if two solutions are equal for all criteria then their ordering does not matter.

When the number of objectives is two, we use the method as shown in Algorithm 2. The idea is to sort the solutions based on the first criterion (good ones first). The first solution must be Pareto optimal. Then, we go through the list of solutions sequentially and look for solutions with better second objective than the last non-dominated solution.

When the number of objectives $d \geq 3$, we use our divide-and-conquer based recursive algorithm shown in Algorithm 3. The first step is to split the set of solutions S into two sets A and B of approximately the same size based on the last criterion, so that solutions in A have better last criterion than solutions in B (line 8). The splitting procedure is shown in Algorithm 4. Then, we recursively

---

Algorithm 2. Non_Dominated_2D(S): given a set S of 2-dimensional partial solutions, find the set of non-dominated solutions in S.

---

1 Sort solutions in S by their first element, in descending order if we aim to maximize the element, in ascending order otherwise;
2 $P \leftarrow \{S[1]\}$;
3 **foreach** $s \in S[2:]$ **do**
4 | **if** *s is not dominated by the last element of P* **then**
5 | | Append s to P;
6 | **end**
7 **end**
8 **return** P;

Algorithm 3. Non_Dominated(S): given a set S of d-dimensional partial solutions ($d \geq 2$), find the set of non-dominated solutions in S.

---

1 $d \leftarrow$ dimensionality of solutions in S;
2 $n \leftarrow$ number of solutions in S;
3 **if** $n = 1$ **then**
4 | **return** S
5 **else if** $d = 2$ **then**
6 | **return** Non_Dominated_2D(S)
7 **else**
8 | $A, B \leftarrow$ Split(S, d);
9 | $A' \leftarrow$ Non_Dominated(A); // solutions in A' are non-dominated in S.
10 | $B' \leftarrow$ Non_Dominated(B);
11 | **return** $A' \cup$ Marry($A', B', d - 1$);
12 **end**

---

identify the non-dominated solutions from A and B (line 9 and 10). We know that the non-dominated solutions from set A' are also non-dominated in S, but the same statement may not be true for non-dominated solutions from B'. Thus, the last step is to find the solutions from set B' that are not dominated by solutions from A' (line 11). Note that we already know that the dth objective of solutions in A' are better than the dth criterion of solutions in B', so we only need to consider the first $d - 1$ criteria. To find non-dominated solutions in B', we introduce a slightly modified divide-and-conquer procedure shown in Algorithm 5.

The algorithm Marry(A, B, d') shown in Algorithm 5 returns the set of all solutions in B that are not dominated by any solution in A considering only the first d' criteria. The inputs A and B must be disjoint and no two solutions from the same set dominate one another. Let n be the total number of solutions in $A \cup B$. We split the set of solutions $A \cup B$ into two sets X and Y of approximately the same size based the d'th criterion, so that solutions in X have better d'th criterion than solutions in Y (line 7). Next, we consider the four disjoint subsets $X \cap A$, $X \cap B$, $Y \cap A$, $Y \cap B$. Note that they cover all solutions in $A \cup B$, and

---

Algorithm 4. Split(S, d): given a set S of partial solutions, split S into two disjoint sets of roughly the same size based on the dth criterion of each solution.

---

1 Sort solutions in S by their dth criterion, in descending order if we aim to maximize the criterion, in ascending order otherwise;

2 $A \leftarrow S[1 : \lfloor n/2 \rfloor]$;

3 $B \leftarrow S - A$;

4 return A, B;   // A and B are disjoint and solutions in A have better dth criterion.

Algorithm 5. Marry(A, B, d'): consider only the first d' elements in each solution, return the set of all solutions in B that are not dominated by any solutions in A. A and B must be disjoint and no two solutions from the same set dominate one another.

---

1 $n \leftarrow$ number of solutions in $A \cup B$;

2 **if** $d' = 2$ **then**

3 | return $B \cap$ Non_Dominated_2D($A \cup B$);   // a base case of recursion

4 **else if** $A = \emptyset$ or $B = \emptyset$ **then**

5 | return B ;   // also a base case of recursion

6 **else**

7 | $X, Y \leftarrow$ Split$(A \cup B, d')$;

8 | $B_x \leftarrow$ Marry$(X \cap A, X \cap B, d')$;   // n reduce in half

9 | $B_y \leftarrow$ Marry$(Y \cap A, Y \cap B, d')$;   // n reduce in half

10 | $B'_y \leftarrow$ Marry$(X \cap A, B_y, d' - 1)$;   // d' reduce by 1

11 | return $B_x \cup B'_y$

12 **end**

$|(X \cap A) \cup (X \cap B)| \approx n/2 \approx |(Y \cap A) \cup (Y \cap B)|$. In line 8 and 9, we recursively call Marry on half-sized problems. Similarly as before, we know that the solutions in B_x are non-dominated in $A \cup B$, but we need to figure out which solutions in B_y are non-dominated in $A \cup B$. Solutions in B_y can only be dominated by solutions in $X \cap A$ since solutions inside B cannot dominate each other, so we only need to recursive call Marry on $X \cap A$ and B_y. Note that solutions in $X \cap A$ have better d'th criterion than solutions in B_y, so we only need to consider the first $d' - 1$ objectives. Finally, we return $B_x \cup B'_y$.

3.2   Runtime Analysis

For the runtime analysis, we assume that we use a sorting algorithm based on comparison, so the time complexity of Non_Dominated(S) is $O(n(\log n)^{d-1})$.

Proposition 1. *Given a set S of n 2-dimensional solutions, Non_Dominated_2D (S) runs in $O(n \log n)$ time.*

This is because the sorting step takes $O(n \log n)$ time and the for-loop takes $O(n)$ time.

Proposition 2. *Given a set S containing n solutions, $\texttt{Split}(S,d)$ runs in $O(n \log n)$ time.*

This is also because the sorting step takes $O(n \log n)$ time.

Proposition 3. *Given two disjoint sets A and B such that no two solutions from the same set dominate one another and that $A \cup B$ contains n solutions, $\texttt{Marry}(A, B, d')$ runs in $O(n(\log n)^{d'-1})$ time.*

Proof: We denote the runtime of $\texttt{Marry}(A, B, d')$ as $t(n, d')$. For the base case $d' = 2$, the proposition obviously holds. For $d'_0 \geq 3$, assume the proposition holds for $d' < d'_0$, which means that $t(n, d'_0 - 1) = O(n(\log n)^{d'_0 - 1 - 1}) = O(n(\log n)^{d'_0 - 2})$ for any positive integer n. Now we consider cases where $n = 2^k$ for some positive integer k and $d' = d'_0$. The major components of $\texttt{Marry}$ are: a $\texttt{Split}$ step ($O(n \log n)$ time), two half sized $\texttt{Marry}$ steps ($2\, t(n/2, d'_0)$ time), and a $\texttt{Marry}$ step with dimension reduced by one ($t(n, d'_0 - 1)$ time). With induction, we know that $t(n, d'_0 - 1) = O(n(\log n)^{d'_0 - 2})$. For any positive integer k and $n = 2^k$, we have

$$t(2^k, d'_0) = O(2^k \log(2^k)) + 2 \cdot t((2^k)/2, d'_0) + t(2^k, d'_0 - 1)$$
$$= 2 \cdot t(2^{k-1}, d'_0) + O(2^k \cdot k^{d'_0 - 2}).$$

Then, by induction on k, we can prove the following statement

$$t(2^k, d'_0) = O(n(\log n)^{d'_0 - 1}).$$

Since the runtime of $\texttt{Marry}$ increases monotonically with n, the proposition also holds when n is not a power of 2. Hence, $\texttt{Marry}(A, B, d')$ runs in $O(n(\log n)^{d'-1})$ time.

Proposition 4. *Given a set S containing n d-dimensional solutions ($d \geq 3$), $\texttt{Non\_Dominated}(S)$ runs in $O(n(\log n)^{d-1})$ time.*

Proof: When $d \leq 2$, the proposition clearly holds. For $d \geq 3$, we denote the runtime as $T(n, d)$. Similarly as in the proof of Proposition 3, we have

$$T(2^k, d) = 2 \cdot T(2^{k-1}, d) + O(2^k \cdot k^{d-2}).$$

Then, by induction on k, we get

$$T(2^k, d) = O(n(\log n)^{d-1}).$$

Since the runtime of $\texttt{Non\_Dominated}(S)$ increases monotonically with n, the proposition also holds when n is not a power of 2. Hence, $\texttt{Non\_Dominated}(S)$ runs in $O(n(\log n)^{d-1})$ time.

3.3   Implementation Notes

Split: The split procedure shown in Algorithm 4 can also be implemented using an $O(n)$ find median algorithm. However, the numerous steps of copying arrays and creating new arrays in the $O(n)$ find median algorithm are hard to implement and perform poorly in practice. Hence, we chose to use sorting to work "in-place" on the sets of solutions. Each time we drop a dimension we must create a new array sorted based on that dimension and then in the recursive process we simply keep track of the location within the array that we are working on. We found that in practice sorting and working in place give us much better performance.

Batching: Our new divide-and-conquer algorithm for pruning dominated solutions considerably speeds up the DP algorithm and allow us to solve problems on much larger networks, with higher precision, and with more objectives. However, the number of solutions to evaluate grows exponentially with the number of objectives and memory soon becomes a problem. For example, for the entire Amazon basin, for four criteria, with a precision of $\epsilon = 0.01$, the algorithm has to evaluate $144,823,974,336$ partial solutions at a single node of the tree, which is way beyond the memory available. To circumvent this problem, we introduced a batching process: at each tree node, instead of evaluating all possible solutions at once, we feed them to Non_Dominated in smaller batches of size $K = 10^7$. Then, we run Non_Dominated on the set of all non-dominated solutions from each batch. In practice, this batching routine actually also speeds up the DP algorithm. In the future we plan to consider different batching strategies and also parallel batching, which can be done in a straightforward way.

4   MIP-Based Pareto Frontier

We also proposed a MIP formulation (see Fig. 3) and a scheme for ϵ-approximating the Pareto-frontier of a multi-objective optimization problem in [18]. The key idea is to divide the space of objectives into small hyper-rectangles and query whether there exists a **feasible** solution in each hyper-rectangle. Then, from each feasible hyper-rectangle, we find one solution and form a set S of all the solutions we find. Under the condition that for each dimension, the upper bound of each hyper-rectangle is $(1 + \epsilon)$ of the lower bound, the set of non-dominated solutions from S forms an ϵ-approximate Pareto-frontier [9].

   In this paper we exploit restart strategy and introduce a new scheme to reduce the number of MIPs to solve. We first **optimize** for one of the objectives. We divide the space of the remaining objectives into small hyper-rectangles. Specifically, the hyper-rectangles are designed to satisfy the condition that, for each dimension, the upper bound is $(1 + \epsilon)$ of the lower bound (assuming the objectives are always positive values). For each cell, we formulate a MIP to find the solution in that cell that **optimizes** the target objective if a feasible solution exists. We form a set S of all the solutions found by MIP. Under the assumption

Minimize: d

subject to: $d = 0$ (d is a dummy variable)

$$\hat{C} \le C \le (1 + \epsilon)\hat{C}; \ \hat{S}_d \le S_d \le (1 + \epsilon)\hat{S}_d$$

$$\hat{E} \le E \le (1 + \epsilon)\hat{E}; \ \hat{S}_s \le S_s \le (1 + \epsilon)\hat{S}_s$$

$$C = \sum_{v \in V} n_v c_v; \ S_s = \sum_{e \in E} \pi_e r_e$$

$$E = \sum_{e \in E} \pi_e h_e; \ S_d = \sum_{v \in V} y_v s_v$$

$$n_v \in \{0, 1\}, \forall v \in V; \pi_e \in \{0, 1\}, \forall e \in E$$

$$n_s = 1; n_u \ge n_v, \forall (u, v) \in E$$

$$n_v \le 1 - \pi_{u,v}, \forall (u, v) \in E$$

$$n_v \ge n_u - \pi_{u,v}, \forall (u, v) \in E$$

$$y_s = 1$$

$$y_v \le y_u \text{ and } y_v \ge (1 - p_e) y_u, \forall (u, v) \in E$$

$$y_v \le (1 - p_e) y_u + (1 - \pi_e), \forall e = (u, v) \in E$$

$$y_v \ge y_u - \pi_e, \forall e = (u, v) \in E$$

Fig. 3. MIP formulation of the dam placement problem for four criteria. $\hat{C}, \hat{E}, \hat{S}_d, \hat{S}_s$: bounds on the objectives. V: the set of all nodes; E: the set of all edges (dams); s: the root of the tree; $e = (u, v)$: u is downstream of v; c_v, s_v, r_e, and h_e: connectivity value, sediment production, seismic risk, and hydropower associated with each hypernode or dam, respectively; p_e: percentage of sediment trapped by dam e; π_e: indicator variable of whether the dam will be built; n_v: indicator variable of whether node v can be reached from the river mouth without passing a dam; y_v: continuous variable representing the percentage of the sediment produced at the node v not trapped by dams.

that we solve the MIPs optimally, the set of non-dominated solutions from S forms an ϵ-approximate Pareto-frontier. In practice, we repeat the above scheme as many times as the number of criteria, cycling through every objective as target objective to get better coverage and a more accurate approximation. See details of the MIP formulation in [18]. A key difference in this new scheme is that we always optimize for the target objective, instead of solving decision problems.

Theorem: Let P be the set of all solutions on the Pareto frontier. Let $\bar{P}$ be the set of non-dominated solutions from S. Then, $\bar{P}$ ϵ-approximates P.

Proof: Assume that we are optimizing for k objectives $O_1, O_2, O_3, \cdots$, and O_k where k is greater or equal to 2. Without loss of generality, assume we aim to maximize O_1. For any $\pi \in P$, assume $(O_2(\pi), O_3(\pi), \cdots, O_k(\pi))$ lies in the rectangular cell $[\hat{O}_2, (1 + \epsilon)\hat{O}_2] \times [\hat{O}_3, (1 + \epsilon)\hat{O}_3] \times \cdots \times [\hat{O}_k, (1 + \epsilon)\hat{O}_k]$. Since there is already a solution π in the rectangular cell, MIP can find a solution π' in the same cell that optimizes O_1, which means that $O_1(\pi') \ge O_1(\pi)$ and π' ϵ-dominates π. If $\pi' \notin \bar{P}$, then there exists a $\pi'' \in \bar{P}$ that dominates π' and consequently ϵ-dominates π. Hence, $\bar{P}$ ϵ-approximates P.

Table 1. Sample of runtimes and number of solutions for the different methods. A (Amazon); WA (Western Amazon); and M (Marañon); (E energy; C connectivity; S_d sediment; S_e seismic risk). *mem* denotes memory limit. N/A denotes MIP cannot produce the exact Pareto frontier. We bolded several entries to highlight performance improvements.

B	Criteria	ϵ	DP orig (secs)	DP new (secs)	MIP orig (secs)	MIP new (secs)	DP #Sols	MIP #Sols
A	E, C	exact	*18291*	*254*	N/A	N/A	*39841*	N/A
A	E, C	0.001	72	14	*6432*	*48*	3020	894
M	E, S_d	exact	2077	64	N/A	N/A	25732	N/A
M	E, S_d	0.001	4	2	1day+	1day+	318	—
WA	E, S_d	exact	*46291*	*924*	N/A	N/A	*58808*	N/A
WA	E, S_d	0.001	30	13	1day+	2187	2668	1671
A	E, S_d	exact	mem	15153	N/A	N/A	177490	N/A
A	E, S_d	0.001	2368	226	1day+	1day+	7973	—
A	E, S_d	0.1	0.1	0.1	297	3359	83	24
A	E, S_e	exact	54581	335	N/A	N/A	72591	N/A
A	E, S_e	0.001	2471	83	*35050*	*31*	8737	1558
M	E, C, S_d	exact	*2day+*	*526*	N/A	N/A	*283898*	N/A
M	E, C, S_d	0.001	630	32	1day+	1day+	5563	—
WA	E, C, S_d	exact	mem	90251	N/A	N/A	3267859	N/A
WA	E, C, S_d	0.001	mem	1120	1day+	1day+	88710	—
WA	E, C, S_d	0.0025	1667	269	1day+	1day+	28804	—
WA	E, C, S_d	0.005	254	69	1day+	65638	12655	4129
A	E, C, S_d	0.005	mem	32175	1day+	1day+	758462	—
A	E, C, S_d	0.025	70348	607	1day+	1day+	48381	—
A	E, C, S_d	0.05	1680	58	1day+	1day+	12866	—
A	E, C, S_d	0.1	40	6	1day+	4503	4724	62
A	E, C, S_d	0.15	11	2	1day+	6025	2493	43
A	E, C, S_e	0.005	*mem*	*88246*	*1day+*	*4809*	*2274168*	*40981*
A	E, C, S_e	0.05	109910	2121	238	51	47978	581
M	E, C, S_d, S_e	exact	*mem*	*763150*	N/A	N/A	*23364120*	N/A
M	E, C, S_d, S_e	0.001	mem	53620	1day+	1day+	1479660	—
M	E, C, S_d, S_e	0.02	*278310*	*649*	1day+	1day+	15961	—
M	E, C, S_d, S_e	0.1	886	28	1day+	15484	1406	773
WA	E, C, S_d, S_e	0.01	mem	695153	1day+	1day+	3540829	—
WA	E, C, S_d, S_e	0.1	47704	1154	1day+	1day+	107087	—
WA	E, C, S_d, S_e	0.15	11712	424	1day+	13692	69422	296
A	E, C, S_d, S_e	0.05	mem	*437271*	1day+	1day+	*2193314*	—
A	E, C, S_d, S_e	0.1	mem	19510	1day+	1day+	491578	—
A	E, C, S_d, S_e	0.15	mem	7471	1day+	1day+	198772	—
A	E, C, S_d, S_e	0.2	74940	1333	1day+	1day+	47059	—
A	E, C, S_d, S_e	0.25	11358	410	1day+	1505	23019	95

We observed fat and heavy-tailed behavior in the MIP runtime distributions [7]. To improve performance, we run the MIP solver with a cutoff, using a geometric restart strategy that doubles the cutoff time in every run [7,15]. Our experiments show that the restart strategy significantly boosts performance.

5   Experimental Results

To test the performance of the new methods at different scales, we used three datasets: the Marañon, Western Amazon, and Amazon basins, with 107, 219, and 467 hypernodes, respectively (corresponding to 128801, 455156 and 4083059

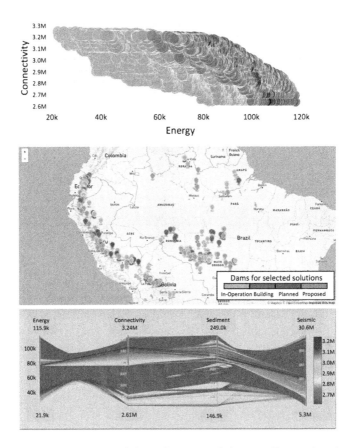

Fig. 4. Top: The visualization of the 4-dimensional Amazon Pareto frontier ($\epsilon = 0.4$). X axis: energy; Y axis: connectivity; Marker size: sediment; Color: seismic risk. **Middle:** The dam placement of a particular Pareto solution. **Bottom:** Parallel coordinate plot for the Amazon Pareto frontier ($\epsilon = 0.4$). The four axes are hydropower, connectivity, sediment and seismic risk. The color of each solution is based on its hydropower output. The plot displays only **1440** solutions due to the bounding of the objectives (pink lines on the axes). (Color figure online)

river segments, respectively). We compare the performance of the new DP and MIP methods with the methods in [18] and see significant improvements in both **speed** and **accuracy**. See Fig. 1 and Table 1 for a summary of results.

Specifically, in terms of accuracy, the new MIP approach is substantially better than the previous MIP approach. As shown in Fig. 1, in the 2-dimensional case, the solutions produced by the new MIP approach are on the exact Pareto frontier, slightly better than the DP approximation for the same $\epsilon = 0.1$.

The new DP approach has the same high level of accuracy as the previous DP approach and still produces more solutions than the MIP approaches.

In terms of speed, our experiments show that the new DP approach is up to three orders of magnitude faster than the original DP and scales to significantly larger instances and more criteria. The batching technique also solves the issue of hitting the memory limit when computing for three or more objectives. The new MIP approach is faster and can now solve larger problems.

The DP and MIP methods are complementary since in practice our new MIP scheme provides solutions closer to the exact Pareto frontier (for a given ϵ) and it provides more flexibility for considering additional constraints for what-if analyses, which is important to decision makers.

We are developing a web-based visualization tool for policy makers to explore the Pareto frontier interactively. For example, Fig. 4 displays: (1) the Pareto frontier for four criteria for the entire Amazon ($\epsilon = 0.4$); (2) the placement of the selected dams for a particular Pareto solution; and (3) a parallel coordinate plots to visualize the solutions, in which each axis represents an objective, and each line across the different axes represents a solution. We can bound each objective (pink lines on the axes) and only show solutions that satisfy the bounds. By bounding each objective appropriately, we notably decrease the number of solutions to consider.

6   Conclusions

We introduced new DP and MIP approaches that significantly boost the efficiency and accuracy of computing the exact Pareto frontier and its approximation with guarantees on tree-structured networks. Our DP and MIP approaches show complementary strengths and are now able to scale up to much larger real-world problems. We are developing interactive tools for what-if analyses and visualizations for policy makers. The overall goal of this project is to assist policy makers in making informed decisions when planning hydropower dams in the Amazon Basin. Our methods are general and can be adapted to other multi-objective optimization problems on tree-structured networks.

Acknowledgments. This work was supported by NSF Expedition awards for Computational Sustainability (CCF-1522054 and CNS-0832782), NSF CRI (CNS-1059284) and Cornell University's Atkinson Center for a Sustainable Future.

References

1. Altwaijry, N., EI Bachir Menai, M.: Data structures in multi-objective evolutionary algorithms. J. Comput. Sci. Technol. **27**(6), 1197–1210 (2012)
2. Deb, K., Pratap, A., Agarwal, S., Meyarivan, T.: A fast and elitist multiobjective genetic algorithm: NSGA-II. IEEE Trans. Evol. Comput. **6**(2), 182–197 (2002)
3. Ehrgott, M., Gandibleux, X.: A survey and annotated bibliography of multiobjective combinatorial optimization. OR Spectrum **22**(4), 425–460 (2000)
4. Finer, M., Jenkins, C.N.: Proliferation of hydroelectric dams in the Andean Amazon and implications for Andes-Amazon connectivity. PLoS One **7**(4), e35126 (2012)
5. Gavanelli, M.: An algorithm for multi-criteria optimization in CSPs. In: Proceedings of the 15th European Conference on Artificial Intelligence, ECAI, pp. 136–140 (2002)
6. Gomes, C.P.: Computational sustainability: computational methods for a sustainable environment, economy, and society. Bridge **39**(4), 5–13 (2009)
7. Gomes, C.P., Selman, B., Crato, N., Kautz, H.: Heavy-tailed phenomena in satisfiability and constraint satisfaction problems. J. Auto. Reason. **24**(1), 67–100 (2000)
8. Neumann, F.: Expected runtimes of a simple evolutionary algorithm for the multi-objective minimum spanning tree problem. Eur. J. Oper. Res. **181**(3), 1620–1629 (2007)
9. Papadimitriou, C.H., Yannakakis, M.: On the approximability of trade-offs and optimal access of web sources. In: Proceedings of the 41st Annual Symposium on Foundations of Computer Science, FOCS 2000 (2000)
10. Qian, C., Tang, K., Zhou, Z.-H.: Selection hyper-heuristics can provably be helpful in evolutionary multi-objective optimization. In: Handl, J., Hart, E., Lewis, P.R., López-Ibáñez, M., Ochoa, G., Paechter, B. (eds.) PPSN 2016. LNCS, vol. 9921, pp. 835–846. Springer, Cham (2016). https://doi.org/10.1007/978-3-319-45823-6_78
11. Qian, C., Yu, Y., Zhou, Z.-H.: Pareto ensemble pruning. In: Proceedings of the Twenty-Ninth AAAI Conference on Artificial Intelligence, AAAI 2015, pp. 2935–2941 (2015)
12. Schaus, P., Hartert, R.: Multi-objective large neighborhood search. In: Schulte, C. (ed.) CP 2013. LNCS, vol. 8124, pp. 611–627. Springer, Heidelberg (2013). https://doi.org/10.1007/978-3-642-40627-0_46
13. Sheng, W., Liu, Y., Meng, X., Zhang, T.: An improved strength pareto evolutionary algorithm 2 with application to the optimization of distributed generations. Comput. Math. Appl. **64**(5), 944–955 (2012)
14. Terra-Neves, M., Lynce, I., Manquinho, V.: Introducing pareto minimal correction subsets. In: Gaspers, S., Walsh, T. (eds.) SAT 2017. LNCS, vol. 10491, pp. 195–211. Springer, Cham (2017). https://doi.org/10.1007/978-3-319-66263-3_13
15. Walsh, T.: Search in a small world. In: Proceedings of the 16th International Joint Conference on Artificial Intelligence, IJCAI 1999, San Francisco, CA, USA, vol. 2, pp. 1172–1177. Morgan Kaufmann Publishers Inc. (1999)
16. Wiecek, M.M., Ehrgott, M., Fadel, G., Figueira, J.R.: Multiple criteria decision making for engineering (2008)
17. Winemiller, K.O., McIntyre, P.B., Castello, L., Fluet-Chouinard, E., Giarrizzo, T., Nam, S., Baird, I.G., Darwall, W., Lujan, N.K., Harrison, I., et al.: Balancing hydropower and biodiversity in the Amazon, Congo, and Mekong. Science **351**(6269), 128–129 (2016)

18. Wu, X., Gomes-Selman, J.M., Shi, Q., Xue, Y., Garcia-Villacorta, R., Sethi, S., Steinschneider, S., Flecker, A., Gomes, C.P.: Efficiently approximating the pareto frontier: hydropower dam placement in the Amazon basin. In: AAAI (2018)
19. Yukish, M.: Algorithms to identify Pareto points in multi-dimensional data sets. Ph.D. thesis (2004)
20. Yukish, M., Simpson, T.W.: Analysis of an algorithm for identifying pareto points in multi-dimensional data sets. In: 10th AIAA/ISSMO Multidisciplinary Analysis and Optimization Conference, p. 4324 (2004)
21. Zarfl, C., Lumsdon, A.E., Berlekamp, J., Tydecks, L., Tockner, K.: A global boom in hydropower dam construction. Aquat. Sci. **77**(1), 161–170 (2015)
22. Ziv, G., Baran, E., Nam, S., Rodríguez-Iturbe, I., Levin, S.A.: Trading-off fish biodiversity, food security, and hydropower in the Mekong River Basin. Proc. Nat. Acad. Sci. **109**(15), 5609–5614 (2012)

Bandits Help Simulated Annealing to Complete a Maximin Latin Hypercube Design

Christian Hamelain, Kaourintin Le Guiban$^{(\boxtimes)}$, Arpad Rimmel, and Joanna Tomasik

LRI, CentraleSupélec, Université Paris Saclay, 91405 Orsay Cedex, France
kaourintin.leguiban@lri.fr

Abstract. Simulated Annealing (SA) is commonly considered as an efficient method to construct Maximin Latin Hypercube Designs (LHDs) which are widely employed for Experimental Design. The Maximin LHD construction problem may be generalized to the Maximin LHD completion problem in an instance of which the measurements have already been taken at certain points.

As the Maximin LHD completion is NP-complete, the choice of SA to treat it shows itself naturally. The SA performance varies greatly depending on the mutation used. The completion problem nature changes when the number of given points varies. We thus provide SA with a mechanism which selects an appropriate mutation. In our approach the choice of a mutation is seen as a bandit problem. It copes with changes in the environment, which evolves together with the thermal descent.

The results obtained prove that the bandit-driven SA adapts itself on the fly to the completion problem nature. We believe that other parametrized problems, where SA can be employed, may also benefit from the use of a decision-making algorithm which selects the appropriate mutation.

1 Introduction

Maximin Latin Hypercube Designs are widely used in the operations research field to sample complex systems, where experiments are costly, in order to build a model for the system. Each point of the design represents an experiment and the coordinates of the points represent the parameters chosen for the experiment. Two properties of the design are needed to obtain a good quality for the model: the design should be **non collapsing**, which means that no parameter value can be chosen more than once, and they should be **space filling**, which means that the points of the designs should be evenly spread in space. The first property is equivalent to the Latin constraint. To satisfy the second one, we build an LHD with a **separation distance** $D_{\min}$, which signifies the minimal distance between any pair of points, as high as possible.

The complexity of the Maximin LHD construction problem is still unknown. The authors of [5] defined and studied its generalization: the completion of a

© Springer International Publishing AG, part of Springer Nature 2018
W.-J. van Hoeve (Ed.): CPAIOR 2018, LNCS 10848, pp. 280–288, 2018.
https://doi.org/10.1007/978-3-319-93031-2_20

Maximin LHD. Instead of building up a Maximin LHD from scratch, certain of its points are fixed beforehand. The problem consists in adding points with respect to the constraints until the design is completed. This problem models practical situations when certain experiments have already been concluded and we want to choose the best settings for the following ones. The completion problem becomes a construction problem when there are no points fixed in advance. The theoretical analysis of the problem of filling up a partial LHD [5] shows that it is NP-complete and inapproximable with a constant factor. We therefore devise a heuristic algorithm taking as a base the Simulated Annealing (SA) method. Section 2 gives its description together with the rules of its usage to construct a Maximin LHD.

The contributions we made are twofold. First, we designed a mutation dedicated to the completion problem for its instances with a relatively small number of fixed points. This mutation, defined in Sect. 3 and called *cmpl1D-move*, allows SA to avoid being trapped within a small set of configurations and to pursue the descent towards a global optimum.

Second, we turned to our advantage the fact that completion instance changes its nature in function of the number of preset points. In Sect. 4 we proposed a method, called *BDM* and based upon the bandit principle, to make SA choose itself a mutation for a given instance.

The experimental results (Sect. 5) confirm the capacity of SA, assisted by *BDM*, to adapt the search space exploration on-the-fly, without any supplementary numerical effort, to the aspect of an instance treated. We believe that the decision-making algorithm we proposed may be used in other evolutionary algorithms to solve problems the instances of which require mutation tailoring.

2   State of the Art for SA Solving Maximin LHDs

Simulated Annealing is used with success for the construction of Maximin LHDs as reported in [7]. Its principal components are: a function E that evaluates configurations, called **energy**, and a randomized operation on a configuration that modifies it slightly, called **a mutation**. Each SA iteration consists in applying the mutation on a current configuration ω to get a new configuration ω'. If the mutation causes the energy descent, ω' it is accepted, otherwise ω' it is accepted with the probability p given by the Metropolis-Hastings formula [4]:

$$p = \exp\left(\frac{E(\omega) - E(\omega')}{kT}\right), \tag{1}$$

where T is the temperature which is set at the beginning to a high value and decreases as the system cools down with the increase of the number of iterations and k is the Boltzmann constant (in practice k is typically set to one). Probability p computed according to Eq. (1) ensures the iteration convergence.

The article [7] is the most recent survey of algorithms based upon the local search which construct Maximin LHDs. All the mutations used in the construction, which it enumerates, need a notion of a **critical point** which is a point whose distance to another point is equal to the minimal distance. Put differently, a critical point is responsible for establishing $D_{\min}$ of the design. The mutations we need for the purpose of this study are:

m1: select one critical point and any other point at random, exchange their coordinates on a random number of dimensions,
m2: select one critical point and any other point at random, exchange their coordinates on a single dimension randomly chosen,
m3: select one critical point and any other point at random, exchange the coordinates on the dimension which ensures the greatest energy descent.

All these mutations may affect two critical points as "any other point" taken as the second one can be critical. The most efficient mutation according to [3], *1D-move*, targets the Latin constraint. It involves two points $p^{(1)}, p^{(2)}$ which are **neighbors**. This means there is a dimension on which their coordinates differ by one: there is $j \in [[1; k]]$ such that $\left| p_j^{(1)} - p_j^{(2)} \right| = 1$. **1D-move** is specified as:

1. Select one critical point and any of its neighbors at random.
2. Determine the dimension which makes these points be neighbors.
3. Exchange their coordinates on this dimension.

This mutation is particularly efficient for a local search because it makes it possible to follow a step-by-step path on the energy surface of an LHD without jumping over possible minima. We also emphasize that both the points involved in it may be critical.

The evaluation function designed to express the energy (corresponding to E in Eq. (1)) of the Maximin LHD construction is ϕ described in [6]:

$$\phi = \left(\sum_{d \in D} \frac{1}{|d|^p} \right)^{1/p},$$

where D is the set containing all distances between each pair of points in the hypercube and p is a parameter (typically set to 10).

3 Mutation Targeting "Relatively Empty" Hypercubes

The mutations conceived for the construction problem may be reused in the completion context. The single difference is that only the points which have not been fixed in advance may be involved in them. In the remainder we use the term **authorized points** to refer to points that can be modified.

The choice of an appropriate mutation is conditioned by the number of points set *a priori*. When the design is entirely empty, the completion becomes a construction and *1D-move* is a natural choice as, according to [3], it is the best mutation to this day.

The mutation *cmpl1D-move*, illustrated in Fig. 1, is a variant of *1D-move*. The major difference with the previous mutations is that it selects an **oriented**

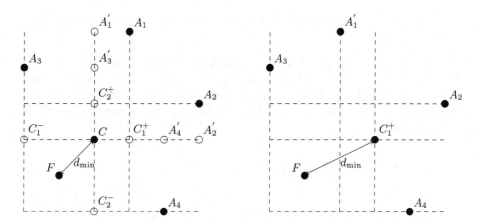

Fig. 1. On the left, the points marked with solid black circles form the initial configuration for this example. Authorized points are noted A_1, A_2, A_3, A_4, and C, the latter being the only authorized critical point. The smallest distance between lines is equal to one. Point F is a fixed point. Mutation *cmpl1D-move* selects an authorized critical point. In this example, it can only select C. It then chooses one dimension (either the first or second in this example) and one sign (either + or −). If the + sign and first dimension have been selected, *cmpl1D-move* exchanges the first coordinate of C and A_1, resulting in points C_1^+ and A_1'. The other possibilities of *cmpl1D-move* result in points either (C_2^+ and A_2') or (C_1^- and A_3') or (C_2^- and A_4'). The dashed lines represent all the coordinates that C can take after the mutation. The figure on the right represents the configuration after the mutation, if the sign + and first dimension have been selected.

dimension, which means that it takes into account a direction when looking for another point to exchange coordinates. Once the direction has been fixed, this mutation searches for the authorized point with the minimal distance on the dimension and direction selected. Thus, considering the direction in which we look for the authorized point beforehand ensures that, given a critical point and a dimension, *cmpl1D-move* will have two outcomes (in the two opposite directions, possibly not symmetrical one to another with reference to the critical point on the dimension selected). This is necessary to avoid situations where the mutation will only have one possible outcome in each dimension, leading to an impasse where the algorithm is stuck with a current solution alternating between a few solutions. ***cmpl1D-move*** in summary:

1. Select one critical point $p^{(1)}$, one dimension j and the sign +/− of the coordinate difference at random.
2. Find the authorized $p^{(2)}$ such that $p_j^{(2)} > p_j^{(1)}$ (if the + sign has been selected) or $p_j^{(2)} < p_j^{(1)}$ (if the − sign has been selected) the coordinate difference $\left| p_j^{(1)} - p_j^{(2)} \right|$ is minimal.
3. Exchange the coordinates of $p^{(1)}$ and $p^{(2)}$ on the dimension j.

4. If there is no authorized point on the dimension and direction chosen, perform mutation *m2*.

As the restriction of the search space, induced by the fixed points, gets stronger when the number of these points increases, the mutation *cmpl1D-move* would not look for a solution sufficiently far from a current configuration. Its interest is therefore limited to instances in which a relatively small number of points is fixed in advance.

4   Bandit-Driven Mutation

At this stage we hypothesize that among the considered mutations there is no single mutation which could cope with the entire spectrum of incomplete hypercubes, from almost empty to almost filled up. We know, however, that *cmpl1D-move* works well for the first category of instances while *m2* is well suited to explore the search space of the second one. As we do not want to change the mutation depending on the instance, we would like to create a decision-making algorithm for choosing the mutation on-the-fly. In order to do this, we consider that at each iteration of SA, we have at our disposal several mutations and that choosing a mutation is a multi-armed bandit (MAB) problem as described in [1].

The MAB problem can be defined by several random variables X_i where i is the index of a gambling machine. At each step, a machine j is chosen and a reward which is the realization of X_j is given. The goal is to maximize the sum of the rewards. In our case, the choice of the mutation would correspond to the choice of the machine and the reward would correspond to the evaluation of the new individual, *i.e.* the absolute energy variation in the SA context.

However, in our problem, the choice of a mutation changes the state of the system. We make the hypothesis that this is equivalent to a small modification of the random variables behind each machine. With this hypothesis, our problem can be seen as a classical variant of the MAB problem: the non-stationary multi-armed bandit (NSMAB) problem [2].

We propose the following algorithm to solve this problem in our context. It is based on a classical compromise between exploration and exploitation. For the exploitation part, we compute the average reward for each mutation. However, this average is computed only on the last w results in order to take into account the fact that the problem is non-stationary.

The exploration part of our algorithm is handled in two different ways. First, the choice of a mutation is based on a draw according to a softmax distribution of the exploitation terms, so each mutation has a chance to be selected. The second aspect is based on a characteristic of our problem. As SA converges to a local minimum, the rewards decrease over time on average. As the exploitation is based only on the last w rewards, the exploitation term of a mutation will decrease when a mutation is selected.

Here is a formal description of our algorithm. Let us note W_i the set of size w of the last rewards x_i for mutation i. At each step, we select the next mutation according to the probability p_i computed as:

$$p_i = \frac{\exp(\overline{X_i})}{\sum_j \exp(\overline{X_j})}, \quad \text{where} \quad \overline{X_i} = \frac{1}{w} \sum_{x_i \in W_i} x_i. \tag{2}$$

In the following, we apply this algorithm to the mutations $m2$ and $cmpl1D$-$move$ and name the resulting mutation Bandit Driven Mutation (BDM). We note that it can be applied to any number of mutations. **BDM** is as follows:

1. Compute the probability of $m2$ and $cmpl1D$-$move$ according to Eq. (2).
2. Randomly select the mutation to be used based on these probabilities.
3. Apply the chosen mutation.

5 Numerical Experiments

Our experiments are conducted for the problem of dimension 4 and size 75 as done in [7]. They are performed for the hypercube taken from spacefilling-designs.nl, with the distance expressed by the Euclidean norm $\mathcal{L}_2$. According to the site mentioned, this hypercube has $D_{\min}$ of 867 (we obtained, however, a hypercube with a score of 889 during our experiments).

Unless it is stated otherwise, the number of mutations of the annealing used for the experiments is 10^5. We remark that for mutation $m3$, experiments are realized with four times less mutations than for the others, because this mutation evaluates the $D_{\min}$ of the configuration once per dimension of the instance, thus we limited the total number of mutations performed in order to fairly compare the different types of mutations. Also, the temperature follows a linear cooling scheme that decreases the temperature every 100 mutations, from an initial acceptance probability of 0.4 down to 0.

Incomplete hypercube instances are obtained by removing points according to a uniform distribution. The scores are averages over 200 incomplete instances except for some points in Table 2 and Fig. 2 where a significant difference was

Table 1. The mean score in function of the number of deleted points, for an instance of size 75, in four dimensions, with $D_{\min} = 867$

Deleted points	Mutations				
	m1	m2	m3	1D-Move	Cmpl1D-Move
5	846 ± 49	$\mathbf{864 \pm 14}$	858 ± 32	331 ± 54	853 ± 32
15	810 ± 55	$\mathbf{854 \pm 20}$	832 ± 42	173 ± 30	710 ± 57
25	745 ± 36	769 ± 28	$\mathbf{791 \pm 35}$	137 ± 28	732 ± 32
35	735 ± 14	728 ± 14	752 ± 13	126 ± 25	$\mathbf{767 \pm 15}$
45	735 ± 11	724 ± 10	745 ± 8	134 ± 24	$\mathbf{780 \pm 9}$
55	740 ± 10	722 ± 10	747 ± 8	173 ± 31	$\mathbf{799 \pm 7}$
65	747 ± 11	724 ± 10	753 ± 7	325 ± 53	$\mathbf{818 \pm 6}$
75	751 ± 12	730 ± 10	761 ± 8	$\mathbf{844 \pm 5}$	844 ± 6

Table 2. The mean score of *BDM* and its components: *m2* and *cmpl1D-move*

Deleted points	Mutations		
	m2	cmpl1D-move	BDM
5	**864 ± 11**	858 ± 18	863 ± 12
15	**853 ± 14**	709 ± 41	830 ± 25
20	**819 ± 21**	713 ± 31	811 ± 24
25	766 ± 19	736 ± 22	**794 ± 20**
30	736 ± 11	750 ± 14	**780 ± 14**
35	727 ± 8	763 ± 10	**773 ± 9**
40	725 ± 7	774 ± 7	**775 ± 6**
45	724 ± 6	**782 ± 6**	779 ± 5
55	723 ± 6	**798 ± 4**	791 ± 4
65	726 ± 6	**818 ± 3**	807 ± 4
75	731 ± 6	**843 ± 3**	830 ± 3

not observed and therefore the average over 2000 runs was used. The same set of 200 (2000, respectively) instances of hypercubes with fixed points randomly generated with 200 (2000, respectively) different seeds were used as starting hypercubes across all algorithms.

We first compare, in Table 1, the performance of existing mutations with the one we designed for the completion problem: *cmpl1D-move* (Sect. 3) in function of the number of deleted points in the hypercube.

The results in bold type correspond to the best average for a given number of deleted points. The last line (75 deleted points) of this table corresponds to the construction problem.

We note that the mutation *1D-move* which gives the best results for the construction problem performs very poorly for the completion problem, which was explained in Sect. 3. The mutation we proposed: *cmpl1D-move* is successful as it obtains the best average for 35 to 75 deleted points. However, it is interesting to observe that this is not the case for a smaller number of deleted points where mutations *m2* and *m3* produce better scores. This confirms that the problem behaves very differently depending on the number of deleted points.

We will now present in Table 2 and Fig. 2 the results of the comparison of *BDM* with mutations *m2* and *cmpl1D-move* of which *BDM* is composed. The size of the sliding window w discussed in Sect. 4 is arbitrarily set for our experiments to 100 (for a total of 10^5 mutation steps during an annealing process).

We see that for the extreme cases (a large or a small number of deleted points), *BDM* obtains scores close to those produced by the best performing mutation. Our goal of having a mutation that can handle both cases has therefore been achieved. On top of that, for intermediate values of deleted points, *BDM* reaches significantly better results than both other mutations. This shows that a dynamic choice of a mutation during a single run based on a bandit formula

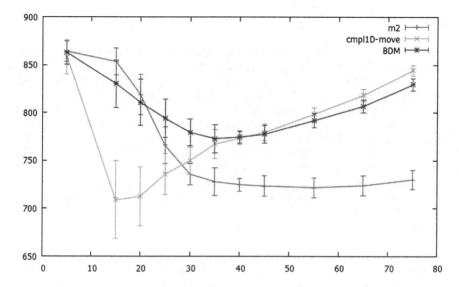

Fig. 2. Mean $D_{\min}$ with confidence interval for mutations: *m2*, *cmpl1D-move*, and *BDM* in function of the number of deleted points

can not only perform as well as each individual mutation but also combine the advantages of the different mutations to obtain even a better result. This seems promising and we hope to generalize our approach to more mutations and to use it to solve other problems.

6   Conclusion and Future Work

We proposed an algorithm based on SA that produces satisfying results for the problem of completing Latin hypercubes. We showed that this problem behaves differently when numerous points should be added to complete a design and when there are only a few of them. For the former case, an existing mutation *m2* can be used, for the latter case we proposed *cmpl1D-move* adapted to the problem. Finally, we proposed a method to dynamically chose the appropriate mutation during the execution of the annealing process based upon a bandit algorithm. This approach allows us to perform the mutation choice automatically and to obtain even better results than each independent mutation for the most typical instances.

There are two possible directions for future work. Firstly, we plan to further improve the results of the algorithm on this problem by trying new mutations and other evaluation functions. Secondly, we believe that the approach of using a bandit algorithm to choose between several mutations can be generalized and we will apply this principle to other problems where it may be relevant.

References

1. Auer, P., Cesa-Bianchi, N., Fischer, P.: Finite-time analysis of the multiarmed bandit problem. Mach. Learn. **47**(2–3), 235–256 (2002)
2. Auer, P., Cesa-Bianchi, N., Freund, Y., Schapire, R.E.: The nonstochastic multi-armed bandit problem. SIAM J. Comput. **32**(1), 48–77 (2002)
3. Bergé, P., Guiban, K.L., Rimmel, A., Tomasik, J.: Search space exploration and an optimization criterion for hard design problems. In: Proceedings of GECCO (compagnon), pp. 43–44, July 2016
4. Hastings, W.K.: Monte Carlo sampling methods using Markov chains and their applications. Biometrika **57**(1), 97–109 (1970)
5. Le Guiban, K., Rimmel, A., Weisser, M.A., Tomasik, J.: Completion of partial Latin Hypercube Designs: NP-completeness and inapproximability. J. Theor. Comput. Sci. Sect. A **715**, 1–20 (2018)
6. Morris, M.D., Mitchell, T.J.: Exploratory designs for computational experiments. J. Stat. Plann. Infer. **43**(3), 381–402 (1995)
7. Rimmel, A., Teytaud, F.: A survey of meta-heuristics used for computing maximin Latin hypercube. In: Blum, C., Ochoa, G. (eds.) EvoCOP 2014. LNCS, vol. 8600, pp. 25–36. Springer, Heidelberg (2014). https://doi.org/10.1007/978-3-662-44320-0_3

A Dynamic Discretization Discovery Algorithm for the Minimum Duration Time-Dependent Shortest Path Problem

Edward He$^{(\boxtimes)}$, Natashia Boland$^{(\boxtimes)}$, George Nemhauser,
and Martin Savelsbergh

H. Milton Steward School of Industrial and Systems Engineering,
Georgia Institute of Technology, 765 Ferst Dr, Atlanta, GA 30332, USA
`edwardhe@gatech.edu`,
{`natashia.boland,george.nemhauser,martin.savelsbergh`}`@isye.gatech.edu`

Abstract. We present an exact algorithm for the Minimum Duration Time-Dependent Shortest Path Problem with piecewise linear arc travel time functions. The algorithm iteratively refines a time-expanded network model, which allows for the computation of a lower and an upper bound, until - in a finite number of iterations - an optimal solution is obtained.

1 Introduction

Finding a shortest path between two locations in a network is a critical component of many algorithms for solving transportation problems. There is a growing interest in the setting where the travel time along an arc in the network is a function of the time the arc is entered. Time-dependent travel times are typically a result of congestion. We refer to these problems as Time-dependent Shortest Path Problems (TDSPPs). It is commonly assumed that travel times on arcs satisfy the First-In First-Out (FIFO) property, i.e., it is impossible to arrive at the end of the arc earlier by entering the arc later. Given a departure time at the source, the standard approach for finding a path that reaches the sink as early as possible is detailed in [1]. For an overview of other methods, see [2]. In this paper, we consider the problem of finding a path such that the difference between the departure time at the source and the arrival time at the sink is as small as possible. We call it the Minimum Duration Time-Dependent Shortest Path Problem (MD-TDSPP), sometimes referred to as the least travel time TDSPP or the minimum delay TDSPP. The MD-TDSPP arises in many contexts. It has been studied, for example, in the context of path planning in traffic networks [3], and it has even arisen in the analysis of social networks [4].

We present an efficient dynamic discretization discovery algorithm for the variant of MD-TDSPP in which travel times on the arcs are given by piecewise linear functions. It was established only recently that an algorithm polynomial in the number of travel time function breakpoints exists [5]. Our key contribution

© Springer International Publishing AG, part of Springer Nature 2018
W.-J. van Hoeve (Ed.): CPAIOR 2018, LNCS 10848, pp. 289–297, 2018.
https://doi.org/10.1007/978-3-319-93031-2_21

is the development of an algorithm that, in practice, investigates only a small fraction of the travel time function breakpoints in the search for an optimal path and the proof of its optimality. In Sect. 2, we formally introduce MD-TDSPP and briefly discuss the relevant literature. In Sect. 3, we describe our algorithm and illustrate it on a small instance. In Sect. 4, we present the results of a small computational study.

2   Problem Description

We are given a directed network $D = (N, A)$ with $N = \{1, 2, \ldots, n\}$ and $A \subseteq N \times N$, a time interval $[0, T]$, and piecewise linear travel times $c_{i,j}(t)$ for $t \in [0, T]$ satisfying the FIFO property for arcs $(i, j) \in A$. Without loss of generality, we let 1 be the source and n be the sink. Satisfying the FIFO property, in this case, is equivalent to having the slopes of the linear pieces being at least -1. (Note the FIFO property implies that waiting anywhere except at the source is sub-optimal, since it is always better to depart immediately.)

The MD-TDSPP is to find a starting time $0 \leq \tau \leq T$ and a time-dependent path $P(\tau) = (t_1, a_1, t_2, a_2, \ldots, a_{m-1}, t_m)$, which is a path $P = (a_1, a_2, \ldots, a_{m-1})$ from 1 to n in D and a set of associated departure times $(t_1 = \tau, t_2, \ldots, t_{m-1})$ and arrival time t_m at node n, where the departure times satisfy $t_k + c_{a_k}(t_k) = t_{k+1}$ for all $k = 1, \ldots, m-1$, meaning that the arrival time for one arc is the departure time of the next arc. Among all possible paths and starting times, $P(\tau)$ minimizes the duration, which is given by $t_m - t_1$. Furthermore, we require that $t_m \leq T$. We characterize time-dependent paths (TDPs) by their starting times as the problem of finding the minimum duration given a starting time is a TDSPP, which can be solved easily.

The MD-TDSPP has attracted much attention since the early work of Orda and Rom [6]. There are two classes of approaches: discrete and continuous. In the discrete approaches, a time-expanded network (TEN) is formed and the problem can be solved using the same method as for TDSPP. The DOT algorithm presented in [1] solves the TDSPP with complexity $\mathcal{O}(SSP + nM + mM)$, where SSP is the cost of solving a static SP, n is the number of nodes, m is the number of arcs, and M is the size of the time discretization. Discrete approaches are inexact and rely heavily on the quality of the discretization. A denser discretization leads to a better approximation, but an increase in computation time. Continuous methods, such as the Dijkstra's algorithm variants [7,8], and the A* algorithm variant [9], create and update arrival time functions at each node and are exact. The complexity analysis of these methods has relied on being able to store and manipulate such functions efficiently and is given in terms of these operations, which are hard to quantify. Even for continuous piecewise linear functions, it was only recently that an algorithm that is polynomial in the total number of breakpoints (in the piecewise linear functions) was proposed [5]. The authors show that there is an optimal path that contains an arc (i, j) where the departure time occurs exactly at a breakpoint. Their algorithm investigates all arcs (i, j) and all its breakpoints t, solves the TDSPP from i to n starting at time t and the TDSPP from 1 to i ending at time t. The latter is done by pre-computing the inverse costs (given an arc (i, j) and an arrival time t, what is the latest time to depart i so that we arrive at time t) so that we can solve the TDSPP from 1 to i ending at time t. If we let K be the total number of breakpoints in the network, then the complexity is $\mathcal{O}(K \times SSP)$. Such an app-roach performs many extraneous calculations due to its brute force nature. Our algorithm very significantly reduces the number of breakpoints investigated.

3 Dynamic Discretization Discovery Algorithm

Our algorithm is inspired by [10] and dynamically updates the discretization of a TEN. Any TEN allows the computation of lower and upper bounds on the duration of an optimal path. The lower and upper bounds are used to determine whether a minimum duration path has been found and, if not, for which parts of the TEN the time discretization should be refined.

We illustrate our ideas using the network in Fig. 1 with travel time functions as given in Table 1. The time interval is $[0, 5]$, breakpoints for each arc are at every integer point, with the exception of arc $(3, 4)$ which only has breakpoints at $0, 1, 2, 5$. These values have been chosen to increase the visibility of the algorithm progression and to reduce the number of iterations.

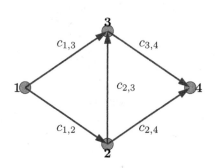

Fig. 1. Network D

Table 1. Arc travel times at each breakpoint (BP)

BP time	Arc travel times				
	$(1, 2)$	$(1, 3)$	$(2, 3)$	$(2, 4)$	$(3, 4)$
0	1.34	2.85	1.99	1.29	0.61
1	0.66	2.95	1.82	1.02	0.73
2	0.14	3.00	1.51	1.63	0.83
3	0.01	2.98	1.10	2.57	—
4	0.35	2.90	0.67	3.00	—
5	1.00	2.76	0.30	2.54	1.00

We maintain a set of *Arc-completed Backwards Shortest Path Trees* (ABSPTs), each denoted by $\mathcal{D}^{(k,t_k)}$, where $k \in N$ and $t_k \in [0, T]$, and is created by the following procedure. First, find a TDSP from (k, t_k) to n to obtain an arrival time t_n at n, see Fig. 2a. Then, compute a time-dependent *backwards shortest path tree* (BSPT), giving node-time pairs (i, t_i) for each node in N, see Fig. 2b. Finally, "arc-complete" the tree by adding an arc $((i, t_i), (j, t_j))$ for each arc $(i, j) \in A$ which is missing, see Fig. 2c. Note that $\mathcal{D}^{(k,t_k)}$ can also be identified as $\mathcal{D}^{(i,t_i)}$, for any $(i, t_i) \in \mathcal{D}^{(k,t_k)}$, since the procedure starting at either (k, t_k)

and (i, t_i) generates the same ABSPT. In particular, it is convenient to identify an ABSPT by its departure time at 1. By the FIFO property, the ABSPTs in a TEN have a natural chronological order (the ABSPTs can be sorted in nondecreasing order of their departure time at 1). The benefit of working with ABSPTs is that any possible path can be represented in an ABSPT and an ABSPT can be used to compute a lower and upper bound on the duration of an optimal path.

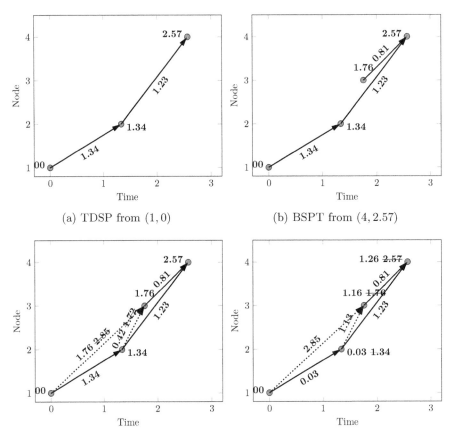

(a) TDSP from $(1, 0)$

(b) BSPT from $(4, 2.57)$

(c) Arc-completed BSPT (actual travel times in red)

(d) Arc-completed BSPT with underestimated travel times (using $\mathcal{D}^{(4,5)}$) and node times (actual node times in red)

Fig. 2. Procedure to generate the ABSPT corresponding to $(1, 0)$ (departure times in blue; travel times in black). (Color figure online)

Suppose we have at least two ABSPTs in a TEN. Let (i, t_i^1) be a node-time pair in one ABSPT and (i, t_i^2) be the node-time pair for i in the (chronologically) next ABSPT. Instead of actual travel times $c_{ij}(t_i^1)$ on $\mathcal{D}^{(i, t_i^1)}$, use the *underestimated travel times* (UTTs) given by $\underline{c}_{ij}(t_i^1) = \min_{t'}\{c_{ij}(t') \mid t_i^1 \leq t' \leq t_i^2\}$, see

Fig. 2d. It is easy to see that a shortest path from $(1, t_1^1)$ to (n, t_n^1) using the underestimated travel times gives a lower bound on a minimum duration path departing in $[t_1^1, t_1^2)$, and that a shortest path from $(1, t_1^1)$ to (n, t_n^1) using the actual travel times gives an upper bound. The last ABSPT, which will always be $\mathcal{D}^{(n,T)}$ in our algorithm, is treated separately. Let $(1, t_1)$ be the node-time pair in $\mathcal{D}^{(n,T)}$ for 1. By the construction of $\mathcal{D}^{(n,T)}$, it is not possible to depart later than t_1 and arrive at n by time T, hence, we do not need to use underestimated travel times and instead keep the actual travel times for this particular ABSPT.

Given an ordered set L of ABSPTs, $L = (\mathcal{D}^{(1,t_1^1)}, \ldots, \mathcal{D}^{(1,t_1^p)})$, as well as associated lower and upper bounds, suppose that $\mathcal{D}^{(1,t_1^k)}$ contains the smallest lower bound. We choose to refine our time discretization by exploring the gap in the TEN between ABSPTs $\mathcal{D}^{(1,t_1^k)}$ and $\mathcal{D}^{(1,t_1^{k+1})}$. Adding an ABSPT corresponding to any node-time pair (i, t) such that $t_i^k < t < t_i^{k+1}$ and updating the underestimated travel times for $\mathcal{D}^{(1,t_1^k)}$ (since the next ABSPT is no longer $\mathcal{D}^{(1,t_1^{k+1})}$) may improve the lower bound, since the interval used to calculate the underestimated travel times has shortened.

The concepts presented so far can be used to devise an algorithm that converges to an optimal path, but not enough to ensure finite termination. Finite termination can be achieved by exploiting the fact that there exists an optimal path that contains a departure at a node i at time t for some arc (i, j) that has a breakpoint at time t (see [5]). Therefore, we only create ABSPTs that contain at least one node-time pair (i, t) corresponding to a breakpoint. Since there are a finite number of breakpoints, this ensures finite termination. To achieve efficiency, we exploit the fact that the arrival time function at n for departures between t_1^1 and t_1^2 is concave if no shortest path that departs between t_1^1 and t_1^2 contains a breakpoint (also shown in [5]). This situation occurs when there are no more breakpoints remaining in the gap between two ABSPTs $\mathcal{D}^{(1,t_1^k)}$ and $\mathcal{D}^{(1,t_1^{k+1})}$, and we know that the minimum duration path departing between t_1^k and t_1^{k+1} departs at either t_1^k or t_1^{k+1}, both of which have already been calculated as an upper bound, hence the lower bound for $\mathcal{D}^{(1,t_1^k)}$ can be updated to one of these upper bounds and thus no longer needs to be considered. This gives us an additional termination criterion: we can terminate when the smallest lower bound among the ABSPTs still being under consideration is larger than the best upper bound obtained so far. A high-level overview of our algorithm can be found in Algorithm 1.

The algorithm explores breakpoints. We choose to look for a breakpoint t in the travel time function of arc (i, j) such that i is minimized, then j is minimized, and t is the median among the breakpoints. The performance of the algorithm depends greatly on being able to efficiently compute the minimum arc travel time in a departure time interval. This is accomplished by (efficiently) pre-computing a look-up table that gives the next local minimum for any breakpoint.

Next, we illustrate the algorithm on the example; see Fig. 3. The algorithm is initialized with $\mathcal{D}^{(1,0)}, \mathcal{D}^{(4,5)}$, which are generated by the paths $P = ((1,2), (2,4))$ and $P = ((1,2), (2,3), (3,4))$, respectively.

Algorithm 1. Dynamic Discretization Discovery (DDD) Algorithm.

input : $G = (N, A)$, $c_{i,j}(t)$, T
output: minimum duration shortest path
$L \leftarrow (\mathcal{D}^{(1,0)}, \mathcal{D}^{(n,T)})$;
$UB \leftarrow \min\{computeUB(\mathcal{D}^{(1,0)}), computeUB(\mathcal{D}^{(n,T)})\}$;
$LB \leftarrow computeLB(\mathcal{D}^{(1,0)})$;
$\mathcal{D}^{(1,t_1^k)} \leftarrow \mathcal{D}^{(1,0)}$;
while $(LB < UB)$ **do**

> **if** *there is a breakpoint* (j, τ) *between* $\mathcal{D}^{(1,t_1^k)}$ *and* $\mathcal{D}^{(1,t_1^{k+1})}$ **then**
>> **if** $computeUB(\mathcal{D}^{(j,\tau)}) < UB$ **then**
>>> $UB \leftarrow UB(\mathcal{D}^{(j,\tau)})$
>>
>> **end**
>> $recomputeLB(\mathcal{D}^{(1,t_1^k)})$;
>> $computeLB(\mathcal{D}^{(j,\tau)})$; $insert(L, \mathcal{D}^{(j,\tau)})$;
>
> **else**
>> $LB(\mathcal{D}^{(1,t_1^k)}) = UB(\mathcal{D}^{(1,t_1^k)})$;
>
> **end**
> $LB \leftarrow updateLB(L)$;
> $\mathcal{D}^{(1,t_1^k)} \leftarrow getBestLB(L)$;

end

In Iteration 1, since $\mathcal{D}^{(1,0)}$ gives the lower bound, look at the section of the TEN succeeding $\mathcal{D}^{(1,0)}$, and observe that arc $(1,2)$ has a breakpoint at $t = 1$, so we add $\mathcal{D}^{(1,1)}$. It turns out that $\mathcal{D}^{(1,1)}$ has UB= 2.0804 and LB= 1.4470. Since $LB < UB$, we continue with the algorithm. We proceed to add $\mathcal{D}^{(1,2)}$ and $\mathcal{D}^{(2,2)}$. In Iteration 4, $\mathcal{D}^{(1,1)}$ contains the LB, however, since there are no breakpoints in the succeeding section, we replace the LB of $\mathcal{D}^{(1,1)}$ with its UB. The new LB is contained in $\mathcal{D}^{(2,2)}$. We proceed to replace the LB of $\mathcal{D}^{(2,2)}$ and $\mathcal{D}^{(1,2)}$ with their UB due to the lack of breakpoints in the succeeding sections, at which stage UB= 1.8886 is less than LB= 1.8888 and hence the algorithm terminates. The optimal path is the one that corresponds to UB, which was found when $\mathcal{D}^{(1,2)}$ was created and is $((1,2),(2,4))$ starting at time 2.

4   Computational Study

To analyze the performance of the DDD algorithm, we apply it to several randomly generated instances. In particular, we solve 10 instances with $n = 20$ and $T = 200$ and 10 instances with $n = 30$ and $T = 200$. The instances are generated as follows. The arc set A consists of all pairs of nodes (i,j) where $i < j$. The travel time on arc (i,j) is the piecewise linear interpolant of the function $f_{i,j}(t) = (j-i) + sin(b_{i,j} \times t)$ at each integer point from 0 to T, so that there are $T + 1$ breakpoints, where $b_{i,j}$ is a random number between 0 and 1 generated using a pre-specified random seed and the Matlab 'twister' RNG. The function

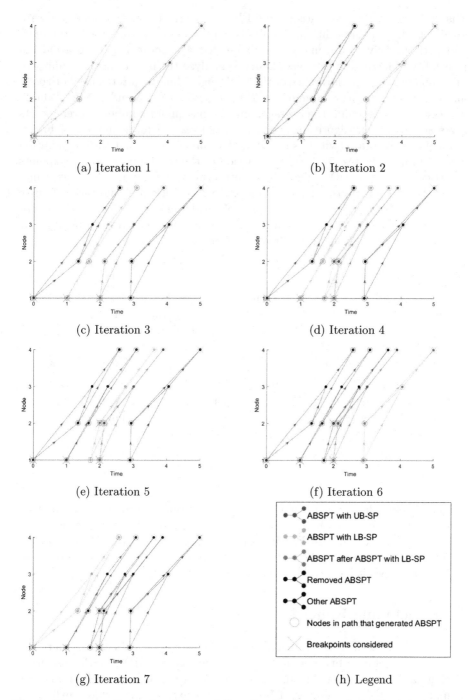

(a) Iteration 1

(b) Iteration 2

(c) Iteration 3

(d) Iteration 4

(e) Iteration 5

(f) Iteration 6

(g) Iteration 7

(h) Legend

Fig. 3. Time-expanded network in each iteration for the example in Fig. 1 (note that a ABSPT may satisfy multiple criteria in the legend).

f is designed so that $c_{i,j}$ satisfies the FIFO property (since the slope is always greater than -1). In addition, due to the additional constant $(j-i)$, the optimal path is likely to use many arcs. Note that we choose $T \geq n$ to avoid the possibility of having no feasible TDSP. To analyze the impact of the number of breakpoints on the performance of the DDD algorithm, we stretch the horizon T and the function $f_{i,j}(t)$; we multiply T by a factor $S = 2.5$ and $= 5$. In Table 2, we report the minimum, the average, and the maximum number of breakpoints investigated by our algorithm over the instances, the total number of breakpoints in the instance (which is the number of breakpoints that the enumeration algorithm investigates [5]), and the fraction of the total number of breakpoints investigated by the DDD algorithm. Furthermore, we report the solve times of the DDD algorithm and the enumeration algorithm and the ratio of these solves times. Note that the total number of breakpoints is $|N-1| \times (T+1)$ not $|A| \times (T+1)$, as in [5], since for arcs with a common tail node and breakpoint, we only need to investigate the breakpoint once.

Table 2. Computational results.

n	S	BP			Total #BP	%BP	Avg. time DDD	Avg. time enum.	Ratio
		Min.	Avg.	Max.					
20	1	112	169.7	224	3800	4.47	103.4	55.8	1.86
	2.5	127	175.2	229	9500	1.84	118.3	218.9	0.54
	5	139	185.0	242	19000	0.97	136.2	690.0	0.20
30	1	182	229.4	268	5800	3.96	290.6	170.4	1.71
	2.5	165	243.9	303	14500	1.68	376.7	774.9	0.49
	5	179	261.8	328	29000	0.90	439.5	2493.0	0.18

The results show clearly that the DDD algorithm investigates only a small fraction of the total number of breakpoints. In addition, the fraction decreases when both n and T increase. For finer discretizations of time, the DDD algorithm significantly outperforms the enumeration algorithm in terms of solve times as well even though it has not been optimized for efficiency.

5   Concluding Remarks

We have shown that dynamic discretization discovery concepts can dramatically reduce the number of breakpoints explored when solving MD-TDSPP instances. Preliminary computational results show that our method scales well in both the number of nodes and number of breakpoints. Next, we will explore extending these ideas to other types of transportation problems, e.g., the Time-Dependent Traveling Salesman Problem.

Acknowledgements. This material is based upon work supported by the National Science Foundation under Grant No. 1662848.

References

1. Chabini, I.: Discrete dynamic shortest path problems in transportation applications: complexity and algorithms with optimal run time. Transp. Res. Rec. **1645**, 170–175 (1998)
2. Dean, B.C.: Shortest paths in FIFO time-dependent networks: theory and algorithms. Rapport technique, Massachusetts Institute of Technology (2004)
3. Demiryurek, U., Banaei-Kashani, F., Shahabi, C., Ranganathan, A.: Online computation of fastest path in time-dependent spatial networks. In: Pfoser, D., Tao, Y., Mouratidis, K., Nascimento, M.A., Mokbel, M., Shekhar, S., Huang, Y. (eds.) SSTD 2011. LNCS, vol. 6849, pp. 92–111. Springer, Heidelberg (2011). https://doi.org/10.1007/978-3-642-22922-0_7
4. Gunturi, V.M., Joseph, K., Shekhar, S., Carley, K.M.: Information lifetime aware analysis for dynamic social networks. Technical report, University of Minnesota (2012)
5. Foschini, L., Hershberger, J., Suri, S.: On the complexity of time-dependent shortest paths. Algorithmica **68**(4), 1075–1097 (2014)
6. Orda, A., Rom, R.: Shortest-path and minimum-delay algorithms in networks with time-dependent edge-length. J. ACM (JACM) **37**(3), 607–625 (1990)
7. Nachtigall, K.: Time depending shortest-path problems with applications to railway networks. Eur. J. Oper. Res. **83**(1), 154–166 (1995)
8. Ding, B., Yu, J.X., Qin, L.: Finding time-dependent shortest paths over large graphs. In: Proceedings of the 11th International Conference on Extending Database Technology: Advances in Database Technology, pp. 205–216. ACM (2008)
9. Kanoulas, E., Du, Y., Xia, T., Zhang, D.: Finding fastest paths on a road network with speed patterns. In: Proceedings of the 22nd International Conference on Data Engineering, ICDE 2006, p. 10. IEEE (2006)
10. Boland, N., Hewitt, M., Marshall, L., Savelsbergh, M.: The continuous-time service network design problem. Oper. Res. **65**(5), 1303–1321 (2017)

Observations from Parallelising Three Maximum Common (Connected) Subgraph Algorithms

Ruth Hoffmann[1], Ciaran McCreesh[2(✉)], Samba Ndojh Ndiaye[3],
Patrick Prosser[2], Craig Reilly[2], Christine Solnon[4], and James Trimble[2]

[1] University of St Andrews, St Andrews, UK
[2] University of Glasgow, Glasgow, Scotland
ciaran.mccreesh@glasgow.ac.uk
[3] Université Lyon 1, LIRIS, UMR5205, 69621 Villeurbanne, France
[4] INSA-Lyon, LIRIS, UMR5205, 69621 Villeurbanne, France

Abstract. We discuss our experiences adapting three recent algorithms for maximum common (connected) subgraph problems to exploit multi-core parallelism. These algorithms do not easily lend themselves to parallel search, as the search trees are extremely irregular, making balanced work distribution hard, and runtimes are very sensitive to value-ordering heuristic behaviour. Nonetheless, our results show that each algorithm can be parallelised successfully, with the threaded algorithms we create being clearly better than the sequential ones. We then look in more detail at the results, and discuss how speedups should be measured for this kind of algorithm. Because of the difficulty in quantifying an average speedup when so-called anomalous speedups (superlinear and sublinear) are common, we propose a new measure called *aggregate speedup*.

1 Introduction

Finding a maximum common subgraph is the key step in measuring the similarity or difference between two graphs [3,12,19]. Because of this, maximum common subgraph problems frequently arise in biology and chemistry [10,14,33] where graphs represent molecules or reactions, and also in computer vision [5,7], computer-aided manufacturing [23], the analysis of programs and malware [13,31], crisis management [8], and social network analysis [11].

A *subgraph isomorphism* is an injective mapping from a *pattern* graph to a *target* graph which *preserves adjacency*—that is, it maps adjacent vertices to adjacent vertices. The isomorphism is *induced* if additionally it maps non-adjacent vertices to non-adjacent vertices, preserving non-adjacency as well.

C. McCreesh, P. Prosser, C. Reilly and J. Trimble—This work was supported by the Engineering and Physical Sciences Research Council [grant numbers EP/K503058/1, EP/M508056, and EP/P026842/1].

S. N. Ndiaye and C. Solnon—This work was supported by the ANR project SoLStiCe (ANR-13-BS02-0002-01).

W.-J. van Hoeve (Ed.): CPAIOR 2018, LNCS 10848, pp. 298–315, 2018.
https://doi.org/10.1007/978-3-319-93031-2_22

When working with labelled graphs, a subgraph isomorphism must preserve labels, and on directed graphs, it must preserve orientation. A *common induced subgraph* of two graphs G and H is a pair of induced subgraph isomorphisms from a pattern graph P, one to G and one to H. A *maximum common induced subgraph* is one with as many vertices as possible. (The *maximum common partial subgraph* problem is non-induced, with as many edges as possible; this paper discusses only induced problems.) A common variant of the problem requires a largest *connected* subgraph [10,23,33,36].

Although both the connected and non-connected variants are NP-hard, recently progress has been made towards solving the problem in practice. This paper looks at three branch and bound algorithms for maximum common (connected) induced subgraph problems, each of which is the state of the art for certain classes of instance. We discuss our experiences in adding parallel tree-search to these three algorithms. In each case, our results show that the parallel version of the algorithm is clearly better than the sequential version, although a closer look at the results shows many nuances. Thus this paper focusses primarily on presenting and interpreting the experimental data, rather than heavy implementation details, in the hopes that the lessons we learned are helpful to other practitioners—in particular, we introduce a new measure called *aggregate speedup* which is suitable for determining speedups for decision problems or optimisation problems where anomalous speedups are common.

2   Sequential Algorithms

There are three competitive approaches for the maximum common subgraph problem, each being the strongest on certain classes of instance. The first involves a reduction to the maximum clique problem, whilst the other two approaches are inspired by constraint programming.

2.1   Reduction to Maximum Clique

A *clique* in a graph is a subgraph where every vertex is adjacent to every other. There is a well-known reduction from the maximum common subgraph problem to the problem of finding a maximum clique in an *association graph* [21,25,33]; this reduction resembles the microstructure encoding [17] of the constraint programming approach described below. When combined with a modern maximum clique solver [35], this is the current best approach for solving the problem on labelled graphs [25]. A modified clique-like algorithm can also be used to solve the maximum common connected subgraph problem, by ensuring connectedness during search [25]; again, this is the best known way of solving the problem on labelled graphs. However, the association graph encoding is extremely memory-intensive, limiting its practical use to pairs of graphs with no more than a few hundred vertices.

2.2   Constraint Programming

The maximum common induced subgraph problem may be reformulated as a constraint optimisation problem, as follows. Observe that an equivalent definition of a common subgraph of graphs G and H is an injective *partial* mapping from G to H which preserves both adjacency and non-adjacency. Hence we pick whichever input graph has fewer vertices, and call it the *pattern*; the other graph is called the *target*. The model then follows from this new definition: for each vertex in the pattern, we create a variable, whose domain ranges over each vertex in the target graph, plus an additional value $\perp$ representing an unmapped vertex. We then have three sets of constraints. The first set says that for each pair of adjacent vertices in the pattern (that is, for each edge in the pattern), if neither of these vertices are mapped to $\perp$ then these vertices must be mapped to an adjacent pair of target vertices. The second set is similar, but looks at non-adjacent pairs (or non-edges). Finally, the third set ensures injectivity, by enforcing that the variables must be all different except when using $\perp$. This final set of constraints may either be implemented using binary constraints between all pairs of variables, or a special global "all different except $\perp$" propagator [32]. The objective is simply to find an assignment of values to variables, maximising the number of variables not set to $\perp$. The state of the art for this technique is a dedicated (non-toolkit) implementation of a forward-checking branch and bound search over this model [25,30].

Two approaches exist for ensuring connectedness: either a conventional global constraint and propagator can be used [25], or a special branching rule can enforce connectedness during search [36]. The two techniques are broadly comparable performance-wise [25], but the branching rule is simpler to implement.

2.3   Domain Splitting (McSplit and McSplit↓)

McCreesh et al. [28] observe that due to the special structure of the maximum common subgraph problem, the following property holds throughout the search process using the constraint programming model: any two variables either have domains with no values in common (with the possible exception of $\perp$), or have identical domains. The McSplit algorithm exploits this property. It explores essentially the same search tree as the basic forward-checking constraint programming approach, but using different supporting algorithms and data structures. Rather than storing a domain for each vertex in the pattern graph, equivalence classes of vertices in both graphs are stored in a special data structure which is modified in-place and restored upon backtracking. This enables fast propagation of the constraints and smaller memory requirements. In addition, this data structure enables stronger branching heuristics to be calculated cheaply. The McSplit algorithm effectively dominates conventional constraint programming approaches, being consistently over an order of magnitude faster.

The McSplit↓ algorithm is a variant designed for instances where we expect nearly all of the smaller graph to be found. It branches first on result size, from largest possible result downwards.

2.4 k-Less Subgraph Isomorphism

A different take on the constraint programming approach is presented by Hoffmann et al. [16]. They approach maximum common subgraph via the subgraph isomorphism problem, asking the question "if a pattern graph cannot be found in the target, how much of the pattern graph can be found?". The $k\downarrow$ algorithm tries to solve the subgraph isomorphism problem first for $k = 0$ (asking whether the whole pattern graph can be found in the target). Should that not be satisfiable, it tries to solve the problem for $k = 1$ (one vertex cannot be matched), and should that also not be satisfiable, it iteratively increases k until the result is satisfiable. This approach exploits strong invariants using paths and the degrees of vertices to prune large portions of the search space.

This algorithm is aimed primarily at large instances, where the two graphs are of different orders, and where it is expected that the solution will involve most of the smaller graph (that is, k is expected to be low). The sequential implementation we start with does not support labels or the connected variant.

3 Benchmark Instances

Most of the benchmark instances we will use come from a standard database for maximum common subgraph problems [6,34]. This benchmark set can be used in a number of ways, for different variants of the problem. Following other recent work [16,25,28], we use it to create five families of instances, as follows:

Unlabelled undirected instances, by selecting the first ten members of each parameter class where the graphs have up to 50 vertices each—this gives us a total of 4,110 instances.

Vertex labelled undirected instances, by selecting the first ten members of each parameter class (and so graphs have up to 100 vertices each), using the 33% labelling scheme [34] for vertices only. This gives 8,140 instances.

Both labelled, directed instances, by selecting the first ten members of each parameter class, and applying the 33% labelling scheme [34] to both vertices and edges. Again, this gives 8,140 instances.

Unlabelled, connected instances, as per the *unlabelled* case.

Both labelled, connected instances, starting in the same way as the *both labelled, directed* case. These are then converted to undirected graphs by treating edges as undirected, picking the label of the lower-numbered edge.

Following Hoffmann et al. [16], we also work with the 5,725 **Large** instances originally introduced for studying portfolios of subgraph isomorphism algorithms [18]. These graphs are unlabelled and undirected, and can include up to 6,671 vertices. We do not use the clique encoding on these instances due to its memory requirements.

4   Parallel Search

The clique and $k\!\downarrow$ algorithms already make use of fine-granularity bit-parallelism. To introduce coarse-grained thread parallelism, we will parallelise search: viewing backtracking search as forming a tree, we can explore different portions of the tree using different threads. We use a shared incumbent, so better solutions found by one thread can be used by others immediately. In this paper we use C++11 native threads, and so only support shared memory systems.

Parallel tree-search has a long history [1]. Of particular interest to us are so-called *anomalies* [2,20,22]: because we are not performing a fixed amount of work, we should have no expectation of a linear speedup, and instead we could see a sublinear speedup (much less than n from n processors, if speculative work turns out to be wasted) or a superlinear speedup (much more than n from n processors, if a strong incumbent is found more quickly). An absolute slowdown (a speedup much less than 1) is also possible when using some parallelisation techniques.

We stress that these anomalies are due to changes in the amount of work done, and are not due to work balance problems (although work balance is *also* unusually difficult for this problem). Anomalies can have a very strong effect on these algorithms, and we will therefore try to mitigate them as far as possible. In the evaluation of their "embarrassingly parallel search" technique, Malapert et al. [24] "consider unsatisfiable, enumeration and optimization [problem] instances", and "ignore the problem of finding a first feasible solution because the parallel speedup can be completely uncorrelated to the number of workers, making the results hard to analyze". They do "consider optimization problems for which the same variability can be observed, but at a lesser extent because the optimality proof is required". Unfortunately, many of the instances we consider behave more like decision problem instances than optimisation instances: due to the combination of a low solution density, good value-ordering heuristics, and a strong bound function in cases where the optimal solution is relatively large, it is often the case that the runtime is determined almost entirely by how long it takes to find an optimal solution, with the proof of optimality being nearly trivial. Indeed, attempts to parallelise the basic constraint programming approach by static decomposition have had limited success [29].

4.1   Parallel Maximum Clique

Thread-parallel versions of state-of-the-art maximum clique algorithms already exist. McCreesh et al. [27] compare several of these approaches, and make an important observation: although work balance is a problem due to the irregularity of the search tree, often the interaction between search order and parallel work decomposition is the dominating factor in determining speedups. They explain why anomalies are in fact common in practice: many clique problem instances benefit immensely from having found a strong incumbent, but have solutions which are either unique or rare, and are hard to find. They propose a work splitting mechanism which offsets anomalies, guaranteeing reproducibility (two runs with the same instance on the same hardware will give similar runtimes), scalability (increasing the number of cores cannot make things worse), and no absolute slowdowns. Additionally, this mechanism explicitly offsets the commitment to early branching choices, where search ordering heuristics are most likely to be inaccurate [4,15], making superlinear speedups common.

We will use this mechanism for our experiments. The clique-based maximum common subgraph algorithm effectively differs only in the preprocessing stage, and the clique-inspired connected algorithm described by McCreesh et al. [25] is sufficiently similar that it may be parallelised in exactly the same way. Based upon preliminary experiments, we set the mechanism's splitting depth limit parameter to be five rather than the original three, since maximum common subgraph instances appear to give even more irregular search trees than normal clique problem instances.

4.2   Parallel Constraint-Based Search

A similar approach may be used for the $k\downarrow$ algorithm. Although it is not quite a conventional branch and bound algorithm, each individual k pass is a treesearch, and may be parallelised. For each pass, we use the same work splitting mechanism as in the clique algorithm, starting by splitting only at the top level of search to explicitly introduce diversity, and then iteratively increasing the splitting depth as additional work is needed (up to a limit of five levels deep). Because the $k\downarrow$ algorithm uses a conventional constraint programming domain store, there is no need to use recomputation; the state is naturally copied at each branching point.

In principle the McSplit algorithm may be parallelised in exactly the same way. However, this algorithm makes heavy use of an in-place, backtrackable data structure, which is not copied for recursive calls. In order to introduce the *potential* for parallelism, we must make copies of the state data structure. Implemented naïvely, this can give an order of magnitude slowdown to the sequential algorithm, which can be hard to recover using parallelism. To lessen the effects, rather than copying state for each recursive call, we copy once before the main branching loop, and then copy that copy in each "helper" thread, replaying the branching loop without making duplicate recursive calls. (We believe a better approach using partial recomputation may be possible, and intend to investigate this further in the future.)

5   Empirical Evaluation

We perform our experiments on systems with dual Intel Xeon E5-2697A v4 processors and 512 GBytes RAM, running Ubuntu 17.04, with GCC 6.3.0 as the compiler. Each machine has a total of thirty-two cores. We run all our experiments with a one thousand second timeout for each instance. All of our sequential runtimes are from optimised implementations by their original authors which were not designed with parallelism in mind—that is, speedups from parallelism are genuine improvements over the state of the art.

5.1 Parallel Search Is Better Overall

In Fig. 1 we plot empirical cumulative distribution functions showing the number of instances solved over time, for both sequential (solid lines) and parallel (dotted lines) versions of each algorithm. To read these plots, make a choice of timeout along the x-axis (which uses a log scale). The y value at that point shows the number of instances whose runtime (individually) is at most x, for a particular algorithm. In other words, at any given x value, the highest line shows which algorithm is able to solve the largest number of instances using a per-instance timeout of that x value, bearing in mind that the actual sets of instances solved by each algorithm may be completely different.

With one exception, each plot gives the same conclusion: if we are working with a solving time of at least 100 ms, then for any problem family and any sequential algorithm, if given the option of switching to the corresponding parallel algorithm, then we should do so. For the McSplit algorithm on both labelled, connected instances, the parallel algorithm does not quite catch up to the sequential algorithm.

Although good at showing general trends, cumulative plots can hide interesting details. We therefore now take a closer look at each of the three algorithms in turn.

5.2 Clique Results in Depth

In the first column of Fig. 2, we see scatter plots comparing the sequential and parallel runtimes of the clique algorithm on an instance by instance basis, using a log-log plot. Each point represents one instance, with the x-axis being the sequential runtime and the y-axis the parallel runtime. Instances which timed out using one algorithm but not the other are shown as points along the outer borders. Points below the $x-y$ diagonal line represent speedups. The colour of the points indicates the relative size of the solution—darker points represent instances where the solution uses most of the vertices of the input graphs. (We use these conventions for scatter plots throughout this paper.)

Broadly speaking, the results are similar on each of the five families. For runtimes below 100 ms, overheads and the preprocessing step dominate, and we are usually only able to achieve a small speedup. At higher runtimes, most speedups appear to be between ten and thirty, except on the final family of both labelled connected instances, where they are mostly between five and ten. For a few instances, the speedups are lower (but they are still clearly speedups), whilst in the first four families, we also see evidence of superlinear speedups being relatively common.

However, attempting to determine a speedup by staring at a scatter plot is not particularly quantitative. We *could* attempt to find a best fit line through these points, pretending that the superlinear speedups are outliners. We might perhaps get away with this if outliers were rare enough, but in practice we are not expecting linear speedups (and for the other two algorithms, we will see that superlinear speedups are even more common). Alternatively, we could rig

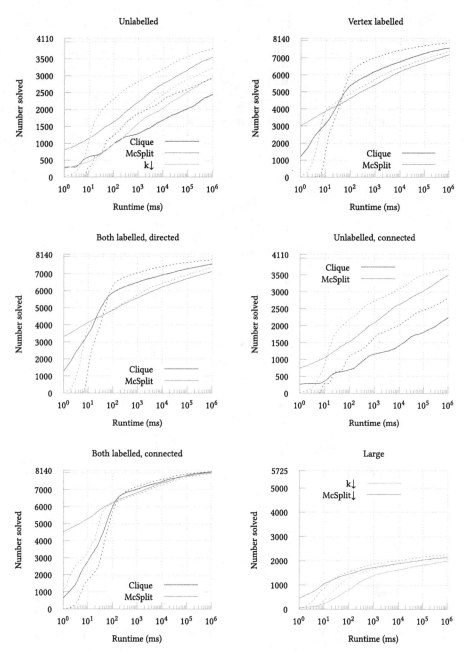

Fig. 1. The cumulative number of instances solved over time. Except in the bottom left plot, the 32 threaded parallel versions (shown using dotted lines) are always better in aggregate than the sequential versions (shown using solid lines).

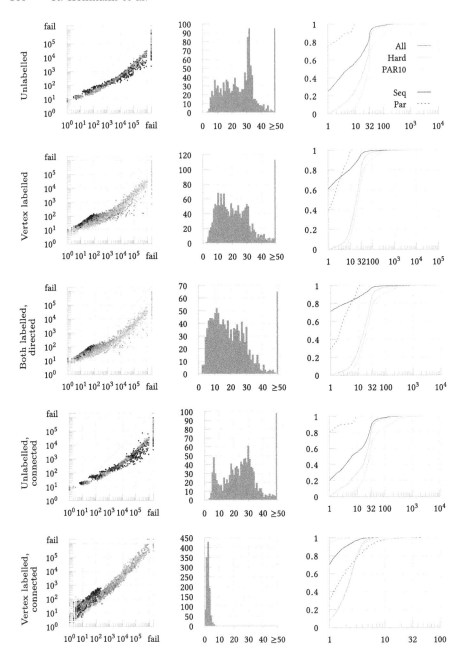

Fig. 2. In the left column, per-instance speedups, using the clique algorithm. The x-axis is sequential performance and the y-axis is 32 threaded performance. In the centre, histograms plotting the distribution of speedups for instances whose sequential runtime was at least 500 ms, and below the timeout. On the right, performance profiles.

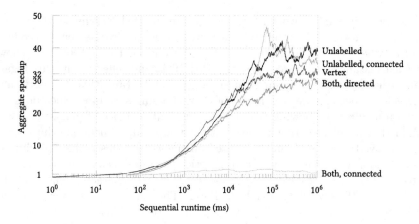

Fig. 3. Aggregate speedups from 32 threads, shown as a function of sequential runtime, for each family supported by the clique algorithm.

our experiments to remove anomalies, by priming search with a known-optimal solution; however, since the time to find an optimal solution (but not prove its optimality) is so important, we do not consider this to be a fair measure of algorithm performance [27].

A more principled approach is given in the second column of Fig. 2. For instances where the sequential run both succeeded and took at least 500 ms, we plot the distribution of speedups obtained. These histograms confirm our informal observations. However, these plots are still not especially satisfactory: in order to calculate a speedup, we can only consider instances where the sequential algorithm succeeded, and so these plots underestimate superlinear speedups. The choice of a 500 ms minimum sequential runtime is also rather arbitrary, and is acceptable only if we expect the parallel algorithms will only be used on relatively hard instances.

In the third column we show performance profiles [9]. A performance profile is a cumulative plot of how many times worse the performance of an algorithm is relative to the virtual best algorithm. Each plot shows three options as different lines. The 'all' lines include easy instances whose sequential runtime is below 500 ms, whilst the other two lines exclude them. The 'hard' line treats sequential timeouts as having been solved at the time limit, whilst the 'PAR10' line treats timeouts as taking ten times longer than the timeout (this convention is common in portfolios [37]). The solid lines show the sequential algorithms, whilst the dotted lines show the parallel algorithms. (There are no dotted lines on the top four plots for the 'hard' and 'PAR10' cases, since the parallel algorithm always beats the sequential algorithm in these cases.) We have normalised the y-axis to the number of counted instances in a given class.

Unfortunately, these three lines can paint very different pictures. For example, for unlabelled instances on the top row, if we include easy instances, it appears that the parallel algorithm can be up to ten times worse, whereas if we exclude them, it is never worse. If we do not use the PAR10 scheme, the performance profile also suggests that there are around twenty-five percent of the hard instances where the speedup is below 10, whilst using PAR10 correctly shows that such instances are rare. However, PAR10 is only effective in this regard because the "typical" speedup is in the region of 10 (and this is a particular inconvenience because we seek a way of characterising speedups which does not rely upon us already knowing that 10 is a reasonable choice of penalty).

A further problem is that to deal with the large superlinear speedups sometimes observed, a log scale must be used on the x-axis; this makes speedups of 10 and 30 look very similar, whilst in practice the difference is important.

To avoid these weaknesses, we propose a new way of characterising speedups. Refer back to the cumulative plots in Fig. 1. The usual way of comparing two algorithms on these plots is by measuring the vertical difference between lines, which would tell us how many more instances the parallel algorithm can solve than the sequential algorithm can with a particular choice of timeout. However, measuring the *horizontal* distance between lines also conveys information. Suppose the sequential algorithm can solve y instances with a selected timeout of s. By moving to the left on a cumulative plot, we can find the timeout p required for the parallel algorithm to solve the same number of instances, bearing in mind that *the two sets of instances could have completely different members*. We define the *aggregate speedup* to be s/p; this can be expressed as a function of time (i.e. s) or of the number of instances solved (y).

We plot aggregate speedups as a function of time in Fig. 3. For a sequential timeout of one thousand seconds, we get speedups of thirty to forty in the unlabelled, vertex labelled, and both labelled, directed cases. In the unlabelled cases, our aggregate speedup are over thirty-two, which is superlinear. With some detailed knowledge of the underlying sequential algorithm, this should perhaps not surprise us: for instances with a large solution, once we have found that solution, a proof of optimality is relatively easy. However, finding that solution can be unusually hard, particularly since the branching strategy for the connected constraint necessarily interferes with the tailored search order used by modern clique algorithms. In contrast, for the both labelled connected case, our aggregate speedup is barely larger than one. A closer inspection of the results shows that the search tree is unusually narrow and deep for these instances, making work balance harder and contention higher.

What about scalability and reproducibility? The first plot in Fig. 4 shows the effects of going from sequential to threaded with two cores, and the next four plots show the effects of doubling the number of threads each time. These plots show that most of the superlinear effects occur with fairly small numbers of threads, with nearly all of the benefits of increased diversity in search being obtained once eight threads are used. As expected, in no case does increasing the number of threads make things substantially worse. The final plot in Fig. 4 shows that runtimes are reproducible: running the same instance on the same hardware twice takes almost exactly the same amount of time.

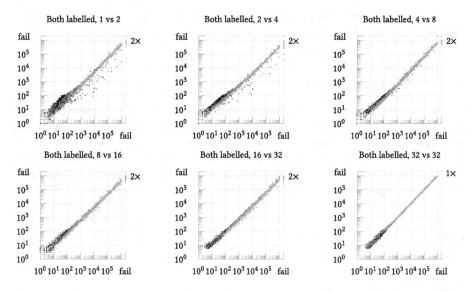

Fig. 4. Per-instance speedups from the clique algorithm on vertex- and edge-labelled, directed instances, when going from sequential to two threads in the first plot, then increasing the number of threads in subsequent plots. The final plot shows 32 threads versus a repeated run also with 32 threads.

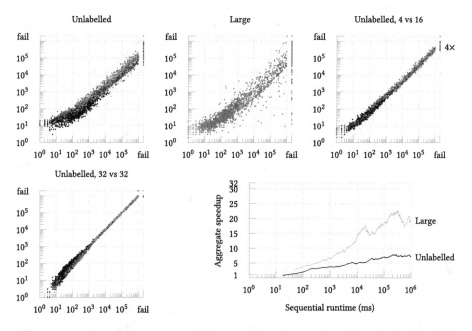

Fig. 5. In the first two plots, per-instance speedups, using the $k\downarrow$ algorithm. The x-axis is sequential performance and the y-axis is 32 threaded performance. Next, scalability and reproducibility, and finally, aggregate speedups for both families.

These results are comforting: they show that anomalies can be controlled, and that switching to a parallel algorithm is not only better, but also safe from a scientific reproducibility perspective.

5.3 $k\!\downarrow$ Results in Depth

In Fig. 5 we show per-instance and aggregate speedups for the $k\!\downarrow$ algorithm. On unlabelled instances, we see a range of speedups between 0.9 and ten, with an aggregate speedup of seven. These results are not as good as with the clique algorithm. Profiling suggests memory allocation problems: although the amount of work done would suggest good parallelism, the time taken to perform each domain copy operation increases as the number of threads increases. Unlike the clique algorithm, which has very small, cache-friendly data structures which are modified in-place, the state for the $k\!\downarrow$ algorithm is large and much of the runtime is spent copying data structures. (Our hardware is a dual multi-core processor configuration, and each core has its own low-level cache, but memory bandwidth is shared. Interestingly, on older Xeon E5 v2 systems, this problem is much more pronounced.)

For the large instances, our aggregate speedup is higher, at around twenty. This has two causes: for larger graphs, the computational effort per recursive call increases by more than the amount the memory copying does, reducing the memory problem slightly, and additionally a much larger number of superlinear speedups occurred with this family of instances. We could perhaps anticipate this latter effect: in many of these instances the maximum common subgraph covers all or nearly all of the smaller of the two graphs, and so once it is found, the proof of optimality is trivial. However, finding a witness can be difficult. We should also expect value-ordering heuristics in these algorithms to be weak at the top of search (they are based upon degree, and many graphs do not have a large degree spread), and so the benefits of high-up diversity can be extremely large [4,15,27]. Indeed, similar results were seen with a parallel version of the subgraph isomorphism algorithm upon which $k\!\downarrow$ is based [26].

The third and fourth plots in Fig. 5 show that as with the clique algorithm, this parallelism is reproducible, and that runtimes do not get worse when the number of threads is increased. (Although not shown, we also tried to parallelise $k\!\downarrow$ using randomised work-stealing from Intel Cilk Plus. Doing so gives generally reasonable results on average, as it does for the clique algorithm [27], but now repeat runtimes can differ by more than an order of magnitude.)

5.4 McSplit Results in Depth

Finally, we look at our attempts to parallelise the McSplit algorithm. Recall that doing so required heavy modifications to the implementation, introducing significant amounts of speculative copying of a data structure that is usually backtrackable and modified in-place.

For unlabelled, unlabelled connected, and large instances, Fig. 6 shows a particularly high proportion of strongly superlinear speedups. This is because the

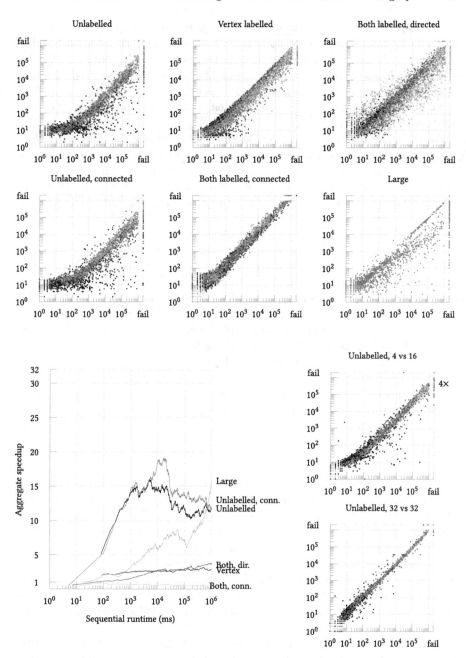

Fig. 6. On the first two rows, per-instance speedups, using McSplit. Below, aggregate speedups on the left, and on the right, scalability and reproducibility.

McSplit algorithm is focussed upon exploring the search space very quickly, and its branching heuristics do not have the advantage of the domain filtering performed by $k\downarrow$, or the rich inter-domain knowledge coming from the combination of the association graph encoding and the colour ordering used by clique algorithms. Thus making a correct value-ordering choice at the top of search is harder for McSplit than for other algorithms, and so increased diversity can be particularly beneficial.

For the large instances, we see evidence of work balance problems. McSplit's use of a "smallest domain first" variable-ordering heuristic, combined with the presence of $\perp$ in domains, tends to produce narrow (nearly binary) and deep search trees. These balance problems are even more evident in the labelled cases (where following a guessed assignment, many domains are left with only two values), and often lead to little to no speedup being obtained. Indeed, for the labelled, connected case, we see a slight aggregate *slowdown*.

The scatter plots also show occasional large absolute slowdowns, sometimes by over an order of magnitude. These are due to the changes which had to be made to the sequential algorithm (and because we are benchmarking against the sequential algorithm, not a parallel algorithm with one thread), rather than search order effects. In cases where parallelism cannot be exploited, the cost of speculatively copying domains at each level of search can dominate the runtimes. Because of this, fixing work balance problems by increasing the splitting depth typically makes matters much worse, not better.

What about scalability and reproducibility? Figure 6 presents a less ideal picture than for the previous two algorithms—again, this is due to speculative overheads that fail to pay off, rather than being anomalies in the classical sense.

6   Conclusion

We have parallelised three state-of-the-art maximum common (connected) subgraph algorithms with a reasonable degree of success by using dynamic work-splitting. Despite having a branch and bound flavour, all three sequential algorithms had their own difficulties and performance characteristics which prevented them from cleanly fitting into common abstraction frameworks. Nonetheless, our results show that the parallel algorithms are not just better in aggregate, but also preserve the desirable reproducibility properties of sequential algorithms. A large part of our success was down to using parallelism to explicitly introduce diversity into the search process, offsetting weak early value-ordering branching choices.

There is room for improvement, particularly with respect to work balance. However, improvements to work balance must not come at the expense of the search order properties, nor at the cost of increased overheads.

More generally, we introduced the idea of *aggregate speedups*, to deal with measuring a speedup in the presence of anomalies. This measure gives sensible answers even when working with instances which behave like decision problems. Aggregate speedups informed part of our analysis, but our results highlight the importance of viewing results in multiple ways, and in using large families of instances with different characteristics when evaluating parallel search algorithms—had we looked only at unlabelled instances, or only at labelled connected instances, our conclusion would be very different.

References

1. Bader, D.A., Hart, W.E., Phillips, C.A.: Parallel algorithm design for branch and bound. In: Greenberg, H.J. (ed.) Tutorials on Emerging Methodologies and Applications in Operations Research. ISOR, vol. 76, pp. 1–44. Springer, New York (2005). https://doi.org/10.1007/0-387-22827-6_5
2. de Bruin, A., Kindervater, G.A.P., Trienekens, H.W.J.M.: Asynchronous parallel branch and bound and anomalies. In: Ferreira, A., Rolim, J. (eds.) IRREGULAR 1995. LNCS, vol. 980, pp. 363–377. Springer, Heidelberg (1995). https://doi.org/10.1007/3-540-60321-2_29
3. Bunke, H.: On a relation between graph edit distance and maximum common subgraph. Pattern Recogn. Lett. **18**(8), 689–694 (1997)
4. Chu, G., Schulte, C., Stuckey, P.J.: Confidence-based work stealing in parallel constraint programming. In: Gent, I.P. (ed.) CP 2009. LNCS, vol. 5732, pp. 226–241. Springer, Heidelberg (2009). https://doi.org/10.1007/978-3-642-04244-7_20
5. Combier, C., Damiand, G., Solnon, C.: Map edit distance vs. graph edit distance for matching images. In: Kropatsch, W.G., Artner, N.M., Haxhimusa, Y., Jiang, X. (eds.) GbRPR 2013. LNCS, vol. 7877, pp. 152–161. Springer, Heidelberg (2013). https://doi.org/10.1007/978-3-642-38221-5_16
6. Conte, D., Foggia, P., Vento, M.: Challenging complexity of maximum common subgraph detection algorithms: a performance analysis of three algorithms on a wide database of graphs. J. Graph Algorithms Appl. **11**(1), 99–143 (2007)
7. Cook, D.J., Holder, L.B.: Substructure discovery using minimum description length and background knowledge. J. Artif. Intell. Res. **1**, 231–255 (1994)
8. Delavallade, T., Fossier, S., Laudy, C., Lortal, G.: On the challenges of using social media for crisis management. In: Rogova, G., Scott, P. (eds.) Fusion Methodologies in Crisis Management, pp. 137–175. Springer, Cham (2016). https://doi.org/10.1007/978-3-319-22527-2_8
9. Dolan, E.D., Moré, J.J.: Benchmarking optimization software with performance profiles. Math. Program. **91**(2), 201–213 (2002)
10. Ehrlich, H.C., Rarey, M.: Maximum common subgraph isomorphism algorithms and their applications in molecular science: a review. Wiley Interdisc. Rev.: Comput. Mol. Sci. **1**(1), 68–79 (2011)
11. Fang, M., Yin, J., Zhu, X., Zhang, C.: Trgraph: cross-network transfer learning via common signature subgraphs. IEEE Trans. Knowl. Data Eng. **27**(9), 2536–2549 (2015)
12. Fernández, M., Valiente, G.: A graph distance metric combining maximum common subgraph and minimum common supergraph. Pattern Recogn. Lett. **22**(6/7), 753–758 (2001)
13. Gao, D., Reiter, M.K., Song, D.: BinHunt: automatically finding semantic differences in binary programs. In: Chen, L., Ryan, M.D., Wang, G. (eds.) ICICS 2008. LNCS, vol. 5308, pp. 238–255. Springer, Heidelberg (2008). https://doi.org/10.1007/978-3-540-88625-9_16

14. Gay, S., Fages, F., Martinez, T., Soliman, S., Solnon, C.: On the subgraph epimorphism problem. Discret. Appl. Math. **162**, 214–228 (2014)
15. Harvey, W.D., Ginsberg, M.L.: Limited discrepancy search. In: Proceedings of the Fourteenth International Joint Conference on Artificial Intelligence, IJCAI 95, Montréal Québec, Canada, 20–25 August 1995, vol. 2, pp. 607–615. Morgan Kaufmann (1995)
16. Hoffmann, R., McCreesh, C., Reilly, C.: Between subgraph isomorphism and maximum common subgraph. In: Singh, S.P., Markovitch, S. (eds.) Proceedings of the Thirty-First AAAI Conference on Artificial Intelligence, 4–9 February 2017, San Francisco, California, USA, pp. 3907–3914. AAAI Press (2017)
17. Jégou, P.: Decomposition of domains based on the micro-structure of finite constraint-satisfaction problems. In: Fikes, R., Lehnert, W.G. (eds.) Proceedings of the 11th National Conference on Artificial Intelligence, Washington, DC, USA, 11–15 July 1993, pp. 731–736. AAAI Press/The MIT Press (1993)
18. Kotthoff, L., McCreesh, C., Solnon, C.: Portfolios of subgraph isomorphism algorithms. In: Festa, P., Sellmann, M., Vanschoren, J. (eds.) LION 2016. LNCS, vol. 10079, pp. 107–122. Springer, Cham (2016). https://doi.org/10.1007/978-3-319-50349-3_8
19. Kriege, N.: Comparing graphs. Ph.D. thesis, Technische Universität Dortmund (2015)
20. Lai, T., Sahni, S.: Anomalies in parallel branch-and-bound algorithms. Commun. ACM **27**(6), 594–602 (1984)
21. Levi, G.: A note on the derivation of maximal common subgraphs of two directed or undirected graphs. CALCOLO **9**(4), 341–352 (1973)
22. Li, G., Wah, B.W.: Coping with anomalies in parallel branch-and-bound algorithms. IEEE Trans. Comput. **35**(6), 568–573 (1986)
23. Luo, C., Wang, X., Su, C., Ni, Z.: A fixture design retrieving method based on constrained maximum common subgraph. IEEE Trans. Autom. Sci. Eng. **PP**(99), 1–13 (2017)
24. Malapert, A., Régin, J., Rezgui, M.: Embarrassingly parallel search in constraint programming. J. Artif. Intell. Res. **57**, 421–464 (2016)
25. McCreesh, C., Ndiaye, S.N., Prosser, P., Solnon, C.: Clique and constraint models for maximum common (connected) subgraph problems. In: Rueher, M. (ed.) CP 2016. LNCS, vol. 9892, pp. 350–368. Springer, Cham (2016). https://doi.org/10.1007/978-3-319-44953-1_23
26. McCreesh, C., Prosser, P.: A parallel, backjumping subgraph isomorphism algorithm using supplemental graphs. In: Pesant, G. (ed.) CP 2015. LNCS, vol. 9255, pp. 295–312. Springer, Cham (2015). https://doi.org/10.1007/978-3-319-23219-5_21
27. McCreesh, C., Prosser, P.: The shape of the search tree for the maximum clique problem and the implications for parallel branch and bound. TOPC **2**(1), 8:1–8:27 (2015)
28. McCreesh, C., Prosser, P., Trimble, J.: A partitioning algorithm for maximum common subgraph problems. In: Proceedings of the Twenty-Sixth International Joint Conference on Artificial Intelligence, IJCAI 2017, Melbourne, Australia, 19–25 August 2017 (2017, to appear)
29. Minot, M., Ndiaye, S.N., Solnon, C.: A comparison of decomposition methods for the maximum common subgraph problem. In: 27th IEEE International Conference on Tools with Artificial Intelligence, ICTAI 2015, Vietri sul Mare, Italy, 9–11 November 2015, pp. 461–468. IEEE Computer Society (2015)

30. Ndiaye, S.N., Solnon, C.: CP models for maximum common subgraph problems. In: Lee, J. (ed.) CP 2011. LNCS, vol. 6876, pp. 637–644. Springer, Heidelberg (2011). https://doi.org/10.1007/978-3-642-23786-7_48
31. Park, Y.H., Reeves, D.S., Stamp, M.: Deriving common malware behavior through graph clustering. Comput. Secur. **39**, 419–430 (2013)
32. Petit, T., Régin, J.-C., Bessière, C.: Specific filtering algorithms for over-constrained problems. In: Walsh, T. (ed.) CP 2001. LNCS, vol. 2239, pp. 451–463. Springer, Heidelberg (2001). https://doi.org/10.1007/3-540-45578-7_31
33. Raymond, J.W., Willett, P.: Maximum common subgraph isomorphism algorithms for the matching of chemical structures. J. Comput. Aided Mol. Des. **16**(7), 521–533 (2002)
34. Santo, M.D., Foggia, P., Sansone, C., Vento, M.: A large database of graphs and its use for benchmarking graph isomorphism algorithms. Pattern Recogn. Lett. **24**(8), 1067–1079 (2003)
35. Segundo, P.S., Matía, F., Rodríguez-Losada, D., Hernando, M.: An improved bit parallel exact maximum clique algorithm. Optim. Lett. **7**(3), 467–479 (2013)
36. Vismara, P., Valery, B.: Finding maximum common connected subgraphs using clique detection or constraint satisfaction algorithms. In: Le Thi, H.A., Bouvry, P., Pham Dinh, T. (eds.) MCO 2008. CCIS, vol. 14, pp. 358–368. Springer, Heidelberg (2008). https://doi.org/10.1007/978-3-540-87477-5_39
37. Xu, L., Hoos, H., Leyton-Brown, K.: Hydra: automatically configuring algorithms for portfolio-based selection. In: Fox, M., Poole, D. (eds.) Proceedings of the Twenty-Fourth AAAI Conference on Artificial Intelligence, AAAI 2010, Atlanta, Georgia, USA, 11–15 July 2010. AAAI Press (2010)

Horizontally Elastic Not-First/Not-Last Filtering Algorithm for Cumulative Resource Constraint

Roger Kameugne[1,2]([✉]), Sévérine Betmbe Fetgo[3], Vincent Gingras[4],
Yanick Ouellet[4], and Claude-Guy Quimper[4]

[1] University of Maroua, Maroua, Cameroon
[2] University of Bamenda, Bamenda, Cameroon
rkameugne@gmail.com
[3] University of Dschang, Dschang, Cameroon
betmbe200@yahoo.fr
[4] Université Laval, Québec, QC, Canada
{vincent.gingras.5,yanick.ouellet.2}@ulaval.ca,
Claude-Guy.Quimper@ift.ulaval.ca

Abstract. Fast and powerful propagators are the main key to the success of constraint programming on scheduling problems. It is, for example, the case with the CUMULATIVE constraint, which is used to model tasks sharing a resource of discrete capacity. In this paper, we propose a new not-first/not-last rule, which we call the *horizontally elastic not-first/not-last*, based on strong relaxation of the earliest completion time of a set of tasks. This computation is obtained when scheduling the tasks in a horizontally elastic way. We prove that the new rule is sound and is able to perform additional adjustments missed by the classic not-first/not-last rule. We use the new data structure called *Profile* to propose a $\mathcal{O}(n^3)$ filtering algorithm for a relaxed variant of the new rule where n is the number of tasks sharing the resource. We prove that the proposed algorithm still dominates the classic not-first/not-last algorithm. Experimental results on highly cumulative instances of resource constrained project scheduling problems (RCPSP) show that using this new algorithm can substantially improve the solving process of instances with an occasional and marginal increase of running time.

1 Introduction

Cumulative scheduling is the allocation of a scarce resource to tasks over time. It appears in many real-world problems such as university timetable, ship loading, employee scheduling, bridge or building constructions. The challenging part of this problem comes from the resource constraint. To solve it efficiently with a constraint programming solver, it is important to have fast and efficient propagators for the CUMULATIVE [1] constraint. The CUMULATIVE constraint models

---

This work was partially supported by a grant from the Niels Henrik Abel board and the University Laval.

the problem where a limited number of tasks can be executed simultaneously. In a cumulative scheduling problem (CuSP), a set of tasks T has to be executed on a resource of capacity C. Each task $i \in T$ is executed without interruption during p_i time units and uses $c_i \leq C$ units of resource. For a task $i \in T$, the earliest start time est_i and the latest completion time lct_i are specified. A solution to a CuSP instance is an assignment of valid start time s_i to each task $i \in T$ such that the resource constraint is satisfied, i.e.,

$$\forall i \in T, \qquad est_i \leq s_i \leq s_i + p_i \leq lct_i \qquad (1)$$

$$\forall \tau, \qquad \sum_{i \in T,\ s_i \leq \tau < s_i + p_i} c_i \leq C \qquad (2)$$

The inequalities in (1) ensure that each task is assigned a feasible start and end time, while (2) enforces the resource constraint. Each task $i \in T$ has an energy $e_i = c_i \cdot p_i$, an earliest completion time $ect_i = est_i + p_i$ and a latest start time $lst_i = lct_i - p_i$. These notations can be extended to non-empty sets of tasks as follows:

$$e_\Omega = \sum_{j \in \Omega} e_j, \quad est_\Omega = \min_{j \in \Omega} est_j, \quad lct_\Omega = \max_{j \in \Omega} lct_j, \quad ECT_\Omega = \min_{j \in \Omega} ect_j. \qquad (3)$$

By convention, for empty sets we have: $e_\emptyset = 0$, $est_\emptyset = +\infty$, $lct_\emptyset = -\infty$ and $ECT_\emptyset = +\infty$. Throughout the paper, we assume that for any task $i \in T$, $ect_i \leq lct_i$ and $c_i \leq C$, otherwise the problem has no solution. We let $n = |T|$ denotes the number of tasks, $k = |\{c_i, i \in T\}|$ denotes the number of distinct capacity requirements of tasks. $H = \{ect_i, i \in T\}$ denotes the set of distinct earliest completion times of tasks with $|H| = h$. The global constraint CUMULATIVE removes inconsistent values from the domain of starting time variable $s_i \in [est_i, lst_i]$. Since the CuSP is a NP-Hard problem [2], it is NP-Hard to remove all such values. Polynomial time algorithms only exist for relaxations of the problem.

The global constraint CUMULATIVE integrates many filtering algorithms which perform different pruning and are sometimes combined for more pruning (depending on the characteristics of the instance) to reduce the search space and thus the running time [3]. Each of these filtering algorithms can be called thousands of times during the search. Therefore, it is important for them to be fast, exact and efficient. Among these filtering algorithms, edge-finding [4,7] and timetabling [5] are the most used, but there exists many other filtering algorithms such as not-first/not-last [6,9,10], energetic reasoning [3,12], and more recently, horizontally elastic edge-finder [8]. For the remainder of this paper we focus solely on the algorithm for updating the earliest starting times, as the latest completion time algorithm is both symmetric and easily derived.

In this paper, we propose a new formulation of the not-first/not-last rule based on a strong relaxation of the earliest completion time of a set of tasks. The novel formula, which we call *horizontally elastic not-first/not-last* subsumes the classic not-first/not-last rule. With the data structure *Profile* from [8], we propose a $\mathcal{O}(n^3)$ relaxation of the new rule. Experimental results on highly cumulative instances of resource constrained project scheduling problems (RCPSP) from suites benchmarks of libraries BL [11], Pack [13] and KSD15_D [14] highlight that using this new algorithm reduces the number of backtracks for a large majority of instances with an occasional and marginal increase of the running time.

The rest of the paper is organized as follows. Section 2 presents the classic not-first rule for cumulative resource constraint and Sect. 3 defines the novel function for computing the earliest completion time of a set of tasks with its corresponding algorithm as it is formulated and presented in [8]. In Sect. 4, we propose a new formulation of the not-first rule based on the earliest completion time of a set of tasks which subsumes the classic not-first/not-last rule. Section 5 focuses on the presentation of a $\mathcal{O}(n^3)$ not-first algorithm for the horizontally elastic not-first rule where n is the number of tasks being scheduled on the resource, while Sect. 6 presents a relaxation of the above algorithm with the same complexity. Section 7 shows that the relaxed horizontally elastic not-first algorithm dominates the classic not-first algorithm. Finally, in Sect. 8, the empirical evaluation of the new algorithm on highly cumulative instances of the RCPSP is presented while Sect. 9 concludes the paper.

2   Classic Not-First Rule

The not-first/not-last rule detects tasks that cannot run first/last relatively to a set of tasks and prunes their time bounds. If a task i cannot be the first to be executed in $\Omega \cup \{i\}$ then the earliest start time of task i is updated to the minimum earliest completion time of the set. The not-first filtering rule is formalized as follows:
$\forall \Omega \subset T, \ \forall i \in T \setminus \Omega$

$$\begin{cases} \mathrm{est}_i < \mathrm{ECT}_\Omega \\ e_\Omega + c_i(\min(\mathrm{ect}_i, \mathrm{lct}_\Omega) - \mathrm{est}_\Omega) > C(\mathrm{lct}_\Omega - \mathrm{est}_\Omega) \end{cases} \Rightarrow \mathrm{est}_i \geq \mathrm{ECT}_\Omega. \qquad (\mathrm{NF})$$

Recently, in [9] the authors proposed a quadratic not-first algorithm using the Timeline data structure. Some $\mathcal{O}(n^2 \log n)$ algorithms proposed for this rule can be found in [6,10].

3   Function of the Earliest Completion Time

We present a function to compute the earliest completion time of a set of tasks as in [8]. We use the notation ect_Ω^F to denote the fully-elastic earliest completion time of a set of tasks Ω and it is computed by spending a maximum amount of energy as early as possible without any regards to the resource required of the tasks using the following formula [7].

$$\mathrm{ect}_\Omega^F = \left\lceil \frac{\max\{C\mathrm{est}_{\Omega'} + e_{\Omega'} | \Omega' \subseteq \Omega\}}{C} \right\rceil. \qquad (4)$$

This value is a relaxation of the real earliest completion time ect_Ω of the set Ω. Note that ect_Ω is NP-hard to compute [2]. A stronger relaxation for the function ect_Ω called horizontally elastic earliest completion time and noted ect_Ω^H is introduced in [8].

During the computation of this value, any task i consumes e_i units of resource within the interval $[est_i, lct_i)$ and is allowed to consume at any time $t \in [est_i, lct_i)$, between 0 and c_i units of resource. Given a set of tasks Ω, ect_Ω^H is computed using the following functions.

- $c_{max}(t)$ the amount of resource that can be allocated to the tasks in Ω at time t, i.e.,

$$c_{max}(t) = \min \left(\sum_{i \in \Omega | est_i \leq t < lct_i} c_i, C \right) \tag{5}$$

- $c_{req}(t)$ the amount of resource required at time t by the tasks in Ω if they were all starting at their earliest starting times, i.e.,

$$c_{req}(t) = \sum_{i \in \Omega | est_i \leq t < ect_i} c_i \tag{6}$$

- $ov(t)$ called overflow is the energy from $c_{req}(t)$ that cannot be executed at time t due to the limited capacity $c_{max}(t)$.
- $c_{cons}(t)$ the amount of resource that is actually consumed at time t, i.e.,

$$c_{cons}(t) = \min(c_{req}(t) + ov(t-1), c_{max}(t)) \tag{7}$$
$$ov(t) = ov(t-1) + c_{req}(t) - c_{cons}(t) \tag{8}$$
$$ov(est_\Omega) = 0 \tag{9}$$

The horizontally elastic earliest completion time occurs when all tasks are completed, i.e.,

$$ect_\Omega^H = \max\{t | c_{cons}(t) > 0\} + 1. \tag{10}$$

For a set of tasks, it is proven in [8] that the horizontally elastic earliest completion time is a relaxation of the earliest completion time and is stronger than the fully-elastic one.

Theorem 1 [8]. *For all $\Omega \subseteq T$, $ect_\Omega^F \leq ect_\Omega^H \leq ect_\Omega$.*

The computation of ect^H is done with the Profile data structure [8] that stores the resource utilization over time. The tuples $\langle time, cap, \delta_{max}, \delta_{req} \rangle$ (where $time$ is the start time, cap is the remaining capacity of the resource at the start time, δ_{max} and δ_{req} are two quantities initialized to zero) are stored in a sorted linked list whose nodes are called time points. The Profile is initialized with a time point of capacity C for every distinct value of est, ect, and lct. A sufficiently large time point is added to act as a sentinel. While initializing the data structure, pointers are kept so that t_{est_i}, t_{ect_i} and t_{lct_i} return the time point associated to est_i, ect_i, and lct_i. The algorithm *ScheduleTasks* computes the functions $c_{req}(t)$, $c_{max}(t)$, $c_{cons}(t)$ and $ov(t)$ to schedule a set of tasks Θ on the profile P.

Algorithm 1. *ScheduleTasks*(Θ, C) [8]

Input: Θ a set of tasks and C the capacity of the resource
Output: A lower bound ect_Θ^H of the set Θ
1 **for** *all time point t* **do**
2 $\quad\lfloor\; t.\delta_{\max} \leftarrow 0$ and $t.\delta_{req} \leftarrow 0$

3 **for** $i \in \Theta$ **do**
4 $\quad\mid\;$ Increment $t_{\text{est}_i}.\delta_{\max}$ and $t_{\text{est}_i}.\delta_{req}$ by c_i
5 $\quad\lfloor\;$ Decrement $t_{\text{lct}_i}.\delta_{\max}$ and $t_{\text{ect}_i}.\delta_{req}$ by c_i

6 $t \leftarrow P.first$, $ov \leftarrow 0$, $\text{ect} \leftarrow -\infty$, $S \leftarrow 0$, $c_{req} \leftarrow 0$
7 **while** $t.time \neq \text{lct}_\Theta$ **do**
8 $\quad\mid\;$ $t.ov \leftarrow ov$, $l \leftarrow t.next.time - t.time$, $S \leftarrow S + t.\delta_{\max}$
9 $\quad\mid\;$ $c_{max} \leftarrow \max(S, C)$
10 $\quad\mid\;$ $c_{req} \leftarrow c_{req} + t.\delta_{req}$
11 $\quad\mid\;$ $c_{cons} \leftarrow \min(c_{req} + ov, c_{max})$
12 $\quad\mid\;$ **if** $0 < ov < (c_{cons} - c_{req}) \cdot l$ **then**
13 $\quad\mid\quad\lceil\;$ $l \leftarrow \max\left(1, \left\lfloor \frac{ov}{c_{cons} - c_{req}} \right\rfloor\right)$
14 $\quad\mid\quad\lfloor\;$ $t.insertAfter(t.time + l, t.cap, 0, 0)$
15 $\quad\mid\;$ $ov \leftarrow ov + (c_{req} - c_{cons}) \cdot l$
16 $\quad\mid\;$ $t.cap \leftarrow C - c_{cons}$
17 $\quad\mid\;$ **if** $t.cap < C$ **then** $\text{ect} \leftarrow t.next.time$
18 $\quad\lfloor\;$ $t \leftarrow t.next$

19 **return** ect

The interesting properties of this data structure come from the number of time points and the linear-time algorithm *ScheduleTasks*.

Proposition 1 ([8]). *The Profile contains at most $4n + 1$ time points and the algorithm ScheduleTasks runs in $\mathcal{O}(n)$ time where n is the number of tasks.*

4 New Formulation of the Not-First Rule

Before generalizing the not-first rule, let us state the classic not-first using the earliest completion time of a set of tasks. Let $\Omega \subset T$ be a set of tasks and $i \in T \setminus \Omega$ be a task. From task i and set of tasks Ω, a new task i' can be derived with the following attributes: $\text{est}_{i'} = \text{est}_\Omega$, $\text{lct}_{i'} = \min(\text{ect}_i, \text{lct}_\Omega)$, $p_{i'} = \min(\text{ect}_i, \text{lct}_\Omega) - \text{est}_\Omega$ and $c_{i'} = c_i$. Substituting this into rule (NF) leads to:

$$\begin{cases} \text{est}_i < \text{ECT}_\Omega \\ e_\Omega + e_{i'} > C(\text{lct}_\Omega - \text{est}_{\Omega \cup \{i'\}}) \end{cases} \Rightarrow \text{est}_i \geq \text{ECT}_\Omega \qquad (11)$$

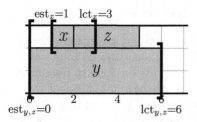

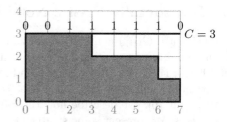

Fig. 1. A CuSP problem of 3 tasks sharing a resource of capacity $C = 3$.

Fig. 2. The resource utilization profile of tasks $\{x, y, z'\}$ where z' is derived from z and $\Omega = \{x, y\}$ and whose parameters $\langle \text{est}_{z'}, \text{lct}_{z'}, p_{z'}, c_{z'} \rangle$ are $\langle 0, 3, 3, 1 \rangle$.

The rule (11) is equivalent to

$$\begin{cases} \text{est}_i < \text{ECT}_\Omega \\ \text{ect}^F_{\Omega \cup \{i'\}} > \text{lct}_\Omega \end{cases} \Rightarrow \text{est}_i \geq \text{ECT}_\Omega \tag{12}$$

The horizontally elastic not-first rule is obtained from (12) by replacing $\text{ect}^F_{\Omega \cup \{i'\}}$ by $\text{ect}^H_{\Omega \cup \{i'\}}$ and is given by the formula:
$\forall \Omega \subset T, \ \forall i \in T \setminus \Omega,$

$$\begin{cases} \text{est}_i < \text{ECT}_\Omega \\ \text{ect}^H_{\Omega \cup \{i'\}} > \text{lct}_\Omega \end{cases} \Rightarrow \text{est}_i \geq \text{ECT}_\Omega \tag{HNF}$$

where i' is a task derived from task i whose parameters $\langle \text{est}_{i'}, \text{lct}_{i'}, p_{i'}, c_{i'} \rangle$ are $\langle \text{est}_\Omega, \min(\text{ect}_i, \text{lct}_\Omega), \min(\text{ect}_i, \text{lct}_\Omega) - \text{est}_\Omega, c_i \rangle$.

Theorem 2. *The not-first rule* (NF) *is subsumed by the horizontally elastic not-first rule* (HNF).

Proof. Since $\text{ect}^F_\Omega \leq \text{ect}^H_\Omega$ for all Ω (Theorem 1) and from the equivalence of rules (NF) and (12), it follows that condition $e_\Omega + c_i(\min(\text{ect}_i, \text{lct}_\Omega) - \text{est}_\Omega) > C(\text{lct}_\Omega - \text{est}_\Omega)$ implies the condition $\text{ect}^H_{\Omega \cup \{i'\}} > \text{lct}_\Omega$. Therefore, all the adjustments performed by rule (NF) are also done by rule (HNF).

Consider the CuSP instance of Fig. 1 where three tasks share a resource of capacity $C = 3$. The resource utilization profile of tasks $\{x, y, z'\}$ is given in Fig. 2 where z' is derived from z and $\Omega = \{x, y\}$ whose parameters $\langle \text{est}_{z'}, \text{lct}_{z'}, p_{z'}, c_{z'} \rangle$ are $\langle 0, 3, 3, 1 \rangle$. The numbers above the bold line of capacity limit are overflow units of energy remaining at each time point i.e., $ov(0) = 0$, $ov(1) = 0$, $ov(2) = 1$, and so forth. The horizontally elastic earliest completion time of the tasks set $\{x, y, z'\}$ is 7 which is greater than $\text{lct}_{\{x, y, z'\}} = 6$. When we apply the not-first rule (NH) with $\Omega = \{x, y\}$ and $i = z$, it appears that $0 = \text{est}_z < 2 = \text{ECT}_\Omega$ but $\text{ect}^F_{\Omega \cup \{z'\}} = 3 \leq \text{lct}_\Omega$. Therefore, no detection is found and consequently no adjustment follows. But the horizontally elastic not-first rule (HNF) applied to the same instance gives $\text{ect}^H_{\Omega \cup \{z'\}} = 7 > \text{lct}_\Omega$ and the earliest start time of task z is updated to 2. $\qquad\square$

5 Horizontally Elastic Not-First Algorithm

We present a $\mathcal{O}(n^3)$ cumulative horizontally elastic not-first algorithm where n is the number of tasks sharing the resource. The new algorithm is sound as in [6,10] i.e., the algorithm may take additional iterations to perform maximum adjustments and uses the concept of the left cut of the set of tasks T by a task j as in [6]. We describe how the left cut can be used to check the not-first conditions. We present some strategies to reduce the practical complexity of the algorithm and to fully utilize the power of the the Profile data structure.

Definition 1 [6]. *Let i and j be two different tasks. The left cut of T by task j relatively to task i is the set of tasks $LCut(T, j, i)$ defined as follows:*

$$LCut(T, j, i) = \{k \mid k \in T \wedge k \neq i \wedge est_i < ect_k \wedge lct_k \leq lct_j\}. \tag{13}$$

Using this set, we have the following new rule:
    For all $i, j \in T$ with $i \neq j$,

$$ect^H_{LCut(T,j,i)\cup\{i'\}} > lct_j \Rightarrow est_i \geq ECT_{LCut(T,j,i)} \tag{HNF'}$$

where i' is a task derived from task i whose parameters $\langle est_{i'}, lct_{i'}, p_{i'}, c_{i'} \rangle$ are $\langle est_{LCut(T,j,i)}, \min(ect_i, lct_j), \min(ect_i, lct_j) - est_{LCut(T,j,i)}, c_i \rangle$.

Theorem 3. *For a task i, at most $h-1$ iterative applications of the rule (HNF') achieve the same filtering as one application of the rule (HNF) where h is the number of different earliest completion time of tasks.*

Proof. Let Ω be the set which induces the maximum change of the value est_i by the rule (HNF). Let $j \in \Omega$ be a task with $lct_j = lct_\Omega$. Until the same value of est_i is reached, in each iteration of the rule (HNF') holds that $\Omega \subseteq LCut(T, j, i)$. Indeed, because $est_i < ECT_\Omega$ and $i \notin \Omega$, it follows that for all $k \in \Omega$, $k \neq i$, $est_i < ect_k$ and $lct_k \leq lct_j$. From the inclusion $\Omega \subseteq LCut(T, j, i)$, it follows that $ect^H_{LCut(T,j,i)\cup\{i'\}} \geq ect^H_{\Omega\cup\{i'\}} > lct_j$ and the rule (HNF') holds and propagates.
    After each successful application of the rule (HNF'), the value est_i is increased. This removes all the tasks from the set $LCut(T, j, i)$ having the earliest completion time $ECT_{LCut(T,j,i)}$. Therefore the final value of est_i must be reached after at most $h - 1$ iterations and it is the same as for the rule (HNF). □

Example 1 ([6]). Consider the CuSP instance of Fig. 3 where four tasks share a resource of capacity $C = 3$. The not-first rule (14) holds for task d and the set $LCut(T, a, d) = \{a, b, c\}$. Indeed, $est_d < ECT_{\{a,b,c\}}$ and $ect^H_{LCut(T,a,d)\cup\{d'\}} = 6 > lct_a$. Hence, $est_d = 3$. After this adjustment, we have $LCut(T, a, d) = \{a\}$, $est_d < ECT_{\{a\}}$, and $ect^H_{LCut(T,a,d)\cup\{d'\}} = 6 > lct_a$ which leads to $est_d = 5$. The maximum adjustment holds.

    To reduce the practical computational complexity of the algorithm, we deduce from the properties of $LCut(T, j, i)$ and the rigidity of task i' some strategies to learn from failures and successes and anticipate the detection of future tasks.

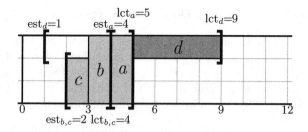

Fig. 3. A scheduling problem of 4 tasks sharing a resource of capacity $C = 3$.

5.1   Reducing the Number of Sets $\Theta = \mathrm{LCut}(T, j, i)$ to Consider

To speed-up the algorithm without reducing its filtering power, we can know whether or not the set of tasks $\Theta = \mathrm{LCut}(T, j, i)$ is in conflict with the task i. The detection of the not-first rule of task i with the set of tasks Θ is only possible when Θ is conflicting with i. This happens when $\sum_{k \in \Theta} c_k > C - c_i$ since when the sum of the capacity requirements of the tasks in Θ is less than $C - c_i$, then the set Θ is not conflicting with task i.

5.2   Deduction from Detection Failure of Tasks

Let i be a task such that the not-first detection rule (HNF') fails for all set of tasks $\mathrm{LCut}(T, j, i)$. Then for any other task u such that $u \neq i$, $\mathrm{lct}_u \leq \mathrm{lct}_i$, $\mathrm{est}_u = \mathrm{est}_i$, $c_u \leq c_i$ and $\mathrm{ect}_u \leq \mathrm{ect}_i$, we can deduce that, for all sets of tasks $\mathrm{LCut}(T, j, i)$ the rule (HNF') will also fail with task u. This assertion is formally proven in the following proposition.

Proposition 2. *Let $i \in T$ be a task such that the not-first rule (HNF') fails for all sets of tasks $\mathrm{LCut}(T, j, i)$. Then for any task $u \in T$ such that $u \neq i$, $\mathrm{lct}_u \leq \mathrm{lct}_i$, $\mathrm{est}_u = \mathrm{est}_i$, $c_u \leq c_i$ and $\mathrm{ect}_u \leq \mathrm{ect}_i$, the not-first rule (HNF') also fails with task u for all sets of tasks $\mathrm{LCut}(T, j, u)$.*

Proof. By contradiction, let $u \in T$ be a task with $u \neq i$, $\mathrm{lct}_u \leq \mathrm{lct}_i$, $\mathrm{est}_u = \mathrm{est}_i$, $c_u \leq c_i$ and $\mathrm{ect}_u \leq \mathrm{ect}_i$ such that the not-first rule (HNF') detects and adjusts the earliest start time of task u, i.e., there exists a task $j \in T$ such that $\mathrm{ect}^H_{\mathrm{LCut}(T,j,u) \cup \{u'\}} > \mathrm{lct}_j$ and the task u is updated such that $\mathrm{est}_u \geq \mathrm{ECT}_{\mathrm{LCut}(T,j,u)}$. We distinguish two cases: $\mathrm{lct}_j < \mathrm{lct}_i$ and $\mathrm{lct}_i \leq \mathrm{lct}_j$.

1. If $\mathrm{lct}_j < \mathrm{lct}_i$, then $\mathrm{LCut}(T, j, u) \subseteq \mathrm{LCut}(T, j, i)$ and from the fact that i' is more constrained than u' it follows that $\mathrm{ect}^H_{\mathrm{LCut}(T,j,i) \cup \{i'\}} \geq \mathrm{ect}^H_{\mathrm{LCut}(T,j,i) \cup \{u'\}} \geq \mathrm{ect}^H_{\mathrm{LCut}(T,j,u) \cup \{u'\}} > \mathrm{lct}_j$, which contradicts the non-detection of the not-first rule (HNF') with task i.
2. If $\mathrm{lct}_i \leq \mathrm{lct}_j$, then $\mathrm{LCut}(T, j, u) \subseteq \mathrm{LCut}(T, j, i) \cup \{i\}$. Since the set of tasks $\mathrm{LCut}(T, j, i) \cup \{i'\}$ is more constrained than $\mathrm{LCut}(T, j, u) \cup \{u'\}$ it follows that $\mathrm{ect}^H_{\mathrm{LCut}(T,j,i) \cup \{i'\}} \geq \mathrm{ect}^H_{\mathrm{LCut}(T,j,u) \cup \{u'\}} > \mathrm{lct}_j$, which contradicts the non-detection of the rule (HNF') with task i.   □

5.3   Deduction from Success Detection of Tasks

Let i and j be two tasks such that $i \neq j$, $\mathrm{est}_i < \mathrm{ect}_j$ and the not-first rule (HNF') holds for i and $\mathrm{LCut}(T, j, i)$. Then for any other task u such that $u \neq i$, $\mathrm{lct}_u \leq \mathrm{lct}_i$, $\mathrm{est}_u \leq \mathrm{est}_i$, $c_u \geq c_i$ and $\mathrm{ect}_u \geq \mathrm{ect}_i$, the tasks set $\mathrm{LCut}(T, j, u)$ successfully detected the not-first rule (HNF') with task u, $u \notin \mathrm{LCut}(T, j, i)$. This assertion is formally proven in the following proposition.

Proposition 3. *Let i and j be two tasks such that $i \neq j$, $\mathrm{est}_i < \mathrm{ect}_j$ and the not-first rule* (HNF') *holds for i and $L\mathit{Cut}(T, j, i)$. Then for any other task u such that $u \neq i$, $\mathrm{lct}_u \leq \mathrm{lct}_i$, $\mathrm{est}_u \leq \mathrm{est}_i$, $c_u \geq c_i$ and $\mathrm{ect}_u \geq \mathrm{ect}_i$, if $u \notin L\mathit{Cut}(T, j, i)$ then the not-first rule* (HNF') *holds with u and $L\mathit{Cut}(T, j, u)$.*

Proof. Let i and j be two tasks such that $i \neq j$, $\mathrm{est}_i < \mathrm{ect}_j$ and the rule (HNF') holds for i and $\mathrm{LCut}(T, j, i)$. Let u be a task such that $u \neq i$, $\mathrm{lct}_u \leq \mathrm{lct}_i$, $\mathrm{est}_u \leq \mathrm{est}_i$, $c_u \geq c_i$, $\mathrm{ect}_u \geq \mathrm{ect}_i$ and $u \notin \mathrm{LCut}(T, j, i)$. From $\mathrm{est}_u \leq \mathrm{est}_i$ and $u \notin \mathrm{LCut}(T, j, i)$, it is obvious that $\mathrm{LCut}(T, j, i) \subseteq \mathrm{LCut}(T, j, u)$. Since the set of tasks $\mathrm{LCut}(T, j, u) \cup \{u'\}$ is more constrained than $\mathrm{LCut}(T, j, i) \cup \{i'\}$ it follows that $\mathrm{ect}^H_{\mathrm{LCut}(T,j,u) \cup \{u'\}} \geq \mathrm{ect}^H_{\mathrm{LCut}(T,j,i) \cup \{i'\}} > \mathrm{lct}_j$. $\qquad\square$

To apply these reductions, we start with a set $\Lambda = T$ of tasks sorted by non-decreasing order of lct_j, by non-increasing order of est_j, by non-decreasing order of c_j and by non-decreasing order of ect_j to break ties. If a task $i \in \Lambda$ fails for detection of the rule (HNF'), then we are sure that the detection will fail with all tasks $u \in \Lambda$ such that $u \neq i$, $\mathrm{lct}_u \leq \mathrm{lct}_i$, $\mathrm{est}_u = \mathrm{est}_i$, $c_u \leq c_i$ and $\mathrm{ect}_u \leq \mathrm{ect}_i$. On the other hand, when the rule (HNF') holds with a task i and the set $\mathrm{LCut}(T, j, i)$, if the detection of the rule (HNF') fails with the set $\mathrm{LCut}(T, j, u)$ and the task $u \in \Lambda$ such that $u \neq i$, $\mathrm{lct}_u \leq \mathrm{lct}_i$, $\mathrm{est}_u \leq \mathrm{est}_i$, $c_u \geq c_i$ and $\mathrm{ect}_u \geq \mathrm{ect}_i$, then the task is postponed to the next iteration.

5.4   Horizontally Elastic Not-First Algorithm

In Algorithm 2, we iterate through the set of tasks sorted by non-decreasing order of lct, by non-increasing order of est, by non-decreasing order of c_j and by non-decreasing order of ect_j to break ties (line 5) and for each unscheduled task (line 3), we iterate over the different set $\Theta = \mathrm{LCut}(T, j, i)$ (line 7). For each set Θ satisfying the reduction of Sect. 5.1, the minimum earliest completion time is computed (line 9) during the initialization of the increment values of the function *ScheduleTasks*. The horizontally-elastic earliest completion time of the set of tasks $\Theta \cup \{i'\}$ is computed at line 10 by the function *ScheduleTasks* and if this value is greater than lct_j (line 11), then the adjustment of est_i occurs (line 12). The boolean "detect" of line 4 allows breaking for the while loop of line 5 when a detection is found. The loop of line 15 is used to avoid similar set $\Theta = \mathrm{LCut}(T, j, i)$ since $\mathrm{LCut}(T, j, i) = \mathrm{LCut}(T, j', i)$ if $\mathrm{lct}_j = \mathrm{lct}_{j'}$. The complexity of Algorithm 2 is given in the following theorem.

Algorithm 2. Horizontally elastic Not-First Algorithm in $\mathcal{O}(n^3)$ time.

Input: Λ tasks sorted by non-decreasing lct_j, by non-increasing est_j, by
   non-decreasing c_j and by non-decreasing ect_j to break ties.
Output: A lower bound est'_i for each task i

1 **for** $i \in T$ **do** $est'_i \leftarrow est_i$
2 **for** $i = n$ **to** 1 **do**
3 $\quad$ **if** $ect_i < lct_i$ **then**
4 $\quad\quad$ $detect \leftarrow false,\ j \leftarrow 1,\ t \leftarrow -1$
5 $\quad\quad$ **while** $j \le n \wedge detect = false$ **do**
6 $\quad\quad\quad$ **if** $j \ne i \wedge est_i < ect_j$ **then**
7 $\quad\quad\quad\quad$ $\Theta \leftarrow \mathrm{LCut}(T, j, i)$
8 $\quad\quad\quad\quad$ **if** $\sum_{k \in \Theta} c_k > C - c_i$ **then**
9 $\quad\quad\quad\quad\quad$ $\mathrm{ECT} \leftarrow \mathrm{ECT}_{\mathrm{LCut}(T,j,i)}$
10 $\quad\quad\quad\quad\quad$ $ect^H \leftarrow ScheduleTasks(\Theta \cup \{i'\}, C)$
11 $\quad\quad\quad\quad\quad$ **if** $ect^H > lct_j$ **then**
12 $\quad\quad\quad\quad\quad\quad$ $est'_i \leftarrow \max(est'_i, \mathrm{ECT})$
13 $\quad\quad\quad\quad\quad\quad$ $detect \leftarrow true$
14 $\quad\quad\quad\quad\quad\quad$ $t \leftarrow j$
15 $\quad\quad\quad\quad\quad$ **while** $j + 1 \le n \wedge lct_j = lct_{j+1} \wedge detect = false$ **do**
16 $\quad\quad\quad\quad\quad\quad$ $j \leftarrow j + 1$

17 $\quad\quad\quad$ $j \leftarrow j + 1$

18 $\quad\quad$ **if** $detect = true$ **then**
19 $\quad\quad\quad$ **for** $u = i - 1$ **to** 1 **do**
20 $\quad\quad\quad\quad$ **if** $est_u \le est_i \wedge c_u \ge c_i \wedge ect_u \ge ect_i$ **then**
21 $\quad\quad\quad\quad\quad$ $\Theta \leftarrow \mathrm{LCut}(T, t, u)$
22 $\quad\quad\quad\quad\quad$ $\mathrm{ECT} \leftarrow \mathrm{ECT}_{\mathrm{LCut}(T,t,u)}$
23 $\quad\quad\quad\quad\quad$ $ect^H \leftarrow ScheduleTasks(\Theta \cup \{u'\}, C)$
24 $\quad\quad\quad\quad\quad$ **if** $ect^H > lct_j$ **then**
25 $\quad\quad\quad\quad\quad\quad$ $est'_u \leftarrow \max(est'_u, \mathrm{ECT})$
26 $\quad\quad\quad\quad$ $\Lambda \leftarrow \Lambda \setminus \{u\}$

27 $\quad\quad$ **if** $detect = false$ **then**
28 $\quad\quad\quad$ **for** $u = i - 1$ **to** 1 **do**
29 $\quad\quad\quad\quad$ **if** $est_u = est_i \wedge c_u \le c_i \wedge ect_u \le ect_i$ **then**
30 $\quad\quad\quad\quad\quad$ $\Lambda \leftarrow \Lambda \setminus \{u\}$

31 **for** $i \in T$ **do** $est_i \leftarrow est'_i$

Theorem 4. *Algorithm 2 runs in $\mathcal{O}(n^3)$ time.*

Proof. The linear time algorithm *ScheduleTasks* is called $\mathcal{O}(n^2)$ time for total complexity of $\mathcal{O}(n^3)$. $\qquad\square$

We perform a preliminary comparison of this algorithm with the state of the art algorithms on the resource constrained project scheduling problems (RCPSP) instances of the BL set [3]. It appears that on many instances, when the proposed algorithm is used, the solver spends 1.5–2 more time than with the others not-first/not-last algorithms for a reduction of the number of backtracks of less than 40%. We observe that for a task i, many set $\Theta = \text{LCut}(T, j, i)$ used to check the not-first conditions are fruitless and should be avoided.

6   Relaxation of the Horizontally Elastic Not-First Algorithm

We propose a relaxation of the previous algorithm based on a new criterion used to reduce the number of subsets $\text{LCut}(T, j, i)$ to consider. Without changing the computational complexity, the relaxed horizontally elastic not-first algorithm still dominates the classic not-first algorithm with a good trade-off between the filtering power and the running time. To do so, it is important to have a criteria to select the task j for which the set $\Theta = \text{LCut}(T, j, i)$ has more potential to detect at least the classic not-first conditions.

Definition 2. *Let $i \in T$ be a task. The not-first set of tasks with task i denoted NFSet(T, i) is given by*

$$NFSet(T, i) = \{j, j \in T \wedge j \neq i \wedge est_i < ect_j\}.$$

The set $\text{NFSet}(T, i)$ is the set of tasks conflicting with task i. If a not-first condition is detected with a set Ω i.e., $\text{ect}_{\Omega \cup \{i'\}} > \text{lct}_\Omega$, then $\Omega \subseteq \text{NFSet}(T, i)$. In this condition, the earliest start time of task i' can be replaced by $\text{est}_{\min} = \min\{\text{est}_k, k \in T\}$ since none of the tasks from $\text{NFSet}(T, i)$ starts and ends before est_i. We schedule the tasks from $\text{NFSet}(T, i) \cup \{i'\}$ and compute the overflow energy that cannot be executed at time $t = t_{\text{lct}_j}$ for $j \in \text{NFSet}(T, i)$. The algorithm *ScheduleNFConflictingTasks* is a variant of the algorithm *ScheduleTasks* which schedules the set $\text{NFSet}(T, i) \cup \{i'\}$ and returns the set Δ of task j such that the overflow energy at time point t_{lct_j} is greater than 0.

We use the condition $t_{\text{lct}_j}.ov > 0$ to reduce the number of sets $\text{LCut}(T, j, i)$ to be considered during the detection of the not-first conditions with task i. The above improvements are incorporated in Algorithm 3. The complexity of the resulting algorithm remains $\mathcal{O}(n^3)$ but the condition $t_{\text{lct}_j}.ov > 0$ used to reduce the number of sets $\text{LCut}(T, j, i)$ during the detection considerably reduces the running time, as shown from the experimental results section.

Example 2. Consider the CuSP instance of Fig. 1 with an additional task t where attributes $\langle \text{est}_t, \text{lct}_t, p_t, c_t \rangle$ are $\langle 3, 7, 1, 1 \rangle$. The function *ScheduleNFConflicting-Tasks* return an empty set when the set $\text{NFSet}(T, z) \cup \{z'\}$ is scheduled because the overflow energy will be consumed before the time point 6. Therefore, the relaxed algorithm will miss the adjustment of est_z to 2.

The filtering power of the algorithm is reduced. We prove later that the relaxed horizontally elastic not-first algorithm subsumes the classic not-first algorithm.

Algorithm 3. Relaxation of the horizontally elastic Not-First in $\mathcal{O}(n^3)$.

Input: Λ tasks sorted by non-decreasing lct_j, by non-increasing est_j, by non-decreasing c_j and by non-decreasing ect_j to break ties.

Output: A lower bound est'_i for each task i

```
 1  for i ∈ T do  est'_i ← est_i
 2  for i = n to 1 do
 3      if ect_i < lct_i ∧ i ∈ Λ then
 4          detect ← false, t ← −1
 5          if detect = false then
 6              Δ ← ScheduleNFConflictingTasks(i, C)
 7              j ← |Δ|
 8              while j ≥ 1 ∧ detect = false do
 9                  Θ ← LCut(T, j, i)
10                  ECT ← ECT_LCut(T,j,i)
11                  ect^H ← ScheduleTasks(Θ ∪ {i'}, C)
12                  if ect^H > lct_j then
13                      est'_i ← max(est'_i, ECT)
14                      detect ← true
15                      t ← j
16                  j ← j − 1
17          if detect = true then
18              for u = i − 1 to 1 do
19                  if est_u ≤ est_i ∧ c_u ≥ c_i ∧ ect_u ≥ ect_i then
20                      Θ ← LCut(T, t, u)
21                      ECT ← ECT_LCut(T,t,u)
22                      ect^H ← ScheduleTasks(Θ ∪ {u'}, C)
23                      if ect^H > lct_t then
24                          est'_u ← max(est'_u, ECT)
25                      Λ ← Λ \ {u}
26          if detect = false then
27              for u = i − 1 to 1 do
28                  if est_u = est_i ∧ c_u ≤ c_i ∧ ect_u ≤ ect_i then
29                      Λ ← Λ \ {u}
30  for i ∈ T do  est_i ← est'_i
```

---

7 Properties of the Relaxation of the Horizontally Elastic Not-First Algorithm

In this section, we prove that the relaxation of the horizontally elastic not-first algorithm (Algorithm 3) subsumes the standard not-first algorithm.

Lemma 1. *Let $i \in T$ be a task. If the not-first condition (NF) is detected with the set of tasks Ω, then after a horizontally elastic scheduling of tasks $NFSet(T, i) \cup \{i'\}$, it appears that $t_{lct_j}.ov > 0$ where $lct_j = lct_\Omega$.*

Proof. Let $j \in T$ be a task such that $lct_j = lct_\Omega$. If the not-first conditions (4) are detected with the set of tasks Ω, then $ect^F_{\Omega \cup \{i'\}} > lct_j$ and $\Omega \subseteq LCut(T, j, i)$. Therefore, in the fully elastic schedule of tasks from $\Omega \cup \{i'\}$, the resource is fully used at any time points from est_Ω to lct_Ω with a surplus of energy not executed. Then from $ect^H_{LCut(T,j,i) \cup \{i'\}} \geq ect^F_{LCut(T,j,i) \cup \{i'\}} \geq ect^F_{\Omega \cup \{i'\}} > lct_j$ and $LCut(T, j, i) \subseteq NFSet(T, i)$ it follows that during the scheduling of tasks set $NFSet(T, i) \cup \{i'\}$, $t_{lct_j}.ov > 0$. ☐

Theorem 5. *The relaxation of the horizontally elastic not-first algorithm (Algorithm 3) subsumes the classic not-first algorithm.*

Proof. According to Lemma 1, any detection and adjustment performed by the classic not-first algorithm are also detected and adjusted by the relaxed horizontally elastic not-first algorithm. In the CuSP instance of Example 1, the classic not-first algorithm fails to adjust est_z while the relaxation of the horizontally elastic not-first algorithm succeeds to update est_z to 2. ☐

We know from [3] that the classic not-first/not-last rule is not subsumed by the energetic reasoning rule and vice-versa. According to Theorem 5, we can deduce that the relaxation of the horizontally elastic not-first/not-last rule is not subsumed by the energetic reasoning and vice-versa.

8   Experimental Results

We carry out experimentations on resource-constrained project scheduling problems (RCPSP) to compare the new algorithm of not-first/not-last with the state-of-the art algorithms. A RCPSP consists of a set of resources of finite capacities, a set of tasks of given processing times, an acyclic network of precedence constraints between tasks, and a horizon (a deadline for all tasks). Each task requires a fixed amount of each resource over its execution time. The problem is to find a starting time assignment for all tasks satisfying the precedence and resource capacity constraints, with the least makespan (i.e., the time at which all tasks are completed) at most equals to the horizon.

Tests were performed on benchmark suites of RCPSP known to be highly cumulative [3]. On highly cumulative scheduling instances, many tasks can be scheduled simultaneously as contrary to the highly disjunctive ones. We use the libraries BL [11], Pack [13] and KSD15_D [14]. The data set BL consists of 40 instances of 20 and 25 tasks sharing three resources, Pack consists of 55 instances of 15–33 tasks sharing a resource of capacity 2–5 while the set KSD15_D consists of 480 instances of 15 tasks sharing a resource of capacity 4.

Starting with the provided horizon as an upper bound, we modeled each problem as an instance of Constraint Satisfaction Problem (CSP); variables are start times of tasks and they are constrained by the precedence graph (i.e., precedence relations between pairs of tasks were enforced with linear constraints) and resource limitations (i.e., each resource was modeled with a single CUMULATIVE constraint [1]). We used a branch and bound search to minimize the makespan.

We implemented three different propagators of the global constraint CUMULATIVE in Java using Choco solver 4.0.1 [17].

1. The first CUMULATIVE propagator noted "TT-NF" (for not-first with Θ -tree) is a sequence of two filtering algorithms: the $\mathcal{O}(n^2 \log n)$ not-first algorithm from [6] and timetabling algorithm from [15].
2. The second propagator noted "CHE-NF" (for not-first with complete horizontally elastic) is obtained when replacing in the first propagator the not-first algorithm with Timeline by the complete horizontally elastic not-first algorithm presented in Algorithm 2.
3. The third propagator noted "RHE-NF" (for not-first with relaxed horizontally elastic) is obtained when replacing in the first propagator the not-first algorithm with Timeline by the relaxed horizontally elastic not-first algorithm presented in Algorithm 3.

Branching scheme is another ingredient to accelerate the solving process. The heuristics used to select tasks and values are directly linked to the type of problems and the filtering algorithms considered in the solver. We combine the conflict-ordering search heuristic [16] with the heuristic $minDomLBSearch$ from Choco. During the search, the solver records conflicting tasks and at the backtrack, the last one is selected in priority until they are all instantiated without causing any failure. When no conflicting tasks is recorded, the heuristic $minDomLBSearch$ which consists of selecting the unscheduled tasks with the smallest domain and assigning it to its lower bound is used. Tests were performed on a data center equiped with Intel(R) Xeon X5560 Nehalem nodes, 2 CPUs per node, 4 cores per CPU at 2.4 GHz, 24 GB of RAM per node. Any search taking more than 10 min was counted as a failure.

In Table 1, the columns "solve" report the number of instances solved by each propagator. Columns "time", "backt", and "speedup" denote the average CPU time (in second) used to reach the optimal solution, the average number of backtracks, and the average speedup factor (TT-NF time over new algorithms time) reported on instances solved by "TT-NF" vs. "CHE-NF" (sp_1) and "TT-NF" vs. "RHE-NF" (sp_2) respectively. 527 instances were solved by the three propagators with one instance solved only by "CHE-NF" and "RHE-NF" (pack016) and two instances solved only by "TT-NF" and "RHE-NF" (pack015 and j30_45_2).

The propagator "TT-NF" performs better in average on BL set while "RHE-NF" is the best on Pack and KSD15_D with an average speedup factor of 124.4% and 103.1% wrt. "TT-NF". We observe a reduction of the average number of backtracks from "RHE-NF" on Pack set while "CHE-NF" dominated on BL and KSD15_D. Figure 4 compares the runtimes (a), the number of backtracks (b) and the number of adjustments (propagations) (c) made at the fixed point of the node of the search tree on the 527 instances solved by the three propagators. It appears in (a) that the running time of "RHE-NF" is generally close to "TT-NF" and sometimes less. In (b), the number of backtracks of "RHE-NF" is

Table 1. We report the number of instances solved (solve), the average number of backtracks (backts), the average time in second (time) and the average speedup factor (TT-NF time over new algorithms time) required to solve all instances that are commonly solved by the three propagators on set BL, Pack and KSD15_D.

	TT-NF			CHE-NF			RHE-NF			Speedup (%)	
	Solve	Time	Backts	Solve	Time	Backts	Solve	Time	Backts	sp_1	sp_2
BL	40	**4.497**	32789	40	6.952	**23114**	40	6.616	27193	64.7	68
Pack	**19**	30.524	161467	18	42.608	90663	18	**24.543**	**66154**	71.6	**124.4**
KSD15_D	**471**	0.766	2196	470	1.2	**2008**	**471**	**0.743**	2020	63.8	**103.1**

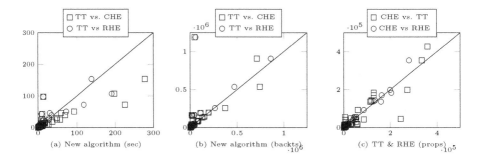

(a) New algorithm (sec)   (b) New algorithm (backts)·10^6 (c) TT & RHE (props)·10^5

Fig. 4. (a) Runtimes comparison of TT-NF vs. CHE-NF and TT-NF vs. RHE-NF, (b) Comparison of the number of Backtracts TT-NF vs. CHE-NF and TT-NF vs. RHE-NF, (c) Comparison of the number of adjustments (Propagations) CHE-NF vs. TT-NF and CHE-NF vs. RHE-NF on instances of BL, Pack and KSD15_D where the three propagators found the best solution.

always less than the number of backtracks of "TT-NF". In (c), the average number of propagations of "RHE-NF" are always less than the average number of propagations of "CHE-NF" when on a few number of instances, the number of propagations of "TT-NF" is less than the number of propagations of "CHE-NF".

9   Conclusion

We proposed a generalization of the not-first/not-last rule for the cumulative resource constraint based on a strong relaxation of the earliest completion time of a set of tasks. A relaxation of the corresponding horizontally elastic not-first/not-last algorithm running in $\mathcal{O}(n^3)$ is also proposed, where n is the number of tasks sharing the resource. The new algorithm is sound and can reach a better fixed point than the state-of-the-art algorithms. The new algorithm is based on the data structure Profile used to compute a strong lower bound on the earliest completion time of a set of tasks. Experimental results demonstrate that the new algorithm has more impact in terms of backtracks reduction and running time on highly cumulative instances of RCPSPs. Future work will focus on finding how to improve the complexity of this algorithm from $\mathcal{O}(n^3)$ to $\mathcal{O}(n^2 \log n)$ and to design a branching scheme more suitable for the new rule.

References

1. Aggoun, A., Beldiceanu, N.: Extending CHIP in order to solve complex scheduling and placement problems. Math. Comput. Model. **17**(7), 57–73 (1993)
2. Garey, M.R., Johnson, D.S.: Computers and Intractability, vol. 29. W. H. Freeman, New York (2002)
3. Baptiste, P., Le Pape, C., Nuijten, W.: Constraint-Based Scheduling: Applying Constraint Programming to Scheduling Problems. Kluwer, Boston (2001)
4. Kameugne, R., Fotso, L.P., Scott, J., Ngo-Kateu, Y.: A quadratic edge-finding filtering algorithm for cumulative resource constraints. Constraints **19**(3), 243–269 (2014)
5. Gay, S., Hartert, R., Schaus, P.: Simple and scalable time-table filtering for the cumulative constraint. In: Pesant, G. (ed.) CP 2015. LNCS, vol. 9255, pp. 149–157. Springer, Cham (2015). https://doi.org/10.1007/978-3-319-23219-5_11
6. Kameugne, R., Fotso, L.P.: A cumulative not-first/not-last filtering algorithm in $\mathcal{O}(n^2 \log(n))$. Indian J. Pure Appl. Math. **44**(1), 95–115 (2013)
7. Vilím, P.: Edge finding filtering algorithm for discrete cumulative resources in $\mathcal{O}(kn \log n)$. In: Gent, I.P. (ed.) CP 2009. LNCS, vol. 5732, pp. 802–816. Springer, Heidelberg (2009). https://doi.org/10.1007/978-3-642-04244-7_62
8. Gingras, V., Quimper, C.-G.: Generalizing the edge-finder rule for the cumulative constraint. In: Proceedings of the 25th International Joint Conference on Artificial Intelligence (IJCAI 2016), pp. 3103–3109 (2016)
9. Fahimi, H., Ouellet, Y., Quimper, C.-G.: Linear-time filtering algorithms for the disjunctive constraint and a quadratic filtering algorithm for the cumulative not-first not-last. Constraints (2018). https://urldefense.proofpoint.com/v2/url?u= https-3A__doi.org_10.1007_s10601-2D018-2D9282-2D9&d=DwIGaQ&c=vh6FgFnd uejNhPPD0fl_yRaSfZy8CWbWnIf4XJhSqx8&r=UyK1_569d50MjVlUSODJYRW2 epEY0RveVNq0YCmePcDz4DQHW-CkWcttrwneZ0md&m=aL081BMc0-Mz9R6 8wFZEUyFJk8ey6WR_yrftmQnZo5M&s=hgOsaJRlHR1tDxzWdCLdLc6yr4SUt5 P6x9Nz5aecTfQ&e
10. Schutt, A., Wolf, A.: A new $\mathcal{O}(n^2 \log n)$ not-first/not-last pruning algorithm for cumulative resource constraints. In: Cohen, D. (ed.) CP 2010. LNCS, vol. 6308, pp. 445–459. Springer, Heidelberg (2010). https://doi.org/10.1007/978-3-642-15396-9_36
11. Baptiste, P., Le Pape, C.: Constraint propagation and decomposition techniques for highly disjunctive and highly cumulative project scheduling problems. Constraints **5**(1–2), 119–139 (2000)
12. Derrien, A., Petit, T.: A new characterization of relevant intervals for energetic reasoning. In: O'Sullivan, B. (ed.) CP 2014. LNCS, vol. 8656, pp. 289–297. Springer, Cham (2014). https://doi.org/10.1007/978-3-319-10428-7_22
13. Carlier, J., Néron, E.: On linear lower bounds for the resource constrained project scheduling problem. Eur. J. Oper. Res. **149**(2), 314–324 (2003)
14. Koné, O., Artigues, C., Lopez, P., Mongeau, M.: Event-based milp models for resource-constrained project scheduling problems. Comput. Oper. Res. **38**(1), 3–13 (2011)
15. Letort, A., Beldiceanu, N., Carlsson, M.: A scalable sweep algorithm for the *cumulative* constraint. In: Milano, M. (ed.) CP 2012. LNCS, pp. 439–454. Springer, Heidelberg (2012). https://doi.org/10.1007/978-3-642-33558-7_33

16. Gay, S., Hartert, R., Lecoutre, C., Schaus, P.: Conflict ordering search for scheduling problems. In: Pesant, G. (ed.) CP 2015. LNCS, vol. 9255, pp. 140–148. Springer, Cham (2015). https://doi.org/10.1007/978-3-319-23219-5_10
17. Prud'homme, C., Fages, J.-G., Lorca, X.: Choco Solver Documentation, TASC, INRIA Rennes, LINA CNRS UMR 6241, COSLING S.A.S. (2016). http://www.choco-solver.org

Soft-Regular with a Prefix-Size Violation Measure

Minh Thanh Khong[1(✉)], Christophe Lecoutre[2(✉)], Pierre Schaus[1(✉)], and Yves Deville[1(✉)]

[1] ICTEAM, Université catholique de Louvain, Louvain-la-Neuve, Belgium
{minh.khong,pierre.schaus,yves.deville}@uclouvain.be
[2] CRIL-CNRS UMR 8188, Université d'Artois, Lens, France
lecoutre@cril.fr

Abstract. In this paper, we propose a variant of the global constraint soft-regular by introducing a new violation measure that relates a cost variable to the size of the longest prefix of the assigned variables, which is consistent with the constraint automaton. This measure allows us to guarantee that first decisions (assigned variables) respect the rules imposed by the automaton. We present a simple algorithm, based on a Multi-valued Decision Diagram (MDD), that enforces Generalized Arc Consistency (GAC). We provide an illustrative case study on nurse rostering, which shows the practical interest of our approach.

1 Introduction

Global constraints play an important role in Constraint Programming (CP) due to their expressiveness and capability of efficiently filtering the search space. Popular global constraints include, among others, allDifferent [1], count [2], element [3], cardinality [4,5], cumulative [6] and regular [7]. The constraint regular imposes a sequence of variables to take their values in order to form a word recognized by a finite automaton. This constraint happens to be useful when modeling various combinatorial problems such as rostering and car sequencing problems.

In many time-oriented problems, such as planning, scheduling, and timetabling, one has to take decisions while paying attention to the future demands, resources, etc. Those are generally obtained from a forecasting model. The standard approach is basically to define an horizon and to try having the problem instances solved over that horizon. Unfortunately, many problem instances are over-constrained [8,9]. Following the approach of [10], some hard-constraints can then be relaxed and replaced by their soft versions. Although attractive, this approach applied to time-oriented problems has a major drawback which is that constraints are equally penalized, whether the violation is about the beginning or the end of the horizon. We claim that the importance of completely satisfying the constraints must decrease with elapsed time. In other words, satisfying the constraints in the near future must be considered as more important than

© Springer International Publishing AG, part of Springer Nature 2018
W.-J. van Hoeve (Ed.): CPAIOR 2018, LNCS 10848, pp. 333–343, 2018.
https://doi.org/10.1007/978-3-319-93031-2_24

satisfying the constraints in the far future (in addition, forecasts may be more or less accurate).

In this paper, we are interested in the relaxation of the global constraint `regular`. Existing violation measures for `soft-regular` are based on the concept of distances in term of variables or edit operations [11–13]. We propose an alternative violation measure that is based on the size of the longest prefix that is consistent with the underlying automaton of the constraint. This violation measure can be useful when first variables of the sequence (scope) are critical. We illustrate our approach on a rostering application.

2   Technical Background

A Constraint Satisfaction Problem (CSP) [14–16] is composed of a set of n variables, $X = \{x_1, \ldots, x_n\}$, and a set of e constraints, $C = \{c_1, \ldots, c_e\}$. On the one hand, each variable x has an associated domain, denoted by $D(x)$, that contains the set of values that can be assigned to x. Assuming that the domain $D(x)$ of a variable x is totally ordered, $min(x)$ and $max(x)$ will respectively denote the smallest value and the greatest value in the domain of x. Note also that d will denote the maximum domain size for the variables in a given CSP. On the other hand, each constraint c involves an ordered set of variables, called the scope of c and denoted by $scp(c)$. Each constraint c is mathematically defined by a relation, denoted by $rel(c)$, which contains the allowed combinations of values for $scp(c)$. The arity of a constraint c is the size of $scp(c)$, and will usually be denoted by r.

Given a sequence $\langle x_1, \ldots, x_i, \ldots, x_r \rangle$ of r variables, an r-tuple τ on this sequence of variables is a sequence of values $\langle a_1, \ldots, a_i, \ldots, a_r \rangle$, where the individual value a_i is also denoted by $\tau[x_i]$ or, when there is no ambiguity, $\tau[i]$. An r-tuple τ is *valid* on an r-ary constraint c iff $\forall x \in scp(c), \tau[x] \in D(x)$. A tuple τ is *allowed* by a constraint c iff $\tau \in rel(c)$; we also say that c accepts τ. A *support* on c is a valid tuple on c that is also allowed by c. A *literal* is a pair (x, a) where x is a variable and a a value, not necessarily in $dom(x)$. A literal (x, a) is *Generalized Arc-Consistent* (GAC) iff it admits a *support* on c, i.e., a valid tuple τ on c such that τ is allowed by c and $\tau[x] = a$. A constraint c is GAC iff $\forall x \in scp(c), \forall a \in D(x), (x, a)$ is GAC.

3   Constraint `soft-regular`^{prx}

Definition 1 (DFA). *A deterministic finite automaton (DFA) is defined by a 5-tuple $(Q, \Sigma, \delta, q_0, F)$ where Q is a finite set of states, Σ is a finite set of symbols called the alphabet, $\delta : Q \times \Sigma \to Q$ is a transition function, $q_0 \in Q$ is the initial state, and $F \subseteq Q$ is the set of final states.*

Given an input string (a finite sequence of symbols taken from the alphabet Σ), the automaton starts in the initial state q_0, and for each symbol in sequence of the string, applies the transition function to update the current state. If the last state reached is a final state then the input string is accepted by the automaton. The set of strings that the automaton accepts constitutes a language, denoted by $L(M)$, which is technically a regular language.

In [7], a global constraint, called **regular**, is introduced: the sequence of values taken by the successive variables in the scope of this constraint must belong to a given regular language. For such constraints, a DFA can be used to determine whether or not a given tuple is accepted. This can be an attractive approach when constraint relations can be naturally represented by regular expressions in a known regular language. For example, in rostering problems, regular expressions can represent valid patterns of activities.

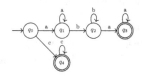

Fig. 1. A DFA with initial state q_0 and final states q_3 and q_4.

Definition 2 (regular). *Let M be a DFA, and $X = \langle x_1, \ldots, x_r \rangle$ be a sequence of r variables. The constraint* **regular**(X, M) *accepts all tuples in* $\{(v_1, \ldots, v_r) \mid v_1 \ldots v_r \in L(M) \wedge v_i \in D(x_i), \forall i \in 1..r\}$.

Example 1. Let us consider a sequence of 4 variables $X = \langle x_1, x_2, x_3, x_4 \rangle$ with $D(x_i) = \{a, b, c\}, \forall i \in 1..4$ and the automaton M depicted in Fig. 1. The tuples (c, c, c, c) and (a, a, b, a) are accepted by the constraint **regular**(X, M) whereas (c, a, a, b) and (c, c, c, a) are not.

Working with constraints defined by a DFA, Pesant's filtering algorithm [7] enforces generalized arc consistency by means of a two-stage forward-backward exploration. This two-stage process constructs a layered directed multi-graph and collects the set of states that support each literal (x, a). The worst-case time and space complexities of this incremental algorithm are both $O(rd|Q|)$. In [17], the authors proposed a filtering algorithm for extended finite automata (with counters) by reformulating them as a conjunction of signature and transition constraints; their method can also be applied to constraints **regular** when there are no counters.

It is worth noting that there is a direct correspondence between MDDs (Multi-valued Decision Diagrams) and DFAs. An acyclic and minimized deterministic finite automaton is equivalent to a reduced (ordered) MDD [18]. This is the reason why we can use an underlying MDD [19, 20] as a basis for filtering constraints **regular**.

When problems are over-constrained, it is sometimes relevant to relax some hard constraints. More specifically, it is possible to soften global constraints by adopting some known violation measures [10, 11]. A violation measure μ is simply a cost function that guarantees that cost 0 is associated with, and only with, any tuple that fully satisfies the constraint. The violation measure "var" is general-purpose: it measures the smallest number of variables whose values must be changed in order to satisfy the constraint (it basically expresses a Hamming distance).

For the relaxed version of `regular`, we can use "var" as well as the violation measure "edit", defined in [11] and revised in [12], which stands for the smallest number of insertions, deletions and substitutions required to satisfy the constraint. In a local search context, it was proposed [13] a violation measure by dividing the sequence of variables into segments that are accepted by the underlying DFA. This measure overestimates the Hamming distance. In this paper, we propose an original violation measure "prx" that is related to the size (length) of the longest prefix of a tuple compatible with a DFA.

Definition 3 (Violation Measure "prx"). *The violation measure μ^{prx} of an r-tuple τ with respect to a DFA M is $\mu^{prx}(M, \tau) = r - k$, where k is the size of the longest prefix of τ that can be extended to an r-tuple in $L(M)$.*

Definition 4 (`soft-regular`^{prx}). *Let M be a DFA, $X = \langle x_1, \ldots, x_r \rangle$ be a sequence of r variables and z be a (cost) variable. The constraint `soft-regular`$^{prx}(X, M, z)$ accepts all tuples in $\{(v_1, \ldots, v_r, d) \mid \mu^{prx}(M, (v_1, \ldots, v_r)) \leq d \wedge v_i \in D(x_i), \forall i \in 1..r \wedge d \in D(z)\}$.*

Note that only the bounds of z are considered. We do not reason about equality, because z is supposed to be minimized. The constraint `soft-regular`$^{prx}(X, M, z)$ supports an r-tuple $\tau = (v_1, \ldots, v_r)$ for X iff τ is valid and $\mu^{prx}(M, \tau) \leq max(z)$, i.e., the first $r - max(z)$ values of τ can be extended to a tuple recognized by M. We also have that a value $d \in D(z)$ is a support of the constraint iff there exists a valid r-tuple τ for X such that $d \geq \mu^{prx}(M, \tau)$.

Example 2. Let us consider the automaton M depicted in Fig. 1, a sequence of 4 variables $X = \langle x_1, x_2, x_3, x_4 \rangle$ with $D(x_i) = \{a, b, c\}, \forall i \in 1..4$ and a cost variable z with $D(z) = \{0, 1, 2\}$. The constraint `soft-regular`$^{prx}(X, M, z)$ supports (c, c, c, a) for X but not (c, a, a, b) because $\mu^{prx}(M, (c, c, c, a)) = 1 \leq max(z)$ while $\mu^{prx}(M, (c, a, a, b)) = 3 > max(z)$. Suppose now that x_3 and x_4 are assigned to c and a, respectively. The r-tuple with the longest prefix consistent with M is $\tau = (c, c, c, a)$ with violation cost $\mu^{prx}(M, \tau) = 1$. Hence, $z = 0$ can not satisfy the constraint and should be removed from $D(z)$.

4   A GAC Algorithm

We now introduce a filtering algorithm to enforce generalized arc consistency on a constraint `soft-regular`^{prx}, i.e., a soft regular constraint, using "prx" as violation measure. As presented in [7,11], the main data structure, the DFA, can be traversed in order to identify the values that are supported by the constraint. An MDD-based algorithm [19] can be applied for this step. Indeed, it is rather immediate to unfold the automaton on r levels, where each level corresponds to a variable in the main sequence X of variables. Arcs labelled with values for the first variable leave the root node (at level 0), whereas arcs labelled with values for the last variable reach the terminal node (at level r). An illustration is given by Fig. 2.

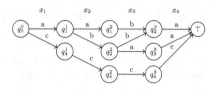

Fig. 2. MDD built for the constraint `soft-regular`^{prx} in Example 2.

The principle of filtering is the following. The MDD is traversed first in order to identify the deepest level for which a consistent prefix exists. With this information, it is then possible to update the min bound, $min(z)$, of the cost variable z. This is exactly the same kind of filtering performed with respect to the variable y in an inequality binary constraint $x \leq y$ when we apply: $min(y) \leftarrow max(min(y), min(x))$.

The value of the max bound, $max(z)$, is also useful for possibly pruning some values for variables in X. More precisely, for the first $r - max(z)$ variables in X, we need to only keep the values that occur in any consistent prefix whose size is at least equal to $r - max(z)$. This is the reason why we use an array *collected* that gives for each variable x_i, such that $i \leq r - max(z)$, the set of values $collected[x_i]$ that respect that condition. Roughly speaking, this is the spirit of the filtering performed with respect to the variable x in an inequality binary constraint $x \leq y$ when we apply: $max(x) \leftarrow min(max(x), max(y))$.

Algorithm 1 can be called to enforce GAC on a specified constraint `soft-regular`$^{prx}(X, M, z)$. We introduce a map, called *explored*, that stores for any processed node the size of the longest consistent prefix that can be reached from it. As usual for a map, we use the operations (i) *clear()*, for reinitializing the map, (ii) *contains()*, for determining if a specified node has already been processed, (iii) *get()*, for getting the size of the longest consistent prefix that can be reached from a specified node, (iv) *put()*, for putting an entry in the map.

Algorithm 1. `soft-regular`$^{prx}(X = \langle x_1, \ldots, x_r \rangle, root, z)$

---

1 $explored.clear()$
2 $collected[x] \leftarrow \emptyset, \forall x \in X$
3 $maxSuccessLevel \leftarrow exploreTree(root)$
4 $min(z) \leftarrow max(min(z), r - maxSuccessLevel)$
5 **if** $D(z) = \emptyset$ **then**
6 $\quad$ **return** Failure
7 **else**
8 $\quad$ **foreach** $i \in 1..r - max(z)$ **do**
9 $\qquad$ $D(x_i) \leftarrow collected[x_i]$

Algorithm 1 works as follows. Data structures are initialized at lines 1 and 2. Then, the exploration of the MDD starts from the root, and the size of the longest found consistent prefix is stored in the variable $maxSuccessLevel$. At this point, it is possible to update both the domain of the cost variable z and the domains of the first $r - max(z)$ variables of X, as explained earlier in this section. Note that if the domain of z becomes empty, a failure is returned. Also, note that no domain wipe-out (failure) can happen when updating the domains of variables in X.

Algorithm 2 makes an exploration of a (sub-)MDD and returns the size of the longest prefix compatible with the DFA that can be reached from the specified node. If this node corresponds to the terminal node (this is identified by a level equal to r) or a node that has already been processed, the algorithm returns the corresponding level. Otherwise, the algorithm explores each child (i.e., node reached from an outgoing edge) such that the value labeling the linking arc is still valid. Some values can then be collected, but note that we collect supported values for only the first $r - max(z)$ variables (lines 10–11). The level of a node is equal to the maximum level reached by its children, which can be expressed by the formula: $maxLevel(node) = max(node.level, max_{n:node.outs}maxLevel(n))$. Finally, Algorithm 2 adds an entry to the map $explored$ and returns the size of the longest consistent prefix.

Algorithm 2. exploreTree($node$) : *Integer*

---

1 **if** $node.level = r$ **then**
2 $\quad$ **return** r

3 **if** $explored.contains(node)$ **then**
4 $\quad$ **return** $explored.get(node)$

5 $x \leftarrow node.var$
6 $maxLevel \leftarrow node.level$
7 **foreach** $arc \in node.outs$ **do**
8 $\quad$ **if** $arc.value \in D(x)$ **then**
9 $\quad\quad$ $currMaxLevel \leftarrow exploreTree(arc.dest)$
10 $\quad\quad$ **if** $node.level \leq r - max(z) \leq currMaxLevel$ **then**
11 $\quad\quad\quad$ $collected[x] \leftarrow collected[x] \cup \{arc.value\}$
12 $\quad\quad$ $maxLevel \leftarrow max(maxLevel, currMaxLevel)$

13 $explored.put(node, maxLevel)$
14 **return** $maxLevel$

Example 3. Let us consider the constraint $\texttt{soft-regular}^{prx}(X, M, z)$ from Example 2. After the execution of Algorithm 1, we have: $collected[x_1] = \{a, c\}$, $collected[x_2] = \{a, b, c\}$, $collected[x_3] = collected[x_4] = \{\}$. Since $max(z) = 2$, only domains of x_1 and x_2 may be updated. Here, b is removed from $D(x_1)$.

Suppose now that x_3 and x_4 are respectively assigned to c and a, and 2 is removed from $D(z)$, i.e., $max(z) = 1$. Since $max(z)$ is now 1, we need to collect values for the first $4 - 1 = 3$ variables x_1, x_2, and x_3. Any value labeling an arc is collected iff this arc can reach at least the level 3. We have: $collected[x_1] = collected[x_2] = collected[x_3] = \{c\}$. Consequently, we now have

$D(x_1) = D(x_2) = D(x_3) = \{c\}$. Note that the deepest reachable consistent level is 3, which implies that $min(z) = 1$.

Proposition 1. *Algorithm 1 enforces GAC on any specified constraint* soft-regular$^{prx}(X, M, z)$.

Proof. After the execution of Algorithm 2, $\forall x_i \in X$, with $i \in 1..r - max(z)$, $\forall v_i \in D(x_i)$, the arc associated with (x_i, v_i) must reach a node with a level at least equal to $r - max(z)$, i.e., there must exist a path from the root to this node whose size is at least $r - max(z)$. Values collected on this path can represent a support for (x_i, v_i). The maximum level reached from the root corresponds to the size of the longest prefix of a tuple that is consistent with the DFA M. Hence, the lower bound of $D(z)$ must be updated using this value. □

Since Algorithm 2 traverses at most one each arc in the corresponding graph, the time complexity of Algorithm 1 is $O(r|\delta|)$ where $|\delta|$ is the number of transitions in the DFA.

5   Possible Decomposition

The constraint soft-regular^{prx} can be decomposed using cost MDD [21] constraints with the unfolded automaton described in Sect. 4 by adding the following arcs: for each node at level i, add an escape arc to the terminal node with cost $r - i$; each arc in the unfolded automaton has cost 0. The time complexity to filter this constraint is also linear in the number of arcs but it is not straightforward to implement if the cost MDD is not available.

The constraint soft-regular$^{prx}(X, M, z)$ can also be decomposed by using reified table constraints [22] and (ordinary) table constraints as follows. First, we introduce $r + 1$ new variables y_i ($i \in 0..r$) such that $D(y_0) = \{q_0\}$, $D(y_i) = Q, \forall i \in 1..r - 1$ and $D(y_r) = F$. We also introduce r Boolean variables b_i, $i \in 1..r$, for reification purpose. Next, we introduce r reified table constraints $c_i^{reif} : c_i \Leftrightarrow b_i$ where c_i is a classical ternary positive table constraint such that $scp(c_i) = \{y_{i-1}, x_i, y_i\}$ and $rel(c_i) = \delta$ for $i = 1..r$. These constraints reflect the truth of a valid transition. We then introduce r functionality constraints c_i^f ($i = 1..r$) where each constraint c_i^f is a ternary negative table constraint such that $scp(c_i^f) = \{y_{i-1}, x_i, y_i\}$ and $rel(c_i^f) = \{(q_k, v, \neq q_l) | (q_k, v, q_l) \in \delta\}$. Finally, we enforce the prefix restriction by adding the constraints: $(z \leq r - k) \Rightarrow (\sum_{i=1..k} b_i = k)$. Note that when $z \leq r - k$, a prefix of size at least equal to k must be consistent with the underlying DFA, which implies that $b_i = 1$, $i = 1..k$ and this is equivalent to $\sum_{i=1..k} b_i = k$.

6   Experimental Results

We illustrate the practical interest of our approach on a variant of the Nurse Rostering Problem (NRP). This problem has been extensively studied in both

Table 1. Results obtained on NRP instances. Timeout set to 3,600 s.

					Soft-regular		
Instance	#days	#nurses	#shifts	LNS	LNS (1 nurse)		
				Objective	Objective	violHoz
Instance1	14	8	1	607.0	512.0	5.9
Instance2	14	14	2	890.4	837.7	4.0
Instance3	14	20	3	1055.6	1006.2	3.6
Instance4	28	10	2	1732.4	1645.3	12.6
Instance4a	28	10	2	1694.2	1548.1	10.9
Instance4b	28	10	2	1722.4	1672.4	12.1
Instance5	28	16	2	1477.1	1261.9	11.5
Instance5a	28	16	2	1471.2	1390.0	11.7
Instance5b	28	16	2	1382.2	1303.0	11.1
Instance6	28	18	3	2629.1	2498.9	11.2
Instance6a	28	18	3	2539.6	2433.9	7.5
Instance6b	28	18	3	2706.7	2642.1	8.8
Instance7	28	20	3	1756.4	1555.1	9.1
Instance7a	28	20	3	1903.3	1810.6	8.3
Instance7b	28	20	3	1674.4	1485.8	9.4

domains of Operational Research and Artificial Intelligence for more than 40 years due to its importance in real-world hospital contexts [23,24].

The NRP consists of creating a roster by assigning nurses different shift types satisfying some constraints [23]. They are divided in two groups: hard constraints and soft constraints. Hard constraints must be satisfied in order to have a feasible solution whereas soft constraints can be partially violated by introducing some violation costs. The objective is to minimize the sum of these costs.

In real NRP situations, it may happen that one or even several nurses indicate that in a near future (not precisely indicated) they will have to be absent. One possible solution is to relax the hard `regular` constraints corresponding to the regulation rules for these nurses while trying to find a roster satisfying as much as possible the first steps (shifts) of the underlying automata. The constraint `soft-regular`^{prx} can be applied for that. In [8], Schaus proposed to relax the demands instead using a `soft-cardinality` constraint. However, this relaxation does not allow us to optimize the longest feasible prefix easily over a horizon.

We have conducted an experimentation under Linux (CPUs clocked at 2.5 GHz, with 5 GB of RAM). We have been interested in the NRP instances recently proposed in [25]. We have also randomly generated some variants for the last four instances by slightly modifying soft demands of nurses (we use symbols a and b as suffix in their names). For our experiments, a nurse was randomly chosen to be the person who can not totally follow the roster. So, we first minimize the global violation cost (as initially computed for these instances), and then we attempt to maximize the longest prefix satisfying the shift horizon of the "relaxed" nurses. Note that we also force this staff's roster to satisfy regulation rules over at least the first half of the scheduling horizon.

We run our algorithm 10 times on each instance, with a timeout set to 3, 600 s, and we report average results. We chose Large Neighborhood Search (LNS) [26] as a method to solve NRP instances. The variable ordering heuristic selects the variable that admits the highest violation cost over a day while the value ordering heuristic selects the value that reduces the overall cost most. Concerning relaxation, firstly, one nurse is selected randomly. If no solution is found after 10 executions, the number of nurses relaxed will increase by one. It will be set back to one when a solution is found. For each restart, the number of failures is limited to 100K.

Table 1 shows some representative results for only one relaxed nurse. The first 4 columns gives some information about the instances. The 3 last columns present solving results. The first column (out of these 3 last columns) shows the violation cost obtained with LNS for the initial problem (i.e., without any relaxation). The last 2 columns present results for the relaxed problem: the obtained objective cost, and the horizon violated by the roster of the relaxed nurse. Generally speaking, one can observe an improvement on the overall objective while keeping a reasonable roster for the relaxed nurse.

Table 2. Execution time(s) for dedicated and decomposition approaches.

	Instance1	Instance2	Instance3	Instance4	Instance5	Instance6	Instance7
Dedicated	71.7	40.4	58.4	38.0	64.5	98.5	98.6
Decomp	190.7	81.2	122.6	57.9	74.4	110.5	118.9

We also compared our soft constraint $\texttt{soft-regular}^{prx}$ with the decomposition approach. For simplicity, a static branching was used, and the program was stopped when the number of failures reached 500K. The execution times in seconds are reported in Table 2. Clearly, the dedicated approach is more robust than the decomposition one, as it can be twice as fast.

Acknowledgments. The first author is supported by the FRIA-FNRS. The second author is supported by the project CPER Data from the "Hauts-de-France".

References

1. Régin, J.-C.: A filtering algorithm for constraints of difference in CSPs. In: Proceedings of AAAI 1994, pp. 362–367 (1994)
2. Beldiceanu, N., Contejean, E.: Introducing global constraints in CHIP. Math. Comput. Modell. **20**(12), 97–123 (1994)

3. Van Hentenryck, P., Carillon, J.-P.: Generality versus specificity: an experience with AI and OR techniques. In: Proceedings of AAAI 1988, pp. 660–664 (1988)
4. Régin, J.-C.: Generalized arc consistency for global cardinality constraint. In: Proceedings of AAAI 1996, pp. 209–215 (1996)
5. Hooker, J.N.: Integrated Methods for Optimization. Springer, Heidelberg (2012). https://doi.org/10.1007/978-1-4614-1900-6
6. Aggoun, A., Beldiceanu, N.: Extending chip in order to solve complex scheduling and placement problems. Math. Comput. Modell. **17**(7), 57–73 (1993)
7. Pesant, G.: A regular language membership constraint for finite sequences of variables. In: Wallace, M. (ed.) CP 2004. LNCS, vol. 3258, pp. 482–495. Springer, Heidelberg (2004). https://doi.org/10.1007/978-3-540-30201-8_36
8. Schaus, P.: Variable objective large neighborhood search: a practical approach to solve over-constrained problems. In: 2013 IEEE 25th International Conference on Tools with Artificial Intelligence (ICTAI), pp. 971–978. IEEE (2013)
9. van Hoeve, W.: Over-constrained problems. In: van Hentenryck, P., Milano, M. (eds.) Hybrid Optimization, pp. 191–225. Springer, New York (2011). https://doi.org/10.1007/978-1-4419-1644-0_6
10. Petit, T., Régin, J.-C., Bessière, C.: Specific filtering algorithms for over-constrained problems. In: Walsh, T. (ed.) CP 2001. LNCS, vol. 2239, pp. 451–463. Springer, Heidelberg (2001). https://doi.org/10.1007/3-540-45578-7_31
11. van Hoeve, W., Pesant, G., Rousseau, L.-M.: On global warming: flow-based soft global constraints. J. Heuristics **12**(4–5), 347–373 (2006)
12. He, J., Flener, P., Pearson, J.: Underestimating the cost of a soft constraint is dangerous: revisiting the edit-distance based soft regular constraint. J. Heuristics **19**(5), 729–756 (2013)
13. He, J., Flener, P., Pearson, J.: An automaton constraint for local search. Fundamenta Informaticae **107**(2–3), 223–248 (2011)
14. Montanari, U.: Network of constraints: fundamental properties and applications to picture processing. Inf. Sci. **7**, 95–132 (1974)
15. Dechter, R.: Constraint Processing. Morgan Kaufmann, Burlington (2003)
16. Lecoutre, C.: Constraint Networks: Techniques and Algorithms. ISTE/Wiley, Hoboken (2009)
17. Beldiceanu, N., Carlsson, M., Petit, T.: Deriving filtering algorithms from constraint checkers. In: Wallace, M. (ed.) CP 2004. LNCS, vol. 3258, pp. 107–122. Springer, Heidelberg (2004). https://doi.org/10.1007/978-3-540-30201-8_11
18. Hadzic, T., Hansen, E.R., O'Sullivan, B.: On automata, MDDs and BDDs in constraint satisfaction. In: Proceedings of ECAI 2008 Workshop on Inference methods based on Graphical Structures of Knowledge (2008)
19. Cheng, K., Yap, R.: An MDD-based generalized arc consistency algorithm for positive and negative table constraints and some global constraints. Constraints **15**(2), 265–304 (2010)
20. Perez, G., Régin, J.-C.: Improving GAC-4 for table and MDD constraints. In: O'Sullivan, B. (ed.) CP 2014. LNCS, vol. 8656, pp. 606–621. Springer, Cham (2014). https://doi.org/10.1007/978-3-319-10428-7_44
21. Perez, G., Régin, J.-C.: Soft and cost MDD propagators. In: Proceedings of AAAI 2017, pp. 3922–3928 (2017)
22. Khong, M.T., Deville, Y., Schaus, P., Lecoutre, C.: Efficient reification of table constraints. In: 2017 IEEE 29th International Conference on Tools with Artificial Intelligence (ICTAI). IEEE (2017)
23. Burke, E., De Causmaecker, P., Berghe, G.V., Van Landeghem, H.: The state of the art of nurse rostering. J. Sched. **7**(6), 441–499 (2004)

24. Ernst, A., Jiang, H., Krishnamoorthy, M., Sier, D.: Staff scheduling and rostering: a review of applications, methods and models. Eur. J. Oper. Res. **153**(1), 3–27 (2004)
25. Curtois, T., Qu, R.: Computational results on new staff scheduling benchmark instances. Technical report, ASAP Research Group, School of Computer Science, University of Nottingham, 06 October 2014
26. Shaw, P.: Using constraint programming and local search methods to solve vehicle routing problems. In: Maher, M., Puget, J.-F. (eds.) CP 1998. LNCS, vol. 1520, pp. 417–431. Springer, Heidelberg (1998). https://doi.org/10.1007/3-540-49481-2_30

Constraint and Mathematical Programming Models for Integrated Port Container Terminal Operations

Damla Kizilay[1,2(✉)], Deniz Türsel Eliiyi[2], and Pascal Van Hentenryck[1]

[1] University of Michigan, Ann Arbor, MI 48109, USA
dkizilay@umich.edu
[2] Yasar University, 35100 Izmir, Turkey

Abstract. This paper considers the integrated problem of quay crane assignment, quay crane scheduling, yard location assignment, and vehicle dispatching operations at a container terminal. The main objective is to minimize vessel turnover times and maximize the terminal throughput, which are key economic drivers in terminal operations. Due to their computational complexities, these problems are not optimized jointly in existing work. This paper revisits this limitation and proposes Mixed Integer Programming (MIP) and Constraint Programming (CP) models for the integrated problem, under some realistic assumptions. Experimental results show that the MIP formulation can only solve small instances, while the CP model finds optimal solutions in reasonable times for realistic instances derived from actual container terminal operations.

Keywords: Container terminal operations · MIP
Constraint programming

1 Introduction

Maritime transportation has significant benefits in terms of cost and capability for carrying a higher number of cargos. Indeed, sea trade statistics indicate that 90% of global trade is performed by maritime transportation. This has led to new investments in container terminals and a variety of initiatives to improve the operational efficiency of existing terminals. Operations at a container terminal can be classified as quay side and yard side. They handle materials using quay cranes (QC), yard cranes (YC), and transportation vehicles such as yard trucks (YT). QCs load and unload containers at the quay side, while YCs load and discharge containers at the yard side. YTs provide transshipment of the containers between the quay and the yard sides.

This study was supported by a Fulbright Program grant sponsored by the Bureau of Educational and Cultural Affairs of the United States Department of State and administered by the Institute of International Education.

In a typical container terminal, it is important to minimize the vessel berthing time, i.e., the period between the arrival and the departure of a vessel. When a vessel arrives at the terminal, the berth allocation problem selects when and where in the port the vessel should berth. Once a vessel is berthed, its stowage plan determines the containers to be loaded/discharged onto/from the vessel. This provides an input to the *QC assignment and scheduling*, which determines the sequence of the containers to be loaded or discharged from different parts of the vessel by the QCs. In addition, the containers discharged by the QCs are placed onto YTs and transported to the storage area, which corresponds to a *vehicle dispatching problem*. Each discharged container is assigned a storage location, giving rise to a *yard location assignment problem*. Finally, the containers are taken from YTs by YCs and placed onto stacks in storage blocks, specifying a *YC assignment and scheduling problem*.

A container terminal aims at completing the operations of each berthed vessel as quickly as possible to minimize vessel waiting times at the port and thus to maximize the turnover, i.e., the number of handled containers. Optimizing the integrated operations within a container terminal is computationally challenging [16]. Therefore, the optimization problems identified earlier are generally considered separately in the literature, and the number of studies considering integrated operations is rather limited. However, although the optimization of individual problems brings some operational improvements, the main opportunity lies in optimizing terminal operations holistically. This is especially important since the optimization sub-problems have conflicting objectives that can adversely affect the overall performance of the system.

This paper considers the integrated optimization of container terminal operations and proposes MIP and CP formulations under some realistic assumptions. To the best of our knowledge, the resulting optimization problem has not been considered in the literature so far. Experimental results show that the MIP formulation is not capable of solving instances of practical relevance, while the CP model finds optimal solutions in reasonable times for realistic instances derived from real container terminal operations.

The rest of the paper is organized as follows. Section 2 specifies the problem and the assumptions considered in this work. Section 3 provides a detailed literature review for the integrated optimization of container terminal operations. Section 4 presents the MIP model, while Sect. 5 presents the constraint programming model for the same problem. Section 6 presents the data generation procedure, the experimental results, and the comparison of the different models. Finally, Sect. 7 presents concluding remarks and future research directions.

2   Problem Definition

This section specifies the Integrated Port Container Terminal Problem (IPCTP) and its underlying assumptions. The IPCTP is motivated by the operations of actual container terminals in Turkey.

In container terminals, berth allocation assigns a berth and a time interval to each vessel. Literature surveys and interviews with port management officials reveal that significant factors in berth allocation include priorities between customers, berthing privileges of certain vessels in specific ports, vessel sizes, and depth of the water. Because of all these restrictions, the number of alternative berth assignments is quite low, especially in small Turkish ports. As a result, the berthing plan is often determined without the need of intelligent systems and the berthing decisions can be considered as input data to the scheduling of the material-handling equipment. The vessel stowage plan decides how to place the outbound containers on the vessel and is prepared by the shipping company. These two problems are thus separated from the IPCTP. In other words, the paper assumes that vessels are already berthed and ready to be served.

The IPCTP is formulated by considering container groups, called *shipments*. A single shipment represents a group of containers that travel together and belong to the same customer. Therefore, the containers in a single shipment must be stored in the same yard block and in the same vessel bay. In addition, each shipment is handled as a single batch by QCs and YCs.

The IPCTP determines the storage location in the yard for inbound containers. The yard is assumed to be divided into a number of areas containing the storage blocks. Each inbound shipment has a number of possible location points in each area. Each YC is assumed to be dedicated to a specific area of the yard. Note that outbound shipments are at specified yard locations at the beginning of the planning period and hence their YCs are known in advance. In contrast, for inbound shipments, the YC assignment is derived from the storage block decisions. The IPCTP assumes that each yard location can store at most one shipment but there is no difficulty in relaxing that assumption.

The inbound and outbound shipments and their vessel bays are specified in the vessel stowage plan. The IPCTP assigns each shipment to a QC and schedules each QC to process a sequence of shipments. The QC scheduling is constrained by movement restrictions and safety distances between QCs. Two adjacent QCs must be apart from each other by safety distance, so that they can perform their tasks simultaneously without interference as described in [12].

The IPCTP assumes the existence of a sufficient number of YTs so that cranes never wait. This assumption is motivated by observations in real terminals where many YTs are dedicated to each QC in order to ensure a smooth operation. This organization is justified by the fact that QCs are the most critical handling equipment in the terminal. As a result, QCs are very rarely blocked while discharging and almost never starve while loading. After solving the IPCTP, the YT schedule can be found in a post-processing step. Indeed, the YP scheduling problem can be reduced to the tactical fixed job scheduling problem, which is polynomial-time solvable. Therefore, the assumption of having a sufficient number of YTs is realistic and simplifies the IPCTP.

The IPCTP also assumes that the handling equipment (QC, YC, YT) is homogeneous, and their processing times are deterministic and known. Since the QCs cannot travel beyond the berthed vessel bays and must obey a safety distance [14], each shipment can only be assigned to an eligible set of QCs that respect safety distance and non-crossing constraints. These are illustrated in

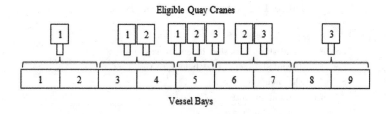

Fig. 1. The vessel bays and their available QCs.

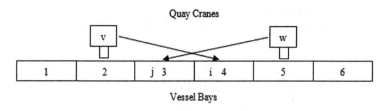

Fig. 2. An example of interference for shipments i, j and quay cranes v, w.

Fig. 1 where berthed vessel bays and QCs are indexed in increasing order and the safety distance is assumed to be 1 bay. For instance, only QC-1 is eligible to service bays 1–2. Similarly, only QC-3 can operate on vessel bays 8–9. In contrast, bays 3–4 can be served by QC-1 and QC-2.

The main objective of a container terminal is to maximize total profit by increasing productivity. Terminal operators try to lower vessel turn times and decrease dwell times. To lower vessel turn times, the crane operations must be well-coordinated and the storage location of the inbound shipments must be chosen carefully, since they impact the distance traveled by the YTs. Therefore, the IPCTP jointly considers the storage location assignment for the inbound shipments from multiple berthed vessels and the crane assignment and scheduling for both outbound and inbound containers. The objective of the problem is to minimize the sum of weighted completion times of the vessels.

The input parameters of the IPCTP are given in Table 1. Most are self-explanatory but some necessitate additional explanation. The smallest distance $\delta_{v,w}$ between quay cranes v and w is given by $\delta_{v,w} = (\delta + 1)\,|v - w|$ where δ is the safety distance. The minimum time between the starting times of shipments i and j when processed by cranes v and w is given by

$$
\Delta_{i,j}^{v,w} = \begin{cases} (b_i - b_j + \delta_{v,w})\,s_{QC} & \text{if } v < w \text{ and } i \neq j \text{ and } b_i > b_j - \delta_{v,w} \\ (b_j - b_i + \delta_{v,w})\,s_{QC} & \text{if } v > w \text{ and } i \neq j \text{ and } b_i < b_j - \delta_{v,w} \\ 0 & \text{otherwise.} \end{cases}
$$

This captures the time needed for a quay crane to travel to a safe distance in case of potential interference. This is illustrated in Fig. 2. If shipments i and j are processed by cranes v and w, then their starting times must be separated by s_{QC} time units in order to respect the safety constraints (assuming that $w = v + 1$).

Table 1. The parameters of the IPCTP.

S	Set of berthed vessels
C_u^s	Set of inbound shipments that belong to vessel $s \in S$
C_l^s	Set of outbound shipments that belong to vessel $s \in S$
C	Set of all shipments
C_u	Set of inbound shipments
C_l	Set of outbound shipments
L_u	Set of available yard locations for inbound shipments
l_i	Yard location of outbound shipment $i \in C_l$
L	Set of all yard locations
QC	Set of QCs
YC	Set of YCs
P	Set of precedence relationships between containers on QCs
B	Set of vessel bays
b_i	Vessel bay position of shipment $i \in C$
$QC(i)$	Set of eligible QCs for shipment $i \in C$
$YC(k)$	The YC responsible for yard location $k \in L$
w_s	Weight (priority) of vessel $s \in S$
Q_i	QC handling time of shipment $i \in C$
Y_i	YT handling time of shipment $i \in C$
tyt_i	YT handling time of outbound shipment $i \in C_l$
tt_k	YT transfer time of inbound shipment to yard location $k \in L_u$
$tyc_{k,l}$	YC travel time between yard locations k and l
$eqc_{i,j}$	QC travel time from shipment $i \in C$ to shipment $j \in C$
$eyc_{i,j}$	YC travel time from yard location $i \in L$ to yard location $j \in L$
s_{QC}	Travel time for unit distance of equipment QC
δ	Safety distance between two QCs
$\delta_{v,w}$	Smallest allowed difference between bay positions of quay cranes v and w
$\Delta_{i,j}^{v,w}$	Minimum time between the starting times of shipments i and j when processed by cranes v and w
Θ	Set of all combinations of shipments and QCs with potential interferences
0	Dummy initial shipment
N	Dummy last shipment
C^0	Set of all shipments including dummy initial shipment $C \cup \{0\}$
C^N	Set of all shipments including dummy last shipment $C \cup \{N\}$
M	A sufficiently large constant integer
$Qtime_i$	QC handling time of a first container in inbound shipment $i \in C_u$
$Ytime_i$	YC handling time of a first container in outbound shipment $i \in C_l$

For instance, if shipment i is processed first, crane v must move to bay 1 before shipment j can be processed. Finally, the set of interferences can be defined by

$$\Theta = \{(i, j, v, w) \in C^2 \times QC^2 \mid i < j \ \& \ \Delta_{i,j}^{v,w} > 0\}.$$

The dummy (initial and last) shipments are only used in the MIP model.

3   Literature Review

Port container terminal operations have received significant attention and many studies are dedicated to the sub-problems described earlier: See [1] for a classification of these subproblems. Recent work often consider the integration of two or three problems but very few papers propose formulations covering all the sub-problems jointly. Some papers give mathematical formulations of integrated problems but only use heuristic approaches given the computational complexity of solving the models. This section reviews recent publications addressing the integrated problems and highlights their contributions.

Chen et al. [3] propose a hybrid flowshop scheduling problem (HFSP) to schedule QCs, YTs, and YCs jointly. Both outbound and inbound operations are considered, but outbound operations only start after all inbound operations are complete. In their mathematical model, each stage of the flowshop has unrelated multiple parallel machines and a tabu-search algorithm is used to address the computational complexity. Zheng et al. [21] study the scheduling of QCs and YCs, together with the yard storage and vessel stowage plans. They consider an automated container handling system, in which twin 40' QCs and a railed container handling system is used. A rough yard allocation plan is maintained to indicate which blocks are available for storing the outbound containers from each bay. No mathematical model is provided, and the yard allocation, vessel stowage, and equipment scheduling is performed using a rule-based heuristic.

Xue et al. [20] propose a mixed integer programming (MIP) model for integrating the yard location assignment for inbound containers, quay crane scheduling, and yard truck scheduling. Non-crossing constraints and safety distances between QCs are ignored, and the assignment of QCs and YTs are predetermined. The yard location assignment considers block assignments instead of container slots. The resulting model cannot be solved for even medium-sized problems. Instead, a two-stage heuristic algorithm is employed, combining an ant colony optimization algorithm, a greedy algorithm, and local search.

Chen et al. [4] consider the integration of quay crane scheduling, yard crane scheduling, and yard truck transportation. The problem is formulated as a constraint-programming model that includes both equipment assignment and scheduling. However, non-crossing constraints and safety margins are ignored. The authors state that large-scale instances are computationally intractable for constraint programming and that even small-scale instances are too time-consuming. A three-stage heuristic algorithm is solved iteratively to obtain solutions for large-scale problems with up to 500 containers.

Wu et al. [17] study the scheduling of different types of equipment together with the storage strategy in order to optimize yard operations. Only loading operations for outbound containers are considered, and the tasks assigned to each QC and their processing sequence are assumed to be known. The authors formulate models to schedule the YCs and automated guided vehicles (AGV), and use a genetic algorithm to solve large-scale problems.

Homayouni et al. [5] study the integrated scheduling of cranes, vehicles, and storage platforms at automated container terminals. A mathematical model of

---

Model 1. The MIP model for the IPCTP: Decision variables

Variables

$x_{i,k} \in \{0,1\}$: inbound shipment i is assigned to yard location k

$z_{i,j}^q \in \{0,1\}$: shipment j is handled immediately after shipment i by QC q

$qz_{i,j} \in \{0,1\}$: shipment j is handled after shipment i by QC

$v_{i,j}^c \in \{0,1\}$: shipment j is handled immediately after shipment i by YC c

$sqc_i \geq 0$: start time of shipment i by its QC

$syc_i \geq 0$: start time of shipment i by its YC

$t_i \geq 0$: travel time of YT for inbound shipment i to assigned yard location k

$sy_{i,j} \geq 0$: travel time of YC from location i to location j

$Cmax_s$: time of the last handled container at vessel s

---

the same problem is proposed in [6]. In these studies, both outbound and inbound operations are considered. The origin and destination points of the containers are assumed to be predetermined and, in addition, empty for inbound containers. The earlier study proposes a simulated annealing (SA) algorithm to solve the problem, whereas the latter proposes a genetic algorithm (GA) outperforming SA under the same assumptions.

Lu and Le [11] propose an integrated optimization of container terminal scheduling, including YTs and YCs. The authors consider uncertainty factors such as YT travel speed, YC speed, and unit time of the YC operations. The assignment of YTs and YCs are not considered, and pre-assignments are assumed. The objective is to minimize the operation time of YCs in coordination with the YTs and QCs. The authors use a simulation of the real terminal operation environment to capture uncertainties. The authors also formulate a mathematical model and propose a particle swarm optimization (PSO) algorithm. As a future study, they indicate that the scheduling for simultaneous outbound and inbound operations should be considered for terminals adopting parallel operations.

Finally, a few additional studies [3, 7–10, 13, 15, 18, 19] integrate different sub-problems, highlighting the increasing attention given to integrated solutions. They propose a wide range of heuristic or meta-heuristic algorithms, e.g., genetic algorithm, tabu search, particle swarm optimization, and rule-based heuristics.

Although most papers in container port operations focus on individual problems, recent developments have emphasized the need and potential for coordinating these interdependent operations. This paper pushes the state-of-the-art further by optimizing all operations holistically and demonstrating that constraint programming is a strong vehicle to address this integrated problem.

4   The MIP Model

The MIP decision variables are presented in Model 1, while the objective function and the constraints are given in Model 2. They use the formulation of QC interference constraints from [2]. The objective function (2-01) minimizes the maximum weighted completion time of each vessel. Constraints (2-02–2-03) compute the weighted completion time of each vessel. Inbound shipments start their operations at a QC and finish at a YC, whereas outbound shipments follow the reverse order. Constraint (2-04) expresses that each available storage block stores at most one inbound shipment. Constraint (2-05) ensures that each inbound shipment is assigned an available storage block. All the containers of a shipment are assigned to the same block. Constraints (2-06–2-07) assigns the first (dummy) shipments to each QC and YC and Constraints (2-08–2-09) do the same for the last (dummy) shipments. Constraint (2-10) states that every shipment is handled by exactly one eligible QC. Constraints (2-11–2-12) ensures that each shipment is handled by a single YC. In Constraint (2-12), yard blocks are known at the beginning of the planning horizon for outbound shipments: They are thus directly assigned to the dedicated YCs. Constraints (2-13–2-14) guarantee that the shipments are handled in well-defined sequences by each handling equipment (QC and YC). Constraint (2-15) defines the YT transportation times for the inbound shipments. Constraints (2-16–2-18) specify the empty travel times of the YCs according to yard block assignments of the shipments. Constraints (2-19–2-22) specify the relationship between the start times of two consecutive shipments processed by the same handling equipment. Constraints (2-23–2-24) are the precedence constraints for each shipment based on the handling time of the first container, which again differ for inbound and outbound shipments. Constraint (2-25) ensures that, if shipment i precedes shipment j on a QC, shipment j cannot start its operation on that QC until shipment i finishes. Constraint (2-26) guarantees that shipments that potentially interfere are not allowed to be processed at the same time on any QC. Constraint (2-27) imposes a minimum temporal distance between the processing of such shipments, which corresponds to the time taken by the QC to move to a safe location. Precedence constraints (2-28) are used to express that shipment j is located under a shipment i and to ensure vessel balance during operations.

There are nonlinear terms in constraint (2-17), which computes the empty travel time of a YC, i.e., when it travels between the destinations of two inbound shipments. These terms can be linearized by introducing new binary variables of the form $\theta_{i,k,j,l}$ to denote whether inbound shipments $i, j \in C_u$ are assigned to yards locations $k, l \in L_u$. The constraints then become:

$$sy_{i,j} = \sum_{k \in L_u} \sum_{l \in L_u} tyc_{k,l}\theta_{i,k,j,l} \quad \forall i,j \in C_u$$

$$\theta_{i,k,j,l} \geq x_{i,k} + x_{j,l} - 1, \forall i,j \in C_u \quad \forall k,l \in L_u, i \neq j, k \neq l$$

$$2 - (x_{i,k} + x_{j,l}) \leq 2(1 - \theta_{i,k,j,l}) \quad \forall i,j \in C_u, \forall k,l \in L_u, i \neq j, k \neq l$$

Model 2. The MIP model for the IPCTP: Objective and constraints

Objective

minimize $\sum_{s \in S} Cmax_s$ $\hspace{2cm}$ (2-01)

Constraints

$Cmax_s \geq w_s \left(sqc_i + Q_i\right)$ $\quad \forall s \in S, \forall i \in C_l^s$ $\hspace{1.5cm}$ (2-02)

$Cmax_s \geq w_s \left(syc_i + Y_i\right)$ $\quad \forall s \in S, \forall i \in C_u^s$ $\hspace{1.5cm}$ (2-03)

$\sum_{i \in C_u} x_{i,k} \leq 1$ $\quad \forall k \in L_u$ $\hspace{1.5cm}$ (2-04)

$\sum_{k \in L_u} x_{i,k} = 1$ $\quad \forall i \in C_u$ $\hspace{1.5cm}$ (2-05)

$\sum_{j \in C^N} z_{0,j}^q = 1$ $\quad \forall q \in QC$ $\hspace{1.5cm}$ (2-06)

$\sum_{j \in C^N} v_{0,j}^c = 1$ $\quad \forall c \in YC$ $\hspace{1.5cm}$ (2-07)

$\sum_{j \in C^0} z_{i,N}^q = 1$ $\quad \forall q \in QC$ $\hspace{1.5cm}$ (2-08)

$\sum_{j \in C^0} v_{i,N}^c = 1$ $\quad \forall c \in YC$ $\hspace{1.5cm}$ (2-09)

$\sum_{q \in QC(i)} \sum_{j \in C^N} z_{i,j}^q = 1$ $\quad \forall i \in C, i \neq j$ $\hspace{1.5cm}$ (2-10)

$\sum_{j \in C^N} v_{i,j}^{YC(k)} = x_{i,k}$ $\quad \forall k \in L_u, \forall i \in C_u, i \neq j$ $\hspace{1.5cm}$ (2-11)

$\sum_{j \in C^N} v_{i,j}^{YC(l_i)} = 1$ $\quad \forall i \in C_l, i \neq j$ $\hspace{1.5cm}$ (2-12)

$\sum_{j \in C^0} z_{j,i}^q - \sum_{j \in C^N} z_{i,j}^q = 0$ $\quad \forall i \in C, \forall q \in QC$ $\hspace{1.5cm}$ (2-13)

$\sum_{j \in C^0} v_{j,i}^c - \sum_{j \in C^N} v_{i,j}^c = 0$ $\quad \forall i \in C, \forall c \in YC$ $\hspace{1.5cm}$ (2-14)

$t_i = \sum_{k \in L_u} \left(tt_k * x_{i,k}\right)$ $\quad \forall i \in C_u$ $\hspace{1.5cm}$ (2-15)

$sy_{i,j} = \sum_{m \in L_u} \left(tyc_{m,l_j} * x_{i,m}\right)$ $\quad \forall i \in C_u, \forall j \in C_l$ $\hspace{1.5cm}$ (2-16)

$sy_{i,j} = \sum_{m \in L_u} \sum_{l \in L_u} \left(tyc_{m,l} * x_{i,m} * x_{j,l}\right)$ $\quad \forall i,j \in C_u$ $\hspace{1.5cm}$ (2-17)

$sy_{i,j} = \sum_{m \in L_u} \left(tyc_{i,m} * x_{j,m}\right)$ $\quad \forall i \in C_l, \forall j \in C_u$ $\hspace{1.5cm}$ (2-18)

$sqc_j + M \left(1 - z_{i,j}^q\right) \geq sqc_i + Q_i + eqc_{i,j}$ $\quad \forall i,j \in C, \forall q \in QC$ $\hspace{1cm}$ (2-19)

$syc_j + M \left(1 - v_{i,j}^c\right) \geq syc_i + Y_i + sy_{i,j}$ $\quad \forall i \in C_u, \forall j \in C, \forall c \in YC$ $\hspace{0.5cm}$ (2-20)

$syc_j + M \left(1 - v_{i,j}^c\right) \geq syc_i + Y_i + sy_{i,j}$ $\quad \forall i \in C_l, \forall j \in C_u, \forall c \in YC$ $\hspace{0.3cm}$ (2-21)

$syc_j + M \left(1 - v_{i,j}^c\right) \geq syc_i + Y_i + eyc_{i,j}$ $\quad \forall i,j \in C_l, \forall c \in YC$ $\hspace{0.5cm}$ (2-22)

$sqc_i \geq syc_i + Ytime_i + tyt_i$ $\quad \forall i \in C_l$ $\hspace{1.5cm}$ (2-23)

$syc_i \geq sqc_i + Qtime_i + t_i$ $\quad \forall i \in C_u$ $\hspace{1.5cm}$ (2-24)

$sqc_i + Q_i - sqc_j \leq M \left(1 - qz_{i,j}\right)$ $\quad \forall i,j \in C$ $\hspace{1.5cm}$ (2-25)

$\sum_{u \in C^0} z_{u,i}^v + \sum_{u \in C^0} z_{u,j}^w \leq 1 + qz_{i,j} + qz_{j,i}$ $\quad \forall (i,j,v,w) \in \Theta$ $\hspace{0.8cm}$ (2-26)

$sqc_i + Q_i + \Delta_{i,j}^{v,w} - sqc_j \leq M \left(3 - qz_{i,j} - \sum_{u \in C^0} z_{u,i}^v - \sum_{u \in C^0} z_{u,j}^w\right)$ $\quad \forall (i,j,v,w) \in \Theta$

$\hspace{12cm}$ (2-27)

$sqc_j \geq sqc_i$ $\quad \forall i,j \in P$ $\hspace{1.5cm}$ (2-28)

5 The Constraint Programming Model

The CP model is presented in Model 3 using the OPL API of CP Optimizer. It uses interval variables for representing the QC handling of all shipments and the YC handling of inbound shipments. In addition, optional interval variables represent the handling of shipment i on QC j and the handling of shipment i at yard location k. The model also declares a number of sequence variables associated with each QC and YC: Each sequence constraint collects all the optional interval variables associated with a specific crane. Finally, the model declares a number of sequences for optional interval variables that may interfere.[1]

Model 3. The CP model for the IPCTP

Variables

qc_i : Interval variable for the QC handling of shipment i

yt_i : Interval variable for the YT handling of inbound shipment i

$aqc_{i,j}$: Optional interval variable for shipment i on QC j with duration Q_i

$ayc_{i,k}$: Optional interval variable for shipment i on YC at yard location k with duration Y_i

qcs_j : Sequence variable for QC j over $\{aqc_{i,j} \mid i \in C\}$

ycs_j : Sequence variable for YC j over $\{ayc_{i,k} \mid i \in C \wedge YC(k) = j\}$

$interfere_{i,v,j,w}$: Sequence variable over $\{aqc_{i,v}, aqc_{j,w}\}$

Objective

$$\text{minimize } \sum_{s \in S} w_s \left(max \left(\max_{i \in C_u^s} \text{ENDOF}(yt_i), \max_{j \in C_l^s} \text{ENDOF}(qc_j) \right) \right) \tag{3-01}$$

Constraints

$$\text{ALTERNATIVE}(qc_i, \text{all}(j \text{ in } QC(i)) \, aqc_{i,j}) \quad \forall i \in C \tag{3-02}$$

$$\text{ALTERNATIVE}(yt_i, \text{all}(k \text{ in } L_u) \, ayc_{i,k}) \quad \forall i \in C_u \tag{3-03}$$

$$\sum_{i \in C_u} \text{PRESENCEOF}(ayc_{i,k}) \leq 1 \quad \forall k \in L_u \tag{3-04}$$

$$\text{PRESENCEOF}(ayc_{i,l_i}) = 1 \quad \forall i \in C_l \tag{3-05}$$

$$\text{NOOVERLAP}(ycs_m, eyc_{i,j}) \quad \forall m \in YC \tag{3-06}$$

$$\text{NOOVERLAP}(qcs_m, eqc_{i,j}) \quad \forall m \in QC \tag{3-07}$$

$$\text{STARTBEFORESTART}(aqc_{i,n}, ayc_{i,k}, Qtime_i + tt_k) \quad \forall i \in C_u, k \in L_u, n \in qc_i \tag{3-08}$$

$$\text{STARTBEFORESTART}(ayc_{i,l_i}, aqc_{i,n}, Ytime_i + tyt_i) \quad \forall i \in C_l, n \in qc_i \tag{3-09}$$

$$\text{STARTBEFORESTART}(aqc_{i,m}, aqc_{j,n}) \quad \forall i,j \in P, m \in qc_i, n \in qc_j \tag{3-10}$$

$$\text{NOOVERLAP}\left(interfere_{i,v,j,w}, \Delta_{i,j}^{v,w}\right) \quad \forall i,j \in C, v \in qc_i, w \in qc_j : \Delta_{i,j}^{v,w} > 0 \tag{3-11}$$

The CP model minimizes the weighted completion time of each vessel by computing the maximum end date of the yard cranes (inbound shipments) and for the quay crane (outbound shipments). Alternative constraints (3-02) ensure

---

[1] Not all such sequences are useful but we declare them for simplicity.

that the QC processing of a shipment is performed by exactly one QC. Alternative constraints (3-03) enforce that each inbound shipment is allocated to exactly one yard location, and hence one yard crane. Constraints (3-04) state that at most one shipment can be allocated to each yard location. Constraints (3-05) fix the yard location (and hence the yard crane) of outbound shipments. Cranes are disjunctive resources and can execute only one task at a time, which is expressed by the NOOVERLAP constraints (3-06–3-07) over the sequence variables associated with the cranes. These constraints also enforce the transition times between successive operations, capturing the empty travel times between yard locations (constraints 3-06) and bay locations (constraints 3-07). Constraints (3-08) impose the precedence constraints between the QC and YC tasks of inbound shipments, while adding the travel time to move the shipment from its bay to its chosen yard location. Constraints (3-09) impose the precedence constraints between the YC and QC operations of outbound shipments, adding the travel time from the fixed yard location to the fixed bay of the shipment. Constraints (3-10) impose the precedences between shipments. Interference constraints for the QCs are imposed by constraints (3-11). These constraints state that, if there is a conflict between two shipments and their QCs, then the two shipments cannot overlap in time, and their executions must be separated by a minimum time. This is expressed by NOOVERLAP constraints over sequences consisting of the pairs of optional variables associated with these tasks.

6   Experimental Results

The MIP and CP models were written in OPL and run on the IBM ILOG CPLEX 12.8 software suite. The results were obtained on an Intel Core i7-5500U CPU 2.40 GHz computer.

6.1   Data Generation

The test cases were generated in accordance with earlier work, while capturing the operations of an actual container terminal. These are the first instances of this type, since the IPCTP has not been considered before in the literature. Travel and processing times in the test cases model those in the actual terminal. Different instances have slightly different times, as will become clear.

Figure 3 depicts the layout of the yard side considered in the experiments, which also models the actual terminal. The yard side is divided into 3 separate fields denoted by A, B, and C. Each field has two location areas and a single YC is responsible for each location area, giving a total of 6 yard cranes. In each location area, there are 2 yard block groups, shown as the dotted region, and traveling between them takes one unit of time. Field C is the nearest to the quay side, and there is a hill from field C to field A. The transportation times for YTs are generated according to these distances. YTs can enter and exit each field from the entrance shown in the figure, so the transfer times between the vessels and the yard blocks close to the entrance take less time. YT transfer times are

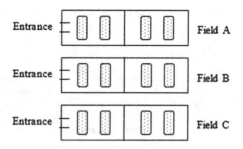

Fig. 3. Layout of the yard side.

generated between [5, 10] considering the position of the yard blocks. At the quay side, the travel of a QC between consecutive vessel bay locations takes 3 units of time.

The processing times of the cranes for a single container are generated uniformly in [2, 5] for YCs and [2, 4] for QCs. The safety margin for the QCs is set to 1 vessel bay. The IPCTP is expressed in terms of shipments and the number of containers in each shipment is uniformly distributed between [4, 40].

The experiments evaluate the impact of the number of shipments, the number of vessel bays, the inbound-outbound shipment ratio, and the number of available yard locations for inbound shipments. The number of shipments varies between 5 and 25, by increments of 5. The instances can thus contain up to 1,000 containers. The number of vessel bays are taken in {4, 6, 8}. The number of QCs depends on the vessel bays due to the QC restrictions: There are half as many QCs as there are vessel bays. The inbound-outbound shipment ratios are 20% and 50%, representing the fraction of inbound shipments over the outbound shipments. Finally, the number of available yard locations (U-L ratio) is computed from the number of inbound shipments: There are 2 to 3 times more yard locations than inbound shipments. For each configuration of the parameters, 5 random instances were generated.

6.2   Computational Results and Analysis

The Results. The results are given in Tables 2 and 3 for each configuration, for a total of 300 instances. Table 2 reports the results for 20% inbound-outbound ratio, and Table 3 for 50%. In the tables, each configuration is specified in terms of the U-L ratio, the number of bays and the number of shipments (Shp.). The average number of containers (Cnt.) in each shipment is also presented. For each such configuration, the tables report the average objective value and the average CPU time for its five instances. The CPU time is limited to an hour (3,600 s). The MIP solver did not always find feasible solutions within an hour for some or all five instances of a configuration. Note that this may result in an average objective value that is lower for the MIP model than the CP model, even when CP solves all instances optimally, since the MIP model may not find a feasible

Table 2. Results for import-export rate 20%

U-L Ratio	# of Bays	# of Shp.	Avg. # of Cont.	MIP			CP		RPD%
				Obj.	CPU (sec.)	GAP%	Obj.	CPU (sec.)	
2	4	5	126.2	292.80	0.40	0.00	292.80	0.05	0.00
		10	224.4	$541.60^{0/5}$	3600.05	0.36	541.60	9.41	0.00
		15	368.8	$739.80^{0/5}$	3601.68	0.69	739.80^{1}	734.91	0.00
		20	459.8	$822.40^{0/5}$	3601.39	0.75	820.00^{1}	788.81	0.00
		25	563.6	$1018.00^{2/3}$	3600.81	0.77	998.40^{2}	1791.34	0.00
	6	5	83.0	179.20	0.16	0.00	179.20	0.14	0.00
		10	223.8	$359.80^{0/1}$	1717.16	0.05	359.80	6.20	0.00
		15	347.8	$544.40^{0/5}$	3600.50	0.58	541.00	36.47	0.00
		20	400.0	$660.75^{1/4}$	3600.75	0.67	647.40	274.15	0.00
		25	577.0	NA	NA	NA	752.20	479.15	0.00
	8	5	127.2	427.00	0.16	0.00	426.80	0.40	0.00
		10	190.0	507.20	98.30	0.00	506.90	10.26	0.00
		15	315.0	$843.60^{0/5}$	3600.26	0.48	841.10	150.59	0.00
		20	427.8	$967.33^{2/3}$	3600.13	0.57	1052.40	542.60	0.00
		25	540.4	$1466.00^{4/1}$	3614.05	0.64	1320.30	1255.55	1.26
3	4	5	98.6	206.60	0.40	0.00	206.60	0.21	0.00
		10	205.8	$425.60^{0/4}$	3297.84	0.31	425.60	6.80	0.00
		15	336.0	$642.60^{0/5}$	3605.24	0.65	640.60	79.30	0.00
		20	447.6	$864.80^{0/5}$	3600.71	0.75	859.80^{1}	897.30	0.00
		25	506.8	$824.00^{3/2}$	3601.27	0.73	841.60^{2}	1895.17	0.00
	6	5	125.0	247.80	0.16	0.00	247.80	0.15	0.55
		10	237.8	$401.40^{0/1}$	1449.26	0.08	401.40	8.70	0.00
		15	299.0	$482.40^{0/1}$	3600.39	0.50	482.00	539.06	0.00
		20	476.6	$625.50^{1/4}$	3600.50	0.65	613.80^{1}	778.93	0.00
		25	542.6	$988.00^{2/3}$	3600.22	0.74	855.60^{4}	2938.55	0.00
	8	5	112.4	418.20	0.22	0.00	418.20	1.27	0.00
		10	232.4	655.20	449.44	0.00	655.20	26.02	0.00
		15	329.2	$894.60^{0/5}$	3600.45	0.47	881.50	160.79	0.00
		20	422.0	$1272.00^{1/4}$	3611.86	0.63	1112.70^{1}	1524.57	0.38
		25		NA	NA	NA	1381.90^{3}	2681.45	2.87

solution to an instance with a high optimal value. These cases are flagged by superscripts of the form x/y, where x is the number of infeasible and y is the number of suboptimal solutions in that average. The superscripts for CP indicate the number of suboptimal solutions. An entry 'NA' in the table means that the MIP cannot find a feasible solution to any of the five instances. For the MIP, the tables also report the optimality gap on termination, i.e., the gap in percentage between the best lower and upper bounds.

For CP, the experiments were also run with a CPU limit of 600 s. The relative percentage deviations (RPD%) from the 1-h runs are listed to assess CP's ability to find high-quality solutions quickly. The RPD is computed as follows:

$$RPD\% = \frac{(\text{Obj. in 600 s} - \text{Obj. in 3600 s}) * 100}{(\text{Obj. in 3600 s})}.$$

Table 3. Results for import-export rate 50%

U-L Ratio	# of Bays	# of Shp.	# of Cont.	MIP			CP		
				Obj.	CPU (sec.)	GAP%	Obj.	CPU (sec.)	RPD%
2	4	5	101.2	217.80	0.43	0.00	217.80	0.07	0.00
		10	214.8	$439.60^{0/5}$	3600.16	0.51	439.60	5.35	0.00
		15	349.0	$711.40^{0/5}$	3601.40	0.68	710.80	723.95	0.00
		20	394.2	$801.00^{3/2}$	3600.28	0.75	743.60^2	1997.88	0.00
		25	552.0	NA	NA	NA	1030.60^4	3368.38	0.00
	6	5	103.8	236.00	0.40	0.00	236.00	0.20	0.00
		10	228.4	$364.00^{0/5}$	3600.26	0.30	363.40	3.62	0.00
		15	331.4	$513.00^{0/5}$	3600.72	0.58	512.40	143.53	0.00
		20	428.2	$684.67^{2/3}$	3601.16	0.69	698.00^2	1554.96	0.00
		25	523.8	$1341.00^{4/1}$	3600.61	0.82	762.40^4	2953.04	0.00
	8	5	107.0	400.40	0.31	0.00	400.00	1.40	0.00
		10	229.6	$719.40^{0/2}$	1702.43	0.08	719.10	21.13	0.00
		15	318.0	$870.75^{1/4}$	3600.88	0.49	861.40	96.71	0.00
		20	457.6	$1315.50^{3/2}$	3600.31	0.60	1125.10	384.26	0.00
		25	570.2	$4250.00^{4/1}$	3601.03	0.91	1373.10^1	1692.20	2.51
3	4	5	110.8	255.00	0.36	0.00	255.00	0.42	0.00
		10	218.8	$390.20^{0/5}$	3603.10	0.47	390.20	6.56	0.00
		15	333.2	$746.00^{0/5}$	3600.78	0.76	743.60	123.28	0.00
		20	462.0	$1355.00^{3/2}$	3600.50	0.83	836.60^1	1025.25	0.00
		25	526.6	NA	NA	NA	857.20^4	3103.01	0.00
	6	5	118.4	244.80	0.46	0.00	244.80	0.08	0.00
		10	226.0	$401.60^{0/4}$	3491.75	0.34	401.00	6.73	0.00
		15	303.2	$481.60^{0/5}$	3600.90	0.55	481.00	129.67	0.00
		20	444.6	$25557.00^{3/2}$	3600.53	0.90	622.40	363.12	0.00
		25	571.6	NA	NA	NA	798.40^4	3073.44	0.00
	8	5	107.2	348.00	0.20	0.00	347.80	0.56	0.00
		10	215.2	$659.20^{0/2}$	2262.06	0.09	659.20	33.83	0.00
		15	329.4	$771.00^{1/4}$	3601.29	0.41	776.10	204.93	0.00
		20	472.4	$2430.67^{2/3}$	3600.60	0.80	1184.80^1	1527.23	0.57
		25	544.4	$3798.00^{4/1}$	3600.61	0.91	1315.20^1	2226.52	1683.57

MIP Versus CP. The experimental results indicate that CP is orders of magnitude more efficient than MIP on the IPCTP. This is especially remarkable since this paper compares two black-box solvers. Overall, within the time limit, the MIP model does not find feasible solutions for 71 out of 300 instances and cannot prove optimality for 214 instances. On all but the smallest instances, the MIP solver cannot prove optimality for all five instances of the same configuration. In almost all configurations with 20 or more shipments, the MIP solver fails to find feasible solutions on at least one of the instances. In contrast, the CP model always find feasible solutions and proves optimality for 260 instances out of 300 instances. CP proves optimality on all but 16 instances in Table 3, and all but 24 in Table 2. On instances where both models find optimal solutions, the CP model is almost always 1–3 orders of magnitude faster (except for the smallest instances). Of course, the benefits are even more substantial when the MIP fails to find feasible and/or optimal solutions. For instance, on instance 3/6/15, CP proves optimality on all five instances in 2 min, while the MIP cannot find the optimal solution to any of them in an hour. Finally, the CP model always dominates the MIP model: It proves optimality every time the MIP does.

Short Runs. On all configurations but one (3/8/15), the CP model finds optimal, or near optimal, solutions within 10 min, which is significant in practice.

Sensitivity Analysis. The sensitivity analysis is restricted to the CP model for obvious reasons. The sensitivity of each factor is analyzed by comparing their respective run times and objective values. In general, the effect of the number of bays on the solution values and on CPU times tend to be small. In contrast, increasing the U-L ratio from 2 to 3 gives inbound shipments more alternatives for yard locations, which typically increases CPU times. Increasing the ratio of inbound-outbound containers also increases problem difficulty. This is not a surprise, since inbound shipments are more challenging, as they require a yard location assignment, while outbound shipments have both their yard locations and vessel bays fixed. Nevertheless, the CP model scales reasonably well when this ratio is increased. These analyses indicate that the number of shipments/containers is by far the most important element in determining the computing times in the IPCTP: The other factors have a significantly smaller impact, which is an interesting result in its own right.

7   Conclusion

This paper introduced the Integrated Port Container Terminal Problem (IPCTP) which, to the best of our knowledge, integrates for the first time, a wealth of port operations, including the yard location assignment, the assignment of quay and yard cranes, and the scheduling of these cranes under realistic constraints. In particular, the IPCTP considers empty travel time of the equipment and interference constraints between the quay cranes. The paper proposed both an MIP and a CP model for the IPCTP of a configuration based on an actual container terminal, which were evaluated on a variety of configurations regarding the number of vessel bays, the number of yard locations, the ratio of inbound-outbound shipments, and the number of shipments/containers. Experimental results indicate that the MIP model can only be solved optimally for small instances and often cannot find feasible solutions. The CP model finds optimal solutions for 87% of the instances and, on instances where both models can be solved optimally, the CP model is typically 1–3 orders of magnitude faster and proves optimality each time the MIP does. The CP model scales reasonably well with the number of vessel bays and yard locations, and the ratio of inbound-outbound shipments. It also solves large realistic instances with hundreds of containers. These results contrast with the existing literature which typically resort to heuristic or meta-heuristic algorithms, with no guarantee of optimality.

Future work will be devoted to capturing a number of additional features, including operator-based processing times, the stacking of inbound containers using re-shuffling operations, the scheduling of the yard trucks, and larger container operations.

References

1. Bierwirth, C., Meisel, F.: A follow-up survey of berth allocation and quay crane scheduling problems in container terminals. Eur. J. Oper. Res. **244**, 675–689 (2015)
2. Bierwirth, C., Meisel, F.: A fast heuristic for quay crane scheduling with interference constraints. J. Sched. **12**(4), 345–360 (2009)
3. Chen, L., Bostel, N., Dejax, P., Cai, J., Xi, L.: A tabu search algorithm for the integrated scheduling problem of container handling systems in a maritime terminal. Eur. J. Oper. Res. **181**, 40–58 (2007)
4. Chen, L., Langevin, A., Lu, Z.: Integrated scheduling of crane handling and truck transportation in a maritime container terminal. Eur. J. Oper. Res. **225**(1), 142–152 (2013)
5. Homayouni, S.M., Vasili, M.R., Kazemi, S.M., Tang, S.H.: Integrated scheduling of SP-AS/RS and handling equipment in automated container terminals. In: Proceedings of International Conference on Computers and Industrial Engineering, CIE, vol. 2 (2012)
6. Homayouni, S.M., Tang, S.H., Motlagh, O.: A genetic algorithm for optimization of integrated scheduling of cranes, vehicles, and storage platforms at automated container terminals. J. Comput. Appl. Math. **270**(Supplement C), 545–556 (2014)
7. Homayouni, S., Tang, S.: Multi objective optimization of coordinated scheduling of cranes and vehicles at container terminals (2013)
8. Kaveshgar, N., Huynh, N.: Integrated quay crane and yard truck scheduling for unloading inbound containers. Int. J. Prod. Econ. **159**(Supplement C), 168–177 (2015)
9. Lau, H.Y., Zhao, Y.: Integrated scheduling of handling equipment at automated container terminals. Int. J. Prod. Econ. **112**(2), 665–682 (2008)
10. Liang, L., Lu, Z.Q., Zhou, B.H.: A heuristic algorithm for integrated scheduling problem of container handling system. In: 2009 International Conference on Computers Industrial Engineering, pp. 40–45. IEEE (2009)
11. Lu, Y., Le, M.: The integrated optimization of container terminal scheduling with uncertain factors. Comput. Ind. Eng. **75**(Supplement C), 209–216 (2014)
12. Moccia, L., Cordeau, J.F., Gaudioso, M., Laporte, G.: A branch-and-cut algorithm for the quay crane scheduling problem in a container terminal. Naval Res. Logist. (NRL) **53**(1), 45–59 (2006)
13. Niu, B., Xie, T., Tan, L., Bi, Y., Wang, Z.: Swarm intelligence algorithms for yard truck scheduling and storage allocation problems. Neurocomputing **188**(Supplement C), 284–293 (2016)
14. Sammarra, M., Cordeau, J.F., Laporte, G., Monaco, M.F.: A tabu search heuristic for the quay crane scheduling problem. J. Sched. **10**(4), 327–336 (2007)
15. Tang, L., Zhao, J., Liu, J.: Modeling and solution of the joint quay crane and truck scheduling problem. Eur. J. Oper. Res. **236**(3), 978–990 (2014)
16. Vis, I., de Koster, R.: Transshipment of containers at a container terminal: an overview. Eur. J. Oper. Res. **147**, 1–16 (2003)
17. Wu, Y., Luo, J., Zhang, D., Dong, M.: An integrated programming model for storage management and vehicle scheduling at container terminals. Res. Transp. Econ. **42**(1), 13–27 (2013)
18. Xin, J., Negenborn, R.R., Corman, F., Lodewijks, G.: Control of interacting machines in automated container terminals using a sequential planning approach for collision avoidance. Transp. Res. Part C: Emerg. Technol. **60**(Supplement C), 377–396 (2015)

19. Xin, J., Negenborn, R.R., Lodewijks, G.: Energy-aware control for automated container terminals using integrated flow shop scheduling and optimal control. Transp. Res. Part C: Emerg. Technol. **44**(Supplement C), 214–230 (2014)
20. Xue, Z., Zhang, C., Miao, L., Lin, W.H.: An ant colony algorithm for yard truck scheduling and yard location assignment problems with precedence constraints. J. Syst. Sci. Syst. Eng. **22**(1), 21–37 (2013)
21. Zheng, K., Lu, Z., Sun, X.: An effective heuristic for the integrated scheduling problem of automated container handling system using twin 40' cranes. In: 2010 Second International Conference on Computer Modeling and Simulation, pp. 406–410. IEEE (2010)

Heuristic Variants of A* Search for 3D Flight Planning

Anders N. Knudsen, Marco Chiarandini, and Kim S. Larsen$^{(\boxtimes)}$

Department of Mathematics and Computer Science,
University of Southern Denmark, Campusvej 55, 5230 Odense M, Denmark
{andersnk,marco,kslarsen}@imada.sdu.dk

Abstract. A crucial component of a flight plan to be submitted for approval to a control authority in the pre-flight phase is the prescription of a sequence of airways and airway points in the sky that an aircraft has to follow to cover a given route. The generation of such a path in the 3D network that models the airways must respect a number of constraints. They generally state that if a set of points or airways is visited then another set of points or airways must be avoided or visited. Paths are then selected on the basis of cost considerations. The cost of traversing an airway depends, directly, on fuel consumption and on traversing time, and, indirectly, on weight and on weather conditions.

Path finding algorithms based on A* search are commonly used in automatic planning. However, the constraints and the dependency structure of the costs invalidate the classic domination criterion in these algorithms. A common approach to tackle the increased computational effort is to decompose the problem heuristically into a sequence of horizontal and vertical route optimizations. Using techniques recently designed for the simplified 2D context, we address the 3D problem directly. We compare the direct approach with the decomposition approach. We enhance both approaches with ad hoc heuristics that exploit the expected appeal of routes to speed-up the solution process. We show that, on data resembling those arising in the context of European airspaces, the direct approach is computationally practical and leads to results of better quality than the decomposition approach.

1 Introduction

The Flight Planning Problem (FPP) aims at finding, for a given aircraft, a sky trajectory and an initial fuel load, minimizing the total cost determined by fuel consumption and travel time. The sky is subdivided into airspaces where airway points and airways between them are predefined. Thus, a route, denoted by origin and destination airports, corresponds to a path in a network that models the 3D space. The path starts at the origin airport, climbs to a favorable altitude and finally descends to the destination airport while satisfying different types

K. S. Larsen—Supported in part by the Independent Research Fund Denmark, Natural Sciences, grant DFF-7014-00041.

© Springer International Publishing AG, part of Springer Nature 2018
W.-J. van Hoeve (Ed.): CPAIOR 2018, LNCS 10848, pp. 361–376, 2018.
https://doi.org/10.1007/978-3-319-93031-2_26

of constraints. There are clear operational, financial, and environmental motivations for aiming at feasible and cost-optimal paths. However, the size of airway networks, the nature of a large number of the constraints, and the dependencies among different factors that affect costs make the problem substantially more difficult to solve compared with classic shortest path problems in road networks. In addition, some of the constraints from central control authorities (Eurocontrol in Europe or FAA in USA) are issued to take rapidly changing traffic conditions into account. Thus, they are updated frequently, and airlines, and to an even larger extent small private airplanes, tend to determine the precise flight route only a few hours before take-off, when more constraints are available. Although the route can be adjusted during operation, doing so will lead to a suboptimal route compared with a complete route determined offline. As take-off approaches, available processing time will be on the order of a few seconds.

The cost of flying through an airway depends on the time when the traversal occurs and the fuel consumed so far. There is a direct dependency between the two, because airlines calculate the total cost as a weighted sum of total travel time and total fuel consumption. There is also an indirect dependency, because the time spent and amount of fuel consumed on an arc depend on the performance of the aircraft, which is influenced by: (i) the weather conditions, which in turn depend on the time when the arc is traversed, and (ii) the weight, which in turn depends on the fuel consumed so far. A consequence of these dependencies is that the cost of each arc of the airway network is not statically given but becomes known only when the path to a node is determined. A further complication is that while the departure time is fixed, the initial amount of fuel is to be decided. This cost structure leads to issues similar to time dependencies in shortest path on road networks [1].

The indirect cost dependencies together with the impossibility of waiting at the nodes, even at the departure airport, mean that it might be a disadvantage to arrive cheaply at a node (and hence earlier or lighter than other alternatives) since it may preclude obtaining high savings on the remaining path, due to favorable weather conditions developing somewhere at a later time. In other words, the so-called First-In-First-Out (FIFO) property cannot be assumed to hold. For labeling algorithms typically used to approach routing problems, Dijkstra [2] and A* [3], this means that a total ordering of the labels at the nodes is not available, which has a tremendous impact on efficiency.

Further complexity is added by the quantity and type of constraints. The most challenging ones are Route Availability Document (RAD) constraints. They include local constraints affecting the availability of airways and airspaces at certain times, but most of them are conditional constraints. For example, if the route comes from a given airway, then it must continue through another airway. Or some airways can only be used if coming from, or arriving to, certain airspaces. Or flights between some locations are not allowed to cross certain airspaces at certain flight levels. Or short-haul and long-haul flights are segregated in congested zones.

Due to increasing strain on the European airspace, the networks become more heavily constrained to ensure safe flights. There are typically more than 16,000 constraints in the European network and they can be updated several times a day. Some of the RAD constraints are generalizations of *forbidden pairs*, which make the problem at least as hard as the *path avoiding forbidden pairs* problem, shown to be NP-hard [4]. Additionally, RAD constraints invalidate the FIFO assumption, with far-reaching effects in terms of complexity. Therefore, a common approach in industry is to implement a first stage, where the algorithms ignore some constraints that are difficult to formalize and check during route construction, and enforce them later by somehow repairing the route. All constraints are verified by the control authority when a route is submitted for approval.

A common approach in industry is to solve the FPP in a decomposed form: find a horizontal 2D route and then determine a vertical profile for it. While this reduces the computation cost considerably, it does not guarantee optimality. However, due to the practical relevance, recent research has focused on these two simplified problems. In [5], the authors address the horizontal 2D flight planning problem without constraints and study methods for calculating weather-dependent estimates for the heuristic component of A* algorithms. They show that with these estimates, A* achieves better experimental results in both preprocessing and query time than contraction hierarchies [6], a technique from state-of-the-art algorithms for shortest path on road networks. In [7], we proposed a framework for handling the RAD constraints in the horizontal 2D problem. We introduced a tree representation for constraints to maintain them during path construction, and compared lazy techniques for efficiency. Experimental work suggested that the best running times were obtained using an approach similar to a logic-based Benders decomposition with no-good cuts, in which we ignore constraints in the initial search and only add those that are violated in the subsequent searches. In [8], we focused on the impact of the FIFO assumption on the vertical route optimization problem, constrained to a given horizontal route, but not considering constraints. We showed that wrt. cost considerations, heuristically assuming the FIFO property in vertical routing is unlikely to lead to suboptimal solutions.

We extend our work on constraint handling to the 3D context to show that an A* approach can be practical in solving the FPP directly in a 3D network. With respect to our 2D work [7], the networks we consider increase in size from 11,000 nodes and 1,000,000 arcs to 200,000 nodes and 124,000,000 arcs. We design heuristics to improve the efficiency of the algorithms and empirically study the quality loss that they imply. We focus on two types of heuristics: those affecting the estimate of the remaining cost from a node to the goal in the A* framework and those using the desired shape of the vertical profile to prune label expansion in the A* algorithm. We present the comparison of the most relevant among these heuristics and combinations. We include the comparison with the obvious consequent outcome from our previous research, namely a two-phase approach with constraints handled lazily by restart as in [7] and costs as in [8]. Our results demonstrate that a direct 3D approach leads to better results in terms of quality, with computation times that remain suitable for practical needs.

There are elements from real-life problems we ignore: First, we do not consider the initial fuel as a decision variable. We assume this value is given on the basis of historical data. In practice, we would approach the problem using a line-search method as in [8]. Second, we do not consider (possibly non-linear) overflight costs associated with some airspaces. We refer to [9] for a treatment of these issues; authors suggest a cost projection method to anticipate the cost incurred by the overflight and then eliminate potential paths. Finally, not all airspaces have predefined airway points; free route airspaces lead to algorithms exploiting geometric properties, and they have been shown to be more efficient than graph-based ones [10] in those scenarios. These algorithms could in principle be integrated with the framework presented here, calling them as subroutines when an entry point to a free route airspace is expanded.

This work is in collaboration with an industrial partner. Many of its customers are owners of private planes who plan their flights shortly before departure and expect almost immediate answers to the portable device. Thus, a query should be answered in a few seconds.

2   The Flight Planning Problem

The 3D airspace is represented by *waypoints* that can be traversed at different altitudes (*flight levels*). Waypoints are connected by *airways*. This gives rise to a network in the form of a directed graph $D = (V, A)$. The nodes in V represent waypoints at different latitude, longitude, and altitude. The arcs in A connect all nodes that can be traced back to two different waypoints connected by an airway and whose implied difference in altitude can be operated by the aircraft. Each arc has associated resource consumption and costs. The resource consumption for flying via an arc $a \in A$ is defined by a pair $\tau_a = (\tau_a^x, \tau_a^t) \in \mathbb{R}_+^2$, where the superscripts x and t denote the fuel and time components, respectively. The cost c_a is a function of the resource consumption, i.e., $c_a = f(\tau_a)$.[1] A 3D (*flying*) *route* is an (s, g)-path in D, represented by n nodes, a departure node (source) s, and an arrival node (goal) g, that is, $P = (s, v_1, \ldots, v_n, g)$, with $s, v_i, g \in V$ for $i = 1...n$, $v_i v_{i+1} \in A$ for $i = 1...n-1$, and $sv_1, v_n g \in A$. The cost of the route P is defined as $c_P = c_{sv_1} + \sum_{i=1...n-1} c_{v_i v_{i+1}} + c_{v_n g}$.

The route must satisfy a set $\mathcal{C}$ of *constraints* imposed on the path. These constraints are of the following type: if a set of nodes or arcs A is visited, then another set of nodes or arcs B must be avoided or visited. The visit or avoidance of the sets A and B can be further specified by restrictions on the order of nodes in the route, on the time window, and on the flight level range.

Definition 1 (Flight Planning Problem (FPP)). *Given a network $N = (V, A, \tau, c)$, a departure node s, an arrival node g, and a set of constraints $\mathcal{C}$,*

---

[1] The total cost is calculated as a weighted sum of time and fuel consumed. In our specific case, we have used 1.5\$ per gallon of fuel and 1000\$ per hour.

find an (s,g)-path P in $D = (V,A)$ that satisfies all constraints in C and that minimizes the total cost, c_P.

All constraints can be categorized into two classes: *forbidden* and *mandatory*. Constraints consist of an *antecedent* and a *consequent* expression, p and q. A constraint of the mandatory type is *satisfied* when $p \to q$ and *violated* otherwise; one of the forbidden type is *satisfied* when $p \to \neg q$ and *violated* otherwise. The expressions p and q give Boolean values. They contain terms that express possible path choices, such as passing through a node (representing a specific flight level at an airway point), traversing an arc (possibly belonging to an airspace), departing or arriving at a node, or visiting a set of nodes in a given order.

3   A* Search with Constraints

We use A* search as our base algorithm, and the following strategy, detailed and shown effective in [7], to handle the numerous constraints: First, we find a path using A*, ignoring all constraints. Then, the path obtained is checked against all constraints. If no constraint is violated, the path is feasible and the procedure terminates. Otherwise, the violating constraints are added to the input and a new search is initiated. The new path will not violate any of the included constraints but it could violate others. Thus, the procedure is repeated until a feasible path is found. In the iterations that follow the initial one, the search must handle the constraints during the construction. A template for A* modified to handle constraints is given in Algorithm 1.

In Algorithm 1, the function FINDPATH takes the initial conditions as input: a network $N = (V, A, \boldsymbol{\tau}, \boldsymbol{c})$ built using information from the airspace, aircraft performance data, and weather conditions; the query from s to g; the initial fuel load τ_s^x and the departure time τ_s^t; an array $\boldsymbol{h}$ of values $h(v)$ for every node $v \in V$, indicating the precomputed, static, *estimated cost* of flying from that node v to the destination node g; the set of constraints (see below). The time and fuel consumptions for each arc $\boldsymbol{\tau}$ depend on: (i) the flight level, (ii) the weight, (iii) international standard atmosphere deviation (temperature), (iv) the wind component, and (v) the cost index.[2] Inputs (ii), (iii), and (iv) depend on the time of arrival at the arc and hence on the partial path. We can regard these values as retrieved from data tables but we refer to Sect. 5 for more details.

We represent partial paths under construction in Algorithm 1 by labels. A *label* ℓ is associated with a node $\phi(\ell) = u \in V$ and contains information about a partial path from the departure node $s \in V$ to the node u, that is, $P_\ell = (s, ..., u)$.

Labels also have constraints associated. We represent constraints as trees, where leaves are terms and internal nodes are logical operators (AND, OR, NOT). Constraint trees are, potentially, associated with each label. Initially, all constraint terms are in an unknown state and labels have no constraints associated. Then, if during the extension of a path of a label, a term of a constraint

---

[2] The cost index is an efficiency ratio between the time-related cost and the fuel cost, decided upon at a strategic level and unchangeable during the planning phase.

```
 1  Function FINDPATH(N = (V, A, τ, c), s, g, (τ_s^x, τ_s^t), h, Γ)
 2  │   initialize the open list Q by inserting ℓ_s = ((s), 0, {})
 3  │   initialize ℓ_r = ((), ∞, {})
 4  │   while Q is not empty do
 5  │   │   ℓ ← retrieve and remove the cheapest label from Q
 6  │   │   if (c_ℓ + h(φ(ℓ))) > c_{ℓ_r}) then break                    ▷ termination criterion
 7  │   │   if (φ(ℓ) = g) and (c_ℓ < c_{ℓ_r}) then
 8  │   │   │   ℓ_r ← ℓ
 9  │   │   │   continue
10  │   │   foreach node v such that uv in A do
11  │   │   │   ℓ' ← label at v expanded from ℓ
12  │   │   │   evaluate constraint trees in Δ_{ℓ'}
13  │   │   │   if one or more constraints in Δ_{ℓ'} are violated then
14  │   │   │   │   continue
15  │   │   │   INSERT(ℓ',Q)
16  │   return P_{ℓ_r} and c_{ℓ_r}

17  Function INSERT(ℓ', Q)
18  │   foreach label ℓ ∈ Q with φ(ℓ) = φ(ℓ') do
19  │   │   if (c_ℓ > c_{ℓ'}) then
20  │   │   │   if (Δ_ℓ is implied by Δ_{ℓ'}) then                    ▷ ℓ is dominated
21  │   │   │   │   remove ℓ from Q
22  │   │   else if (c'_ℓ > c_ℓ) then
23  │   │   │   if (Δ_ℓ is implied by Δ_{ℓ'}) then return              ▷ ℓ' is dominated
24  │   insert ℓ' in Q
25  │   return
```

Algorithm 1. An A^* search template for solving FPP.

is resolved, the corresponding constraint becomes *active* for that label. After resolution, the term is removed from the constraint tree and the truth value is propagated upwards. The evaluation of the constraint terminates when the root node is resolved. The constraint is satisfied if the root node evaluates to false.

In practice, we translate the set of constraints $\mathcal{C}$ into a dictionary of constraint trees Γ with constraint identifiers as keys and the corresponding trees as values. For a constraint $\gamma \in \Gamma$, we let $\iota(\gamma)$ denote the constraint identifier and $T(\gamma)$ the corresponding tree. Then, for each node, $v \in V$, and each arc, $uv \in A$, we maintain a set of identifiers of the constraints that have those nodes or arcs, respectively, as leaves in the corresponding tree. We denote these sets E_v and E_{uv}, with $E_v = \{\iota(\gamma) \mid \gamma \in \Gamma, v \text{ appears in } \gamma\}$ and E_{uv} defined similarly.

We denote by Δ_ℓ the set of constraint trees copied from Γ to a label ℓ when they became active for ℓ. These constraint trees may be reduced immediately after copying because some terms have been resolved. Let $\rho(\gamma, P_\ell)$ be the tree $T(\gamma)$ after propagation of the terms in P_ℓ. Then, we can formally define, $\Delta_\ell = \{\rho(\gamma, P_\ell) \mid \iota(\gamma) \in E_u \cup E_{uv}, u, v \in P_\ell\}$.

In Algorithm 1, each label ℓ is thus the information record made of $(P_\ell, c_\ell, \Delta_\ell)$. We maintain all labels that are created in a structure $\mathcal{Q}$, called the *open list*. At each iteration of the while loop in Lines 5–15 of FINDPATH, a label ℓ is selected for extraction from the open list if its evaluation, given by $c(\ell) + h(\phi(\ell))$, is the smallest among the labels in $\mathcal{Q}$. Successively, the label ℓ with $\phi(\ell) = u \in V$ is *expanded* along each arc $uv \in A$, and new labels $\ell' = ((s, \ldots, u, v), c_\ell + c_{uv}, \Delta_{\ell'})$ are created (Lines 10–12).[3] The new set of constraint trees $\Delta_{\ell'}$ is obtained by copying the trees from Δ_ℓ, and the trees from Γ identified by E_v and E_{uv}. While performing these operations, the trees are reduced based on the resolution of the terms u and/or uv. If the root of a constraint tree in $\Delta_{\ell'}$ evaluates to true, then the label ℓ' is deleted, because the corresponding path became infeasible (Lines 13–14). On the other hand, if a root evaluates to false, then the corresponding constraint tree is resolved but is kept in $\Delta_{\ell'}$ to prevent re-evaluating it if, at a later stage, one of the terms that were logically deduced appears in the path. Formally, $\Delta_{\ell'} = \{\rho(\gamma, P_{\ell'}) \mid \gamma \in \Delta_\ell\} \cup \{\rho(\gamma, P_{\ell'}) \mid \iota(\gamma) \in E_{\phi(\ell')} \cup E_{\phi(\ell)\phi(\ell')}\}$.

If a new label ℓ' is not deleted, it is proposed for insertion in $\mathcal{Q}$ to the function INSERT. The function INSERT takes care of checking the domination between the proposed label ℓ' and the other labels in $\mathcal{Q}$ at the same node. The domination criterion plays a crucial role in the efficiency of the overall algorithm.

The loop in the function FINDPATH terminates on Line 6 when the goal g has been reached and the incumbent best path to g has total cost less than or equal to the evaluation of all labels in $\mathcal{Q}$. Under some well known conditions for h that we describe in Sect. 4.1, the solution returned is optimal.

Label Domination: We motivate our specific choice for the definition of the domination criterion using the following simple example taken from [7]. Let $C_1 = ((a \lor b) \land c)$ be the only constraint present. Let $\ell_1 = ((s, a, x), 3, \{C_1\})$, $\ell_2 = ((s, d, x), 4, \varnothing)$ and $\ell_3 = ((s, b, x), 2, \{C_1\})$ be the only three labels at x. Consider the labels ℓ_3 and ℓ_2. Although ℓ_3 is cheaper than ℓ_2, ℓ_2 has not activated C_1 and hence its path ahead is less constrained than ℓ_3. Indeed, ℓ_2 can use the cheapest route to g, while ℓ_3 must avoid c. It is therefore good not to discard ℓ_2, in spite of being more expensive. The labels ℓ_1 and ℓ_3 are identical with regards to constraints. However, the cost of ℓ_1 is greater than the cost of ℓ_3. Both labels will continue selecting the path through the arc with cost 5, and ℓ_3 will end up being the cheapest. The only way for ℓ_1 to recover and become the cheapest would be if the time it arrives at x is so different from the time of ℓ_3 that a drastic change in the weather conditions could be experienced. This seems unlikely, and the results of [8] seem to suggest that it is safe to assume the FIFO property for costs and not to assume it for constraints.

Further, it is possible to define a partial order relation between the constraint sets associated with labels. Intuitively, if ℓ_a has activated the same constraints as ℓ_b but ℓ_b has activated fewer terms than ℓ_a, then we can say that the constraints

---

[3] Although D contains cycles and although, theoretically, the cycles could be profitable because of the time dependency of costs, we do not allow labels to expand to already visited vertices because routes with cycles would be impractical.

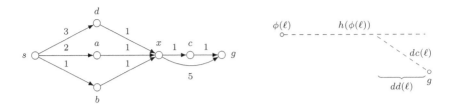

Fig. 1. Left, a domination example. The labels are $\ell_1 = ((s, a, x), 3, \{C_1\})$, $\ell_2 = ((s, d, x), 4, \varnothing)$ and $\ell_3 = ((s, b, x), 2, \{C_1\})$. Right, the setup for the SingleDescent estimate.

of the label ℓ_b are *implied* by those of ℓ_a. We refer to [7] for a precise definition of constraint implication and for implementation details.

Finally, we can state the domination criterion implemented in Lines 19–23. A label ℓ_a is *dominated* by another label ℓ_b if $\phi(\ell_a) = \phi(\ell_b)$, $c_{\ell_a} > c_{\ell_b}$ and Δ_{ℓ_b} is implied by Δ_{ℓ_a}. A label that is dominated is removed from (Line 21) or not inserted into (Line 23) the open list. If $\phi(\ell_a) = \phi(\ell_b)$, $c_{\ell_a} > c_{\ell_b}$ but Δ_{ℓ_b} is *not* implied by Δ_{ℓ_a}, then we say that ℓ_a is only *partially dominated* by ℓ_b. Partially dominated labels cannot be removed from the open list.

4   Heuristics

We present novel heuristics designed to cope with the large sizes of networks representing the 3D problem. We distinguish between two kinds of heuristics: those that implement the estimation of the remaining cost from a node to the goal, a crucial component of A* algorithms, and those that use the desired shape of the vertical profile to add constraints and prune the search space.

Further, we describe a two-phase approach, inspired by the current practice in industry, and we will use that as a benchmark for the heuristic 3D algorithms.

4.1   Remaining Cost Estimation in A*

The estimated cost h is said to be *admissible* if, for any node $u \in V$, the value $h(u)$ does not overestimate the final cost from u to the goal. Further, it is said to be *consistent* if for every node $u \in V$, $h(u)$ is at most the cost of getting to a successor $v \in V$ plus $h(v)$. Consistency can be shown to be the stronger property as it also implies admissibility. If both properties hold, then Algorithm 1 is optimal. If a heuristic is admissible, but not consistent, an optimal solution can be guaranteed if the algorithm allows for more than one label to be expanded from the same node. In standard A* search algorithms, labels expanded from a node are moved to a closed list, which is used to avoid expanding labels again from the same node. Instead, our Algorithm 1 allows more than one label to be expanded from a node in order to handle the constraint-induced lack of FIFO property. The use of an inconsistent rather than a consistent estimate increases the risk of re-expansions occurring and thus lessens the efficiency of the algorithm.

However, guaranteeing consistency, by ensuring that for all arcs $uv \in A$, $h(u) \leq h(v) + c_{uv}$, can be challenging in the FPP setting. First, it is not possible to compute a time-dependent estimate for each node of the network considering all its three coordinates. This would require a full 3D backwards search, which would be computationally heavy, quickly invalidated by the change of data, and pose memory problems. Second, restricting to considering only the 2D positions of the waypoints, say, the projection $\pi(u)$ onto the Earth for any node $u \in V$, one could define an estimate based on the great circle distance from $\pi(u)$ to the goal. But choosing the additional weather conditions would be more challenging: for example, assuming tailwind at cruise level everywhere would not only yield quite optimistic (i.e., loose) estimates but also would not guarantee consistency due to the ignored, possibility of cheaper descent of some levels. Other attempts at defining the flying conditions might lead to violation of admissibility as well.

In this light, with the goal of allowing precomputation of the estimate when a query is received and before the A* algorithm is started, we explore two directions: making the heuristic consistent in spite of worsening its tightness wrt. true cost, and giving up consistency while trying to improve the tightness. In both cases, we strive to maintain estimate admissibility, without which Algorithm 1 would not find the optimal solution. Both estimate procedures contain the following elements. They consider a 2D network obtained by the projection on the ground of all points from V, they associate time and weight independent costs with each arc, and they compute a cheapest path using a backwards Dijkstra from the goal to each node of the 2D network. The estimate $h(v)$ of each node $v \in V$ is then equal to the estimate of the projected node, i.e., $h(v) = h(\pi(v))$. The two procedures differ in the way the resource independent cost is determined.

All Descents: We set the cost of an arc between any two projected nodes to be equal to the cheapest possible cost of going between any pair of nodes that project on those nodes. More formally, let x and y be two nodes in the 2D network and let $\pi^{-1}(x)$ and $\pi^{-1}(y)$ be the set of nodes from $V(D)$ that project to x and y, respectively. We set the cost δ_{xy} between x and y in the 2D network as follows: $\delta_{xy} = \min\{c_{uv} \mid u \in \pi^{-1}(x), v \in \pi^{-1}(y), uv \in A\}$. Clearly, this is the most conservative choice; it guarantees that the derived estimates are consistent for all $u, v \in V$. Indeed, since $h(x) - h(y) = \delta_{xy}$, for any pair of points $u, v \in V$ that project to x and y, respectively, the inequality $h(u) - h(v) \leq c_{uv}$ will be satisfied. We refer to this estimate as AllDescents. The algorithm described in Sect. 3 with the AllDescents estimates is optimal, in the very likely case that the FIFO assumption on costs holds.

Single Descent: While AllDescents is consistent and admissible, its estimates are unrealistically cheaper than the real costs. Indeed, the cheapest arc between any two connected waypoints is very likely a descent from the top-most allowed flight level to the lowest flight level the aircraft can reach. The estimate associated with a node at a given altitude then becomes the cost of the aircraft descending between every two nodes in the shortest path to the destination.

We design an estimate that is closer to the real cost by considering only cruises between nodes. More specifically, between any pair of nodes x and y in the projected network, we consider the cost $\delta_{xy} = \min\{c_{uv} \mid u \in \pi^{-1}(x), v \in \pi^{-1}(y), uv \in A, uv$ is a cruise$\}$. It is realistic in most scenarios as the cruising phase is by far the most important and longest phase of a flight and thus the estimated value will often be consistent. However, there are scenarios where it is optimal for the aircraft to descend early on the route, which could lead to inconsistent estimates. Unfortunately, the estimates could also be inadmissible as the aircraft will have to descend at some point to reach the destination airport. To decrease the chance of this occurring, we correct the estimate by including a single direct descent to the goal when calculating the shortest path to the goal. We achieve this by updating the estimate when it is needed. Thus, whenever a label ℓ at $\phi(\ell)$ needs to be evaluated, we subtract from its estimate $h(\phi(\ell))$ the difference between the cost of a descent to the ground and a cruise of the same length using the great circle distance of the potential descent. We retrieve the cost of descending to the ground using the weather and weight conditions of ℓ. We denote this cost $dc(\ell)$ and the distance required for the descent by $dd(\ell)$. From the tables, we then retrieve the cost of cruising for $dd(\ell)$ under the same weather and weight conditions of ℓ. We denote this cost $\delta_{dd(\ell)}$. See Fig. 1, right. Thus, formally, the updated estimate h' for ℓ at $\phi(\ell)$ becomes $h'(\ell) = h(\pi(\phi(\ell)) - (\delta_{dd(\ell)} - dc(\ell))$. We refer to this estimate as SingleDescent.

4.2   Pruning Heuristics

Descent Disregard: Every aircraft has a flight level at which, under average weather conditions, it is optimal for it to cruise. In industry, this flight level is often enforced. We use the information on this *desired flight level* (DFL) heuristically to avoid expanding unpromising labels at low flight levels, thus reducing the search space of the A* search algorithm. We achieve this as follows.

Normally A* tries any possible climb, descent, and cruise along an arc when expanding a label, but many labels created (especially descents and cruises at low and inefficient flight levels) will not lead to optimal solutions. We aim for the algorithm to reduce the number of irrelevant descent and cruise expansions. Using the aircraft performance data we can determine the greatest descent distance (GDD), which is the greatest distance required to be able to reach the ground level from any flight level under any weight and weather conditions. If the distance to the destination airport is greater than the GDD, we do not need to worry about being able to reach the ground in time and can focus on remaining at an efficient flight level. We could therefore disallow descent and cruise expansions for labels not yet at the DFL and outside of the GDD. We still need to worry about constraints that could be blocking the DFL, so we must allow some deviation. We therefore define a threshold for flight levels where, if the aircraft is below this threshold and the distance to the destination is greater than the GDD, descents and cruises are not allowed. The value of the threshold determines a trade-off between missed optimality and speed-up. We have chosen to set the threshold to 3/4 of the DFL, as experiments indicated that this value yields a large decrease in run-time and only a small decrease in solution quality. We refer to this heuristic as DD.

Climb Disregard: Flight routes generally consist of three phases: climb, cruise, and descent. It is desirable for passengers' comfort to have only one of each of these phases. Our algorithm does not take this into account as it allows for unlimited climbs and descents in order to avoid constraints or to find better weather conditions. Limiting the number of times the aircraft can alternate between climbing and descending will result in more comfortable routes and speed up the algorithm. On the other hand, as the routes will be more constrained, it will lead to an increase in the total cost of the path.

We implement the rule that allows switching only once from climbing to descending, so once a descent has been initiated, no further climbs are allowed. We still allow paths to have a staircase shape, having climbs and descents interleaved with cruising arcs. This could be advantageous when further climbing becomes appealing only after some weight has been lost by consuming fuel.

To monitor the switches we equip the labels with a binary information indicating whether or not a descent has been performed. The label ℓ_a of a path that has already made a descent will be more constrained than the label ℓ_b of a path that has not. Thus, ℓ_a should not be allowed to dominate ℓ_b, even if ℓ_b is more expensive. The decreased effect of domination implies an increase in run-time that may outweigh the reduction of labels expanded due to the restriction. In order to assess the effect of this lack of domination experimentally, we include two configurations in our tests: one where we ignore the switch constraint in domination (thus, ℓ_a would have dominated ℓ_b), and one in which we treat the switch constraint as any other constraint (thus, ℓ_a would be only partially dominating ℓ_b and hence ℓ_b is not discarded). We refer to the former version as CD and to the latter as CDS.

4.3   Two-Phase Approach

In this approach to the FPP, we first find a path through airway points, solving a horizontal 2D routing problem, and then decide the vertical profile of the route by solving a vertical path finding problem.

Solving the horizontal 2D problem, considering only a given altitude at every airway point, does not work well with the second phase, because for many of the routes found, it is then impossible to find a feasible vertical profile. Instead, we "simulate" a horizontal 2D path finding problem by only considering one arc between two airway points when expanding a label, namely the feasible arc whose arrival node is closest to the DFL for the aircraft. Thus, for example, the label at the departure node only expands through arcs going from 0 to 200 nominal altitude; at the next node a label only expands on arcs going from 200 to 400 nominal altitude; and, from there, if 400 is the DFL, the label only expands through arcs to nodes at the same flight level. We only consider descents if it is the only option. Thus, we consider any node, whose projection is the destination airport, as goal nodes.

Climbing and descending decisions are handled in the vertical routing, where we restrict ourselves to the path through the airway points found in the previous phase but reintroduce freedom to determine the altitude. During the search, all possible climbs and descents from each node are allowed.

In both phases, we make use of the same constraint framework as in the 3D solution, iterating the search until a feasible path is found. However, limiting the search in the first phase to only one arc per pairs of waypoints can lead to a route that cannot be made feasible in the second phase, because, for example, of missing links for the descent phase. Hence, when the second phase fails to find a feasible vertical profile, we analyze which combination of arcs caused the problem, introduce a forbidden constraint containing those arcs, and restart the first phase. This can lead to many restarts and even unsolvable instances, as the constraints we introduce may be more severe than they are in reality.

We also include versions of the two-phase approach that use the DD and the CD heuristics. These heuristics address the vertical profile and hence only apply in the vertical optimization phase. We do not vary the estimate heuristics because the conditions here are different from the 3D case: in the first phase, we use an estimate heuristic similar to SingleDescent but without update. In the second phase we do not use any estimate and hence the search corresponds to a Dijkstra search. We were not able to come up with an A* estimate that could be helpful because, with the main focus in this phase being on altitude, estimates tend to be loose and pose computational problems due to the need of calculating these values on the 3D network (or recalculating them every time a new horizontal route is found).

5   Experimental Results

We have conducted experiments on real-life data to compare combinations of the elements presented here. All algorithms used the template of Algorithm 1. In the instantiated versions, the algorithms use the AllDescents or the SingleDescent estimate heuristics combined with reasonable pruning heuristics: DD, CD, CDS, DD + CD, DD + CDS. We also include the versions where AllDescents and SingleDescent are not combined with any pruning heuristic. Finally, we compare with the two-phase approach, which is closest to methods used in practice.

The real-life data is provided by our industrial partner. This data consists of aircraft performance data, weather forecast data in standardized GRIB2 format, and a navigation database containing the information for the graph. The graph consists of approximately 200,000 nodes and 124,000,000 arcs. The aircraft performance data refers to one single aircraft. The data for the weather forecast is given at intervals of three hours on specific grid points that may differ from the airspace waypoints. When necessary, we interpolate both in space and time.

In the calculation of the AllDescents and SingleDescent estimates, we assume piece-wise linearity of the consumptions τ on the arcs. Consequently, the cheapest cost of an arc can be determined by looking only at the points where measurements are available (e.g., at time intervals of three hours). Under this assumption, the AllDescents estimate is admissible. Note however that, as shown in [5], the travel time function between data points is not piece-wise linear and so neither is our cost function that includes fuel consumption. To get as close as possible to having an optimal algorithm as *baseline* for the comparison, we therefore included an algorithm without pruning heuristics and with the AllDescents estimate derived from costs on arcs given by a piece-wise linear function, whose pieces in the time scale are reduced to five minutes intervals. This approach would not be feasible in practice, as it took more than 40 h in our computational environment (see below) to precompute the estimates.

A test instance is specified by a departure airport and time, and a destination airport. A set of 13 major airports in Europe was selected uniformly at random to explore a uniform coverage of the constraints in the network. Among the 156 possible pairings, 16 were discarded because of short distances, resulting in 140 pairs that were used as queries. The great circle distances of these instances range from 317 to 1682 nautical miles. All algorithms were implemented in C# and the tests were conducted on a Dell XPS 15 laptop with an Intel Core i7 6700HQ at 2.6 GHz with 16 GB of RAM. Each algorithm was run 3 times on each instance and only the fastest run was recorded. All runs had a time limit of one hour.

We visualize the results in Fig. 2. The plots are disposed such that column-wise we distinguish the estimate heuristic and row-wise the performance measure considered. The pruning heuristics are represented along the x-axis of each plot. The quality of a route is its monetary cost. We show the percentage gap with respect to the results found by the baseline algorithm (5 min pieces). The run-time is the time spent for preprocessing (i.e., calculating the estimates) plus the total time spent searching. The preprocessing times for any configuration (apart from the baseline algorithm, which is not shown) are almost identical, so we do not distinguish. Time is measured in seconds and a logarithmic transformation is applied. We did not observe a clear dependency of the time on the mile distance of the instances, hence this latter is not visualized. Points represent the results of the selected runs. Points of results attained on the same instance are linked by a gray line. The boxes show the first, second (median) and third quartile of the distribution.

Comparison: The first observation is that the gap of AllDescents without pruning heuristic is different from zero only in a couple of instances and only in one making the gap worse than 1.5%. Hence, in practice assuming piece-wise linearity of consumptions seldom deteriorates the results. However, this observation might be dependent on our setting, as Blanco et al. [5] do show a relevant impact. Comparing AllDescents with SingleDescent, we observe that the inadmissibility of SingleDescent does not worsen the solution quality in any instance while wrt. run-time, SingleDescent leads to a considerable save for most instances. Tighter estimates allow for more restrictive search; a narrower area around the optimal route. We give evidence of this in Fig. 3, where we show the search results for one instance using the two estimate heuristics. The figure depicts the horizontal section of the 3D network restricted to the arcs that were actually expanded

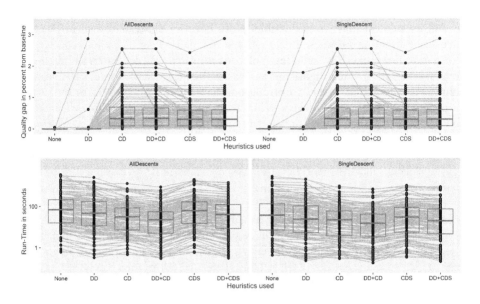

Fig. 2. Solutions quality (top) and run-time (bottom) results on 140 queries.

during the search. SingleDescent is much better at narrowing the search compared with AllDescents. However, the figure does not show the vertical dimension and the several flight levels expanded from nodes nor the arcs that were expanded more than once in SingleDescent. There are 14 instances where the extra expansions to cope with the inconsistency outweigh the expansions saved from the tighter estimate and make SingleDescent slower than AllDescents.

Considering pruning heuristics, DD leads to worse solution quality in only 5 instances for both estimate heuristics while improving the run-time for both. As expected, solutions become more costly when including heuristics that restrict the vertical profiles, CD and CDS. However, because these routes are more stable, they are more comfortable, a quality component not included in the monetary cost otherwise considered. If we include correct domination, as done in CDS, the solution quality improves slightly over algorithm versions that use CD, but considering the run-time, CDS is computationally more demanding than CD. In general, profile-based heuristics have considerable impact on the run-time. The CD and DD + CD heuristics contribute most to the speed-up. In a few instances, however, they can perform worse than the variants with no pruning heuristic. In these cases, finding a different solution (which might violate a constraint) can cause the algorithm using CD to do extra passes. Finally, the configuration using both DD and CD is the fastest. It has a median run-time slightly above 14 s and a worst-case performance of 14 min.

Comparison with the Two-Phase Approach: The two-phase approach with no pruning heuristics has a median quality gap of 1.9%, which is significantly higher than any of the 3D algorithms. There are also a few heavy outliers with

Fig. 3. Comparison of arcs expanded by AllDescents (left) and SingleDescent (right) with no heuristics on a route from Innsbruck to Nantes.

5 instances having a gap over 10%. In these instances, the two-phase approach returns a path through arcs where flying at the desired flight level is forbidden by some constraints and thus the vertical optimization will not be able to repair and obtain a good quality solution. There are also two instances where the two-phase approach fails to find any solution at all. As far as run-time is concerned, the two-phase approach does entail faster run-times than any of the 3D algorithms with a median of 10 s and a worst-case performance of 47 min. In general, there are some outliers caused by instances where the vertical optimization has trouble finding a feasible solution causing a large number of restarts. Using the DD heuristic does not yield any relevant effect, while using the CD heuristic does decrease run-time but also worsens the cost of the final path further. CD also increases the number of unsolved instances to 10.

6   Concluding Remarks

We found the SingleDescent estimate together with the DD heuristic particularly interesting. With respect to the near optimal solution of the baseline algorithm, this algorithm leads to deterioration of solution quality in only a few instances while it provides a considerable decrease in run-time. If further speed-up is needed, the CD heuristic could be added at the cost of only a slight increase in deterioration. The routes attained adopting the CD heuristic might be preferable in practice anyway due to the more stable vertical profile.

Perhaps our most important contribution is that we demonstrate the practicability of a direct 3D flight planning approach. Our comparison against the two-phase approach, more commonly used in practice, shows that although median results for the latter are better in terms of run-time, the solution quality and the outliers in run-time and unsolved instances are considerably worse than the results attainable with direct 3D approaches. The 3D approach exhibits fewer outliers, terminates on every instance, and is therefore more robust. If run-time is really an issue, we have shown that by combining the SingleDescent, the DD, and the CD heuristics, competitive time performance can be achieved, while retaining superior robustness and smaller reduction in solution quality relative to the two-phase approach. The ideas tested here will be included in future releases of software from our industrial partner.

References

1. Batz, G.V., Geisberger, R., Sanders, P., Vetter, C.: Minimum time-dependent travel times with contraction hierarchies. ACM J. Exp. Algorithmics **18**(1), 1.4:1–1.4:43 (2013). Article no. 1.4
2. Dijkstra, E.W.: A note on two problems in connexion with graphs. Numerische Mathematik **1**(1), 269–271 (1959)
3. Hart, P.E., Nilsson, N.J., Raphael, B.: A formal basis for the heuristic determination of minimum cost paths. IEEE Trans. Syst. Sci. Cybern. **4**(2), 100–107 (1968)
4. Yinnone, H.: On paths avoiding forbidden pairs of vertices in a graph. Discrete Appl. Math. **74**(1), 85–92 (1997)
5. Blanco, M., Borndörfer, R., Hoang, N.-D., Kaier, A., Schienle, A., Schlechte, T., Schlobach, S.: Solving time dependent shortest path problems on airway networks using super-optimal wind. In: 16th Workshop on Algorithmic Approaches for Transportation Modelling, Optimization, and Systems (ATMOS). OpenAccess Series in Informatics (OASIcs), vol. 54, pp. 12:1–12:15. Schloss Dagstuhl-Leibniz-Zentrum für Informatik (2016)
6. Bast, H., Delling, D., Goldberg, A., Müller-Hannemann, M., Pajor, T., Sanders, P., Wagner, D., Werneck, R.F.: Route planning in transportation networks (2015). arXiv:1504.05140 [cs.DS]
7. Knudsen, A.N., Chiarandini, M., Larsen, K.S.: Constraint handling in flight planning. In: Beck, J.C. (ed.) CP 2017. LNCS, vol. 10416, pp. 354–369. Springer, Cham (2017). https://doi.org/10.1007/978-3-319-66158-2_23
8. Knudsen, A.N., Chiarandini, M., Larsen, K.S.: Vertical optimization of resource dependent flight paths. In: 22nd European Conference on Artificial Intelligence (ECAI). Frontiers in Artificial Intelligence and Applications, vol. 285, pp. 639–645. IOS Press (2016)
9. Blanco, M., Borndörfer, R., Dung Hoàng, N., Kaier, A., Casas, P.M., Schlechte, T., Schlobach, S.: Cost projection methods for the shortest path problem with crossing costs. In: D'Angelo, G., Dollevoet, T., (eds.) 17th Workshop on Algorithmic Approaches for Transportation Modelling, Optimization, and Systems (ATMOS). OpenAccess Series in Informatics (OASIcs), vol. 59, pp. 15:1–15:14. Schloss Dagstuhl-Leibniz-Zentrum für Informatik (2017)
10. Jensen, C.K., Chiarandini, M., Larsen, K.S.: Flight planning in free route airspaces. In: D'Angelo, G., Dollevoet, T., (eds.) 17th Workshop on Algorithmic Approaches for Transportation Modelling, Optimization, and Systems (ATMOS). OpenAccess Series in Informatics (OASIcs), vol. 59, pp. 14:1–14:14. Schloss Dagstuhl-Leibniz-Zentrum für Informatik (2017)

Juniper: An Open-Source Nonlinear Branch-and-Bound Solver in Julia

Ole Kröger[✉], Carleton Coffrin, Hassan Hijazi, and Harsha Nagarajan

Los Alamos National Laboratory, Los Alamos, NM, USA
o.kroeger@wikunia.de

Abstract. Nonconvex mixed-integer nonlinear programs (MINLPs) represent a challenging class of optimization problems that often arise in engineering and scientific applications. Because of nonconvexities, these programs are typically solved with global optimization algorithms, which have limited scalability. However, nonlinear branch-and-bound has recently been shown to be an effective heuristic for quickly finding high-quality solutions to large-scale nonconvex MINLPs, such as those arising in infrastructure network optimization. This work proposes JUNIPER, a Julia-based open-source solver for nonlinear branch-and-bound. Leveraging the high-level Julia programming language makes it easy to modify JUNIPER's algorithm and explore extensions, such as branching heuristics, feasibility pumps, and parallelization. Detailed numerical experiments demonstrate that the initial release of JUNIPER is comparable with other nonlinear branch-and-bound solvers, such as BONMIN, MINOTAUR, and KNITRO, illustrating that JUNIPER provides a strong foundation for further exploration in utilizing nonlinear branch-and-bound algorithms as heuristics for nonconvex MINLPs.

1 Introduction

Many of the optimization problems arising in engineering and scientific disciplines combine both nonlinear equations and discrete decision variables. Notable examples include the blending/pooling problem [1,2] and the design and operation of power networks [3–5] and natural gas networks [6]. All of these problems fall into the class of mixed-integer nonlinear programs (MINLPs), namely,

$$\text{minimize: } f(x, y)$$
$$\text{s.t.}$$
$$g_c(x, y) \leq 0 \;\; \forall c \in \mathcal{C} \tag{MINLP}$$
$$x \in \mathbb{R}^m, y \in \mathbb{Z}^n$$

where f and g are twice continuously differentiable functions and x and y represent real and discrete valued decision variables, respectively [7]. Combining nonlinear functions with discrete decision variables makes MINLPs a broad and challenging class of mathematical programs to solve in practice. To address this

© Springer International Publishing AG, part of Springer Nature 2018
W.-J. van Hoeve (Ed.): CPAIOR 2018, LNCS 10848, pp. 377–386, 2018.
https://doi.org/10.1007/978-3-319-93031-2_27

challenge, algorithms have been designed for special subclasses of MINLPs, such as when f and g are convex functions [8,9] or when f and g are nonconvex quadratic functions [10,11]. For generic nonconvex functions, global optimization algorithms [12–15] are required to solve MINLPs with a proof of optimality. However, the scalability of such algorithms is limited and remains an active area of research. Although global optimization algorithms have been widely successful at solving industrial MINLPs with a few hundred variables, their limited scalability precludes application to larger real-world problems featuring thousands of variables and constraints, such as AC optimal transmission switching [16].

One approach to addressing the challenge of solving large-scale industrial MINLPs is to develop heuristics that attempt to quickly find high-quality feasible solutions without guarantees of global optimality. To that end, it has been recently observed that nonlinear branch-and-bound (NLBB) algorithms can be effective heuristics for the nonconvex MINLPs arising in infrastructure systems [4–6] and that they present a promising avenue for solving such problems on real-world scales. To the best of our knowledge, BONMIN and MINOTAUR are the only open-source solvers that implement NLBB for the most general case of MINLP, which includes nonlinear expressions featuring transcendental functions. Both BONMIN and MINOTAUR provide optimized high-performance C++ implementations of NLBB with a focus on convex MINLPs.

The core contribution of this work is JUNIPER, a minimalist implementation of NLBB that is designed for rapid exploration of novel NLBB algorithms. Leveraging the high-level Julia programming language makes it easy to modify JUNIPER's algorithm and explore extensions, such as branching heuristics, feasibility pumps, and parallelization. Furthermore, the solver abstraction layer provided by JuMP [17] makes it trivial to change the solvers used internally by JUNIPER's NLBB algorithm. Detailed numerical experiments on 300 challenging MINLPs are conducted to validate JUNIPER's implementation. The experiments demonstrate that the initial release of JUNIPER has comparable performance to other established NLBB solvers, such as BONMIN, MINOTAUR, and KNITRO, and that JUNIPER finds high-quality solutions to problems that are challenging for global optimization solvers, such as COUENNE and SCIP. These results illustrate that JUNIPER's minimalist implementation provides a strong foundation for further exploration of NLBB algorithms for nonconvex MINLPs.

The rest of the paper is organized as follows. Section 2 provides a brief overview of the NLBB algorithms. Section 3 introduces the JUNIPER NLBB solver. The experimental validation is conducted in Sect. 4, and Sect. 5 concludes the paper.

2   The Core Components of Nonlinear Branch-and-Bound

To provide context for JUNIPER's implementation, we begin by reviewing the core components of an NLBB algorithm. NLBB is a natural extension of the well-known branch-and-bound algorithm for mixed-integer linear programs (MIPs) to MINLPs. The algorithm implicitly represents all possible discrete variable assignments in a MINLP by a decision tree that is exponential in size. The algorithm then searches through this implicit tree (i.e., *branching*), looking for the best assignment of the discrete variables and keeping track of the best feasible solution found thus far, the so-called incumbent solution. At each node in the tree, the partial assignment of discrete variables is fixed and the remaining discrete variables are relaxed to continuous variables, resulting in a nonlinear program (NLP) that can be solved using an established solver, such as IPOPT [18]. If the solution to this NLP is globally optimal, then it provides a lower bound to the MINLP's objective function, $f(x, y)$. Furthermore, if this NLP *bound* is worse than the best solution found thus far, then the NLP relaxation proves that the children of the given node can be ignored. If this algorithm is run to completion, it will provide the globally optimal solution to the MINLP. However, if the MINLP includes nonconvex constraints, the NLP solver provides only local optimality guarantees, and the NLBB algorithm will be only a heuristic for solving the MINLP. The key to designing this kind of NLBB algorithm is to develop generic strategies that find feasible solutions quickly and direct the tree search toward higher-quality solutions. We now briefly review some of the core approaches to achieve these goals.

Branching Strategy: In each node of the search tree, the branching strategy defines the order in which the children (i.e., variable/value pairs) of that node should be explored. The typical branching strategies are (1) *most infeasible*, which branches on the variables that are farthest from an integer value in the NLP relaxation; (2) *pseudo cost*, which tracks how each variable affects the objective function during search and then prioritizes variables with the best historical record of improving the objective value [19]; (3) *strong*, which tests all branching options by brute-force enumeration and then takes the branch with the most promising NLP relaxation [20]; and (4) *reliability*, which uses a threshold parameter to limit strong branching to a specified amount of times for each variable [21].

Traversal Strategy: At any point during the tree search there are a number of *open* nodes that have branches that remain to be explored. The traversal strategy determines how the next node will be selected for exploration. The typical traversal strategies include (1) *depth first*, which explores the most recent open node first; and (2) *best first*, which explores the open node with the best NLP bound first. The advantage of depth first search is that it only requires a memory overhead that is linear in the number of discrete variables. In contrast, best first search results in the smallest number of nodes explored but can consume an exponential amount of memory.

Incumbent Heuristics: In some classes of MINLPs, finding an initial feasible solution can be incredibly difficult, and the NLBB algorithm can spend a prohibitive amount of time in unfruitful parts of the search tree. Running dedicated feasiblity heuristics at the root of the search tree is often effective in mitigating this issue. The most popular such heuristic is the *feasibility pump*, which is a fixed-point algorithm that alternates between solving an NLP relaxation of the MINLP for assigning the continuous variables and solving a MIP projection of the NLP solution for assigning the discrete variables [22,23].

Code Block 1 Installing and Solving a MINLP with JuMP and JUNIPER

```
Pkg.add("JuMP"); Pkg.add("Ipopt"); Pkg.add("Cbc"); Pkg.add("Juniper")
using JuMP, Ipopt, Cbc, Juniper

ipopt = IpoptSolver(print_level=0); cbc = CbcSolver()
m = Model(solver=JuniperSolver(ipopt, mip_solver=cbc))

v = [10,20,12,23,42]; w = [12,45,12,22,21]
@variable(m, 0 <= x[1:5] <= 10, Int)

@objective(m, Max, dot(v,x))
@constraint(m, sum(x[i] for i=1:5) <= 6)
@NLconstraint(m, sum(w[i]*x[i]^2 for i=1:5) <= 300)

status = solve(m); getvalue(x)
```

Relaxation Restarts: In traditional branch-and-bound algorithms, the continuous relaxation is convex and guaranteed to converge to the global optimum or prove that the relaxation is infeasible. However, in the case of nonconvex MINLPs, a local NLP solver provides no such guarantees. Thus, it can be advantageous to restart the NLP solver from a variety of different starting points in the hopes of improving the lower bound or finding a feasible solution [8,24].

3   The Juniper Solver

The motivation for developing JUNIPER [25] is to provide relatively simple and compact implementation of NLBB so that a wide variety of algorithmic modifications can be explored in the pursuit of developing novel heuristics for nonconvex MINLPs. To that end, Julia is a natural choice for the implementation for two reasons: (1) Julia provides high-level programming, similar to Matlab and Python, that is preferable for rapid prototyping; and (2) the mathematical programming package JuMP [17] provides an AMPL-like modeling layer, which makes it easy to state MINLP problems, and a solver abstraction layer, which makes a wide range of NLP and MIP solvers available for use in Juniper. To demonstrate these properties, Code Block 1 provides a simple Julia v0.6 example illustrating the software installation, stating a JuMP v0.18 MINLP model, and solving it with JUNIPER. In this example, the NLP solver IPOPT is used for solving the continuous relaxation subproblems and the MIP solver CBC is used in the feasibility pump heuristic.

From Code Block 1, it is clear how JUNIPER can be reconfigured to use different NLP and MIP solvers at runtime. As is typical for solvers, JUNIPER also features a wide variety of parameters for augmenting the NLBB algorithm. These include options for selecting the branching strategy, tree traversal strategy, feasibility pump, parallelized tree search, and numerical tolerances, among others. A complete list of algorithm parameters is available in JUNIPER's documentation. After rigorous testing on hundreds of MINLP problems, the following default settings were identified: Strong branching is performed at the root node, and pseudo-cost branching is used afterward. Typically, complete strong branching is conducted; however, if the NLP runtime combined with the number of branches will require more than 100 s, the number of branches explored is reduced to meet this time limit. If the NLP relaxation fails in the root node, it will be restarted up to three times. Best first search is used for exploring the decision tree, and the runtime of the feasibility pump is limited to 60 s.

4   Experimental Evaluation

This section conducts a detailed numerical study of JUNIPER's performance under a variety of configurations and compares its performance to established MINLP solvers. Five points of comparison were considered for solving MINLPs. BONMIN v1.8 [8], MINOTAUR v0.2 [26], and KNITRO v10.3 [24] were included as alternative NLBB implementations, whereas COUENNE v0.5 [13] and SCIP v5.0 [10, 27] were used for a global optimization reference. All of the open-source solvers utilize IPOPT v3.12 [18] compiled with HSL [28] for solving NLP sub-problems and their respective default LP and MIP solvers. All of the solvers, except JUNIPER, were accessed through their AMPL NL file interface. All of the computations were conducted on a cluster of HPE ProLiant XL170r servers featuring two Intel 2.10 GHz 16 Core CPUs and 128 GB of memory. All solvers were configured with an optimality gap of 0.01% and a runtime limit of 1 h. It is important to note that Julia's JIT takes around 3–10 s the first time JUNIPER is run; this time is not reflected in the runtime results.

MINLP Problem Selection: The first step in performing this evaluation is to select an appropriate collection of MINLP test problems. We began with 1500 MINLP problems from MINLPLIB2 [29], which are available in Julia via the MINLPLibJuMP package [30]. Second, all of the problems with no discrete variables or fewer than ten constraints were eliminated, resulting in about 700 problems that focus on the constrained mixed-integer problems that JUNIPER is intended for. Through a preliminary study, it was observed that more than half of these cases are solved to global optimality or are proven to be infeasible by SCIP or COUENNE in less than 60 s, suggesting that these are relatively easy cases for state-of-the-art global optimization methods and that they are not of interest to this work. The final collection of test problems consists of 298 MINLPs that are challenging for both NLBB and global optimization solvers.

Solver Comparison: The first and foremost goal is to demonstrate that JUNIPER has comparable computational performance to BONMIN and MINOTAUR. Figure 1 (top) provides an overview of the runtime for each solver to

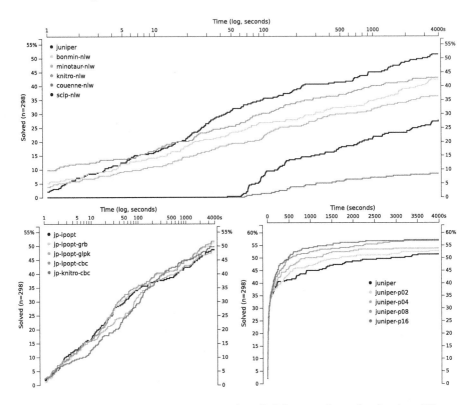

Fig. 1. Runtime profiles on 298 instances for all different solvers (top) using different MIP solvers (bottom left) and parallelized tree search (bottom right).

complete its tree search procedure. This figure highlights two key points: (1) JUNIPER is slower for small models that can be solved in less than 30 s; however, it consistently solves more models after 30 s; and (2) the search completes in no more than 50% of the cases considered, demonstrating that the selected MINLP instances present challenging tree search problems for both the NLBB and global optimization solvers.

Table 1 provides further details on the performance of each solver, including problem sizes, objective gaps from the best-known solution, and runtime results. The table begins with summary statistics. The first row shows the number of feasible solutions found by each solver as well as the number of test cases where the runtime limit was reached. The following three rows show the average optimality gaps and runtime for each solver. The first average is for all instances where the specific solver was able to find a feasible solution. The second average is for instances where all six solvers were able to find feasible solutions. The third average is for instances where all four NLBB solvers were able to find feasible solutions. These summary results indicate two key points: (1) JUNIPER is one of the most robust solvers (only SCIP had a higher feasible solution count); and (2) for cases where all NLBB algorithms have feasible solutions, BONMIN has the

Table 1. Quality and runtime results for various instances

Instance	$\|V\|$	$\|C\|$	$\|I\|$	$\|NC\|$	best obj.	Gap (%)						Runtime (seconds)					
						juniper	bon	minot	knitro	coue	scip	juniper	bon	minot	knitro	coue	scip
Feasible instances / Time limit reached						228	196	193	187	183	257	114	105	134	144	247	215
Average solver feasible						324.51	6.34	25.52	1365.77	395.68	1e+04	1425	1672	1789	1226	3285	2853
Average all solvers feasible (n=85)						21.43	0.86	3.64	24.57	16.19	2e+04	929	834	1558	1002	2788	2661
Average all NLBB solvers feasible (n=113)						18.48	0.94	11.19	20.69	-	-	918	845	1420	1112	-	-
oil2	937	927	2	284	-0.73	0.00	0.00	0.00	0.00	-	-	7	3	2	8	-	-
bchoco08	169	191	9	79	0.98	-	-	0.00	0.71	-	-	-	-	46	153	-	-
squfl015-080persp	2416	2481	15	1200	402.48	0.00	0.00	4.61	0.00	0.93	0.00	27	23	259	23	T.L	94
transswitch0014p	139	278	20	121	8082.58	0.00	0.00	0.00	0.00	0.00	0.00	2	14	<1	32	T.L	-
squfl030-100persp	6031	6101	30	3000	363.09	0.00	0.00	43.71	0.00	137.24	0.00	683	919	T.L	5	T.L	489
FLay05M	63	66	40	5	64.50	0.00	0.00	0.00	0.00	0.00	0.00	1283	561	1213	1679	117	71
ndcc13	631	255	42	42	84.63	5.80	0.00	0.00	-	-	3.30	T.L	1721	1172	T.L	-	T.L
CLay0205H	261	366	50	40	8092.50	0.00	0.00	891.18	1.61	-	0.00	2548	1091	T.L	T.L	-	71
procurement1mot	785	750	60	1	291.54	0.00	0.00	0.00	0.00	8.19	1.95	445	204	385	2345	T.L	T.L
fo9_ar2_1	181	436	72	18	32.62	-	-	115.90	-	56.21	0.00	-	-	T.L	-	T.L	3454
crudeoil_pooling_ct1	311	566	80	37	210537.49	43.82	0.00	3.06	-	2.34	0.00	T.L	T.L	T.L	-	T.L	T.L
multiplants_mtg1a	194	257	93	28	391.61	0.00	0.00	3.08	-	2.46	0.00	909	901	T.L	-	T.L	997
multiplants_mtg2	230	307	112	37	7099.19	0.00	0.00	0.14	-	0.00	27.54	3390	1877	T.L	-	T.L	T.L
crudeoil_pooling_ct3	730	1199	128	211	287000.00	32.69	-	-	-	16.82	0.00	T.L	-	-	-	T.L	T.L
pooling_epa3	1105	1718	150	274	-14948.12	0.00	-	-	0.17	-	-	T.L	-	-	T.L	-	-
transswitch0118r	1240	1764	179	1311	129457.85	0.45	-	-	0.00	-	-	181	-	-	491	-	-
multiplants_stg1	415	262	198	34	355.09	24.36	0.00	1.03	-	-	-	T.L	T.L	T.L	-	-	-
qap	226	31	225	1	389718.00	0.50	0.38	6.79	0.00	10.94	6.53	1404	428	T.L	137	T.L	T.L
csched2	401	138	308	1	-166102.00	-	0.00	4.16	-	14.05	6.69	-	18	T.L	-	T.L	T.L
edgecross22-048	463	6161	462	1	84.00	0.00	0.00	0.00	0.00	0.00	0.00	12	121	98	23	162	2201
crudeoil_pooling_dt1	3713	5779	916	570	209585.27	368.25	-	4.59	-	-	0.00	T.L	-	T.L	-	T.L	3429

highest solution quality, and JUNIPER, MINOTAUR, and KNITRO have similar quality, on average. The remaining rows in the table provide a representative sample of the 298 problems considered. The first five columns describe each problem by name, number of variables V, number of constraints C, number of discrete variables I, and number of nonlinear constraints NC. The general trends are summarized as follows: (1) there is a great diversity among which solver is the best on the MINLP instances considered; and (2) in most cases, the solutions found by the NLBB solvers tie or improve those found by the global solvers; however, there are a few notable cases where global solvers find the best solutions. Overall, these results indicate that JUNIPER in its default configuration is comparable with the NLBB solvers considered here.

Subsolver Selection: One of the key features of JUNIPER is that it can use different solvers for the NLP relaxation and for the MIP aspect of the feasibility pump heuristic. Figure 1 (bottom left) shows a performance profile for a number of subsolver variants of JUNIPER, both with and without a MIP solver (i.e., GLPK [31], CBC [32], GUROBI [33]), as indicated by JUNIPER-ipopt. JUNIPER-KNITRO-CBC shows the result of using KNITRO as the NLP solver instead of IPOPT. The runtime difference between using the feasibility pump and using no heuristic is quite notable in some cases; however, given sufficient time, JUNIPER solves a similar number of cases even without a feasibility pump. To our surprise, there was little difference in using CBC as the MIP solver compared to using GUROBI, suggesting that CBC is a suitable default solver. We also observed that GLPK is not a suitable choice because it was typically unable to terminate in less than 60 s, which is the preferred feasibility pump time limit.

Parallel Tree Search: A key feature of Julia is native and easy-to-use support for parallel processing. JUNIPER leverages this capability to implement a parallel tree search algorithm. Figure 1 (bottom right) illustrates the benefits from this simple parallelization of the algorithm, where the first thread orchestrates the computation and all additional worker threads process open nodes in the search tree. The figure indicates that having two worker threads (instead of using the sequential algorithm) is about 1.7 times faster and having four worker threads is about 3.3 times faster. The difference between eight and sixteen worker threads is not that notable (both increase speed by around 5.8 times).

5   Conclusion

This work has highlighted the potential for leveraging NLBB algorithms as heuristics for solving challenging nonconvex MINLPs. To assist in the design of such algorithms, a new Julia-based solver, JUNIPER, is proposed as the base implementation for future exploration in this area. A detailed experimental study demonstrated that, despite its minimalist implementation, JUNIPER performs comparably to established NLBB solvers on the class of MINLPs for which it was designed. We hope that JUNIPER will provide the community with a valuable reference implementation for collaborative open-source research on heuristics for large-scale nonconvex MINLPs.

References

1. Audet, C., Brimberg, J., Hansen, P., Digabel, S.L., Mladenovic, N.: Pooling problem: alternate formulations and solution methods. Manage. Sci. **50**(6), 761–776 (2004)
2. Trespalacios, F., Kolodziej, S.P., Furman, K.C., Sawaya, N.W.: Multiperiod blend scheduling problem. Cyber Infrastructure for MINLP, June 2013. www.minlp.org/library/problem/index.php?i=168
3. Jabr, R.A.: Optimization of AC transmission system planning. IEEE Trans. Power Syst. **28**(3), 2779–2787 (2013)
4. Coffrin, C., Hijazi, H.L., Lehmann, K., Hentenryck, P.V.: Primal and dual bounds for optimal transmission switching. In: 2014 Power Systems Computation Conference, pp. 1–8, August 2014
5. Coffrin, C., Hijazi, H.L.: Heuristic MINLP for optimal power flow problems. In: 2014 IEEE Power & Energy Society General Meetings (PES) Application of Modern Heuristic Optimization Algorithms for Solving Optimal Power Flow Problems Competition (2014)
6. Borraz-Sanchez, C., Bent, R., Backhaus, S., Hijazi, H., Hentenryck, P.V.: Convex relaxations for gas expansion planning. INFORMS J. Comput. **28**(4), 645–656 (2016)
7. Belotti, P., Kirches, C., Leyffer, S., Linderoth, J., Luedtke, J., Mahajan, A.: Mixed-integer nonlinear optimization. Acta Numer. **22**, 1–131 (2013)
8. Bonami, P., Biegler, L.T., Conn, A.R., Cornuéjols, G., Grossmann, I.E., Laird, C.D., Lee, J., Lodi, A., Margot, F., Sawaya, N., Wächter, A.: An algorithmic framework for convex mixed integer nonlinear programs. Discrete Optim. **5**(2), 186–204 (2008). In memory of George B. Dantzig
9. Lubin, M., Yamangil, E., Bent, R., Vielma, J.P.: Extended formulations in mixed-integer convex programming. In: Louveaux, Q., Skutella, M. (eds.) IPCO 2016. LNCS, vol. 9682, pp. 102–113. Springer, Cham (2016). https://doi.org/10.1007/978-3-319-33461-5_9
10. Achterberg, T.: SCIP: solving constraint integer programs. Math. Program. Comput. **1**(1), 1–41 (2009)
11. Bonami, P., Gunluk, O., Linderoth, J.: Solving box-constrained nonconvex quadratic programs (2016). http://www.optimization-online.org/DB_HTML/201 6/06/5488.html
12. Ryoo, H., Sahinidis, N.: A branch-and-reduce approach to global optimization. J. Global Optim. **8**(2), 107–138 (1996)
13. Belotti, P.: Couenne: user manual (2009). https://projects.coin-or.org/Couenne/. Accessed 04 Oct 2015
14. Nagarajan, H., Lu, M., Wang, S., Bent, R., Sundar, K.: An adaptive, multivariate partitioning algorithm for global optimization of nonconvex programs. arXiv preprint arXiv:1707.02514 (2017)
15. Nagarajan, H., Lu, M., Yamangil, E., Bent, R.: Tightening McCormick relaxations for nonlinear programs via dynamic multivariate partitioning. In: Rueher, M. (ed.) CP 2016. LNCS, vol. 9892, pp. 369–387. Springer, Cham (2016). https://doi.org/10.1007/978-3-319-44953-1_24
16. Sahraei-Ardakani, M., Korad, A., Hedman, K.W., Lipka, P., Oren, S.: Performance of AC and DC based transmission switching heuristics on a large-scale polish system. In: 2014 IEEE PES General Meeting—Conference Exposition, pp. 1–5, July 2014

17. Dunning, I., Huchette, J., Lubin, M.: JuMP: a modeling language for mathematical optimization. SIAM Rev. **59**(2), 295–320 (2017)
18. Wächter, A., Biegler, L.T.: On the implementation of a primal-dual interior point filter line search algorithm for large-scale nonlinear programming. Math. Program. **106**(1), 25–57 (2006)
19. Benichou, M., Gauthier, J.M., Girodet, P., Hentges, G., Ribiere, G., Vincent, O.: Experiments in mixed-integer linear programming. Math. Program. **1**(1), 76–94 (1971)
20. Applegate, D., Bixby, R., Chvatal, V., Cook, B.: Finding cuts in the TSP (a preliminary report). Technical report (1995)
21. Achterberg, T., Koch, T., Martin, A.: Branching rules revisited. Oper. Res. Lett. **33**(1), 42–54 (2005)
22. Fischetti, M., Glover, F., Lodi, A.: The feasibility pump. Math. Program. **104**(1), 91–104 (2005)
23. D'Ambrosio, C., Frangioni, A., Liberti, L., Lodi, A.: A storm of feasibility pumps for nonconvex MINLP. Math. Program. **136**(2), 375–402 (2012)
24. Byrd, R.H., Nocedal, J., Waltz, R.A.: KNITRO: an integrated package for nonlinear optimization. In: Di Pillo, G., Roma, M. (eds.) Large-Scale Nonlinear Optimization. Nonconvex Optimization and Its Applications, vol. 83, pp. 53–59. Springer, Boston (2006). https://doi.org/10.1007/0-387-30065-1_4
25. Kröger, O., Coffrin, C., Hijazi, H., Nagarajan, H.: Juniper (2017). https://github.com/lanl-ansi/Juniper.jl. Accessed 14 Dec 2017
26. Mahajan, A., Leyffer, S., Linderoth, J., Luedtke, J., Munson, T.: Minotaur: a mixed-integer nonlinear optimization toolkit (2017)
27. Gleixner, A., Eifler, L., Gally, T., Gamrath, G., Gemander, P., Gottwald, R.L., Hendel, G., Hojny, C., Koch, T., Miltenberger, M., Müller, B., Pfetsch, M.E., Puchert, C., Rehfeldt, D., Schlösser, F., Serrano, F., Shinano, Y., Viernickel, J.M., Vigerske, S., Weninger, D., Witt, J.T., Witzig, J.: The SCIP optimization suite 5.0. Technical report 17–61, ZIB, Takustr. 7, 14195 Berlin (2017)
28. Research Councils UK: The HSL mathematical software library. http://www.hsl.rl.ac.uk/. Accessed 30 Oct 2017
29. Vigerske, S.: MINLP Library 2 (2017). http://www.gamsworld.org/minlp/minlplib2/html/. Accessed 17 Dec 2017
30. Wang, S.: MINLPLibJuMP (2017). https://github.com/lanl-ansi/MINLPLibJuMP.jl. Accessed 14 Dec 2017
31. Free Software Foundation Inc.: GNU linear programming kit (2017). https://www.gnu.org/software/glpk/
32. The COIN-OR Foundation: COIN-OR CBC (2017). https://projects.coin-or.org/Cbc
33. Gurobi Optimization Inc.: Gurobi optimizer reference manual (2014). http://www.gurobi.com

Objective Landscapes for Constraint Programming

Philippe Laborie[(✉)]

IBM, 9 rue de Verdun, 94250 Gentilly, France
`laborie@fr.ibm.com`

Abstract. This paper presents the concept of *objective landscape* in the context of Constraint Programming. An objective landscape is a light-weight structure providing some information on the relation between decision variables and objective values, that can be *quickly* computed *once and for all* at the beginning of the resolution and is used to *guide* the search. It is particularly useful on decision variables with large domains and with a continuous semantics, which is typically the case for time or resource quantity variables in scheduling problems. This concept was recently implemented in the automatic search of CP Optimizer and resulted in an average speed-up of about 50% on scheduling problems with up to almost 2 orders of magnitude for some applications.

Keywords: Constraint Programming · Scheduling · Search
Optimization

1 Introduction

Motivations. A recognized weakness of Constraint Programming (CP) for solving combinatorial optimization problems (when compared to Mixed Integer Linear Programming for instance) is the lack of a global vision of the problem and in particular of the influence of decision variables on the objective function.

This paper presents the concept of *objective landscape* in the context of CP. The purpose of *objective landscapes* is to capture some information on the relation between decision variables and objective values. They help answering questions like: How much does a given decision variable contribute to the cost? What is the impact on the cost of modifying the value of a variable? What is the ideal value of a variable with respect to the cost function? An objective landscape is a light-weight structure that can be *quickly* computed *once and for all* at the beginning of the resolution by exploiting constraint propagation and is used to *guide* the search. It is particularly useful on decision variables with large domains and with a continuous semantics, which is typically the case for time or resource quantity variables in scheduling problems.

Although validated mostly on scheduling problems, the concept is generic and in this paper we consider a general combinatorial optimization problem defined as:

© Springer International Publishing AG, part of Springer Nature 2018
W.-J. van Hoeve (Ed.): CPAIOR 2018, LNCS 10848, pp. 387–402, 2018.
https://doi.org/10.1007/978-3-319-93031-2_28

$$\text{minimize } f(x)$$
$$\text{subject to } c(x, y)$$

Variables $x = [x_1, ..., x_n]$, $y = [y_1, ..., y_m]$ are **decision variables**. The objective function f functionally depends on a subset of decision variables x that we will call **objective decision variables**.[1]

Comparison with Existing Approaches. The idea of measuring the interactions between variables and objective function is not new in CP. Impact-based search (IBS) [12] incrementally maintains during the search an impact measurement for each possible variable assignment ($x_i = a$) that estimates the impact of the assignment on the other variables domain. In Activity-based search (ABS) [8], the activity of an assignment ($x_i = a$) estimates the number of affected variables when it is propagated. As we will see, *objective landscapes* share a common feature with the above techniques in that they are computed by exploiting constraint propagation. But there is also a number of differences:

– In IBS, *impacts* on cost are measured by modifying the domain of a variable x_i and measuring its impact on the cost whereas *objective landscapes* work the other way round: it modifies the bounds of the cost function and measures the impact on the variables x_i.
– *Impacts* and *activity* measurements are mostly designed for discrete variables with reasonably small domains whereas *objective landscapes* were initially designed for continuous domains: their complexity does not depend on domain size.
– *Impacts* are continuously updated during the search and are based on some *averaging* assumption whereas *objective landscapes* are computed once and for all at the beginning of the search and make some kind of *optimistic* assumption about the values of variables.

Objective landscapes differ from the *cost-based solution densities* introduced in [11] in that (1) they are parameter-free (no parameter ϵ related with the optimality corridor), (2) they do not require specific (and potentially complex) computation algorithms for each constraint, (3) they can scale to large domain size with almost no computational overhead and (4) they estimate the impact of a variable's value on the cost rather than on the number of solutions.

Some approaches in CP use a linear relaxation of the problem to guide the search with a more global view on the objective. In [1], a linear relaxation (LR) of a sub-problem using only *objective decision variables* is solved at each search node and the solution of this relaxation is used to guide the search. In [9] the LR is computed with specific algorithms and integrated in a global constraint. The automatic search of CP Optimizer also uses a LR of the problem at the root node of each Large Neighborhood Search move [7]. As we will see, *objective landscapes* differ from LR in several points:

---

[1] The distinction between objective decision variables x and the other variables y clearly depends on the formulation of the problem and may be considered as a bit arbitrary. This will be discussed in Sect. 7.

- LR does not give the contribution of different values of a variable to the cost whereas *objective landscapes* do (in a certain sense).
- LR heavily exploits the constraints of the problem (their linearization) and is usually computed several times during the search whereas *objective landscapes* focus on the cost function only and are computed once and for all.
- LR needs to *convexify* the objective function $f(x)$ whereas *objective landscapes* make less assumption about the convexity of objective terms and are related to the more general notion of *quasiconvexity*.
- LR needs building and solving a linear program or running specific algorithms whereas *objective landscapes* are much lighter to compute and exploit.

Example. All along the paper we will use a flow-shop scheduling problem with earliness-tardiness cost inspired from [10] for illustration. Using the scheduling concepts of CP Optimizer, an instance of this problem for n jobs and m machines can be formulated with a set of interval variables $o_{i,j}$ of fixed length (processing time), one for each operation, as follows:

$$
\begin{aligned}
\text{minimize} \qquad & \sum_{i=1}^{n} \text{endEval}(o_{i,m}, ET_i) \\
\text{subject to} \qquad & \text{endBeforeStart}(o_{i,j}, o_{i,j+1}) && \forall i \in [1, n], \forall j \in [1, m-1] \\
& \text{noOverlap}(\{o_{i,j}\}_{i \in [1,n]}) && \forall j \in [1, m]
\end{aligned}
$$

In the formulation above, ET_i is a piecewise linear function representing the earliness-tardiness cost for finishing job i at a date t as illustrated on Fig. 1. In this example, *objective decision variables* are the end values of interval variables $o_{i,m}$ representing the last operation of job i.

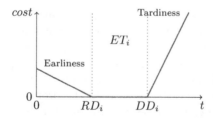

Fig. 1. Example of earliness-tardiness cost function

Paper Organization. Objective landscapes are formally defined in Sect. 2 whereas Sect. 3 presents some of their properties. From a more practical point of view, Sects. 4 and 5 respectively describe how landscapes can be computed and exploited to guide the search. Objective landscapes have been implemented in CP Optimizer (in version 12.7.1). Section 6 presents an experimental study on a large number of scheduling problems showing the practical interest of landscapes.

2   Objective Landscape Definition

We consider a general combinatorial optimization problem defined as:

$$\text{minimize } z = f(x)$$
$$\text{subject to } c(x, y)$$

With $x = [x_1, ..., x_n]$, $y = [y_1, ..., y_m]$. Let x_i be a decision variable and S a feasible solution to the problem, we denote:[2]

- D_i the initial domain of variable x_i
- $D = D_1 \times ... \times D_n$
- X_i^L a lower bound on the value of variable x_i
- X_i^U an upper bound on the value of variable x_i
- X_i^S the value of variable x_i in solution S
- Z^L a lower bound on the objective function f
- Z^U an upper bound on the objective function f.

Let x_i be an objective variable, we define function $f_i^* : D_i \to \mathbb{R}$ such that $f_i^*(v)$ denotes the optimal objective value one can obtain when $x_i = v$ if one only considers the objective function f. That is:

$$f_i^*(v) = \min_{x \in D \text{ s.t. } x_i = v} f(x)$$

Value $f_i^*(v)$ gives, in a certain sense, the contribution of the assignment $x_i = v$ to the cost under the optimistic assumption that all the other variables $x_{j,j \neq i}$ are fixed to values minimizing their overall contribution to the cost.

The idea of objective landscapes is to build, for each decision variable x_i, a function that approximates f_i^* and that we will call the objective landscape of variable x_i.

We suppose the existence of a given propagation algorithm $\mathcal{P}$ that is able to propagate constraints like $f(x) \leq z$ in order to reduce the domain of variables x_i. It is important to note that in the context of landscapes, propagation algorithm $\mathcal{P}$ ignores the constraints of the problem $c(x, y)$. We suppose that propagation algorithm $\mathcal{P}$ is *monotonous* (if $z' < z$, the propagation of $f(x) \leq z'$ does not lead to larger domains for variables x_i than the propagation of $f(x) \leq z$) and, of course, that it is *sound* (it does not remove feasible values). Let x_i be an objective variable, we define:[3]

- Z^L the smallest value of $z \in \mathbb{R}$ such that the propagation of objective cut $f(x) \leq z$ does not fail
- $X_i^L(z)$ for $z \geq Z^L$ as the lower bound on variable x_i obtained after the propagation of an objective cut $f(x) \leq z$

---

[2] Through the paper, we use lower cases for variables and upper cases for constants.
[3] By abuse of notation, as $\mathcal{P}$ is supposed to be given, we do not use it in the notations of these bounds although they clearly depend on the propagation algorithm.

– $X_i^U(z)$ for $z \geq Z^L$ as the upper bound on variable x_i obtained after the propagation of an objective cut $f(x) \leq z$.

Clearly, by monotonicity of the propagation, for any $z \geq Z^L$, $X_i^L(z)$ (resp. $X_i^U(z)$) is a non-increasing (resp. non-decreasing) function of z and $X_i^L(z) \leq X_i^U(z)$. An example of these two functions is shown on the left part of Fig. 2.

Informally speaking, the objective landscape function of an objective variable x_i is a function L_i whose graph is the 90° rotate of the union of the graphs of the two functions $X_i^L(z)$ and $X_i^U(z)$, as illustrated on the right part of Fig. 2.

Objective landscapes can now be defined more formally.

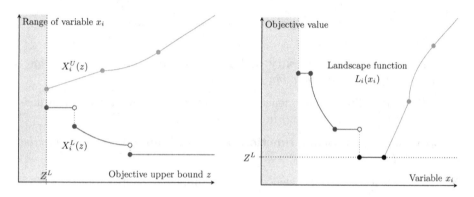

Fig. 2. Left: decision variable ranges as a function of objective upper-bound z. Right: objective landscape function of an objective variable x_i

Definition 1 (Objective landscape). *Given a propagation algorithm $\mathcal{P}$, the* **objective landscape function** *of an objective variable x_i is a function L_i : $D_i \rightarrow [Z^L, +\infty)$ defined as follow:*

– *For $v \in [X_i^L(Z^L), X_i^U(Z^L)] : L_i(v) = Z^L$*
– *For $v < X_i^L(Z^L) : L_i(v) = \min\{z \mid X_i^L(z) \leq v\}$*
– *For $v > X_i^U(Z^L) : L_i(v) = \min\{z \mid X_i^U(z) \geq v\}$.*

3   Objective Landscape Properties

We first recap the notion of a quasiconvex function [3].

Definition 2 (Quasiconvex function). *A function $f : S \rightarrow \mathbb{R}$ defined on a convex subset S of a real vector space is* **quasiconvex** *if for all $x, y \in S$ and $\lambda \in [0, 1]$ we have $f(\lambda x + (1 - \lambda)y) \leq \max\{f(x), f(y)\}$.*

Informally, along any stretch of the curve the highest point is one of the endpoints. Examples of quasi convex and non-quasiconvex functions are shown on Fig. 3.

We can now present some properties of objective landscapes.

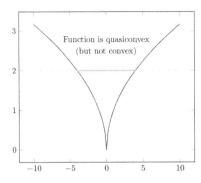 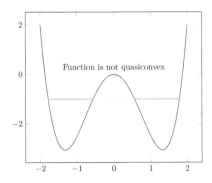

Fig. 3. Examples of quasi convex and non-quasiconvex functions

Proposition 1 (Quasiconvexity of landscape functions). *For any objective variable* x_i, *landscape function* L_i *is quasiconvex.*

Proof. Proof is a direct consequence of the fact $X_i^L(z)$ (resp. $X_i^U(z)$) is a non-increasing (resp. non-decreasing) function of z and $X_i^L(z) \leq X_i^U(z)$. □

Proposition 2 (Landscape functions as lower bounds). *For any objective variable* x_i, $L_i \leq f_i^*$.

Proof. Proof is a direct consequence of the soundness of propagation. □

Definition 3 (Exact landscape function). *A landscape function* L_i *for an objective variable* x_i *is said to be* **exact** *if and only if* $L_i = f_i^*$.

The proposition below gives a sufficient condition for a landscape function to be exact.

Proposition 3. *If function* f_i^* *is quasiconvex and if the propagation of* $f(x) \leq z$ *performs bound-consistency on variables* x *then* L_i *is exact.*

Proof. By Property 2, we know $L_i \leq f_i^*$. The proof that $f_i^* \leq L_i$ exploits bound-consistency (and quasiconvexity of f_i^* where bounding functions $X_i^L(z)$ and $X_i^U(z)$ are discontinuous). See Appendix for details. □

Proposition 3 is interesting because its condition holds in many practical cases. In particular, it holds as soon as the objective function is an aggregation $\Theta_{i=1}^n(f_i(x_i))$ (where Θ is a *sum, min, max,* or *product of non-negative terms*) of individual quasiconvex terms $f_i(x_i)$ as it is easy for these expressions to provide a bound-consistent propagation algorithm. The flow-shop scheduling problem with earliness-tardiness cost presented in the introduction is clearly an example as soon as the earliness-tardiness cost function is quasiconvex like the one on Fig. 1. Note that on this example, as the minimal value of the cost functions ET_i is 0, the landscape function L_i of a job i is exactly its earliness-tardiness cost function ET_i.

In fact we can also show that the condition of Proposition 3 is a necessary condition leading to the following theorem that characterizes exact landscapes.

Theorem 1 (Characterization of exact landscapes). *A landscape function* L_i *is exact if and only if function* f_i^* *is quasiconvex and the propagation of* $f(x) \leq z$ *performs bound-consistency on variables* x.

Proof. Proof combines Propositions 1 and 3 and the fact that if propagation does not perform bound-consistency, one can easily exhibit cases where $L_i(v) > f_i^*(v)$ for a given v.   □

It is important to keep in mind that objective landscapes are used as a heuristic to guide the search so in practice, even if the conditions of Theorem 1 are not met, the landscape functions can still convey some relevant information and be useful. This is in particular the case when the same variable x_i appears several time in the formulation of objective function f so that the propagation algorithm is not guaranteed to achieve bound consistency.

4   Objective Landscape Computation

Principles of the Computation. Given a range $[Z^L, Z^U]$ for the objective function, coming for instance from the propagation at the root node, a simple way of computing landscape functions consists in discretizing the objective values by selecting m values in the domain $z_1 = Z^U > z_2 > \cdots > z_j > \cdots > z_{m-1} > z_m = Z^L$ and recording the bounds $X_i^L(z_j)$ and $X_i^U(z_j)$ for each variable x_i as shown in Algorithm 1. Note that the objective bounds z_j are traversed by decreasing value so, by propagation monotony, the domains of variables x_i will only decrease and no reversibility/backtrack is needed.

Algorithm 1. Objective landscapes basic computation

1: **for** j in 1...m **do**
2:     propagate($f(x) \leq z_j$)
3:     **for** i in 1...n **do**
4:         $X_i^L(z_j) \leftarrow$ current lower bound of x_i
5:         $X_i^U(z_j) \leftarrow$ current upper bound of x_i

---

The discretized values of the objective landscape functions computed in Algorithm 1 can be interpolated by a linear approximation or by a lower bounding step function that would preserve the lower bound property of Proposition 2.

It is of course important to select some objective values z_j that are representative of the values explored during the search. In a minimization problem, we can expect that small objective values are more useful than large ones, this is why in our implementation in CP Optimizer we decided to use a logarithmic scheme. More precisely, before the model is actually solved (see [6] for an overview of the automatic search), the landscape computation step extracts only the objective function $f(x)$ and performs m_{LOG} logarithmic steps of Algorithm 1 using $z_j = \frac{Z^U - Z^L}{2^{j-1}} + Z^L$. It records the first objective value $Z = z_j$ in the descent such that the next one (z_{j+1}) has an impact on the bounds of some variable x_i and then performs a new run of Algorithm 1, this time with a linear sampling with m_{LIN} steps of interval $[Z^L, Z]$. This process is illustrated on Fig. 4.

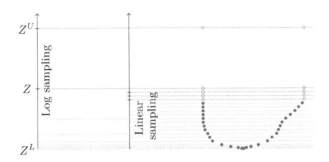

Fig. 4. Objective landscape computation

Complexity. In practice, just a few tens of points (value m in Algorithm 1) are needed to represent a landscape function L_i with sufficient precision. The landscape function structure can be as simple as an array of pairs $[v_j, L_i(v_j)]_{j \in [1...m]}$ and estimating its value $L_i(v)$ for a given v can be performed in $O(\log(m))$ with a binary search. The algorithmic cost of computing the landscapes is in $O(m(n + p))$ if n is the number of objective variables and p the cost of propagating an objective upper bound $f(x) \leq z$. This is negligible compared to the typical resolution time of the optimization problem.

Examples. Some examples of computed objective landscapes using a linear interpolation are illustrated on Figs. 5, 6, 7 and 8. On these figures, each function describes the landscape of a decision variable of the problem. Note that in these problems, objective decision variables are time variables (start/end of a time interval or temporal delay on a precedence constraint).

– On Fig. 5 the problem is a one-machine problem with V-shape earliness-tardiness costs and a common due date for all activities [2]. We see that the common due date (actual value is 2171 on this instance) as well as the different weights for the earliness-tardiness costs are effectively captured by the landscapes.
– On Fig. 6 the problem is the flow-shop scheduling with earliness-tardiness cost of our example with some particular V-shape costs, we observe in particular the different due-dates of the jobs.
– Fig. 7 illustrates a semiconductor manufacturing scheduling problem described in [4] whose objective function is the weighted sum of two types of cost: some weighted completion times of jobs (the linear functions) and some delays on precedence constraints between operations that incur a large quadratic cost.

- On Fig. 8 the problem is a resource-constrained project scheduling problem with maximization of net present value [13]. The landscape functions clearly distinguish between the tasks with positive cash flow (increasing landscape function) and the ones with negative one (decreasing landscape function) and reveal the exponential nature of the objective terms $(e^{-\alpha t})$.

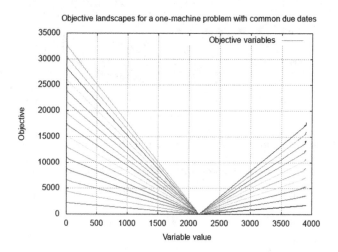

Fig. 5. Objective landscapes for a problem with common due dates [2]

5 Objective Landscape Exploitation

Once available, the objective landscape functions could be used in several ways during the search. In this article we focus on their use in the context of Large Neighborhood Search (LNS) but they could be used in variable/value ordering heuristic in a classical CP search tree, and more generally in Constrained Optimization Problems. In an LNS framework as the one used in the automatic search of CP Optimizer [6], landscapes can be used:

- To prevent cost degradation for some variables in LNS moves
- To define new types of neighborhoods
- To select variables and values in LNS completion strategies.

At the root node of an LNS move, if we are trying to improve an incumbent solution S of cost Z, once the selected fragment has been relaxed and given that constraint $f(x) \leq Z$ has been propagated, there are three values of interest for a given objective variable x_i:

- X_i^L the current lower bound of the variable
- X_i^U the current upper bound of the variable
- X_i^S the value of the variable in solution S.

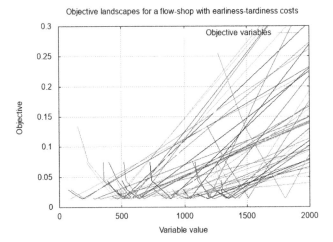

Fig. 6. Objective landscapes for a flow-shop with earliness-tardiness cost [10]

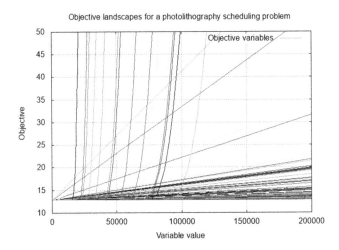

Fig. 7. Objective landscapes for a problem in semiconductor manufacturing [4]

We have of course $X_i^L \leq X_i^S \leq X_i^U$ as solution S is consistent with the current bounds. Thanks to the landscape function L_i of the variable, we can compute some interesting indicators illustrated on Fig. 9:

- $l_i = L_i(X_i^S)$ is the objective landscape value of variable x_i at solution S.
- As function L_i is quasiconvex, the set $\{v \in [X_i^L, X_i^U] | L_i(v) \leq l_i\}$ is a convex interval denoted $[X_i^{S-}, X_i^{S+}]$, one of its endpoint (X_i^{S-} or X_i^{S+}) being equal to X_i^S. This interval represents the values v in the current domain of x_i that do not degrade the landscape function value compared to solution S.

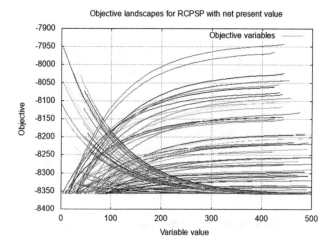

Fig. 8. Objective landscapes for an RCPSP with net present value [13]

- $o_i = l_i - \min_{v \in [X_i^L, X_i^U]} L_i(v)$ is an optimistic bound on how much the objective landscape value of the variable could be improved by choosing the best value in the current domain.
- $p_i = \max_{v \in [X_i^L, X_i^U]} L_i(v) - l_i$ is a pessimistic bound on how much the objective landscape value of the variable could be degraded by choosing the worse value in the current domain.
- ω_i is the derivative of landscape function L_i evaluated at value X_i^S in direction of the improvement of the landscape value. It measures how much a small change from the solution value could improve the landscape function.
- π_i is the derivative of landscape function L_i evaluated at value X_i^S in direction of the degradation of the landscape value. It measures how much a small change from the solution value could degrade the landscape function.

We are far from having investigated all the above indicators and, more generally, all the potentialities of landscapes during the search. In our initial implementation in CP Optimizer 12.7.1, landscapes are exploited as follows.

- At a given LNS move, the landscape strategy described below is used with a certain probability that is learned online as described in [5]
- Landscape strategy:
  - A certain number of objective variables are randomly selected. The ratio of selected variables is learned online from a set of possible ratio values in $[0, 1]$.
  - For each selected objective variable x_i, a constraint $v \in [X_i^{S-}, X_i^{S+}]$ is added to ensure that the landscape value of these variables is not degraded during the LNS iteration.

This type of strategy focuses the search on the improvement of a subset of the objective terms, it is particularly useful for objective expressions that are not

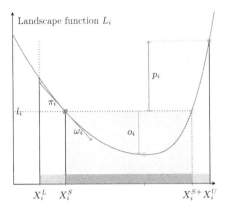

Fig. 9. Some objective landscape indicators

well exploited by constraint propagation, typically like *sum* expressions which are extremely common in scheduling problems.

In the flow-shop scheduling example, when the landscape strategy is used at an LNS move, it will randomly select a subset of jobs and for each selected job i, post the additional constraint on the job end time that it must not degrade its earliness-tardiness cost ET_i compared to what it was in the incumbent solution.

6   Results

In this section, we compare the performance of CP Optimizer V12.7.1 (containing the implementation of landscapes as described in Sects. 4 and 5) with the same version that do not compute and use landscapes[4]. The comparison was performed on the IBM scheduling benchmark which, as of today, consists of 135 different families of scheduling problems collected from different sources (classical instances for famous scheduling problems like job-shop or RCPSP, benchmarks proposed in academic papers, industrial scheduling problems modeled by our consultants, partners or end-users, problems submitted on our Optimization forums, *etc.*). The performance comparison between the two versions of the automatic search is measured in terms of resolution time *speed-up*. As the engine is not able to solve all instances to optimality in reasonable time, the time speed-up estimates how many times faster the default version of the automatic search (using landscapes) is able to reach similar quality solutions to the version without landscapes. Results are summarized on Fig. 10. The 135 different families are sorted on the x-axis by increasing speed-up factor. Each point is the average speed-up over the different instances of the family. The geometrical mean of the speed-up over the 135 families is around 1.5. On Table 1, the families are classified by objective types, depending on the aggregation function (usually, a *sum* or a *max*) and the nature of the aggregated terms. These objective types are also

---

[4] A specific parameter can be used to switch off objective landscapes.

shown with different marks on Fig. 10. As expected, the objective landscapes have a strong added value in the case of sums of terms involving expressions with large domains (like start/end of intervals with an average speed-up of 2.62, interval lengths with a speed-up larger than 3 or resource capacities (height)). As an illustration, for the flow-shop scheduling example with earliness-tardiness cost described in the introduction, the average speed-up factor is larger than 18. Note that for some families of scheduling problems the speed-up factor is close to two orders of magnitude. The speed-up is much more modest for objectives like makespan (max of ends) or total resource allocation costs (sum of presence of intervals).

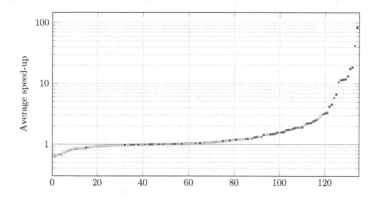

Fig. 10. Average speed-up when using objective landscapes measured on 135 different families of scheduling problems.

Table 1. Average speed-up by objective type

Aggregation type	Main variables type	Number of families	Average speed-up	Mark on Fig. 10
Max	Start/end	55	1.05	○
Sum	Start/end	40	2.62	●
Sum	Presence	29	0.98	■
Sum	Length	7	3.31	◆
Sum	Height	4	1.88	▲

7   Conclusion and Discussion

This paper introduces the notion of *objective landscapes* in CP. Objective landscapes are light-weight structures that can be fast computed before the search by exploiting constraint propagation on the objective function expression. We show some formal properties of landscapes and give some results about their implementation in the automatic search of CP Optimizer: an average speed-up of 50% on a large panel of scheduling problems, with up to almost 2 orders of magnitude for some applications.

We have only explored a small part of the potentials of objective landscapes and there are still many interesting open questions:

- Landscapes are currently defined by only considering the objective expression $f(x)$ and the objective variables x. The definition of objective variables clearly depend on how the problem is formulated. There may be many different subsets of decision variables x such that the objective functionally depends on x. Some may be more interesting than others. There is here an evident link with the notion of *functional dependency* heavily studied in relational databases but much less in CP [14].
- In fact functional dependency is not strictly required. We could also compute some landscapes on non-objective variables by considering the constraints $c(x, y)$ of the problem. But this messes up the theoretical framework about exact landscape computations mostly because achieving bound-consistency is in general impossible on the whole problem. It also results in a phenomenon that tends to hide the lowest part of landscape functions L_i as, because of constraint propagation, the model including constraints $c(x, y)$ will start to fail for larger values of z resulting in a larger value for Z^L.
- Objective landscapes can be extended without too much difficulty to handle holes in the domains. The landscape structure would be slightly more complex, looking like a tree describing how values or intervals of values are removed from the domain of a variable x_i when the objective z is decreasing. This type of landscape could be useful for variables with a discrete semantics.
- As mentioned in Sect. 5, we only explored the tip of the iceberg as about how to exploit objective landscapes to guide the search.

Appendix: Proof of Proposition 3

Let $v \in D_i$, we want to prove that $L_i(v) = f_i^*(v)$. We distinguish 3 cases depending on the position of v with respect to $X_i^L(Z^L)$ and $X_i^U(Z^L)$.

Case 1: $v \in [X_i^L(Z^L), X_i^U(Z^L)]$

By definition of the landscape function we have $L_i(v) = Z^L$.

We first note that by definition of Z^L, for every objective value $z < Z^L$ as the propagation of $f(x) \leq z$ fails, it means that Z^L is a lower bound on $Z^* = \min_{x \in D} f(x)$.

As by definition of function f_i^* we have $\forall v \in D_i, Z^* \leq f_i^*(v)$. We can deduce: $L_i(v)(= Z^L) \leq f_i^*(v)$.

We will now use the bound-consistency property of propagation to show that $f_i^*(X_i^L(Z^L)) \leq Z^L$ and $f_i^*(X_i^U(Z^L)) \leq Z^L$. By definition $X_i^L(Z^L)$ is the minimal value in the domain of x_i after propagation of $f(x) \leq Z^L$. Because propagation is bound-consistency on x_i, it means that there exist some x with $x_i = X_i^L(Z^L)$ such that $f(x) \leq Z^L$ (otherwise, this value would have been filtered from the domain). Furthermore, by definition of function f_i^*, we have $f_i^*(X_i^L(Z^L)) \leq f(x)$, thus $f_i^*(X_i^L(Z^L)) \leq Z^L$. The proof that $f_i^*(X_i^U(Z^L)) \leq Z^L$ is symmetrical. Finally, the quasiconvexity of f_i^* implies that for all $v \in [X_i^L(Z^L), X_i^U(Z^L)]$, $f_i^*(v) \leq Z^L(= L_i(v))$.

Case 2: $v < X_i^L(Z^L)$

By definition of the landscape we have $L_i(v) = \min\{z|X_i^L(z) \leq v\}$.

We first prove that $f_i^*(v) \geq L_i(v)$. By contradiction, suppose that $f_i^*(v) < L_i(v)$ and let $z' = f_i^*(v)$. This would mean that there exist x with $x_i = v$ such that $f(x) = z'$. For such a z' we would have $X_i^L(z') \leq v$ because x supports objective value z' and thus value v cannot be removed from the domain of x_i. Such a value $z' < L_i(v)$ such that $X_i^L(z') \leq v$ contradict the fact $L_i(v)$ is the smallest z satisfying $X_i^L(z) \leq v$. This proves $f_i^*(v) \geq L_i(v)$.

For proving that $f_i^*(v) \leq L_i(v)$, we distinguish 2 sub-cases depending on whether or not there exist an objective value z such that $X_i^L(z) = v$.

Case 2.1: There exist a value z such that $X_i^L(z) = v$ so that $L_i(v) = z$.

By definition of X_i^L, this means that v is the minimal value in the domain of x_i after propagating $f(x) \leq z$. Because propagation is bound-consistent on x_i, it means that there exist some x with $x_i = v$ such that $f(x) \leq z$ (otherwise, this value would have been filtered from the domain). Furthermore, by definition of function f_i^*, we have $f_i^*(v) \leq f(x)$, thus $f_i^*(v) \leq z(= L_i(v))$.

Case 2.2: There does not exist any value z such that $X_i^L(z) = v$.

This is the case of a discontinuity of function $X_i^L(z)$. Let $v^+ = X_i^L(L_i(v))$ denote the value of X_i^L at the discontinuity. By definition of L_i, we have $L_i(v) = L_i(v^+)$. Furthermore, as v^+ satisfies case 2.1 above, we also know that $f_i^*(v^+) \leq L_i(v^+)$. We have seen in case 1 that $f_i^*(X_i^L(Z^L)) \leq Z^L$ and, of course, $Z^L \leq L_i(v^+)$. To summarize: $v \in [X_i^L(Z^L), v^+]$, $f_i^*(X_i^L(Z^L)) \leq L_i(v^+)$ and $f(v^+) \leq L_i(v^+)$. Thus, by quasiconvexity of f_i^*: $f(v) \leq L_i(v^+)(= L_i(v))$.

Case 3: $v > X_i^U(Z^L)$

This case is the symmetrical of Case 2. $\square$

References

1. Beck, J.C., Refalo, P.: A hybrid approach to scheduling with earliness and tardiness costs. Ann. Oper. Res. **118**(1–4), 49–71 (2003)
2. Biskup, D., Feldmann, M.: Benchmarks for scheduling on a single machine against restrictive and unrestrictive common due dates. Comput. Oper. Res. **28**(8), 787–801 (2001)
3. Greenberg, H., Pierskalla, W.: A review of quasi-convex functions. Oper. Res. **19**(7), 1553–1570 (1971)
4. Knopp, S., Dauzère-Pérès, S., Yugma, C.: Modeling maximum time lags in complex job-shops with batching in semiconductor manufacturing. In: Proceedings of the 15th International Conference on Project Management and Scheduling (PMS 2016), pp. 227–229 (2016)

5. Laborie, P., Godard, D.: Self-adapting large neighborhood search: application to single-mode scheduling problems. In: Baptiste, P., Kendall, G., Munier-Kordon, A., Sourd, F. (eds.) Proceedings of the 3rd Multidisciplinary International Conference on Scheduling: Theory and Applications (MISTA), Paris, France, pp. 276–284, 28–31 Aug 2007

6. Laborie, P., Rogerie, J., Shaw, P., Vilím, P.: IBM ILOG CP optimizer for scheduling. Constraints J. **23**(2), 210–250 (2018)

7. Laborie, P., Rogerie, J.: Temporal linear relaxation in IBM ILOG CP optimizer. J. Sched. **19**(4), 391–400 (2016)

8. Michel, L., Van Hentenryck, P.: Activity-based search for black-box constraint programming solvers. In: Beldiceanu, N., Jussien, N., Pinson, É. (eds.) CPAIOR 2012. LNCS, vol. 7298, pp. 228–243. Springer, Heidelberg (2012). https://doi.org/10.1007/978-3-642-29828-8_15

9. Monette, J., Deville, Y., Hentenryck, P.V.: Just-in-time scheduling with constraint programming. In: Proceedibgs of the 19th International Conference on Automated Planning and Scheduling (ICAPS 2009) (2009)

10. Morton, T., Pentico, D.: Heuristic Scheduling Systems. Wiley, New York (1993)

11. Pesant, G.: Counting-based search for constraint optimization problems. In: Proceedings of the 13th AAAI Conference on Artificial Intelligence (AAAI 2016), pp. 3441–3447 (2016)

12. Refalo, P.: Impact-based search strategies for constraint programming. In: Wallace, M. (ed.) CP 2004. LNCS, vol. 3258, pp. 557–571. Springer, Heidelberg (2004). https://doi.org/10.1007/978-3-540-30201-8_41

13. Vanhoucke, M.: A scatter search heuristic for maximising the net present value of a resource-constrained project with fixed activity cash flows. Int. J. Prod. Res. **48**, 1983–2001 (2010)

14. Vardi, M.Y.: Fundamentals of dependency theory. Technical report RJ4858, IBM Research (1985)

An Update on the Comparison of MIP, CP and Hybrid Approaches for Mixed Resource Allocation and Scheduling

Philippe Laborie[(⊠)]

IBM, 9 rue de Verdun, 94250 Gentilly, France
laborie@fr.ibm.com

Abstract. We consider a well known resource allocation and scheduling problem for which different approaches like mixed-integer programming (MIP), constraint programming (CP), constraint integer programming (CIP), logic-based Benders decompositions (LBBD) and SAT-modulo theories (SMT) have been proposed and experimentally compared in the last decade. Thanks to the recent improvements in CP Optimizer, a commercial CP solver for solving generic scheduling problems, we show that a standalone tiny CP model can out-perform all previous approaches and close all the 335 instances of the benchmark. The article explains which components of the automatic search of CP Optimizer are responsible for this success. We finally propose an extension of the original benchmark with larger and more challenging instances.

Keywords: Constraint programming · Resource allocation
Scheduling · CP Optimizer

1 Introduction

We consider the well known resource allocation and scheduling problem proposed in [6]. After recapping the problem definition (Sect. 2) and giving an overview of the state-of-the-art methods that have been proposed for solving it (Sect. 3), we present a very concise formulation of the problem in CP Optimizer (Sect. 4). Experimental results (Sect. 5) show that, using this model together with a parameter focusing the search on optimality proofs, CP Optimizer 12.7.1 outperforms all existing approaches and closes all the 335 instances of the benchmark. We finally propose an extension of the original benchmark (Sect. 6) with larger and more challenging instances.

2 Problem Definition

The problem proposed in [6] is defined by a set of jobs $\mathcal{J}$ and a set of facilities $\mathcal{I}$. Each job $j \in \mathcal{J}$ must be assigned to a facility $i \in \mathcal{I}$ and scheduled to start after

© Springer International Publishing AG, part of Springer Nature 2018
W.-J. van Hoeve (Ed.): CPAIOR 2018, LNCS 10848, pp. 403–411, 2018.
https://doi.org/10.1007/978-3-319-93031-2_29

its release date r_j, end before its due date d_j, and execute for p_{ij} consecutive time units. Each job j has a facility assignment cost f_{ij} and a resource requirement c_{ij} when allocated to facility i. Each facility $i \in \mathcal{I}$ has a capacity C_i and the constraint that the resource capacity must not be exceeded at any time. The problem is to minimize the total facility assignment cost.

A time-indexed MIP formulation of the problem where Boolean variable x_{ijt} is one if job j starts at discrete time t on facility i is:

$$\min \sum_{i \in \mathcal{I}} \sum_{j \in \mathcal{J}} \sum_{t \in \mathcal{T}} f_{ij} x_{ijt}$$

$$\text{s.t.} \sum_{i \in \mathcal{I}} \sum_{t \in \mathcal{T}} x_{ijt} = 1, \qquad \forall j \in \mathcal{J}$$

$$\sum_{j \in \mathcal{J}} \sum_{t' \in \mathcal{T}_{ijt}} c_{ij} x_{ijt'} \leq C_i, \quad \forall i \in \mathcal{I}, t \in \mathcal{T}$$

$$x_{ijt} = 0, \qquad \forall i \in \mathcal{I}, j \in \mathcal{J}, t \in \mathcal{T},$$
$$\qquad\qquad t < r_j \text{ or } t > d_j - p_{ij}$$

$$x_{ijt} \in \{0,1\}, \qquad \forall i \in \mathcal{I}, j \in \mathcal{J}, t \in \mathcal{T}.$$

In the above formulation, $\mathcal{T}_{ijt}$ denotes the set of time values t' such that if job j is allocated to resource i and starts at t' then it is executing at time t:

$$\mathcal{T}_{ijt} = \{t' \in \mathcal{T} | t - p_{ij} < t' \leq t\}$$

3   State of the Art

Although the problem can be considered as being *simple* compared to many real-world scheduling problem, it is clearly of interest because it mixes resource allocation and scheduling and it is NP-Hard in the strong sense. This problem has received a lot of attention in the combinatorial optimization community and several approaches have been studied and evaluated on the benchmark proposed in [6,7]. We give a short review of these approaches below.

Mixed Integer Programming (MIP). The time-indexed formulation described in Sect. 2 is often used as a baseline method to compare with [4,7]. As reported in [1], other MIP formulations were tried (like the event based ones, see [13] for a review) but on this particular problem, time-indexed formulations seem to perform better. In [5], a time-indexed MIP model using some redundant formulation was used and the reported results are, to the best of our knowledge, the best MIP results on the benchmark.

Constraint Programming (CP). Results using an OPL model on top of ILOG Scheduler were reported in [6,7]. In [4] the authors used an early version of IBM ILOG CP Optimizer (V2.3). More recently [12] has been using IBM ILOG CP

Optimizer 12.7 but the details of the model are not provided. All the experiments so far suggested that standalone CP is not really competitive with hybrid methods or even a standalone MIP.

Logic-Based Benders Decomposition (LBBD). An LBBD approach was originally proposed in [6,7] and shown to outperform both CP and MIP. In this approach, the master problem consists of the facility allocation problem whereas sub-problems consist in scheduling the jobs given a facility allocation provided by the master problem. A relaxation of the scheduling problem inspired from energetic reasoning [2] is also included in the master problem. In the original work, the scheduling sub-problems were solved using ILOG Scheduler. These results were updated in [1] by using IBM ILOG CP Optimizer (V12.4) for solving the scheduling sub-problems. Other results with LBBD on top of CP (denoted LBBD-CP in this paper) are also reported in [5].

Constraint Integer Programming (CIP). CIP is a resolution paradigm that exploits both the constraint propagation traditionally used in CP and the linear relaxations used in MIP in the same search tree. CIP models were studied in [3–5] showing that they were competitive with LBBD.

Satisfiability Modulo Theories (SMT). More recently, an SMT approach combining SAT with the theory of real arithmetics has been studied in [12] for solving the problem, together with a more efficient hybrid LBBD approach that uses SMT for solving the sub-problems (denoted LBBD-SMT).

4   CP Optimizer Model

CP Optimizer extends classical CP on integer variables with a few mathematical concepts (intervals, functions) that make it easier to model scheduling problems while providing interesting problem structure for its automatic search algorithm [10]. The complete CP Optimizer formulation of the resource allocation and scheduling problem proposed in [6] is shown on Fig. 1 (using OPL).

Lines 2–8 read data using similar notations as the ones introduced in Sect. 2. Line 10 creates one interval variable job[j] for each job j that is constrained to start after r_j and end before d_j. For each possible allocation of a job j on a facility i, line 11 creates an optional interval variable mode[i][j] of size p_{ij}. This interval variable will be present if and only if job j is allocated to facility i. The objective at line 13 is to minimize the weighted sum of the selected modes. Constraints on line 15 state that a job j is allocated to one facility: only one of the interval variables mode[i][j] is present and the start and end value of interval job[j] are the same as the ones of the selected mode. Finally at line 16, a cumul function expression on the left hand side represents the facility usage over time and is constrained to be lower than the facility capacity C_i.

```
1   using CP;
2   tuple ModeData { int p; int c; int f; }
3   tuple JobData { int r; int d; }
4   int n = ...; range J = 1..n;
5   int m = ...; range I = 1..m;
6   ModeData M[I,J] = ...;
7   JobData Job[J] = ...;
8   int C[I] = ...;
9
10  dvar interval job[j in J] in Job[j].r..Job[j].d;
11  dvar interval mode[i in I][j in J] optional size M[i][j].p;
12
13  minimize sum(i in I, j in J) (M[i][j].f * presenceOf(mode[i][j]));
14  subject to {
15    forall(j in J) { alternative(job[j], all(i in I) mode[i][j]); }
16    forall(i in I) { sum(j in J) pulse(mode[i][j],M[i][j].c) <= C[i]; }
17  }
```

Fig. 1. Complete CP Optimizer model

5   Results

5.1   Benchmark Description

The benchmark proposed in [7] is composed of 4 families of instances (c, e, de, df) with slightly different characteristics resulting in a total of 335 instances. For all instances the cost is such that faster facilities tend to be more expensive. In each families there are 5 instances for every combination (n, m) where n is the number of jobs and m the number of facilities.

Results in this section were obtained on an IBM blade with 20 GB RAM and an Intel® Xeon® X5570 2.93 GHz running GNU/Linux with CP Optimizer V12.7.1. Unless stated otherwise, we are using 4 parallel workers (Workers=4).

5.2   Experimental Evaluation of CP Optimizer Search Components

As explained in [10], the automatic search of CP Optimizer consists of two components run in concurrence: (1) a Large Neighborhood Search (LNS) heuristic that tries to improve the current solution by successively unfreezing and re-optimizing a part of the solution [8] and (2) a Failure-directed Search (FDS) that aims at proving optimality of the current solution [15].

CP Optimizer targets in priority industrial scheduling problems that are in general much larger than the ones of the benchmark. This explains why, by default, the automatic search spends more effort on LNS than on FDS but, as explained in [15], this behavior can be changed thanks to a search parameter FailureDirectedSearchEmphasis.

In a preliminary experimental study, using a 1 h time limit, we compare three variants of the CP Optimizer search: the default version (default), a version with a strong focus on FDS with 3.5 out of the 4 workers dedicated to FDS[1] (`fds`) and a version that do not use FDS[2] (`no_fds`). Results are shown on Table 1.

Table 1. Comparison of different CP Optimizer variants.

	no_fds	default	fds
Number of feasible solutions found	335/335	335/335	335/335
Maximal time for finding a feasible solution	0.33 s	0.33 s	0.33 s
Number of optimal solutions found	305/335	320/335	334/335
Maximal gap of best solution found	1.18%	0.92%	0.16%
Number of optimality proofs	143/335	309/335	330/335

First, it is to be noted that, whatever variant is used, CP Optimizer finds an initial feasible solution or prove infeasibility[3] for all 335 instances of the benchmark in less than 0.5 s. This is to be compared with LBBD approaches that, by construction, do not provide any feasible solution before the optimal solution is eventually found. For 309 instances, the `default` version finds the optimal solution and proves optimality, for the remaining 26 instances, the gap with respect to the optimal solution is less than 1%. The main difference between the 3 variants concerns the capability to prove optimality: as expected, the more FDS is used and the more optimality proofs we get. Within the 1 h time limit, the variant with a strong focus on FDS closes all but 5 instances of the benchmark. In next section, we compare the results of this variant with state-of-the-art approaches.

5.3 Comparison with Previous Results

Table 2 compares the results of CP Optimizer 12.7.1 (with a focus on optimality proofs through FDS) on the most challenging instances of the 'c' family using a time limit of 1 h with the best known results compiled for the main four approaches (MIP, LBBD-CP, LBBD-SMT, CIP-CP). For MIP and LBBD-CP, we compare with the results of [5] that are on average better than the ones reported in [1,7,12]. For LBBD-SMT we compare with the results of [12][4] and for CIP-CP with the ones of [5]. As in [5], the columns `geom` denote the shifted geometrical mean with a shift of 10 s over the 5 instances, taking any time lower than 1 s as being equal to 1s. Experiments in [5] were run using a single thread with a 2 h time-limit on a slightly slower machine (2.50 GHz v.s. 2.93GHz) so for a more fair comparison we also provide results of our model using one worker but still using a 1 h time-limit. CP Optimizer easily solves all the instances of the other families.

---

[1] Using `FailureDirectedSearchEmphasis=3.5`.

[2] Using `FailureDirectedSearch=Off`.

[3] All instances of the benchmark are feasible except for 5 instances of the `de` family.

[4] Computed from the detailed results the authors gratefully sent us.

Table 2. Comparison with state-of-the-art approaches.

#I	#J	MIP		LBBD-CP		LBBD-SMT		CIP-CP		CPO (fds) 1W		CPO (fds) 4W	
		opt	geom	opt	geom	opt	geom	opt	geom	opt	geom	opt	geom
2	16	5	8.0	5	**1.0**	5	2.82	5	4.7	5	1.05	5	1.00
	18	5	16.9	5	**1.3**	5	1.64	5	1.7	5	2.72	5	1.60
	20	5	29.0	5	3.7	5	1.47	5	**1.5**	5	2.01	5	1.62
	22	*4*	*812.4*	5	51.4	5	72.06	*3*	*382.5*	5	**7.37**	5	**4.06**
	24	*3*	*883.0*	*4*	*214.8*	5	196.72	*2*	*573.4*	5	**10.17**	5	**6.54**
	26	*4*	*1069.2*	5	209.0	*3*	*554.03*	*4*	*464.9*	5	**22.33**	5	**11.28**
	28	*4*	*378.9*	5	536.5	*4*	*38.58*	*4*	*42.0*	5	48.39	5	**17.00**
	30	*3*	*861.2*	*3*	*401.2*	*1*	*1147.84*	*2*	*587.6*	**5**	122.21	**5**	**92.30**
	32	*3*	*792.1*	*0*	-	*3*	*332.85*	*2*	*1140.5*	**5**	235.81	**5**	**120.14**
	34	*3*	*879.7*	*2*	*1745.1*	*3*	*509.45*	*1*	*1995.3*	*3*	*419.24*	*3*	*253.09*
	36	*2*	*1534.1*	*1*	*4770.2*	*2*	*450.68*	*3*	*548.4*	*3*	*1199.91*	*3*	*491.11*
	38	*2*	*4980.2*	*1*	*5848.7*	*4*	*428.51*	*2*	*1334.0*	*4*	*390.86*	*4*	*127.07*
3	18	5	46.0	5	5.8	5	2.43	5	4.8	5	**2.08**	5	**1.56**
	20	*4*	*98.5*	5	1.5	5	**1.33**	5	6.9	5	2.21	5	1.75
	22	*4*	*554.6*	5	2.3	5	**2.17**	5	6.6	5	4.55	5	2.90
	24	5	304.5	5	**6.7**	5	9.41	5	78.6	5	11.05	5	**6.24**
	26	*3*	*1652.8*	5	**19.8**	5	44.50	5	40.2	5	21.71	5	**10.28**
	28	*3*	*987.6*	5	35.4	5	70.54	*3*	*194.9*	5	**34.14**	5	**15.13**
	30	*3*	*3100.2*	*4*	*178.3*	*3*	*540.18*	*4*	*520.9*	5	158.48	5	**54.17**
	32	*2*	*3601.3*	*4*	*1951.8*	*2*	*665.42*	*3*	*559.0*	5	220.02	5	**117.26**
4	20	5	25.3	5	1.8	5	**1.15**	5	4.3	5	1.65	5	**1.09**
	22	5	60.0	5	3.7	5	**2.48**	5	15.0	5	3.22	5	**2.29**
	24	*4*	*1399.0*	5	12.1	5	19.22	5	42.9	5	**9.58**	5	**4.95**
	26	*3*	*2787.8*	5	**14.9**	5	17.12	5	112.7	5	20.56	5	**12.44**
	28	*3*	*2124.2*	5	**9.6**	5	29.15	5	200.0	5	17.30	5	10.32
	30	*2*	*3253.6*	5	**31.7**	5	110.23	5	581.1	5	153.54	5	**53.10**
	32	*1*	*4691.0*	5	118.3	5	450.84	5	1519.1	5	**93.22**	5	**44.09**

As we see, CP Optimizer generally outperforms existing approaches, both in terms of number of instances solved to optimality and in terms of solve time. With 130 instances solved out of 135, it is better than the virtual best solver of the 4 approaches studied in [5] (127 solved instances). In fact, the 5 remaining open problems could be closed with the same model by using a larger time limit (up to 160 h for the hardest one). Detailed list of optimal costs can be found at http://ibm.biz/AllocSched.

We think that an important ingredient of the success of Failure-Directed Search on these instances is its capability to opportunistically mix allocation and scheduling decisions in the same decision tree based on its decision rating system. In the CP Optimizer model, both allocation decisions (presence status of interval variables) and scheduling decisions (interval variables start and end values) hold on the same decision variables that are efficiently pruned by constraint propagation on optional intervals. The search space explored by the FDS is reduced by using strong pruning (like timetable edge finding [14]) and efficient propagation of conditional bounds on alternative constraints [11].

6   Benchmark Extension

Given the progress achieved until now on the original benchmark, we propose an extension with more challenging instances. In particular we want to address the following limitations:

- The current instances are small (up to 50 jobs and 10 facilities) and not really representative of the typical size of actual scheduling problems. We propose to generate new problems with a size up to 1000 jobs and 20 facilities.
- The time granularity is coarse, as the maximal duration of jobs is less than a few tens of units. We propose a grain 100 times finer.
- On a similar line, the granularity of facilities capacity (maximal capacity is 10) is also made 100 times finer.
- In the current benchmark, a given job j always requires the same quantity c_{ij} of the different facilities i. In our extension they can use different capacity. Instead of being roughly inversely proportional to the job duration, the facility cost is roughly inversely proportional to the job energy (product of the duration by the demanded quantity).
- As in the original version of the benchmark, we generated some precedence constraints between jobs. These precedences can be used optionally. Precedence constraints are ubiquitous in real-life scheduling but, as highlighted in [5], they destroy the independent sub-problem structure that LBBD and, to a lesser extend, the other approaches (CP Optimizer apart) exploit.

In the new instances, the capacity of each facility is 1000. As in the existing instances, we suppose that the facilities get in average slower as their index i increases. The duration p_{ij} is drawn uniformly from $[100\sqrt{i}, 1000\sqrt{i}]$. The required capacity c_{ij} is drawn uniformly from $[1, 1000]$. The cost f_{ij} for executing job j on machine i is roughly proportional to the inverse of the energy $e_{ij} = p_{ij}c_{ij}$ with a variability α_j that depends on the job j and is drawn uniformly from $[0.5, 1]$. More precisely, the cost f_{ij} is drawn from $[\alpha_j F_{ij}, F_{ij}]$ where $F_{ij} = 10^7\sqrt{m}/e_{ij}$. As for the original instances of the 'c' family, the release dates are all 0 and the deadlines are all equal to αL where $\alpha = 1/3$ and $L = \sum_{ij} p_{ij}/m^2$ is the average total processing time per facility. A set of precedence constraints can also optionally be considered: we generated $n/2$ precedence constraints between random pairs of predecessor/successor jobs while ensuring the resulting precedence graph is acyclic. We selected 20 different combinations $(n, m) \in [20, 1000] \times [2, 20]$. As in the original benchmark, 5 instances are generated for each different combinations (n, m) they are denoted fnjmmk.dat for $k \in [1, 5]$, so the benchmark extension consists of 100 new instances. In this paper we only report results without precedence constraints. The new benchmark is available at http://ibm.biz/AllocSched together with the detailed results discussed in this section as well as results with precedence constraints.

We experimented with 3 variants of the CP Optimizer search with a time limit of 1 h: the one with a focus on FDS (col. fds), the default search without any parameter change (col. default) and a version not using the temporal linear relaxation[5] [9] (col. no_tlr). The results are summarized on Table 3. We see that the first feasible solutions are produced, even for the largest instances, in less than 1 s (column fst). Some optimality proofs could be provided only for the smallest instances with 20 or 30 jobs and 2 facilities. The s_gap columns (for 'scheduling gap') measure the average gap of the produced solutions with respect to the lower bound of a MIP formulation of the allocation part of the problem with energetic resource relaxation (this is basically the MIP of the initial iteration of the master problem in LBBD approaches). As we see for problems with 20 jobs and 2 facilities for which CP Optimizer proves optimality of the global problem, this gap is significant (21.8%) meaning that the allocation part of the problem does not dominate the scheduling part[6]. As expected, when the size of the problem grows, the FDS becomes less useful and the performance is better when not using it[7]. Interestingly, for these problems, the temporal linear relaxation does not seem to be very helpful and, for the largest instances, it is penalized by the cost of running the LP relaxation at the root node of LNS moves. This would deserve a more detailed analysis.

Table 3. Comparison of CP Optimizer variants on the new instances.

#J	#I	fst	fds		default	no_tlr	#J	#I	fst	fds		default	no_tlr
		(s)	opt	s_gap	s_gap	s_gap			(s)	opt	s_gap	s_gap	s_gap
20	2	0.0	5	21.8%	21.8%	21.8%	200	10	0.1	0	20.4%	18.2%	**16.8%**
30	2	0.0	1	**10.6%**	11.0%	11.3%	200	20	0.2	0	25.2%	23.3%	**20.2%**
40	2	0.0	0	10.7%	8.9%	**8.7%**	500	2	0.1	0	**12.8%**	13.2%	14.1%
50	2	0.0	0	11.3%	9.8%	**9.0%**	500	5	0.1	0	**19.8%**	22.6%	20.3%
50	5	0.1	0	15.8%	**14.4%**	14.5%	500	10	0.2	0	27.0%	26.3%	**23.7%**
100	2	0.1	0	10.9%	8.5%	**8.1%**	500	20	0.4	0	33.2%	35.0%	**24.4%**
100	5	0.1	0	14.7%	13.1%	**13.0%**	1000	2	0.2	0	18.8%	19.2%	**17.8%**
100	10	0.1	0	18.5%	16.3%	**15.3%**	1000	5	0.3	0	30.6%	28.1%	**22.8%**
200	2	0.1	0	14.5%	9.8%	**9.3%**	1000	10	0.5	0	35.4%	36.2%	**33.4%**
200	5	0.1	0	17.2%	14.4%	**13.5%**	1000	20	1.0	0	35.7%	35.1%	**35.0%**

7   Conclusion

This paper provides an update on the comparison of different approaches for solving a well known allocation and scheduling problem. We show that with the recent advances in the automatic search algorithm of CP Optimizer, a standalone simple CP model outperforms all existing approaches. This simple declarative

---

[5] Using TemporalRelaxation=Off.

[6] As a comparison, this scheduling gap is only 0.77% in average for the instances of the 'c' family with 20 jobs and 2 facilities.

[7] Note that FDS is automatically switched off for large problems. Here, it is not being used for problems with 500 and 1000 jobs.

model allows to close the benchmark. We proposed an extension of the original benchmark with larger and more challenging instances for future research and provide a preliminary analysis of the results of CP Optimizer on the new instances.

References

1. Ciré, A., Çoban, E., Hooker, J.N.: Logic-based benders decomposition for planning and scheduling: a computational analysis. Knowl. Eng. Rev. **31**(5), 440–451 (2016)
2. Erschler, J., Lopez, P.: Energy-based approach for task scheduling under time and resources constraints. In: Proceedings of the 2nd International Workshop on Project Management and Scheduling, pp. 115–121 (1990)
3. Heinz, S., Beck, J.C.: Solving resource allocation/scheduling problems with constraint integer programming. In: Proceedings of the Workshop on Constraint Satisfaction Techniques for Planning and Scheduling Problems (COPLAS 2011), pp. 23–30 (2011)
4. Heinz, S., Beck, J.C.: Reconsidering mixed integer programming and MIP-based hybrids for scheduling. In: Beldiceanu, N., Jussien, N., Pinson, É. (eds.) CPAIOR 2012. LNCS, vol. 7298, pp. 211–227. Springer, Heidelberg (2012). https://doi.org/10.1007/978-3-642-29828-8_14
5. Heinz, S., Ku, W.-Y., Beck, J.C.: Recent improvements using constraint integer programming for resource allocation and scheduling. In: Gomes, C., Sellmann, M. (eds.) CPAIOR 2013. LNCS, vol. 7874, pp. 12–27. Springer, Heidelberg (2013). https://doi.org/10.1007/978-3-642-38171-3_2
6. Hooker, J.N.: A hybrid method for planning and scheduling. In: Wallace, M. (ed.) CP 2004. LNCS, vol. 3258, pp. 305–316. Springer, Heidelberg (2004). https://doi.org/10.1007/978-3-540-30201-8_24
7. Hooker, J.N.: Planning and scheduling by logic-based benders decomposition. Oper. Res. **55**(3), 588–602 (2007)
8. Laborie, P., Godard, D.: Self-adapting large neighborhood search: application to single-mode scheduling problems. In: Baptiste, P., Kendall, G., Munier-Kordon, A., Sourd, F. (eds.) Proceedings of the 3rd Multidisciplinary International Conference on Scheduling: Theory and Applications (MISTA 2007), pp. 276–284. Paris, France, 28–31 Aug 2007
9. Laborie, P., Rogerie, J.: Temporal linear relaxation in IBM ILOG CP optimizer. J. Sched. **19**(4), 391–400 (2016)
10. Laborie, P., Rogerie, J., Shaw, P., Vilím, P.: IBM ILOG CP optimizer for scheduling. Constraints J. **23**(2), 210–250 (2018)
11. Laborie, P., Rogerie, J.: Reasoning with conditional time-intervals. In: Proceedings of the 21th International Florida Artificial Intelligence Research Society Conference (FLAIRS 2008), pp. 555–560 (2008)
12. Mistry, M., D'Iddio, A.C., Huth, M., Misener, R.: Satisfiability modulo theories for process systems engineering. Optimization Online (2017)
13. Tesch, A.: Compact MIP models for the resource-constrained project scheduling problem. Master's thesis, Technische Universität Berlin (2015)
14. Vilím, P.: Timetable edge finding filtering algorithm for discrete cumulative resources. In: Achterberg, T., Beck, J.C. (eds.) CPAIOR 2011. LNCS, vol. 6697, pp. 230–245. Springer, Heidelberg (2011). https://doi.org/10.1007/978-3-642-21311-3_22
15. Vilím, P., Laborie, P., Shaw, P.: Failure-directed search for constraint-based scheduling. In: Michel, L. (ed.) CPAIOR 2015. LNCS, vol. 9075, pp. 437–453. Springer, Cham (2015). https://doi.org/10.1007/978-3-319-18008-3_30

Modelling and Solving the Senior Transportation Problem

Chang Liu[✉], Dionne M. Aleman, and J. Christopher Beck[✉]

Department of Mechanical and Industrial Engineering, University of Toronto,
Toronto, ON M5S 3G8, Canada
{cliu,aleman,jcb}@mie.utoronto.ca

Abstract. This paper defines a novel transportation problem, the
Senior Transportation Problem (STP), which is inspired by the elderly
door-to-door transportation services provided by non-profit organizations. Building on the vehicle routing literature, we develop solution
approaches including mixed integer programming (MIP), constraint programming (CP), two logic-based Benders decompositions (LBBD), and
a construction heuristic. Empirical analyses on both randomly generated
datasets and large real-life datasets are performed. CP achieved the best
results, solving to optimality 89% of our real-life instances of up to 270
vehicles with 385 requests in under 600 s. The best LBBD model can
only solve 17% of those instances to optimality. Further investigation of
this somewhat surprising result indicates that, compared to the LBBD
approaches, the pure CP model is able to find better solutions faster
and then is able to use the bounds from these sub-optimal solutions to
reduce the search space slightly more effectively than the decomposition
models.

1 Introduction

As the world population ages, there is an increasing demand for transit options
for elderly people who have difficulties accessing the regular public transit system but yet do not have disabilities that qualify them for specialized transit
services. As a consequence, there are non-profit organizations that provide such
"senior transportation" services in many communities. However, the resources
for these services are often limited and many elders are put on waiting lists.
Furthermore, due to lack of expertise and decision support tools, the schedules
assigned to the drivers are often sub-optimal as many vehicles do not operate at
full capacity. Therefore, finding optimal schedules is crucial for organizations to
meet increasing demands.

The Senior Transportation Problem (STP) is a static optimization problem in
which a fixed fleet of heterogeneous vehicles from multiple depots must satisfy
as many door-to-door transportation requests as possible within a fixed time
horizon. Due to the limited resources, not all requests can be met within the given
time and, therefore, the problem is to select a subset of requests such that the
total weight of all served requests is maximized. As some of the drivers operate on

© Springer International Publishing AG, part of Springer Nature 2018
W.-J. van Hoeve (Ed.): CPAIOR 2018, LNCS 10848, pp. 412–428, 2018.
https://doi.org/10.1007/978-3-319-93031-2_30

a volunteer basis, the problem includes characteristics such as multiple depots, heterogeneous vehicles, and time windows on both locations and vehicles.

Our primary contributions are to formally define the STP and to provide solution techniques for the STP. Four exact methods based on mixed integer programming (MIP), constraint programming (CP), and two logic-based Benders decomposition (LBBD) models plus a construction heuristic are developed. We define and present detailed experimental results and analyses for each approach. On real-world data from a non-profit organization, CP performs substantially better than the other approaches, solving over 89% of the problems to optimality.

2   Problem Definition

Let $G = (\mathcal{V}, \mathcal{A})$ be a directed complete graph with vertex set $\mathcal{V} = \mathcal{D} \cup \mathcal{N}$, where $\mathcal{D}$ represents the depot vertices and $\mathcal{N}$ represents the client vertices. Each vertex $i \in \mathcal{V}$ is associated with a time window $[E_i, L_i]$ and a service duration S_i corresponding to the time to be spent at location i. Each arc $(i, j) \in \mathcal{A}$ has a non-negative routing time $T_{i,j}$ satisfying the triangular inequality.

Let $\mathcal{K} = \{1, \ldots, |\mathcal{K}|\}$ represent the set of vehicles. Each vehicle $k \in \mathcal{K}$ is associated with a starting and ending depot $i_{k^+}, i_{k^-} \in \mathcal{D}$ where the vehicle must start and end, respectively. Multiple vehicles can share a depot but relocation of vehicles between depots is not allowed. Each vehicle also specifies its availability via time windows: $[E_{i_{k^+}}, L_{i_{k^+}}]$ and $[E_{i_{k^-}}, L_{i_{k^-}}]$. If the vehicle is used, it must leave its starting depot during the first interval, perform all pickup and delivery requests assigned to it, and arrive at its ending depot during the second interval. Furthermore, vehicles differ in capacity, with each vehicle $k \in \mathcal{K}$ associated with a maximum capacity C_k.

Let $\mathcal{R} = \{1, \ldots, |\mathcal{R}|\}$ represent the set of requests. Each request r is paired with a positive weight, W_r, denoting its importance. The total weight of served requests is the basis of the objective function. A request $r \in \mathcal{R}$ has an associated pickup location $i^+ \in \mathcal{N}$ and a delivery location $i^- \in \mathcal{N}$. In addition, each client is restricted to a maximum ride time, F, on any vehicle. The time horizon is denoted by Z. The load size is positive for a pickup location vertex and negative for a delivery vertex, $Q_i = -Q_{|\mathcal{R}|+i}, \forall i \in \mathcal{R}^+$.

A *route* for vehicle k is a sequence of vertices, $[i_{k^+}, \ldots, i_{k^-}]$ and a request is *served* when it is part of a route. The set of routes must satisfy the following constraints:

1. The pickup and delivery vertices of any request must be on the same route;
2. The pickup vertex must precede the delivery vertex;
3. A vertex is visited by at most one vehicle;
4. The load of a vehicle k cannot exceed its maximum capacity C_k at any point;
5. A route must start and end within the vehicle's availability window;
6. No sub-tours are allowed in any route;
7. The ride time of a client cannot exceed the maximum ride time F;
8. All pickups and deliveries must be served within their time windows.

3   Related Work

There are three levels of decisions in the STP: the selection of requests, the assignment of vehicles to requests, and the routing of vehicles. Each decision problem is a well-studied problem on its own.

The selectivity and routing aspects of STP can be viewed as a Team Orienteering Problem (TOP) [10]. Alternatively, the routing and assignment of requests can be seen as a Pickup and Delivery Problem with Time Windows (PDPTW) [5,6] or a Dial-a-Ride Problem (DARP) [2]. In addition to minimizing total travel cost in the classical DARP, Cordeau and Laporte [2] noted that there can be other objectives, such as maximizing the number of fulfilled demands or overall quality of service, but did not provide any formulation or references. The PDPTW has been solved to optimality for loosely constrained instances of sizes up to 100 requests [9] while the DARP has only been solved to optimality for problems with 24 requests [2]. The most common solution approaches are heuristic.

The combination of the three decisions has only been looked at by two groups. Baklagis et al. [1] proposed a branch-and-price framework to tackle this problem and Qiu et al. [7] investigated a graph search and a maximum set packing formulation specially tailored for homogeneous fleets. These works however are missing three components that are critical to the STP: multiple depots, maximum ride times for clients, and heterogeneous fleets.

4   Models for the Senior Transportation Problem

We present four exact methods (MIP, CP, and two LBBD approaches) and one heuristic to solve the STP. Both LBBD approaches employ a CP sub-problem while they use MIP and CP for the master problem, respectively.

4.1   Mixed Integer Programming

In Fig. 1, we present a MIP formulation adapted from the PDPTW formulation of Ropke and Cordeau [9]. The formulation uses three variables: a binary variable $x_{k,i,j}$ and two continuous variables $u_{k,i}$ and $v_{k,i}$. $x_{k,i,j} = 1$ if vehicle k visits location j immediately after visiting location i and 0 otherwise. $u_{k,i}$ indicates the time when vehicle k leaves location $i \in \mathcal{V}$. It is non-negative and less than or equal to the maximum time horizon Z. Variables $v_{k,i}$ indicate the load of vehicle k after visiting location $i \in \mathcal{V}$. They are non-negative and less than or equal to the vehicle capacity C_k.

The objective function (1) maximizes the sum of the weights of served requests. Constraints (2) and (3) ensure that each vehicle leaves from its starting depot and ends at its ending depot. Constraint (4) allows for the selectivity of requests. Constant flow is enforced with Constraint (5). Constraint (6) specifies that the pickup and delivery locations of a request must be visited by the

$$\max \sum_{k \in \mathcal{K}} \sum_{r \in \mathcal{R}} \sum_{j \in \mathcal{V}} \left(W_r \times x_{k,i_{r+},j} \right) \tag{1}$$

$$\text{s.t.} \quad \sum_{j \in \mathcal{N}^+} x_{k,i_{k+},j} + x_{k,i_{k+},i_{k-}} = 1 \qquad \forall k \in \mathcal{K} \tag{2}$$

$$\sum_{i \in \mathcal{N}^-} x_{k,i,j_{k-}} + x_{k,i_{k+},i_{k-}} = 1 \qquad \forall k \in \mathcal{K} \tag{3}$$

$$\sum_{k \in \mathcal{K}} \sum_{j \in \mathcal{V}} x_{k,i_{r+},j} \leq 1 \qquad \forall r \in \mathcal{R} \tag{4}$$

$$\sum_{j \in \mathcal{V}} (x_{k,i,j} - x_{k,j,i}) = 0 \qquad \forall k \in \mathcal{K}, i \in \mathcal{N} \tag{5}$$

$$\sum_{j \in \mathcal{V}} \left(x_{k,i_{r+},j} - x_{k,j,i_{r-}} \right) = 0 \qquad \forall k \in \mathcal{K}, r \in \mathcal{R} \tag{6}$$

$$u_{k,j} \geq (u_{k,i} + T_{i,j} + S_j) - M \times (1 - x_{k,i,j}) \qquad \forall k \in \mathcal{K}, i, j \in \mathcal{V} \tag{7}$$

$$u_{k,i} \geq E_i - M \times \left(1 - \sum_{j \in \mathcal{V}} x_{k,i,j} \right) \qquad \forall k \in \mathcal{K}, i \in \mathcal{V} \tag{8}$$

$$u_{k,i} \leq L_i - S_i + M \times \left(1 - \sum_{j \in \mathcal{V}} x_{k,i,j} \right) \qquad \forall k \in \mathcal{K}, i \in \mathcal{V} \tag{9}$$

$$u_{k,i_{r+}} \leq u_{k,i_{r-}} \qquad \forall k \in \mathcal{K}, r \in \mathcal{R} \tag{10}$$

$$\left(u_{k,i_{r-}} - u_{k,i_{r+}} \right) \leq F \qquad \forall k \in \mathcal{K}, r \in \mathcal{R} \tag{11}$$

$$v_{k,j} \geq (v_{k,i} + Q_i) - M \times (1 - x_{k,i,j}) \qquad \forall k \in \mathcal{K}, i, j \in \mathcal{V} \tag{12}$$

$$x_{k,i,j} \in \{0,1\} \qquad \forall k \in \mathcal{K}, (i,j) \in \mathcal{A} \tag{13}$$

$$0 \leq u_{k,i} \leq Z \qquad \forall k \in \mathcal{K}, i \in \mathcal{V} \tag{14}$$

$$0 \leq v_{k,i} \leq C_k \qquad \forall k \in \mathcal{K}, i \in \mathcal{V} \tag{15}$$

Fig. 1. MIP model for the Senior Transportation Problem.

same vehicle. In Constraint (7), the travel time and service time of visited locations are enforced. Constraints (8) and (9) make sure that each location that is visited must be visited within its time window. Constraint (10) imposes that pickup locations must precede delivery locations. Constraint (11) enforces that each ride does not exceed the maximum ride time. Constraint (12) keeps track of the load of each vehicle after visiting the location.

4.2 Constraint Programming

The CP formulation (Fig. 2) employs optional interval variables [4] that are linked using cumulative functions and sequence expressions. Each location $i \in \mathcal{N}$ is an optional interval variable x_i that is bounded by its time windows and the length of its service time. We assume that each vehicle visits its depot locations

$$\max \sum_{r \in \mathcal{R}} \left(W_r \times \texttt{PresenceOf}(x_{i_{r+}}) \right) \tag{16}$$

$$\text{s.t. } \texttt{Alternative}(x_i, X_i) \qquad\qquad\qquad \forall i \in \mathcal{N} \tag{17}$$

$$\texttt{Before}(\bar{X}_{k,i_{r+}}, \bar{X}_{k,i_{r-}}) \qquad\qquad\qquad \forall k \in \mathcal{K}, \forall r \in \mathcal{R} \tag{18}$$

$$\texttt{PresenceOf}(\bar{X}_{k,i_{r+}}) = \texttt{PresenceOf}(\bar{X}_{k,i_{r-}}) \qquad \forall k \in \mathcal{K}, \forall r \in \mathcal{R} \tag{19}$$

$$\texttt{StartOf}(x_{i_{r-}}) - \texttt{EndOf}(x_{i_{r+}}) \leq F \qquad\qquad \forall r \in \mathcal{R} \tag{20}$$

$$v_{k,i} = \texttt{StepAtStart}(\bar{X}_{k,i}, Q_i) \qquad\qquad \forall k \in \mathcal{K}, \forall i \in \mathcal{N} \tag{21}$$

$$\sum_{i \in \mathcal{N}} v_{k,i} \leq C_k \qquad\qquad\qquad\qquad \forall k \in \mathcal{K} \tag{22}$$

$$\texttt{First}(u_k, x_{i_{k+}}) \qquad\qquad\qquad\qquad \forall k \in \mathcal{K} \tag{23}$$

$$\texttt{Last}(u_k, x_{i_{k-}}) \qquad\qquad\qquad\qquad \forall k \in \mathcal{K} \tag{24}$$

$$\texttt{NoOverlap}(u_k, T) \qquad\qquad\qquad\qquad \forall k \in \mathcal{K} \tag{25}$$

Fig. 2. CP model for the Senior Transportation Problem.

regardless of whether it is assigned requests or not. The presence of x_i in the final solution implies that the location is visited by a vehicle. Auxiliary interval variables $X_{i,k}$ and $\bar{X}_{k,i}$ are transpositions of each other (i.e., $X_{i,k} = \bar{X}_{k,i}$), and link the x_i variables to vehicles through the use of the **Alternative** constraint. The presence of $X_{i,k}$ and $\bar{X}_{k,i}$ indicates that location i is visited by vehicle k. Cumulative functions $v_{k,i}$ are expressions that model the load of vehicle k after visiting location i. Finally, each route is modelled by a sequence variable u_k whose value is a permutation of locations visited by vehicle k.

The objective function (16) maximizes the total weight of fulfilled requests. The **Alternative** constraint in Constraint (17) indicates that if a variable (x_i) is present, then exactly one variable in the set of variables X_i (a vector of variables $X_{i,k}$) can be present, ensuring that at most one vehicle can visit location i. In Constraint (18), the **Before** constraint ensures that each pickup location is visited before its corresponding delivery location. Constraint (19) enforces that if the pickup location is served by vehicle k, then its associated delivery location must also be served by the same vehicle k. The difference between the end time of a delivery location variable and the start time of the respective pickup location variable must be less than the maximum ride time and is enforced through Constraint (20). In Constraint (21), the *cumul* function $v_{k,i}$ is defined such that for each vehicle k, the variable changes by the load size of location i, Q_i, at the start of the location variable of vehicle k ($\bar{X}_{k,i}$) where the size is positive for a pickup and negative for a delivery. Constraint (22) enforces that the sum of the load variables does not exceed the capacity of the vehicle. Constraints (23) and (24) indicate that each route must start at its associated start depot and end at its associated end depot. The CP model uses the **NoOverlap** global constraint (25) to prevent sub-tours on each route; it specifies that all present

$$\max \sum_{r \in \mathcal{R}} \sum_{k \in \mathcal{K}} (W_r \times \varphi_{k,r}) \tag{26}$$

$$\text{s.t.} \sum_{k \in \mathcal{K}} y_{k,i} \leq 1 \qquad\qquad\qquad \forall i \in \mathcal{N} \quad (27)$$

$$\zeta_r = S_{i_{r+}} + T_{i_{r+},j_{r-}} + S_{i_{r-}} \qquad\qquad \forall r \in \mathcal{R} \quad (28)$$

$$Q_r \times \varphi_{k,r} \leq P_k \qquad\qquad \forall k \in \mathcal{K}, r \in \mathcal{R} \quad (29)$$

$$(E_{i_{r+}} + \zeta_r) \times \varphi_{k,r} \leq L_{i_{k-}} \qquad\qquad \forall k \in \mathcal{K}, r \in \mathcal{R} \quad (30)$$

$$(E_{i_{k+}} + \zeta_r) \times \varphi_{k,r} \leq L_{i_{r-}} \qquad\qquad \forall k \in \mathcal{K}, r \in \mathcal{R} \quad (31)$$

$$\sum_{i \in \mathcal{N}} (y_{k,i} \times T_i + S_i) + T_{i_{k+}} + S_{i_{k+}}$$

$$\leq L_{i_{k-}} - E_{i_{k+}} \qquad\qquad \forall k \in \mathcal{K} \quad (32)$$

$$y_{k,r} = y_{k,r+|\mathcal{R}|} = \varphi_{k,r} \qquad\qquad \forall k \in \mathcal{K}, r \in \mathcal{R} \quad (33)$$

$$y_{k,i}, \varphi_{k,r} \in \{0,1\} \qquad\qquad \forall k \in \mathcal{K}, i \in \mathcal{N}, r \in \mathcal{R} \quad (34)$$

$$\textit{Benders Cuts}$$

Fig. 3. A MIP model for the LBBD master problem of the STP.

interval variables on the sequence variable u_k must not overlap in operation times while considering the transition time between all locations defined through the transition distance matrix T.

4.3   Logic-Based Benders Decompositions

For the LBBD approaches [3], we decompose the STP into a relaxed master problem and a number of sub-problems. The master problem finds the optimal relaxed assignment of requests to vehicles. Each sub-problem is, then, an optimization problem to find the optimal route given the assigned requests. If the optimal route for each sub-problem satisfies all requests assigned to it, then the global optimal solution has been found, otherwise, a Benders cut is produced. The LBBD models find a feasible global solution at every iteration since the route found in each sub-problem is feasible when the master objective value is ignored. We present one MIP and one CP formulation of the master problem and a single CP model for the sub-problem.

MIP Master Problem. The master problem assigns each request into a vehicle using integer decision variables $\varphi_{k,r}$ which equal 1 if request r is assigned to vehicle k and 0 otherwise, and $y_{k,i}$ which equal 1 if location i is visited by vehicle k and 0 otherwise. Instead of modelling the exact travel distance between consecutive locations, we compute the minimum travel time from each location i to any other, denoted with T_i. The sum of minimum travel time of all locations assigned to a vehicle must be less than or equal to the time availability of the vehicle. All other routing constraints are ignored in the master problem.

max Objective (16)

s.t. Constraints (17), (19)

$$\texttt{EndBeforeStart}(x_{i_{k^+}}, X_{i,k}) \hspace{3cm} \forall k \in \mathcal{K}, i \in \mathcal{N} \hspace{1cm} (35)$$

$$\texttt{EndBeforeStart}(X_{i,k}, x_{i_{k^-}}) \hspace{3cm} \forall k \in \mathcal{K}, i \in \mathcal{N} \hspace{1cm} (36)$$

$$\texttt{EndBeforeStart}(x_{i_{k^+}}, x_{i_{k^-}}) \hspace{3cm} \forall k \in \mathcal{K} \hspace{1cm} (37)$$

$$\sum_{i \in \mathcal{N}} (\texttt{PresenceOf}(X_{i,k}) \times \mathcal{T}_i + S_i)$$

$$+ \mathcal{T}_{i_{k^+}} + S_{i_{k^+}} \le L_{i_{k^-}} - E_{i_{k^+}} \hspace{2cm} \forall k \in \mathcal{K} \hspace{1cm} (38)$$

Benders Cuts

Fig. 4. A CP model for the LBBD master problem of the STP.

Figure 3 presents the model. The objective function (26) maximizes the total weight of all the requests served. Constraint (27) ensures all locations are visited at most once. The approximate length of a request, ζ_r is modelled in Constraint (28) and Constraints (29)–(31) remove all infeasible requests from a specific vehicle. The relaxed total travel time for a vehicle is restricted to the time availability of the vehicle through Constraint (32). The relationship between the $y_{k,i}$ and $\varphi_{k,r}$ variables is established in Constraint (33) which also specifies that a corresponding pickup and delivery must be served by the same vehicle.

CP Master Problem. The CP formulation presented in Fig. 4 uses significantly fewer of variables than the full STP model in Fig. 2. Since we are relaxing all the temporal constraints in the master problem, there is no need for sequence variables. In this CP formulation of the LBBD master problem, we only employ interval variables x_i and $X_{i,k}$ as defined in Sect. 4.2.

The objective and a number of constraints remain the same as in the full STP model (Fig. 2). Since sequences are relaxed, no sequence variables are modelled but Constraints (35)–(37) ensure that each vehicle visits its starting depot and ending depot first and last, respectively. Finally, the distance relaxation is the same as in the MIP master problem represented by Constraint (38).

CP Sub-problem. After the master problem allocates the requests, a sub-problem is created for each vehicle with at least two assigned requests.[1] Each sub-problem is a single vehicle STP maximizing the total weight of served requests of those assigned by the master problem. If the sub-problem is able to schedule all the requests given to it, then the vehicle has a feasible assignment. Otherwise, the requests assigned to the vehicle are not feasible and the solution of the sub-problem is the optimal assignment for a proper subset of the assigned requests. The objective value of the sub-problem is then used in a Benders cut. With

---

[1] The master problem guarantees a solution for a vehicle with only one request.

$$\max \sum_{r \in \mathcal{R}^*} \left(W_r \times \texttt{PresenceOf}(x_{i_{r+}}) \right) \tag{39}$$

s.t.

$$\texttt{Before}(u, x_{i+}, x_{i-}) \qquad\qquad \forall i \in \mathcal{R}^* \tag{40}$$

$$\texttt{StartOf}(x_{i-}) - \texttt{EndOf}(x_{i+}) \leq F \qquad\qquad \forall i \in \mathcal{R}^* \tag{41}$$

$$v_i = \texttt{StepAtStart}(v_i, Q_i) \qquad\qquad \forall i \in \mathcal{N}^* \tag{42}$$

$$0 \leq \sum_{i \in \mathcal{N}^*} v_i \leq C_{k^*} \tag{43}$$

$$\texttt{First}(u, x_{i_{k^*+}}) \tag{44}$$

$$\texttt{Last}(u, x_{i_{k^*-}}) \tag{45}$$

$$\texttt{NoOverlap}(u) \tag{46}$$

Fig. 5. A CP model for the LBBD sub-problem of the STP.

optimization sub-problems, at each iteration of the LBBD, the algorithm finds a globally feasible solution.

Let k^* represent the vehicle and $\mathcal{R}^*$ the subset of requests assigned to k^* by the master problem. The CP formulation of the sub-problem uses three decision variables. For each location $i \in \mathcal{V}^*$, the optional interval variable x_i represents the time interval in which location i is served and is not present if it is not visited. This variable is bounded by the time window of the specific location. Cumulative functions y_i represent the load of the vehicle after visiting location i. Finally, a sequence variable u represents the sequence of visits of the vehicle.

Figure 5 presents the CP model for the subproblems. The objective function (39) maximizes the sum of weights of served requests. Constraint (40) makes sure that the pickup location is visited before the delivery location. The maximum ride time is enforced through Constraint (41). Constraints (42) and (43) keep track of the load of the vehicle after visiting each location and make sure that the load does not exceed the capacity of the vehicle at any location. Constraints (44) and (45) force the start and end of the sequence to be at the starting and ending depot, respectively. Finally, Constraint (46) takes into account the travel distances between locations for the sequence and eliminates sub-tours.

Benders Cut. If a sub-problem schedules all the requests assigned to it, then it is feasible. Otherwise, an optimality cut is returned to the master problem. The cut specifies that, given the subset of requests $\mathcal{R}^*$ to vehicle k^* in iteration h, denoted by $\mathcal{J}_{h,k}$, the objective value cannot be larger than the sub-problem's optimal value denoted by z^*. This cut is modelled in a MIP formulation as in Inequality (47) and in a CP formulation as in Inequality (48).

$$\sum_{r \in \mathcal{J}_{h,k}} (\varphi_{k,r} \times W_r) \leq z^* \qquad \forall k \in \mathcal{K}, h \in \{1, ..., H-1\} \quad (47)$$

$$\sum_{r \in \mathcal{J}_{h,k}} \big(\texttt{PresenceOf}(X_{i_{r+},k} \times W_r)\big) \leq z^* \quad \forall k \in \mathcal{K}, h \in \{1, ..., H-1\} \quad (48)$$

4.4   A Construction Heuristic

We designed a simple heuristic for the STP. It is used both as a basis of comparison with and as a warm-start solution for the exact techniques.

Since the objective function is to maximize the weight of served requests, it is reasonable to first schedule the requests that have the highest ratio of weight to length (i.e., W_r/ζ_r with ζ_r as defined in Constraint (28)). Furthermore, vehicles are sorted in ascending order of the size of their interval of availability so that requests are spread out amongst all vehicles and not concentrated on a single vehicle with a large time window. The algorithm schedules the highest weight ratio request to the first vehicle that can perform the request. If no currently available vehicle can satisfy a request, then the request is not scheduled. The algorithm is outlined in Algorithm 1.

Algorithm 1. Construction Heuristic for the STP

Data: Set of requests $\mathcal{R}$ and set of vehicles $\mathcal{K}$
Result: A set of scheduled routes
1 Sort $\mathcal{R}$ based on descending order of $\frac{W_r}{\zeta_r}$;
2 Sort $\mathcal{K}$ based on ascending time window size;
3 **for** *all requests r in $\mathcal{R}$* **do**
4 　 **for** *all vehicles v in $\mathcal{K}$* **do**
5 　　 **if** *r can be served by v* **then**
6 　　　 assign r to v and set start time of r as earliest start time on r that is after the earliest pickup time of r;
7 　　　 split v into 2 vehicle pieces, v_1 and v_2;
8 　　　 set start and end locations and start and end times for v_1 and v_2;
9 　　　 insert v_1 and v_2 into $\mathcal{K}$ based on the new time window sizes;
10 　　　 break;
11 　 **end**
12 　 **end**
13 **end**
14 Regroup all pieces of the same vehicle to make scheduled routes;

---

5   Experimental Results

In this section, we discuss the datasets used in our experiments and present the performance of the five approaches proposed above, including using the construction heuristic to provide a starting solution for the exact techniques. All approaches are coded using IBM's CPLEX Studio 12.7 in C++. The experiments are run on a computer with Intel Xeon E3-1226 v3 @ 3.30 GHz, 16 GB RAM using a single thread and a 600 s runtime limit. The CP Optimizer solver is set to use its default search.

5.1 Datasets

We generated 75 random datasets and extracted 280 problem instances from real world data provided by a partnering organization. In the generated problem instances, we varied the number of requests and vehicles, and the sizes of the time window (TW) of each request and vehicle. We also experimented with three different time window sizes: big, normal and small. All other characteristics are generated randomly following normal distributions. Table 1 outlines the lower and upper bounds of each characteristic.

Table 1. Bounds on problem characteristics for generated datasets.

	Characteristic	Lower bound	Upper bound
Vehicle	Number of vehicles	2	20
	Capacity	2	6
	Start and end depot service time	2	16
Request	Number of requests	6	50
	Size	1	3
	Weight	1	5
	Pickup and delivery location service time	2	16
	Travel time	1	60
Time windows	Small	80	180
	Normal	180	360
	Big	600	900

From the historical records of our partnering organization, we extracted 72,883 requests and 54,494 vehicle records over 280 operating days from January 2015 to January 2016. A total of 280 datasets were created. On average, there are 260 requests per day and the maximum number of requests per day is 554. There are on average 187 vehicles available each day.

5.2 Results

Table 2 summarizes the results of all approaches on the generated datasets. CP solved all 75 (100%) instances to optimality in an average of 1.02 s, MIP/CP LBBD solved 71 (95%) instances with an average runtime of 21.78 s, CP/CP

LBBD solved 49 (65%) instances with an average runtime of 110.14 s, and MIP solved 35 (48%) instances with an average of 90.00 s. The heuristic was able to find, but of course not prove, the optimal solution for 45 (60%) of the instances. In terms of relative solution quality compared to the optimal solutions, CP is again the best performer with the heuristic finding, on average, better solutions than both the MIP and CP/CP LBBD models.

Each of the four exact methods were then run with the heuristic solution as a warm start. Both MIP and CP/CP LBBD have a substantial improvement with the heuristic start. However, the only additional instances that they were able to solve to optimality were those for which the heuristic found an optimal solution. The heuristic start only improves MIP/CP LBBD a little while it has very minimal effects on CP. CP/CP LBBD exhibits lower solution quality than the heuristic, even when warm-started. Recall that the relaxed master problem often has better (but globally infeasible) solutions and so the warm start solution is replaced by a better master problem incumbent before the subproblems are solved.

Table 2. Number of instances solved to optimality, average runtime, and average optimality gap for generated datasets. The '*' indicates the heuristic found but did not prove optimal solutions.

Approach	# instances solved to optimality	% solved to optimality	Average runtime	Average optimality gap
Heuristic	45*	60.00%*	**0.01**	4.13%
MIP	35	46.67%	90.00	30.68%
MIP_H	52	69.33%	22.36	1.80%
CP	**75**	100.00%	1.02	**0.00%**
CP_H	**75**	100.00%	2.38	**0.00%**
MIP/CP LBBD	71	94.67%	21.78	0.15%
MIP/CP LBBD_H	73	97.33%	2.54	0.09%
CP/CP LBBD	49	65.33%	110.14	18.95%
CP/CP LBBD_H	61	81.33%	42.58	10.85%

Given the good performance of CP and MIP/CP LBBD, we apply them to the real world datasets. As shown in Table 3, out of 280 instances, 250 instances are solved to optimality with an average of 126.74 s using the pure CP model while the MIP/CP LBBD could only solve 47 instances in 331.31 s.

The evolution of runtime of the CP model as the problem sizes of the real instances increase is shown in Fig. 6. It can be observed that there is an approximately linear increase in runtime up to about 400 nodes (with some outliers) but with larger problems, the runtime substantially increases.

We also ran CP with an 8-h time limit. An additional 21 instances were solved to optimality but nine instances are still open. Thus Table 4 reports the

Table 3. Number of instances solved to optimality, average runtime, and average optimality gap for real world datasets.

Approach	# instances solved to optimality	% solved to optimality	Average runtime	Average optimality gap
CP	250	89.29%	126.74	3.03%
MIP/CP LBBD	47	16.79%	331.31	18.38%

Table 4. Average optimality gap summary for CP and MIP/CP LBBD on CHATS instances.

Instances	CP avg. gap	MIP/CP LBBD avg. gap
All 280 instances	**5.25%**	11.84%
233 instances not solved by MIP/CP LBBD	**6.31%**	14.23%
30 instances not solved by CP	49.02%	**18.38%**

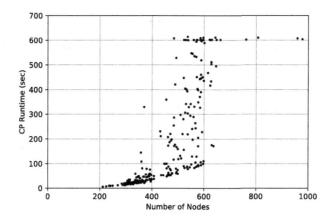

Fig. 6. CP runtime of real world instances over number of vertices.

solution quality relative to the best known solution for the real world datasets. The overall mean optimality gap for CP is 5.25% and 11.84% for MIP/CP LBBD.

6   Analysis

The strong results for CP compared to the LBBD approaches differ from much of the literature. Here, we explore three, non-mutually exclusive, hypotheses.

1. The default search of CP Optimizer is particularly suited to our problems.

2. The first feasible solutions found by the CP model are better and found more quickly than those found by the LBBD approaches.
3. Good solutions result in strong back-propagation from the lower-bound on the objective function, creating greater impact of search space reduction [8].

6.1   CP and Depth First Search

CP Optimizer's default search employs a combination of Large Neighbourhood Search (LNS) and Failure-directed Search (FDS) [11]. To observe its impact, we ran CP on the generated dataset using depth-first search (DFS). All instances were solved to optimality by DFS with an increase in the average runtime from 1.016 s to 1.873 s, a *decrease* in the average optimality gap of the first feasible solution from 29.14% to 24.14%, and an increase on the mean time to find the first solution from 0.163 s to 0.207 s.

The difference when using DFS appears marginal, perhaps due to using a single thread in all experiments. However, it does not appear that we can attribute the strong performance of our CP model, relative to the LBBD approaches, to the sophisticated default search of CP Optimizer.

6.2   First Solution Quality and Time

We recorded the time to find the first feasible solution and its quality for the CP model and both LBBD approaches on the generated dataset. The objective value, z', is compared to the known optimal solution, z^* via the optimality gap computed as $(z^* - z')/z^*$. Figures 7 and 8, respectively, compare the MIP/CP LBBD approach and the CP/CP LBBD approach to the CP model.

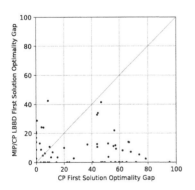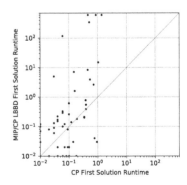

Fig. 7. First solution quality of MIP/CP LBBD compared to CP.

For the LBBD approaches, the first feasible solution is often the actual optimal solution and therefore is usually better than the CP model. However, the time to find these solutions for the LBBD approaches is much longer.

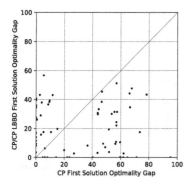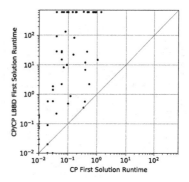

Fig. 8. First solution quality of MIP/CP LBBD compared to CP.

To further analyze the effect of the first solution, we used the first solution found in CP as a starting solution for the better performing LBBD approach, MIP/CP LBBD. We then let the algorithm run and report the change of runtime with and without the warm start. The warm start solution consists of an assignment of requests to vehicles which is a solution to the master problem of the MIP/CP LBBD but it does not contain any temporal information. For this experiment, the runtime does not include the time to compute the warm start solution. The results are shown in Fig. 9.

For big time windows, some instances are solved more quickly with the warm start solution. However, on average, the run-times with or without the warmstart are the same. As with the CP/CP LBBD_H results in Table 2, in many cases, the warm start solution provided by the CP model is not as good as the first master problem solution and thus is discarded.

We conducted the inverse experiment, inserting the MIP/CP LBBD first solution into the CP model as a warm start with the results shown in Fig. 10. In most cases, given the assignment of the warm start solution, the CP model performs slightly slower. Examination of the vehicle assignments of the first solutions showed that MIP/CP LBBD's assignment often clusters requests onto few vehicles. When CP is warm-started with such solutions, it needs to backtrack and reassign many requests to different vehicles in order improve the solution and/or prove optimality.

6.3   Search Space Reduction

The next set of experiments measures the impact of search space reduction of artificial lower bounds. If we denote the set of possible values that a variable x_i can take as D_{x_i}, then the logarithm of the size of the search space $\log(|P|)$ is computed as in Eq. (49) [8].

$$\log(|P|) = \log(|D_{x_i}|) + \ldots + \log(|D_{x_n}|) \tag{49}$$

For interval variables, the domain size is simply the size of the interval minus the duration of the variable, or $|D_{x_i}| = L_i - E_i - S_i + 1$. For optional interval

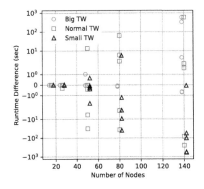

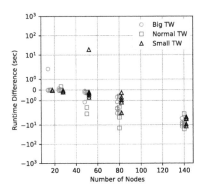

Fig. 9. Runtime difference of pure MIP/CP LBBD minus MIP/CP LBBD with CP starting solution.

Fig. 10. Runtime difference of pure CP minus CP with MIP/CP LBBD starting solution.

variables, there is an additional boolean value to represent the presence of the variable, thus the domain size is multiplied by 2. We focus on the CP/CP LBBD model so as to not conflate the comparison with fundamentally different problem solving bases (e.g., back-propagation is less important for MIP solving).

From the known optimal solutions, we compute five different lower bounds for each dataset that are 100%, 80%, 60%, 40%, and 20% of the optimal solution. Note that since we are maximizing, a lower bound on the objective function still results in a feasible solution. We then add this lower bound as a constraint on the objective function for both the CP model and the CP/CP LBBD approach. The search space is calculated before and after propagation of the root node. Table 5 presents how many instances show search space reduction and the average percentage reduction *over those instance which showed non-zero reduction*, given the different lower bounds for both CP and CP/CP LBBD.

Table 5. Number of instances (out of 25 in each row) that show a reduction in search space and the average percentage reduction after applying the artificial lower bound. The average only includes instances with non-zero reduction.

	TW Type	Lower Bound Percentage				
		100%	80%	60%	40%	20%
CP	small	8 (9.53%)	3 (23.56%)	2 (34.55%)	0 (-)	0 (-)
	normal	3 (1.50%)	1 (0.30%)	0 (-)	0 (-)	0 (-)
	big	0 (-)	0 (-)	0 (-)	0 (-)	0 (-)
CP/CP LBBD	small	1 (5.61%)	1 (5.61%)	0 (-)	0 (-)	0 (-)
	normal	0 (-)	0 (-)	0 (-)	0 (-)	0 (-)
	big	0 (-)	0 (-)	0 (-)	0 (-)	0 (-)

There are several instances that show a search space reduction for CP, after applying a lower bound, indicating back-propagation. Only one instance demonstrated search space reduction for the CP/CP LBBD showing a poor propagation of the first solution quality to the entire search space.

The average percentage reduction should be interpreted carefully. Since we are taking the mean the over instances with non-zero reduction, it may increase even when the lower bound decreases due to fewer problems showing any reduction (i.e., a smaller denominator).

7   Conclusion

Inspired by a real-world problem, we define the Senior Transportation Problem (STP), a problem encountered by organizations responsible for providing elder transportation. We show that it is a challenging combination of Pickup-and-Delivery with Time Windows, the Dial-a-Ride Problem, and the Team Orienteering Problem. In this paper, a formal problem definition for the STP was proposed, illustrating multiple constraints in real life problems.

Five different approaches using mixed integer programming, constraint programming, logic-based Benders decomposition, and a construction heuristic are developed to solve the STP. Each method is tested on 75 instances from a generated dataset and 280 real-world instances from our industrial partner. Constraint programming proves to be the best performing approach on both problem sets in terms of the number of instances solved to proven optimality, faster runtime, and solution quality. An LBBD approach combining mixed integer programming and constraint programming achieves the second best performance, though substantially worse than the pure constraint programming model. Our subsequent analysis lends support to the hypotheses that the strong performance of the CP model stems from the ability to quickly find feasible solutions and then to use the bounds on those solutions to reduce the search space.

While our conclusion is that the current CP model is superior, we plan to try to improve the logic-based Benders models in order to further challenge the pure CP approach and, more importantly, develop at a deeper understanding of the problem characteristics that favor CP or decomposition approaches.

References

1. Baklagis, D., Dikas, G., Minis, I.: The team orienteering pick-up and delivery problem with time windows and its applications in fleet sizing. RAIRO-Oper. Res. **50**(3), 503–517 (2016)
2. Cordeau, J.F., Laporte, G.: The dial-a-ride problem: models and algorithms. Ann. Oper. Res. **153**(1), 29 (2007)
3. Hooker, J.N., Ottosson, G.: Logic-based benders decomposition. Math. Program. **96**(1), 33–60 (2003)
4. Laborie, P., Rogerie, J.: Reasoning with conditional time-intervals. In: FLAIRS Conference, pp. 555–560 (2008)

5. Parragh, S.N., Doerner, K., Hartl, R.F.: A survey on pickup and delivery models: part I: transportation between customers and depot. J. für Betriebswirtschaft **58**(1), 21–51 (2008)
6. Parragh, S.N., Doerner, K.F., Hartl, R.F.: A survey on pickup and delivery problems: part II: transportation between pickup and delivery locations. J. für Betriebswirtschaft **58**, 81–117 (2008)
7. Qiu, X., Feuerriegel, S., Neumann, D.: Making the most of fleets: a profit-maximizing multi-vehicle pickup and delivery selection problem. Eur. J. Oper. Res. **259**(1), 155–168 (2017)
8. Refalo, P.: Impact-Based Search Strategies for Constraint Programming. In: Wallace, M. (ed.) CP 2004. LNCS, vol. 3258, pp. 557–571. Springer, Heidelberg (2004). https://doi.org/10.1007/978-3-540-30201-8_41
9. Ropke, S., Cordeau, J.F.: Branch and cut and price for the pickup and delivery problem with time windows. Transp. Sci. **43**(3), 267–286 (2009)
10. Vansteenwegen, P., Souffriau, W., Van Oudheusden, D.: The orienteering problem: a survey. Eur. J. Oper. Res. **209**(1), 1–10 (2011)
11. Vilím, P., Laborie, P., Shaw, P.: Failure-Directed Search for Constraint-Based Scheduling. In: Michel, L. (ed.) CPAIOR 2015. LNCS, vol. 9075, pp. 437–453. Springer, Cham (2015). https://doi.org/10.1007/978-3-319-18008-3_30

Solver Independent Rotating Workforce Scheduling

Nysret Musliu[1]([⊠]), Andreas Schutt[2,3]([⊠]), and Peter J. Stuckey[2,3]([⊠])

[1] TU Wien, Vienna, Austria
[2] Data61, CSIRO, Docklands, VIC, Australia
andreas.schutt@data61.csiro.au
[3] University of Melbourne, Parkville, VIC, Australia

Abstract. The rotating workforce scheduling problem aims to schedule workers satisfying shift sequence constraints and ensuring enough shifts are covered on each day, where every worker completes the same schedule, just starting at different days in the schedule. We give two solver independent models for the rotating workforce scheduling problem and compare them using different solving technology, both constraint programming and mixed integer programming. We show that the best of these models outperforms the state-of-the-art for the rotating workforce scheduling problem, and that solver independent modeling allows us to use different solvers to achieve different aims: *e.g.*, speed to solution or robustness of solving (particular for unsatisfiable problems). We give the first complete method able to solve all of the standard benchmarks for this problem.

1 Introduction

Rotating workforce scheduling is a specific personnel scheduling problem arising in many spheres of life such as, *e.g.*, industrial plants, hospitals, public institutions, and airline companies. Table 1 shows a workforce schedule for 7 employees during one week, in which a row represents the weekly schedule of one employee. There are three shifts: day shift (D), afternoon shift (A), and night shift (N). The first employee works from Monday till Thursday in the afternoon shift and has days-off in the remaining week. The second employee has a day-off on Thursday and Friday and works in the day shift in the other days. The last employee starts the week with 3 night shifts, then rests for two days, and ends the week with 2 day shifts. A schedule must meet many constraints such as workforce requirements for shifts and days, minimal and maximal length of shifts, and shift transition constraints, which are described in detail in the next section. For rotating workforce scheduling, the schedule is rotating (or cyclic), *i.e.*, the ith employee has the schedule of the $(((i-1+k) \bmod n)+1)^{th}$ employee after the kth week, where n is the number of employees. Due to that, no personal preferences of employees can be considered. The aim is to find a schedule satisfying all the constraints. Rotating workforce scheduling problems are NP-complete [7].

© Springer International Publishing AG, part of Springer Nature 2018
W.-J. van Hoeve (Ed.): CPAIOR 2018, LNCS 10848, pp. 429–445, 2018.
https://doi.org/10.1007/978-3-319-93031-2_31

Table 1. A typical week schedule for 7 employees.

Employee	Mon	Tue	Wed	Thu	Fri	Sat	Sun
1	A	A	A	A	-	-	-
2	D	D	D	-	-	D	D
3	D	-	-	N	N	N	N
4	-	-	-	-	A	A	A
5	D	D	D	D	D	-	-
6	N	N	N	N	N	-	-
7	N	N	N	-	-	D	D

According to [1,26], the problem studied can be characterized as a single-activity tour scheduling problem with non-overlapping shifts and rotation constraints.

Many practical real-life rotating workforce scheduling problems have been solved by complete techniques [2,12,19,22,28] and heuristic algorithms [20,21], but solving large problems is still a challenging task. Balakrishnan and Wong [2] formulate the problem as a network flow problem. Laporte [18] proposes a Integer Linear Programming approach. Methods based on Constraint Programming (CP) techniques are studied in [19,22,28]. Recently, Erkinger and Musliu [12] propose a Satisfiability Modulo Theory (SMT) approach, which—to our best knowledge—defines the state of the art complete method. All methods have been evaluated on the benchmark set with 20 instances [20,21], which are based on real life problems from different business areas, or on a sub-set of them. The state of the art complete method [12] was able to solve 18 of them, whereas the state-of-the-art heuristic approach [20,21] based on min-conflicts heuristic and tabu search (MC-T) found a solution for all of them. Other research studies focus at the creation of efficient rotation schedules by hand [17], and the design and the analysis of rotating schedules with an algebraic computational approach [13].

There are various variants of personnel scheduling (see, *e.g.*, the surveys [4, 5]), one group of them is the (multi-activity) shift scheduling problem, which is generally concerned about a finer schedule of the shifts over a planning horizon of one day considering, *e.g.*, meal breaks and more workforce regulations. Because of the finer nature, there has been a great deal of work in formalizing languages and their automata or network flows for capturing most of the regulations (see, *e.g.*, [9,11,16,25,27]). Beside the technologies mentioned in the previous paragraph, researchers has been investigating in Column Generation based methods (see, *e.g.*, [15,26]) to tackle large personnel scheduling problems.

We define two solver-independent models for rotating workforce scheduling, and compare them using a CP and a Mixed-Integer Programming (MIP) solver. The first model is rather direct, where each constraint is separately stated. The second model attempts to model as much as possible of the regulations using a single `regular` constraint [24], which models the regulations as a deterministic finite automaton. We consider redundant constraints, and symmetry breaking constraints that can be added to the model to possibly improve them.

We compare the variations of the models experimentally using the two solvers, and explore good search strategies to be used with the CP solver. Moreover, we generated 1980 additional instances and show that our two best methods outperform the state of the art approaches [12,20,21], on both the standard 20 benchmark instances and the extended 2000 instances.

2 The Rotating Workforce Scheduling Problem

We focus on a specific variant of a general workforce scheduling problem, which we formally define in this section. The following definition is from Musliu *et al.* [22] and proved to be able to satisfactorily handle a broad range of real-life scheduling instances in commercial settings. A rotating workforce scheduling problem as discussed in this paper consists of:

- n: Number of employees.
- $\mathbf{A}$: Set of m shifts (activities). There is also a "day-off" activity denoted O. We let $\mathbf{A}^+ = \mathbf{A} \cup \{O\}$.
- w: Length of the schedule. A typical value is $w = 7$, to assign one shift type for each day of the week to each employee. The total length of a planning period is $n \times w$ due to the schedule's cyclicity as discussed below.
- R: Temporal requirements matrix, an $m \times w$-matrix where each element $R_{i,j}$ shows the required number of employees that need to be assigned shift type $i \in \mathbf{A}$ during day j. The number o_j of day-off "shifts" for a specific day j is implicit in the requirements and can be computed as $o_j = n - \sum_{i=1}^{m} R_{i,j}$. In an abuse of notation we let $R_{O,j} = o_j$.
- Sequences of shifts not permitted to be assigned to employees. We consider two kinds of forbidden sequences: length 2 sequences, *e.g.*, ND (Night Day): after working in the night shift, it is not allowed to work the next day in the day shift; and length 3 sequences, *e.g.*, DON (Day Off Night): after working a day shift and then having a day off, it is not allowed to work the next day in the night shift. In length 3 sequences the middle shift is always O (Off). A typical rotating workforce instance forbids several shift sequences, often due to legal reasons and safety concerns. These two kinds are sufficient for all the cases we have met in practice. We represent the forbidden sequence as two sets of pairs $(sh_1, sh_2) \in F_2$ if it is forbidden to take shift sh_2 directly after sh_1; and $(sh_1, sh_2) \in F_3$ if it is forbidden to take shift sh_2 directly after a single O shift after sh_1.
- l_s and u_s: Each element of these vectors shows, respectively, the required minimal and permitted maximal length of periods of consecutive shifts $s \in \mathbf{A}^+$ of the same type.
- l_w and u_w: Minimal and maximal length of blocks of consecutive work shifts. This constraint limits the number of consecutive days on which the employees can work without having a day off.

The task in rotating workforce scheduling is to construct a cyclic schedule, which we represent as an $n \times w$ matrix $S_{i,j} \in \mathbf{A}^+, 1 \leq i \leq n, 1 \leq j \leq w$. Each element $S_{i,j}$ denotes the shift that employee i is assigned during day j in the first period of the cycle, or whether the employee has time off on that day. In a cyclic schedule, the schedule for one employee consists of a sequence of all rows of the matrix S.

The task is called rotating or cyclic scheduling because the last element of each row is adjacent to the first element of the next row, and the last element of the matrix is adjacent to its first element. Intuitively, this means that employee i $(i < n)$ assumes the place (and thus the schedule) of employee $i + 1$ after each week, and employee n assumes the place of employee 1. This cyclicity must be taken into account for the last three constraints above.

In the present paper, we consider the satisfaction problem satisfying all constraints given in the problem definition, which is usually sufficient in practice. This means the generation of one schedule is sufficient. The commercial software FCS [14,22] uses the same constraints for generating rotating workforce schedules. This system has been used since 2000 in practice by many companies in Europe and the scheduling variant we discuss in this paper proved to be sufficient for a broad range of uses.

3   Direct Model

The direct model of the problem asserts each of the constraints individually. To make it easy to handle the cyclic nature of the schedule we define a new view on the schedule $T_k = S_{k \div w + 1, k \mod w + 1}, 0 \le k \le n \times w - 1$ which simply maps the days of the schedule to a list of length $n \times w$ indexed from $TT = \{0, \ldots, n \times w - 1\}$. Let $t(x) = x \mod (n \times w)$ be a map from days to indexes of the list. We can then assert the constraints individually

$$\sum\nolimits_{k \in 0}^{u_w} (T_{t(j+k)} = O) > 0, \quad j \in TT \tag{1}$$

$$\sum\nolimits_{k \in 1}^{l_w} (T_{t(j+k)} = O) = 0, \quad j \in TT, T_j = O \wedge T_{t(j+1)} \ne O \tag{2}$$

$$\sum\nolimits_{k \in 0}^{u_O} (T_{t(j+k)} \ne O) > 0, \quad j \in TT \tag{3}$$

$$\sum\nolimits_{k \in 1}^{l_O} (T_{t(j+k)} \ne O) = 0, \quad j \in TT, T_j \ne O \wedge T_{t(j+1)} = O \tag{4}$$

$$\sum\nolimits_{k \in 0}^{u_{sh}} (T_{t(j+k)} \ne sh) > 0, \quad j \in TT, sh \in \mathbf{A} \tag{5}$$

$$\sum\nolimits_{k \in 1}^{l_{sh}} (T_{t(j+k)} \ne sh) = 0, \quad j \in TT, sh \in \mathbf{A}, T_j \ne sh \wedge T_{t(j+1)} = sh \tag{6}$$

$$T_j = sh_1 \rightarrow T_{t(j+1)} \ne sh_2, \quad j \in TT, (sh_1, sh_2) \in F_2 \tag{7}$$

$$T_j = sh_1 \wedge T_{t(j+1)} = O \rightarrow T_{t(j+2)} \ne sh_2, \quad j \in TT, (sh_1, sh_2) \in F_3 \tag{8}$$

Constraint (1) enforces there are no sequences of length $u_w + 1$ with no O shift, *i.e.*, the maximum length of a work block. Constraint (2) enforces there are no sequences of length less than l_w of work shifts, *i.e.*, the minimum length of a work block. Constraint (3) enforces there are no sequences of length $u_O + 1$ of just

O shifts, *i.e.*, the maximum length of an off block. Constraint (4) enforces there are no sequences of length less than l_O of off shifts, *i.e.*, the minimum length of an off block. Constraint (5) enforces there are no sequences of length $u_{sh} + 1$ of just sh shifts, *i.e.*, the maximum length of an sh block. Constraint (6) enforces there are no sequences of length less than l_{sh} of sh shifts, *i.e.*, the minimum length of an sh block. Constraint (7) enforces no forbidden sequences of length 2. Constraint (8) enforces no forbidden sequences of length 3.

To complete the model, we enforce that each day has the correct number of each type of shift.

$$\sum_{i\in 1..n}(S_{i,j} = sh) = R_{sh,j}, \quad j \in 1..w, sh \in \mathbf{A} \tag{9}$$

We can do the same for the off shifts as follows. Note that it is a redundant constraint.

$$\sum_{i\in 1..n}(S_{i,j} = O) = o_j, \quad j \in 1..w \tag{10}$$

Note that this model appears to consist entirely of linear constraints (at least once we use 01 variables to model the decisions $S_{i,j} = sh, sh \in \mathbf{A}^+$). This is misleading, since Eqs. (2), (4) and (6) are all contingent on variable conditions. The entire model can be easily expressed with linear constraints, and (half-)reified linear constraints.

4 Alternative Model Choices

The direct model (1–10) described in the previous section simply uses linear-styled constraints. However, there are alternative ways to model the shift transitions and the temporal requirements using global constraints. In addition, we can add more redundant constraints and symmetry breaking constraints to the model. In this section, we look at these choices except for the shift transition, for which we devote a separate section after this one.

4.1 Temporal Requirements

Instead of using the linear constraints in (9) and (10), we can respectively use these global cardinality constraints for each week day $j \in 1..w$.

$$gcc\_low\_up([S_{i,j}|i \in 1..n], \mathbf{A}, [R_{sh,j}|sh \in \mathbf{A}], [R_{sh,j}|sh \in \mathbf{A}]) \tag{11}$$
$$gcc\_low\_up([S_{i,j}|i \in 1..n], \mathbf{A}^+, [R_{sh,j}|sh \in \mathbf{A}^+], [R_{sh,j}|sh \in \mathbf{A}^+]) \tag{12}$$

They state that the number of shifts of each type in $sh \in \mathbf{A}$ (or $sh \in \mathbf{A}^+$) occurring in each day j must exactly equal the requirement $R_{sh,j}$.

4.2 Redundant Constraints

In a cyclic schedule, we know that there are equal numbers of work blocks and off blocks. We exploit this knowledge to create redundant constraints for each week by ensuring the lower bounds and upper bounds of these blocks do not cross.

Let $twl = \sum_{sh \in A} \sum_{j=1}^{n} R_{sh,j}$ be the total workload over the planning period. Then we can define the number of days-off ow_i at the end of the week i from the beginning of the schedule as

$$ow_i = \begin{cases} 0 & i = 0 \\ ow_{i-1} + \sum_{j \in 1..w}(S_{i,j} = O) & i \in 1..n-1 \\ n \times w - twl & i = n \end{cases}$$

Define ro_i to be the number of days-off remaining after the end of week i, and similarly rw_i to be the number of work days remaining after the end of week i

$$ro_i = n \times w - twl - ow_i, \qquad rw_i = twl - w \times i + ro_i.$$

We can determine a lower bound lo_i for the number of remaining off blocks starting from week i, and similarly an upper bound uo_i for the number of remaining off blocks as:

$$lo_i = \lceil ro_i/u_O \rceil - (S_{1,1} \neq O \wedge S_{i+1,1} = O)$$
$$uo_i = \lfloor ro_i/l_O \rfloor + (S_{1,1} = O \wedge S_{i,w} = O)$$

Note that the potential additional minus and plus one from the evaluation of the conjunction accounts for the fact that the number of work and off blocks can differ by one in the remaining schedule. For the lower bound, when the schedule starts with a work day and the week after the week i with a day-off then there might be one off block more in the remaining schedule than the number of remaining work blocks. For the upper bound, if the schedule starts with an off day and the week i ends with a day-off then there might be one off block less in the remaining schedule than the number of remaining work blocks.

Similarly, we can compute a lower bound lw_i for the number of the remaining work blocks after the end of week i, and similarly an upper bound uw_i for the number of remaining work blocks as:

$$lw_i = \lceil rw_i/u_w \rceil - (S_{1,1} = O \wedge S_{i+1,1} \neq O)$$
$$uw_i = \lfloor rw_i/l_w \rfloor + (S_{1,1} \neq O \wedge S_{i,w} \neq O)$$

Finally, we constrain these bounds to agree.

$$lo_i \leq uw_i \wedge lw_i \leq uo_i, \qquad i \in 1..n \qquad (13)$$

4.3   Symmetry Breaking Constraints

Given a schedule S a symmetric solution can easily be obtained by shifting the schedule by any number of weeks. If there must be at least one off day at the end of the week then we impose that the last day in the schedule is an off day. Note that it happens for all instances used.

$$ow > 0 \rightarrow S_{n,w} = O \qquad (14)$$

Note that we could have chosen any day and possible shift for breaking this symmetry, but we choose this one because it aligns with our other model choice for the shift transitions described in the next section.

Another symmetry occurs when all temporal requirements are the same for each day and each shift. In this case, a symmetric schedule can be obtained by shifting the schedule by any number of days. Thus, we can enforce that the work block starts at the first day in the schedule. The same constraint can be enforced if the number of working days in the first day is greater than in the last day of the week, because there must be at least one work block starting at the first day.

$$\left(\forall sh \in \mathbf{A}, \forall j \in 1..w-1 : R_{sh,j} = R_{sh,j+1} \vee \sum_{sh \in \mathbf{A}} R_{sh,1} > \sum_{sh \in \mathbf{A}} R_{sh,w}\right)$$
$$\rightarrow S_{1,1} \neq O \wedge S_{n,w} = O \quad (15)$$

In comparison to (14), (15) also enforces an off day on the last day and thus it is stronger symmetry breaking constraint, but less often applicable. Note that there might be further symmetries, especially instance specific ones, but here we focus on more common symmetries.

5   Automata Based Model

The automata-based model attempts to capture as much of the problem as possible in a single **regular** [24] constraint.

In order to enforce the forbidden sequences constraints we need to keep track of the last shift taken, and if the last shift was an O shift then the previous shift before that. In order to track the lower and upper bounds for each shift type, we need to track the number of consecutive shifts of a single type (including O). In order to track the lower and upper bounds for consecutive work shifts, we need to track the number of consecutive work shifts.

We define an automata M with $Q = m + u_o + \sum_{sh \in \mathbf{A}}(u_w \times u_{sh})$ states. We bracket state names to avoid ambiguity with shift types. They represent in sequence: an artificial start state [start]; states for the first O shift in a sequence, recording the type of the previous work shift [sO]; states for 2 or more O shifts in sequence ($2 \leq i \leq u_o$) [O^i], states encoding that the last $1 \leq j \leq u_s$ consecutive shifts are type $s \in \mathbf{A}$ directly after a sequence of $0 \leq j < u_w$ consecutive work shifts (not O) [$w^i s^j$]. Note each state effectively records a sequence of previous shifts. Note that some of the states may be useless, since, e.g., a state encoding 3 consecutive D shifts after a sequence of 4 other works shifts with $o_w = 6$ is not possible (it represents 7 consecutive work shifts).

The transition function d for the states is defined as follows (missing transitions go to an error state):

- [start]: on $sh \in \mathbf{A}$ goto [sh], on O goto [OO]. Note that transitions assume that the previous shift was O.
- [sO]: on O goto [OO] (assuming $u_O \geq 2$), on $sh \in A$ goto [sh] unless $s \, O \, sh$ is forbidden ((s, sh) $\in F_3$) or $l_O > 1$.

– $[O^i]$, $2 \leq i \leq u_O$: on O goto $[O^{i+1}]$ unless $i = u_O$, on $sh \in \mathbf{A}$ goto $[sh]$ unless $i < l_O$.

– $[w^i s^j]$, $0 \leq i \leq u_w - 1$, $1 \leq j \leq u_s$: on O goto $[sO]$ unless $j < l_s$, on s goto $[w^i s^{j+1}]$ unless $j = u_s$ or $i + j \geq u_w$, on $sh \in \mathbf{A} - \{s\}$ goto $[w^{i+j} sh]$ unless $s\ sh$ is forbidden $((s, sh) \in F_2)$ or $i + j \geq u_w$ or $j < l_s$.

Each state is accepting in this automata.

An example automata with two shifts D (Day) and N (Night) with forbidden sequences ND and DON and limits $l_D = 2$, $u_D = 3$, $l_N = 1$, $u_N = 2$, $l_O = 1$, $u_O = 3$ and $l_w = 2$, $u_w = 4$ is shown in Fig. 1. Unreachable states are shown dotted, and edges from them are usually omitted, except horizontal edges which do not break the total work limit. Edges for D shifts are full, N are dashed and O are dotted.

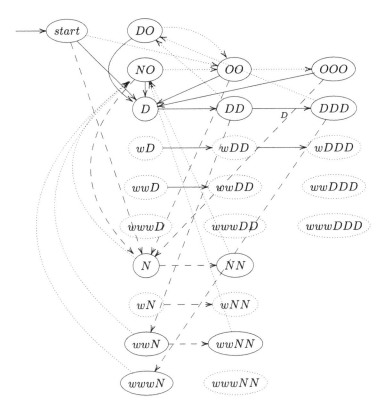

Fig. 1. The automaton capturing correct shift sequences for a problem with work shifts D and N, and forbidden sequences ND and DON. D shifts are indicated by full arrows, N shifts by dashed arrows, and O shifts by dotted arrows.

In order to define a cyclic schedule the `regular` constraint is applied on a sequence that duplicates the first w shifts at the end. This is safe assuming that $u_w < w$, which occurs in all our examples.

[start] D	[D]	D	[DD]	N	[wwN]	O	[NO]	O	[OO]	N	[NO]	N
[NN] O	[NO]	D	[D]	D	[DD]	D	[DDD]	O	[DO]	O	[OO]	D
[D]	D	[DD]	D	[DDD]	N	[wwwN]	O	[NO]	O	[OO]	N	[NO] N

Fig. 2. A two week sequence $DDNOONNODDDOOD$ with the first week repeated illustrating how the **regular** constraint can be in different states in the two copies of the first week.

The remaining constraints simply enforce the correct number of each shift type on each day. In summary the total model consists of either Eqs. (11) or (12) together with

$$\texttt{regular}([S_{1,1}, \ldots, S_{n,w}, S_{1,1}, \ldots, S_{1,w}], Q, m+1, d, [start], 1..Q) \qquad (16)$$

that is a **regular** constraint over the $n + 1$ week extended shift list, using an automata defined by Q states, $m + 1$ shift possibilities, transition function d, start state $[start]$ will all states being final states.

Note that the states in the first week, and the copy need not be the same. For example, given the automaton of Fig. 1, Fig. 2 shows a two week schedule, with the first week repeated, giving the state of the automata across the schedule. The two bold subsequences show where the states are different for the two copies of the first week. The sequence is accepted.

A simpler model would be possible if we had a **regular** constraint that included start and end states as variable arguments. We could then simply constrain the original array of shifts (with no duplication of the first week) and constrain the start and end states to be identical. Current solvers do not support such a **regular** constraint.

Note that while the start state ambiguity means that we make assumptions about the previous state when evaluating the automata on the states until we reach a first O shift, because the constraints are applied twice on the first week of shifts the proper constraints are satisfied. Note also that the **regular** constraint may actually remove solutions, since the first shift is assumed to be the first after an off, for the problems we tackle this simply removes symmetric solutions. This can be incorrect for corner cases, *e.g.*, when $\sum_{s \in \mathbf{A}} R_{s,w} = n$ (so there can be no O shift in the last day of the week). In these cases we can rotate the shift requirements for the days (effectively starting the cycle on a different day) to overcome the problem (it does not occur in any of the benchmarks).

We tried an alternate automata M' which ignores constraints until it can be sure of which state it is in, *i.e.*, a work shift followed by an O shift, or two consecutive O shifts. This proved to be terrible since it allowed erroneous schedules in the first week which then only detected as unsatisfiable when we finish labeling the last week.

6   Search Strategies

Beside the solver's default search strategy, we tested several others for the CP solver used, which are briefly described in this section.

The strategies studied consist of a variable and value selection part, which can be combined freely.

6.1 Variable Selection

Variable selection is critical for reducing the search space for any combinatorial problem. We need to balance the criteria of driving quickly towards failure, with getting the most possible inference from the solver. The key decisions of the model are the schedule variables $S_{i,j}$. We define our variable selection over these variables unless stated otherwise. Ties are broken by input order.

default: Solver's default selection.

random: Randomly select a variable.

worker: Select the variables in the chronological order over the planning horizon, i.e., $S_{1,1}, \ldots, S_{1,w}, S_{2,1}, \ldots, S_{2,w}, \ldots, S_{n,1}, \ldots, S_{n,w}$.

day: Select the variables in the following order the first day of the week from the first to the last week in the planning horizon and then the next day and so on, i.e., $S_{1,1}, \ldots, S_{n,1}, S_{1,2}, \ldots, S_{n,2}, \ldots, S_{1,w}, \ldots, S_{n,w}$.

wd: Select the variables of the first day in the weeks in the chronological order, i.e., $S_{1,1}, \ldots, S_{n,1}$, and then use the variable selection **worker**.

ff: Select the variable with the smallest domain (first fail).

first: Create new Boolean variables $b_k \leftrightarrow T_{t(k-1)} \neq T_{t(k)}$, $k \in TT$. These represent where a change of shift type occurs. Select the Boolean variables in chronological order and assign the value $true$ at first. Then use the variable selection **worker**.

s1: Create new Boolean variables $b_k \leftrightarrow (T_k = O)$, $k \in TT$. These represent where an off shift occurs. Select the Boolean variables in chronological order and assign the value $true$ at first. Then use the variable selection **worker**.

6.2 Value Selection

All instances in the benchmark set have these shift types (D) day, (A) afternoon, and (N) night, as well as the day-off (O) shift. Hence we could consider any of the 24 different static value ordering amongst these four. We consider 4 static orderings: $DANO$, default ordering of the model indomain_min; $ODAN$, off shifts first; $ONAD$, reversed default ordering indomain_max; $NADO$, reversed ordering of shifts.

We also consider static orderings that are computed from features of the instance to be solved. Let $maxb$ be the maximal number of work or off blocks, calculated as

$$maxb = \max(\lceil (\sum\nolimits_{sh \in \mathbf{A}, j \in 1..n} R_{sh,j})/l_w \rceil, \lceil (\sum\nolimits_{j \in 1..n} R_{O,j})/l_O \rceil).$$

We consider two variants of ordering shift types in terms of the tightness of the number of shifts required compared to the minimum number required to be scheduled. Both variants are in ascending order.

slack1: For each shift in $\mathbf{A}^+$, we compute the slack between maximal available space, *i.e.*, number of shift blocks times the maximal shift length, and the required workload or "days-off-load", *i.e.*, $sl_O^1 = u_O \times maxb - \sum_{j=1}^{w} o_j$, and $sl_{sh}^1 = \min(maxb, \sum_{j=1}^{w} R_{sh,j}/l_{sh}) \times u_{sh} - \sum_{j=1}^{w} R_{sh,j}$, $sh \in \mathbf{A}$.

slack2: This variant refines **slack1** for the work shifts. In addition, it considers when the maximal space is restricted by the high block requirement of the other work shifts, *i.e.*, $sl_O^2 = sl_O^1$, and $sl_{sh}^2 = \min(sl_{sh}^1, ti_{sh})$, $sh \in \mathbf{A}$ where $ti_{sh} = u_w \times (maxb - \sum_{z \in A \setminus \{sh\}} \lceil \sum_{j=1}^{w} R_{z,j}/u_z \rceil) - \sum_{j=1}^{w} R_{sh,j}$.

7 Experiments

To evaluate our methods, we took all 20 instances from a standard benchmark set[1] and further 30 hard instances from 1980 additional generated instances, for which the heuristic MC-T [20,21] either did not find a solution or required a long time for it. The instances generated consist of 9 to 51 employees, 2 to 3 shift types, 3 to 4 minimal and 5 to 7 maximal length of work blocks, 1 to 2 minimal and 2 to 4 maximal length of days-off blocks, and minimal and maximal length of periods of consecutive shifts (D: 2 to 3 and 5 to 7, A: 2 to 3 and 4 to 6, N: 2 to 3 and 4 to 5). The same forbidden sequences as for real-life examples are used. Initially the temporal requirements for shifts are distributed randomly between shifts based on the total number of working days and days-off (the number of days-off is set to $\lfloor n \times w \times 0.2857 \rfloor$). With probability 0.3 the temporal requirements during weekend are changed (half of these duties are distributed to the temporal requirements of the weekdays).

Experiments were run on Dell PowerEdge M630 machines having Intel Xeon E5-2660 V3 processors running at 2.6 GHz with 25 MB cache, unless otherwise stated. We imposed a runtime limit of one hour and a memory limit of 16 GB, unless otherwise stated. We tested Gurobi as a MIP solver and Chuffed [6] as a CP solver.[2] The development version of MiniZinc was used for modeling.

Table 2 compares the impact of the different model combinations for the temporal requirement and shift transition constraints for Gurobi and Chuffed using the solvers' default search strategy. The columns show in this order the solver, the constraints used (model), the total number of solved instances (#tot), the average number of nodes (avg. nd), the average runtime (avg. rt) (including time-outs for unsolved instances), the number of instances for which the solver found a solution (#sat), the average runtime of feasible instances (avg. rt), the number of instances for which the solver proved infeasibility (#uns), and the average runtime of infeasible instances (avg. rt). The results for Gurobi are clear. The shift transition constraints are the key constraints for its performance. Using the `regular` constraint (16), it solves all 50 instances regardless of the constraints for the temporal requirements. Its superiority over the direct representation results because MiniZinc transforms the `regular` constraint into

---

[1] Available at http://www.dbai.tuwien.ac.at/staff/musliu/benchmarks/.

[2] We also tried Gecode as a constraint programming solver, but it was not competitive.

Table 2. Results on different constraint choices.

Solver	Model	#tot	avg. nd	avg. rt	#sat	avg. rt	#uns	avg. rt
Gurobi	(1–8), (9)	46	3.2 k	780.6 s	40	587.5 s	6	1794.2 s
Gurobi	(1–8), (9, 10)	44	2.4 k	789.2 s	39	533.5 s	5	2131.6 s
Gurobi	(1–8), (11)	42	3.3 k	891.8 s	38	629.7 s	4	2268.0 s
Gurobi	(1–8), (12)	43	3.5 k	841.0 s	38	639.7 s	5	1897.7 s
Gurobi	(16), (9)	**50**	177	50.1 s	**42**	53.6 s	**8**	32.0 s
Gurobi	(16), (9, 10)	**50**	471	92.7 s	**42**	105.9 s	**8**	**23.2 s**
Gurobi	(16), (11)	**50**	177	49.5 s	**42**	52.8 s	**8**	32.1 s
Gurobi	(16), (12)	**50**	113	**45.0 s**	**42**	**48.6 s**	**8**	26.3 s
Chuffed	(1–8), (9)	33	9.1 m	1323.8 s	33	890.1 s	0	3600.6 s
Chuffed	(1–8), (9,10)	44	2.8 m	505.3 s	39	342.5 s	5	1360.3 s
Chuffed	(1–8), (11)	37	9.9 m	1070.6 s	36	644.2 s	1	3308.9 s
Chuffed	(1–8), (12)	44	4.9 m	482.8 s	39	315.5 s	5	1361.2 s
Chuffed	(16), (9)	30	5.7 m	1539.2 s	30	1146.5 s	0	3600.8 s
Chuffed	(16), (9,10)	44	1.6 m	521.1 s	39	303.2 s	5	1665.6 s
Chuffed	(16), (11)	34	4.3 m	1304.5 s	34	867.2 s	0	3600.4 s
Chuffed	(16), (12)	42	1.8 m	615.5 s	37	469.2 s	5	1383.7 s

a network flow for mixed-integer solvers [3, 10] and hence almost the entire model is totally unimodular. The overall best combination is achieved with the global cardinality constraint (12).

By contrast, Chuffed is not able to solve all instances in any combination and there are two important model choices. As opposed to Gurobi, Chuffed performs better when using the direct constraints (1–8) for the shift transition constraints, even though weaker propagation is achieved. The average number of nodes indicate that a stronger propagation of the `regular` does not convert into runtime savings. This probably results from the fact that the direct constraints introduce intermediate variables which are valuable for learning, whereas the `regular` constraint introduces no intermediate variables. For Chuffed, it is also important to choose temporal constraints covering the days-off. The overall best combination are the direct constraints (1–8) for the shift transition constraints and the global cardinality constraint (12).

Table 3 shows the impact on the performance of Gurobi and Chuffed when adding the redundant and symmetry breaking constraints to the best model combination. Gurobi's performance drastically deteriorate when using the redundant constraints. This is unsurprising since these are mainly linear combinations of constraints the solver already has. Its performance significantly improves when using the symmetry breaking constraints, but only for infeasible instances, which is expected because it makes the search space smaller. For feasible instances, it does not have any impact. For Chuffed, both set of constraints are important

Table 3. Results with redundant (13) and symmetry breaking (14, 15) constraints.

Solver	Model	(13)	(14, 15)	#tot	avg. nd	avg. rt	#sat	avg. rt	#uns	avg. rt
Gurobi	(16), (12)			**50**	113	45.0 s	**42**	48.6 s	**8**	26.3 s
Gurobi	(16), (12)		×	**50**	84	**41.4 s**	**42**	48.4 s	**8**	**4.5 s**
Gurobi	(16), (12)	×		**50**	271	93.5 s	**42**	108.2 s	**8**	16.0 s
Gurobi	(16), (12)	×	×	49	374	107.6 s	41	127.3 s	**8**	4.5 s
Chuffed	(1–8), (12)			44	4.9 m	482.8 s	39	315.5 s	5	1361.2 s
Chuffed	(1–8), (12)		×	44	5.1 m	489.1 s	39	323.9 s	5	1356.2 s
Chuffed	(1–8), (12)	×		48	1.6 m	190.1 s	**42**	51.9 s	6	915.4 s
Chuffed	(1–8), (12)	×	×	48	1.8 m	172.5 s	**42**	**32.2 s**	6	909.2 s

Table 4. Results on best value and variable selections for the search strategies.

Solver	Search	#tot	avg. nd	avg. rt	#sat	avg. rt	#uns	avg. rt
Chuffed	**default**+*DANO*	**48**	1.8 m	172.5 s	**42**	32.2 s	**6**	909.2 s
Chuffed	**worker**+*NADO*	**48**	1.4 m	169.7 s	**42**	28.4 s	**6**	911.1 s
Chuffed	**ff**+*NADO*	**48**	1.4 m	**166.7 s**	**42**	**25.1 s**	**6**	909.8 s
Chuffed	**s1**+*DANO*	47	1.3 m	236.4 s	41	109.0 s	**6**	**905.2 s**

to increase its performance. Still it cannot solve all instances in the given run-time limit, but could find a solution for all feasible instances. Interestingly, the average runtime over all feasible instances is lower than Gurobi's best time.

Table 4 shows the best pairing of value and variable selections for the search strategies for Chuffed.[3] Chuffed alternates between the given search strategy and its default one on each restart. This is important since it allows the powerful default activity based search to be utilized.

Using a search strategy was beneficial for the performance, but either for feasible or infeasible instances, and not both together. For feasible instances, the variable selections **ff** and **worker** were the best two in combination with a value ordering *NADO*, because first fail makes the search space small and worker explores it in chronological order of the schedule allowing removal of impossible values for the next decision. The value ordering *NADO* works best, because of the structure of the instances. The night shift is the most restricted one for the shift transitions and then the afternoon. In addition, the temporal requirements tends to be the least for the night shift and then the afternoon shift. For infeasible instances, deciding work and days-off at first and then which work shift is performed using the value ordering *DANO* performed best.

Because the instances all have very similar relations between the different types of shifts, we took the 50 instances and reversed the order of the forbidden sequences to check that the value ordering *NADO* is not necessarily the best one.

---

[3] We ran preliminary experiments on all possible combinations with a five minutes runtime limit.

Table 5. Results on instances with reversed forbidden sequences (runtime limit 300 s).

Solver	Search	#tot	avg. nd	avg. rt	#sat	avg. rt	#uns	avg. rt
Chuffed	ff+*ODAN*	45	304 k	49.2 s	34	33.7 s	11	**85.4 s**
Chuffed	ff+*DANO*	45	303 k	40.5 s	34	21.0 s	11	86.1 s
Chuffed	ff+*ONAD*	44	277 k	45.1 s	33	27.5 s	11	86.1 s
Chuffed	ff+*NADO*	45	260 k	40.1 s	34	20.6 s	11	85.6 s
Chuffed	ff+slack1	45	244 k	40.1 s	34	20.4 s	11	85.8 s
Chuffed	ff+slack2	45	235 k	**37.9 s**	34	**17.0 s**	11	86.9 s

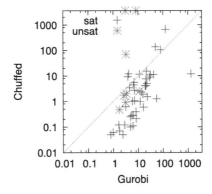

Fig. 3. Runtime comparison between Chuffed (y-axis) and Gurobi (x-axis).

Table 6. Comparison to the state of the art methods on all 2000 instances (runtime limit 200 s).

Solver	#fastest	#tot	avg. rt	#sat	avg. rt	#uns	avg. rt
Gurobi	31	**1988**	**10.3 s**	1320	13.0 s	**668**	**4.9 s**
Chuffed	513	1845	19.4 s	**1322**	5.0 s	523	47.8 s
MathSAT - BV	3	1470	63.8 s	1198	32.1 s	272	126.7 s
MC-T	**781**	1212	82.7 s	1212	23.3 s	0	200 s

We set a runtime limit to 300 s for this experiment. Table 5 clearly shows that the dynamic criteria **slack2** performs the best, which looks at the tightness of the temporal requirements for each shift including the days-off. In comparison to **slack2**, Chuffed is about 20% slower when using *NADO*, which confirms our previous observation that this order is well-suited for the instances in the benchmark library, due to the fact that the night shifts and then the afternoon shifts are normally the most-restrictive ones.

Figure 3 shows the runtime comparison between Chuffed and Gurobi on the 50 instances. Runtimes are given in seconds and the axis use a logarithmic scale. Points below the diagonal line express that Chuffed solved the instance quicker than Gurobi, and vice-versa for points above the line. Except for 5 feasible and 3 infeasible instances, Chuffed solved the instances in a similar speed or faster, even by order of a magnitude for many instances. However, Gurobi is more robust, especially for infeasible instances, which were solved within 10 s.

Table 6 compares the best Chuffed and Gurobi outcome on all 2000 instances to the state-of-the-art heuristic approach based on min-conflicts heuristic and tabu search (MC-T) [20,21] and the state-of-the-art complete SMT approach [12] using bit vectors for modeling and the SMT solver MathSAT 5.5.1 [8] for solving. Note that the results of MC-T reported in this paper were obtained on a Lenovo T440s machine having Intel(R) Core(TM) i5-4200U CPU @ 1.60 GHz 2.30 GHz with 8 GB RAM. To be conservative, we consider that these machines are twice as slow as the machines on which Gurobi, Chuffed and MathSAT were executed. For this comparison we set the runtime limit to 200 s and recorded the number of instances (#fastest), for that each method was the fastest, we halved the runtime of MC-T for computing #fastest. Chuffed and Gurobi significantly outperform the state-of-the-art complete methods in all aspects. The heuristic solver performs better than the SMT method on feasible instances, even though it runs on a slower machine. It is the fastest solver on almost 59% of the feasible instances followed by Chuffed with 39%. However, it cannot compete with Chuffed and Gurobi in term of the number of solved feasible instances. Chuffed and Gurobi were respectively able to solve more feasible instances within 10 s and 30 s than MC-T within 200 s. On top of that, our methods are able to detect infeasibility. Thus, both our methods are more robust than MC-T, whereas Gurobi is the most robust one. Together, both our methods could solve all instances within 200 s, except two instances.

8   Conclusion

We investigated different solver-independent models for solving the rotating workforce scheduling problem using MiniZinc [23]. Surprisingly the best model combination resulted when using **regular** constraints with Gurobi, even though the **regular** constraint is native to constraint programming. This shows the advantages of solver-independent modeling, where we do not commit to a single solver. While **regular** and network-flow models have been used previously for this problem, they made use of multiple **regular** constraints instead of one large **regular**. The advantage of using a high level modeling language was that we could generate a complex automata fully automatically that encoded almost all of the problem. Indeed a simple first version of the automata based model was constructed in under an hour. We tested our approaches on the standard benchmark set and created more challenging instances for our evaluation. Interestingly, Gurobi and Chuffed excelled on different model combinations. On the majority of instances Chuffed is the quickest solver, but Gurobi the most robust one, because of its superiority in proving infeasibility.

Acknowledgments. This work was partially supported by the Asian Office of Aerospace Research and Development grant 15-4016 and by the Austrian Science Fund (FWF): P24814-N23.

References

1. Baker, K.R.: Workforce allocation in cyclical scheduling problems: a survey. J. Oper. Res. Soc. **27**(1), 155–167 (1976)
2. Balakrishnan, N., Wong, R.T.: A network model for the rotating workforce scheduling problem. Networks **20**(1), 25–42 (1990)
3. Belov, G., Stuckey, P.J., Tack, G., Wallace, M.G.: Improved linearization of constraint programming models. In: Rueher, M. (ed.) Principles and Practice of Constraint Programming - CP 2016, pp. 49–65. Springer International Publishing, Cham (2016)
4. Van den Bergh, J., Beliën, J., De Bruecker, P., Demeulemeester, E., De Boeck, L.: Personnel scheduling: a literature review. Eur. J. Oper. Res. **226**(3), 367–385 (2013)
5. Burke, E.K., De Causmaecker, P., Berghe, G.V., Van Landeghem, H.: The state of the art of nurse rostering. J. Sched. **7**(6), 441–499 (2004)
6. Chu, G.: Improving combinatorial optimization. Ph.D. thesis, The University of Melbourne (2011). http://hdl.handle.net/11343/36679
7. Chuin Lau, H.: On the complexity of manpower shift scheduling. Comput. Oper. Res. **23**(1), 93–102 (1996)
8. Cimatti, A., Griggio, A., Schaafsma, B.J., Sebastiani, R.: A modular approach to MaxSAT modulo theories. In: Järvisalo, M., Van Gelder, A. (eds.) SAT 2013. LNCS, vol. 7962, pp. 150–165. Springer, Heidelberg (2013). https://doi.org/10.1007/978-3-642-39071-5_12
9. Côté, M.C., Gendron, B., Quimper, C.G., Rousseau, L.M.: Formal languages for integer programming modeling of shift scheduling problems. Constraints **16**(1), 54–76 (2011)
10. Côté, M.-C., Gendron, B., Rousseau, L.-M.: Modeling the regular constraint with integer programming. In: Van Hentenryck, P., Wolsey, L. (eds.) CPAIOR 2007. LNCS, vol. 4510, pp. 29–43. Springer, Heidelberg (2007). https://doi.org/10.1007/978-3-540-72397-4_3
11. Côté, M.C., Gendron, B., Rousseau, L.M.: Grammar-based integer programming models for multiactivity shift scheduling. Manag. Sci. **57**(1), 151–163 (2011)
12. Erkinger, C., Musliu, N.: Personnel scheduling as satisfiability modulo theories. In: International Joint Conference on Artificial Intelligence - IJCAI 2017, Melbourne, Australia, 19–25 August 2017, pp. 614–621 (2017)
13. Falcón, R., Barrena, E., Canca, D., Laporte, G.: Counting and enumerating feasible rotating schedules by means of Gröbner bases. Math. Comput. Simul. **125**, 139–151 (2016)
14. Gärtner, J., Musliu, N., Slany, W.: Rota: a research project on algorithms for workforce scheduling and shift design optimization. AI Commun. **14**(2), 83–92 (2001)
15. Hashemi Doulabi, S.H., Rousseau, L.M., Pesant, G.: A constraint-programming-based branch-and-price-and-cut approach for operating room planning and scheduling. INFORMS J. Comput. **28**(3), 432–448 (2016)
16. Kadioglu, S., Sellmann, M.: Efficient context-free grammar constraints. In: AAAI, pp. 310–316 (2008)

17. Laporte, G.: The art and science of designing rotating schedules. J. Oper. Res. Soc. **50**, 1011–1017 (1999)
18. Laporte, G., Nobert, Y., Biron, J.: Rotating schedules. Eur. J. Oper. Res. **4**(1), 24–30 (1980)
19. Laporte, G., Pesant, G.: A general multi-shift scheduling system. J. Oper. Res. Soc. **55**(11), 1208–1217 (2004)
20. Musliu, N.: Combination of local search strategies for rotating workforce scheduling problem. In: International Joint Conference on Artificial Intelligence - IJCAI 2005, Edinburgh, Scotland, UK, 30 July - 5 August 2005, pp. 1529–1530 (2005). http://ijcai.org/Proceedings/05/Papers/post-0448.pdf
21. Musliu, N.: Heuristic methods for automatic rotating workforce scheduling. Int. J. Comput. Intell. Res. **2**(4), 309–326 (2006)
22. Musliu, N., Gärtner, J., Slany, W.: Efficient generation of rotating workforce schedules. Discrete Appl. Math. **118**(1–2), 85–98 (2002)
23. Nethercote, N., Stuckey, P.J., Becket, R., Brand, S., Duck, G.J., Tack, G.: MiniZinc: towards a standard CP modelling language. In: Bessière, C. (ed.) CP 2007. LNCS, vol. 4741, pp. 529–543. Springer, Heidelberg (2007). https://doi.org/10.1007/978-3-540-74970-7_38
24. Pesant, G.: A regular language membership constraint for finite sequences of variables. In: Wallace, M. (ed.) CP 2004. LNCS, vol. 3258, pp. 482–495. Springer, Heidelberg (2004). https://doi.org/10.1007/978-3-540-30201-8_36
25. Quimper, C.G., Rousseau, L.M.: A large neighbourhood search approach to the multi-activity shift scheduling problem. J. Heuristics **16**(3), 373–392 (2010)
26. Restrepo, M.I., Gendron, B., Rousseau, L.M.: Branch-and-price for personalized multiactivity tour scheduling. INFORMS J. Comput. **28**(2), 334–350 (2016)
27. Salvagnin, D., Walsh, T.: A hybrid MIP/CP approach for multi-activity shift scheduling. In: Milano, M. (ed.) CP 2012. LNCS, pp. 633–646. Springer, Heidelberg (2012). https://doi.org/10.1007/978-3-642-33558-7_46
28. Triska, M., Musliu, N.: A constraint programming application for rotating workforce scheduling. In: Mehrotra, K.G., Mohan, C., Oh, J.C., Varshney, P.K., Ali, M. (eds.) Developing Concepts in Applied Intelligence. SCI, vol. 363, pp. 83–88. Springer, Heidelberg (2011). https://doi.org/10.1007/978-3-642-21332-8_12

Greedy Randomized Search for Scalable Compilation of Quantum Circuits

Angelo Oddi[(✉)] and Riccardo Rasconi

Institute of Cognitive Sciences and Technologies (ISTC-CNR),
Via S. Martino della Battaglia, 44, 00185 Rome, Italy
{angelo.oddi,riccardo.rasconi}@istc.cnr.it
http://www.istc.cnr.it

Abstract. This paper investigates the performances of a greedy randomized algorithm to optimize the realization of *nearest-neighbor* compliant quantum circuits. Current technological limitations (*decoherence* effect) impose that the overall duration (makespan) of the quantum circuit realization be minimized. One core contribution of this paper is a lexicographic two-key ranking function for quantum gate selection: the first key acts as a *global* closure metric to minimize the solution makespan; the second one is a *local* metric acting as "tie-breaker" for avoiding cycling. Our algorithm has been tested on a set of quantum circuit benchmark instances of increasing sizes available from the recent literature. We demonstrate that our heuristic approach outperforms the solutions obtained in previous research against the same benchmark, both from the CPU efficiency and from the solution quality standpoint.

Keywords: Quantum computing · Optimization · Scheduling
Planning · Greedy heuristics · Random algorithms

1 Introduction

In this work, we investigate the performances of greedy randomized search (GRS) techniques [1–3] to the problem of compiling quantum circuits to emerging quantum hardware. Quantum Computing represents the next big step towards power consumption minimization and CPU speed boost in the future of computing machines. The impact of quantum computing technology on theoretical/applicative aspects of computation as well as on the society in the next decades is considered to be immensely beneficial [4].

While classical computing revolves around the execution of logical gates based on two-valued *bits*, quantum computing uses *quantum gates* that manipulate multi-valued bits (*qubits*) that can represent as many logical states (*qstates*) as are the obtainable linear combinations of a set of basis states (state *superpositions*). A quantum circuit is composed of a number of qubits and by a series of quantum gates that operate on those qubits, and whose execution realizes a specific quantum algorithm. Executing a quantum circuit entails the chronological

© Springer International Publishing AG, part of Springer Nature 2018
W.-J. van Hoeve (Ed.): CPAIOR 2018, LNCS 10848, pp. 446–461, 2018.
https://doi.org/10.1007/978-3-319-93031-2_32

evaluation of each gate and the modification of the involved qstates according to the gate logic.

Current quantum computing technologies like ion-traps, quantum dots, super-conducting qubits, etc. limit the qubit interaction distance to the extent of allowing the execution of gates between adjacent (i.e., *nearest-neighbor*) qubits only [5–7]. This has opened the way to the exploration of possible techniques and/or heuristics aimed at guaranteeing nearest-neighbor (NN) compliance in any quantum circuit through the addition of a number of so-called *swap* gates between adjacent qubits. The effect of a swap gate is to mutually exchange the qstates of the involved qubits, thus allowing the execution of the gates that require those qstates to rest on adjacent qubits. However, according to [8], adding swap gates also introduces a time overhead in the circuit execution, which depends on the quantum chip's topology (see Fig. 2 for an example of three different 2D topologies). In addition, the Achilles' heel of quantum computational hardware is the problem of *decoherence*, which degrades the performance of quantum programs over time. In order to minimize the negative effects of decoherence and guarantee more stability to the computation, it is therefore essential to produce circuits whose overall duration (i.e., *makespan*) is minimal.

In this work, we present a GRS procedure that synthesizes NN-compliant quantum circuits realizations, starting from a set of benchmark instances of different size belonging to the *Quantum Approximate Optimization Algorithm* (QAOA) class [9, 10] tailored for the MaxCut problem, to be executed on top of a hardware architecture proposed by Rigetti Computing Inc. [11]. We demonstrate that the meta-heuristic we present outperforms the approach used in previous research against the same benchmark, both from the CPU efficiency and from the solution quality standpoint.

The paper is organized as follows. Section 2 provides some background information. Section 3 proposes a formal statement of the solved problem, whereas subsequent Sects. 4 and 5 describe the proposed heuristic solving algorithm and the Greedy Randomized Search approach, respectively. Finally, an empirical validation based on the results proposed in [12] and some conclusions close the paper.

2   Background

Quantum computing is based on the manipulation of qubits rather than conventional bits; a quantum computation is performed by executing a set of quantum operations (called *gates*) on the qubits. A gate whose execution involves k qubits is called k-*qubit quantum gate*. In this work we will focus on 1-qubit and 2-qubit quantum gates. In order to be executed, a quantum circuit must be mapped on a quantum chip which determines the circuit's hardware architecture specification [13]. The chip can be generally seen as an undirected weighted multigraph whose nodes represent the qubits (quantum physical memory locations) and whose edges represent the types of gates that can be physically implemented between adjacent qubits of the physical hardware (see Fig. 1 as an example of three chip

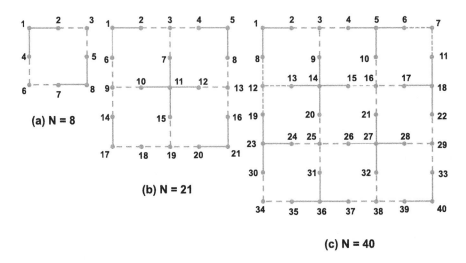

Fig. 1. Three quantum chip designs characterized by an increasing number of qubits ($N = 8, 21, 40$) inspired by Rigetti Computing Inc. Every qubit is located at a different location (node), and the integers at each node represent the qubit's identifier. Two qubits connected by an edge are adjacent, and each edge represents a 2-qubit gate (p-s or $swap$) that can be executed between those qubits (see Sect. 3.1). p-s gates executed on continuous edges have duration $\tau_{p\text{-}s} = 3$, while p-s gates executed on dashed edges have duration $\tau_{p\text{-}s} = 4$. Swap gates have duration $\tau_{swap} = 2$.

topologies of increasing size). Since a 2-qubit gate requiring two specific qstates can only be executed on a pair of adjacent (NN) qubits, the required qstates must be conveyed on such qubit pair prior to gate execution. NN-compliance can be obtained by adding a number of $swap$ gates so that every pair of qstates involved in the quantum gates can be eventually made adjacent, allowing all gates to be correctly executed. Figure 2 shows an example of quantum circuit that only uses the first three qubits of the chip ($N = 8$) of Fig. 1, which assumes that qstates q_1, q_2 and q_3 are initially allocated to qubits n_1, n_2 and n_3. The circuit is composed of four generic 2-qubit gates and one generic 1-qubit gate. Note that the circuit is not NN-compliant as the last gate involves two qstates belonging to two non-adjacent qbits (n_1 and n_3). The right side of Fig. 2(b) shows the same circuit made NN-compliant through the insertion of a swap gate.

According to the authors of [12], the problem of finding a sequence of gates that efficiently realizes an NN-compliant quantum circuit fits perfectly into a temporal planning problem, and their solution consists in modelling quantum gates as PDDL2.1 durative actions [14], enabling domain independent temporal planners [15] to find a quantum circuit realization in terms of a *parallel sequence* of conflict-free operators (i.e., the gates), characterized by minimum completion time (makespan). In this work, we tackle the same problem following a scheduling-oriented formulation, as described in the next sections. In particular, our approach is related to a body of heuristic efforts available in the current literature, see [16,17] for two recent and representative works. Despite these papers

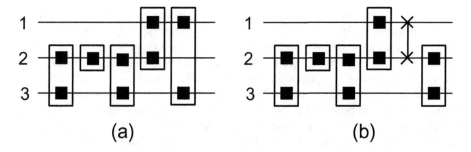

Fig. 2. Example of quantum circuit: (a) not NN-compliant; (b) NN-compliant through the insertion of a swap gate between qbits 1 and 3 just before the last gate, which exchanges their respective qstates. It is implicitly supposed that at the beginning, the i-th qubit is characterized by the i-th qstate.

pursue the same objective, i.e., optimizing the realization of *nearest-neighbor* compliant quantum circuits, they focus on quantum circuits characterized by pre-ordered non-commutative gates. On the contrary, our approach leverages the parallel nature of the considered planning/scheduling problem, and proposes a greedy randomized algorithm that exploits an original lexicographic double-key ranking function for quantum gate selection.

3   Problem Definition

The problem tackled in this work consists in compiling a given quantum circuit on a specific quantum hardware architecture. To this aim, we focus on the same framework used in [12]: (i) the class of *Quantum Approximate Optimization Algorithm* (QAOA) circuits [9,10] to represent an algorithm for solving the MaxCut problem (see below); (ii) a specific hardware architecture inspired by the one proposed by Rigetti Computing Inc. [11]. The QAOA-based benchmark problems are characterized by a high number of commuting quantum gates (i.e., gates among which no particular order is superimposed) that allow for great flexibility and parallelism in the solution, which makes the corresponding optimization problem very interesting and guarantees greater makepan minimization potential for decoherence minimization [12]. Moreover, the Rigetti hardware architecture (see Fig. 1) allows for two types of nearest-neighbor relations characterized by different durations, making the temporal planning/scheduling problem even more challenging. The rest of this section is devoted to: (i) describing the Max-Cut problem and (ii) providing a formulation of the Quantum Gate Compilation Problem (QGCP).

3.1   The MaxCut Problem

Given a graph $G(V, E)$ with $n = |V|$ nodes and $m = |E|$ edges, the objective is to partition the node set V in two subsets V_1 and V_2 such that the number of edges that connect every node pair $\langle n_i, n_j \rangle$ with $n_i \in V_1$ and $n_j \in V_2$ is maximized. The following formula: $U = \frac{1}{2}\sum_{(i,j)\in E}(1 - s_i s_j)$ describes a quadratic objective function U for the MaxCut problem, where s_i is a binary variable corresponding to the i-th node v_i of the graph G, that takes the value $+1$ if $v_i \in V_1$ or -1 if $v_i \in V_2$ at the end of the partition operated by the algorithm.

The compilation process of the MaxCut problem on QAOA circuits is rather simple, and is composed of a *phase separation* (P-S) step and a *mixing step* (MIX), which entails the execution of a set of identical 2-qubit gates (one 2-qubit gate for each quadratic term in the above objective function U) called *p-s* gates, followed by the execution of a set of 1-qubit gate for each node of the graph G, called *mix* gates [9]. Figure 3 (left side) shows an example of the graph G upon which the MaxCut problem is to be executed, corresponding to the problem instance #1 of the $N8\_u0.9\_P2$ benchmark set analyzed in the experimental section of this paper (Sect. 6). The list of *p-s* and *mix* quantum gates that must be executed during the compilation procedure is shown on the right side of the figure. The compilation problem depicted in the figure requires the execution of two phase ($p = 2$) separation steps (*P-S 1* and *P-S 2*), interleaved by two mixing steps (*MIX 1* and *MIX 2*).

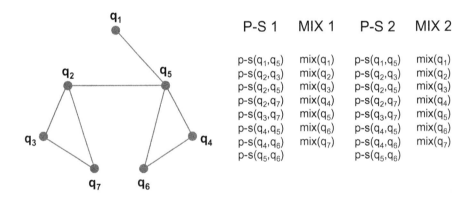

Fig. 3. MaxCut problem instance on a graph with 7 nodes. Each node is associated with a particular qstate q_i and such associations define the compilation objectives as a set of *p-s* and *mix* gates to be planned for and executed (note that the qstate q_8 does not appear in this instance, and therefore it will not participate to any gate). On the right side, the list of *p-s* gates under the *P-S 1* and *P-S 2* labels correspond to the *phase separation* steps, while the list of *mix* gates under the *MIX 1* and *MIX 2* labels correspond to the *mixing* steps.

3.2   Quantum Gate Compilation Problem

Formally, the Quantum Gate Compilation Problem (QGCP) is a tuple $P = \langle C_0, L_0, QM \rangle$, where C_0 is the input quantum circuit, representing the execution of the MaxCut algorithm, L_0 is the initial assignment of the i-th *qstate* q_i to the i-th qubit n_i, and QM is a representation of the quantum hardware as a multigraph.

– The input quantum circuit is a tuple $C_0 = \langle Q, P\text{-}S, MIX, \{g_{start}, g_{end}\}, TC_0 \rangle$, where $Q = \{q_1, q_2, \ldots, q_N\}$ is the set of qstates which, from a planning & scheduling perspective (see for example [15], Chap. 15) represent the *resources* necessary for each gate's execution. $P\text{-}S$ and MIX are, respectively, the set of *p-s* and *mix* gate *operations* such that: (i) every $p\text{-}s(q_i, q_j)$ gate requires two qstates for execution; (ii) every $mix(q_i)$ gate requires one qstate only. g_{start} and g_{end} are two fictitious reference gate operations requiring no qstates. The execution of every quantum gate requires the uninterrupted use of the involved qstates during its processing time, and each qstate q_i can process at most one quantum gate at a time. Finally, TC_0 is a set of simple precedence constraints imposed on the $P\text{-}S$, MIX and $\{g_{start}, g_{end}\}$ sets, such that: (i) each gate in the two sets $P\text{-}S$, MIX occurs after g_{start} and before g_{end}; (ii) according to the total order imposed among the steps $P\text{-}S_1, MIX_1, P\text{-}S_2, MIX_2, \ldots, P\text{-}S_p, MIX_p$ (see the example in Fig. 3 with $p = 2$), all the gates belonging to the step $P\text{-}S_k$ (MIX_k) involving a specific qstate q_i must be executed before all the gates belonging to the next step MIX_k ($P\text{-}S_{k+1}$) involving the same qstate q_i, for $k = 1, 2, \ldots, p$ (for $k = 1, 2, \ldots, (p-1)$).
– L_0 is the initial assignment at the time origin $t = 0$ of qstates q_i to qubits n_i.
– QM is a representation of the quantum hardware as an undirected multi-graph $QM = \langle V_N, E_{p\text{-}s}, E_{swap}, \tau_{mix}, \tau_{p\text{-}s}, \tau_{swap} \rangle$, where $V_N = \{n_1, n_2, \ldots, n_N\}$ is the set of qubits (nodes), $E_{p\text{-}s}$ (E_{swap}) is a set of undirected edges (n_i, n_j) representing the set of *adjacent* locations the qstates q_i and q_j of the gates $p\text{-}s(q_i, q_j)$ ($swap(q_i, q_j)$) can potentially be allocated to. In addition, the *labelling* functions $\tau_{p\text{-}s} : E_{p\text{-}s} \to \mathbb{Z}^+$ and $\tau_{swap} : E_{swap} \to \mathbb{Z}^+$ respectively represent the durations of the gate operations $p\text{-}s(q_i, q_j)$ and $swap(q_i, q_j)$ when the qstates q_i and q_j are assigned to the corresponding adjacent locations. Similarly, the *labelling* function $\tau_{mix} : V \to \mathbb{Z}^+$ represents the durations of the *mix* gate (which can be executed at any node n_i). Figure 1 shows an example of quantum hardware with gate durations.

A feasible solution is a tuple $S = \langle SWAP, TC \rangle$, which extends the initial circuit C_0 with: (i) a set $SWAP$ of additional $swap(q_i, q_j)$ gates added to guarantee the adjacency constraints for the set of $P\text{-}S$ gates, and (ii) a set TC of additional simple precedence constraints such that:

– for each qstate q_i, a total order $\preceq_i$ is imposed among the set Q_i of operations requiring q_i, with $Q_i = \{op \in P\text{-}S \cup MIX \cup SWAP : op \text{ requires } q_i\}$.
– all the $p\text{-}s(q_i, q_j)$ and $swap(q_i, q_j)$ gate operations are allocated on adjacent qubits in QM.

Given a solution S, the makespan $mk(S)$ corresponds to the maximum completion time of the gate operations in S. A *path* between the two fictitious gates g_{start} and g_{end} is a sequence of gates $g_{start}, op_1, op_2, \ldots, op_k, g_{end}$, with

---

Algorithm 1. Find Feasible Plan

Require: A problem $P = \langle C_0, L_0, QM \rangle$
  $S \leftarrow \text{INITSOLUTION}(P)$;
  while not all the $P\text{-}S$ and MIX operations are inserted in S **do**
    $op \leftarrow \text{SELECTEXECUTABLEOPERATION}(P, S)$;
    $S \leftarrow \text{INSERTOPERATION}(op, S)$;
  end while
  return S

$op_j \in P\text{-}S \cup MIX \cup SWAP$, such that $g_{start} \preceq op_1, op_1 \preceq op_2, \ldots, op_k \preceq g_{end} \in (TC_0 \cup TC)$. The length of the path is the sum of the durations of all the path's gates and $mk(S)$ is the length of the longest path from g_{start} to g_{end}. An optimal solution S^* is a feasible solution characterized by the minimum makespan.

4   A Greedy Procedure

Algorithm 1 takes in input a QGCP problem $P = \langle C_0, L_0, QM \rangle$, and proceeds by *chronologically* inserting in the *partial solution* S one gate operation at a time until all the gates in the set $P\text{-}S \cup MIX$ are in S.

As stated earlier, let $op \in Q_i$ be a general gate operation that involves qstate q_i, and let us define $n(op)$ as the QM node at which gate op terminates its execution. Let us also define a *chain* $ch_i = \{op \in Q_i : op \in S\}$ as the set of gates involving q_i and currently present in the partial solution S, among which a total order is imposed (see Fig. 4 for a graphical representation of a complete solution composed of a set of chains, one for each qstate q_i). Let us now define $last(ch_i)$ as the last operation in the chain ch_i according to the imposed total order. Finally, let us define the *current state* L_S of a partial solution S as the tuple $L_S = \langle n(last(ch_1)), n(last(ch_2)), \ldots, n(last(ch_N)) \rangle$ containing the N last gate operations according to each chain ch_i ordering.

As a first step, Algorithm 1 initialises the partial solution S; in particular, it sets the current state L_S to the init value L_0 by initializing the locations of every qstate q_i (i.e., for every chain ch_i) at the time origin $t = 0^1$. The core of the algorithm is the function $\text{SELECTEXECUTABLEOPERATION}()$, which returns at each iteration either one of the gates in the set $P\text{-}S \cup MIX$ or a $swap(q_i, q_j)$ gate necessary to guarantee NN-compliance as described in the previous Sect. 3. It should be noted that $p\text{-}s(q_i, q_j)$ and $mix(q_i)$ gates leave unchanged the locations of the qstates q_i and q_j, whereas $swap(q_i, q_j)$ gate swaps their locations. We remark that the duration of the $p\text{-}s(q_i, q_j)$ gates change depending on the particular quantum chip edge they are executed on (see Fig. 1).

Then, the algorithm proceeds as follows: starting form the current state $L_S = L_0$, Algorithm 1 ranks the set of potentially executable gates according to a determined evaluation function (described below). The gate op with the lowest

---

[1] It is implicitly supposed that at the beginning, the i-th qstate is initialized at the i-th location.

(minimal) evaluation is selected and inserted in the partial solution S as the last operation of the chains[2] relative to the qstates involved in op: $last(ch_i) \leftarrow op$; subsequently, the state L_S of the partial solution is updated accordingly. The process continues iteratively until all the $P\text{-}S$ and MIX gates are inserted in S. At the end, the produced solution will contain a set of additional $swap(q_i, q_j)$ gates necessary to satisfy the NN-constraints.

Given the multi-graph QM introduced in Sect. 3.2, we consider the distance graph $G_d(V, E_{p\text{-}s})$, so as to contain an undirected edge $(n_i, n_j) \in E_{p\text{-}s}$ when QM can execute a $p\text{-}s$ gate on the pair (n_i, n_j). In the graph G_d, an undirected path p_{ij} between a node n_i and a node n_j is the list of edges $p_{ij} = ((n_i, n_{j1}), (n_{j1}, n_{j2}), \ldots, (n_{jk}, n_j))$ connecting the two nodes n_i and n_j and its lenght l_{ij} is the number of edges in the path p_{ij}. Let d_{ij} represent the minimal length among the set of all the paths between n_i and n_j. The distance d_{ij} between all nodes is computed only once at the beginning, by means of all-pairs shortest path algorithm, whose complexity is $O(|V|^3)$ in the worst case [18]. The distance d^{L_S} associated to a given $p\text{-}s(q_i, q_j)$ gate that requires two qstates q_i and q_j w.r.t. the state L_S of the partial solution S is defined as:

$$d^{L_S}(p\text{-}s(q_i, q_j)) = d(n(last(ch_i)), n(last(ch_j))) \tag{1}$$

Two qstates q_i and q_j are in adjacent locations in the state L^S if $d^{L_S}(p\text{-}s$ $(q_i, q_j)) = 1$. Intuitively, given a $p\text{-}s(q_i, q_j)$ gate and a partial solution S, the value $d^{L_S}(p\text{-}s(q_i, q_j))$ yields the minimal number of swaps (in excess of 1) for moving the two qstates q_i and q_j to adjacent locations on the machine QM. The concept of distance defined on a single gate operation $p\text{-}s(q_i, q_j)$ can be extended to a set of gate operations. In particular, let S be a partial solution and $\overline{P\text{-}S}^S$ the set of $p\text{-}s(q_i, q_j)$ gates that are not yet scheduled in S and such that all predecessors according to the temporal order imposed by the set TC_0 have already been scheduled. We propose two different functions to measure the distance separating the set $\overline{P\text{-}S}^S$ from the *adjacent state*. The first sums the set of the distances $d^{L_S}(p\text{-}s(q_i, q_j))$:

$$D^S_{sum}(\overline{P\text{-}S}^S) = \sum_{p\text{-}s \in \overline{P\text{-}S}^S} d^{L_S}(p\text{-}s(q_i, q_j)) \tag{2}$$

The second returns the minimal value of the distance $d^{L_S}(p\text{-}s(q_i, q_j))$ in the set $\overline{P\text{-}S}^S$:

$$D^S_{min}(\overline{P\text{-}S}^S) = MIN_{p\text{-}s \in \overline{P\text{-}S}^S} d^{L_S}(p\text{-}s(q_i, q_j)) \tag{3}$$

Given the functions (2) and (3), we can now evaluate the impact of the selection of a gate operation op on the whole solution S by means of the following ranking function that aggregates them lexicographically:

$$\Delta(S, op, \overline{P\text{-}S}^S) = \begin{cases} (D^{S'}_{sum}(\overline{P\text{-}S}^S \setminus \{op\}), 1) & op \text{ is a p-s;} \\ (D^{S'}_{sum}(\overline{P\text{-}S}^{S'}), 1) & op \text{ is a mix;} \\ (D^{S'}_{sum}(\overline{P\text{-}S}^{S'}), D^{S'}_{min}(\overline{P\text{-}S}^{S'})) & op \text{ is a swap.} \end{cases} \tag{4}$$

---

[2] Note that in general, a k-qubit gate occupies k chains.

where S' is the new partial solution after the addition of the selected gate operation op. In particular, the function SELECTEXECUTABLEOPERATION() returns one gate operation op (p-s, mix or $swap$) that minimises the $\Delta(S, op, \overline{P\text{-}S}^S)$ value, using the resource (i.e., the qstates) lowest indexes as a tie-break criterium. When all the P-S and MIX operations are inserted in S a full solution is returned.

We observe that the two-dimension distance function used for the gate selection ranking has a twofold role. The D_{sum} component acts as a *global* closure metric; by evaluating the overall distance left to be covered by all the qstates still involved in gates yet to be executed, it guides the selection towards the gate that best favours the efficient execution of the remaining gates. Conversely, the D_{min} component acts as a *local* closure metric, in that it favours the mutual approach of the closest qstates pairs. In more details, the role played by the D_{min} component is essential as a "tie-breaker", to avoid the selection of swap gates that may induce deadlock situations (cycles), which are possible in case we based our ranking solely on the D_{sum} component.

The overall time complexity of Algorithm 1 is polynomial. In fact, we can *always* generate a solution by incrementally inserting each gate operation P-$S \cup MIX$ according to any total ordering consistent with the input quantum circuit C_0 constraints, after making each p-$s(q_i, q_j)$ gate NN-compliant through the insertion of the necessary swap gates. If d_{max} is the maximal shortest path length in the distance graph $G_d(V, E_{p\text{-}s})$, the number of added swap gates is clearly bounded by the value $d_{max} - 1$; if n_g is the total size of the two sets P-$S \cup MIX$, it can be proved that the overall time complexity is $\mathcal{O}(n_g^3)$.

5   A Randomized Approach

To provide a capability for expanding the search cases without incurring the combinatorial overhead of a conventional backtracking search, we now define a random counterpart of our conflict selection heuristic (in the style of [1–3]), and embed the result within an iterative random sampling search framework. Note that randomization is also beneficial in systematic and complete approaches, see for instance [19] where randomized restart in backtracking is used. In order to make FINDFEASIBLEPLAN() suitable to random greedy restart, the SELECTEXECUTABLEOPERATION() function is modified according to the following rationale: (1) at each solution step, a set of "equivalent" gate operations (p-s, mix or $swap$) are first identified, and then (2) one of these is randomly selected. As in the deterministic variant, the selected gate operation op is then inserted in the current partial solution. The set of equivalent operations is created by identifying one operation op^* associated with the minimal lexicographic value $\Delta(S, op^*, \overline{P\text{-}S}^S) = (D^*_{sum}, D^*_{min})$ and by considering equivalent to op^* all the operations op such that $\Delta(S, op, \overline{P\text{-}S}^S) = (D_{sum}, D_{min})$ with $D_{sum} = D^*_{sum}$ and $D_{min} = D^*_{min}$. Subsequently, the operation to be inserted in the partial solution is randomly selected from this set, allowing a non-deterministic yet heuristically-biased choice. Successive calls to FINDFEASIBLEPLAN() are intended to explore

Algorithm 2. Greedy Random Sampling

Require: An proplem P, stop criterion
  $S_{best} \leftarrow$ FINDFEASIBLEPLAN(P);
  while (stopping criterion not satisfied) **do**
    $S \leftarrow$ FINDFEASIBLEPLAN(P)
    if (makespan(S) < makespan(S_{best})) **then**
      $S_{best} \leftarrow S$;
    end if
  end while
  return (S_{best})

Table 1. Aggregated results obtained from the **N8_u*_P2**, **N21_u*_P2**, and **N40_u*_P1** sets

Benchmark	Planner	# Imprvd	% Avg. Δ (MK)	# Unchgd	# Unprvd	% Avg. Δ (MK)	CPU (mins)
N8_u0.9_P2	TFD	37/50	9.76	10/50	3/50	−3.43	1 Vs. 10
N8_u1.0_P2	TFD	41/50	9.95	7/50	2/50	−4.95	1 Vs. 10
N21_u0.9_P2	TFD	46/50	10.83	1/50	3/50	−1.87	15 Vs. 60
N21_u1.0_P2	TFD	40/50	13.82	3/50	7/50	−7.13	15 Vs. 60
N40_u0.9_P1	SGPlan	50/50	38.56	0/50	0/50	0.0	1 Vs. 60
	LPG	10/10	18.88	0/50	0/50	0.0	
N40_u1.0_P1	SGPlan	50/50	35.42	0/50	0/50	0.0	1 Vs. 60
	LPG	14/14	16.7	0/50	0/50	0.0	

heuristically equivalent paths through the search space. Algorithm 2 depicts the complete iterative sampling algorithm for generating an optimized solution, which is designed to invoke the FINDFEASIBLEPLAN() procedure until a stop criterion is satisfied.

6 Experiments

In this section, we present the results obtained with our GRS procedure against the same quantum circuit benchmark set utilized in [12]. The benchmark is composed of instances of three different sizes, based on quantum chips with $N = 8, 21$ and 40 qubits, respectively (see Fig. 1). In [12], the authors base their experimentation on two problem classes for each chip size, depending on the number of passes ($p = 1$ or $p = 2$) to be performed during circuit execution. In this work, we will mostly focus on the $p = 2$ problem class because it is the most computationally challenging; we will analyze the $p = 1$ case for the $N = 40$ qubit size only, exclusively for comparison purposes with the results obtained in [12].

The utilized benchmark[3] contains 100 different problem instances for each chip size, where each instance is representative of a graph G to be partitioned by the *MaxCut* procedure to be realized (an example of graph is given in Fig. 3). The benchmark instances are divided in two subsets composed of 50 instances each, depending on the "utilization level" (u) of the available qstates over the circuit. In particular, the 50 instances characterized by $u = 0.9$ are built randomly choosing 90% of the available qstates to allocate over the N edges of the instance graph G, while the other 50 instances ($u = 1.0$) are built by possibly allocating *all* the qstates over the N edges of the graph G. Larger sizes and higher p values will lead to more complex problem instances.

Our experimental campaign is organized as follows. As anticipated earlier, we will mainly focus on the complete benchmark instances, i.e., those characterized by two compilation passes ($p = 2$), as they represent the hardest computational challenge. In particular, in our analysis we tackle 8 instance sets labelled **N**x_**u**y_**P**z, each composed of 50 instances, where **N** marks the number of qubits, **u** marks the occupation level, and **P** marks the number of compilation passes. The first 6 sets are characterized by $x \in [8, 21, 40]$, $y \in [0.9, 1.0]$, and $z \in [2]$, while the remaining 2 sets are characterized by $x \in [40]$, $y \in [0.9, 1.0]$ and $z \in [1]$.

Given that the previous sets are characterized by problem instances of increasing size, we have allotted different CPU time limits for each set, as follows. All runs relatively to the $N8^*$ and $N40\_u * \_P1$ sets are limited to max 60 s each; all runs relatively to the remaining $N21$ and $N40$ sets are limited to max 900 s each. All experiments have been performed on a 64-bit Windows10 O.S. running on Intel(R) Core(TM)2 Duo CPU E8600 @3.33 GHz with 8GB RAM.

6.1   Results

Table 1 exhibits the performances obtained with our GRS procedure. The results are presented in aggregated form for reasons of space; however, the complete set of makespan values, together with the complete set of solutions are available at http://pst.istc.cnr.it/~angelo/qc/. In the table, each row compares the results returned by the GRS procedure using the benchmark set highlighted in the *Benchmark* column, with the results returned by the planner(s) that achieved best performance among those reported in [12] (*Planner* column). The *# Imprvd* column shows the number of improved solutions obtained by our GRS w.r.t. the set of solutions solved in [12]. Similarly, the *# Unchgd* column shows the number of unmodified solutions, while the *# Unprvd* column shows the number of solutions that GRS was not able to improve. The *% Avg.Δ(MK)* columns show the percentage of the average makespan difference computed over the solution set of interest (i.e., the improved and the unimproved set). Note that such percentage is negative when computed on the unimproved solutions set. Lastly, the *CPU (mins)* column reports the maximum CPU time (in minutes) allotted to our GRS procedure Vs. the competitor procedure, for each run.

---

[3] The benchmark is available at: https://ti.arc.nasa.gov/m/groups/asr/planning-and-scheduling/VentCirComp17_data.zip.

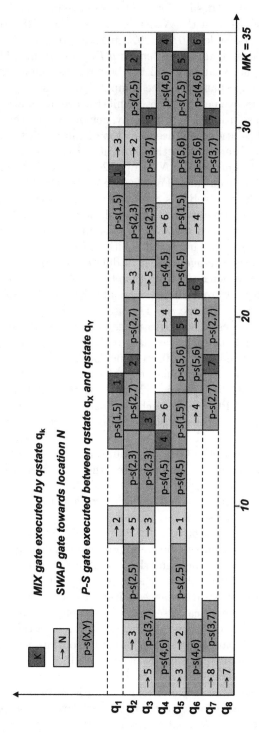

Fig. 4. Gantt representation of the solution relative to the problem instance no.1 belonging to the $N8\_u0.9\_P2$ benchmark set. Each q_i row represents a *chain*, i.e., the set of activities executed by the i-th qstate. At the beginning of the execution, it is assumed that the i-th qstate starts from the i-th location. The *swap* gates convey the information relative to the destination locations only, being the start locations already known. Note that the *swap* gates are always executed in pairs, exactly as the p-s gates. Note also that the p-s gates are characterized by two different durations ($\tau_{p\text{-}s} = 3$ or $\tau_{p\text{-}s} = 4$) depending on the location pair on which they are executed. For instance, the gates p-$s(2,5)$ involving qstates q_2 and q_5 have duration equal to 4 in the first pass (the gates are executed on locations n_3 and n_2), and duration equal to 3 in the second pass (the gates are executed on locations n_2 and n_1). All *mix* gates have duration equal to 1. Lastly, note that the depicted solution is *complete* in the sense that it contains also the *mix* gates pertaining to the second pass ($p = 2$).

The results clearly show a marked superiority of our procedure w.r.t. the counterparts, for all benchmark sets. In the $N8\_u0.9\_P2$ case, the GRS procedure improved 37/50 solutions (74%), left 10/50 (20%) unchanged, and was outperformed on 3/50 (6%) solutions, against the *Temporal FastDownward* (TFD) planner (see [20]) i.e., the planner that returned the best results among those reported in [12]. The average makespan difference computed over the improved solutions (average makespan improvement) is 9.86%, while the average makespan difference over the unimproved solutions is -3.43% only, demonstrating that the quality of the few unimproved solutions is indeed not too distant from the best results. By inspecting the subsequent row in the table, it is clear that equally strong results have also been obtained in the $N8\_u1.0\_P2$ case.

The same kind of analysis has been carried out for the $N21\_u0.9\_P2$ and the $N21\_u1.0\_P2$ benchmark sets, respectively. As a summary of the performances, in the $u = 0.9$ case the GRS procedure improved 46/50 solutions (92%), left 1/50 (2%) unchanged, and was outperformed on 3/50 (6%) solutions. The average makespan difference computed over the improved solutions (average makespan improvement) is 10.83%, while the average makespan difference over the unimproved solutions is -1.87%, again demonstrating that the quality of the unimproved solutions is not too distant from the best results. Equally convincing results can also be easily appreciated in the $u = 1.0$ case.

In order to test the performances of our GRS procedure against the results obtained in [12] for the $N40$ instances, the last two rows of Table 1 show the results for the $N40\_u0.9\_P1$ and the $N40\_u1.0\_P1$ benchmark sets respectively (i.e., relative to the $p = 1$ case only). In particular, the table compares our results with those obtained with the SGPlan planner [21,22] and the LPG planner [23], respectively. Note that the LPG planner achieves better results over the SGPlan planner, even though it succeeds in solving only a strict minority of the instances (see # *Imprvd* column). However, as a remarkable result, all the 50 instances have been improved by the GRS procedure, in both the $u = 0.9$ and the $u = 1.0$ case. To summarize the performances, the average makespan improvement obtained by GRS over SGPlan is 38.56% and 35.42% for $u = 0.9$ and $u = 1.0$ respectively, while the average makespan improvement obtained by GRS over LPG (computed on the subset of instances solved by LPG) is 18.88% and 16.7% for $u = 0.9$ and $u = 1.0$ respectively.

As opposed to the previous experiments, no comparative analysis is possible in the $N40\_u0.9\_P2$ and $N40\_u1.0\_P2$ benchmark case, as no planner in [12] succeeded in solving any instance within the max allotted time of 60 min. However, our GRS procedure solved all such instances within a max allowed time of 15 min. In order to provide a numerical assessment of the efficacy of GRS's random optimization procedure, we compared the makespan values of the initial solutions found against the makespan values at the end of the optimization process, acknowledging an average makespan improvement of 21.15% and 19.25%, in the $u = 0.9$ and $u = 1.0$ case, respectively.

Before concluding, we present an example of solution in Fig. 4. In particular, the figure shows a Gantt representation of the solution relative to the problem instance no.1 belonging to the $N8\_u0.9\_P2$ benchmark set. Each q_i row represents the set of activities executed by the i-th qstate (i.e., the chain ch_i). At the beginning of the execution, it is assumed that the i-th qstate starts from the i-th location. The *swap* gates convey the information relative to the destination locations only (i.e., $\rightarrow i$ is to be intended as "towards location n_i"), since the start locations are implicitly known as those in which the previous activity on the same chain has terminated. Note that the both the *swap* gates and the *p-s* gates are always executed in pairs. Note also that the *p-s* gates are characterized by two different durations ($\tau_{p\text{-}s} = 3$ or $\tau_{p\text{-}s} = 4$) depending on the location pair on which they are executed. For instance, the gates *p-s*$(2,5)$ involving qstates q_2 and q_5 have $\tau_{p\text{-}s} = 4$ in the first pass (the gates are executed on locations n_3 and n_2), and $\tau_{p\text{-}s} = 3$ in the second pass (the gates are executed on locations n_2 and n_1). The previous durations can be checked by visual inspection of Fig. 1. All *mix* gates have duration equal to 1. Lastly, note that the depicted solution is *complete* in the sense that it contains also the *mix* gates pertaining to the second pass ($p = 2$).

7 Conclusions

In this work we propose a greedy random search heuristic to solve the quantum circuit compilation problem, where the objective is essentially to synthesize a quantum gate execution plan characterized by a minimum makespan. We test our procedure against a number of instances from a benchmark repository publicly available, and compare our results with those obtained in a recent work where the same problem is solved by means of PDDL-based planning technology, showing that our procedure is more performing in the vast majority of cases. Despite the very good results, we consider our present contribution to the quantum compilation problem complementary to the PDDL approach. The take-home message from this comparison can in fact be wrapped up as follows. On the one hand, it is confirmed that tackling a problem with a general technique such as PDDL can be less rewarding in terms of overall solution quality than employing heuristics more tailored on the problem; such heuristics, though very simple, can remain extremely efficient when the problem size scales up significantly. On the other hand, it remains true that in more complex domains, a general approach such as PDDL-based planning could still represent a winning factor, also considering that some of the solutions we compared against have demonstrated to be of very high quality. Our conclusion is that an integration between the two techniques might be beneficial in order to appreciate the representational generality of PDDL planning, without renouncing the exploration/exploitation power of state-of-the-art constraint-based metaheuristics.

References

1. Hart, J., Shogan, A.: Semi-greedy heuristics: an empirical study. Oper. Res. Lett. **6**, 107–114 (1987)
2. Resende, M.G., Werneck, R.F.: A hybrid heuristic for the p-median problem. J. Heuristics **10**(1), 59–88 (2004)

3. Oddi, A., Smith, S.: Stochastic procedures for generating feasible schedules. In: Proceedings 14th National Conference on AI (AAAI-1997), pp. 308–314 (1997)
4. Nielsen, M.A., Chuang, I.L.: Quantum Computation and Quantum Information: 10th Anniversary Edition, 10th edn. Cambridge University Press, New York (2011)
5. Cirac, J.I., Zoller, P.: Quantum computations with cold trapped ions. Phys. Rev. Lett. **74**, 4091–4094 (1995)
6. Herrera-Martí, D.A., Fowler, A.G., Jennings, D., Rudolph, T.: Photonic implementation for the topological cluster-state quantum computer. Phys. Rev. A **82**, 032332 (2010)
7. Yao, N.Y., Gong, Z.X., Laumann, C.R., Bennett, S.D., Duan, L.M., Lukin, M.D., Jiang, L., Gorshkov, A.V.: Quantum logic between remote quantum registers. Phys. Rev. A **87**, 022306 (2013)
8. Brierley, S.: Efficient implementation of quantum circuits with limited qubit interactions. arXiv preprint arXiv:1507.04263, September 2016
9. Farhi, E., Goldstone, J., Gutmann, S.: A quantum approximate optimization algorithm. arXiv preprint arXiv:1411.4028, November 2014
10. Guerreschi, G.G., Park, J.: Gate scheduling for quantum algorithms. arXiv preprint arXiv:1708.00023, July 2017
11. Sete, E.A., Zeng, W.J., Rigetti, C.T.: A functional architecture for scalable quantum computing. In: 2016 IEEE International Conference on Rebooting Computing (ICRC), pp. 1–6, October 2016
12. Venturelli, D., Do, M., Rieffel, E., Frank, J.: Temporal planning for compilation of quantum approximate optimization circuits. In: Proceedings of the Twenty-Sixth International Joint Conference on Artificial Intelligence, IJCAI-2017, pp. 4440–4446 (2017)
13. Maslov, D., Falconer, S.M., Mosca, M.: Quantum circuit placement: optimizing qubit-to-qubit interactions through mapping quantum circuits into a physical experiment. In: Proceedings of the 44th Annual Design Automation Conference, DAC 2007, pp. 962–965. ACM, New York (2007)
14. Fox, M., Long, D.: PDDL2.1: An extension to PDDL for expressing temporal planning domains. J. Artif. Int. Res. **20**(1), 61–124 (2003)
15. Nau, D., Ghallab, M., Traverso, P.: Automated Planning: Theory & Practice. Morgan Kaufmann Publishers Inc., San Francisco (2004)
16. Kole, A., Datta, K., Sengupta, I.: A heuristic for linear nearest neighbor realization of quantum circuits by swap gate insertion using n-gate lookahead. IEEE J. Emerg. Sel. Top. Circ. Syst. **6**(1), 62–72 (2016)
17. Kole, A., Datta, K., Sengupta, I.: A new heuristic for n-dimensional nearest neighbor realization of a quantum circuit. IEEE Trans. Comput.-Aided Des. Integr. Circ. Syst. **37**(1), 182–192 (2018)
18. Cormen, T.H., Leiserson, C.E., Rivest, R.L., Stein, C.: Introduction to Algorithms, 2nd edn. MIT Press, Cambridge (2001)
19. Gomes, C.P., Selman, B., Kautz, H.: Boosting combinatorial search through randomization. In: Proceedings of the Fifteenth National/Tenth Conference on Artificial Intelligence/Innovative Applications of Artificial Intelligence. AAAI 1998/IAAI 1998, pp. 431–437. American Association for Artificial Intelligence, Menlo Park (1998)
20. Eyerich, P., Mattmüller, R., Röger, G.: Using the context-enhanced additive heuristic for temporal and numeric planning. In: Proceedings of the 19th International Conference on Automated Planning and Scheduling, ICAPS 2009, Thessaloniki, Greece, 19–23 September 2009 (2009)

21. Wah, B.W., Chen, Y.: Subgoal partitioning and global search for solving temporal planning problems in mixed space. Int. J. Artif. Intell. Tools **13**(04), 767–790 (2004)
22. Chen, Y., Wah, B.W., Hsu, C.W.: Temporal planning using subgoal partitioning and resolution in SGPlan. J. Artif. Int. Res. **26**(1), 323–369 (2006)
23. Gerevini, A., Saetti, A., Serina, I.: Planning through stochastic local search and temporal action graphs in LPG. J. Artif. Int. Res. **20**(1), 239–290 (2003)

A Comparison of Optimization Methods for Multi-objective Constrained Bin Packing Problems

Philippe Olivier[1,2(✉)], Andrea Lodi[1,2], and Gilles Pesant[1]

[1] École Polytechnique de Montréal, Montreal, Canada
{philippe.olivier,andrea.lodi,gilles.pesant}@polymtl.ca
[2] CERC, Montreal, Canada

Abstract. Despite the existence of efficient solution methods for bin packing problems, in practice these seldom occur in such a pure form but feature instead various considerations such as pairwise conflicts or profits between items, or aiming for balanced loads amongst the bins. The Wedding Seating Problem is a combinatorial optimization problem incorporating elements of bin packing with conflicts, bin packing with profits, and load balancing. We use this representative problem to present and compare constraint programming, integer programming, and metaheuristic approaches.

1 Introduction

In the optimization version of the classical bin packing problem, a set of items of various weights must be packed into as few bins of limited capacities as possible. Despite the existence of efficient solution methods for bin packing problems, in practice these seldom occur in such a pure form. They instead feature various considerations such as pairwise conflicts or profits between items, or aiming for balanced loads amongst the bins. The objective then becomes to minimize some scoring function by selecting an optimal distribution of items in the available bins.

In our representative problem, the Wedding Seating Problem (WSP) [1], groups of guests of different sizes must be seated at tables of limited capacities. Some of these groups may or may not like each other, thus some relation is defined over each pair of them. Pairs of groups whose relation is *definitely apart* can never be seated at the same table. While not strictly necessary, pairs of groups whose relation is either *rather together* or *rather apart* should, if possible, be seated together or apart, respectively. Pairs which have no specific relation are *indifferent*. Note that an implicit relation, *definitively together*, is baked into the problem as *groups of guests*, the smallest indivisible entity that can be assigned to a table.

Section 2 gives a formal definition of our problem, and Sect. 3 reviews current methods of solving the WSP and similar problems. Sections 4, 5 and 6 introduce, respectively, our constraint programming (CP) model as well as our

© Springer International Publishing AG, part of Springer Nature 2018
W.-J. van Hoeve (Ed.): CPAIOR 2018, LNCS 10848, pp. 462–476, 2018.
https://doi.org/10.1007/978-3-319-93031-2_33

two integer programming (IP) models. Sections 7 and 8 present the results of our experiments.

2   Description of the Problem

Let

- $\mathcal{I} = \{1, \ldots, n\}$ be the index set of items,
- $\mathcal{B} = \{1, \ldots, m\}$ be the index set of bins,
- ℓ and u be, respectively, the lower and upper bounds on the load of a bin,
- w_i denote the weight of item i with $w = \sum_{i \in \mathcal{I}} w_i$ representing the combined weight of all the items,
- c_{ij} the cost incurred if items i and j are packed into the same bin.

Entries in the cost matrix C can take any integer value. Namely,

$$c_{ij} \begin{cases} = \infty, & \text{if } i \text{ and } j \text{ are in conflict and must be packed into separate bins,} \\ = 0, & \text{if } i \text{ and } j \text{ have no cost,} \\ < 0, & \text{if } i \text{ and } j \text{ should rather be packed into the same bin,} \\ > 0, & \text{if } i \text{ and } j \text{ should rather be packed into separate bins.} \end{cases}$$

Since a conflict is expressed as being a prohibitive cost, the initial cost matrix can be enhanced by adding this prohibitive cost for each pair of items whose combined weights is greater than u, since they can never be packed together.

The problem consists of packing all items into the available bins such that conflicting items are packed into separate bins, while optimizing cost and balance objectives. The cost objective f is to minimize the combined cost of all available bins. The balancing objective is to maximize the balance of loads amongst the bins, which is somewhat abstract as balancing loads with different norms will yield incomparable results. The L_0-norm will minimize the number of values different from the mean bin load (which is not useful unless the mean is an integer). The L_∞-norm will minimize the maximum deviation from the mean. In this paper, we will focus on the L_1- and L_2-norms which will, respectively, minimize the sum of deviations and the sum of squared deviations from the mean.

The correct way to compare solutions with one another is debatable. Both objectives use different units, so at the very least they should be weighted in order to obtain some sort of solution ranking. In this paper, we have instead opted to construct a Pareto set using the ϵ-constraint method [2]: We bounded the cumulative deviation at successively higher values, each time solving the problem by minimizing objective f. The correspondence between multiple objectives is often not directly proportional in practice. As such, presenting the balancing objective with a Pareto set has the advantage of offering decision-makers multiple optimal solutions to choose from depending on their perception of the trade-offs amongst them.

The problem of packing items into bins is the same as that of seating groups of guests at tables, as described by Lewis and Carroll. It has been shown to be $\mathcal{NP}$-hard as it generalizes two other $\mathcal{NP}$-hard problems: the k-partition problem and the graph k-coloring problem [3].

3   Related Work and Existing Methods

The problem of constructing seating plans was originally introduced by Bellows and Peterson [4]. The authors used an IP model to solve various instances of this problem for their own wedding. They were able to solve small instances of 17 guests in a few seconds. For a larger instance of 107 guests, however, no optimal solution could be found in reasonable time.

This seating assignment problem was later formally described as the *Wedding Seating Problem* by Lewis [1]. The author solved the problem with a metaheuristic model based on a two-stage tabu search. In a further paper [3], Lewis and Carroll devised their own quadratic IP model (of which our close variant is discussed in Sect. 5) to be compared with the metaheuristic approach. This latter approach outperformed their IP model both in solution quality and in running time in most cases. The authors also reimplemented the IP model of Bellows and Peterson, which they found performed poorly.

A fairly recent survey of CP work on bin packing and load balancing can be found in [5]. Most research on bin packing with conflicts, such as [6], focuses on minimizing the number of bins used as in the classical problem (albeit with the added conflict dimension). In contrast, the WSP uses a fixed number of bins of dynamic capacities, the objective being to optimize a scoring function subject to additional balancing constraints. The notion of pairwise costs between items used in the WSP is somewhat unconventional, but a similar idea can be found in some bin packing *games* where selfish agents (items) strive to maximize their payoffs by packing themselves into the most profitable bin, which is determined by its item composition [7].

4   CP Model

For each item i we define a decision variable b_i whose value is the bin in which the item will be packed. For each bin k we define an auxiliary variable o_k whose value is the load of the bin. To pack the items into the bins, the model uses a binpacking constraint [8]. The balancing objective represented by variable σ is taken care of by a balance constraint [9], which can handle L_1- and L_2-norms.

$$\texttt{binpacking}\,(\langle b_i \rangle, \langle w_i \rangle, \langle o_k \rangle) \tag{1}$$

$$\texttt{balance}\,(\{o_k\}, w/m, \sigma) \tag{2}$$

$$\ell \leq o_k \leq u \tag{3}$$

$$b_i \in \mathcal{B}, \quad i \in \mathcal{I} \tag{4}$$

$$\ell, u, o_k \in \mathbb{N}, \quad k \in \mathcal{B} \tag{5}$$

The model uses a conflict graph to infer `alldifferent` constraints, similar to what has been used by Gualandi and Lombardi in their decomposition of the `multibin packing` constraint [10]. In a conflict graph, each item is represented by a vertex, and an edge joins two vertices if they are conflicting. By extracting all the maximal cliques of this graph, it is possible to add `alldifferent` constraints to the model for each one of those cliques. Furthermore, the maximum clique of an instance (the largest of all the maximal cliques) determines a lower bound on the number of bins necessary to find a feasible solution to that instance. The Bron-Kerbosch algorithm is an exact method which can be used to find all the maximal cliques of the conflict graph [11]. Let $\mathcal{M}$ be the set of all maximal cliques (maximal clique x being, for example, $\mathcal{M}_x = \{b_1^x, \ldots, b_k^x\}$):

$$\texttt{alldifferent}\left(\{\mathcal{M}_x\}\right), \quad \forall x \in \{1, \ldots, |\mathcal{M}|\} \tag{6}$$

While we have not explored all specific edge cases in this paper, if we were to solve highly constrained instances of the problem (i.e., with a dense conflict graph) these could be intractable for the Bron-Kerbosch algorithm. A heuristic for finding cliques could instead be applied, or simple binary disequality constraints could be used in lieu of the conflict graph.

Some symmetry is broken by fixing items of weights strictly greater than $u/2$ to separate bins. In theory, better symmetry breaking could be achieved first by fixing each item in the maximum clique of the conflict graph to a separate bin, and then by forcing an order on the loads of the remaining bins. In practice, however, symmetry breaking for the CP model is tricky as it interferes with the branching heuristic, whose strength lies in finding very good solutions very quickly. While the overall time needed to solve an instance to optimality decreases with the use of symmetry breaking, the downside is that early solutions will be worse with it than without. Without symmetry breaking, the branching heuristic basically packs the heaviest items at the bottom of the bins (i.e., they are assigned to a bin near the top of the tree). The top items of the bins are thus of lighter weights, and it is naturally less constraining to swap them around and pack them more profitably. Symmetry breaking forces some packings of lighter items at the bottom of the bins, constraining the swapping of items at the top of the bins.

Finally the objective is to minimize f subject to a constraint bounding the value of the cumulative deviation of the bins. Considering disjoint intervals of deviation ensures that the trees explored in each step of the construction of the Pareto set are nonoverlapping, preventing identical solutions from being found in different steps. Let $d_{\min}$ and $d_{\max}$ be, respectively, the lower and upper bounds on the cumulative deviation of a solution:

$$d_{\min} < \sigma \leq d_{\max} \tag{7}$$

$$\min \sum_{i=1}^{n} \sum_{j=1}^{n} (b_i = b_j)\, c_{ij} \tag{8}$$

The model branches on the decision variables according to a heuristic which was inspired by the standard best fit decreasing strategy used to solve the bin packing problem. It first chooses the item of the greatest weight yet unassigned and will pack it into an empty bin, if one is available. Otherwise, the heuristic will pack the item into the bin which would most increase the f value. This heuristic tends to find a good first solution very early in the search.

We have also further tested the CP model by including a large neighborhood search (LNS) strategy [12]. The model finds a good initial solution, after which the LNS strategy takes over. The LNS will iteratively freeze different parts of the solution and search afresh from there for a set amount of time. About a third of the bins are frozen as such, with the most profitable bins having the most chance of being frozen.

5   IP Model A

The first IP model (IP_A) is a generalization of the natural quadratic IP model proposed by Lewis and Carroll [3]. A $n \times m$ matrix of decision variables x represents packing assignments of items into bins (9)

$$x_{ik} := \begin{cases} 1, & \text{if item } i \text{ is packed in bin } k, \\ 0, & \text{otherwise.} \end{cases} \tag{9}$$

$$\min \quad \sum_{k \in \mathcal{B}} \sum_{i=1}^{n-1} \sum_{j=i+1}^{n} x_{ik} x_{jk} c_{ij} \tag{10}$$

s.t.

$$\sum_{k \in \mathcal{B}} x_{ik} = 1 \qquad \forall i \in \mathcal{I} \tag{11}$$

$$x_{ik} + x_{jk} \leq 1 \qquad \forall i,j \in \mathcal{I} : c_{ij} = \infty, \quad \forall k \in \mathcal{B} \tag{12}$$

$$\sum_{i \in \mathcal{I}} x_{ik} w_i \geq \ell \qquad \forall k \in \mathcal{B} \tag{13}$$

$$\sum_{i \in \mathcal{I}} x_{ik} w_i \leq u \qquad \forall k \in \mathcal{B} \tag{14}$$

$$\sum_{i \in \mathcal{I}} x_{ik} w_{ik} - w/m \leq o_k \qquad \forall k \in \mathcal{B} \tag{15}$$

$$\sum_{i \in \mathcal{I}} x_{ik} w_{ik} - w/m \geq -o_k \qquad \forall k \in \mathcal{B} \tag{16}$$

$$\sum_{k \in \mathcal{B}} o_k \geq d_{\min} \tag{17}$$

$$\sum_{k \in \mathcal{B}} o_k \leq d_{\max} \tag{18}$$

$$x_{ik} = 0 \qquad \forall i \in \mathcal{I}, \quad \forall k \in \{i+1, \ldots, m\} \tag{19}$$

$$x_{ik} \in \{0,1\} \qquad \forall i \in \mathcal{I}, \quad \forall k \in \mathcal{B} \tag{20}$$

$$o_k \in \{\ell, \ldots, u\} \qquad \forall k \in \mathcal{B} \tag{21}$$

This model minimizes the sum of all pairwise costs between items in each bin (10). The items are required to be packed into one and only one bin (11), and conflicting items may not be packed into the same bin (12). Constraints (13) and (14) require that the load of every bin be within bounds ℓ and u. This model computes the deviation according to an L_1-norm; Constraints (15)−(16) emulate an absolute value function and constraints (17)−(18) stipulate that the cumulative deviation of the solution must be bounded by $d_{\min}$ and $d_{\max}$ (since we are constructing a Pareto set). Some symmetry breaking is achieved with constraints (19).

This model has been constructed to handle deviation according to an L_1-norm. Integrating convex relaxation with McCormick envelopes [13] to the model would allow the use of the L_2-norm.

6   IP Model B

Our second IP model (IP_B) is based on the definition of an exponential-size collection of possible bin compositions as for the classical bin packing problem. Indeed, as for bin packing, the resulting formulation can be solved by column generation with either a set covering (SC) or set partitioning (SP) model. Let $\mathbb{S}$ be the collection of all subsets of items that can be packed into the same bin

$$\mathbb{S} := \left\{ S \subseteq \{1, \ldots, n\} : \ell \leq \sum_{i \in S} w_i \leq u, \quad \forall i, j \in S : c_{ij} \neq \infty \right\}.$$

We can observe that it may not be possible to *only* construct maximal sets with regards to the bin capacity due to conflicts between items and the constraints enforcing them. There is a binary variable for each subset $S \in \mathbb{S}$ representing a combination of items, or *pattern*, to be packed into the same bin

$$x_S := \begin{cases} 1, & \text{if pattern } S \text{ is selected,} \\ 0, & \text{otherwise.} \end{cases} \tag{22}$$

The sum of all pairwise costs for the items of a pattern and the deviation of the weight of that pattern from the mean bin load are represented by α and β, respectively. In regards to the balancing objective, using a fixed number of bins has two major advantages over using a variable number of bins. First, the values of α and β need to be computed only once per pattern (when it is generated) and remain constant throughout the process. Second and more importantly, since β is computed outside of the program, the norm according to which it is computed does not complexify the problem (i.e., the program remains linear even when balance is computed according to an L_2-norm).

While the solution of a SP model is always directly feasible, that of a SC model must be transformed in order to be feasible for our problem (i.e., we must remove all duplicate items from the bins). This has the unfortunate effect of potentially worsening the objective funtion. Example 1 illustrates the underlying issue of using a SC model to solve our problem.

Example 1. Assume an instance of the problem with 2 bins and 4 items A, B, C, and D. The pairwise costs of AB, BC, CD, and DA are 0, while the pairwise cost of AC is 1 and that of BD is -2. We also have $\ell = 2$ and $u = 3$. The two most profitable maximal subsets are ABD and BCD which both have a value of -2 and cover all the items. The initial solution of the SC model would be $(-2) + (-2) = -4$, which must be rendered feasible for our problem by removing B from one bin and D from the other. This modified solution would have a value of $(0) + (0) = 0$, while the solution of a SP model would be to pack AC and BD in separate bins, for a value of $(1) + (-2) = -1$.

The master problem (MP) is thus based on a SP model.

$$\min \quad \sum_{S \in \mathbb{S}} \alpha_S x_S \tag{23}$$

s.t.

$$\sum_{S \in \mathbb{S}: i \in S} x_S = 1 \qquad \forall i \in \mathcal{I} \tag{24}$$

$$\sum_{S \in \mathbb{S}} x_S = m \tag{25}$$

$$\sum_{S \in \mathbb{S}} \beta_S x_S \geq d_{\min} \tag{26}$$

$$\sum_{S \in \mathbb{S}} \beta_S x_S \leq d_{\max} \tag{27}$$

$$x_S \in \{0, 1\} \qquad \forall S \in \mathbb{S} \tag{28}$$

The MP minimizes the combined costs of all bins (23), under the conditions that each item be packed into one and only one bin (24), that a total of m bins be used (25), and that the cumulative deviation of a solution be within bounds $d_{\min}$ and $d_{\max}$ (26)–(27). In order to begin solving the problem, the column generation algorithm needs m columns making up an initial feasible solution of the problem and of the continuous relaxation obtained by replacing constraints (28) with $x_S \geq 0, \forall S \in \mathbb{S}$. These columns are generated by a compact CP model defined by (1)–(5), (7), and

$$b_i \neq b_j, \quad b_i, b_j \in \mathcal{B}, \quad \forall i, j \in \mathcal{I} : c_{ij} = \infty \tag{29}$$

which searches for the first feasible solution satisfying these constraints. The m columns found by this CP model make up the initial restricted master problem (RMP). The dual of the MP is

$$\max \quad \sum_{i=1}^{n} y_i + m\zeta + d_{\min}\gamma + d_{\max}\delta \tag{30}$$

s.t.

$$\sum_{i \in S} y_i + \zeta + \beta_S(\gamma + \delta) \leq \alpha_S \qquad \forall S \in \mathbb{S} \tag{31}$$

$$y_i \text{ free} \qquad\qquad \forall i \in S \qquad (32)$$
$$\zeta \text{ free} \qquad\qquad\qquad (33)$$
$$\gamma \geq 0 \qquad\qquad\qquad (34)$$
$$\delta \leq 0 \qquad\qquad\qquad (35)$$

which is all that is needed for the column generation algorithm to take over:

1. Solve the continuous relaxation of the RMP to get the dual values.
2. Solve the subproblem, or pricing problem (PP), to generate $S^* \subseteq \{1, \ldots, n\}$ (the most promising new column). Let

$$z_i := \begin{cases} 1, & \text{if item } i \text{ is packed into the new bin/pattern,} \\ 0, & \text{otherwise.} \end{cases} \qquad (36)$$

$$\max \sum_{i=1}^{n} y_i^* z_i + \zeta^* + \beta(\gamma^* + \delta^*) - \sum_{i=1}^{n-1}\sum_{j=i+1}^{n} c_{ij} z_i z_j \quad \text{if } \gamma^* + \delta^* < 0 \qquad (37)$$

$$\max \sum_{i=1}^{n} y_i^* z_i + \zeta^* - \beta(\gamma^* + \delta^*) - \sum_{i=1}^{n-1}\sum_{j=i+1}^{n} c_{ij} z_i z_j \quad \text{if } \gamma^* + \delta^* > 0 \qquad (38)$$

s.t.

$$z_i + z_j \leq 1 \qquad\qquad \forall i, j \in \mathcal{I} : c_{ij} = \infty \quad (39)$$

$$\sum_{i=1}^{n} w_i z_i \geq \ell \qquad\qquad\qquad (40)$$

$$\sum_{i=1}^{n} w_i z_i \leq u \qquad\qquad\qquad (41)$$

$$\sum_{i=1}^{n} w_i z_i - w/m \leq \beta \qquad\qquad\qquad (42)$$

$$\sum_{i=1}^{n} w_i z_i - w/m \geq -\beta \qquad\qquad\qquad (43)$$

$$z_i \in \{0, 1\} \qquad\qquad \forall i \in \mathcal{I} \quad (44)$$

The PP minimizes the value of a bin (37)/(38) under the conditions that no conflicting items be packed into it (39), and that its load be within bounds ℓ and u (40)–(41). Constraints (42)–(43) work in the same manner as constraints (15)–(16) and ensure that deviation β is always positive.

3. Determine if S^* should be added to the RMP. If the inequality $\sum_{k \in S^*} y_k^* + \zeta^* + \beta(\gamma^* + \delta^*) > \sum_{i \in S^*}\sum_{j \in S^*} c_{ij}$ (or $\sum_{k \in S^*} y_k^* + \zeta^* - \beta(\gamma^* + \delta^*) > \sum_{i \in S^*}\sum_{j \in S^*} c_{ij}$, alternatively) is true, compute α_{S^*} and β_{S^*}, and add column S^* to the RMP before going back to step 1. Otherwise, the current solution of the continuous relaxation of the RMP is the lower bound of the initial problem.

The optimal solution of the continuous relaxation of the RMP provides a lower bound to the problem. We subsequently enforce constraints (28) and solve the RMP with the columns which were previously generated by the algorithm in order to find an integral solution. We know such an integral solution exists since we started off with one with our initial columns. While this final integral solution offers no proof of optimality, it is most often relatively close to the lower bound.

Depending on the norm used to balance the bins and on the value of bound d_{max}, we can make arithmetic deductions to determine the optimal ℓ and u bounds which should help prematurely prune nodes leading to infeasible solutions, without eliminating any feasible solution:

$$\ell \geq \max\{0, \lceil w/m - d_{max}/2 \rceil\} \tag{45}$$

$$u \leq \lfloor w/m + d_{max}/2 \rfloor \tag{46}$$

$$\ell \geq \max\{0, \left\lceil w/m - \sqrt{d_{max} \times (m-1)/m} \right\rceil\} \tag{47}$$

$$u \leq \left\lfloor w/m + \sqrt{d_{max} \times (m-1)/m} \right\rfloor \tag{48}$$

For the L_1-norm, in the worst case a single bin can account for at most half of the deviation, since the cumulative weight in excess of the mean bin load will always be equal to the cumulative weight short of the mean bin load (45)−(46). For the L_2-norm we must also take the number of bins into account in order to tightly bound the most deviative bin (47)−(48). This offline optimization of bin load bounds makes a noticeable difference for both IP models, cutting the execution time by half on average. This optimization is of no use for the CP model, as the balancing constraint already ensures that bin loads be consistent with d_{max}.

7   Benchmark Results

Our testbed includes instances of 25 and 50 items, of weights chosen uniformly at random from 1 to 8 (in a similar fashion as Lewis [1]). Costs and conflicts are introduced according to probability p of a pairwise negative cost, probability q of a pairwise positive cost, and probability r of a conflict (i.e., $p+q+r \leq 1$, with the remainder being the probability that the pair incurs no cost). Costs range from −5 to 5, and the number of bins has been chosen so that the mean load is closest to 10.

The experiments were performed on dual core AMD 2.2 GHz processors with 8 GB of RAM running CentOS 7, and all models were constructed with IBM ILOG CPLEX Optimization Studio v12.5. Pareto sets are constructed for the smallest integral ranges of deviations (i.e., min/max deviations of 0/1, 1/2, ..., 19/20). Time limits cover only one range of deviations, meaning that the results for one method, for one figure, involve solving 20 independent problems. This also means that construction of the Pareto sets is easily parallelizable and can be scaled to limited resources by modifying its resolution (e.g., if 4 instances can be solved in parallel, a lower-resolution Pareto set can be constructed with min/max deviations of 0/5, 6/10, 11/15, 16/20). Each data point shows the average results of 5 different cost matrices applied to the same set of items. All figures show the results for instances of 25 or 50 items with the deviations computed according to an L_1-norm. For Figs. 1, 2 and 3 the cost probabilities are $p = 0.25$, $q = 0.25$, and $r = 0.25$.

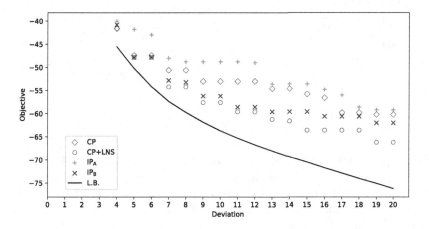

Fig. 1. Instances of 25 items with a time limit of 600 s.

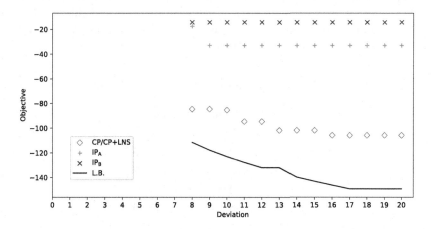

Fig. 2. Instances of 50 items with a time limit of 6 s.

We can observe in Fig. 2 that the CP model finds the best early solutions of all methods. However it quickly reaches a plateau from which it is hard to further improve (notice the similarity between the CP solutions of Fig. 2 with a time limit of 6 s, and those of Fig. 3 with a time limit of 600 s). The introduction

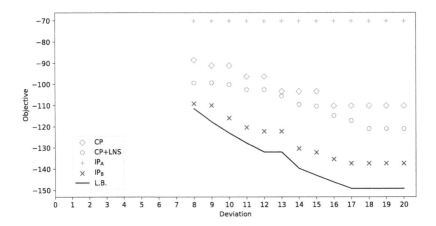

Fig. 3. Instances of 50 items with a time limit of 600 s.

of LNS for the CP model always improves the solution quality. For the small instances of Fig. 1, the CP model with LNS does better than IP_B since the latter only solves the relaxation of the RMP to optimality and tries to get the best integral solution from these limited columns, with no proof of optimality. IP_A does particularly poorly compared with the other models. The results of IP_B are usually off to a slow start, but given enough time this model does better than both previous ones. Similar to the CP model, this model reaches a plateau of its own before the 600 s mark. Further improvements could be achieved via branch and price. Average computation times are shown in Table 1 (owing to details in the implementation of the models, the execution time may be slightly higher than the time limit in some cases).

Table 1. Average computation times for Figs. 1, 2 and 3

	25 items	50 items	
	600 s	6 s	600 s
CP	510.00	3.90	390.00
CP + LNS	508.49	3.90	379.02
IP_A	589.06	6.96	602.04
IP_B	8.35	8.32	178.46

For the CP and IP_A models, infeasibility or optimality can sometimes be proven when the maximum cumulative deviation is low enough. IP_B does well especially when the time limit is high, although its solutions cannot be proven optimal without a branch-and-price scheme (but the CP model generating its initial columns can determine if an instance is infeasible). We have further tried solving instances with conflicts only ($p = 0$, $q = 0$, $r = 0.25$) and every method could find a solution within the time limit. In Fig. 4 we show the results of instances without conflicts and only with costs ($p = 0.25$, $q = 0.25$, $r = 0$). Table 2 shows the computation times for instances with either only conflicts or only costs.

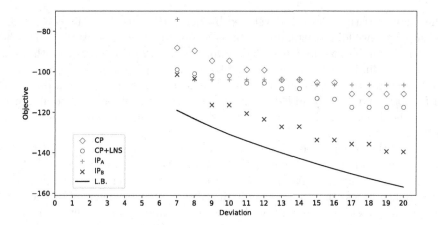

Fig. 4. Instances of 50 items with a time limit of 600 s. (costs only)

Table 2. Average computation times

	50 items	
	Conflicts only	Costs only
CP	0.03	420.00
CP + LNS	0.03	390.78
IP$_A$	601.11	600.62
IP$_B$	4.64	28.65

It is interesting to notice that with only conflicts, the CP model very easily proves optimality for all instances in a fraction of a second, whereas both IP models are orders of magnitude behind it. The weakness of the CP model lies in the optimization of objective f, even with LNS. IP$_A$ appears to generally do better without conflicts, while the performance and results of IP$_B$ are largely unaffected by the parameters of the instances.

A CP/IP$_B$ hybrid could be constructed: The CP part would generate the initial columns, those columns being the current CP solution once the model reaches its plateau; From there, the IP part would take over and improve this solution by bringing it to near-optimality. Experimentation would be necessary to see if such a model would be an improvement over current methods. We have not integrated the CP and IP$_B$ models into a hybrid in this paper, as our objective was to compare individual methods to each other. Furthermore, our simple approach to the generation of the Pareto sets for both IP models could be improved [14].

8   Practical Applications

The metaheuristic approach developed by Lewis [1] is used on the commercial website www.weddingseatplanner.com as a tool to generate seating plans. The problem is similar to ours which is described in Sect. 2 of this paper, with a few exceptions: Bin loads are unbounded, negative costs are always equal to -1 and positive costs are always equal to 1, and the deviation is computed according to an L_1-norm and directly added to the objective (i.e., the objective is to minimize f plus the deviation). The objective functions of our models have been adapted for these tests.

One of the design goals of the website was to solve the problem *quickly* since their clients could easily grow impatient after waiting just a few seconds in front of a seemingly unresponsive browser window. As such, their tool usually solves the problem in less than 5 s, a time limit which cannot be specified by the user. Because of this short time limit, we have bounded the maximum deviation of our models at 20 in order to find better results. For the CP model, since the balancing constraint and the branching heuristic are very effective, we have further decided to solve the instances two more times by bounding the maximum deviation at 10 and 5, respectively (since we are solving thrice as many problems, we have also divided the time limit by three). We will be considering only the best of those three solutions.

Due to an error in the website's implementation of the algorithm, all negative costs are considered to be conflicts. In order to provide a fair comparison, the instances we have generated for these tests do not include negative costs. The cost probabilities are thus $p = 0$, $q = 0.\bar{3}$, and $r = 0.\bar{3}$ (which implies that the probability of a cost of 0 is also $0.\bar{3}$). Since the website always solves the instances

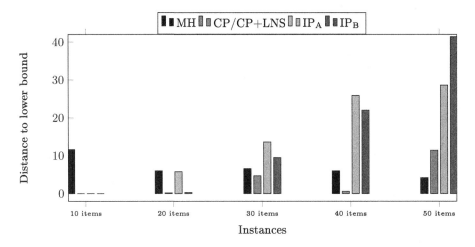

Fig. 5. All methods compared with a time limit of 5 s.

in under 5 s, this is what we have chosen as our time limit. Tests have been run with 10, 20, 30, 40, and 50 items, and are again averaged for five cost matrices. The histograms shown in Fig. 5 represent the distance in score of a solution from the lower bound.

When solving small instances, exact methods have a distinct advantage over metaheuristics, often proving optimality. Both IP models scale poorly with an increasing number of items as they usually require some time to find decent solutions. While it can find the best solutions given enough time, IP_B does particularly badly with a short time limit as the quality of its solutions improves relatively slowly. The metaheuristic model scales very well, with its solution quality being constant with a varying number of items. The CP model does well all-around, proving optimality for small instances as well as having good solutions for all instances.

9   Conclusion

In this paper we have compared how various methods can be used to solve multi-objective constrained bin packing problems with an aspect of load balancing. A metaheuristic model can find good solutions in a short time and scales well to an increasing number of items but will most likely not find optimal solutions. A CP model can also find good solutions quickly, but for large instances it will not reach the best solutions in reasonable time even with the help of LNS. A natural IP model is probably not the best choice, as it scales poorly while its strenghts can also be found in other models. An IP model using column generation does very well given enough time but is not a good contender to solving instances quickly. It would be interesting to see if a CP/IP hybrid using column generation and branch and price could prove optimality in reasonable time for larger instances of the problem.

Acknowledgements. Financial support for this research was provided by NSERC Discovery Grant 218028/2017 and CERC, École Polytechnique de Montréal.

References

1. Lewis, R.: Constructing wedding seating plans: a tabu subject. In: CSREA Press, pp. 24–32 (2013)
2. Miettinen, K.: Nonlinear Multiobjective Optimization. Springer, Heidelberg (1998). https://doi.org/10.1007/978-1-4615-5563-6
3. Lewis, R., Carroll, F.: Creating seating plans: a practical application. J. Oper. Res. Soc. **67**(11), 1353–1362 (2016)
4. Bellows, M.L., Peterson, J.D.L.: Finding an optimal seating chart for a wedding. https://www.improbable.com/2012/02/12/finding-an-optimal-seating-chart-for-a-wedding. Accessed 23 May 2018
5. Schaus, P.: Solving balancing and bin-packing problems with constraint programming. Ph.D thesis, Université catholique de Louvain (2009)

6. Sadykov, R., Vanderbeck, F.: Bin packing with conflicts: a generic branch-and-price algorithm. INFORMS J. Comput. **25**(2), 244–255 (2013)

7. Wang, Z., Han, X., Dósa, G., Tuza, Z.: Bin packing game with an interest matrix. In: Xu, D., Du, D., Du, D. (eds.) COCOON 2015. LNCS, vol. 9198, pp. 57–69. Springer, Cham (2015). https://doi.org/10.1007/978-3-319-21398-9_5

8. Shaw, P.: A constraint for bin packing. In: Wallace, M. (ed.) CP 2004. LNCS, vol. 3258, pp. 648–662. Springer, Heidelberg (2004). https://doi.org/10.1007/978-3-540-30201-8_47

9. Pesant, G.: Achieving domain consistency and counting solutions for dispersion constraints. INFORMS J. Comput. **27**(4), 690–703 (2015)

10. Gualandi, S., Lombardi, M.: A simple and effective decomposition for the multidimensional binpacking constraint. In: Schulte, C. (ed.) CP 2013. LNCS, vol. 8124, pp. 356–364. Springer, Heidelberg (2013). https://doi.org/10.1007/978-3-642-40627-0_29

11. Bron, C., Kerbosch, J.: Algorithm 457: finding all cliques of an undirected graph. Commun. ACM **16**(9), 575–577 (1973)

12. Shaw, P.: Using constraint programming and local search methods to solve vehicle routing problems. In: Maher, M., Puget, J.-F. (eds.) CP 1998. LNCS, vol. 1520, pp. 417–431. Springer, Heidelberg (1998). https://doi.org/10.1007/3-540-49481-2_30

13. Mitsos, A., Chachuat, B., Barton, P.I.: Mccormick-based relaxations of algorithms. SIAM J. Optim. **20**(2), 573–601 (2009)

14. Boland, N., Charkhgard, H., Savelsbergh, M.: A criterion space search algorithm for biobjective integer programming: the balanced box method. INFORMS J. Comput. **27**(4), 735–754 (2015)

A $O(n \log^2 n)$ Checker and $O(n^2 \log n)$ Filtering Algorithm for the Energetic Reasoning

Yanick Ouellet and Claude-Guy Quimper$^{(\boxtimes)}$

Université Laval, Québec City, Canada
yanick.ouellet.2@ulaval.ca, claude-guy.quimper@ift.ulaval.ca

Abstract. Energetic reasoning is a strong filtering technique for the CUMULATIVE constraint. However, the best algorithms process $O(n^2)$ time intervals to perform the satisfiability check which makes it too costly to use in practice. We present how to apply the energetic reasoning by processing only $O(n \log n)$ intervals. We show how to compute the energy in an interval in $O(\log n)$ time. This allows us to propose a $O(n \log^2 n)$ checker and a filtering algorithm for the energetic reasoning with $O(n^2 \log n)$ average time complexity. Experiments show that these two algorithms outperform their state of the art counterparts.

1 Introduction

There exist many filtering rules for the CUMULATIVE constraint. Among them, the energetic reasoning rule [3,7,14] dominates the overload check [8,23], the time-tabling [4], the edge-finder [15], and the time-tabling-edge-finder [22]. To apply the energetic reasoning, one needs to process $O(n^2)$ time intervals, which might be too slow in practice.

We introduce a technique based on Monge matrices to explicitly process only $O(n \log n)$ of the $O(n^2)$ intervals. The remaining intervals are processed implicitly. This allows us to propose the first subquadratic checker for the energetic reasoning, with a $O(n \log^2 n)$ running time. We also propose a new filtering algorithm that filters allsks with an average running time complexity of $O(n^2 \log n)$ and a worst case running time complexity of $O(n^2 \log^2 n)$. However, we do not know whether the bound $O(n^2 \log^2 n)$ is tight as we did not succeed in finding an instance requiring that much time to filter.

The next section formally presents the CUMULATIVE constraint and the energetic reasoning rule. Section 3 presents some algorithmic background, including the Monge matrices that we use to design our algorithms. Section 4 introduces an adaptation of the range trees that is used in Sect. 5 to compute the energy in a time interval. Section 6 presents the subquadratic checker, and Sect. 7, the $O(n^2 \log n)$ filtering algorithm. Section 8 shows the performance of these algorithms on classic benchmarks.

© Springer International Publishing AG, part of Springer Nature 2018
W.-J. van Hoeve (Ed.): CPAIOR 2018, LNCS 10848, pp. 477–494, 2018.
https://doi.org/10.1007/978-3-319-93031-2_34

2   Scheduling Background

Let $\mathcal{I} = \{1, \ldots, n\}$ be the set of task indices. Each task i is defined with four integer parameters: the *earliest starting time* est_i, the *latest completion time* lct_i, the processing time p_i, and the resource consumption rate h_i. From these parameters, one can compute the *earliest completion time* $\text{ect}_i = \text{est}_i + p_i$, the *latest starting time* $\text{lst}_i = \text{lct}_i - p_i$, and the energy $e_i = p_i h_i$ of the task. The horizon spans from time $\text{est}_{\min} = \min_{i \in \mathcal{I}} \text{est}_i$ to time $\text{lct}_{\max} = \max_{i \in \mathcal{I}} \text{lct}_i$. A task i starts executing at time S_i and executes for p_i units of time without interruption. The starting time S_i is unknown but must belong to the time interval $\text{dom}(S_i) = [\text{est}_i, \text{lst}_i]$. The task is necessarily executing during the time interval $[\text{lst}_i, \text{ect}_i)$ if $lst_i < ect_i$. This time interval is called the *compulsory part*. A cumulative resource can simultaneously execute multiple tasks as long as the total resource consumption rate of the tasks executing at any time t does not exceed the capacity C of the resource. The constraint CUMULATIVE [1] ensures that the starting times of the tasks do not overload the capacity of the resource.

$$\textsc{Cumulative}([S_1, \ldots, S_n], [p_1, \ldots, p_n], [h_1, \ldots, h_n], C) \iff \forall t \sum_{i: S_i \leq t < S_i + p_i} h_i \leq C$$

Deciding whether the constraint CUMULATIVE is satisfiable is NP-Complete. For that reason, there are no polynomial time filtering algorithms that can achieve bounds consistency for this constraint. However, there exist multiple filtering rules that partially remove the inconsistent values from the domains of the starting variables. The overload check [8,23], the time-tabling [4], the edge-finder [15], the time-table-edge-finder [22], and the not-first not-last [10,17,18] are popular filtering rules. With the exception of not-first not-last, all these rules are dominated by the energetic reasoning that detects more inconsistencies and filters more values [3]. The energetic reasoning is incomparable to the not-first not-last.

A task i that starts at its earliest starting time spends, during the time interval $[l, u)$, an amount of energy equal to $LS(i, l, u) = h_i \max(\min(\text{ect}_i - l, p_i, u - l), 0)$. This energy is called the *left-shift*. If task i starts at its latest completion time, it spends, during the time interval $[l, u)$, an amount of energy equal to $RS(i, l, u) = h_i \max(\min(u - \text{lst}_i, p_i, u - l), 0)$. This energy is called the *right-shift*. Finally, regardless of its starting time, a task i must spend during the time interval $[l, u)$ an amount of energy called *left-shift/right-shift* and denoted $LSRS(i, l, u)$.

$$LSRS(i, l, u) = \min(LS(i, l, u), RS(i, l, u)) \tag{1}$$

$$= h_i \max(\min(\text{ect}_i - l, u - \text{lst}_i, p_i, u - l), 0) \tag{2}$$

By abuse of notation, we define the *left-shift/right-shift* for a set of tasks Θ: $LSRS(\Theta, l, u) = \sum_{i \in \Theta} LSRS(i, l, u)$. The *slack* $S(\Theta, l, u)$ is the amount of remaining energy, for a cumulative resource of capacity C over an interval $[l, u)$, after spending the left-shift/right-shift of a set of tasks Θ.

$$S(\Theta, l, u) = C \cdot (u - l) - LSRS(\Theta, l, u) \tag{3}$$

The energetic reasoning tests [3], for every time interval $[l, u)$, whether the slack is non-negative: $S(\mathcal{I}, l, u) \geq 0$.

Baptiste et al. [3] showed that not all time intervals $[l, u)$ need to be tested. Let O_1, O_2, and $O(t)$ be such that

$$O_1 = \{\text{est}_i \mid i \in \mathcal{I}\} \cup \{\text{ect}_i \mid i \in \mathcal{I}\} \cup \{\text{lst}_i \mid i \in \mathcal{I}\}$$
$$O_2 = \{\text{lct}_i \mid i \in \mathcal{I}\} \cup \{\text{ect}_i \mid i \in \mathcal{I}\} \cup \{\text{lst}_i \mid i \in \mathcal{I}\}$$
$$O(t) = \{\text{est}_i + \text{lct}_i - t \mid i \in \mathcal{I}\}$$

Only the time intervals $[l, u)$ that fall into one of these three situations are considered of interest: (1) $l \in O_1$ and $u \in O_2$; (2) $l \in O_1$ and $u \in O(l)$; (3) $l \in O(u)$ and $u \in O_2$.

The energetic reasoning filtering consists in increasing est_i and decreasing lct_i to ensure that the energetic reasoning check would pass if the tasks was executed at its earliest starting time or latest starting time. The filtering rule for the est states that if the left-shift of the task i is greater than the slack of the remaining tasks in an interval $[l, u)$, est_i must be adjusted to $\left\lceil u - \frac{S(\mathcal{I} \setminus \{i\}, l, u)}{h_i} \right\rceil$.

$$S(\mathcal{I} \setminus \{i\}, l, u) < LS(i, l, u) \implies \text{est}_i' = \left\lceil u - \frac{S(\mathcal{I} \setminus \{i\}, l, u)}{h_i} \right\rceil \tag{4}$$

Derrien and Petit [7] show that it is sufficient to test a subset of the intervals of Baptiste et al. to reach the fix point. To filter the est of a task i, one has to apply the filtering rule on all intervals in $O_C \cup L_i$. The set O_C contains at most two intervals for each pair of tasks and thus has a cardinality in $O(n^2)$. The set L_i contains $2n + 1$ intervals. Similarly, to filter the lct of a task, one has to apply the filtering rules on intervals in $O_C \cup R_i$, where R_i is symmetric to L_i. The definitions of O_C, L_i, and R_i are based on a long enumeration of cases, but are straightforward to compute. By lack of space, we refer the reader to [7] for a complete definition of these sets.

While the energetic reasoning achieves a strong level of filtering, it is not commonly used in practice due to its slow computation time. Baptiste et al. [3] proposed a checker with a running time complexity of $O(n^2)$. Their algorithm asymptotically remains the fastest checker in the literature. Nevertheless, Derrien and Petit [7] reduced the number of intervals to check by a constant. This improvement led to a checker with equivalent running time complexity, but faster in practice.

Baptiste et al. [3] also present an algorithm in $O(n^3)$ to filter the constraint. Bonifas [5] introduced an algorithm that filters at least one task in $O(n^2 \log n)$ time. Tesch [21] presents an algorithm that achieves a weaker level of filtering in $O(n^2 \log n)$ time which is later improved to perform an exact filtering in $O(n^2 \log^2 n)$ time [20].

3   Algorithmic Background

We present algorithms and data structures that will be used to design a sub-quadratic checker for the Energetic Reasoning.

3.1   Partial Sums

Let $A[1..n]$ be an array of n integers. A partial sum query is defined such as

$$\texttt{Partial-Sum}(A, i, j) = \sum_{k=i}^{j} A[k] \tag{5}$$

To efficiently answer such a query, one preprocesses in $O(n)$ time the array A by creating an array $B[0..n]$ such that $B[0] = 0$ and $B[i] = B[i-1] + A[i]$. $\texttt{Partial-Sum}(A, i, j)$ returns $B[j] - B[i-1]$ in constant computation time.

3.2   Range Trees

Consider a set of n weighted points P in a two-dimensional Cartesian plan. Each point i has two coordinates, x_i and y_i, and a weight w_i. A sum query $Q_{points}(\chi, \gamma, P)$ computes the weighted sum of all points delimited by the quater-plane $\chi \leq x$ and $y \leq \gamma$.

$$Q_{points}(\chi, \gamma, P) = \sum_{\{i \in P \mid \chi \leq x_i \wedge y_i \leq \gamma\}} w_i \tag{6}$$

Such queries can be answered by two-dimensional range-trees [6]. If the fractional cascading technique is used, each query can be answered online in $O(\log |P|)$ time after a $O(|P| \log |P|)$ pre-processing time of P is completed.

Each node of a range tree is associated to a set of points that serves as its label. The root P of a range tree contains all the points in P. The set of points of a node v is partitioned into two subsets left(v) and right(v), one for the left subtree and one for the right subtree. For each node v, the points contained in the left child have an abscissa smaller than or equal to the abscissa of the points contained in the right child: $i \in$ left$(v) \wedge j \in$ right$(v) \Rightarrow x_i \leq x_j$. Each node v of a range-tree has an attribute x_v^{mid} such $x_v^{mid} = \max_{i \in \text{left}(v)} x_i$ is the largest abscissa of a point in the left subtree.

Each node v of the range-tree has a vector Y_v of dimension $|v|$ which contains the ordinates y_i of the points $i \in v$ sorted in non-decreasing order. Three other vectors characterize the nodes. The vector W_v is a partial sum such that $W_v[i]$ is the sum of the weights of the i points in v with the smallest ordinates. The vector L_v and R_v link the points in v with the points in the left and right subtrees. There are $L_v[i]$ points in left(v) whose ordinate is no greater than $Y_v[i]$. Similarly, there are $R_v[i]$ points in right(v) whose ordinate is no greater than $Y_v[i]$.

If v is a leaf, v contains a single point i. Thus, the vectors Y_v, W_v, L_v, and R_v have length 1. The vectors $Y_v = [y_i]$ and $W_v = [w_i]$ contain the ordinate and the weight of that point. The vectors $L_v = R_v = [0]$ are the null vectors. The value x_v^{mid} is undefined.

Range-trees are built using a bottom-up approach similar to the merge sort. By definition, the leaves of range-tree are sorted in non-decreasing order of abscissa x_i. Therefore, one sorts the points in P by abscissa which gives the leaves of the tree. Then the upper level is computed by merging the vectors Y_v, W_v, L_v, and R_v from the lower level. Since creating the node v can be done in $O(|v|)$ time, building the range-tree is done in $O(|P| \log |P|)$ times.

The query $Q_{points}(\chi, \gamma, P)$ can be answered by traversing the tree from the root to a leaf. Let v be the current node initialized to the root. Let i be an index such that $Y_v[i] \leq \gamma < Y_v[i+1]$. This index is initialized by doing a binary search over the vector Y_P. If $\chi > x_v^{mid}$, then $Q_{points}(\chi, \gamma, v) = Q_{points}(\chi, \gamma, \text{right}(v))$. The current node v becomes $\text{right}(v)$ and the index i becomes $R_v[i]$. If $\chi \leq x_v^{mid}$, $Q_{points}(\chi, \gamma, v) = Q_{points}(\chi, \gamma, \text{left}(v)) + W_{\text{right}(v)}[R_v[i]]$. The current node v becomes $\text{left}(v)$ and the index i becomes $L_v[i]$. When the current node $v = \{j\}$ is a leaf, we return w_j if $x_j \geq \chi$ and $y_j \leq \gamma$ and zero otherwise. This computation is done in $O(\log |P|)$ time.

3.3 Monge Matrices

A *Monge matrix* M is an $n \times m$ matrix such that for any pair of rows $1 \leq i_1 < i_2 \leq n$ and any pair of columns $1 \leq j_1 < j_2 \leq m$, the inequality (7) holds.

$$M[i_2, j_2] - M[i_2, j_1] \leq M[i_1, j_2] - M[i_1, j_1] \qquad (7)$$

An *inverse Monge matrix* satisfies the opposite inequality: $M[i_2, j_2] - M[i_2, j_1] \geq M[i_1, j_2] - M[i_1, j_1]$.

Consider the functions: $f_i(x) = M[i, x]$. Inequality (7) imposes the slopes of these functions to be monotonic. By choosing $i_1 = i$, $i_2 = i + 1$, $j_1 = x$, and $j_2 = x + 1$ and substituting in (7), one can observe the monotonic behavior of the slopes.

$$\frac{f_{i+1}(x+1) - f_{i+1}(x)}{(x+1) - x} \leq \frac{f_i(x+1) - f_i(x)}{(x+1) - x}. \qquad (8)$$

It follows that the functions of two distinct rows of a Monge matrix cross each other at most once. Monge matrices satisfy many more properties [19].

Property 1. The submatrix obtained from a subset of rows and columns of a Monge matrix is a Monge matrix.

Property 2. The transpose of a Monge matrix is a Monge matrix.

Property 3. If M is a Monge matrix, v and w are two vectors, then $M'[i, j] = M[i, j] + v[i] + w[j]$ is a Monge matrix.

Properties 1 to 3 also hold for inverse Monge matrices.

Property 4. M is a Monge matrix if and only if $-M$ is an inverse Monge matrix.

Sethumadhavan [19] presents a survey about Monge matrices.

The *envelope* of a Monge matrix M is a function $l^*(j) = \arg \min_i M[i, j]$ that returns the row i on which appears the smallest element on column j. The envelope $l^*(j)$ of a (inverse) Monge matrix is non-increasing (non-decreasing).

A *partial Monge matrix* is a Monge matrix with empty entries. Empty entries are not subject to the inequality (7) and are ignored when computing the envelope. In this paper, we only consider partial (inverse) Monge matrices where $M[i, j]$ is empty if and only if $i > j$. The envelope of such an $n \times m$ partial (inverse) Monge matrix is non-increasing (non-decreasing) on the interval $[1, i]$ and non-decreasing (non-increasing) on the interval $[i, m]$ where $M[i, l^*(i) - 1]$ is empty. Kaplan et al. [11] compute the envelope of a $n \times m$ partial (inverse) Monge matrix in $O(n \log m)$ time. Their algorithm uses binary searches over the columns to find the intersection of the row functions $f_i(x)$.

4   Adapting the Range-Trees

We adapt the range-tree data structure to perform a query on a set of weighted segments S instead of a set of points. Such an adaptation is required for the algorithms we present in the next sections. A segment i of weight w_i, noted $\langle x_i, x'_i, y_i, w_i \rangle$, spans from coordinates x_i to x'_i on the abscissa, at y_i on the ordinate. The query $Q_{segments}(\chi, \gamma, S)$ computes the weighted sum of all segment parts inside the quater-plane of the query. A segment accounts for its weight times the length of the sub-segment that is within the query range.

$$Q_{segments}(\chi, \gamma, S) = \sum_{\substack{\langle x_i x'_i, y_i, w_i \rangle \in S \\ y_i \leq \gamma}} w_i(\max(x'_i - \chi, 0) - \max(x_i - \chi, 0))$$

We can simplify the problem by replacing each segment by two rays i' and i'': $\langle -\infty, x'_i, y_i, w_i \rangle$ and $\langle -\infty, x_i, y_i, -w_i \rangle$. Since ray i'' cancels ray i' when i begins, the result of the query is unchanged.

We adapt the range-tree data structure to answer queries on weighted rays instead of weighted points. We add an attribute $\underline{x}_v$ to each node v of the tree that represents the smallest abscissa of a ray $\langle -\infty, x_i, y_i, w_i \rangle$ in v. In other words, each ray in v ends at $\underline{x}$ or after. We also add a vector Σ_v to each node v such that $\Sigma_v[i] = Q_{segments}(\underline{x}_v, Y_v[i], v)$ is a precomputed result of a query. If rays in v are sorted by ordinates, we have $\Sigma_v[i] = \Sigma_v[i - 1] + w_i(x_i - \underline{x}_v)$.

Similar to the original range-tree, the query $Q_{segments}(\chi, \gamma, S)$ is computed by traversing the tree from the root S to a leaf. Let v be the current node initialized to the root. Let i be an index such that $Y_v[i] \leq \gamma < Y_v[i + 1]$. This index is initialized by doing a binary search over the vector Y_S. If $\chi > x_v^{mid}$, then $Q_{segments}(\chi, \gamma, v) = Q_{segments}(\chi, \gamma, \text{right}(v))$. The current node v becomes $\text{right}(v)$ and the index i becomes $R_v[i]$. If $\chi \leq x_v^{mid}$, $Q_{segments}(\chi, \gamma, v) = Q_{segments}(\chi, \gamma, \text{left}(v)) + \Sigma_v[R_v[i]] + (\underline{x} - \chi) \cdot W_v[R_v[i]]$. The current node v becomes $\text{left}(v)$ and the index i becomes $L_v[i]$. When the current node $v = \{j\}$ is a leaf, we return $(x_j - \chi) \cdot w_j$ if $y_j \leq \gamma$ and 0 otherwise. This computation is done in $O(\log |S|)$ time.

5   Computing the Left-Shift Right-Shift in $O(\log n)$ Time

We want to preprocess n tasks in $O(n \log n)$ time in order to compute the left-shift/right-shift $LSRS(\mathcal{I}, l, u)$ of any time interval $[l, u)$, upon request, in $O(\log n)$ time. We decompose the problem as follows. For every task i, we let $c_i = \max(0, ect_i - lst_i)$ be the length of the task's compulsory part. For every task i, we define p_i weighted semi-open intervals partitioned into two sets: the c_i compulsory intervals CI_i that lie within the compulsory part $[lst_i, ect_i)$ of the task and the $p_i - c_i$ free intervals FI_i that embed the compulsory part. The weight of all intervals is h_i.

$$CI_i = \{\langle [lst_i + k, lst_i + k + 1), h_i \rangle \mid 0 \le k < c_i\} \tag{9}$$

$$FI_i = \{\langle [est_i + k, lct_i - k), h_i \rangle \mid 0 \le k < p_i - c_i\} \tag{10}$$

For a set of tasks, we have: $CI_\Omega = \bigcup_{i \in \Omega} CI_i$ and $FI_\Omega = \bigcup_{i \in \Omega} FI_i$. Computing $LSRS(\mathcal{I}, l, u)$ consists of counting the weight of the intervals nested in $[l, u)$.

$$LSRS(\mathcal{I}, l, u) = \sum_{\substack{\langle [a,b), w \rangle \in CI_\mathcal{I} \\ [a,b) \subseteq [l,u)}} w + \sum_{\substack{\langle [a,b), w \rangle \in FI_\mathcal{I} \\ [a,b) \subseteq [l,u)}} w \tag{11}$$

Figure 1a shows two tasks and their intervals as well as a request interval $[l, u)$ shown in gray. The number of intervals nested in $[l, u)$ gives the amount of processing time the tasks must spend executing in $[l, u)$. When the number of intervals is weighted by the task heights, we obtain the left-shift/right-shift. Compulsory intervals have length 1 since the time at which compulsory energy is spent is known. Free intervals are longer and nested into $[l, u)$ only if the unit of processing time it corresponds to belongs both to the left-shift and the right-shift of the task. Equation (11) counts all intervals nested in $[l, u)$ and weighted by the task heights.

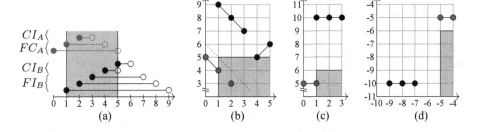

Fig. 1. (a) The intervals of two tasks: task A in gray with $est_A = 0$, $lct_A = 5$, $p_A = 3$, $h_A = 1$ and task B in black with $est_B = 1$, $lct_B = 9$, $p_B = 5$, $h_B = 1$. (b) The representation of the intervals on a Cartesian plan with lower bounds on the abscissa and upper bounds on the ordinate. (c) First transformation. (d) Second transformation. In all figures, the gray rectangle represents the query $[l, u) = [1, 5)$ that contains exactly 3 intervals of weight 1, hence $LSRS(\{A, B\}, 1, 5) = 3$.

Figure 1b represents an interval $[a, b)$ by a point (a, b) on the Cartesian plane with the corresponding weight. The sums in (11) can be computed with the queries $Q_{points}(l, u, CI_{\mathcal{I}})$ and $Q_{points}(l, u, FI_{\mathcal{I}})$ as explained in Sect. 3.2. These queries are represented as gray rectangles on Figs. 1a and 1b. Since the points associated to the intervals in FI_i form segments with a slope of -1 and that points in CI_i form a segment on the line $y = x + 1$, we can design efficient algorithms to compute these queries with $O(n \log n)$ processing time and $O(\log n)$ query time. Section 5.1 shows how to achieve these time bounds when computing the first summation in (11) that we call the *compulsory energy* while Sect. 5.2 shows how to compute the second summation that we call the *free energy*.

5.1 Computing the Compulsory Energy

We compute the compulsory energy that lies within an interval $[l, u)$ as follows. Let $T = \{\text{ect}_i \mid i \in \mathcal{I}\} \cup \{\text{lst}_i \mid i \in \mathcal{I}\}$ be the sorted time points where the compulsory energy can increase or decrease over time. Let Y_t be the amount of compulsory energy spent at time T_t. We compute Y_t using an intermediate vector Y'_t initialized to zero. For each task i, we increase by h_i the component Y'_t such that $T_t = \text{lst}_i$ and decrement by h_i the component Y'_t such that $T_t = \text{ect}_t$. We obtain these relations:

$$Y'_t = \sum_{i \in \mathcal{I} \mid \text{lst}_i = T_t} h_i - \sum_{i \in \mathcal{I} \mid \text{ect}_i = T_t} h_i, \qquad Y_0 = Y'_0, \qquad Y_t = Y_{t-1} + Y'_t.$$

Let Z_t be the amount of compulsory energy in the time interval $[T_t, T_{t+1})$, i.e. $Z_t = Y_t(T_{t+1} - T_t)$. Let t_1 and t_2 be such that $T_{t_1-1} < l \le T_{t_1}$ and $T_{t_2} \le u < T_{t_2+1}$. The amount of compulsory energy within the time interval $[l, u)$ is given below.

$$Q_{points}(l, u, CI_{\mathcal{I}}) = \sum_{\substack{\langle [a,b), w \rangle \in CI_{\mathcal{I}} \\ [a,b) \subseteq [l,u)}} w = Y_{t_1-1}(T_{t_1} - l) + \sum_{t=t_1}^{t_2} Z_t + Y_{t_2}(u - T_{t_2}) \quad (12)$$

Once the tasks are sorted, in $O(n \log n)$ time, by ect and lst, the vector Y' can be computed in linear time. The vectors Y, and Z can also be computed in linear time. The vector Z is preprocessed as a partial sum in linear time (see Sect. 3.1). Overall, the preprocess time is $O(n \log n)$.

To answer a query $Q_{points}(l, u, CI_{\mathcal{I}})$, a binary search finds the indices t_1 and t_2 in $O(\log n)$ time. Equation (12) is computed in constant time since the vector Z was preprocessed as a partial sum. Overall, the query time is $O(\log n)$.

5.2 Computing the Free Energy

We use our adaptation of range trees to answer the query $Q_{points}(l, u, FI_{\mathcal{I}})$. Since range trees process segments that are parallel to the x-axis, we use a geometric transformation to align the points in a set FC_i with the x-axis.

We transform an interval $[a, b) \in FI_{\mathcal{I}}$ into an interval $[a, a + b)$. The weights of the intervals remain unchanged. The intervals in $FI_{\mathcal{I}}$, when transformed, form the following weighted segments.

$$T_{\mathcal{I}}^1 = \{ \langle \text{est}_i, \text{est}_i + p_i - c_i, \text{est}_i + \text{lct}_i, h_i \rangle \mid i \in \mathcal{I} \}$$

We proceed to a second transformation where each interval $[a, b) \in FI_{\mathcal{I}}$ is transformed into $[-b, -a - b)$. The intervals from $FI_{\mathcal{I}}$ become the following weighted segment.

$$T_{\mathcal{I}}^2 = \{ \langle -\text{lct}_i, -\text{lct}_i + p_i - c_i, -\text{est}_i - \text{lct}_i, h_i \rangle \mid i \in \mathcal{I} \}$$

Figures 1c and d show the first and second transformations.

Lemma 1. $Q_{points}(l, u, FI_{\mathcal{I}}) = Q_{segments}(l, l + u, T_{\mathcal{I}}^1) + Q_{segments}(-u, -l - u - 1, T_{\mathcal{I}}^2)$.

Proof. Sketch: $Q_{segments}(l, l + u, T_{\mathcal{I}}^1)$ computes the weights of the points inside the gray box and under the dotted line in Fig. 1b or inside the gray box in Fig. 1c. The query $Q_{segments}(-u, -l - u - 1, T_{\mathcal{I}}^2)$ computes weights of the points inside the gray box and above the dotted line in Fig. 1b or inside the gray box in Fig. 1d.

From Lemma 1, it follows that two range trees can be constructed in $O(n \log n)$ time with the segments in $T_{\mathcal{I}}^1$ and $T_{\mathcal{I}}^2$. Computing the free energy within the interval $[l, u)$ is performed online in $O(\log n)$ time.

6   A Checker that Analyzes $O(n \log n)$ Time Intervals

We show that even though the energetic reasoning can fail in any of the $O(n^2)$ time intervals mentioned in Sect. 2, this number of intervals can be reduced to $O(n \log n)$ during the online computation. After analyzing a subset of $O(n \log n)$ intervals, it is safe to conclude whether the check passes for all $O(n^2)$ intervals.

We define the matrix $\mathbf{E}$ such that $\mathbf{E}[l, u]$ is the left-shift/right-shift energy contained in the interval $[l, u)$ for $\text{est}_{min} \le l < u \le \text{lct}_{max}$. If the interval $[l, u)$ is reversed ($l > u$), the left-shift/right-shift is null. The matrix $\mathbf{S}[l, u]$ contains the slack for the interval $[l, u)$, i.e. the remaining amount of energy in that interval.

$$\mathbf{E}[l, u] = \sum_{i \in \mathcal{I}} LSRS(i, l, u) \qquad \mathbf{S}[l, u] = C \cdot (u - l) - \mathbf{E}[l, u]$$

Theorem 1. *The matrix $\mathbf{E}$ is a Monge matrix.*

Proof. Let $\text{est}_{min} \le l_1 < l_2 \le \text{lct}_{max}$ and $\text{est}_{min} \le u_1 < u_2 \le \text{lct}_{max}$. The quantity $\mathbf{E}[l_2, u_2] - \mathbf{E}[l_2, u_1]$ is the amount of left-shift/right-shift energy that we gain by enlarging the interval $[l_2, u_1)$ to $[l_2, u_2)$. By analyzing (2), we deduce that the quantities $u - \text{lst}_i$ and $u - l$ increase at the same rate when enlarging $[l_2, u_1)$ to $[l_2, u_2)$ than when enlarging $[l_1, u_1)$ to $[l_1, u_2)$. However, the terms $\text{ect}_i - l$ and p_i

might prevent the left-shift/right-shift to increase when the interval is enlarged. It turns out that $\text{ect}_i - l$ increases and p_i remains constant as l decreases. Consequently, the increase of energy from $[l_1, u_1)$ to $[l_1, u_2)$ is less limited than when enlarging $[l_2, u_1)$ to $[l_2, u_2)$. Hence $\mathbf{E}[l_2, u_2] - \mathbf{E}[l_2, u_1] \leq \mathbf{E}[l_1, u_2] - \mathbf{E}[l_1, u_1]$. □

Corollary 1. *The matrix S is an inverse Monge matrix.*

Proof. Follows from $\mathbf{S}[l, u] = v[u] - v[l] - \mathbf{E}[l, u]$ where $v[i] = iC$, Theorem 1, Property 4, and Property 3. □

The energetic reasoning test fails if and only if there exists a non-empty interval $[l, u]$ such that $\mathbf{S}[l, u] < 0$. Inspired from [11], we design an algorithm that finds the smallest entry $\mathbf{S}[l, u]$ for any $l < u$ by checking only $O(n \log n)$ entries in $\mathbf{S}$. The algorithm assumes that the matrix $\mathbf{S}$ is not precomputed, but that any entry can be computed upon request. Since in Sect. 5, we show how to compute $\mathbf{S}[l, u]$ for any interval $[l, u]$ in $O(\log n)$ time, we obtain an algorithm with a running time complexity of $O(n \log^2 n)$. Moreover, we do not need to process the whole matrix $\mathbf{S}$ but submatrices containing a subset of rows and columns from $\mathbf{S}$. These submatrices contain all intervals of interest described in Sect. 2 and by Property 1, are inverse Monge matrices.

We need to execute Algorithm 1 twice to correctly apply the energetic reasoning rule. The first execution processes all intervals of the form $O_1 \times O_2 \cup \bigcup_{l \in O_1} O(l)$. To do so, we call Algorithm 1 with the parameters O_1, O_2, and $O' := O(0) = \{\text{est}_i + \text{lct}_i \mid i \in \mathcal{I}\}$. Moreover, we pass the function $S := (l, u) \mapsto S(\mathcal{I}, l, u)$ that returns the slack in the interval $[l, u]$. The sets O_1, O_2, and O' contain $O(n)$ elements (see Sect. 2) and are computed in linear time. For the second execution, we execute the checker on the reversed problem, processing intervals of the form $O_1 \cup O' \times O_2$. Thus, the algorithm is called with $S := (l, u) \mapsto S(\mathcal{I}, -u, -l)$, $O_1 := \{-u \mid u \in O_2\}$, $O_2 := \{-l \mid l \in O_1\}$, and $O' := \{-(\text{est}_i + \text{lct}_i) \mid i \in \mathcal{I}\}$. If neither execution leads to a failure, the constraint is consistent according to the energetic reasoning rule.

Algorithm 1 is built around the data structure P that encodes the envelope of the inverse Monge matrix $\mathbf{S}[l, u]$. The algorithm proceeds in two phases. The first phase initializes the data structure P while the second phase uses it to perform the check.

Let P be a set of tuples such that $\langle [\underline{u}, \overline{u}], l \rangle \in P$ indicates that the smallest element on any column $u \in [\underline{u}, \overline{u}]$ occurs on row l (we ignore rows greater than or equal to u as they correspond to empty time intervals). The intervals $[\underline{u}, \overline{u}]$ in P are sorted, disjoint, and contiguous. Upon the insertion of a tuple $\langle [\underline{u}, \overline{u}], l \rangle$, the intervals in P must be altered in order to be disjoint from $[\underline{u}, \overline{u}]$. Intervals in P that are nested in $[\underline{u}, \overline{u}]$ must be deleted from P. Intervals that partially overlap with $[\underline{u}, \overline{u}]$ must be shrunk. An interval that embeds $[\underline{u}, \overline{u}]$ needs to be split.

Consider the sequence of tuples $\langle [\underline{u}, \overline{u}], l' \rangle \in P$ sorted by intervals. By property of the envelope of an inverse Monge matrix, the rows l' in the sequence increase up to a maximum and then decrease. We store in a stack P_1 the first tuples of the sequence up to the tuple with the largest row (exclusively). We

---

Algorithm 1. MongeChecker(S, O_1, O_2, O')

$P_1 \leftarrow \emptyset$, $P_2 \leftarrow \{\langle [\min O_2, \max O_2], \min O_1 \rangle\}$, $F \leftarrow \emptyset$;

for $l \in O_1 \setminus \{\min O_1\}$ *in increasing order* **do**

    $\langle [u_1, u_2], l^* \rangle \leftarrow \text{Top}(P_2)$;

1   **while** $S(l, u_2) \leq S(l^*, u_2)$ **do**

2        $\text{Pop}(P_2)$;

        $\langle [u_1, u_2], l^* \rangle \leftarrow \text{Top}(P_2)$;

3    $b \leftarrow \max(\{u \mid u_1 \leq u \leq u_2 \wedge u \in O_2 \wedge S(l, u) < S(l^*, u)\} \cup \{-\infty\})$;

4    $c \leftarrow \max(\{u \mid b \leq u \leq \min(u_2, \text{succ}(b, O_2)) \wedge u + l \in O' \wedge S(l, u) <$
    $S(l^*, u)\} \cup \{-\infty\})$;

    $u \leftarrow \max(b, c)$;

    $u_l \leftarrow \min(\text{succ}(u, O_2), \text{succ}(u + l, O') - l)$;

    $u_{l^*} \leftarrow \min(\text{succ}(u, O_2), \text{succ}(u + l^*, O') - l^*)$;

5    $d \leftarrow$

    $\left\lfloor \frac{(u_l - u)(S(l^*, u) \cdot (u_{l^*} - u) - u(S(l, u_{l^*}) - S(l^*, u))) - (u_{l^*} - u)(S(l, u)(u_l - u) - u(S(l, u_l) - S(l, u)))}{(u_{l^*} - u)(S(l, u_l) - S(l, u)) - (u_l - u)(S(l^*, u_{l^*}) - S(l^*, u))} \right\rfloor$;

    if $d > \min(u_l, u_{l^*})$ **then** $d \leftarrow \max(b, c)$;

    if $d > l + 1$ **then** PushInterval(P_1, P_2, l, d) ;

for $\langle [\underline{u}, \overline{u}], l \rangle \in P_1 \cup P_2$ *in increasing order* **do**

    $u_1 \leftarrow \underline{u}$;

    if $S(l, u_1) < 0$ **then** $F \leftarrow F \cup \{[l, u_1)\}$;

    $u_3 \leftarrow \min\{u \in O_2 \mid u > u_1\}$;

    while $u_3 \leq \overline{u}$ **do**

        if $S(l, u_3) < 0$ **then** $F \leftarrow F \cup \{[l, u_3)\}$;

6   Find u_2 such that $u_2 + l \in O'$, $u_1 < u_2 < u_3$, and
        $S(l, u_2 - 1) \geq S(l, u_2) < S(l, u_2 + 1)$;

        if *such a* u_2 *exists and* $S(l, u_2) < 0$ **then** $F \leftarrow F \cup \{[l, u_2)\}$;

        $u_1 \leftarrow u_3$;

        $u_3 \leftarrow \min\{u \in O_2 \mid u > u_1\}$;

if $F = \emptyset$ **then** **return** (Success, $\emptyset$) **else** **return** (Fail, F) ;

---

store in a stack P_2 the remaining tuple, i.e. the decreasing slice of the sequence. The ends of the sequence are at the bottom of the stacks, the tuple with the largest row l' is at the top of P_2 and the tuple before is at the top of P_1. Algorithm 2 details the process of inserting an interval in the data structure while maintaining the invariant. Lines 1–2 move intervals smaller than the current row l from P_2 to P_1. Lines 3–4 remove overlapping intervals in P_2. The remainder of the algorithm splits the top interval of P_2 and insert the new interval between it. A tuple is always pushed onto P_2 before being moved to P_1 and is never moved once in P_1. Since Algorithm 2 pushes two tuples onto P_2, it has a constant time amortized complexity.

The intervals inserted into P are computed as follows. First, all columns of $\mathbf{S}$ are associated to the first row $\min O_1$. Therefore, P is initialized to $\{\langle [\text{est}_{\min}, \text{lct}_{\max}], \min O_1 \rangle\}$. We process the next rows in increasing order in the first for loop. Each time we process a row, we update the envelope function l^* encoded with the data structure P. Let $f_l(x) = S(\mathcal{I}, l, x)$ and $f_{l^*}(x) = S(\mathcal{I}, l^*(x), x)$ be two functions. Because $\mathbf{S}$ is an inverse Monge matrix, we know that these two functions intersect at most once. We search for the greatest value d where $f_l(d) < f_{l^*}(d)$. Once the value d is computed for a row l, if $d > l + 1$, we insert the tuple $\langle [l+1, d], l \rangle$ in P. If $d \leq l+1$, the functions do not intersect or intersect on an empty interval $[l, d)$ which is not of interest.

We compute d by proceeding in four steps. The while loop on line 1 searches the tuple $\langle [\underline{u}, \overline{u}], l^* \rangle$ in P_2 such that $f_l(x)$ and $f_{l^*}(x)$ intersect in $[\underline{u}, \overline{u}]$. This tuple can not be in P_1 because l is greater than all intervals in P_1 and we want d to be greater than l. The intervals in which $f_l(x)$ is smaller than the functions of the previous rows are removed from P on line 2. On line 3, we perform a binary search over the elements of O_2 within $[\underline{u}, \overline{u}]$ to find the greatest column $b \in O_2$ for which $f_l(b) < f_{l^*}(b)$. Let $\text{succ}(a, A) = \min\{a' \in A \mid a' > a\}$ be the successor of $a \in A$ when A is sorted in increasing order. Once b is found, we narrow the search for the intersection of the functions $f_l(x)$ and $f_{l^*}(x)$ to the interval $[b, \min(u_2, \text{succ}(b, O_2))]$. On line 4 we find the greatest column $c \in O(l)$ that lies in $[b, \min(u_2, \text{succ}(b, O_2))]$ where $f_l(c) < f_{l^*}(c)$. In order not to compute $O(l)$ for each row l, we perform the search in $O' = O(0)$. We have that $c \in O(l)$ if and only if $c + l \in O'$. Therefore, rather than searching for the greatest $c \in O(l)$, line 4 searches for the greatest $c + l \in O'$.

Using the values b and c, we find the value d where $f_l(x)$ and $f_{l^*}(x)$ intersect. The function $f_l(x)$ is piecewise linear with inflection points in $O_2 \cup O(l)$. The function $f_{l^*}(x)$ is piecewise linear with inflections points in $O_2 \cup O(l^*)$. Let $\bar{d}_l = \text{succ}(\max(b, c), O_2 \cup O(l))$ and $\bar{d}_{l^*} = \text{succ}(\max(b, c), O_2 \cup O(l^*(\max(b, c)))$. We know that $\max(b, c) \leq d \leq \min(\bar{d}_l, \bar{d}_{l^*})$, that $f_l(x)$ is linear over the interval $[\max(b, c), \bar{d}_l)$, and that $f_{l^*}(x)$ is linear over $[\max(b, c), \bar{d}_{l^*})$. We let d be the intersection point of these segments, or let $d = \max(b, c)$ if that intersection point does not satisfy $\max(b, c) \leq d \leq \min(\bar{d}_l, \bar{d}_{l^*})$. Once d is computed, we insert $\langle [l+1, d], l \rangle$ into P.

In the second part of the algorithm, we iterate on each tuple $\langle [a, d], l \rangle \in P$ found in the first part. The while loop processes the columns in O_2 that are within the current interval. After checking that the slack of two consecutive columns in O_2, u_1 and u_3, does not yield a negative slack, we try to find a column $u_2 \in O(l)$ such that u_2 is between u_1 and u_3 and has a negative slack.

Derrien and Petit [7] showed that the slope of the slack increases at elements in $O(l)$. Therefore, there is a global minimum between u_1 and u_3 that can be found using a binary search. If $\mathbf{S}[l, t] < \mathbf{S}[l, t+1]$, the global minimum is before t. Otherwise, it is at t or after t. Hence, the binary search finds the element in $O(l)$ between u_1 and u_3 where the slope of the slack shift from negative to positive. If all columns in each interval are processed without causing a failure, we return success.

Lines 3, 4, and 6 are executed $O(n)$ times and perform binary searches over $O(n)$ columns of the matrix leading to $O(n \log n)$ comparisons. Each comparison requires the computation of two entries of the matrix $\mathbf{S}$ which is done in $O(\log n)$

Algorithm 2. PushInterval(P_1, P_2, l, d)

   $\langle [\underline{u}, \overline{u}], l^* \rangle \leftarrow \text{Top}(P_2)$;

1 **while** $l + 1 > \overline{u}$ **do**

      | $\text{Push}(P_1, \text{Pop}(P_2))$;

2 | $\langle [\underline{u}, \overline{u}], l^* \rangle \leftarrow \text{Top}(P_2)$;

   if $d > \overline{u}$ **then**

3 | $u' \leftarrow \underline{u}$;

      | $l' \leftarrow l^*$;

      | **while** $d > \overline{u}$ **do**

      | | $\text{Pop}(P_2)$;

4 | | $\langle [\underline{u}, \overline{u}], l^* \rangle \leftarrow \text{Top}(P_2)$;

      | $\text{Push}(P_1, \langle [u', l], l' \rangle)$;

   else

      | $\langle [\underline{u}, \overline{u}], l^* \rangle \leftarrow \text{Pop}(P_2)$;

      | $\text{Push}(P_1, \langle [\underline{u}, l], l^* \rangle)$;

   $\text{Push}(P_2, \langle [d + 1, \overline{u}], l^* \rangle)$;

   $\text{Push}(P_2, \langle [l + 1, d], l \rangle)$;

---

times (see Sect. 5). This leads in an $O(n \log^2 n)$ running time complexity. The space complexity of the algorithm is dominated by the complexity of the range-trees, which is $O(n \log n)$.

7 Filtering Algorithm

We present a filtering algorithm for the energetic reasoning with average complexity of $O(n^2 \log n)$ based on the checker we presented in Sect. 6 and inspired by Derrien and Petit's filtering algorithm [7]. We only show how to filter the est of the tasks. To filter the lct, one can create the reversed problem by multiplying by -1 the values in the domains. Filtering the est in the reversed problem filters the lct in the original problem. We define the function $S_i^1(l, u)$ to be the amount of slack in the interval $[l, u)$ if i was assigned to its earliest starting time, i.e. $S_i^1(l, u) = S(\mathcal{I}, l, u) + LSRS(i, l, u) - LS(i, l, u)$. Similarly, $S_i^2(l, u) = S_i^1(-u, -l)$ represents the same concept on the reversed problem.

Algorithm 3 filters each task i by running the checker presented in Sect. 6 with S_i^1 and S_i^2 rather than S. We execute Algorithm 1 twice for each task. The first execution handles intervals of the form $O_1 \times O_2 \cup \bigcup_{l \in O_1} O(l)$ and the second, intervals of the form $O_1 \cup \bigcup_{u \in O_2} O(u) \times O_2$. On line 1, for each interval $[l, u)$ whose slack is negative when i starts at its earliest starting time, we filter est$_i$ using the energetic reasoning rule.

Algorithm 3 reaches the same fix point as [7]. A task i needs to be filtered if $S(\mathcal{I} \setminus \{i\}, l, u) - LS(i, l, u)$ is negative on an interval $[l, u)$. If such intervals exist, our algorithm finds at least one since it necessarily processes the interval with the minimum value. If not all negative intervals are found or if the filtering creates new negative intervals, our algorithm processes them at the next iteration.

Algorithm 3. MongeFilter($\mathcal{I}$)

$\text{est}'_i \leftarrow \text{est}_i \ \forall i \in \mathcal{I}$;

for $i \in \mathcal{I}$ **do**

$\quad S^1_i \leftarrow (l, u) \mapsto S(\mathcal{I}, l, u) + LSRS(i, l, u) - LS(i, l, u)$;

$\quad S^2_i \leftarrow (l, u) \mapsto S(\mathcal{I}, -u, -l) + LSRS(i, -u, -l) - LS(i, -u, -l)$;

$\quad (r_1, F_1) \leftarrow \text{MongeChecker}(S^1_i, O_1, O_2, \{\text{est}_i + \text{lct}_i \mid i \in \mathcal{I}\})$;

$\quad (r_2, F_2) \leftarrow \text{MongeChecker}(S^2_i, \{-u \mid u \in O_2\}, \{-l \mid l \in O_1\}, \{-(\text{est}_i + \text{lct}_i) \mid i \in \mathcal{I}\})$;

1 $\quad$ **for** $[l, u) \in F_1 \cup \{[l, u) \mid [-u, -l) \in F_2\}$ **do**

$\quad\quad \text{est}'_i \leftarrow \min(\text{est}'_i, \lceil u - \frac{S(\mathcal{I}, l, u) + LSRS(i, l, u)}{h_i} \rceil)$

7.1 Running Time Analysis

Algorithm 3 makes $2n$ calls to the checker that each makes $O(n \log n)$ slack queries answered in $O(\log n)$ time, hence a running time in $O(n^2 \log^2 n)$. However, we use a memorization based on virtual initialization [13] with a space complexity of $O(\max_i \text{lct}^2_i)$. A hash table could also work with $O(n \log n)$ space. When $S(\mathcal{I}, l, u)$ is computed, we store its value so that further identical queries get answered in $O(1)$. There are $O(n^2)$ intervals of interest. On line 5 of Algorithm 1, the slack for intervals that are not of interest is computed $O(n)$ times in the checker, hence $O(n^2)$ in the filtering algorithm. For these $O(n^2)$ intervals, a total of $O(n^2 \log n)$ time is spent thanks to memorization.

Line 4 performs a binary search over the elements $u \in O(l) \cap [b, \text{succ}(b, O_2)]$ and computes the slack $S(l^*, u)$ which might not be an interval of interest. The binary search could perform up to $O(\log n)$ slack evaluations and therefore lead to $O(n^2 \log n)$ distinct evaluations in the filtering algorithm. However, having n elements in $O(l)$ that occurs between two consecutive elements in O_2 (namely b and $\text{succ}(b, O_2)$) seldomly happens. Suppose that $|O_2| = 3n$ and $|O(l)| = n$ and that the time points in these sets are evenly spread. We obtain an average of $\frac{1}{3}$ elements in $O(l)$ between two consecutive elements in O_2. This number can vary depending on the cardinalities of O_2 and $O(t)$, but in practice, we rather observe an average of 0.141 elements when $n = 16$ and this number decreases as n increases. To reach the worst time bound of $O(n^2 \log^2 n)$, one would need to construct an instance that triggers $O(n^2)$ binary searches on line 4, each time over $O(n)$ elements. We did not succeed to construct such an instance. Assuming that the number of elements in the search is bounded, in average, by a constant, we obtain an average running time of $O(n^2 \log n)$. Under these assumptions, the binary search might as well be substituted by a linear search.

7.2 Optimization

As Derrien and Petit in [7], we improve the practical performance of Algorithm 3 by processing, for each task i, only intervals in O_C and L_i. We partition O_C into two sets: O^1_C contains the intervals in O_C with lower bound in O_1 and O^2_C

contains the intervals in O_C with lower bound in $\bigcup_{u \in O_2} O(u)$. Let $lb^s(A) = \{s \cdot a \mid [a, b] \in A\}$ and $ub^s(A) = \{s \cdot b \mid [a, b] \in A\}$ be the lower bound and upper bounds of the intervals in A, multiplied by s. We execute Algorithm 3 as usual, but the first call to the checker is done with parameters $O_1 := lb^1(O_C^1)$ and $O_2 := up^1(O_C^1)$ while the second call is made with parameters $O_1 := ub^{-1}(O_C^2)$ and $O_2 := lb^{-1}(O_C^2)$. The set O' is empty and S is unchanged. Since we do not process all elements in O_2, we can't use a binary search at line 6 of Algorithm 1 anymore and we must search linearly leading to a worse case complexity of $O(n^3 \log n)$ when there are $O(n^2)$ upper bounds in O_C. However, this seldom happens and the reduced constant outweighs the increased complexity.

To process intervals in L_i for $i \in \mathcal{I}$, we do as Derrien and Petit [7] by applying the filtering rule on each of the $O(n)$ intervals in L_i. Since we compute the slack in $O(\log n)$, we obtain a running time of $O(n^2 \log n)$ for all tasks.

8   Experiments

We implemented the algorithms in the solver Choco 4 [16] with Java 8. We ran the experiments on an Intel Core i7-2600 3.40 GHz. We used both BL [2] and PSPLIB [12] benchmarks of the Resources Constrained Project Scheduling Problems (RCPSP). An instance consists of tasks, subject to precedences, that need to be simultaneously executed on renewable resources of different capacities. We model this problem using one starting time variable S_i for each task i and a makespan variable. We add a constraint of the form $S_i + p_i \leq S_i$ to model a precedence and use one CUMULATIVE constraint per resource. For these constraints, we do not use other filtering algorithms than the algorithms being tested. We minimize the makespan. All experiments are performed using the Conflict Ordering Search [9] search strategy with a time limit of 20 min.

Figure 2 shows the time to optimally solve instances of benchmark BL using the checker from Sect. 6 (on the y axis) and Derrien and Petit's checker [7] (on

Table 1. Percentage of the time taken by Monge Filter to optimally solve instances of n tasks, compared to Derrien and Petit's filtering algorithm.

Benchmark	n	Instances solved	% time
BL	20	20	0.92
BL	25	20	0.89
PSPLIB	30	445	0.94
PSPLIB	60	371	0.57
PSPLIB	90	369	0.44
PSPLIB	120	197	0.48

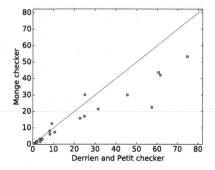

Fig. 2. Comparison of the time to optimally solve instances of the benchmark BL. Derrien et al.'s checker vs the Monge Checker.

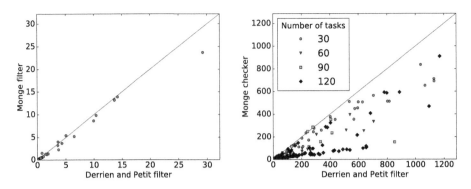

Fig. 3. Comparison of the time to solve to optimality instances of BL (left) and PSPLIB (right) benchmarks for Derrien and Petit's filtering algorithm and Monge Filter.

the x axis). The Monge Checker is faster for most instances and takes, in average, 77% of the time required by Derrien and Petit's checker to solve an instance to optimality.

Figure 3 shows the time to optimally solve instances using the filtering algorithm from Sect. 7 and Derrien and Petit's filtering algorithm [7], for BL (left) and PSPLIB (right) benchmarks. While our algorithm is only marginally faster on BL instances, where the number of tasks is small (between 20 and 25), the difference significantly increases as the number of tasks increases, as shown in Table 1. This shows the impact of decreasing the complexity of the energetic reasoning from $O(n^3)$ to $O(n^2 \log n)$.

9   Conclusion

We introduced a new method to explicitly process only $O(n \log n)$ intervals for the energetic reasoning using Monge matrices. We showed how to compute the energy in an interval in $O(\log n)$. We proposed a checker in $O(n \log^2 n)$ and a filtering algorithm in $O(n^2 \log n)$. Experiments showed that these algorithms are faster in theory and in practice. Future work will focus on extending these algorithms to produce explanations.

Acknowledgment. In memory of Alejandro López-Ortiz (1967–2017) who introduced me to research, algorithm design, and even Monge matrices.   – C.-G. Q

References

1. Aggoun, A., Beldiceanu, N.: Extending chip in order to solve complex scheduling and placement problems. Math. Comput. Model. **17**(7), 57–73 (1993)
2. Baptiste, P., Le Pape, C.: Constraint propagation and decomposition techniques for highly disjunctive and highly cumulative project scheduling problems. Constraints **5**(1–2), 119–139 (2000)

3. Baptiste, P., Le Pape, C., Nuijten, W.: Constraint-Based Scheduling. Kluwer Academic Publishers, Dordrecht (2001)
4. Beldiceanu, N., Carlsson, M.: A new multi-resource *cumulatives* constraint with negative heights. In: Van Hentenryck, P. (ed.) CP 2002. LNCS, vol. 2470, pp. 63–79. Springer, Heidelberg (2002). https://doi.org/10.1007/3-540-46135-3_5
5. Bonifas, N.: A $O(n^2 \log(n))$ propagation for the energy reasoning. In: ROADEF 2016 (2016)
6. de Berg, M., Cheong, O., van Kreveld, M.J., Overmars, M.H.: Computational Geometry: Algorithms and Applications, 3rd edn. Springer, Heidelberg (2008). https://doi.org/10.1007/978-3-540-77974-2
7. Derrien, A., Petit, T.: A new characterization of relevant intervals for energetic reasoning. In: O'Sullivan, B. (ed.) CP 2014. LNCS, vol. 8656, pp. 289–297. Springer, Cham (2014). https://doi.org/10.1007/978-3-319-10428-7_22
8. Fahimi, H., Quimper, C.G.: Linear-time filtering algorithms for the disjunctive constraint. In: AAAI, pp. 2637–2643 (2014)
9. Gay, S., Hartert, R., Lecoutre, C., Schaus, P.: Conflict ordering search for scheduling problems. In: Pesant, G. (ed.) CP 2015. LNCS, vol. 9255, pp. 140–148. Springer, Cham (2015). https://doi.org/10.1007/978-3-319-23219-5_10
10. Kameugne, R., Fotso, L.P.: A cumulative not-first/not-last filtering algorithm in $O(n^2 log(n))$. Indian J. Pure Appl. Math. **44**(1), 95–115 (2013)
11. Kaplan, H., Mozes, S., Nussbaum, Y., Sharir, M.: Submatrix maximum queries in Monge matrices and partial Monge matrices, and their applications. ACM Trans. Algorithms (TALG) **13**(2), 338–355 (2017)
12. Kolisch, R., Sprecher, A.: PSPLIB-a project scheduling problem library: OR software-orsep operations research software exchange program. Eur. J. Oper. Res. **96**(1), 205–216 (1997)
13. Levitin, A.: Introduction to the Design & Analysis of Algorithms, 3rd edn. Pearson Education Inc., Boston (2012)
14. Lopez, P., Esquirol, P.: Consistency enforcing in scheduling: a general formulation based on energetic reasoning. In: 5th International Workshop on Project Management and Scheduling (PMS 1996) (1996)
15. Mercier, L., Van Hentenryck, P.: Edge finding for cumulative scheduling. INFORMS J. Comput. **20**(1), 143–153 (2008)
16. Prud'homme, C., Fages, J.-G., Lorca, X.: Choco Documentation. TASC, LS2N, CNRS UMR 6241 and COSLING S.A.S. (2017)
17. Schutt, A., Wolf, A.: A new $\mathcal{O}(n^2 \log n)$ not-first/not-last pruning algorithm for cumulative resource constraints. In: Cohen, D. (ed.) CP 2010. LNCS, vol. 6308, pp. 445–459. Springer, Heidelberg (2010). https://doi.org/10.1007/978-3-642-15396-9_36
18. Schutt, A., Wolf, A., Schrader, G.: Not-first and not-last detection for cumulative scheduling in $\mathcal{O}(n^3 \log n)$. In: Umeda, M., Wolf, A., Bartenstein, O., Geske, U., Seipel, D., Takata, O. (eds.) INAP 2005. LNCS (LNAI), vol. 4369, pp. 66–80. Springer, Heidelberg (2006). https://doi.org/10.1007/11963578_6
19. Sethumadhavan, S.: A survey of Monge properties. Master's thesis, Cochin University of Science and Technology (2009)
20. Tesch, A.: Exact energetic reasoning in $O(n^2 \log^2 n)$. Technical report, Zuse Institute Berlin (2016)
21. Tesch, A.: A nearly exact propagation algorithm for energetic reasoning in $\mathcal{O}(n^2 \log n)$. In: Rueher, M. (ed.) CP 2016. LNCS, vol. 9892, pp. 493–519. Springer, Cham (2016). https://doi.org/10.1007/978-3-319-44953-1_32

22. Vilím, P.: Timetable edge finding filtering algorithm for discrete cumulative resources. In: Achterberg, T., Beck, J.C. (eds.) CPAIOR 2011. LNCS, vol. 6697, pp. 230–245. Springer, Heidelberg (2011). https://doi.org/10.1007/978-3-642-21311-3_22

23. Wolf, A., Schrader, G.: $\mathcal{O}(n \log n)$ overload checking for the cumulative constraint and its application. In: Umeda, M., Wolf, A., Bartenstein, O., Geske, U., Seipel, D., Takata, O. (eds.) INAP 2005. LNCS (LNAI), vol. 4369, pp. 88–101. Springer, Heidelberg (2006). https://doi.org/10.1007/11963578_8

The WEIGHTEDCIRCUITSLMAX Constraint

Kim Rioux-Paradis and Claude-Guy Quimper[(⊠)]

Université Laval, Québec City, Canada
kim.rioux-paradis.1@ulaval.ca, claude-guy.quimper@ift.ulaval.ca

Abstract. The travelling salesman problem is a well-known problem that can be generalized to the m-travelling salesmen problem with min-max objective. In this problem, each city must be visited by exactly one salesman, among m travelling salesmen. We want to minimize the longest circuit travelled by a salesman. This paper generalizes the CIRCUIT and WEIGHTEDCIRCUIT constraints and presents a new constraint that encodes m cycles all starting from the same city and whose lengths are bounded by a variable L_{max}. We propose two filtering algorithms, each based on a relaxation of the problem that uses the structure of the graph and the distances between each city. We show that this new constraint improves the solving time for the m travelling salesmen problem.

1 Introduction

Constraint programming offers a large number of constraints to model and to solve combinatorial problems [1]. These constraints serve two purposes: they facilitate the modelling of a problem and they offer strong filtering algorithms that reduce the search space. When solving an optimization problem with a Branch & Bound approach, computing a tight bound on the objective function is crucial to limit the size of the search tree. Global constraints can help compute the bound by filtering the objective variable with respect to a large number of variables in the problem.

We propose a new global constraint that helps to model and solve the m-travelling salesman problem with a min-max objective. This problem consists in planning the routes of m salesmen that start at the depot D, visit $n - 1$ cities exactly once, and return to the depot. We want to minimize the longest route. While there exist several global constraints that can be used to encode this problem, few include an optimization criterion, and none minimize the length of the longest cycle. With the new global constraint WEIGHTEDCIRCUITSLMAX, one can easily model the m-travelling salesman problem and compute tight bounds on the objective function.

The filtering algorithm we propose uses two relaxations both inspired by the 1-tree relaxation [2] used for the WEIGHTEDCIRCUIT constraint. We name these relaxations the 1-forest relaxation and the cluster relaxation. For each relaxation, we develop a filtering algorithm that takes into account the minimization of the longest cycle. We compare a model that uses this new constraint against a model that does not use it and we show its efficiency.

© Springer International Publishing AG, part of Springer Nature 2018
W.-J. van Hoeve (Ed.): CPAIOR 2018, LNCS 10848, pp. 495–511, 2018.
https://doi.org/10.1007/978-3-319-93031-2_35

Section 2 describes the problem and the data structures used to design new filtering algorithms. Section 3 presents the new constraint WEIGHTEDCIRCUIT-SLMAX. Sections 4 and 5 each present a relaxation and a filtering algorithm. Section 6 explains how to detect situations where the constraint is consistent and filtering algorithms do not need to be executed. Finally, Sect. 7 presents the experimental results.

2   Background

We present the m-travelling salesman problem. We follow with the presentation of data structures, graph theory, and algorithms that we later use as tools to design our filtering algorithms. We also present global constraints useful to model the travelling salesman problem and its variants.

2.1 m-TSP

The m-travelling salesmen problem, denoted m-TPS, is a well-known generalization of the travelling salesman problem (TSP) [3]. The problem's input is a graph $G = (V, E, d)$ where V is a set of vertices (or cities), E a set of edges, and $d(i, j) = d(j, i)$ is a symmetric function that returns the distance between the vertices i and j. The city $n - 1$ is called the depot $D = n - 1$. The goal is to find m disjoint routes for the m salesmen leaving and returning to the depot. All cities have to be visited exactly once by only one salesman. The objective is to minimize the total distance. The min-max m-TSP [4–7] is the same problem but the objective is to minimize the longest route. L_z is the sum of the distance travelled by the zth salesman and L_{max} the maximum of the L_z. Minimizing the longest circuit is often sufficient to balance the workload between each salesman.

There are different ways to solve a m-TSP exactly. Exact methods based on branch and bound [8,9] and integer programming are often used [10]. One can also reduce the m-TSP to a TSP problem with one salesman and use exact methods for this problem [11,12]. There are many heuristics that have been developed to efficiently obtain good solutions for the m-TSP problem, without providing guaranties about the optimality of the solution. We recommend a survey by Bektas [13] for an overview of these techniques.

2.2   The Constraints CIRCUIT and CYCLES

Laurière [14] introduces the CIRCUIT($[X_1, \ldots, X_n]$) constraint. This constraint is satisfied if the directed graph composed of the vertices $V = \{1, \ldots, n\}$ and the arcs $E = \{(i, X_i) \mid i \in V\}$ contains exactly one cycle. The variable X_i is the node that follows i on the cycle. This constraint models the Hamiltonian cycle problem. Let $G = \langle V, E \rangle$ be an undirected graph. For every node i, one defines a variable X_i with a domain containing the nodes adjacent to i. The constraint CIRCUIT($[X_1, \ldots, X_n]$) is satisfiable if and only if G contains a cycle visiting each node exactly once, i.e. a Hamiltonian cycle. Therefore, it is NP-Hard to perform a complete filtering on this constraint. However, many algorithms [15–17] perform a partial filtering in polynomial time.

Beldiceanu and Contejean [18] generalize the CIRCUIT constraint into the CYCLES($[N_1, \ldots, N_n], m$) constraint to impose exactly m cycles rather than one. The cycles must be disjoint, the salesmen do not leave from the same depot. They generalize CYCLES further by proposing CYCLES($[X_1, \ldots, X_n]$, $m, [W_1, \ldots, W_n], \min, \max$) where W_i is a weight assigned to node i, not to an edge. The variables min and max are weights of the smallest and largest cycles. Further generalization can force nodes to belong to different cycles and control the length of individual cycles. These constraints are primarily introduced as powerful modelling tools.

Benchimol et al. [19] present several filtering algorithms for the WEIGHTEDCIRCUIT($[X_1, \ldots, X_n], d, L$) constraint. Focacci et al. [20,21] also study this constraint that is satisfied when the constraint CIRCUIT($[X_1, \ldots, X_n]$) is satisfied and when the total weight of the selected edges is no more than L, i.e. $\sum_{i=1}^{n} w(X_i, i) \leq L$. They use different relaxations and reduced-cost based filtering algorithms to filter the constraint, including the 1-tree relaxation that we describe in Sect. 2.6.

Without introducing a new constraint, Pesant et al. [22] show how to combine constraint programming and heuristics to solve the travelling salesman problem with and without time windows.

2.3   Disjoints Sets

The algorithms presented in the next sections use the disjoint sets data structure. This data structure maintains a partition of the set $\{0, \ldots, n-1\}$ such that each subset is labelled with one of its elements called the representative. The function FIND(i) returns the representative of the subset that contains i. The function UNION(i, j) unites the subsets whose representatives are i and j and returns the representative of the united subset. FIND(i) runs in $\Theta(\alpha(n))$ amortized time, where $\alpha(n)$ is Ackermann's inverse function. UNION(i, j) executes in constant time [23].

2.4   Minimum Spanning Tree

A weighted tree $T = (V, E, w)$ is an undirected connected acyclic graph. Each pair of vertices is connected by exactly one path. The weight of a tree $w(T) = \sum_{e \in E} w(e)$ is the sum of the weight of its edges. A *spanning tree* of a graph $G = \langle V, G \rangle$ is a subset of edges from E that forms a tree covering all vertices in V. A *minimum spanning tree* $T(G)$ is a spanning tree of G whose weight $w(T(G))$ is minimal [24].

Kruskal's algorithm [25] finds such a tree in $\Theta(|E| \log |V|)$ time. It starts with an empty set of edges S. The algorithm goes through all edges in non-decreasing order. An edge is added to S if it does not create a cycle among the selected edges in S. The disjoint sets are used to verify this condition. The algorithm maintains an invariant where the nodes of each tree form disjoint sets. The nodes that belong to the same sets are connected by a path of edges selected by Kruskal's algorithm. To test whether the addition of an edge (i, j) creates a cycle, the algorithm checks whether the nodes i and j belong to the same disjoint set. If not, the edge is selected and the disjoint sets that contain i and j are united.

Let $T(G)$ be the minimum spanning tree of the graph $G = \langle V, E \rangle$. Let $T_e(G)$ be the minimum spanning tree of G for which the presence of the edge e is imposed. The reduced cost of the edge e, denoted $\widetilde{w}(e)$, is the cost of using e in the spanning tree: $\widetilde{w}(e) = w(T_e(G)) - w(T(G))$. The reduced cost $\widetilde{w}(e)$ of an edge $e = (i, j) \in E$ can be computed by finding the edge s with the largest weight lying on the unique path between nodes i and j in $T(G)$. We obtain $\widetilde{w}(e) = w(e) - w(s)$ [26].

A *spanning forest* of a graph $G = \langle V, E \rangle$ is a collection of trees whose edges belong to E and whose nodes span V. The weight of a forest is the sum of the weights of its trees. A *minimum spanning forest* of m trees is a spanning forest of m trees whose weight is minimum. Kruskal's algorithm can be adapted to find a spanning forest. One simply needs to prematurely stop the algorithm after finding m trees, i.e. after $|V| - m$ unions.

2.5   Cartesian Tree

The Path Maximum Query problem is defined as follows: Given a weighted tree $T = (V, E, w)$ and two nodes $u, v \in V$, find the edge with the largest weight that lies on the unique path connecting u to v in T. This problem can be solved with a simple traversal of the tree in $\mathcal{O}(|E|)$ time. The offline version of the problem is defined as follows. Given a weighted tree $T = \langle V, E, w \rangle$ and a set of queries $Q = \{(u_1, v_1), \ldots, (u_{|Q|}, v_{|Q|})\}$, find for each query $(u_i, v_i) \in Q$ the edge with the largest weight that lies on the unique path connecting u_i to v_i in T. This problem can be solved in $\mathcal{O}(|E| + |Q|)$ time using a Cartesian tree as we explain below.

A Cartesian tree T_C of a tree $T = \langle V, E, w \rangle$ is a rooted binary, and possibly unbalanced, tree with $|V|$ leaves and $|E|$ inner nodes. It can be recursively constructed following Demaine et al. [27]. The Cartesian tree of a tree containing a single node and no edge ($|V| = 1$ and $|E| = 0$) is a node corresponding to the unique node in V. If the tree T has more than one node, the root of its Cartesian tree corresponds to the edge with the largest weight denoted $e_{max} = \arg\max_{e \in E} w(e)$. The left and right children correspond to the Cartesian trees of each of the two trees in $E \setminus \{e_{max}\}$. Finding the largest edge on a path between u_i and v_i in T is equivalent to finding the lowest common ancestor of u and v in the Cartesian tree T_C. The Cartesian tree can be created in $\mathcal{O}(|V|)$ time after sorting the edges in preprocessing [27]. Tarjan's off-line lowest common ancestor algorithm [23] takes as input the Cartesian tree T_C and the queries Q and returns the lowest common ancestor of each query in Q, i.e. the edge with largest weight lying on the path from u_i to v_i in T for each i. When implemented with the disjoint set data structure by Gabow and Tarjan [28], the algorithm has a complexity of $\mathcal{O}(|V| + |Q|)$.

2.6   1-Tree Relaxation

Held and Karp [2] introduce the 1-tree relaxation to solve the travelling sales-man problem that Benchimol et al. [19] use to filter the WEIGHTEDCIRCUIT con-straint. The relaxation partitions the edges E into two disjoint subsets: the subset of edges connected to the depot D, denoted $E_D = \{(i, j) \in E \mid i = D \vee j = D\}$, and the other edges denoted $E_O = E \setminus E_D$. A solution to the travelling salesman problem is a cycle and therefore has two edges in E_D and the remaining edges in E_O form a simple path. In order to obtain a lower bound on the weight of this cycle, Held and Karp [2] select the two edges in E_D that have the smallest weights and compute the minimum spanning tree of the graph $G' = \langle V \setminus \{D\}, E_O \rangle$. The weight of the two selected edges in E_D plus the weight of the minimum spanning tree gives a lower bound on the distance travelled by the salesman. This relax-ation is valid since a simple path is a tree and therefore, the minimum spanning tree's weight is no more than the simple path's weight.

3   Introducing WEIGHTEDCIRCUITSLMAX

We introduce the constraint WEIGHTEDCIRCUITSLMAX that encodes the m cir-cuits of the salesman that start from the depot D, visit all cities once, and return to the depot. All circuits have a length bounded by L_{max}.

$$\text{WEIGHTEDCIRCUITSLMAX}([S_0, \ldots, S_{m-1}],$$
$$[N_0, \ldots, N_{n-2}], d[0, \ldots, n-1][0, \ldots, n-1], L_{max}) \tag{1}$$

The variable S_k is the first city visited by the salesman k. The variable N_i is the next city visited after city i. The symmetric matrix parameter d provides the distances between all pairs of cities. The variable L_{max} is an upper bound on the lengths of all circuits. The constraint WEIGHTEDCIRCUITSLMAX is designed to be compatible with the CIRCUIT constraint in order to take advantage of its filtering algorithms. For that reason, the value associated to the depot is duplicated m times. Each salesman returns to a different copy of the depot. The integers from 0 to $n-2$ represent the $n-1$ cities and the integers from $n-1$ to $n+m-2$ represent the depot. If $N_i \leq n-2$, the salesman visit city N_i after city i. If $n-1 \leq N_i$, the salesman returns to the depot after visiting city i. Consequently, we have $\text{dom}(S_k) \subseteq \{0, \ldots, n-2\}$ and $\text{dom}(N_i) \subseteq \{0, \ldots, n+m-2\}$.

While using the constraint WEIGHTEDCIRCUITSLMAXone can post the con-straint CIRCUIT($[N_0, \ldots, N_{n-2}, S_0, \ldots, S_{m-1}]$) to complement the filtering of the WEIGHTEDCIRCUITSLMAX constraint. The filtering algorithms we present in Sects. 4 and 5 for the WEIGHTEDCIRCUITSLMAX constraint are based on the cost L_{max}. On the other hand, the filtering algorithms of CIRCUIT constraints are based on the structure of the graph. The filtering algorithms are complemen-tary.

CIRCUIT is a special case of WEIGHTEDCIRCUITSLMAX where $m = 1$ and $L_{max} = \infty$. Enforcing domain consistency on circuit is NP-Hard [14]. Therefore, enforcing domain consistency on WEIGHTEDCIRCUITSLMAX is NP-Hard.

4   1-Forest Relaxation

We now describe a relaxation of the WEIGHTEDCIRCUITSLMAX constraint. We introduce two relaxations to the WEIGHTEDCIRCUITSLMAX constraint. These relaxations are used to compute a bound on the length of the longest cycle and to filter the starting variables S_i and the next variables N_i. As seen in the previous section, it is possible to use the WEIGHTEDCIRCUITSLMAX constraint in conjunction with the CIRCUIT constraint. The filtering algorithm that we present is added to the filtering that is already done by the CIRCUIT constraint.

4.1   Relaxation

The domains of the variables S_i and N_i encode the following graph that we denote G. Let the vertices be $V = \{0, \ldots, n-1\}$ where the depot is the node $D = n - 1$. Let the edges $E = E_D \cup E_O$ be partitioned into two sets. The set of edges adjacent to the depot $E_D = \{(D, i) \mid \max(\mathrm{dom}(N_i)) \geq n - 1\} \cup \bigcup_{k=0}^{m-1} \mathrm{dom}(S_k)$ and the edges that are not adjacent to the depot $E_O = \{(i, j) \mid i < j \wedge (i \in \mathrm{dom}(N_j) \vee j \in \mathrm{dom}(N_i))\}$. The weight of an edge is given by the distance matrix $w(i, j) = d[i][j]$.

We generalize the 1-tree relaxation to handle multiple cycles passing by a unique depot D. We call this generalization the 1-forest. Rather than choosing 2 edges in E_D, we choose the m shortest edges $a_0, \ldots, a_{m-1} \in E_D$. We do not choose the $2m$ shortest edges because there could be a salesman that goes back and forth to a node v using twice the edge $e_{v,D}$. Choosing $2m$ different edges is therefore not a valid lower bound. Rather than computing the minimum spanning tree with the edges in E_O, we compute the minimum spanning forest of m trees $T_0, \ldots, T_{m-1}$ using the edges in E_O. Then, L_{max} has to be greater than or equal to the average cost of the trees. The lower bound on L_{max} is given by $c([a_0, \ldots, a_{m-1}], [T_0, \ldots, T_{m-1}])$.

$$c([a_0, \ldots, a_{m-1}], [T_0, \ldots, T_{m-1}]) = \frac{1}{m}\left(2\sum_{i=0}^{m-1} w(a_i) + \sum_{i=0}^{m-1} w(T_i)\right) \quad (2)$$

We show the validity of this relaxation.

Theorem 1. $c([a_0, \ldots, a_{m-1}], [T_0, \ldots, T_{m-1}])$ *is a lower bound on L_{max} for the constraint* WEIGHTEDCIRCUITSLMAX$([S_0, \ldots, S_{m-1}], [N_1, \ldots, N_{n-1}], d[0, \ldots, n - 1][0, \ldots, n - 1], L_{max})$.

Proof. Consider a solution to the constraint where the length of the longest circuit is minimized. Let $E_D^S \subseteq E_D$ be the edges of the circuits connected to the depot and let $E_O^S \subseteq E_O$ be the other edges.

There are 2 edges by circuit going or returning to the depot (with the possibility of duplicates for salesmen visiting a single city). Therefore, there are $2m$ edges in E_D^S, counting duplicates. A lower bound for the cost of the edges in E_D^S is twice the cost of the m shortest edges in E_D.

The edges in E_O^S form a forest of m trees. Hence, a lower bound of the cost of the edges in E_O^S is the cost of a minimum forest of m trees T_i.

$$c([a_0, \ldots, a_{m-1}], [T_0, \ldots, T_{m-1}]) = \frac{1}{m} \left(2 \sum_{i=0}^{m-1} w(a_i) + \sum_{i=0}^{m-1} w(T_i) \right)$$

$$\leq \frac{1}{m} \left(\sum_{e_{D,i} \in E_D^S} w(e_{D,i}) + \sum_{e_{i,j} \in E_O^S} w(e_{i,j}) \right)$$

Since the average cost per salesman is smaller than or equal to the maximum cost, we obtain: $c([a_0, \ldots, a_{m-1}], [T_0, \ldots, T_{m-1}]) \leq L_{max}$.   □

The algorithm to compute $c([a_0, \ldots, a_{m-1}], [T_0, \ldots, T_{m-1}])$ works as follows. First, we pick the m smallest edges in E_D. Then, we compute a forest of m trees with Kruskal's algorithm in $\mathcal{O}(|E| \log |V|)$ time.

4.2 Filtering the Edges in E_D

We want to filter the edges $e = (D, i)$ in E_D that are not selected by the relaxation. We want to know whether $c([a_0, \ldots, a_{m-2}, e], [T_0, \ldots, T_{m-1}])$ is greater than L_{max}. If it is the case, then e cannot be in the solution because the cost of using this edge is too large. The filtering rule (3) removes $i + m$ from the domain of all starting time variable S_k and removes all values associated to the depot from the domain of N_i.

$$c([a_0, \ldots, a_{m-1},], [T_0, \ldots, T_{m-1}]) + \frac{1}{m}(w(e) - w(a_{m-1})) > \max(\text{dom}(L_{max}))$$

$$\implies S_k \neq i \wedge N_i < n - 1 \quad \forall k \in \{0, \ldots, m-1\}$$

(3)

4.3 Filtering the Edges in E_O

We want to decide whether there exists a support for the edge $e = (i, j)$ in E_O. In other words, we want to find a forest of m trees $T_0', \ldots, T_{m-1}'$ that contains e such that $c([a_0, \ldots, a_{m-1},], [T_0', \ldots, T_{m-1}'])$ is no greater than $\max(\text{dom}(L_{max}))$. If e belongs to the trees $T_0, \ldots, T_{m-1}$ computed in Sect. 4.1, e has as support and should not be filtered. Otherwise, there are two possible scenarios: the nodes i and j belong to the same tree in $T_0, \ldots, T_{m-1}$ or they do not.

Same Tree T_β. Given the edge $e = (i, j)$, if nodes i and j belong to a tree T_β, the cost of adding the edge e is equal to its reduced cost $\widetilde{w}(e)$ (see Sect. 2.4), i.e. the weight $w(e)$ minus the largest weight of an edge lying on the unique path in T_β between i and j. If $c([a_0, \ldots, a_{m-1},], [T_0, \ldots, T_{m-1}]) + \widetilde{w}(e)$ is greater than $\max(\text{dom}(L_{max}))$, then edge e must be filtered out from the graph, i.e. i is removed from the domain of N_j and j is removed from the domain of N_i.

To efficiently compute the reduced costs, we construct a Cartesian tree (see Sect. 2.5) for each tree in $T_0, \ldots, T_{m-1}$ in $\mathcal{O}(|V|)$ time. For each edge $e = (i, j)$

such that i and j belong to the same tree, we create a query to find the edge with the largest weight between i and j in the tree. Tarjan's off-line lowest common ancestor algorithm [23] answers, in batches, all queries in time $\mathcal{O}(|V|+Q)$ where Q is the number of queries. For each edge, we compute the reduced cost and check whether filtering is needed.

Different Trees T_ϵ, T_δ. If the nodes i and j of the edge $e = (i, j)$ do not belong to the same tree, adding the edge e to the trees $T_0, \ldots, T_{m-1}$ connects two trees together. In order, to maintain the number of trees to m, one needs to remove an edge from any tree. To minimize the weight of the trees, the edge e' we remove must be the one with the largest weight, i.e. the last edge selected by Kruskal's algorithm. We obtain the reduced cost $\widetilde{w}(e) = w(e) - w(e')$. Using this reduced cost, we filter the variables N_i as we did when the nodes i and j belong to the same tree.

5   Clusters Relaxation

The 1-forest relaxation is fast to compute, but its bound is not always tight. Counting twice the m shortest edges is less effective than counting the $2m$ edges that could be in the circuits. Moreover, for $m = 2$ salesmen, if the solution has a long and a short circuit, the bound lies between the length of both circuits. In that case, the lower bound for the longest circuit is not tight. We refine the 1-forest relaxation to better capture the structure of the graph and to obtain a tight bound on the longest cycle when all variables are assigned.

5.1   Relaxation

We consider the m trees $\{T_0, \ldots, T_{m-1}\}$ as computed in the 1-forest relaxation. Let $C_1, \ldots, C_r$ be a partition of the trees into clusters such that the trees that belong to the same cluster are connected with each other with edges in E_O and are not connected to the trees from the other clusters. Figure 1 shows an example of three trees partitioned into two clusters. Note that the number of clusters r is between 1 and m. In a solution, cities visited by a salesman necessarily belong to the same cluster. We compute a lower bound on L_{max} for each cluster and keep the tightest bound.

We choose two edges $e_1^{C_\alpha}$ and $e_2^{C_\alpha}$ in E_D for each cluster C_α. If the cluster contains a single node v, we choose twice the edge $e_{v,D}$, i.e. $e_1^{C_\alpha} = e_2^{C_\alpha} = e_{v,D}$. If the cluster contains more than one node, we choose the two shortest edges that connect the depot to a node in the cluster. The weight of a cluster C_α, denoted $w(C_\alpha)$, is the weight of the two chosen edges $e_1^{C_\alpha}$ and $e_2^{C_\alpha}$ and the weight of the trees in the cluster.

$$w(C_\alpha) = \sum_{T_\beta \in C_\alpha} w(T_\beta) + w(e_1^{C_\alpha}) + w(e_2^{C_\alpha})$$

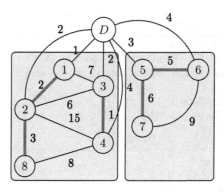

Fig. 1. Three trees grouped into two clusters. Edges that belong to a tree are in bold.

The nodes that belong to a cluster can be visited by a maximum of $\lambda = m - r + 1$ salesmen. In such a case, the average length of a circuit is given by the weight of the cluster divided by λ and it constitutes a valid lower bound of L_{max}.

$$\mu(\{C_1, \ldots, C_r\}) = \frac{1}{\lambda} \max_{C_\alpha} w(C_\alpha) \tag{4}$$

Theorem 2. $\mu(\{C_1, \ldots, C_r\})$ *is a lower bound for L_{max}.*

Proof. Consider a solution to the constraint where the length of the longest circuit is minimized. Let $E_D^S \subseteq E_D$ be the edges of the circuits connected to the depot and let $E_O^S \subseteq E_O$ be the other edges.

$$\mu(\{C_1, \ldots, C_r\}) = \frac{1}{\lambda} \max_{C_\alpha} \left(\sum_{T_\beta \in C_\alpha} w(T_\beta) + w(e_1^{C_\alpha}) + w(e_2^{C_\alpha}) \right) \tag{5}$$

$$\leq \max_{C_\alpha} \frac{1}{\lambda} \left(\sum_{\substack{e_{i,j} \in E_O^S \\ i \in C_\alpha}} w(e_{i,j}) + \sum_{\substack{e_{i,D} \in E_D^S \\ i \in C_\alpha}} w(e_{i,D}) \right) \tag{6}$$

Let c denote the longest circuit. Since the longest circuit is longer than or equal to the average length circuit, we obtain:

$$\leq \sum_{\substack{e_{i,j} \in E_O^S \\ i \in c}} w(e_{i,j}) + \sum_{\substack{e_{i,D} \in E_D^S \\ i \in c}} w(e_{i,D}) \tag{7}$$

$$\leq L_{max} \tag{8}$$

$\square$

To compute the lower bound $\mu(\{C_1, \ldots, C_r\})$, we need to compute the clusters and select edges from E_O. To do so, we adapt Kruskal's algorithm as shown

---

Algorithm 1. ComputeClusters(G,m)

Data: $G = (V, E)$, m, the number of salesmen
Result: num_nodes_Eo, the number of nodes by clusters, max_edge_Eo, the longest edge of each cluster, weight_Eothe weight of each cluster

begin

    trees $\leftarrow$ *DisjointsSets*($|V|$)

    clusters $\leftarrow$ *DisjointsSets*($|V|$)

    num_clusters $\leftarrow |V|$

    for $i = 0..|V| - 1$ **do**

        weight_Eo[i] $\leftarrow 0$

        max_edge_Eo[i] $\leftarrow 0$

        num_nodes_Eo[i] $\leftarrow 1$

    for $e = (i, j) \in E$ *in non-decreasing order of weight* **do**

        $v_i \leftarrow$ clusters.FIND(i)

        $v_j \leftarrow$ clusters.FIND(j)

        if $v_i \neq v_j$ **then**

            if *num_clusters* $< |V| - m$ **then**

                $v_f \leftarrow$ trees.UNION(v_i, v_j)

            $v_f \leftarrow$ clusters.UNION(v_1, v_2)

            num_clusters $\leftarrow$ num_clusters $- 1$

1   weight_Eo[v_f] $\leftarrow$ weight_Eo[v_1] + weight_Eo[v_2]

2   max_edge_Eo[v_f] $\leftarrow w(e)$

            num_nodes_Eo[v_f] $\leftarrow$ num_nodes_Eo[v_1] + num_nodes_Eo[v_2]

    return weight_Eo, max_edge_Eo, num_nodes_Eo

---

in Algorithm 1. Kruskal already uses a disjoint set data structure where the nodes of a tree are grouped into a set. We add another disjoint set data structure where the nodes of a cluster are grouped in a set. We process the edges in non-decreasing order of weight. When processing the edge (i, j), if i and j belong to two distinct clusters, we unite these clusters to form only one. The algorithm also merges the tree that contains i with the tree that contains j, but only if there are more than $|V| - m$ trees in the current forest. While we create each set, we keep three vectors that keep track of: the longest edge of each cluster, the weight of each cluster, and the number of nodes in each cluster. As for Kruskal's algorithm, the running time complexity is dominated by sorting the edges by weight which is done in $\mathcal{O}(|E| \log |V|)$.

Using the clusters computed by Algorithm 1, Algorithm 2 selects the edges from E_D by processing the edges in non-decreasing order of weight. When processing the edge (i, D), the algorithm finds the cluster that contains the node i. It selects the edge (i, D) only if fewer than two edges were selected for the cluster that contains i. Lines 1 and 2 update the sum of the weights of the selected edges for that cluster and the largest edge selected for the cluster. This information will be used later in the filtering algorithm. The second for loop checks whether there are clusters linked to the depot with a single edge. If it is the case and

---

Algorithm 2. ComputeEdgesFromE_D(E_D,num_nodes_Eo)

Data: E_D, num_nodes_Eo
Result: max_weight_Ed, weight_Ed, num_edges_Ed
for $i = 1..n - 1$ **do**
  max_weight_Ed[i] ← 0
  weight_Ed[i] ← 0
  num_edges_Ed[i] ← 0

for $(i, D) \in E_D$ *in non-decreasing order of weight* **do**
  v_i ← clusters.FIND(i)
  if *num_edges_Ed*[v_i] < 2 **then**
    weight_Ed[v_i] ← weight_Ed[v_i] + $w(e)$
    num_edges_Ed[v_i] ← num_edges_Ed[v_i] + 1
    max_weight_Ed[v_i] ← $w(e)$

for $i = 0..n - 1$ **do**
  rep ←clusters.FIND(i)
  if $rep = i$ **then**
    if *num_edges_Ed*[i] = 1 $\wedge$ *num_nodes_Eo*[i] > 1 **then**
      weight_Ed[i] ← 2 × weight_Ed[i]
    else if *num_edges_Ed*[i] < 2 **then** Fail
  return *max_weight_Ed, weight_Ed*

---

the cluster contains only one node, we make that edge count for double. If the cluster has more than one node, the constraint is unsatisfiable.

To compute a lower bound on L_{max}, we go through each cluster C. In (4), the summation is given by the entry of the vector weight_Eo corresponding to the cluster C_α as computed by Algorithm 1. The weight of the edges $w(e_1^{C_\alpha}) + w(e_2^{C_\alpha})$ is given by the entry in the vector weight_Ed corresponding to the cluster C_α.

Overall, the cluster relaxation complements the 1-forest relaxation as follows. At the top of the search tree, when variable domains contain many values, this cluster relaxation is not as tight as the 1-forest relaxation. There are very few clusters and the cluster relaxation only selects two edges per cluster in the set E_D while the 1-forest relaxation selects m edges no matter how many clusters there are. However, as the search progresses down the search tree, there are fewer values in the domains and more clusters. Computing the maximum cycle per cluster becomes more advantageous than computing the average tree of the 1-forest relaxation. The cluster relaxation provides an exact bound when all variables are instantiated, which is not the case for the 1-forest relaxation.

5.2 Filtering the Edges in E_D

We filter the edges in E_D based on the cluster relaxation as follows. Let $C_{\alpha(i)}$ be the cluster that contains the node i. For each edge $e = (D, i) \in E_D$, we check whether $\frac{1}{\lambda} \left(w(C_\alpha) - e_2^{C_{\alpha(i)}} + w(e) \right) \leq L_{max}$. In other words, we check whether substituting the edge $e_2^{C_{\alpha(i)}}$ by e induces a cost that is still below the desired threshold. If not, we remove i from the domain of S_k for all $k \in \{0, \ldots, m-1\}$ and we remove D from the domain of N_i.

5.3   Filtering the Edges in E_O

As for the 1-forest relaxation, we filter edges in E_O differently depending they connect two nodes of the same tree or from different trees.

Same Tree: Consider an edge $e = (i, j)$ such that i and j belong to the same tree and the same cluster that we denote $C_{\alpha(i)}$. We compute the reduced cost $\widetilde{w}(e)$, using a Cartesian tree, exactly as we do for the 1-forest relaxation in Sect. 4.3. If the inequality $w(C_{\alpha(i)}) + \widetilde{w}(e) \leq \max(\mathrm{dom}(L_{max}))$ does not hold, we remove i from the domain of N_j and remove j from the domain of N_i.

Different Trees: Consider an edge $e = (i, j)$ such that i and j do not belong to the same tree. However, by definition of a cluster, i and j belong to the same cluster that we denote $C_{\alpha(i)}$. Let e' be the edge with the largest weight in a tree of the cluster $C_{\alpha(i)}$, i.e. the last edge selected by Algorithm 1. We can substitute e' by e without changing the number of trees. The reduced cost of edge e is $\widetilde{w}(e) = w(e) - w(e')$. If $w(C_{\alpha(i)}) + \widetilde{w}(e) > \max(\mathrm{dom}(L_{max}))$, we remove i from the domain of N_j and remove j from the domain of N_i.

6   Special Filtering Cases

There exist conditions when the filtering algorithm, whether it is based on the 1-forest or the cluster relaxation, does not do any pruning. Some of these conditions are easy to detect and can prevent useless executions of the filtering algorithms.

We consider two consecutive executions of the filtering algorithm. The second execution is either triggered by the instantiation of a variable or by constraint propagation. We check which values are removed from the domains between both executions. The removal of these values can trigger further filtering in two situations:

1. The upper bound of $\mathrm{dom}(L_{max})$ is filtered;
2. An edge selected by the 1-forest or the cluster relaxation is filtered;

In situation 1, the edges selected by the 1-forest and cluster relaxations remain unchanged. However, it is possible that the reduced cost of some edges are too large for the new bound on L_{max} and that these edges need to be filtered. In such a case, we do not need to recompute the relaxation nor the Cartesian trees, but we need to check whether the reduced cost of each edge is too high.

In situation 2, the trees and the clusters must be recomputed and the filtering algorithm needs to be executed entirely. If neither situation 1 nor 2 occurs, for instance if only an edge in E_O that does not belong to a tree is filtered, no filtering needs to be done and the algorithm does not need to be executed.

$$\min L_{max}$$

s.t.

$$S_v \in \{0, \ldots, n-2\} \qquad \forall \, v \in \{0, \ldots, m-1\}$$
$$N_i \in \{0, \ldots, n+m-1\} \setminus \{i\} \qquad \forall i \in \{1, \ldots, n\}$$
$$\text{CIRCUIT}(N_0, \ldots, N_{n-1}, S_0, \ldots, S_{m-1})$$
$$D[i] = 0 \qquad \forall \, i \in \{n-1, \ldots, n+m-2\}$$
$$D[i] = d'[n-1][S_{i-m-n+1}] + D[S_{i-m-n+1}] \qquad \forall \, i \in \{n+m-1, \ldots, n+2m-2\}$$
$$D[i] = d'[i][N_i] + D[N_i] \qquad \forall \, i \in \{0, \ldots, n-1\}$$
$$L_{max} \geq D[i] \qquad \forall \, i \in \{n+m-1, \ldots, n+2m-2\}$$

Fig. 2. Model 1 for the m-TSP.

7 Experiments

We solve the m-TSP problem as defined in Sect. 2.1. We compare two different models in our experimentation. The model 1 is presented in Fig. 2. The variable S_v indicates the starting address for the salesman v. The variable N_v indicates the address visited after address v. The integers between 0 and $n-2$ represent the addresses while the integers from $n-1$ to $n+m-2$ represent the arrival at the depot. The variable $D[v]$ encodes the remaining distance that a salesman needs to travel to reach the depot from address v. The constraint ELEMENT connects the distance variables $D[v]$ with the next variables N_v. The model 2 is based on model 1 that we augment with the constraint WEIGHTEDCIRCUITSLMAX$(S_0, \ldots, S_{m-1}, N_1, \ldots, N_n, d', L_{max})$. In the model, we make sure that the distance matrix can report the distances with the duplicates of the depot. We define the matrix $d'[i][j] = d[\min(i, n-1)][\min(j, n-1)]$.

Both models use the CIRCUIT constraint to exploit the graph's structure and to achieve a strong *strucural filtering*. In addition, Model 2 uses the WEIGHTED-CIRCUITSLMAX constraint to perform an *optimality filtering* that is based on the bound of the objective function. Our experiments aim at showing the advantage the new algorithms offer by performing optimality filtering.

We use instances from the m-TSP benchmark developed by Necula et al. [29]. We also generate instances of $\{10, 20, 50\}$ addresses in Quebec City with uniformly distributed longitude in $[-71.20, -71.51]$ and latitude in $[46.74, 46.95]$ with the shortest driving time as the distance between the addresses. Experiments are run on a MacBook Pro with a 2.7 GHz Intel Core i5 processor and 8 Gb of memory with the solver Choco 4.0.6 compiled with Java 8. We select the variable N_i and assign it to value j such that the distance $d(i, j)$ is minimal. However, other heuristics could have been used [30]. We report the best solution found after a 10-min timeout.

7.1 Result and Discussion

The first experiment aims to determine which relaxation should be used: the 1-forest relaxation, the cluster relaxation, or both. We solved random instances with 10 addresses with 1, 2, and 3 salesmen using model 2. Figure 3a shows that the cluster relaxation is better than the 1-forest relaxation. However, all instances except one were solved faster when combining both relaxations. For this reason, for model 2, we combine both relaxations for the rest of the experiments.

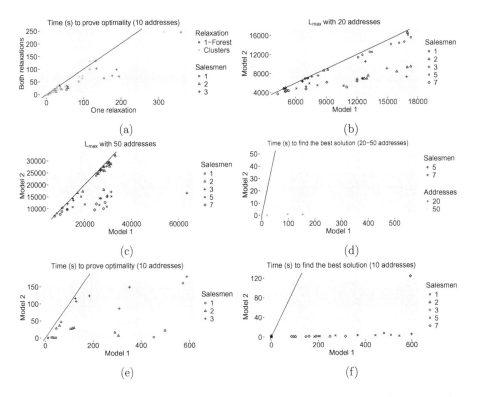

Fig. 3. Comparison of model 1 (without the WEIGHTEDCIRCUITSLMAX) and model 2 (with WEIGHTEDCIRCUITSLMAX) for instances with random addresses in Quebec City.

We compare model 1 against model 2. Figure 3b and c present the objective function obtained by both models after 10 min with instances of 20 and 50 random addresses. Since we minimize, a point under the identity function indicates that model 2 finds a better solution. No solutions were proved optimal. However, we clearly see that model 2 finds better solutions and the gap increases as the number of salesmen increase. As shown on Fig. 3d, for instances where solutions of equivalent quality are returned by both models, model 2 finds the solution faster in most of the cases.

Some instances with 10 addresses are solved to optimality. Figure 3e reports the solving times for these instances and model 2 is clearly faster. For unsolved instances, both models return the same solutions (without proving their optimality). Figure 3f shows that model 2 returns the solution instantaneously.

Table 1 compares models 1 and 2 using a standard benchmark from [29]. Model 2 either finds a better solution or finds an equivalent solution faster for all instances.

Table 1. Result for TSPLIB instances

Instances	n	m	Model 1		Model 2	
			L_{max}	Time last solution	L_{max}	Time last solution
Eil51	51	1	2307	6.0	2307	**1.5**
	51	2	1183	13.9	1183	**1.4**
	51	3	1906	0.5	**1183**	0.1
	51	5	638	0.1	**631**	211.2
	51	7	486	591.2	**457**	130
Berlin52	52	1	41534	3.1	**41276**	175.3
	52	2	32316	599.6	**20979**	164.4
	52	3	14629	599.3	**14457**	40.6
	52	5	14543	599.4	**13986**	42.5
	52	7	9668	347.4	**9668**	0.5
Eil76	76	1	3459	497.1	**3458**	307.0
	76	2	3284	1.8	**3278**	466.7
	76	3	2989	561.4	**2988**	375.2
	76	5	847	585.7	**730**	578.9
	76	7	651	34.9	651	**0.3**
Rat99	99	1	12093	68.8	**12084**	213.9
	99	2	11946	69.1	**11937**	343.4
	99	3	11881	78.2	**11872**	298.6
	99	5	11863	64.9	**11854**	372.9
	99	7	11813	108.4	**11804**	325.9

8 Conclusion

We presented a constraint that models the m-travelling salesmen problem. We proposed two complementary filtering algorithms based on two relaxations. Experiments show that these filtering algorithms improve the solving times and the quality of the solutions. In future work, inspired from [19], we would like to use additive bounding to reinforce the relaxations. We also want to consider the number of salesman m as a variable instead of a parameter.

References

1. Beldiceanu, N., Carlsson, M., Rampon, J.: Global constraint catalog, 2nd edn (revision a). Technical report 03, SICS (2012)
2. Held, M., Karp, R.: The traveling-salesman problem and minimum spanning trees. Oper. Res. **18**(6), 1138–1162 (1970)
3. Laporte, G., Nobert, Y.: A cutting planes algorithm for the m-salesmen problem. J. Oper. Res. Soc. **31**, 1017–1023 (1980)
4. França, P.M., Gendreau, M., Laportt, G., Müller, F.M.: The m-traveling salesman problem with minmax objective. Transp. Sci. **29**(3), 267–275 (1995)
5. Necula, R., Breaban, M., Raschip, M.: Tackling the bi-criteria facet of multiple traveling salesman problem with ant colony systems. In: 2015 IEEE 27th International Conference on Tools with Artificial Intelligence (ICTAI), pp. 873–880. IEEE (2015)
6. Narasimha, K.V., Kivelevitch, E., Sharma, B., Kumar, M.: An ant colony optimization technique for solving min–max multi-depot vehicle routing problem. Swarm Evol. Comput. **13**, 63–73 (2013)
7. Somhom, S., Modares, A., Enkawa, T.: Competition-based neural network for the multiple travelling salesmen problem with minmax objective. Comput. Oper. Res. **26**(4), 395–407 (1999)
8. Ali, A.I., Kennington, J.L.: The asymmetric m-travelling salesmen problem: a duality based branch-and-bound algorithm. Discret. Appl. Math. **13**(2–3), 259–276 (1986)
9. Gromicho, J., Paixão, J., Bronco, I.: Exact solution of multiple traveling salesman problems. In: Akgül, M., Hamacher, H.W., Tüfekçi, S. (eds.) Combinatorial Optimization, pp. 291–292. Springer, Heidelberg (1992). https://doi.org/10.1007/978-3-642-77489-8_27
10. Kara, I., Bektas, T.: Integer linear programming formulations of multiple salesman problems and its variations. Eur. J. Oper. Res. **174**(3), 1449–1458 (2006)
11. Rao, M.R.: A note on the multiple traveling salesmen problem. Oper. Res. **28**(3-part-i), 628–632 (1980)
12. Jonker, R., Volgenant, T.: An improved transformation of the symmetric multiple traveling salesman problem. Oper. Res. **36**(1), 163–167 (1988)
13. Bektas, T.: The multiple traveling salesman problem: an overview of formulations and solution procedures. Omega **34**(3), 209–219 (2006)
14. Lauriere, J.L.: A language and a program for stating and solving combinatorial problems. Artif. Intell. **10**(1), 29–127 (1978)
15. Caseau, Y., Laburthe, F.: Solving small TSPs with constraints. In: Proceedings of the 14th International Conference on Logic Programming (ICLP 1997), pp. 316–330 (1997)
16. Kaya, L.G., Hooker, J.N.: A filter for the circuit constraint. In: Benhamou, F. (ed.) CP 2006. LNCS, vol. 4204, pp. 706–710. Springer, Heidelberg (2006). https://doi.org/10.1007/11889205_55
17. Fages, J., Lorca, X.: Improving the asymmetric TSP by considering graph structure. Technical report 1206.3437, arxiv (2012)
18. Beldiceanu, N., Contejean, E.: Introducing global constraints in chip. Math. Comput. Modell. **20**(12), 97–123 (1994)
19. Benchimol, P., Hoeve, W.J.V., Régin, J.C., Rousseau, L.M., Rueher, M.: Improved filtering for weighted circuit constraints. Constraints **17**(3), 205–233 (2012)

20. Focacci, F., Lodi, A., Milano, M.: Embedding relaxations in global constraints for solving TSP and TSPTW. Ann. Math. Artif. Intell. **34**(4), 291–311 (2002)
21. Focacci, F., Lodi, A., Milano, M.: A hybrid exact algorithm for the TSPTW. INFORMS J. Comput. **14**(4), 403–417 (2002)
22. Pesant, G., Gendreaul, M., Rousseau, J.-M.: GENIUS-CP: a generic single-vehicle routing algorithm. In: Smolka, G. (ed.) CP 1997. LNCS, vol. 1330, pp. 420–434. Springer, Heidelberg (1997). https://doi.org/10.1007/BFb0017457
23. Tarjan, R.E.: Applications of path compression on balanced trees. J. ACM (JACM) **26**(4), 690–715 (1979)
24. Graham, R.L., Hell, P.: On the history of the minimum spanning tree problem. Ann. Hist. Comput. **7**(1), 43–57 (1985)
25. Kruskal, J.B.: On the shortest spanning subtree of a graph and the traveling salesman problem. Proc. Am. Math. Soc. **7**(1), 48–50 (1956)
26. Chin, F., Houck, D.: Algorithms for updating minimal spanning trees. J. Comput. Syst. Sci. **16**(3), 333–344 (1978)
27. Demaine, E.D., Landau, G.M., Weimann, O.: On Cartesian trees and range minimum queries. In: Albers, S., Marchetti-Spaccamela, A., Matias, Y., Nikoletseas, S., Thomas, W. (eds.) ICALP 2009. LNCS, vol. 5555, pp. 341–353. Springer, Heidelberg (2009). https://doi.org/10.1007/978-3-642-02927-1_29
28. Gabow, H.N., Tarjan, R.E.: A linear-time algorithm for a special case of disjoint set union. In: Proceedings of the 15th Annual ACM Symposium on Theory of Computing, pp. 246–251 (1983)
29. Necula, R., Breaban, M., Raschip, M.: Performance evaluation of ant colony systems for the single-depot multiple traveling salesman problem. In: Onieva, E., Santos, I., Osaba, E., Quintián, H., Corchado, E. (eds.) HAIS 2015. LNCS (LNAI), vol. 9121, pp. 257–268. Springer, Cham (2015). https://doi.org/10.1007/978-3-319-19644-2_22
30. Fages, J.G., Prud'Homme, C.: Making the first solution good! In: ICTAI 2017 29th IEEE International Conference on Tools with Artificial Intelligence (2017)

A Local Search Framework for Compiling Relaxed Decision Diagrams

Michael Römer[1,2,3]([✉]), Andre A. Cire[2], and Louis-Martin Rousseau[3]

[1] Institute of Information Systems and OR,
Martin Luther University Halle-Wittenberg, Halle, Germany
`michael.roemer@wiwi.uni-halle.de`
[2] Department of Management, University of Toronto Scarborough, Toronto, Canada
`acire@utsc.utoronto.ca`
[3] CIRRELT, École Polytechnique de Montréal, Montreal, Canada
`louis-martin-rousseau@polymtl.ca`

Abstract. This paper presents a local search framework for constructing and improving relaxed decision diagrams (DDs). The framework consists of a set of elementary DD manipulation operations including a redirect operation introduced in this paper and a general algorithmic scheme. We show that the framework can be used to reproduce several standard DD compilation schemes and to create new compilation and improvement strategies. In computational experiments for the 0–1 knapsack problem, the multidimensional knapsack problem and the set covering problem we compare different compilation methods. It turns out that a new strategy based on the local search framework consistently yields better bounds, in many cases far better bounds, for limited-width DDs than previously published heuristic strategies.

1 Introduction

Relaxed decision diagrams are pivotal components in the use of decision diagrams for optimization [6]. In particular, they provide an adjustable approximation of the solution space of a discrete optimization problem, supplying optimization bounds for combinatorial problems [4,5,8] as well as serving as a constraint store in constraint programming approaches [2,9]. The strength of these bounds, and the speed at which they can be generated, are critical for the success of this area.

Previous methods for compiling relaxed decision diagrams, such as incremental refinement [9] and top-down-merging [8], can be viewed as construction heuristics that stop when a given limit on the size of the diagram is reached. Typically these approaches emphasize the quality of the resulting optimization bound by either (i) determining a good variable ordering [4]; (ii) by heuristically selecting nodes to split [9]; or (iii) by heuristically selecting nodes to merge [8]. Recently, however, Bergman and Cire [7] proposed to consider the problem of compiling a relaxed decision diagram as an optimization problem, specifically by considering a mixed-integer linear programming formulation. While this approach may be useful for benchmarking heuristic compilation methods, its computational costs are too high for any practical application.

Contributions. We present a local search framework that serves as a general scheme for the design of relaxed decision diagram compilation strategies. In particular, we focus on obtaining strong bounds within an acceptable computation time. As in local search methods for combinatorial optimization (see, e.g., [1]), the key ingredients of the framework are a set of "local" operations for obtaining new diagrams from modifications of other diagrams (similar to the concept of *neighborhood*), as well as strategies for guiding the local search. We identify three elementary local operations, from which two (node splitting and merging) have been used in previous works, and one (arc redirection) is introduced in this paper. We demonstrate that several published compilation methods can be cast in our local search framework, and demonstrate how it allows the design of new compilation strategies for relaxed diagrams. In this context, we propose a novel compilation strategy and provide a set of computational experiments with instances of the 0–1 knapsack problem, the multidimensional knapsack problem, and the set covering problem. The new compilation strategy can lead to considerably stronger relaxations than those obtained with standard techniques in the literature, often with faster computational times.

The remainder of the paper is organized as follows. In Sect. 2 we introduce decision diagrams and the notation used in this paper. Section 3 describes a set of local operations, and Sect. 4 presents the generic algorithmic scheme along with a new compilation strategy. Section 5 discusses a preliminary experimental evaluation, and a conclusion is provided in Sect. 6.

2   Preliminaries

A decision diagram (DD) $\mathcal{M} = (\mathcal{N}, \mathcal{A})$ is a layered acyclic arc-weighted digraph with node set $\mathcal{N}$ and arc set $\mathcal{A}$. The paths in $\mathcal{M}$ encode solutions to a discrete optimization problem associated with a maximization objective and an n-dimensional vector of decision variables $x_1, \ldots, x_n \in \mathbb{Z}$. To this end, the node set $\mathcal{N}$ is partitioned into $n+1$ layers $L_1, \ldots, L_{n+1}$, where $L_1 = \{\mathbf{r}\}$ and $L_{n+1} = \{\mathbf{t}\}$ for a *root node* $\mathbf{r}$ and a *terminal node* $\mathbf{t}$. Each node $u \in \mathcal{N}$ belongs to the layer $\ell(u) \in \{1, \ldots, n+1\}$, i.e., $L_i = \{u \mid \ell(u) = i\}$. An arc $a = (u^s(a), u^t(a))$ has a *source* $u^s(a)$ and a *target* $u^t(a)$ with $\ell(u^t(a)) - \ell(u^s(a)) = 1$, i.e., arcs connect nodes in adjacent layers. Each arc a is associated with a value $d(a)$ which represents the assignment $x_{u^s(a)} = d(a)$. Thus, an arc-specified path $p = (a_1, \ldots, a_n)$ starting from $\mathbf{r}$ and ending at $\mathbf{t}$ encodes the solution $x(p) = (d(a_1), \ldots, d(a_n))$. Moreover, each arc a has length $v(a)$; $\sum_{i=1}^{n} v(a_i)$ provides the length of path p.

A DD $\mathcal{M}$ is *relaxed* with respect to a discrete maximization problem if (i) every feasible solution to the problem is associated with some path in $\mathcal{M}$; and (ii) the length of a path p is an upper bound to the objective function value of $x(p)$. The longest path in $\mathcal{M}$ therefore provides an upper bound to the optimal solution value of the discrete optimization problem.

We consider the framework proposed by Bergman et al. [5] for manipulating relaxed DDs. Namely, the discrete problem is formulated as a dynamic program, where each node $u \in \mathcal{N}$ refers to a state $s(u)$, an arc a represents a transition from a source state $s^s(a) := s(u^s(a))$ to a target state $s^t(a) := s(u^t(a))$ according to the action $d(a)$, and an arc length $v(a)$ represents the transition reward of a. If during construction the DD exceeds a maximum number of nodes per layer (i.e., its maximum *width*), nodes are merged and their states according to a problem-specific merge operator $\oplus$. Such an operator must ensure that the resulting $\mathcal{M}$ is a valid relaxation (we refer to examples of operators in [5]).

Example 1. The DDs in Fig. 1 are examples of relaxed DD for the knapsack problem $\max\{4x_1 + 3x_2 + 2x_3 : 3x_1 + 2x_2 + 2x_3 \le 5, x \in \{0,1\}^3\}$. The dashed and solid arcs represent arc values 0 and 1, respectively. Arc lengths are depicted by the number above the arcs; the dashed arcs always have the length 0. The longest paths of the diagrams are represented by the quantity "LP" below each figure. We also depict the state of each node within the circles, in this case the current weight of the knapsack; the merge operator applied in the example is the minimum operator.

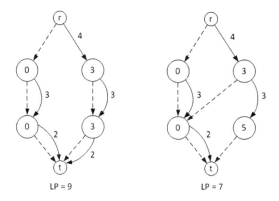

Fig. 1. Example DDs for knapsack problem illustrating operation REDIRECTARC.

3   Local Operations on Decision Diagrams

The standard compilation procedures for relaxed decision diagrams can be alternatively viewed as a sequence of elementary local operations applied to a given DD $\mathcal{M}$. For the purposes of our framework, we define three of such operations below. The first two can be found in the literature, while the third operation is proposed in this paper.

SplitNode. The operation SPLITNODE is key in the incremental refinement algorithm for compiling relaxed DDs proposed in [9]. Such an algorithm starts with a trivial relaxed DD with one node per layer. It then *splits* nodes one by one so as to improve the approximation, observing that the maximum width is satisfied.

Formally, the operation splits a node $u \in \mathcal{N}$ according to a given partition $\{E_1, \ldots, E_m\}$ of the incoming arcs at u. First, $m-1$ new nodes are created on layer $\ell(u)$ of $\mathcal{M}$. Then, the arcs in each of the sets E_2 to E_m are redirected to one of the new nodes; the arcs in E_1 remain pointing to u. Finally, the outgoing arcs of node u are copied to the new nodes, that is, for each out-arc a of u, an out-arc a' is added to each new node u' with $u^t(a') = u^t(a)$ and $v(a') = v(a)$. This operation increases the width of a layer by $m - 1$.

MergeNodes. The operation MERGENODES is key to the top-down compilation algorithm proposed in [8]. Such an algorithm constructs a DD $\mathcal{M}$ one layer at a time, starting at L_1 and expanding layer i before starting layer $i + 1$, for $i = 2, \ldots, n$. If layer i exceeds the maximum width, nodes are merged and their states are replaced according to the operator $\oplus$ as described previously.

Formally, the operation merges a set of nodes $U = \{u_1, \ldots, u_m\}$ on the same layer l by first redirecting all incoming arcs from the nodes $U \setminus u_1$ to node u_1, and by then deleting the nodes in $U \setminus \{u_1\}$ along with their outgoing arcs. The operation MERGENODES reduces the width of a layer by $m - 1$.

RedirectArc. The operation REDIRECTARC changes the target $u^t(a)$ of a given arc a to an existing node u on the same layer, that is, $\ell(u) = \ell(u^t(a))$. The number of nodes in a decision diagram remains unchanged.

Intuitively, redirecting an arc a may strengthen the information at its initial target node since it reduces the number of states that are relaxed in that node. Nonetheless, it may further relax its new target since it adds an incoming arc to u, thereby forcing an additional application of $\oplus$. In general, carefully selecting the target node u for the redirection is crucial for the efficiency of this operation. Given that u is selected from w nodes in the target layer, the complexity of REDIRECTARC is $O(w) *$ COMPSTATES where COMPSTATES is a state comparison operation.

Example 2. Figure 1 depicts an example of arc redirection and its impact on the quality of the bound. In particular, the dashed arc on the right-most node of layer L_2 is redirected, which changes the state of the node and the underlying cost of the longest path.

Some notes concerning the operations above are in order. Applying a local manipulation to a DD induces changes to the states of the involved nodes, as well as on the nodes of subsequent layers. On the one hand, the new target states may become infeasible with respect to the underlying dynamic program, in which case the corresponding arcs are removed from the DD (i.e., *filtered*). On the other hand, the converse is also true: After a REDIRECTARC, arcs that were filtered before may be added back due to the new state of the target node, so as to ensure that the modified DD is still a relaxation. These updating operations may be computationally expensive, thereby implying a trade-off between the frequency of the updates and the quality of the relaxation one obtains.

4   Generic Local Search Scheme

We propose a local search framework for constructing relaxed DDs which alternates local operations with node state updates. Namely, any valid compilation method is defined as a strategy that combines the operations described in Sect. 3 in a systematic way. This is typically written as an iterative procedure that can be repeated until a stopping criterion is reached. To provide concrete examples, we now show how three DD compilation methods from the literature can be reproduced within this perspective. In addition, we present a new compilation strategy using the operation REDIRECTARC that may improve the bound of a given DD without affecting its maximum width.

As a first example, the incremental refinement algorithm presented in [9] is an iterative sequence of SPLITNODES, where the incoming arcs of a node are partitioned according to a *state distance threshold* that decreases in each main iteration. A filtering procedure is called after each sequence of modifications.

As a second example, consider the top-down merging algorithm proposed in [8]. The local search framework can be employed to reproduce its behavior as follows. Starting with a trivial one-width relaxed DD, the algorithm proceeds one layer at a time beginning at L_1. The single node on each layer is first completely split by consecutive applications of the operation SPLITNODE. If the resulting number of nodes exceeds the maximum width, the operation MERGENODES is applied by selecting nodes heuristically. Since this algorithm proceeds top-down, no filtering is needed – a fact contributing to its time efficiency.

As a third example, the longest path trimming algorithm presented in [3] can also be expressed in terms of the local search framework. The procedure starts with a one-width DD. The SPLITNODE operation is then applied on the longest path of the diagram in a top-down order. That is, the procedure partitions the incoming arcs of each node into two sets: The first set contains the incoming arcs that belongs to a longest path, and the second set contains all remaining arcs. Finally, the algorithm removes the last arc on the (possibly infeasible) longest path pointing to the terminal node; if otherwise the longest path is feasible, the associated solution is optimal. No state updates are performed.

An important characteristic of all three algorithms above is that they are *constructive heuristics*, i.e., they are performed until a certain limit on the DD size is reached. In order to obtain an *improvement* heuristic, one needs to design a local search strategy in which the size of the DD does not increase. Given the local operations described above, we present a new compilation strategy in Algorithm 1 that employs all three operations SPLITNODE, MERGENODE, and REDIRECTARC for this purpose.

Algorithm 1 first computes a set U of nodes on the longest path that may be subject to a split operation. A node is u is a candidate for a split operation if splitting results in at least one new node u' for which $s(u') \neq s(u)$. If the longest path does not contain any such node, it defines an optimal solution to the optimization problem. Otherwise, three steps are considered: The application of local operations on the nodes on the longest path, the state updates, and the re-computation of the longest path. In particular, the sequence of local operations can be summarized in terms of the combined operations SPLITANDREDIRECT and MERGESPLITANDREDIRECT, described below.

Algorithm 1. Longest Path Splitting and Redirecting

Data: Decision Diagram $\mathcal{M}$
Result: Manipulated Decision Diagram $\mathcal{M}$

1   Compute the set U of nodes on the longest path to be split
2   **while** U *is not empty and time limit not reached* **do**
3   **for** $u \in U$ *ordered by layer* **do**
4   **if** $width(\ell(u)) < maximum\ width$ **then**
5   SPLITANDREDIRECT$(u, \mathcal{M})$
6   **else**
7   MERGESPLITANDREDIRECT$(u, \mathcal{M})$
8   Update $\mathcal{M}$
9   Compute the set U of nodes on the longest path to be split

---

SplitAndRedirect. This combined operation first applies SPLITNODE to a given node u. Namely, the incoming arcs of u are partitioned into two subsets: One subset defined by the single arc that belongs to the longest path to u, and another subset containing the remaining arcs. For each outgoing arc of the new node u' with the longest-path incoming arc, the operation REDIRECTARC is executed. Different strategies for selecting the new target node of such arc can be applied. For instance, we can consider all nodes for which the addition of the new incoming arc does not modify the associated state. Alternatively, one could also select a node that results in the smallest state change.

The rationale behind this redirection is that, after the split operation, the new node u' traversed by the longest path is likely to have a considerably different state than the original node u. In such a case the target nodes of the outgoing arcs of the original node u may not reflect the states resulting from applying the transitions associated with the outgoing arcs of u', and thus, it is effective to find a target node that "better reflects" those target states.

MergeSplitAndRedirect. The operation MERGESPLITANDREDIRECT is applied when the limit on the width w in layer $\ell(u)$ of a node u has already been reached. In such a case, the procedure searches for a pair of nodes in layer $\ell(u)$ which can be merged without increasing the value of the longest path. If such a pair has been found, these nodes are merged and the operation SPLITANDREDIRECT is applied to node u; otherwise neither the merge nor the split is performed.

Finally, we note that the strategies above can be changed in a straightforward way to consider a maximum number of nodes in $\mathcal{M}$ instead of a maximum width as a stopping criterion.

5   Experimental Results

In this section, we evaluate of our new local search strategy in instances of the 0–1 knapsack problem, multidimensional knapsack, and set covering. All experiments were run on an Intel Core i5 with 12 GB RAM, single core, implemented in C++.

For the 0–1 knapsack problem (KP), we apply it to the same 180 instances used in [7], consisting of 15 and 20 items and varying ratios between the knapsack right-hand side and the sum of item weights. Figure 2 depicts the average percentage optimality gap (i.e., *(upper bound - lower bound)/lower bound*) obtained with different DD construction approaches for a maximum width of 8. The x-axis indicates the knapsack ratio parameter. We compared the IP-based approach (IP) by [7], the proposed local search heuristic (LS), the top-down merging (MinLP) from [8] based on longest paths, and the best bound from 50 random DD constructions (MinRANDOM). The figure suggests that LS yields much stronger bounds than the standard approaches and, in almost all cases, it even results in better bounds than the exact IP approach after 1,800 s. In about 50% of the instances for which the IP was solved to optimality, LS also found an optimal solution. The local search takes less than half a second to be performed.

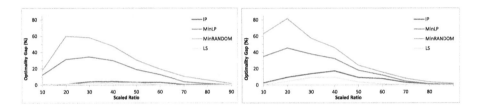

Fig. 2. Percentage gap × Scaled ratio ($r \times 10$) for $|\mathcal{I}| = 15$ (left) and $|\mathcal{I}| = 20$ (right) for a maximum width of 8

We now report results from additional experiments with KP, the multi-dimensional knapsack problem (MKP), and the set covering problem (SCP). For each problem class, the results depicted in Fig. 3 are averages over 15 instances obtained as follows: For the KP, we generated 15 instances uniformly at random with 1,000 items and with three different ratios $(0.25, 0.5, 0.75)$. For the MKP, we used 15 ORlib instances with 100 items, 5 dimensions and with three different ratios $(0.25, 0.5, 0.75)$. For the SCP, we generated 15 instances uniformly at random with 2,000 variables, 150 constraints, and with three different bandwidths. In the experiments, we compare four different compilation strategies using a time limit of two minutes: The top-down merging (TD-M) from [8], a variant of the longest-path-trimming strategy (LP-Trim) proposed in [3], and two variants of the longest-path-splitting-and-redirecting strategy proposed in this paper: The first variant (LP-SR) does not try to improve a DD having reached the maximum width, while the second (LP-MSR) does.

The left-hand side of Fig. 3 shows the percentage optimality gap obtained for different maximum DD widths. For both the KP and the MKP instances, the figure suggests that the bound obtained with the new local search strategies are far superior to the bounds obtained with the existing algorithms, while this difference is much smaller for the SCP instances. The right-hand side shows the development of the bounds over time. The plot illustrates that for the local search strategies, the rate of bound improvement decreases over time, but a significant improvement can be obtained in a short amount of time. In addition, the top-down merging (single blue dot) is fast compared to the strategies based on the longest path, possibly since this strategy does not require any filtering or state update mechanism.

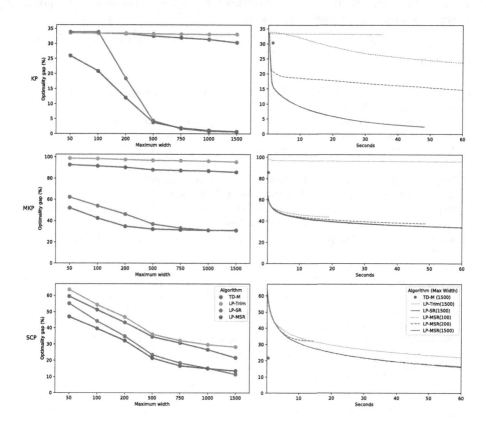

Fig. 3. Bound vs width and time averaged over instance sets for three problem classes (Color figure online)

6   Conclusions and Future Work

This paper presents the first generic framework for designing local search strategies for compiling relaxed DDs. We describe a set of local operations that can be performed within a DD, showing that several strategies from the literature can be perceived as a combination of these different operations. We leverage this new framework to present a new strategy based on an arc redirection operation, also novel to the area. Computational results with three different problem classes show that the new strategy often yields much superior bounds than previously published strategies.

This paper provides the basis for several future research opportunities. Since relaxed DDs are an important component of several constraint programming and optimization approaches, providing better bounds may make these approaches more efficient. Furthermore, the new search strategy proposed in this paper is relatively simple in terms of how the local operations are combined. More sophisticated strategies, e.g., based on meta-heuristics such as variable neighborhood search, are promising research directions.

References

1. Aarts, E., Lenstra, J.K. (eds.): Local Search in Combinatorial Optimization, 1st edn. Wiley, New York (1997)
2. Andersen, H.R., Hadzic, T., Hooker, J.N., Tiedemann, P.: A constraint store based on multivalued decision diagrams. In: Bessière, C. (ed.) CP 2007. LNCS, vol. 4741, pp. 118–132. Springer, Heidelberg (2007). https://doi.org/10.1007/978-3-540-74970-7_11
3. Bergman, D., Cire, A.A.: Theoretical insights and algorithmic tools for decision diagram-based optimization. Constraints **21**, 533–556 (2016)
4. Bergman, D., Cire, A.A., van Hoeve, W.-J., Hooker, J.N.: Variable ordering for the application of BDDs to the maximum independent set problem. In: Beldiceanu, N., Jussien, N., Pinson, É. (eds.) CPAIOR 2012. LNCS, vol. 7298, pp. 34–49. Springer, Heidelberg (2012). https://doi.org/10.1007/978-3-642-29828-8_3
5. Bergman, D., Cire, A.A., van Hoeve, W.J., Hooker, J.N.: Discrete optimization with decision diagrams. INFORMS J. Comput. **28**(1), 47–66 (2016)
6. Bergman, D., Cire, A.A., van Hoeve, W.J., Hooker, J.: Decision Diagrams for Optimization. Artificial Intelligence: Foundations, Theory, and Algorithms. Springer, Cham (2016). https://doi.org/10.1007/978-3-319-42849-9
7. Bergman, D., Cire, A.A.: On finding the optimal BDD relaxation. In: Salvagnin, D., Lombardi, M. (eds.) CPAIOR 2017. LNCS, vol. 10335, pp. 41–50. Springer, Cham (2017). https://doi.org/10.1007/978-3-319-59776-8_4
8. Bergman, D., van Hoeve, W.-J., Hooker, J.N.: Manipulating MDD relaxations for combinatorial optimization. In: Achterberg, T., Beck, J.C. (eds.) CPAIOR 2011. LNCS, vol. 6697, pp. 20–35. Springer, Heidelberg (2011). https://doi.org/10.1007/978-3-642-21311-3_5
9. Hadzic, T., Hooker, J.N., O'Sullivan, B., Tiedemann, P.: Approximate compilation of constraints into multivalued decision diagrams. In: Stuckey, P.J. (ed.) CP 2008. LNCS, vol. 5202, pp. 448–462. Springer, Heidelberg (2008). https://doi.org/10.1007/978-3-540-85958-1_30

Symmetry Breaking Inequalities
from the Schreier-Sims Table

Domenico Salvagnin[(✉)]

Department of Information Engineering (DEI), University of Padova, Padua, Italy
domenico.salvagnin@unipd.it

Abstract. We propose a way to derive symmetry breaking inequalities for a mixed-integer programming (MIP) model from the Schreier-Sims table of its formulation group. We then show how to consider only the action of the formulation group onto a subset of the variables. Computational results show that this can lead to considerable speedups on some classes of models.

1 Motivation

An optimization problem is symmetric if its variables can be permuted without changing the structure of the problem. Even for relatively small cases, symmetric optimization problems can be difficult to solve to proven optimality by traditional enumerative algorithms, as many subproblems in the enumeration tree are isomorphic, forcing a wasteful duplication of effort. Symmetry has long been recognized as a challenge for exact methods in constraint and integer programming, and many different methods have been proposed in the literature, see for example the surveys in [6,14].

A common source of symmetry in a model is the underlying presence of identical objects. Let's consider a MIP model in which a subset of variables, say $x_1, \ldots, x_k$, corresponds to k identical objects, e.g., the k colors in a classical graph coloring model [8]. A natural way to get rid of the symmetry implied by those objects, which is well known and widely used by modelers, is to add to the formulation the following chain of static symmetry breaking inequalities:

$$x_1 \geq x_2 \geq \ldots \geq x_k \qquad (1)$$

The validity of (4) can be easily proved as follows. If the k objects are identical, then G acts on $x_1, \ldots, x_k$ as the full symmetric group S_k, and thus all those variables are in the same orbit. As such, we can always permute the variables such that the first is not less than the others, which means adding the inequalities $x_1 \geq x_i$ for all $i > 1$. This effectively singles out variable x_1. Still, what is left is the full symmetric group S_{k-1} on the $k-1$ variables $x_2, \ldots, x_k$. We can apply the very same reasoning and add the inequalities $x_2 \geq x_i$ for all $i > 2$. Repeating the same argument till the bitter end we get that we added all inequalities of the form $x_i \geq x_j$ for all $i > j$, which is equivalent to the chain above after removing the redundant ones.

© Springer International Publishing AG, part of Springer Nature 2018
W.-J. van Hoeve (Ed.): CPAIOR 2018, LNCS 10848, pp. 521–529, 2018.
https://doi.org/10.1007/978-3-319-93031-2_37

The effectiveness of such a simple symmetry handling technique can vary significantly. In particular, it depends on the distribution of the feasible solutions of the model: if many symmetric solutions lie on the hyperplanes $x_i = x_{i+1}$, then the method is usually rather ineffective, and not competitive with other more elaborate symmetry handling techniques, like isomorphism pruning [12,13] and orbital branching [15]. On the other hand, if there are no solutions with $x_i = x_{i+1}$, e.g. because the variables are linked by some all-different constraint, then (4) is equivalent to a chain of strict inequalities, and all symmetry in the model is effectively broken [17].

Example 1. Let's consider a simple 2D packing model[1], in which we have k squares of size 10×10 and a container of size $10(k - 1) + 1 \times 11$. As no two squares can fit vertically in the container, clearly only $k - 1$ squares can be put into the container. A typical MIP model for this problem has a pair of continuous variables (x_i, y_i) for each square, encoding the coordinates of, say, the lower-left corner of the squares, plus $\Theta(k^2)$ binary variables to encode the non-overlapping constraints among squares. Now, because of the shape of the container, we can assume that w.l.o.g. $y_i = 0$ for all squares in any feasible solution, while clearly $x_i \neq x_j$ for any two squares in any feasible solution. So adding the symmetry breaking inequalities $x_1 \geq x_2 \geq \ldots \geq x_k$ effectively removes all the symmetries from the formulation—and indeed works very well in practice with most solvers— while adding the chain $y_1 \geq y_2 \geq \ldots \geq y_k$ is totally ineffective, and actually harms the solution process as it still destroys the symmetry in the formulation, preventing other symmetry handling methods to kick in.   □

The example above confirms that inequalities (4) might or might not be an effective way to deal with symmetry in MIP. Still, there are cases in which they outperform all other symmetry handling techniques, so the question is: can we derive those static symmetry breaking inequalities from the model *automatically*? The answer is positive, as it turns out that inequalities (4) are a special case of a wider class of symmetry breaking inequalities that can be derived from the so-called *Schreier-Sims table* [19], a basic tool in computational group theory.

The outline of the paper is as follows. In Sect. 2 we review the basic concepts of group theory needed for our discussion, and define the Schreier-Sims representation of a group. In Sect. 3 we show how to use the Schreier-Sims table to derive symmetry breaking inequalities, and present some extensions/improvements over the basic method in Sect. 4. In Sect. 5 we outline an algorithm to actually compute the table. Computational results are given in Sect. 6, with conclusions and future directions of research drawn in Sect. 7.

2   The Schreier-Sims Table

We follow the description of the Schreier-Sims [18,19] representation given in [12]. Let G be a permutation group on the ground set $N = \{1, \ldots, n\}$. A

---

[1] This is a much simplified version of the `pigeon` models [2] in MIPLIB 2010 [9].

permutation $g \in G$ is represented by an n-vector, with $g[i]$ being the image of i under g. Consider the following chain of (nested) subgroups of G:

$$G_0 = G$$
$$G_1 = \{g \in G_0 | \ g[1] = 1\}$$
$$G_2 = \{g \in G_1 | \ g[2] = 2\}$$
$$\dots$$
$$G_n = \{g \in G_{n-1} | \ g[n] = n\} \tag{2}$$

In other words, G_i is the stabilizer of i in G_{i-1}. Note that no such subgroup is empty, as the identity permutation is always contained in all G_i. For each $i = 1, \dots, n$, let $orb(i, G_{i-1}) = \{j_1, \dots, j_p\}$ be the orbit of i under G_{i-1}, i.e., the set of elements onto which i can be mapped according to G_{i-1}. Note that the orbit is never empty, as it always contains at least i. By definition, for each element j_k of the orbit there exists a permutation in G_{i-1} mapping i to j_k, and let h_{i,j_k} be any such permutation. Let $U_i = \{h_{i,j_1}, \dots, h_{i,j_p}\}$ be the set of these permutations, called *coset* representatives. Again, U_i is never empty. We can arrange the permutations in the sets U_i in an $n \times n$ table T, with:

$$T_{i,j} = \begin{cases} h_{i,j} & \text{if } j \in orb(i, G_{i-1}) \\ \emptyset & \text{otherwise} \end{cases} \tag{3}$$

The table T is called the Schreier-Sims representation of G. The most basic property of the table is that the set of permutations stored in the table form a set of strong generators for the group G, i.e., any permutation of $g \in G$ can be expressed as a product of at most n permutations in the set. It is also worth noting that the Schreier-Sims table not only provides a set of strong generators for G but also for all the nested subgroups G_i: indeed, a set of strong generators for G_i is obtained by taking all permutations in the table with row index $k \geq i$.

Example 2. Let's consider the simple permutation group G of the symmetries of the 2×2 square, with cells numbered top to bottom and left to right. G contains 8 permutations. The corresponding Schreier-Sims table is depicted below, where each permutation is written in cycle notation, and i is the identity. Note that G is not equal to S_4 because, e.g., no permutation exists maps cell 2 to cell 4 without affecting cell 1 as well.

	1	2	3	4
1	i	$(1\ 2)(3\ 4)$	$(1\ 3)(2\ 4)$	$(1\ 4)(3\ 2)$
2		i	$(2\ 3)$	
3			i	
4				i

$\square$

Note that the Schreier-Sims table is always upper triangular, and it is fully dense iff $G = S_n$, so constructing the table is sufficient to detect whether a group G is the full symmetric group S_n. Given an arbitrary set of generators for G, the Schreier-Sims table can be constructed in polynomial time (more details are given in Sect. 5).

3   Deriving Symmetry Breaking Inequalities

Consider a MIP model P with n variables and let $G = \langle g_1, \ldots, g_r \rangle$ be its formulation symmetry group, i.e., the group of permutations of the variables that lead to an equivalent formulation, see [11,12] for a formal definition. Let T be the Schreier-Sims representation of G. Then it follows that:

Theorem 1. *The set of symmetry breaking inequalities $x_i \geq x_j$ for all $T_{ij} \neq \emptyset$ is valid for P.*

Proof. The proof is a simple generalization of the argument used to prove the validity of chain (4). Let us consider the orbit O_1 of variable x_1 according to G. By definition we can always permute the variables such that x_1 takes a value which is no less than the values taken by the other variables in the orbit, so the set of inequalities corresponding to the first row of T, namely $x_1 \geq x_j$ $\forall x_j \in O_1$ is valid for P. Now, let's add those inequalities to the model. The formulation group of the resulting model contains G_1, i.e., the stabilizer of x_1 in G, so we can proceed to the second row of the table, which gives exactly the orbit of x_2 in G_1. Thus we can reiterate the argument and conclude that the set of inequalities corresponding to the second row of T is valid for P. By induction we can continue until the very last row of T, which proves the theorem. □

It is worth noting that the addition of symmetry breaking inequalities can in principle result in new symmetries in the formulation, as shown by the following example:

Example 3. Consider the LP:

$$\min\{x_1 + x_2 + x_3 + x_4 : x_3 - x_4 \geq 0\} \tag{4}$$

The corresponding formulation group has only one symmetry, namely $(x_1\ x_2)$. However, adding $x_1 - x_2 \geq 0$ we get the additional symmetry $(x_1\ x_3)(x_2\ x_4)$, while the stabilizer of x_1 according to G would contain the identity permutation only. □

Note that Theorem 1 only proves that we can derive a valid set of symmetry breaking inequalities from the Schreier-Sims table T, but not that the inequalities above are in general sufficient to break all the symmetries in the model. Indeed, the latter statement would be false in general. What we can state however is that (i) adding those inequalities breaks all symmetries in the *original* formulation, and (ii) all solution symmetries of the original formulation are broken if the variables of the model are linked by an all-different constraint, a result already proved in [17]. The fact that in any case formulation symmetries are broken is a double-edged sword: if solution symmetries are also broken then everything is fine, otherwise the addition of those inequalities is not only ineffective but also prevents other methods from being applied, as they would find no (or very little, see Example 3) symmetries to exploit, as shown in Example 1.

4   Improvements

Suppose we are interested in how the formulation group G acts on a subset T of variables of the model. For example, we might want to check whether G acts as $S_{|T|}$ on T, despite G possibly not being S_n. This can easily be achieved by a small extension of the Schreier-Sims construction, in which we do not consider the variables in order from x_1 to x_n when constructing the table, but in a different order, say β, such that the variables in T are considered first. Such order β is called the *base* of the table, and the construction can easily be extended to deal with an arbitrary base. Once the table is constructed, then we can conclude that G acts as $S_{|T|}$ on T iff the upper left $|T| \times |T|$ submatrix of T is (upper triangular) fully dense.

Constructing the complete Schreier-Sims table of order n when we are actually interested only in its upper left corner of size $|T| \times |T|$ can potentially be a big waste of computing resources. For example, in Example 1, the model has size $\Theta(k^2)$, while the continuous variables that encode the placing of each object are only $O(k)$. In general the full computation is needed if the set T has no structure. However, if we assume that T is an orbit according to the original group G, then we have a much better alternative: intuitively, we can project the generators of G and work with a new group G_T whose ground set is just T. Then we can construct the Schreier-Sims table of G_T which is exactly of size $|T| \times |T|$. Let us formalize this argument.

Any generator g of G (as any permutation for that matter), can be written in cycle notation. Because of our choice of T, by construction all cycles in g either move only variables in T or only variables in $N \setminus T$, as there is no permutation in G moving an element from T into $N \setminus T$, otherwise T would not be an orbit.

Define the operator $\varphi : S_n \leftrightarrow S_{|T|}$ as the operator that drops from a permutation written in cycle notation all the cycles not in T. For example, if $g = (13)(25)(789)$ and $T = \{1, 2, 3, 4, 5\}$, then $\varphi(g) = (13)(25)$.

Let $t_1, \ldots, t_r$ be the permutations obtained by applying φ to the generators of G and let $G_T = \langle t_1, \ldots, t_r \rangle$. It is not difficult to prove that φ is a homomorphism from G to G_T: let $a = \gamma_1 \cdots \gamma_k \delta_1 \ldots \delta_p$ and $b = \sigma_1 \cdots \sigma_l \omega_1 \ldots \omega_q$, where we used γ and σ to indicate the cycles moving variables in T and δ and ω to indicate the cycles moving variables in $N \setminus T$ (and we can always write a and b into this form as the cycles can be written down in any order). Then:

$$\varphi(ab) = \varphi(\gamma_1 \cdots \gamma_k \delta_1 \ldots \delta_p \sigma_1 \cdots \sigma_l \omega_1 \ldots \omega_q) \tag{5}$$

$$= \varphi(\gamma_1 \cdots \gamma_k \sigma_1 \cdots \sigma_l \delta_1 \ldots \delta_p \omega_1 \ldots \omega_q) \tag{6}$$

$$= \gamma_1 \cdots \gamma_k \sigma_1 \cdots \sigma_l \tag{7}$$

$$= \varphi(a)\varphi(b) \tag{8}$$

where the first rearrangement of the cycles is allowed because they are disjoint.

In addition, as a homomorphism from G to G_T, φ is surjective. Indeed, let π be a permutation in G_T. By definition it can be obtained by the generators of G_T and their inverses. But for each generator t_i of G_T we know a permutation

h of G such that $\varphi(h) = t_i$, namely $h = g_i$ and the same is true for the inverses, because $\varphi(g_i^-) = t_i^-$. Thus we can always construct a permutation $g \in G$ such that $\varphi(g) = \pi$. For example, let $\pi = t_1 t_2^- t_5$. Then $g = g_1 g_2^- g_5$.

Thus, by working directly with the group G_T we are not introducing (nor removing) any symmetry *among the variables in T* that was not already in G, hence we can use G_T to study how G acts on T. The results still holds if T is not just a single orbit but a union of orbits of G.

5   Constructing the Schreier-Sims Table

A recursive algorithm to compute the Schreier-Sims table is described in [10], and used in [12,13]. However, in our computational experience, we found a different iterative algorithm to perform better in practice. The iterative algorithm constructs the Schreier-Sims table one row at the time, and works as follows. At any given iteration i, the algorithm assumes that a set of generators for G_{i-1} is readily available (this condition is trivially satisfied for the first row, where we can just use the generators of G). Then, it computes the orbit O_i and the set of coset representatives U_i for element i. This is a basic algorithm in computational group theory, called Schreier vector construction [3]. Note that this is enough to fill row i of the table. Then we need to compute the generators for G_i, in order to be ready for the next iteration. This is achieved in two steps:

1. Compute a set of generators for G_i applying the Schreier's lemma. In details, given $G_{i-1} = \langle g_1, \ldots, g_r \rangle$ and the coset representatives $U_i = \{r_1, \ldots, r_k\}$, we can obtain a set of generators for G_i as $\langle r_s^{-1} gr \rangle$, with $g \in G_i$, $r \in U_i$, and r_s chosen such that $(r_s^{-1} gr)[i] = i$.
2. Reduce the set of generators for G_i applying the Sims' filter. This leaves at most $O(n^2)$ generators for G_i. This is needed in order to obtain a polynomial algorithm for the Schreier-Sims table construction. Note that other filters are known, such as for example Jerrum's filter [4]. However, those are more complicated to implement.

The overall complexity of the construction is $O(n^6)$. As noted already in [12], an algorithm with a worst-case complexity of $O(n^6)$ might seem impractical even for reasonable values of n. However, we confirm that those bounds are very pessimistic and that the actual runtime of the algorithm was always negligible w.r.t. to the overall solution process. Still, care must be taken in the implementation, allowing the construction to be interrupted in case it becomes too time consuming.

6   Computational Results

We implemented the separation of the static symmetry breaking inequalities described in Theorem 1 during the development cycle between IBM ILOG CPLEX 12.7.0 and 12.7.1 [7]. In particular, at the end of presolve, we use the generators of the formulation group, freshly computed with AUTOM [16], to construct the Schreier-Sims table on the orbit of continuous variables with largest domain. While the approach can in principle be applied to binary and general integer variables as well, we decided to apply the method very conservatively. The choice of continuous variables with large domains is intuitively justified by the fact that it is "less likely" to have solutions lying on the $x_i = x_{i+1}$ in this case. If the table is sufficiently dense, we add the symmetry breaking inequalities and erase the generators (they are no longer valid), otherwise we forget about the table and continue.

We tested the method on the CPLEX internal testbed, which consists of approximately 3270 models, coming from a mix of publicly available and commercial sources. Tests were executed on a cluster of identical machines, each equipped with two Intel Xeon E5-2667v4 CPUs (for a total of 16 cores) running at 3.2 GHz, and 64 GB of RAM. Each run was given a time limit of 10.000 seconds. To limit the effect of performance variability [5,9], we compared the two methods, namely CPLEX defaults with (symbreak) and without (cpx) the addition of the symmetry breaking inequalities derived from the Schreier-Sims table, with 5 different random seeds. Aggregated results over the 5 seeds are given in Table 1.

The structure of the table is as follows. Instances are divided in different subsets, based on the *difficulty of the models*. To avoid any bias in the analysis, the level of difficulty is defined by taking into account both methods under comparison. The subclasses "[n, 10k}" ($n = 1, 10, 100, 1k$), contain the subset of models for which at least one of the methods took at least n seconds to solve and that were solved to optimality within the time limit by at least one of the methods. Finally, the subclasses "[n, 10k)" ($n = 1, 10, 100, 1k$) contain all models in "[n, 10k}" but considering only the models that were solved to optimality by both methods. The first column of the table identifies the class of models. Then the first group of 5 columns, under the heading "all models", reports results on all instances in the class, while the second group of columns, under the heading "affected", repeats the same information for the subset of models in each class where the two methods took a different solution path. Within each group, column "# models" reports the number of models in the class, columns "#tl" the number of time limits for each method, and columns "time" and "nodes" report the shifted geometric means [1] of the ratios of solution times and number of branch-and-bound nodes, respectively. Ratios $t < 1$ indicate a speedup factor of $1/t$.

According to Table 1, the symmetry breaking inequalities affect only around 2% of the models, which is not unexpected given the conservative criteria that trigger their generation. Still, they are so effective that they produce a non negligible speedup also on the whole testbed, with speedups ranging from 1% to 7% (for the subset of hard models in the "[100, 10k)" bracket). Also the number of time limits is significantly reduced. Aggregated results seed by seed (not reported) also confirm that the improvement is consistent across seeds.

Table 1. Aggregated results.

	all models					affected		
	cpx	symbreak				cpx	symbreak	
class	# models	# tl	# tl	time	nodes	# models	time	nodes
$[0, 10K\}$	16185	127	111	0.99	0.98	317	0.52	0.32
$[1, 10K\}$	9475	82	66	0.98	0.96	303	0.50	0.30
$[100, 10K\}$	2645	79	63	0.93	0.87	159	0.27	0.10
$[0, 1)$	6665	0	0	1.00	1.00	14	1.36	1.94
$[1, 10)$	3905	0	0	1.00	1.00	48	0.97	1.00
$[10, 100)$	2920	0	0	1.00	1.00	96	1.05	1.09
$[100, 1K)$	1765	0	0	0.99	0.97	86	0.73	0.51
$[1K, 10K)$	680	0	0	0.94	0.89	40	0.35	0.14

7   Conclusions

In this paper we investigated computationally the effectiveness of generating static symmetry breaking inequalities from the Schreier-Sims table of the formulation symmetry group. Computational results show that the approach can be extremely effective on some models. The technique is implemented and activated by default in the release 12.7.1 of the commercial solver IBM ILOG CPLEX. Future direction of research include extending the classes of models on which the method is tried, e.g., on pure binary models.

Acknowledgements. The author would like to thank Jean-François Puget for an inspiring discussion about the Schreier-Sims table, and three anonymous reviewers for their careful reading and constructive comments.

References

1. Achterberg, T.: Constraint integer programming. Ph.D thesis. Technische Universität Berlin (2007)
2. Allen, S.D., Burke, E.K., Marecek, J.: A space-indexed formulation of packing boxes into a larger box. Oper. Res. Lett. **40**, 20–24 (2012)
3. Butler, G., Cannon, J.J.: Computing in permutation and matrix groups I: normal closure, commutator subgroups, series. Math. Comput. **39**, 663–670 (1982)
4. Cameron, P.J.: Permutation Groups. London Mathematical Society St. Cambridge University Press, Cambridge (1999)
5. Danna, E.: Performance variability in mixed integer programming. In: MIP 2008 Workshop in New Work (2008). http://coral.ie.lehigh.edu/~jeff/mip-2008/talks/danna.pdf
6. Gent, I.P., Petrie, K.E., Puget, J.-F.: Symmetry in constraint programming. In: Rossi, F., van Beek, P., Walsh, T. (eds.) Handbook of Constraint Programming, pp. 329–376. Elsevier (2006)

7. IBM: ILOG CPLEX 12.7.1 User's Manual (2017)
8. Kaibel, V., Pfetsch, M.: Packing and partitioning orbitopes. Math. Program. **114**, 1–36 (2008)
9. Koch, T., Achterberg, T., Andersen, E., Bastert, O., Berthold, T., Bixby, R.E., Danna, E., Gamrath, G., Gleixner, A.M., Heinz, S., Lodi, A., Mittelmann, H., Ralphs, T., Salvagnin, D., Steffy, D.E., Wolter, K.: MIPLIB 2010 - mixed integer programming library version 5. Math. Program. Comput. **3**, 103–163 (2011)
10. Kreher, D.L., Stinson, D.R.: Combinatorial Algorithms: Generation, Enumeration, and Search. CRC Press, Boca Raton (1999)
11. Liberti, L.: Reformulations in mathematical programming: automatic symmetry detection and exploitation. Math. Program. **131**, 273–304 (2012)
12. Margot, F.: Pruning by isomorphism in branch-and-cut. Math. Program. **94**(1), 71–90 (2002)
13. Margot, F.: Exploiting orbits in symmetric ILP. Math. Program. **98**(1), 3–21 (2003)
14. Margot, F.: Symmetry in integer linear programming. In: Jünger, M., et al. (eds.) 50 Years of Integer Programming, pp. 647–686. Springer, Heidelberg (2009). https://doi.org/10.1007/978-3-540-68279-0_17
15. Ostrowski, J., Linderoth, J., Rossi, F., Smriglio, S.: Orbital branching. Math. Program. **126**(1), 147–178 (2011)
16. Puget, J.-F.: Automatic detection of variable and value symmetries. In: van Beek, P. (ed.) CP 2005. LNCS, vol. 3709, pp. 475–489. Springer, Heidelberg (2005). https://doi.org/10.1007/11564751_36
17. Puget, J.-F.: Breaking symmetries in all different problems. In: Kaelbling, L.P., Saffiotti, A. (eds.) IJCAI, pp. 272–277 (2005)
18. Seress, Á.: Permutation Group Algorithms. Cambridge University Press, Cambridge (2003)
19. Sims, C.C.: Computational methods in the study of permutation groups. In: Computational problems in abstract algebra (Oxford 1967), pp. 169–183. Pergamon Press, Oxford (1970)

Frequency-Based Multi-agent Patrolling Model and Its Area Partitioning Solution Method for Balanced Workload

Vourchteang Sea$^{(\boxtimes)}$, Ayumi Sugiyama, and Toshiharu Sugawara

Department of Computer Science and Communications Engineering,
Waseda University, Tokyo 169-8555, Japan
vourchteang@asagi.waseda.jp,
sugi.ayumi@ruri.waseda.jp, sugawara@waseda.jp

Abstract. Multi-agent patrolling problem has received growing attention from many researchers due to its wide range of potential applications. In realistic environment, e.g., security patrolling, each location has different visitation requirement according to the required security level. Therefore, a patrolling system with non-uniform visiting frequency is preferable. The difference in visiting frequency generally causes imbalanced workload amongst agents leading to inefficiency. This paper, thus, aims at partitioning a given area to balance agents' workload by considering that different visiting frequency and then generating route inside each sub-area. We formulate the problem of frequency-based multi-agent patrolling and propose its semi-optimal solution method, whose overall process consists of two steps – graph partitioning and sub-graph patrolling. Our work improve traditional k-means clustering algorithm by formulating a new objective function and combine it with simulated annealing – a useful tool for operations research. Experimental results illustrated the effectiveness and reasonable computational efficiency of our approach.

Keywords: Frequency-based patrolling · Graph partitioning
Balanced workload · Multi-agent systems · Linear programming
k-means based · Simulated annealing

1 Introduction

Recent advances on autonomous mobile robots have been evident in the last couple of decades. The patrolling problem with a team of agents, in particular, has received much focus. Patrolling refers to the act of continuously walking around and visiting the relevant area or important point of an environment, with some regularity/at regular intervals, in order to protect, navigate, monitor or supervise it. A group of agents is usually required to perform this task efficiently as multi-robot systems are generally believed to hold several advantages over single-robot systems. The most common motivation for developing multi-robot

© Springer International Publishing AG, part of Springer Nature 2018
W.-J. van Hoeve (Ed.): CPAIOR 2018, LNCS 10848, pp. 530–545, 2018.
https://doi.org/10.1007/978-3-319-93031-2_38

system solutions in the real-world applications is that a single robot cannot adequately deal with task complexities [9].

Multi-agent (multi-robot) patrolling, however, is not limited to patrolling real-world areas, but they can be found in applications on several domains, such as continuous sweeping, security patrolling, surveillance systems, network security systems and games. In other word, patrolling can be useful in any domain characterized by the need of systematically visiting a set of predefined points [17]. For instance, in many cases of real police works, there are services with human such as electronic security services [22]. The benefits of those systems are the cost-effectiveness against labor costs, and because it is monitored by sensors, visual overlook and human error are less likely to occur [20]. However, most of current studies assume that the frequency of visit to each node/location is uniform, yet in the realistic applications, the frequencies of visit differ; for example, in security patrolling, each location has different visitation requirement or risk status according to the required security level.

We divide multi-agent patrolling task into three steps: how to partition the work into a number of sub-works, how to allocate the individual sub-task to one of the agents and how to select the visiting sequence for each agent. We call this the *partition, allocation* and *sequencing* problem respectively. In this paper, we assume *homogeneous* agents that have the same capability and use the same algorithms. This assumption makes the allocation problem trivial, and thus, we only consider the algorithms for partitioning and sequencing. The combination of the partition algorithm and the sequencing algorithm is referred as a *strategy*.

In this paper, we will model the problem of patrolling as a problem of visiting vertices in a graph with visitation requirement by dividing it into a number of clusters. Then, after clustering nodes in this graph, each agent is responsible for patrolling the allocated cluster, and its nodes must be visited to meet the visitation requirement, that is, *frequency of visit*. In the partitioning step, we applied k-means based algorithm as a clustering algorithm by modifying its objective function and the initialization of centroids so as to make it fit to our problem. Our goal in this step is to cluster a given graph so that the potential workloads of individual clusters are balanced, which means trying to balance the workload amongst all agents. Moreover, the sequencing step addressed how to select the route (sequence of nodes) for each agent in its allocated cluster with a minimized cost. We used the *simulated annealing* (SA) here as a sequencing algorithm because our problem is similar to the multiple traveling salesman problem (mTSP), which is a generalization of the well-known traveling salesman problem (TSP) as mentioned in [2], and SA is often used to find the acceptable solutions due to the fact that SA is considered to be a flexible meta-heuristic method for solving a variety of combinatorial optimization problems. The difference between our problem and mTSP is that in mTSP, a number of cities have to be visited by m-salesman whose objective is to find m tours with minimum total travel, where all the cities must be visited exactly once, while in our problem, all the locations in a patrolled area must be visited to meet the required frequency of visit. We believe that our model of partitioning and sequencing with the frequency of visit to each node is more fit to realistic environment.

The contributions of our paper are three folds. First, we introduced the model of a frequency-based balanced patrolling problem for multi-agent systems to clarify our problem and requirement. Then, we developed an effective and scalable clustering algorithm based on the visitation requirement of each location by formulating a new k-means based approach for multi-agent patrolling systems; our main objective is to balance the workload amongst all patroller agents. Finally, we generated the route for each agent to patrol in its allocated region, in which the cost of visiting all nodes is minimized by taking into account the difference in each node's frequency of visit. We also demonstrated the computational efficiency of our proposed method that could be run in a short amount of time.

The remainder of this paper is organized as follows. We describe related work in the next section and introduce our problem formulation in Sect. 3. Section 4 explains our proposed method where agents firstly cluster a given graph/area by taking into account the non-uniform visitation requirement, and then find a route for patrolling with a minimized cost. We then show our experimental results in Sect. 5 indicating that agents with the proposed method achieves a computationally efficient and effective clustering in term of balancing the workload for multi-agent patrolling systems, and thus state our conclusion in Sect. 6.

2   Related Work

Multi-agent patrolling problem has been investigated and studied by many researchers. Initial researches [1,18,19] presented a theoretical analysis of various strategies for multi-agent patrolling systems and an overview of the recent advances in patrolling problems. Portugal and Rocha [10] proposed a multi-robot patrolling algorithm based on balanced graph partition, yet this paper did not consider when the required frequency of visit is not uniform. The same author, then, addressed a theoretical analysis of how two classical types of strategies, graph partition and cyclic-based techniques, perform in generic graphs [11]. A survey of multi-agent patrolling strategies can be found in [12], where strategies are evaluated based on robot perception, communication, coordination and decision-making capabilities.

I-Ming et al. [5] presented a heuristic for the team orienteering problem in which a competitor starts at a specified control point trying to visit as many other control points as possible within a fixed amount of time, and returns to a specified control point. The goal of orienteering is to maximize the total score of each control point, while in our patrolling problem, the main goal is to minimize the difference in workload amongst all patroller agents. Sak et al. [17] proposed a centralized solution for multi-agent patrolling systems by presenting three new metrics to evaluate the patrolling problem. Mihai-Ioan et al. [7] addressed the problem of multi-agent patrolling in wireless sensor networks by defining and formalizing the problem of vertex covering with bounded simple cycles (CBSC). This approach consequently considered polynomial-time algorithms to offer solutions for CBSC. Tao and Laura [16] investigated multi-agent frequency based patrolling in undirected circle graphs where graph nodes have non-uniform visitation requirements, and agents have limited communication.

Elor and Bruckstein [3] introduced a novel graph patrolling algorithm by integrating the ant pheromone and balloon models, where the region is segmented into sub-regions that are individually assigned to a certain agent. However, this method partitioned the region into equal-size sub-regions. As the characteristic of the area is not always uniform, equal-size sub-areas are inappropriate. Yehuda et al. [21] proposed a centralized algorithm which guarantees optimal uniform frequency, i.e., all cells are visited with maximal and uniform frequency in a non-uniform, grid environment. However, grid-based representation has a limitation in handling partially occluded cells or cover areas close to the boundaries in continuous spaces.

Sea et al. [13,14] proposed a decentralized coordinated area partitioning method by autonomous agents for continuous cooperative tasks. Agents in this approach could learn the locations of obstacles and the probabilities of dirt accumulation and could divide the area in a balanced manner. However, these papers considered the grid environment and mainly focused on specific application in cleaning/sweeping domain. Sugiyama et al. [15] also introduced an effective autonomous task allocation method that can achieve efficient cooperative work by enhancing divisional cooperation in multi-agent patrolling tasks. This paper addressed the *continuous cooperative patrolling problem* (CCPP), in which agents move around a given area and visit locations with the required and different frequencies for given purposes. However, this paper did not consider area partitioning and was implemented in a 2-dimensional grid space.

The most relevant work to ours is the work of Jeyhun and Murat [6], which introduced a new hybrid clustering model for k-means clustering, namely HE-kmeans, to improve the quality of clustering. This proposed model integrated *particle swarm optimization, scatter search* and *simulated annealing* to find good initial centroids for k-means. Another relevant work is from Elth et al. [4], which proposed a decentralized clustering method by extending the traditional k-means in a grid pattern. These two approaches could produce a good quality of clustering. However, they did not consider when the frequencies of visit to each location are different. As the frequencies of visit in the real-world environment are not always uniform which makes the clustering imbalanced, a clustering method that can take into account the non-uniform frequency of visit and at the same time tries to balance the workload amongst all patroller agents is preferable for realistic applications. Our proposed method, thus, aims at dealing with these requirements.

3   Problem Formulation

This paper aims at proposing solutions for multi-agent patrolling under frequency constraints, while trying to balance the workload amongst all patroller agents and then minimize the cost for patrolling. First, we formulate our problem in this section.

Let $G = (V, E)$ be a complete graph that can be embedded in $\mathbb{R}^2$, where $V = \{v_1, v_2, \ldots, v_n\}$ is a set of nodes, and $E = \{(v_i, v_j) : v_i, v_j \in V, i \neq j\}$ is a set of edges. The patrolled area is described as a graph G, where a location $v_i \in V$ is represented by its (x, y) coordinates in the 2D plane, and E contains $\frac{n \times (n-1)}{2}$ edges. In our patrolling problem, a node represents a location to be patrolled/visited, and an edge represents a path between nodes along which agents move. Let $A = \{1, 2, \ldots, m\}$ be a set of agents, and $m = |A|$ denotes the number of agents patrolling graph G, where $m < |V|$.

Each edge in G has its associated cost which is a traveling distance. Because G is embedded in $\mathbb{R}^2$, the distance between a pair of nodes is the Euclidean distance between two spatial coordinates $v_i \in V$ and $v_j \in V$ denoted by $\|v_i - v_j\| = \sqrt{(x_i - x_j)^2 + (y_i - y_j)^2}$, where (x_i, y_i) and (x_j, y_j) are the coordinates of nodes v_i and v_j respectively.

In the general multi-agent patrolling problem, a team of m agents patrols an area represented by a complete graph $G = (V, E)$. Thus, there are n nodes to be patrolled and $|E|$ possible paths for m agents to move. Our multi-agent frequency-based patrolling problem, however, consists of two main steps: graph partitioning and sub-graph patrolling. Each node in graph G has its associated visitation requirement, simply called *frequency of visit*. Let $f(v_i) \in \mathbb{Z}^+$ be the frequency of visit to each location in G.

Firstly, we partition a patrolled area represented by a graph G into k disjoint clusters, $C = \{C_1, \ldots, C_m\}$, and then allocate cluster C_i to agent i. The main goal is to cluster G by taking into account the required frequency of visit to each node in a balanced manner, such that the expected workload of each cluster is not much different from one another.

Let W_{C_s} be an *expected workload* of each agent in its allocated cluster, denoted by:

$$W_{C_s} = \sum_{v_i, v_j \in C_s} \frac{f(v_i)\|v_i - v_j\|}{|C_s| - 1}, \tag{1}$$

where $|C_s|$ is the number of nodes in each cluster. The expected workload here refers to an estimated amount of work a patroller agent has to do if they generate the shortest (or near-shortest) path, which is the estimated total cost/length agent i has to patrol in its allocated cluster/region, not the actual cost. We used this as a metric to evaluate the clustering performance of our proposed method in Sect. 4. If the value of W_{C_s} for all patroller agents are not much different from one another, we can conclude that the overall workload amongst all agents is considered to be balanced.

After obtaining clusters from the first step, the next goal is to generate a route for each agent to patrol in its allocated cluster based on the required frequency of visit to each node. Let route $s = \langle v_1, v_2, \ldots, v_\ell \rangle, \forall v_i \in V$ be a sequence of nodes agent j has to visit in C_j. In patrolling process, an agent tries to find a route with a minimum cost. The *route* is defined as the selected path in which an agent patrols in its allocated region. Then, the length of route s is denoted by:

$$len(s) = \sum_{i=1}^{\ell-1} \|v_i - v_{i+1}\| \tag{2}$$

For all agents in A, let $O(s, v_i)$ be the number of occurence of node v_i in route s, where $O(s, v_i)$ is the number of nodes v_i appear/exist in route s. Thus, the following condition is satisfied.

$$\begin{cases} O(s, v_i) > 0, & \text{if } v_i \in s \\ O(s, v_i) = 0, & \text{otherwise} \end{cases} \tag{3}$$

Then, the route s must satisfy, $\forall v_i \in C_i$, the following conditions:

$$O(s, v_i) \geq f(v_i) \tag{4}$$
$$min_{v_i \in V} f(v_i) = 1, \tag{5}$$

because clusters $(C_1, \ldots, C_m)$ are disjoint.

Let $S = \{s_1, \ldots, s_m\}$ be a set of routes, and thus m routes must be generated for all m agents to patrol G. Then, the multi-agent patrolling problem is to find m routes, such that each node is visited at least $f(v_i)$ times and that the length of total routes is the shortest. Thus, the objective function, R, is to minimize the sum of all routes, denoted by:

$$R(s_1, \ldots, s_m) = min \sum_{i=1}^{m} len(s_i)$$
$$\text{subject to: } \sum_{i=1}^{m} O(s_i, v_j) \geq f(v_j), \forall v_j \in V \tag{6}$$

Because C_i is disjoint and independent, and the shortest route in C_i is generated independently so that it meets the requirement of frequency of visit, the cost $R(s_1, \ldots, s_m)$ in Eq. 6 is identical to the sum of the cost of routes, $(s_1, \ldots, s_m)$. Therefore, our goal is to minimize:

$$R(s_1, \ldots, s_m) = \sum_{i=1}^{m} min \ len(s_i)$$
$$\text{subject to: } \sum_{i=1}^{m} O(s_i, v_j) \geq f(v_j), \forall v_j \in V \tag{7}$$

4   Proposed Method

Our proposed method is divided into two main steps: graph partitioning and sub-graph patrolling. Because we improved the well-known unsupervised traditional k-means clustering algorithm by introducing a new k-means based approach for clustering a given graph by taking into account the non-uniform visitation requirement for each location, we called our proposed method an *improved frequency-based k-means*, namely IF-k-means.

4.1 Graph Partitioning

Clustering refers to the process of partitioning or grouping a given set of patterns into disjoint clusters, $P = \{P_1, P_2, \ldots, P_{|P|}\}$. This step describes how agent could cluster a given graph, G, by taking into account the different frequency of visit to each node as well as balancing the workload of each cluster. We implemented k-means based clustering algorithm by modifying its objective function and centroids initialization so as to make it suit our problem. Each data point is interpreted as a node in a complete graph G, where $V = \{v_1, v_2, \ldots, v_n\}$ is a set of nodes as mentioned in Sect. 3. The main goal is to partition V into k disjoint clusters by taking into account the required frequency of visit to each node. We denote $C = \{C_1, C_2, \ldots, C_k\}$ as its set of clusters, and $c = \{c_1, c_2, \ldots, c_k\}$ as a set of corresponding centroids.

Simply speaking, k-means clustering is an algorithm to classify or to group the objects based on attributes/features into k number of group, where k is a positive integer number. The grouping is done by minimizing the sum of square of distances between data points and the corresponding cluster centroids [8].

The traditional k-means clustering algorithm aims at minimizing the following objective function, which is a squared error function denoted by:

$$J = min \sum_{i=1}^{n} \sum_{s=1}^{k} \sum_{v_i \in C_s} \|v_i - c_s\|^2$$

$$\text{subject to: } C_1 \cup \ldots \cup C_m = C$$
$$C_i \cap C_j = \emptyset, \forall\, 1 \leq i, j \leq m,\ i \neq j,$$

$$\text{where } c_s = \frac{1}{|C_s|} \sum_{v_i \in C_s} v_i,$$

k is the number of clusters, and c_s is the corresponding cluster centroid.

We modified the objective function of the above traditional k-means so as to apply our problem framework with frequency of visit. This method is called IF-k-means, and its objective function is denoted by:

$$Q = min \sum_{i=1}^{n} \sum_{s=1}^{k} \sum_{v_i \in C_s} f(v_i)\|v_i - c_s\|^2$$

$$\text{subject to: } C_1 \cup \ldots \cup C_m = C$$
$$C_i \cap C_j = \emptyset, \forall\, 1 \leq i, j \leq m,\ i \neq j, \tag{8}$$

$$\text{where } c_s = \frac{\sum_{v_i \in C_s} v_i \cdot f(v_i)}{\sum_{v_i \in C_s} f(v_i)},$$

$f(v_i)$ is the frequency of visit to node v_i, and $\|v_i - c_s\|$ is the Euclidean distance between v_i and c_s.

Algorithm 1. Improved frequency-based k-means (IF-k-means)

Input : $G = (V, E)$ and $f(v_i)$
   k (number of clusters), where $k = |A|$
Output: $C = \{C_1, C_2, \ldots, C_k\}$

1: Sort $|V|$ with $f(v_i)$ in descending order
2: Add them into a list N
3: $time = 1$
4: Select first k nodes from $N[k(time - 1) + 1]$ to $N[time * k]$
5: Place k initial centroids on selected k nodes in G
6: **repeat**
7: | Assign each node to the cluster having the closest centroid
8: | Recalculate(centroids)
9: **until** *centroids no longer move*
10: **foreach** *cluster* **do**
11: | Calculate expected workload, W_{C_s}
12: | Calculate difference in workload, T_{diff}
13: | **if** T_{diff} satisfies condition (10) **then**
14: | | Accept(clusters)
15: | **else**
16: | | Go to step(4)
17: | | $time + +$
18: | **end**
19: **end**

The traditional k-means method has been shown to be effective in producing good clustering results for many practical applications. Although it is one of the most well-known clustering algorithm and is widely used in various applications, one of its drawbacks is the highly sensitive to the selection of the initial centroids, which means the result of clustering highly depends on the selection of initial centroids. Therefore, proper selection of initial centroids is necessary for a better clustering. Thus, instead of placing the initial centroids randomly as in the traditional k-means, we place them on the nodes with the highest frequency of visit, $f(v_i)$, because a node with higher frequency of visit should have a shorter distance from its corresponding centroid than the node with lower frequency of visit to make the cluster balanced.

The difference between our IF-k-means and the classical k-means is that we incorporate $f(v_i)$ to both objective function and its constraint of the classical k-means in order to make the cluster balanced. Adding $f(v_i)$ to the objective function of the classical k-means makes the distance between node v_i and centroid c_s change causing the different size of clusters based on the visiting frequency to each node. It is also important to incorporate $f(v_i)$ into the calculation of the centroids to generate the weighted centroids function for producing a better centroids location for each cluster. By implementing our IF-k-means, the clusters having more nodes with high frequency of visit tend to have smaller size comparing to those with lower frequency of visit. At each step of the clustering, the centroids move close to high-frequency nodes after the repeated calculation using our modified centroids function. Thus, without incorporating $f(v_i)$ to both objective function and its constraint, the inefficient clustering would happen due to the inefficient centroids placement.

Let T_{diff} be the *difference in workload* amongst all agents, where we define T_{diff} as follows:

$$T_{diff} = \frac{1}{k(k-1)} \sum_{i=1}^{k} \sum_{j=1}^{k} |W_{C_i} - W_{C_j}|, \ i \neq j \tag{9}$$

The workload amongst all agents is considered to be balanced if it satisfies the following condition:

$$T_{diff} \leq M, \tag{10}$$

where $M \in \mathbb{Z}^+$ is not so large positive integer.

Algorithm 2. Pseudocode for constructing initial solution for SA

Input : $V = \{v_{k_1}, v_{k_2}, \ldots, v_{k_L}\}$ for cluster C_k and $f(v_i)$
Output: $s_0 = \{v_{k_1}, v_{k_2}, \ldots, v_{k_L}\}$ based on conditions (4) and (5), such that $k_i \neq k_{i+1}$

* function **distMatrix** return Euclidean distance between two nodes.
* k_i is the index of node v_i in cluster C_k.

1: $s_0 = \emptyset$
2: Select a current node, $curNode$, randomly from V
3: Add $curNode$ into s_0
4: **while** *(s_0 is not filled up)* and $(V \neq \emptyset)$ **do**
5:     Find the shortest distance from $curNode$ to another node in V:
     $shortestDist = min(\textbf{distMatrix}[curNode][j]$ for j in $V)$
6:     $curNode = \textbf{distMatrix}[curNode]$.index($shortestDist$)
     if $k_i \neq k_{i+1}$ is not satisfied **then**
       | Regenerate new $curNode$
7:     **end**
8:     Let $O(s_0, curNode)$ be an occurence of new $curNode$ in s_0
9:     **if** $O(s_0, curNode) < f(v_i)$ **then**
10:     | Keep new $curNode$ in V
11:     **else if** $O(s_0, curNode) \geq f(v_i)$ and $O(s_0, curNode) \leq 2.f(v_i)$ **then**
12:     | Remove new $curNode$ from V
13:     **end**
14:     Add new $curNode$ to s_0
15: **end**
16: return s_0

In this partitioning process of our proposed work, we calculated the expected workload of each cluster, W_{C_s}, by using Eq. 1. Then, we computed the difference in each workload, T_{diff}, by implementing the formula in Eq. 9. The process of our proposed IF-k-means algorithm is described in Algorithm 1.

4.2   Sub-graph Patrolling

This step presents how agent selected the best route for patrolling in its allocated sub-region with the shortest length by taking into account the required frequency of visit to each location. The goal of this step aims at finding the shortest route for each patroller agent in its allocated cluster with a semi-optimal solution. Because our multi-agent frequency-based patrolling problem is considered to be one of the combinatorial optimization problems and our main purpose is to partition a given area so as to balance the workload amongst all patroller agents, the optimal solution for the cost of visiting all nodes with their required frequency of visit is difficult due to a trade-off between balancing the workload and optimizing the cost of route, and thus, a semi-optimal solution is accepted in our work as a reasonable solution.

Algorithm 3. Pseudocode for route generation using SA

Input : Initial temperature, $T_0 = 1e + 10$
Final temperature, $T_f = 0.0001$
Cooling parameter, $\alpha = 0.95$

Output: s_{best}

1: Obtain initial solution $s_0 = \{v_{k_1}, v_{k_2}, \ldots, v_{k_L}\}$ from Algorithm 2
2: Set initial temperature: $T = T_0$
3: Cost function $C(s)$ is defined as $len(s)$ in Eq. 2, where $C(s) = len(s)$
4: Let current solution $s_{cur} = s_0$ whose cost is $C(s_{cur})$, and the best solution $s_{best} = s_0$ whose cost is $C(s_{best})$
5: **repeat**
6: $\quad$ Generate new solution s_{new} by randomly swapping two nodes in s_0 and get its cost $C(s_{new})$
$\quad$ **if** $k_i \neq k_{i+1}$ is not satisfied **then**
$\quad\quad$ | Regenerate s_{new} and $C(s_{new})$
7: $\quad$ **end**
8: $\quad$ Compute relative change in cost: $\delta = C(s_{new}) - C(s_{cur})$
9: $\quad$ Acceptance probability: $P(\delta, T) = exp(-\delta/T)$, where $T > 0$
10: $\quad$ **if** $\delta < 0$ or $P(\delta, T) > rand(0, 1)$ **then**
11: $\quad\quad$ | $s_{cur} = s_{new}$ and $C(s_{cur}) = C(s_{new})$
12: $\quad$ **else if** $C(s_{new}) < C(s_{best})$ **then**
13: $\quad\quad$ | $s_{best} = s_{new}$ and $C(s_{best}) = C(s_{new})$
14: $\quad$ **end**
15: $\quad$ Compute new temperature: $T = \alpha \times T$
16: **until** $T < T_f$

We use a simulated annealing here as a sequencing algorithm to find the shortest route, s_i, for patrolling. As our problem is a multi-agent patrolling problem, m routes will be generated in this step where $m = |A|$. However, in the multiple traveling salesman problem, in order to solve it in an easier and simpler way, a heuristic is formed to transform mTSP to TSP and then optimize the tour of each individual salesman. Because our problem is similar to the mTSP as mentioned in Sect. 1, we did the same by applying SA to each cluster to find the best route for each patroller agent in order to make the problem simpler.

Although the SA algorithm has been widely used in mTSP, we have modified and adapted it to our model with non-uniform frequency of visit to each node. The classical SA algorithm in mTSP generates the best solution/route such that each node must be visited exactly once, while our modified SA algorithm constructs the best route for each patroller agent based on the required frequency of visit, where each node is visited at least $f(v_i)$ times by taking into account the conditions from Eqs. 4 and 5. Furthermore, we have also modified the process of computing an initial solution in the SA by implementing a greedy approach instead of random approach to find an initial feasible solution. The process of how we applied SA to our model with non-uniform frequency of visit to find the shortest route is described in Algorithm 3, and the computation of an initial feasible solution with the implementation of greedy strategy is also described in Algorithm 2.

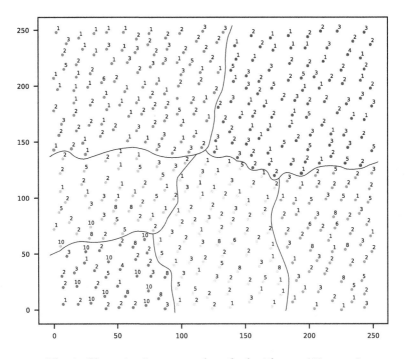

Fig. 1. Clustering by proposed method with $n = 400, m = 6$

Table 1. Numerical results with $n = 400$ and $m = \{6, 10\}$

400 nodes				
No. of agent (m)	Agent (A)	Exp. workload (W_{C_s})	Cost of route $(\ell en(s_i))$	Difference (T_{diff})
6	1	11635	11640	4.80
	2	11630	11638	
	3	11632	11641	
	4	11640	11647	
	5	11634	11642	
	6	11629	11635	
10	1	7136	7142	5.31
	2	7130	7140	
	3	7135	7139	
	4	7125	7134	
	5	7137	7145	
	6	7136	7147	
	7	7127	7138	
	8	7132	7141	
	9	7139	7146	
	10	7134	7143	

5   Experimental Evaluation

The proposed algorithms have been implemented in Python 3.5. All computational results are the averages of 20 trials, and are obtained on a personal computer with Intel(R) Core(TM) i5-6200U CPU @2.30 GHz processor and 8 GB RAM running on Windows 10 64-bit. To run experiments, we generated the coordinates of all nodes and their corresponding frequencies of visit $f(v_i)$, which are randomly distributed in the Euclidean space. We have tested our proposed method with different number of nodes and number of agents to see how well our algorithms can work when the number of nodes and agents increase respectively. In this work, we had run our experiments with 5 different number of nodes, $n = |V|$ is 200, 400, 600, 800 and 1000. We had also tried these with different number of agents, $m = |A|$ is 4, 6, 8 and 10. Moreover, we set $M = 10$ in our experiments. From the best of our knowledge, if M is too small, the solution may not exist, and if it is too large, the solution is not acceptable because agents' works are imbalanced. Therefore, we have to define M according to the problem setting.

Table 2. Numerical results with $n = 600$ and $m = \{6, 10\}$

| 600 nodes | | | | |
No. of agent (m)	Agent (A)	Exp. workload (W_{C_s})	Cost of route $(\ell en(s_i))$	Difference (T_{diff})
6	1	18825	18834	5.86
	2	18823	18832	
	3	18827	18838	
	4	18829	18839	
	5	18838	18845	
	6	18826	18835	
10	1	14326	14330	7.02
	2	14328	14337	
	3	14333	14340	
	4	14336	14342	
	5	14324	14335	
	6	14338	14348	
	7	14323	14332	
	8	14329	14336	
	9	14321	14329	
	10	14334	14345	

Table 3. Numerical results with $n = 1000$ and $m = \{6, 10\}$

| 1000 nodes | | | | |
No. of agent (m)	Agent (A)	Exp. workload (W_{C_s})	Cost of route $(\ell en(s_i))$	Difference (T_{diff})
6	1	30824	30834	7.40
	2	30835	30840	
	3	30841	30849	
	4	30837	30848	
	5	30840	30851	
	6	30842	30853	
10	1	26265	26273	8.02
	2	26273	26278	
	3	26268	26276	
	4	26255	26266	
	5	26260	26269	
	6	26256	26264	
	7	26254	26262	
	8	26268	26275	
	9	26264	26273	
	10	26270	26281	

After running 20 experiments, we randomly plot the result of one experiment as shown in Fig. 1. Figure 1 presents the result of that plot amongst 20 plots obtained from graph clustering using our proposed IF-k-means with $n = 400$ and $m = 6$, where the number on each node represents its required frequency of

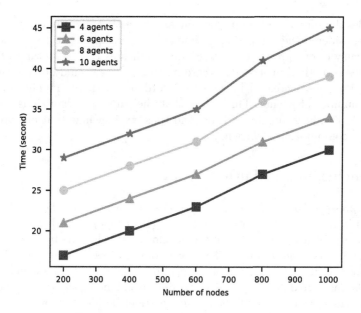

Fig. 2. Computation time of proposed method

visit. According to Fig. 1, we could see that the sizes of all clusters are varied in accordance with the values of $f(v_i)$ in each cluster. Some clusters tend to have small size due to the existence of many values of high visiting frequency in their clusters, while others seem to have bigger size because there are less high frequency of visit in their clusters comparing to those with smaller size. This kind of phenomenon happened because we tried to balance the workload of each cluster. We, thus, say that our proposed clustering algorithm could effectively partition a given graph in a balanced manner.

To evaluate the effectiveness and performance of our proposed work, the expected workload (W_{C_s}), the cost of route $(\ell en(s_i))$ in each cluster and the difference in workload (T_{diff}) are listed in Tables 1, 2 and 3. These tables show the numerical results with the number of agents, $m = 6$ and $m = 10$ for 400, 600 and 1000 nodes respectively. All these tables demonstrate that the difference in workload always satisfied the condition in Eq. 10, where $T_{diff} \leq M$ and $M = 10$. Thus, if IF-k-means cannot find the route whose T_{diff} is less than 10, no solution is generated.

Furthermore, the results from all the tables also clarified that the cost of patrolling in each cluster, $\ell en(s_i)$, has the value which is not much different from its corresponding expected workload, W_{C_s}. This means that the sequencing algorithm in Sect. 4.2 produced a good result in term of generating the route for patrolling and minimizing the cost of each route. Therefore, we conclude that our proposed algorithms not only could balance the workload amongst all agents, but also could generate the patrolling route with a reasonable cost. We only show the results in Tables 1, 2 and 3 when $m = 6$ and 10 because other results exhibit the similar features.

Besides the effectiveness of performing area partition and sub-area patrolling, we also considered the computation time as a significant factor to demonstrate the efficiency of our proposed work. Figure 2 indicates the computation time of our proposed method in second. According to Fig. 2, we could observe that the computation time increased linearly in accordance with the number of nodes and the number of agents. This shows that the proposed algorithms could be computed in a short amount of time, and thus, we conclude that our proposed method is computational efficiency.

6   Conclusion and Future Work

We have presented a new frequency-based area partitioning method for balanced workload in multi-agent patrolling systems. This proposed work considered the non-uniform visitation requirement for each location, where its frequency of visit is high or low depending on the level of importance of that location. Because non-uniform visiting frequencies of all locations could affect the quality of clustering, the main goal of this paper, thus, aims at balancing the workload of each cluster/agent so as to improve the workers morale. Besides the balance in workload, we also believe that computational cost plays a significant role in proving the effectiveness and computational efficiency of the proposed work. Experimental results demonstrated that our proposed method could effectively generate clusters of a given area regarding the non-uniform visitation requirements in a balanced manner and in a satisfied short amount of time.

The study of this problem in a more realistic environment will be considered in our future work. Moreover, we attempt to further extend our work in the future by taking into account the corresponding minimum time interval between the visits to a node that needs frequent patrolling. Also, we intend to incorporate the penalty function into our method in order to prevent the patroller agent from visiting nodes too often or too seldom.

Acknowledgments. This work is partly supported by JSPS KAKENHI grant number 17KT0044.

References

1. Almeida, A., Ramalho, G., Santana, H., Tedesco, P., Menezes, T., Corruble, V., Chevaleyre, Y.: Recent advances on multi-agent patrolling. In: Bazzan, A.L.C., Labidi, S. (eds.) SBIA 2004. LNCS (LNAI), vol. 3171, pp. 474–483. Springer, Heidelberg (2004). https://doi.org/10.1007/978-3-540-28645-5_48
2. Bektas, T.: The multiple traveling salesman problem: an overview of formulations and solution procedures. Omega **34**(3), 209–219 (2006)
3. Elor, Y., Bruckstein, A.M.: Multi-a(ge)nt graph patrolling and partitioning. In: 2009 IEEE/WIC/ACM International Conference on Web Intelligence and Intelligent Agent Technology - Workshops, pp. 52–57 (2009)
4. Elth, O., Benno, O., Maarten van, S., Frances, B.: A method for decentralized clustering in large multi-agent systems. In: AAMAS 2003, pp. 789–796 (2003)

5. I-Ming, C., Bruce, L.G., Edward, A.W.: The team orienteering problem. Eur. J. Oper. Res. **88**(3), 464–474 (1996)
6. Jeyhun, K., Murat, O.: Clustering quality improvement of k-means using a hybrid evolutionary model. Procedia Comput. Sci. **61**, 38–45 (2015)
7. Mihai-Ioan, P., Hervé, R., Olivier, S.: Multi-robot patrolling in wireless sensor networks using bounded cycle coverage. In: 2016 IEEE 28th International Conference on Tools with Artificial Intelligence (ICTAI), pp. 169–176 (2016)
8. Nallusamy, R., Duraiswamy, K., Dhanalaksmi, R., Parthiban, P.: Optimization of non-linear mutiple traveling salesman problem using k-means clustering, shrink wrap algorithm and meta-heuristics. Int. J. Non Linear Sci. **9**(2), 171–177 (2010)
9. Fazli, P., Alireza, D., Alan, K.M.: Multi-robot repeated area coverage. Auton. Robot **34**, 251–276 (2013)
10. Portugal, D., Rocha, R.: MSP algorithm: muti-robot patrolling based on territory allocation using balanced graph partitioning. In: SAC 2010, pp. 1271–1276 (2010)
11. Portugal, D., Pippin, C., Rocha, R.P., Christensen, H.: Finding optimal routes for multi-robot patrolling in generic graphs. In: 2014 IEEE/RSJ International Conference on Intelligent Robots and Systems (IROS), pp. 363–369 (2014)
12. Portugal, D., Rocha, R.: A survey on multi-robot patrolling algorithms. In: Camarinha-Matos, L.M. (ed.) DoCEIS 2011. IAICT, vol. 349, pp. 139–146. Springer, Heidelberg (2011). https://doi.org/10.1007/978-3-642-19170-1_15
13. Sea, V., Sugawara, T.: Area partitioning method with learning of dirty areas and obstacles in environments for cooperative sweeping robots. In: 2015 IIAI 4th International Congress on Advanced Applied Informatics (IIAI-AAI), pp. 523–529 (2015)
14. Sea, V., Kato, C., Sugawara, T.: Coordinated area partitioning method by autonomous agents for continuous cooperative tasks. J. Inf. Process. (JIP) **25**, 75–87 (2017)
15. Sugiyama, A., Sea, V., Sugawara, T.: Effective task allocation by enhancing divisional cooperation in multi-agent continuous patrolling tasks. In: 2016 IEEE 28th International Conference on Tools with Artificial Intelligence (ICTAI), pp. 33–40 (2016)
16. Tao, M., Laura, E.R.: Frequency-based patrolling with heterogeneous agents and limited communication. arXiv preprint arXiv: 1402.1757 (2014)
17. Sak, T., Wainer, J., Goldenstein, S.K.: Probabilistic multiagent patrolling. In: Zaverucha, G., da Costa, A.L. (eds.) SBIA 2008. LNCS (LNAI), vol. 5249, pp. 124–133. Springer, Heidelberg (2008). https://doi.org/10.1007/978-3-540-88190-2_18
18. Yann, C., Francois, S., Geber, R.: A theoretical analysis of multi-agent patrolling strategies. In: Third International Joint Conference on Autonomous Agents and Multiagent Systems (AAMAS), pp. 1524–1525 (2004)
19. Yann, C.: Theoretical analysis of the multi-agent patrolling problem. In: IEEE/WIC/ACM International Conference on Intelligent Agent Technology (IAT), pp. 302–308 (2004)
20. Yasuyuki, S., Hirofumi, O., Tadashi, M., Maya, H.: Cooperative capture by multi-agent using reinforcement learning application for security patrol systems. In: 2015 10th Asian Control Conference (ASCC), pp. 1–6 (2015)
21. Yehuda, E., Noa, A., Gal, A.K.: Multi-robot area patrol under frequency constraints. In: 2007 IEEE International Conference on Robotics and Automation, pp. 385–390 (2007)
22. Definition of Facilities Security Work. https://www.security-law.com/security-services-act/kikai.html

Algorithms for Sparse k-Monotone Regression

Sergei P. Sidorov$^{(\boxtimes)}$, Alexey R. Faizliev, Alexander A. Gudkov, and Sergei V. Mironov

Saratov State University, Saratov, Russian Federation
sidorovsp@info.sgu.ru

Abstract. The problem of constructing k-monotone regression is to find a vector $z \in \mathbb{R}^n$ with the lowest square error of approximation to a given vector $y \in \mathbb{R}^n$ (not necessary k-monotone) under condition of k-monotonicity of z. The problem can be rewritten in the form of a convex programming problem with linear constraints. The paper proposes two different approaches for finding a sparse k-monotone regression (Frank-Wolfe-type algorithm and k-monotone pool adjacent violators algorithm). A software package for this problem is developed and implemented in R. The proposed algorithms are compared using simulated data.

Keywords: Greedy algorithms · Pool-adjacent-violators algorithm
Isotonic regression · Monotone regression · Frank-Wolfe type algorithm

1 Introduction

Let $z = (z_1, \ldots, z_n)^T$ be a vector from $\mathbb{R}^n$, $n \in \mathbb{N}$, and let Δ^k be the finite difference operator of order k, $k \in \mathbb{N} \cup \{0\}$, defined by

$$\Delta^k z_i = \Delta^{k-1} z_{i+1} - \Delta^{k-1} z_i, \quad \Delta^0 z_i = z_i, \quad 1 \le i \le n - k.$$

A vector $z = (z_1, \ldots, z_n)^T \in \mathbb{R}^n$ is said to be k-monotone if and only if $\Delta^k z_i \ge 0$ for each $1 \le i \le n - k$. A vector $z = (z_1, \ldots, z_n)^T \in \mathbb{R}^n$ is called 1-monotone (or monotone) if $z_{i+1} - z_i \ge 0$, $i = 1, \ldots, n - 1$, and 2-monotone vectors are just convex (see Fig. 1).

The recent years have seen an increasing interest in shape-constrained estimation in statistics [4,11,13,16,30,37]. One of such problems is the problem of constructing k-monotone regression which is to find best fitted k-monotone vector to a given vector. The review of results on 1-monotone regression can be found in the book by Robertson and Dykstra [42]. The papers of Barlow and Brunk [5], Dykstra [20], Best and Chakravarti [6], Best [7] consider the

The work was supported by RFBR (grant 18-37-00060).

problem of finding monotone regression in quadratic and convex programming frameworks. Using mathematical programming approach, the works [1,28,49] have recently provided some new results on the topic. The papers [12,21] extend the problem to particular orders defined by the variables of a multiple regression. The recent paper [13] proposes and analyzes a dual active-set algorithm for regularized monotonic regression.

k-monotone regression (and 1-monotone regression in particular) found its applicability in many different areas: in non-parametric mathematical statistics [3,16], in smoothing of empirical data [2,19,25,26,29,36,38,48], in shape-preserving dynamic programming [14,15,32], in shape-preserving approximation [10,18,24,43,46]. Moreover, k-monotone sequences and vectors have broad applications in solving different problems in many mathematical areas [8,9,33,35,39–41,47,51].

Denote Δ_k^n the set of all vectors from $\mathbb{R}^n$, which are k-monotone. The problem of constructing k-monotone regression is to find a vector $z \in \mathbb{R}^n$ with the lowest square error of approximation to the given vector $y \in \mathbb{R}^n$ (not necessary k-monotone) under condition of k-monotonicity of z:

$$(z - y)^T (z - y) = \sum_{i=1}^{n} (z_i - y_i)^2 \to \min_{z \in \Delta_k^n} . \tag{1}$$

In this paper we present two different approaches for finding sparse k-monotone regression. First, following [22,27] in Sect. 2.2 we propose a simple greedy algorithm which employs the well-known Frank-Wolfe type approach. We show that the algorithm should carry out $O(n^{2k})$ iterations to find a solution with error $O(n^{-1/2})$. Finally, extending the ideas of [34] in Sect. 2.3 we propose the k-monotone pool-adjusted-violators algorithm. A software package was developed and implemented in R. The proposed methods are compared using simulated data. To the best of our knowledge, both of proposed algorithms are the first algorithms for the construction of k-monotone regression in case $k > 1$.

2   Algorithms for Monotone Regression

2.1   Preliminary Analysis

The problem (1) can be rewritten in the form of a convex programming problem with linear constraints as follows

$$F(z) = \frac{1}{2} z^T z - y^T z \to \min, \tag{2}$$

where minima is taken over all $z \in \mathbb{R}^n$ such that

$$g_i(z) := -\Delta^k z_i \leq 0, \ 1 \leq i \leq n - k. \tag{3}$$

The problem (2)–(3) is a quadratic programming problem and is strictly convex, and therefore it has a unique solution.

Let $\hat{z}$ be a (unique) global solution of (2)–(3), then there is a Lagrange multiplier $\lambda' = (\lambda'_1, \ldots, \lambda'_{n-k})^T \in \mathbb{R}^{n-k}$ such that

$$\nabla F(\hat{z}) + \sum_{i=1}^{n-k} \lambda'_i \nabla g_i(\hat{z}) = 0, \tag{4}$$

$$g_i(\hat{z}) \le 0, \ \lambda'_i \ge 0, \ 1 \le i \le n-k, \tag{5}$$

$$\lambda'_i g_i(\hat{z}) = 0, \ 1 \le i \le n-k, \tag{6}$$

where ∇g_i denotes the gradient of g_i. The equations (4)–(6) are Karush–Kuhn–Tucker optimality conditions which can be reduced to

$$\hat{z} - y = \sum_{i=1}^{n-k} \lambda_i \sum_{j=0}^{k} \binom{k}{j} (-1)^{k+j+1} e_{i+j}, \tag{7}$$

$$g_i(\hat{z}) \le 0, \ \lambda_i \le 0, \ 1 \le i \le n-k. \tag{8}$$

$$\lambda_i (\Delta^{k-1} \hat{z}_{i+1} - \Delta^{k-1} \hat{z}_i) = 0, \ 1 \le i \le n-k, \tag{9}$$

where e_s, $1 \le s \le n$, are unit standard basis vectors of the Euclidean space $\mathbb{R}^n$, and $\lambda = -\lambda'$.

Preliminary analysis of (7)–(9) shows that the $(k-1)$-th differences of the optimal solution $\hat{z}$ should be sparse, i.e. the sequence $\Delta^{k-1}\hat{z}_i$, $1 \le i \le n-k$, should have many zeroes. For example, if $k = 1$ then the optimal solution should be on a piecewise constant function, and if $k = 2$ then optimal points should lie on a piecewise linear function (see Fig. 1).

2.2 Frank-Wolfe Type Greedy Algorithm

For computational convenience of the problem (1), we moved from points z_i to increments $x_{k+i} = \Delta^{k-1} z_{i+1} - \Delta^{k-1} z_i$, $i = 1, \ldots, n-k$. It was shown in [50] that $z \in \Delta_k^n$ if and only if there exists a vector $x = (x_1, \ldots, x_n)^T \in \mathbb{R}^n$ such that z_i, $1 \le i \le n-k$, can be represented as

$$z_i = \sum_{j_1=1}^{i} \sum_{j_2=1}^{j_1} \cdots \sum_{j_{k-1}=1}^{j_{k-2}} \sum_{j_k=1}^{j_{k-1}} x_{j_k}, \tag{10}$$

where $x_j \ge 0$ for all $k + 1 \le j \le n$. It was proved in [50] by induction from the simple observation that if $z \in \mathbb{R}^n$ is k-monotone then there is a vector $x = (x_1, \ldots, x_n)^T \in \mathbb{R}^n$ with the property that $x = (x_2, \ldots, x_n)^T \in \mathbb{R}^{n-1}$ is $(k-1)$-monotone and such that $z_i = \sum_{j=1}^{i} x_j$.

Then the problem (1) can be rewritten as follows:

$$E(x) := \sum_{i=1}^{n} \left(\sum_{j_1=1}^{i} \sum_{j_2=1}^{j_1} \cdots \sum_{j_{k-1}=1}^{j_{k-2}} \sum_{j_k=1}^{j_{k-1}} x_{j_k} - y_i \right)^2 \to \min_{x \in S}, \tag{11}$$

where S denotes the set of all $x = (x_1, x_2, \ldots, x_n) \in \mathbb{R}^n$ such that $x_1, \ldots, x_k \in \mathbb{R}$, $(x_{k+1}, \ldots, x_n) \in \mathbb{R}_+^{n-k}$ and $\sum_{j=k+1}^n x_j \le \max \Delta^{k-1} y_i - \min \Delta^{k-1} y_i$.

Let $\nabla E(x) = \left(\frac{\partial E}{\partial x_1}, \frac{\partial E}{\partial x_2}, \ldots, \frac{\partial E}{\partial x_n} \right)^T$ be the gradient of function E at a point x. It was shown in [50] that if $z \in \Delta_k^n$ then there is a vector $x = (x_1, \ldots, x_n)^T$, $x_j \ge 0$ for $j = k+1, \ldots, n$, such that $z_i = \sum_{j=1}^i c_{ik}(j) x_j$, $1 \le i \le n$, where $c_{ik}(j)$ are defined by

$$c_{ik}(j) := \begin{cases} \binom{i-1}{j-1}, & \text{if } 1 \le i \le k-1, \\ \binom{k-1}{j-1}, & \text{if } k \le i \le n \text{ and } 1 \le j \le k-1, \\ \binom{i+k-j-1}{k-1}, & \text{if } k \le i \le n \text{ and } k \le j \le i. \end{cases}$$

Then

$$\frac{\partial E}{\partial x_s} = 2 \sum_{i=s}^n c_{ik}(s) \left(\sum_{j=1}^i c_{ik}(j) x_j - y_i \right). \tag{12}$$

For larger-scale problems obtaining the solution of the problem (11) could be computationally quite challenging. In this regard, the present study proposes to use the following Frank-Wolfe type greedy algorithm (k-FWA, Algorithm 1) for finding an approximate sparse solution to the problem (11). The rate of convergence for Algorithm 1 is estimated in Theorem 1.

Denote $\text{reg}_m(\xi)$ the best fitted polynomial regression of order m to the values $\xi = (\xi_{s_1}, \ldots, \xi_{s_2})$ at integer points $s_1, \ldots, s_2$, and we will write

$$z = \text{reg}_m(\xi), \tag{13}$$

where $z = (z_{s_1}, \ldots, z_{s_2})$ are the values predicted by the regression at the same points $s_1, \ldots, s_2$.

Algorithm 1. k-FWA

· Let $y = (y_1, \ldots, y_n)^T$ be the input vector and N be the number of iteration;
begin
   · Let $z^0 = \text{reg}_{k-1}(y)$;
   · Let $x^0 = (x_1^0, \ldots, x_n^0)^T$ be the start point obtained by (10) from z^0;
   · Let the counter $t = 0$;
   · **while** $t < N$ **do**
      · Calculate the gradient $\nabla E(x^t)$ at the current point x^t, using (12);
      · Let $\widetilde{x}^t$ be the solution of the linear optimization problem
      $\langle \nabla E(x^t)^T, x \rangle \underset{x \in S}{\rightarrow} \min$, where $\langle \cdot, \cdot \rangle$ is a scalar product of two vectors;
      · Let $x^{t+1} = x^t + \alpha_t(\widetilde{x}^t - x^t)$, $\alpha_t = \frac{2}{t+2}$, $t := t+1$;
   · Recover the k-monotone sequence $z = (z_1, \ldots, z_n)$ from the vector x^N;
end

Theorem 1. *Let $\{x^t\}$ be generated according to Algorithm 1. Then there is a positive $c(k, y)$ not depending on n such that for all $t \geq 2$*

$$E(x^t) - E^* \leq \frac{c(k,y)n^{2k-\frac{1}{2}}}{t+2}, \tag{14}$$

where E^* is the optimal solution of (11).

Proof. It is know [23] that for all $t \geq 2$

$$E(x^t) - E^* \leq \frac{2L(\mathrm{Diam}(S))^2}{t+2},$$

where L is the Lipschitz constant of E and $\mathrm{Diam}(S)$ is the diameter of S. It is easy to prove that $\mathrm{Diam}(S)$ is bounded. Let $\nabla^2 E(x) := \left(\frac{\partial^2 E}{\partial x_1^2}, \frac{\partial^2 E}{\partial x_2^2}, \ldots, \frac{\partial^2 E}{\partial x_n^2} \right)^T$. It is well known that if ∇E is differentiable, then its Lipschitz constant L satisfies the inequality $L \leq \sup_x \|\nabla^2 E(x)\|_2$. It follows from (12) that $\frac{\partial^2 E}{\partial x_s^2} = 2\sum_{i=s}^n c_{ik}^2(s)$. Then

$$L \leq \sup_x \sqrt{\sum_{s=1}^n \left(\frac{\partial^2 E}{\partial x_s^2} \right)^2} = 2\sqrt{\sum_{s=1}^n \left(\sum_{i=s}^n c_{ik}^2(s) \right)^2}$$

$$\leq 2\sqrt{\sum_{s=1}^n \left(\sum_{i=s}^n \binom{i+k-s-1}{k-1}^2 \right)^2} = 2\sqrt{\sum_{s=1}^n \left(\sum_{i=1}^{n-s+1} \binom{i+k-2}{k-1}^2 \right)^2}$$

$$\leq 2\sqrt{\sum_{s=1}^n \left(\sum_{i=1}^{n-s+k-1} \left(\frac{i^{k-1}}{(k-1)!} \right)^2 \right)^2} = \frac{2}{(k-1)!^2} \sqrt{\sum_{s=1}^n \left(\sum_{i=1}^{n-s+k-1} i^{2(k-1)} \right)^2}. \tag{15}$$

It follows from the inequality $\sum_{i=1}^n i^l \leq \frac{1}{l} n^{l+1}$ that

$$L \leq \frac{2}{(k-1)!^2} \sqrt{\sum_{s=1}^n (n-s+k-1)^{4k-2}} \leq \frac{2}{(k-1)!^2} \sqrt{\sum_{s=1}^{n+k-2} s^{4k-2}}$$

$$\leq \frac{2}{(k-1)!^2} \sqrt{(n+k-2)^{4k-1}} \tag{16}$$

The disadvantages of k-FWA are the slow convergence and the dependence of the theoretical degree of convergence on the dimensionality of the problem. Theorem 1 shows that we need $O(n^{2k})$ iterations of Algorithm 1 to obtain a solution with error $O(n^{-1/2})$. Another drawback of Algorithm 1 is that we do not know the exact number of iterations that is necessary to achieve a desirable accuracy. It may be helpful to follow the ideas of papers [31, 44, 45] and use the values of duality gap as the stopping criterion for k-FWA.

2.3 k-Monotone Pool-Adjusted-Violators Algorithm

Simple iterative algorithm for solving the problem (1) with $k = 1$ is called Pool-Adjacent-Violators Algorithm (PAVA) [17,34]. The work [6] examined the generalization of this algorithm. The paper [52] studied this problem as the problem of identifying the active set and proposed a direct algorithm of the same complexity as the PAVA (the dual algorithm). PAVA computes a non-decreasing sequence of values $z_1, \ldots, z_n$ such that the problem (1) with $k = 1$ is optimized.

In this section we propose the extension of PAVA for finding k-monotone regression which we called k-monotone pool-adjacent-violators algorithm (k-PAVA). Note that 1-PAVA (k-PAVA with $k = 1$) coincides with the PAVA.

Let $y = (y_1, \ldots, y_n)^T$ be the input vector. At the zero iteration ($t = 0$) we assign $z^{[0]} := y$. The algorithm we propose is a simple iterative algorithm, where at each step of the iteration t we bypass the points $z^{[t]}$ from left to right. The result of each iteration t is a vector $z^{[t]} = (z_1^{[t]}, \ldots, z_n^{[t]})^T$. The algorithm finishes its work on the iteration t^* in which $z^{[t^*]} \in \Delta_k^n$.

Let $t = 1$ (the iteration counter), $l = 1$ (the point counter), $j = 0$ (the index of a point at a block).

We calculate $\Delta^k z_l^{[t-1]}$, the finite difference of order k at current point $z_l^{[t-1]}$. If $\Delta^k z_l^{[t-1]} \geq 0$ then we let $z_l^{[t]} := z_l^{[t-1]}$, and move on to the next point $z_{l+1}^{[t-1]}$.

If $\Delta^k z_l^{[t-1]} < 0$ then using the procedure (13) we get $Z_{l,l+k}^{[t]} := \text{reg}_{k-1}$ $\left(Z_{l,l+k}^{[t-1]} \right)$, where $Z_{l,l+k}^{[t]} := \left\{ z_l^{[t]}, \ldots, z_{l+k}^{[t]} \right\}$ denotes the block of points. Then we calculate $\Delta^k z'_{l+1}$, where $z'_{l+1} = (z_{l+1}^{[t]}, \ldots, z_{l+k}^{[t]}, z_{l+k+1}^{[t-1]})^T$, and if $\Delta^k z'_{l+1} \geq 0$ then we assign $z_{l+1}^{[t]} := z_{l+1}^{[t-1]}$, $j := 0$, $l := l+1$, and return to the first step of the algorithm. If $\Delta^k z'_{l+1} < 0$ then we let $j := j+1$ and add the point $z_{l+k+j}^{[t-1]}$ to the block $Z_{l,l+k+j-1}^{[t]}$, and then we calculate $Z_{l,l+k+j}^{[t]} := \text{reg}_{k-1} \left(Z_{l,l+k+j-1}^{[t]} \cup z_{l+k+j}^{[t-1]} \right)$.

Having passed all the points $z_l^{[t-1]}$, $l = 1, \ldots, n - k$, it may be turn out that all the points are from the block $Z_{1,n}^{[1]}$. In this case the resulting vector $z^{[t]} = (z_1^{[t]}, \ldots, z_n^{[t]})^T$ is our k-monotone solution. Otherwise, we obtain separate blocks $Z_{l_1,l_1+k+j_1}^{[t]}, Z_{l_2,l_2+k+j_2}^{[t]}, \ldots$, each of which is k-monotone.

Then we set $t := t + 1$ and we repeat the process described above until $\Delta^k z_l^{[t]} \geq 0$ for all $l = 1, \ldots, n - k$. When all the finite differences become non-negative, we get the desired k-monotone sequence.

The proposed Pool-Adjacent-Violators Algorithm for finding k-monotone solution of the problem (1) (k-PAVA) is described below (Algorithm 2). The resulting vector $z^{[t]}$ of length n will be k-monotone.

3   Empirical Result

The algorithms have been implemented in R. We compared the performance of the Frank-Wolfe type algorithm (k-FWA, Algorithm 1) with the performance of

---

Algorithm 2. k-PAVA

---

begin

    · Let $y = (y_1, \ldots, y_n)^T \in \mathbb{R}^n$ be the input vector and let $z^{[0]} := y$;

    · Let $t := 1$ and let k be the desirable order of monotonicity;

    repeat

        · Let $j := 0$, $l := 1$;

        · **while** $l \leq n - k$ **do**

            · **if** $\Delta^k z_l^{[t-1]} < 0$ **then**

                · $Z_{l,l+k}^{[t]} := \mathrm{reg}_{k-1}\left(Z_{l,l+k}^{[t-1]}\right)$;

                · **if** $\Delta^k(z_{l+1}^{[t]}, \ldots, z_{l+k}^{[t]}, z_{l+k+1}^{[t-1]})^T \geq 0$ **then**

                    $z_{l+1}^{[t]} := z_{l+1}^{[t-1]}$, $j := 0$ and $l := l + 1$;

                · **else** $j := j + 1$ and $Z_{l,l+k+j}^{[t]} := \mathrm{reg}_{k-1}\left(Z_{l,l+k+j-1}^{[t]} \cup z_{l+k+j}^{[t-1]}\right)$;

            · **else** $z_l^{[t]} := z_l^{[t-1]}$, $l := l + 1$;

        · $t := t + 1$;

    until $\Delta^k z_l^{[t]} \geq 0$ *for all* $l = 1, \ldots, n - k$;

    return $z^{[t]} = (z_1^{[t]}, \ldots, z_n^{[t]})$

end

---

the pool-adjacent-violators algorithm for finding k-monotone solution (k-PAVA, Algorithm 2) using simulated data sets.

Results show that the speed of convergence for k-PAVA is higher than for k-FWA. Table 1 presents empirical results for k-PAVA and k-FWA for simulated sets of points. In the case $k = 1$ the simulated points were obtained as the values of logarithmic function with added normally distributed noise: $y_i = \ln(1+i) + \varphi_i$, $\varphi_i \sim \mathcal{N}(0,1)$, $i = 1, \ldots, 10000$. If $k = 2$ then the simulated points were taken as $y_i = 0.001(i - 500)^2 + \varphi_i$, $\varphi_i \sim \mathcal{N}(0,1)$, $i = 1, \ldots, 1000$. The table contains mean errors $\frac{1}{n}\sum_{i=1}^{n}(z_i - y_i)^2$, the cardinality (the number of nonzero increments $x_{k+i} = \Delta^{k-1} z_{i+1} - \Delta^{k-1} z_i$, $i = 1, \ldots, n - k$), the number of iteration and CPU time (in seconds).

The results show that errors of k-FWA are getting closer to the errors of k-PAVA with increased number of iterations for k-FWA. While k-PAVA is better than greedy algorithm in terms of errors, the solutions of k-FWA have a better sparsity for $k = 1$. Both algorithms obtain sparse solutions, but we can control the cardinality in k-FWA as opposed to k-PAVA. Generally, k-FWA's cardinality increases by one at each iteration. Consequently, we should limit the number of iterations to obtain sparser solutions.

Figure 1 ($k = 1$) shows simulated points ($N = 100$) with logarithm structure and isotonic regressions, obtained by 1-FWA (green line) and 1-PAVA (red line). 1-FWA gives a solution with 14 jumps, and 1-PAVA with 16 jumps. Since the solutions of the 1-FWA are sparser, the Frank-Wolfe type algorithm error is slightly higher than the 1-PAVA. Figure 1 ($k = 2$) shows simulated points ($N = 100$) drowned from a quadratic function by adding normally distributed noise and two 2-monotone regressions, obtained by 2-FWA (green line) and 2-PAVA (red line). The solution obtained by 2-FWA after 152 iterations has the cardinality 11 and the value of mean error 1.295. The solution obtained by 2-PAVA after 23 iterations has the cardinality 7 and the value of mean error 1.241.

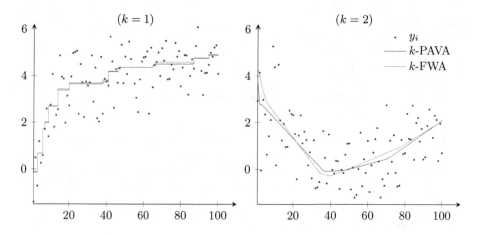

Fig. 1. Solutions obtained by k-PAVA and k-FWA (Color figure online)

Table 1. Comparison of algorithms k-PAVA and k-FWA on the simulated data

Algorithm	$k = 1$				$k = 2$			
	# of iterations	Error	Card.	CPU time	# of iterations	Error	Card.	CPU time
k-PAVA		0.994	82	0.73	98	0.928	32	4.9
k-FWA	100	0.999	41	0.71	100	2.514	14	22.1
k-FWA	200	0.996	57	1.48	150	1.252	23	44.2
k-FWA	500	0.995	76	3.49	250	1.092	31	88.4
k-FWA	1000	0.994	79	6.99	450	1.009	62	176
k-FWA	2000	0.994	82	14.9	650	0.963	88	353

4   Conclusion

Our research proposes two algorithms (k-FWA and k-PAVA) for solving the problem of constructing the best fitted k-monotone regression. One of the main contributions of the paper is Theorem 1, which provides a theoretical convergence result for the Frank-Wolfe type algorithm. We compared the performance of k-FWA with the performance of k-PAVA using simulated data sets. The software was implemented in R. While k-PAVA gives smaller errors than k-FWA, the Frank-Wolfe type algorithm obtains sparser solutions. One of the advantages of the Frank-Wolfe type algorithm is the potential for controlling cardinality of the solution. On the other hand, the drawbacks of k-FWA are the slow convergence and the dependence of the theoretical degree of convergence on the dimensionality of the problem.

References

1. Ahuja, R., Orlin, J.: A fast scaling algorithm for minimizing separable convex functions subject to chain constraints. Oper. Res. **49**(1), 784–789 (2001)
2. Altmann, D., Grycko, E., Hochstättler, W., Klützke, G.: Monotone smoothing of noisy data. Diskrete Mathematik und Optimierung. Technical report feu-dmo034.15. Fern Universität in Hagen, Fakultät für Mathematik und Informatik (2014)
3. Bach, F.: Efficient algorithms for non-convex isotonic regression through submodular optimization (2017), Working paper or preprint
4. Balabdaoui, F., Rufibach, K., Santambrogio, F.: Least-squares estimation of two-ordered monotone regression curves. J. Nonparametr. Stat. **22**(8), 1019–1037 (2010)
5. Barlow, R., Brunk, H.: The isotonic regression problem and its dual. J. Am. Stat. Assoc. **67**(337), 140–147 (1972)
6. Best, M.J., Chakravarti, N.: Active set algorithms for isotonic regression: a unifying framework. Math. Progr.: Ser. A B **47**(3), 425–439 (1990)
7. Best, M., Chakravarti, N., Ubhaya, V.: Minimizing separable convex functions subject to simple chain constraints. SIAM J. Optim. **10**(3), 658–672 (2000)
8. Bor, H.: A study on local properties of Fourier series. Nonlinear Anal.: Theory Methods Appl. **57**(2), 191–197 (2004)
9. Bor, H.: A note on local property of factored Fourier series. Nonlinear Anal.: Theory Methods Appl. **64**(3), 513–517 (2006)
10. Boytsov, D.I., Sidorov, S.P.: Linear approximation method preserving k-monotonicity. Sib. Electron. Math. Rep. **12**, 21–27 (2015)
11. Brezger, A., Steiner, W.J.: Monotonic regression based on bayesian P-splines. J. Bus. Econ. Stat. **26**(1), 90–104 (2008)
12. Burdakov, O., Grimvall, A., Hussian, M.: A generalised PAV algorithm for monotonic regression in several variables. In: Antoch, J. (ed.) Proceedings of the 16th Symposium in Computational Statistics, COMPSTAT, vol. 10, no. 1, pp. 761–767 (2004)
13. Burdakov, O., Sysoev, O.: A dual active-set algorithm for regularized monotonic regression. J. Optim. Theory Appl. **172**(3), 929–949 (2017)
14. Cai, Y., Judd, K.L.: Shape-preserving dynamic programming. Math. Methods Oper. Res. **77**, 407–421 (2013)
15. Cai, Y., Judd, K.L.: Advances in numerical dynamic programming and new applications. In: Handbook of Computational Economics, vol. 3, pp. 479–516. Elsevier (2014)
16. Chen, Y.: Aspects of shape-constrained estimation in statistics. Ph.D. thesis. University of Cambridge (2013)
17. Chepoi, V., Cogneau, D., Fichet, B.: Polynomial algorithms for isotonic regression. Lect. Notes-Monogr. Ser. **31**(1), 147–160 (1997)

18. Cullinan, M.P.: Piecewise convex-concave approximation in the minimax norm. In: Demetriou, I., Pardalos, P. (eds.) Abstracts of Conference on Approximation and Optimization: Algorithms, Complexity, and Applications, Athens, Greece, 29–30 June 2017, p. 4. National and Kapodistrian University of Athens (2017)
19. Diggle Peter, M.S., Tony, M.J.: Case-control isotonic regression for investigation of elevation in risk around a point source. Stat. Med. **18**(1), 1605–1613 (1999)
20. Dykstra, R.: An isotonic regression algorithm. J. Stat. Plan. Inference **5**(1), 355–363 (1981)
21. Dykstra, R., Robertson, T.: An algorithm for isotonic regression for two or more independent variables. Ann. Stat. **10**(1), 708–719 (1982)
22. Faizliev, A.R., Gudkov, A.A., Mironov, S.V., Levshunov, M.A.: Greedy algorithm for sparse monotone regression. In: CEUR Workshop Proceedings, vol. 2018, pp. 23–31 (2017)
23. Frank, M., Wolfe, P.: An algorithm for quadratic programming. Nav. Res. Logist. Q. **3**(1–2), 95–110 (1956)
24. Gal, S.G.: Shape-Preserving Approximation by Real and Complex Polynomials. Birkhäuser, Boston (2008)
25. Gorinevsky, D.: Monotonic regression filters for trending deterioration faults. In: Proceedings of the American Control Conference, vol. 6, pp. 5394–5399 (2004)
26. Gorinevsky, D.: Efficient filtering using monotonic walk model. In: Proceedings of the American Control Conference, pp. 2816–2821. IEEE (2008)
27. Gudkov, A.A., Mironov, S.V., Faizliev, A.R.: On the convergence of a greedy algorithm for the solution of the problem for the construction of monotone regression. Izv. Sarat. Univ. (N. S.) Ser. Math. Mech. Inform. **17**(4), 431–440 (2017)
28. Hansohm, J.: Algorithms and error estimations for monotone regression on partially preordered sets. J. Multivar. Anal. **98**(5), 1043–1050 (2007)
29. Hastie, T., Tibshirani, R., Wainwright, M.: Statistical Learning with Sparsity. Chapman and Hall/CRC, New York (2015)
30. Hazelton, M., Turlach, B.: Semiparametric regression with shape-constrained penalized splines. Comput. Stat. Data Anal. **55**(10), 2871–2879 (2011)
31. Jaggi, M.: Revisiting Frank-Wolfe: projection-free sparse convex optimization. In: Proceedings of the 30th International Conference on Machine Learning, ICML 2013, pp. 427–435 (2013)
32. Judd, K.: Numerical Methods in Economics. The MIT Press, Cambridge (1998)
33. Latreuch, Z., Belaïdi, B.: New inequalities for convex sequences with applications. Int. J. Open Probl. Comput. Math. **5**(3), 15–27 (2012)
34. Leeuw, J., Hornik, K., Mair, P.: Isotone optimization in R: pool-adjacent-violators algorithm (PAVA) and active set methods. J. Stat. Softw. **32**(5), 1–24 (2009)
35. Leindler, L.: A new extension of monotone sequences and its applications. J. Inequal. Pure Appl. Math. **7**(1), 7 (2006). Paper No. 39 electronic only. http://eudml.org/doc/128520
36. Leitenstorfer, F., Tutz, G.: Generalized monotonic regression based on B-splines with an application to air pollution data. Biostatistics **8**(3), 654–673 (2007)
37. Lu, M.: Spline estimation of generalised monotonic regression. J. Nonparametr. Stat. **27**(1), 19–39 (2014)
38. Gorinevsky, D., Kim, S.J., Beard, S., Boyd, S., Gordon, G.: Optimal estimation of deterioration from diagnostic image sequence. IEEE Trans. Signal Process. **57**(3), 1030–1043 (2009)
39. Marshall, A.W., Olkin, I., Arnold, B.C.: Inequalities: Theory of Majorization and Its Applications. Springer, New York (2011). https://doi.org/10.1007/978-0-387-68276-1

40. Milovanović, I.Z., Milovanović, E.I.: Some properties of l_p^k-convex sequences. Bull. Int. Math. Virtual Inst. **5**(1), 33–36 (2015)
41. Niezgoda, M.: Inequalities for convex sequences and nondecreasing convex functions. Aequ. Math. **91**(1), 1–20 (2017)
42. Robertson, T., Wright, F., Dykstra, R.: Order Restricted Statistical Inference. Wiley, New York (1988)
43. Shevaldin, V.T.: Local approximation by splines. UrO RAN, Ekaterinburg (2014)
44. Sidorov, S.P., Mironov, S.V.: Duality gap analysis of weak relaxed greedy algorithms. In: Battiti, R., Kvasov, D.E., Sergeyev, Y.D. (eds.) LION 2017. LNCS, vol. 10556, pp. 251–262. Springer, Cham (2017). https://doi.org/10.1007/978-3-319-69404-7_18
45. Sidorov, S.P., Mironov, S.V., Pleshakov, M.G.: Dual convergence estimates for a family of greedy algorithms in Banach spaces. In: Nicosia, G., Pardalos, P., Giuffrida, G., Umeton, R. (eds.) MOD 2017. LNCS, vol. 10710, pp. 109–120. Springer, Cham (2018). https://doi.org/10.1007/978-3-319-72926-8_10
46. Sidorov, S.: On the saturation effect for linear shape-preserving approximation in Sobolev spaces. Miskolc Math. Notes **16**(2), 1191–1197 (2015)
47. Sidorov, S.P.: Linear k-monotonicity preserving algorithms and their approximation properties. In: Kotsireas, I.S., Rump, S.M., Yap, C.K. (eds.) MACIS 2015. LNCS, vol. 9582, pp. 93–106. Springer, Cham (2016). https://doi.org/10.1007/978-3-319-32859-1_7
48. Siem, A.Y.D., den Hertog, D., Hoffmann, A.L.: Multivariate convex approximation and least-norm convex data-smoothing. In: Gavrilova, M., Gervasi, O., Kumar, V., Tan, C.J.K., Taniar, D., Laganá, A., Mun, Y., Choo, H. (eds.) ICCSA 2006. LNCS, vol. 3982, pp. 812–821. Springer, Heidelberg (2006). https://doi.org/10.1007/11751595_86
49. Stromberg, U.: An algorithm for isotonic regression with arbitrary convex distance function. Comput. Stat. Data Anal. **11**(1), 205–219 (1991)
50. Toader, G.: The representation of n-convex sequences. L'Anal. Numér. et la Théorie de L'Approx. **10**(1), 113–118 (1981)
51. Wu, S., Debnath, L.: Inequalities for convex sequences and their applications. Comput. Math. Appl. **54**(4), 525–534 (2007)
52. Wu, W.B., Woodroofe, M., Mentz, G.: Isotonic regression: another look at the changepoint problem. Biometrika **88**(3), 793–804 (2001)

Revisiting the Self-adaptive Large Neighborhood Search

Charles Thomas[(✉)] and Pierre Schaus

ICTEAM institute, Universite catholique de Louvain, Louvain-la-Neuve, Belgium
{charles.thomas,pierre.schaus}@uclouvain.be

Abstract. This paper revisits the Self-Adaptive Large Neighborhood Search introduced by Laborie and Godard. We propose a variation in the weight-update mechanism especially useful when the LNS operators available in the portfolio exhibit unequal running times. We also propose some generic relaxations working for a large family of problems in a black-box fashion. We evaluate our method on various problem types demonstrating that our approach converges faster toward a selection of efficient operators.

1 Introduction

Back in 2004, Puget [1] said that CP technology was too complex to use and more research efforts should be devoted to make it accessible to a broader audience. A lot of research effort has been invested to make this vision become true. Efficient black-box complete search methods have been designed [2–8] and techniques such as the embarrassingly parallel search are able to select the best search strategy with almost no overhead [9]. For CP, Puget argued that the model-and-run approach should become the target to reach. The improvements went even beyond that vision since for some applications, the model can be automatically derived from the data [10,11].

This work aims at automating the CP technology in the context of Large Neighborhood Search (LNS) [12]. This technique consists in iteratively applying a partial relaxation followed by a reconstruction in order to gradually improve the solution to the problem. The relaxation determines constraints to impose to restrict the problem based on the current best solution. Then, the reconstruction (or search) heuristic guides the search in the resulting search space by assigning values to the remaining variables in order to find one or more new solution(s).

Example 1. For example, a *random* relaxation heuristic selects randomly a percentage of the variables to relax and fix the other ones to their assignment in the current best solution. This heuristic can be parametrized by choosing the percentage to relax in a set of values such as $\{10\%, 20\%, 50\%\}$. A *first fail* heuristic with a fixed limit on the number of backtracks can be used as a reconstruction heuristic which can also be parametrized by choosing a limit on the number of backtracks in a set of values such as $\{50\,bkts, 500\,bkts, 5000\,bkts\}$.

© Springer International Publishing AG, part of Springer Nature 2018
W.-J. van Hoeve (Ed.): CPAIOR 2018, LNCS 10848, pp. 557–566, 2018.
https://doi.org/10.1007/978-3-319-93031-2_40

The relaxation and reconstruction process continues until some limit in terms of iterations or time is reached. From a local search point of view, CP is thus used as a slave technology for exploring a (large) neighborhood around the current best solution. LNS has been successfully used on various types of problems: bin-packing [13,14], vehicle-routing [15,16], scheduling [17–19], etc. Designing good relaxation and reconstruction heuristics with the right parameters is crucial for the efficiency of LNS. Unfortunately this task requires some experience and intuition on the problem to solve.

In order to design an automated LNS, two approaches can be envisioned. A first one would be to recognize the structure of the model in order to select the most suited heuristic from a taxonomy of heuristics described in the literature. This approach, which is used in [20] for scheduling problems, has two disadvantages: (1) some problems are hybrids and thus difficult to classify or recognize, (2) it requires a lot of effort and engineering to develop the problem inspector and to maintain the taxonomy of operators. Therefore, we follow a different approach called Adaptive LNS (ALNS) introduced in [21] which uses a portfolio of heuristics and dynamically learns on the instance which ones are the most suitable. At each iteration, a pair of relaxation and reconstruction heuristics is selected and applied on the current best solution. The challenge is to select the pair having the greatest gradient of the objective function over time (evaluated on the current best solution) based solely on the past executions.

We expand the usage of the Self Adaptive LNS (SA-LNS) framework proposed in [22] on different optimization problems by considering the model as a black-box. Our solver uses a set of generic preconfigured methods (operators) that hypothesize specificities in the problem and leverage them in order to efficiently perform LNS iterations. Given that the operators available in the portfolio are well diversified, we hope to provide a simple to use yet efficient framework able to solve a broad range of discrete optimization problems.

Our contributions to the ALNS framework are: (1) An adaptation of the weight update mechanism able to better cope with unequal running times of the operators. (2) A portfolio of operators easy to integrate and implement in any solver for solving a broad range of problems.

We first explain in Sect. 2 the principles of the ALNS framework. Then, in Sect. 3 we present the heuristics implemented as part of our ALNS portfolio. We present the experiments that we conducted and their results in Sect. 4. Finally, we provide a few concluding remarks and evoke our further research prospects in Sect. 5.

2   Adaptive Large Neighbourhood Search

Each ALNS operator as well as its possible parameters is associated to a weight. These weights allow to dynamically reward or penalize the operators and their parameters along the iterations to bias the operator selection strategy. Algorithm 1 describes the pseudo-code for an ALNS search. $\Delta c \geq 0$ is the objective improvement and Δt is the time taken by the operator.

Algorithm 1. Adaptive Large Neighborhood Search For a minimization problem

$s^* \leftarrow$ feasible solution
repeat
    $relax \leftarrow$ select relaxation operator
    $search \leftarrow$ select search operator
    $(s', \Delta t) \leftarrow search(relax(s^*))$
    $\Delta c \leftarrow cost(s^*) - cost(s')$
    $weight_{relax} \leftarrow updateWeight(relax)$
    $weight_{search} \leftarrow updateWeight(search)$
    if $\Delta c > 0$ **then**
        $s^* \leftarrow s'$
    end if
until stop criterion met
return s^*

Roulette Wheel Selection. We use the Roulette Wheel selection mechanism as in [22,23]. It consists in selecting the operators with probabilities proportional to their weight. The probability $P(i)$ of selecting the i-th operator o_i with a weight w_i among the set of all operators O is $P(i) = \frac{w_i}{\sum_{k=1}^{|O|} w_k}$

Weight Evaluation. In [22], the authors evaluate the operators ran at each iteration using an efficiency ratio r defined as: $r = \frac{\Delta c}{\Delta t}$. This ratio is then balanced with the previous weight of the operator $w_{o,p}$ using a reaction factor $\alpha \in [0,1]$: $w_o = (1-\alpha) \cdot w_{o,p} + \alpha \cdot r$. While the reaction factor is important to accommodate the evolving efficiency of the operators during the search, this method does not cope well with operators having different running times. Indeed, operators with a small execution time will evolve faster as they will be evaluated more often. This can lead less efficient operators to be temporally considered better as their weight will decrease slower.

Example 2. Let us consider two operators A and B with running times of respectively 2 and 4 s. Both operators start with an efficiency ratio of 10 but after some time in the search, A has an efficiency of $\frac{1}{2}$ and B of $(\frac{1}{4})$. If each operator is separately run for 4 s, under a reaction factor of $\alpha = 0.9$; as A will be evaluated twice, its weight will decrease to 0.595 $(0.1 \cdot (0.1 \cdot 10 + 0.9 \cdot \frac{1}{2}) + 0.9 \cdot \frac{1}{2})$. Over the same duration B would be evaluated once and its weight would become 1.225 $(0.1 \cdot 10 + 0.9 \cdot \frac{1}{4})$. While both operators will eventually converge towards their respective efficiency, for a short amount of time, B will have a higher score than A and thus a higher probability to be selected.

This induces a lack of reactivity in the operator selection. In the following, we propose a variation of the weight update rule, more aligned with the expected behavior in case of different runtimes among the operators.

Evaluation Window. We evaluate the operator based on its performances obtained in a sliding evaluation window $[t^* - w, now]$ where t^* is the time at which the last best solution was found and w is the window size meta-parameter. The window thus adapts itself in case of stagnation to always include a fixed part of the search before the last solution was found. This ensures that the operator(s) responsible for finding the last solution(s) will not have their score evaluated to 0 after a while in case of stagnation.

For each LNS iteration i, we record the operator used o_i, the time t_i at which it was executed, the difference Δc_i of the objective and the duration of execution Δt_i. We define the local/total efficiency ratio $L(o)/T(o)$ of an operator and the local/total efficiency L/T of all the operators as:

$$L(o) = \frac{\sum_{i|o_i=o \wedge t_i \in [t^* - w, now]} \Delta c_i}{\sum_{i|o_i=o \wedge t_i \in [t^* - w, now]} \Delta t_i} \qquad T(o) = \frac{\sum_{i|o_i=o \wedge t_i \in [0, now]} \Delta c_i}{\sum_{i|o_i=o \wedge t_i \in [0, now]} \Delta t_i} \qquad (1)$$

$$L = \frac{\sum_{i|t_i \in [t^* - w, now]} \Delta c_i}{\sum_{i|t_i \in [t^* - w, now]} \Delta t_i} \qquad T = \frac{\sum_{i|t_i \in [0, now]} \Delta c_i}{\sum_{i|t_i \in [0, now]} \Delta t_i} \qquad (2)$$

Intuitively, the local efficiency corresponds to estimating the gradient of the objective function with respect to the operator inside the evaluation window. If the operator was not selected during the window, its local efficiency is 0 which might be a pessimistic estimate. Therefore we propose to smooth the estimate by taking into account $T(o)$ normalized by the current context ratio L/T. The evaluation of an operator o is computed as:

$$weight(o) = (1 - \lambda) \cdot L(o) + \lambda \cdot \frac{L}{T} \cdot T(o) \qquad (3)$$

with $\lambda \in [0, 1]$ a balance factor between the two terms. As we desire to evaluate the operator mainly based on its local efficiency, we recommend that $\lambda < 0.5$.

3   Operator Portfolio

In this section, we present the relaxation and search operators that we propose to be part of the portfolio. All of them operate on a vector of integer decision variables. This list is based on our experience and the features available in the solver used for our experiments. Therefore it should not be considered as exhaustive.

Relaxation Heuristics

- Random: Relaxes randomly k variables by fixing the other ones to their value in the current best solution. This heuristic brings a good diversification and was demonstrated to be good despite its simplicity [24].
- Sequential: Relaxes randomly n sequences of k consecutive variables in the vector of decision variables. This heuristic should be efficient on problems where successive decision variables are related to each other, for instance in Lot sizing problems [25,26].

- Propagation Guided and Reversed Propagation Guided: Those heuristics are described in [27]. They consist of exploiting the amount of propagation induced when fixing a variable to freeze together sets of variables whose values are strongly dependent on each other.
- Value Guided: This heuristic uses the values assigned to the variables. We have five different variants: (1) Random Groups: Relaxing together groups of variables having the same value. This variant should be efficient on problems where values represent resources shared between variable such as bin-packing problems. (2) Max Groups: This variant relaxes the largest groups of variables having the same values. It can be useful for problems such as BACP or Assembly line balancing [28,29]. (3) Min Groups: This method relaxes the smallest groups of variables having the same value (which can be single variables). (4) Max Values: It consists in relaxing the k variables having the maximum values. We expect this heuristic to be efficient with problems involving a makespan minimization. (5) Min Values: This heuristic relaxes the k variables having the minimum values. It should be efficient in case of maximization problems.
- K Opt: This heuristic makes the hypothesis that the decision variables form a predecessor/successor model (where variable values indicate the next or previous element in a circuit). It is inspired by the k-opt moves used in local search methods for routing problems. The principle is to relax k edges in the circuit by selecting k variables randomly. The remaining variables have their domain restricted to only their successor and their predecessor in the current best solution in order to allow inversions of the circuit fragments.
- Precedency Based: This relaxation is useful for scheduling problems and hypothesizes that the decision variables corresponds to starting times of activities. It imposes a partial random order schedule as introduced in [17].
- Cost Impact: This operator was described in [24]. The heuristic consists in relaxing the variables that impact most the objective function when fixed.

Search Heuristics. A search heuristic explores the search space of the remaining unbounded variables by iteratively selecting a variable and one of its values to branch on. They can be separated into two components: a variable heuristic and a value heuristic. Here are the variable heuristics used:

- FirstFail tries first variables that have the most chances to lead to failures in order to maximize propagation during the search.
- Conflict Ordering proposed in [4] reorders dynamically the variables to select first the ones having led to the most recent conflicts.
- Weighted Degree introduced in [30] associates a weight to each variable. This weight is increased each time a constraint involving that variable fails.

In combination with these variable heuristics, we used different value heuristics which select the minimum/maximum/median/random value in the domain plus the *value sticking* [31] which remembers the last successful assigned values. We also permit a binary split of the domain into $\leq, >$ branching decisions.

4   Experiments

As in [9] we use an oracle baseline to compare with ALNS. Our baseline consists of a standard LNS with for each instance the best combination of operators (the one that reached the best objective value in the allocated time), chosen a posteriori. Notice that this baseline oracle is not the best theoretical strategy since it sticks with the same operator for all the iterations.

We implemented our framework in the OscaR constraint programming solver [32] where it is available in open-source. We tested our framework on 10 different constraint optimization problems with two arbitrarily chosen medium-sized instances per problem. We compare: (1) The original implementation from [22] (denoted *Laborie* here after) with a reaction factor α of 0.9. (2) The variant of [22] proposed in this article (denoted *Eval window*) with a sliding window $w = 10$ seconds and a balance factor λ of 0.05. (3) The oracle baseline.

Each approach was tested from the same initial solution (found for each instance using a first-fail, min-dom heuristic) with the same set of operators. We used relaxation sizes of $\{10\%, 30\%, 70\%\}$ and backtracks limits of $\{50\,bkts,$ $500\,bkts, 5000\,bkts\}$. We generated a different operator for each parameter(s) value(s) combination but kept the relaxation and reconstruction operators separated. We have 30 relaxation and 36 reconstruction operators, which yields a total of 1080 possible combinations to test for the baseline. Each ALNS variant was run 20 times with different random seeds on each instance for 240 s. We report our results in terms of cost values of the objective function for each instance. In order to compare the anytime behavior of the approaches, we define the *relative distance* of an approach at a time t as the current distance from the best known objective (BKO) divided by the distance of the initial solution: $(objective(t) - BKO)/(objective(0) - BKO)$. A relative distance of 0 thus indicates that the optimum has been reached.

We report the final results in Table 1. For each instance, we indicate the best known objective (BKO) and the results of the evaluated approaches after 240 s of LNS. For each approach, we report the average objective value (obj), the standard deviation (std) if applicable and the relative distance to the best known solution (rdist). The best results between the two evaluated approaches are displayed in bold. Figure 2 plots the average relative distance to the best known solution in function of the search time.

The results seem to indicate (at least on the tested instances) that the weight estimation based on an evaluation window tends to improve the performances of the original ALNS as described in [22]. The average relative distance to the best known solution is of 0.12 at the end of the search using the evaluation window, while it is of 0.18 using our implementation of [22]. None of the ALNS approaches is able to compete with the baseline (except on a few instances), but they obtain reasonably good solutions in a short amount of time. Furthermore, their any-time behavior is good when compared to the baseline and tends to get closer towards the end of the search.

Figure 1 shows a heat map of the relative selection frequency of the relaxation operators for each problem in the Eval window approach. The darker an entry,

Table 1. Experimental results

Instance	Problem	BKO	Baseline obj rdist		Eval window obj	std rdist	Laborie obj	std rdist
la13	JobShop	1150.00	1150.00	0.00	**1157.65**	12.06 **0.00**	1195.65	156.4 0.01
la17	(Lawrence-84)	784.00	784.00	0.00	784.05	0.22 0.00	**784.00**	0.00 **0.00**
chr22b	QAP	6194.00	6292.00	0.01	**6517.80**	115.16 **0.04**	6626.60	135.71 0.06
chr25a	(Christofides-89)	3796.00	3874.00	0.00	**4682.00**	393.52 **0.04**	4982.30	342.88 0.06
j120_11_3	RCPSP	188.00	228.00	0.08	420.75	129.61 0.48	**399.55**	62.88 **0.43**
j120_7_10	(Kolisch-95)	111.00	374.00	0.49	160.60	113.23 0.09	**127.70**	1.42 **0.03**
bench_7_1	Steel	1.00	6.00	0.03	**30.65**	11.69 **0.18**	49.75	8.92 0.29
bench_7_4	(CSPLib)	1.00	9.00	0.04	**24.30**	5.83 **0.12**	42.05	8.05 0.21
kroA200	TSP	29368.00	31466.00	0.01	**50022.35**	11159.91 **0.06**	156239.10	12097.24 0.37
kroB150	(Krolak-72)	26130.00	26141.00	0.00	**28596.20**	872.77 **0.01**	61658.70	6051.92 0.14
C103	VRPTW	82811.00	82814.00	0.00	**105876.10**	5553.99 **0.07**	137749.30	17371.97 0.17
R105	(Solomon-87)	137711.00	137261.00	0.00	**140162.30**	1593.46 **0.02**	144273.05	2051.77 0.05
t09-4	Cutstock	73.00	73.00	0.00	**76.65**	3.38 **0.10**	77.10	1.73 0.11
t09-7	(XCSP)	161.00	161.00	0.00	161.00	0.00 0.00	161.00	0.00 0.00
qwhopt-o18-h120-1	Graph colouring	17.00	17.00	0.00	17.00	0.00 0.00	17.00	0.00 0.00
qwhopt-o30-h320-1	(XCSP)	30.00	30.00	0.00	**541.40**	51.90 **0.59**	653.20	32.46 0.72
PSP_100_4	Lot sizing	8999.00	9502.00	0.03	**11796.85**	1204.18 **0.18**	14682.45	847.30 0.36
PSP_150_3	(Houndji-2014)	14457.00	16275.00	0.23	**18236.15**	569.80 **0.48**	19482.20	339.11 0.63
cap101	Warehouse	804126.00	804126.00	0.00	804599.55	428.34 0.00	**804556.50**	430.50 **0.00**
cap131	(XCSP)	910553.00	910553.00	0.00	**913147.60**	2701.98 **0.00**	913240.40	3197.29 0.00
Average				0.05		1246.05 **0.12**		2156.88 0.18

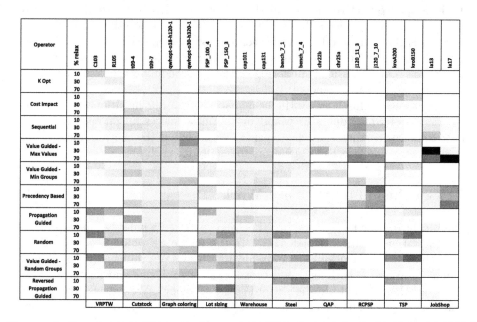

Fig. 1. Heat map of the relaxation operators selection for the eval window approach

the more frequently this operator was selected for the problem instance. Two interesting observations can be made. First, a subset of operators emerges more frequently for most of the problems. Second, this set varies between problems of different types, but is correlated between instances of the same problem. For some problems this set of operators is more uniform than others. For example,

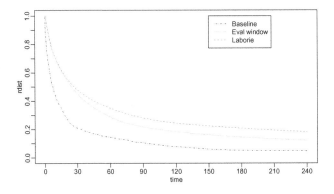

Fig. 2. Average relative distance to BKO during the search

on the warehouse location and the cutting stock problems the operators are selected rather uniformly. The job shop has a strong preference for the max-val and precedency operators. On the contrary, cost-impact is almost useless for the makespan objective of the job shop. Not surprisingly the RCPSP, also a scheduling problem, selects the same two operators as the job shop. The random operator is generally good except for scheduling problems. Due to space limitations, the heat map for Laborie is not given. The selection frequency obtained by the approach of [22] is more uniform, except for scheduling problems for which the same two operators emerge.

The results highlighted by the heat map confirm our intuition and a priori experience, of which operator would be the most successful on each problem. This comforts us that self-adaptive LNS could reach the performances of an expert that would select the operators manually for each problem.

5   Conclusion and Future Work

The weight update mechanism based on an evaluation window seems a promising adaptation for the original ALNS. In the future we would like to continue researching new relaxation operators for other types of problems (time-tabling, planing, etc) and experiment on a broader set of problems and instances. Parallelizing ALNS would also be an interesting challenge. We believe that ALNS would perform well in solver competitions such as [33,34] where the set of problems is very broad.

Acknowledgements. We thank the reviewers for their feedback. This work was funded by the Walloon Region (Belgium) as part of the PRESupply project.

References

1. Puget, J.-F.: Constraint programming next challenge: simplicity of use. In: Wallace, M. (ed.) CP 2004. LNCS, vol. 3258, pp. 5–8. Springer, Heidelberg (2004). https:// doi.org/10.1007/978-3-540-30201-8_2

2. Refalo, P.: Impact-based search strategies for constraint programming. In: Wallace, M. (ed.) CP 2004. LNCS, vol. 3258, pp. 557–571. Springer, Heidelberg (2004). https://doi.org/10.1007/978-3-540-30201-8_41

3. Hebrard, E., Siala, M.: Explanation-based weighted degree. In: Salvagnin, D., Lombardi, M. (eds.) CPAIOR 2017. LNCS, vol. 10335, pp. 167–175. Springer, Cham (2017). https://doi.org/10.1007/978-3-319-59776-8_13

4. Gay, S., Hartert, R., Lecoutre, C., Schaus, P.: Conflict ordering search for scheduling problems. In: Pesant, G. (ed.) CP 2015. LNCS, vol. 9255, pp. 140–148. Springer, Cham (2015). https://doi.org/10.1007/978-3-319-23219-5_10

5. Chu, G., Stuckey, P.J.: Learning value heuristics for constraint programming. In: Michel, L. (ed.) CPAIOR 2015. LNCS, vol. 9075, pp. 108–123. Springer, Cham (2015). https://doi.org/10.1007/978-3-319-18008-3_8

6. Michel, L., Van Hentenryck, P.: Activity-based search for black-box constraint programming solvers. In: Beldiceanu, N., Jussien, N., Pinson, É. (eds.) CPAIOR 2012. LNCS, vol. 7298, pp. 228–243. Springer, Heidelberg (2012). https://doi.org/10.1007/978-3-642-29828-8_15

7. Pesant, G., Quimper, C.G., Zanarini, A.: Counting-based search: branching heuristics for constraint satisfaction problems. J. Artif. Intell. Res. **43**, 173–210 (2012)

8. Vilím, P., Laborie, P., Shaw, P.: Failure-directed search for constraint-based scheduling. In: Michel, L. (ed.) CPAIOR 2015. LNCS, vol. 9075, pp. 437–453. Springer, Cham (2015). https://doi.org/10.1007/978-3-319-18008-3_30

9. Palmieri, A., Régin, J.-C., Schaus, P.: Parallel strategies selection. In: Rueher, M. (ed.) CP 2016. LNCS, vol. 9892, pp. 388–404. Springer, Cham (2016). https://doi. org/10.1007/978-3-319-44953-1_25

10. Picard-Cantin, É., Bouchard, M., Quimper, C.-G., Sweeney, J.: Learning the parameters of global constraints using branch-and-bound. In: Beck, J.C. (ed.) CP 2017. LNCS, vol. 10416, pp. 512–528. Springer, Cham (2017). https://doi.org/10. 1007/978-3-319-66158-2_33

11. Beldiceanu, N., Simonis, H.: A model seeker: extracting global constraint models from positive examples. In: Milano, M. (ed.) CP 2012. LNCS, pp. 141–157. Springer, Heidelberg (2012). https://doi.org/10.1007/978-3-642-33558-7_13

12. Shaw, P.: Using constraint programming and local search methods to solve vehicle routing problems. In: Maher, M., Puget, J.-F. (eds.) CP 1998. LNCS, vol. 1520, pp. 417–431. Springer, Heidelberg (1998). https://doi.org/10.1007/3-540-49481-2_30

13. Malitsky, Y., Mehta, D., O'Sullivan, B., Simonis, H.: Tuning parameters of large neighborhood search for the machine reassignment problem. In: Gomes, C., Sellmann, M. (eds.) CPAIOR 2013. LNCS, vol. 7874, pp. 176–192. Springer, Heidelberg (2013). https://doi.org/10.1007/978-3-642-38171-3_12

14. Schaus, P., Van Hentenryck, P., Monette, J.N., Coffrin, C., Michel, L., Deville, Y.: Solving steel mill slab problems with constraint-based techniques: CP, LNS, and CBLS. Constraints **16**(2), 125–147 (2011)

15. Jain, S., Van Hentenryck, P.: Large neighborhood search for dial-a-ride problems. In: Lee, J. (ed.) CP 2011. LNCS, vol. 6876, pp. 400–413. Springer, Heidelberg (2011). https://doi.org/10.1007/978-3-642-23786-7_31

16. Bent, R., Van Hentenryck, P.: A two-stage hybrid local search for the vehicle routing problem with time windows. Transp. Sci. **38**(4), 515–530 (2004)

17. Godard, D., Laborie, P., Nuijten, W.: Randomized large neighborhood search for cumulative scheduling. In: Biundo, S., et al. (eds.) Proceedings of the International Conference on Automated Planning and Scheduling ICAPS-05, pp. 81–89. Citeseer (2005)

18. Carchrae, T., Beck, J.C.: Principles for the design of large neighborhood search. J. Math. Model. Algorithms **8**(3), 245–270 (2009)

19. Gay, S., Schaus, P., De Smedt, V.: Continuous Casting Scheduling with Constraint Programming. In: O'Sullivan, B. (ed.) CP 2014. LNCS, vol. 8656, pp. 831–845. Springer, Cham (2014). https://doi.org/10.1007/978-3-319-10428-7_59

20. Monette, J.N., Deville, Y., Van Hentenryck, P.: Aeon: synthesizing scheduling algorithms from high-level models. In: Chinneck, J.W., Kristjansson, B., Saltzman, M.J. (eds.) Operations Research and Cyber-Infrastructure. Research/Computer Science Interfaces, vol. 47, pp. 43–59. Springer, Boston (2009). https://doi.org/10.1007/978-0-387-88843-9_3

21. Ropke, S., Pisinger, D.: An adaptive large neighborhood search heuristic for the pickup and delivery problem with time windows. Transp. sci. **40**(4), 455–472 (2006)

22. Laborie, P., Godard, D.: Self-adapting large neighborhood search: application to single-mode scheduling problems. Proceedings MISTA-07, Paris, vol. 8 (2007)

23. Pisinger, D., Ropke, S.: A general heuristic for vehicle routing problems. Comput. Oper. Res. **34**(8), 2403–2435 (2007)

24. Lombardi, M., Schaus, P.: Cost impact guided LNS. In: Simonis, H. (ed.) CPAIOR 2014. LNCS, vol. 8451, pp. 293–300. Springer, Cham (2014). https://doi.org/10.1007/978-3-319-07046-9_21

25. Fleischmann, B.: The discrete lot-sizing and scheduling problem. Eur. J. Oper. Res. **44**(3), 337–348 (1990)

26. Houndji, V.R., Schaus, P., Wolsey, L., Deville, Y.: The stockingcost constraint. In: O'Sullivan, B. (ed.) CP 2014. LNCS, vol. 8656, pp. 382–397. Springer, Cham (2014). https://doi.org/10.1007/978-3-319-10428-7_29

27. Perron, L., Shaw, P., Furnon, V.: Propagation guided large neighborhood search. In: Wallace, M. (ed.) CP 2004. LNCS, vol. 3258, pp. 468–481. Springer, Heidelberg (2004). https://doi.org/10.1007/978-3-540-30201-8_35

28. Monette, J.N., Schaus, P., Zampelli, S., Deville, Y., Dupont, P., et al.: A CP approach to the balanced academic curriculum problem. In: Seventh International Workshop on Symmetry and Constraint Satisfaction Problems, vol. 7 (2007)

29. Schaus, P., Deville, Y., et al.: A global constraint for bin-packing with precedences: application to the assembly line balancing problem. In: AAAI (2008)

30. Boussemart, F., Hemery, F., Lecoutre, C., Sais, L.: Boosting systematic search by weighting constraints. In: Proceedings of the 16th European Conference on Artificial Intelligence, pp. 146–150. IOS Press (2004)

31. Frost, D., Dechter, R.: In search of the best constraint satisfaction search (1994)

32. OscaR Team: OscaR: Scala in OR (2012). https://bitbucket.org/oscarlib/oscar

33. Stuckey, P.J., Feydy, T., Schutt, A., Tack, G., Fischer, J.: The minizinc challenge 2008–2013. AI Mag. **35**, 55–60 (2014)

34. Boussemart, F., Lecoutre, C., Piette, C.: Xcsp3: an integrated format for benchmarking combinatorial constrained problems. arXiv preprint arXiv:1611.03398 (2016)

A Warning Propagation-Based Linear-Time-and-Space Algorithm for the Minimum Vertex Cover Problem on Giant Graphs

Hong Xu$^{(\boxtimes)}$ⓘ, Kexuan Sunⓘ, Sven Koenig, and T. K. Satish Kumar

University of Southern California, Los Angeles, CA 90089, USA
{hongx,kexuansu,skoenig}@usc.edu, tkskwork@gmail.com

Abstract. A vertex cover (VC) of a graph G is a subset of vertices in G such that at least one endpoint vertex of each edge in G is in this subset. The minimum VC (MVC) problem is to identify a VC of minimum size (cardinality) and is known to be NP-hard. Although many local search algorithms have been developed to solve the MVC problem close-to-optimally, their applicability on giant graphs (with no less than 100,000 vertices) is limited. For such graphs, there are two reasons why it would be beneficial to have linear-time-and-space algorithms that produce small VCs. Such algorithms can: (a) serve as preprocessing steps to produce good starting states for local search algorithms and (b) also be useful for many applications that require finding small VCs quickly. In this paper, we develop a new linear-time-and-space algorithm, called MVC-WP, for solving the MVC problem on giant graphs based on the idea of warning propagation, which has so far only been used as a theoretical tool for studying properties of MVCs on infinite random graphs. We empirically show that it outperforms other known linear-time-and-space algorithms in terms of sizes of produced VCs.

1 Introduction

Thanks to the advancement of technologies such as the Internet and database management systems, datasets have been growing tremendously over the past decade. Many of the resulting datasets can be modeled as graphs, such as social networks, brain networks, and street networks. Therefore, it is essential to develop algorithms to solve classical combinatorial problems on giant graphs (with no less than 100,000 vertices).

A *vertex cover* (VC) on an undirected graph $G = \langle V, E \rangle$ is defined as a set of vertices $S \subseteq V$ such that every edge in E has at least one of its endpoint vertices in S. A *minimum VC* (MVC) is a VC on G of minimum size (cardinality), i.e., there exists no VC whose size is smaller than that of an MVC. The MVC problem is to find an MVC on a given graph G. Its decision version is known to be NP-complete [18]. An *independent set* (IS) on G is a set of vertices $T \subseteq V$ such that no two vertices in T are adjacent to each other. The complement of a (maximum)

© Springer International Publishing AG, part of Springer Nature 2018
W.-J. van Hoeve (Ed.): CPAIOR 2018, LNCS 10848, pp. 567–584, 2018.
https://doi.org/10.1007/978-3-319-93031-2_41

IS is a (minimum) VC and vice versa, i.e., for any (maximum) IS T, $V \setminus T$ is always a (minimum) VC.

The MVC problem has been widely used to study various real-world and theoretical problems. For example, in practice, it has been used in computer network security [11], in crew scheduling [24], and in the construction of phylogenetic trees [1]. In theoretical research, it has been used to prove the NP-completeness of various other well-known problems, such as the set cover problem and the dominating set problem [19]. It is also a fundamental problem studied in the theory of fixed-parameter tractability [13].

Various researchers have developed exact solvers [10,21,27,29] for the MVC problem and its equivalents. However, none of these solvers work well for large problem instances of the MVC problem due to its NP-hardness. Furthermore, solving the MVC problem within any approximation factor smaller than 1.3606 is also NP-hard [8].

To overcome the poor efficiency of exact algorithms and the high approximation factor of polynomial-time approximation algorithms, researchers have focused on developing non-exact local search algorithms [2,5,6,22] for solving the MVC problem and its equivalents. These algorithms often require a preprocessing step to construct a VC (usually the smaller the better) before starting the local search. While polynomial-time procedures work well for the preprocessing step on regular-sized graphs, they are prohibitively expensive on giant graphs. On giant graphs, this preprocessing step needs to terminate fast and should use only a moderate amount of memory. Therefore, it is important to develop a linear-time-and-space algorithm to find a small VC.

In addition, many real-world applications on giant graphs require the identification of small VCs but not necessarily MVCs. One example of such applications is the influence-maximization problem in social networks [14]. Here, too, linear-time-and-space algorithms for finding small VCs are important.

In this paper, we develop a new linear-time-and-space algorithm, called MVC-WP, for solving the MVC problem on giant graphs based on the idea of warning propagation, which has so far only been used as a theoretical tool for studying properties of MVCs on infinite random graphs. We then empirically show that MVC-WP has several advantages over other linear-time-and-space algorithms. We also experiment with variants of MVC-WP to empirically demonstrate the usefulness of various steps in it.

2 Background

In this section, we introduce relevant background on random graph models, warning propagation, and existing linear-time-and-space MVC algorithms known to the authors.

2.1 Random Graph Models

The Erdős-Rényi Model. An *Erdős-Rényi model* (ER model) [9] is characterized by two parameters n and p. It generates random graphs with n vertices

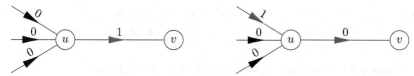

(a) u sends a message of 1 to v since all other incoming messages from its other neighbors are 0.

(b) u sends a message of 0 to v since at least one of its incoming messages from its other neighbors is 1.

Fig. 1. Illustrates the update of a message from $u \in V$ to $v \in V$ in the warning propagation algorithm for the MVC problem on graph $G = \langle V, E \rangle$. Only relevant parts of G are shown, i.e., u, v, and all edges incident to u.

and connects every pair of vertices with probability p. We call a graph generated by an ER model an *ER graph*. The degrees of the vertices of an ER graph follow a Poisson distribution. The average degree of vertices is $c = np$.

The Scale-Free Model. A *scale-free model* (SF model) [4] is characterized by two parameters n and $\lambda > 2$. It generates random graphs whose vertex degree distribution follows a power law, i.e., $P(d) \sim d^{-\lambda}$. The average degree of vertices is therefore

$$c = \sum_{d=1}^{+\infty} P(d)d = \frac{\zeta(\lambda - 1)}{\zeta(\lambda)}, \tag{1}$$

where $\zeta(x) = \sum_{k=1}^{\infty} \frac{1}{k^x}$ is the Riemann zeta function. For notational convenience, we define $Z(\lambda) = \frac{\zeta(\lambda-1)}{\zeta(\lambda)}$. We call a graph generated by an SF model an *SF graph*.

2.2 Warning Propagation

The *warning propagation algorithm* is a specialized *message passing algorithm* where information is processed locally and exchanged between relevant variables [20]. In the warning propagation algorithm, messages can only take one of two values, namely 0 or 1. To analyze properties of MVCs on infinite random graphs, [26] proposed an algorithm that uses warning propagation for solving the MVC problem to help with their theoretical analysis. In their algorithm, messages are passed between adjacent vertices. A message of 1 from $u \in V$ to $v \in V$ indicates that u is not in the MVC and thus it "warns" v to be included in the MVC. Otherwise, if u is in the MVC, this message would be 0. Based on this intuition, the warning propagation algorithm updates messages according to the following rules: A message from u to v is updated to 1 iff all incoming messages to u from its other neighbors equal 0, i.e., no other adjacent vertices of u require u to be in the VC. Otherwise, this message would be 0. Figure 1 illustrates these rules. Upon convergence, vertices with at least one incoming messages equal to 1 are included in the VC, and other vertices are excluded from the VC. The theoretical analysis in [26] mainly focuses on ER graphs. It shows that, on an infinitely large ER graph, a message is 1 with probability $W(c)/c$, where $W(\cdot)$ is the Lambert-W function, i.e., the inverse function of $f(x) = xe^x$.

2.3 Known Linear-Time-and-Space MVC Algorithms

MVC-2. This well-known linear-time-and-space factor-2 approximation algorithm for the MVC problem works as follows [25]: In each iteration, MVC-2 first arbitrarily selects an uncovered edge, then marks it as well as the edges incident to its two endpoint vertices as being covered, and finally adds its endpoint vertices to the VC. It terminates when all edges are marked as covered.

ConstructVC. Serving as a preprocessing step, this algorithm is a greedy linear-time-and-space subroutine in the FastVC solver [5], that constructs a minimal VC[1]. It works as follows: In each iteration, ConstructVC first arbitrarily selects an uncovered edge, then adds its endpoint vertex v with the larger degree to the VC, and finally marks all edges incident to v as being covered. When all edges are marked as covered, it removes all redundant vertices in the VC to construct a minimal VC.

R. This algorithm is used as the preprocessing step to produce a maximal IS (complement of a minimal VC) in the local search algorithm for solving the maximum IS problem developed by [2]. R can be easily adjusted to produce a minimal VC, and the adapted algorithm works as follows: R first adds all vertices into the VC. In each iteration, R randomly removes a vertex v from the VC if it continues to be a VC after the removal. It terminates when the VC is minimal.

MVC-MPL and MVC-L. MVC-MPL is a linear-time-and-space MVC algorithm based on some theoretical results of warning propagation on ER graphs. It works as follows [28]: In each iteration, MVC-MPL first arbitrarily selects a vertex v, then adds it to the IS with probability $(1 - W(c)/c)^{\kappa(v)}$, where $\kappa(v)$ is the degree of v, and otherwise to the VC. It terminates when every vertex has been added to either the VC or the IS. MVC-L is a variant of MVC-MPL with the probability of adding a vertex v to the IS replaced by $1/(\kappa(v) + 1)$ [28].

3 Warning Propagation on Scale-Free Graphs

Assuming that the warning propagation algorithm is applied on an SF graph, we derive the approximate message distribution upon convergence by following a method similar to that in [26, Sect. IV.B]. We use p_0 and p_1 to denote the fractions of all messages with values 0 and 1 upon convergence, respectively. Clearly, we have

$$p_0 + p_1 = 1. \tag{2}$$

---

[1] A minimal VC is a VC such that no proper subset thereof is also a VC.

A message $m_{u \to v}$ from vertex u to vertex v is equal to 1 iff all incoming messages to u from its other neighbors are equal to 0, i.e., $\forall w \in \partial u \setminus v : m_{w \to u} = 0$, where ∂u is the set of vertices adjacent to u. Assuming that all messages incoming to u are independent and using the fact that the probability distribution of the number of such messages follows a power law on an SF graph, we have

$$1 - p_0 = p_1 = \sum_{d=1}^{\infty} \frac{d^{-\lambda}}{\zeta(\lambda)} p_0^{d-1} = \frac{\text{Li}_\lambda(p_0)}{p_0 \zeta(\lambda)}, \tag{3}$$

where $\text{Li}_\lambda(x) = \sum_{k=1}^{\infty} \frac{x^k}{k^\lambda}$ is the polylogarithm function. After making the approximation $\text{Li}_\lambda(p_0) \approx p_0 + \frac{p_0^2}{2^\lambda}$, we solve Eq. (3) for p_0 and have

$$p_0 = \frac{\zeta(\lambda) - 1}{\zeta(\lambda) + \frac{1}{2^\lambda}}, \tag{4}$$

where $\forall \lambda > 2 : 0 \leq p_0 \leq 1$. Therefore, for any $\lambda > 2$, Eq. (4) is always a valid solution for p_0.

4 The Algorithm

Our algorithm MVC-WP (Algorithm 1) is based on the analytical results that govern the warning propagation algorithm for the MVC problem [26]. MVC-WP first uses Algorithm 2, an algorithm that prunes leaves, to identify those vertices that are necessarily in some MVC and modifies the input graph accordingly. It then treats this modified graph as if it were an ER or SF graph and computes p_0 using Algorithm 3. (Although MVC-WP treats the graph as if it were an ER or SF graph, it does not impose any restrictions on the graph.) MVC-WP then assigns each message from vertex u to vertex v to be 1 with probability $p_0^{\kappa(u)-1}$, where $\kappa(u)$ denotes the degree of u. This is done under the assumption that all incoming messages of u have independent probabilities to be 0 or 1. Then, MVC-WP performs warning propagation for M iterations, where M is a given parameter. After M iterations, v is marked as being included in VC if it receives at least one message of 1; otherwise, v is marked as being excluded in VC. If v is excluded, MVC-WP marks all its adjacent vertices as being included in VC. Finally, MVC-WP uses Algorithm 4 to remove redundant vertices from VC to make it a minimal VC. This step is adapted from Lines 6 to 14 of Algorithm 2 in [5].

We note that, a warning propagation iteration in the warning propagation algorithm proposed in [26] is not linear-time due to the requirement of traversing incoming messages of vertex u when updating the message from vertex u to vertex v. To avoid this traversal and thus make each warning propagation iteration linear-time, for each vertex v, MVC-WP keeps track the number of messages incoming to v that are equal to 1 in an array *counter*. This and the value of the message from u to v provide enough information for updating the message.

We also note that, while MVC-WP is based on the analytical results from [26], it differs significantly from the warning propagation algorithm proposed in [26].

Algorithm 1. MVC-WP.

```
 1  Function MVC-WP(G = ⟨V, E⟩, model, M)
        Input: G: The graph to find an MVC for.
        Input: model: The random graph model to use (ER or SF).
        Input: M: Number of iterations of the warning propagation algorithm.
        Output: A minimal VC of G.
 2      VC, IS := Prune-Leaves(G);
 3      p₀ := Compute-p₀(G, model);
 4      Convert G to a directed graph G' = ⟨V, E'⟩ by introducing ⟨u, v⟩ and ⟨v, u⟩ in E' for
        each (u, v) ∈ E;
 5      Build an associative array m for the edges of G' to represent messages;
 6      Build an associative array counter for the vertices of G' to record the number of
        incoming messages that are equal to 1;
 7      Initialize counter to zeros;
 8      • Initialize messages:
 9          for each ⟨u, v⟩ ∈ E' do
10              Draw a random number r ∈ [0, 1];
11              if r ≤ p₀^(κ(u)−1) then
12                  m_{u→v} := 1;
13                  counter(v) := counter(v) + 1;
14              else
15                  m_{u→v} := 0;

16      • Run M iterations of the warning propagation algorithm:
17          for t := 1, ..., M do
18              for each ⟨u, v⟩ ∈ E' do
19                  if counter(u) − m_{v→u} = 0 then
20                      if m_{u→v} = 0 then
21                          m_{u→v} := 1;
22                          counter(v) := counter(v) + 1;
23                      else
24                          if m_{u→v} = 1 then
25                              m_{u→v} := 0;
26                              counter(v) := counter(v) − 1;

27      • Construct a VC:
28          while ∃v ∈ V \ (VC ∪ IS) do
29              v := any vertex in V \ (VC ∪ IS);
30              if counter(v) = 0 then
31                  Add v to IS and all u in ∂v to VC;
32              else
33                  Add v to VC;

34      return Remove-Redundancy(G, VC);
```

MVC-WP introduces preprocessing and postprocessing steps before and after warning propagation iterations. It also initializes messages intelligently. In addition, MVC-WP reduces the time complexity of a warning propagation iteration to linear by using a *counter* array. Most importantly, MVC-WP is specifically designed for being practically run, while the warning propagation algorithm proposed in [26] lacks many algorithmic details, since [26] only uses it as a theoretical tool to study properties of MVCs on ER graphs.

We now formally prove the correctness and time and space complexities of MVC-WP.

Theorem 1. *MVC-WP produces a minimal VC.*

Algorithm 2. Prune leaves.

```
 1  Function Prune-Leaves(G = ⟨V, E⟩)
        Modified: G: The input graph.
 2      Initialize vertex sets VC and IS to the empty set;
 3      for each v ∈ V do
 4          Prune-A-Leaf(G, VC, IS, v);

 5      return VC, IS;

 6  Function Prune-A-Leaf(G = ⟨V, E⟩, VC, IS, v)
        Modified: G: The input graph.
        Modified: VC: The current VC.
        Modified: IS: The current IS.
        Input: v: A vertex in V.
 7      if κ(v) = 1 then
 8          u := the only vertex adjacent to v;
 9          VC := VC ∪ {u};
10          IS := IS ∪ {v};
11          U := ∂u \ {v};
12          Remove v and (u, v) from G;
13          for each w ∈ U do
14              Remove (u, w) from G;
15              Prune-A-Leaf (G, VC, IS, w);

16          Remove u from G;
```

Algorithm 3. Compute p_0 for different random graph models.

```
 1  Function Compute-p₀(G, model)
        Input: G: The input graph.
        Input: model: The random graph model to use (ER or SF).
 2      c := average degree of vertices in G;
 3      if model is ER then
 4          p₀ := 1 − W(c)/c;
 5      else if model is SF then
 6          λ := Z⁻¹(c);
 7          Compute p₀ according to Eq. (4);

 8      return p₀;
```

Proof. Since [5] has proved that `Remove-Redundancy` produces a minimal VC provided that variable VC in Algorithm 1 is a VC, it is sufficient to prove that, right before Line 34 in Algorithm 1, variable IS is an IS, $VC \cup IS = V$, and $VC \cap IS = \emptyset$.

In Algorithm 2, Lines 12 and 14 are the only steps that remove edges. However, these edges are covered by VC as shown on Line 9. In addition, $VC \cup IS$ is the set of all removed vertices and $VC \cap IS = \emptyset$, since each removed vertex is added to either VC or IS. Therefore, IS is an independent set.

In Algorithm 1, message initialization and the M iterations of warning propagation do not change the values of IS and VC.

In Lines 27 and 33 in Algorithm 1, since Line 31 guarantees that no two adjacent vertices are added to IS, IS must be an IS of G. In addition, Line 28 guarantees $IS \cup VC = V$ and Lines 28, 31 and 33 guarantee $IS \cap VC = \emptyset$.

Therefore, this theorem is true.

Theorem 2. *The time complexity of MVC-WP is* $\mathcal{O}(|V| + |E|)$.

Algorithm 4. Remove redundant vertices from a given VC [5].

---

1 **Function** Remove-Redundancy($G = \langle V, E \rangle$, VC)
    Input: G: The input graph.
    Input: VC: A VC of G.
2   Build an associative array *loss* for vertices in VC to record whether they can be removed from VC;
3   Initialize *loss* to zeros;
4   **foreach** $e \in E$ **do**
5   **if** *only one endpoint vertex v of e is in VC* **then**
6             $loss(v) := 1$;

7   **foreach** $v \in VC$ **do**
8   **if** $loss(v) = 0$ **then**
9             $VC := VC \setminus \{v\}$;
10   **foreach** $v' \in \partial v \cap VC$ **do**
11                 $loss(v') := 1$;

12   **return** VC;

Proof. We first prove that Prune-Leaves terminates in $\mathcal{O}(|V| + |E|)$ time by counting the number of times that Prune-A-Leaf is called, since the only loop in Prune-A-Leaf makes only one recursive call in each iteration. Line 4 in Algorithm 2 calls Prune-A-Leaf at most $|V|$ times. Line 15 calls Prune-A-Leaf iff edge (u, w) is removed from G. Therefore, Line 15 calls Prune-A-Leaf at most $|E|$ times.

Obviously, Compute-p_0 terminates in constant time.

In Algorithm 1, Lines 8 to 15 iterate over each edge in G' exactly once, and therefore terminate in $\mathcal{O}(|E|)$ time; Lines 16 to 26 iterate over each edge in G' exactly M times, and therefore terminate in $\mathcal{O}(|E|)$ time; Lines 27 to 33 consider each vertex v in G' at least once and at most $\kappa(v)$ times, and therefore terminate in $\mathcal{O}(|V| + |E|)$ time.

[5] has proved that Remove-Redundancy terminates in $\mathcal{O}(|V| + |E|)$ time.

Combining the results above, MVC-WP terminates in $\mathcal{O}(|V| + |E|)$ time.

Theorem 3. *The space complexity of MVC-WP is $\mathcal{O}(|V| + |E|)$.*

Proof. [5] has proved that Remove-Redundancy uses $\mathcal{O}(|V| + |E|)$ space. The recursive calls of Prune-A-Leaf initiated in Prune-Leaves use $\mathcal{O}(|E|)$ stack space. The remaining steps in MVC-WP require $\mathcal{O}(|E|)$ space to store messages and $\mathcal{O}(|V|)$ space to store *counter* as well as the status of each vertex v, i.e., whether v is in VC, IS or undetermined yet. Therefore, MVC-MP uses $\mathcal{O}(|V| + |E|)$ space.

4.1 Computing Special Functions

In Algorithm 3, we are required to compute a few special functions, namely the Lambert-W function $W(\cdot)$, the Riemann zeta function $\zeta(\cdot)$ and the inverse function of $Z(\cdot)$. For some of these functions, researchers in the mathematics

Table 1. Shows the values of $\zeta(k)$ and $Z(k) = \frac{\zeta(k-1)}{\zeta(k)}$ for $k \in \{1, 2, \ldots, 9\}$. The values of $\zeta(k)$ are taken from [15, Table 23.3], and the values of $Z(k)$ are computed from the values of $\zeta(k)$.

k	1	2	3	4	5	6	7	8	9
$\zeta(k)$	$+\infty$	1.645	1.202	1.082	1.037	1.017	1.008	1.004	1.002
$Z(k)$	-	$+\infty$	1.369	1.111	1.043	1.020	1.009	1.004	1.002

community have already developed various numerical methods [7,16]. However, they are too slow for MVC-WP, which does not critically need this high accuracy. We now present a few new approaches to quickly compute them sufficiently accurately.

4.2   The Lambert-W Function $W(\cdot)$

We approximate $W(\cdot)$ via the first 3 terms of Eq. (4.19) in [7], i.e.,

$$W(c) = L_1 - L_2 + L_2/L_1 + \mathcal{O}\left((L_2/L_1)^2\right), \tag{5}$$

where $L_1 = \log c$ and $L_2 = \log L_1$.

4.3   The Riemann Zeta Function $\zeta(\cdot)$

For the SF model, we need to compute $\zeta(\lambda)$ in Eq. 4 for a given λ. To compute $\zeta(\lambda)$, we approximate $\zeta(\lambda)$ via its first 20 terms, i.e., $\zeta(\lambda) = \sum_{k=1}^{20} \frac{1}{k^\lambda} + O(\frac{1}{21^\lambda})$. This is sufficient, because $\lambda > 2$ always holds in MVC-WP due to Line 6 in Algorithm 3 since $\forall c \geq 1 : Z^{-1}(c) > 2$. In this case, the sum of the remaining terms is sufficiently small and can thus be neglected, because

$$\frac{\sum_{k=21}^{\infty} \frac{1}{k^\lambda}}{\sum_{k=1}^{\infty} \frac{1}{k^\lambda}} \leq \frac{\sum_{k=21}^{\infty} \frac{1}{k^\lambda}}{\sum_{k=1}^{\infty} \frac{1}{k^2}} \approx 0.030. \tag{6}$$

4.4   The Inverse Function of $Z(\cdot)$

- For any $x < 1.002$, we approximate $Z^{-1}(x)$ to be equal to $+\infty$ (and thus approximate p_0 to be equal to 0 in Algorithm 3).
- For any $1.002 \leq x \leq 1.369$, we approximate $Z^{-1}(x)$ via linear interpolation according to Table 1, i.e., we assume $Z^{-1}(x)$ changes linearly between two consecutive entries given in Table 1.
- For any $x > 1.369$, we have $2 < k = Z^{-1}(x) < 3$. In this case, we approximate $\zeta(k)$ via linear interpolation, i.e.,

$$\zeta(k) \approx 1.645 - 0.443 \cdot (k - 2). \tag{7}$$

Table 2. Compares sizes of VCs produced by MVC-WP-ER and MVC-WP-SF, respectively, with those of alternative algorithms. The three numbers in the 3rd to 6th columns represent the numbers of benchmark instances on which MVC-WP-ER and MVC-WP-SF produce smaller/equal/larger VC sizes, respectively. The numbers in parentheses indicate the number of benchmark instances in each benchmark instance set.

Our algorithm	Alternative algorithm	Misc (397)	Web (18)	Street (8)	Brain (26)
MVC-WP-ER	ConstructVC	211/39/147	12/1/5	8/0/0	0/0/26
	MVC-2	241/46/110	16/1/1	8/0/0	26/0/0
	R	376/16/5	17/1/0	8/0/0	26/0/0
	MVC-L	364/19/14	17/1/0	8/0/0	26/0/0
	MVC-MPL	317/18/62	17/1/0	1/0/7	26/0/0
MVC-WP-SF	ConstructVC	209/38/150	11/1/6	8/0/0	0/0/26
	MVC-2	249/45/103	15/1/2	8/0/0	26/0/0
	R	377/15/5	17/1/0	8/0/0	26/0/0
	MVC-L	363/21/13	17/1/0	8/0/0	26/0/0
	MVC-MPL	316/18/63	17/1/0	1/0/7	26/0/0

We approximate $\zeta(k-1)$ via the first three terms of the Laurent series of $\zeta(k-1)$ at $k = 2$, i.e.,

$$\zeta(k-1) = \frac{1}{k-2} + \gamma - \gamma_1(k-2) + \mathcal{O}\left((k-2)^2\right),   \quad (8)$$

where $\gamma \approx 0.577$ is the Euler-Mascheroni constant and $\gamma_1 \approx -0.0728$ is the first Stieltjes constant [12, p. 166]. By plugging these two equations into the definition of $Z(k)$ (i.e., $Z(k) = \frac{\zeta(k-1)}{\zeta(k)} = x$) and solving for k, we have the approximation

$$Z^{-1}(x) \approx \frac{1.645x - \gamma - \sqrt{(1.645x - \gamma)^2 - 4(0.443x - \gamma_1)}}{2 \cdot (0.443x - \gamma_1)} + 2.   \quad (9)$$

5   Experimental Evaluation

In this section, we experimentally evaluate MVC-WP. In our experiments, all algorithms were implemented in C++, compiled by GCC 6.3.0 with the "-O3" option, and run on a GNU/Linux workstation with an Intel Xeon Processor E3-1240 v3 (8 MB Cache, 3.4 GHz) and 16 GB RAM. Throughout this section, we refer to MVC-WP using an ER model and an SF model as MVC-WP-ER and MVC-WP-SF, respectively.

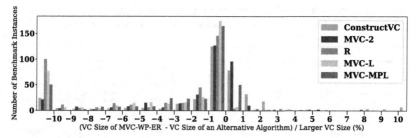

(a) MVC-WP-ER versus ConstructVC/MVC-2/R/MVC-L/MVC-MPL on the misc networks benchmark instance set.

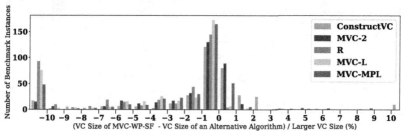

(b) MVC-WP-SF versus ConstructVC/MVC-2/R/MVC-L/MVC-MPL on the misc networks benchmark instance set.

Fig. 2. Compares sizes of VCs produced by MVC-WP-ER, MVC-WP-SF, and alternative algorithms on the misc networks benchmark instance set. The x-axes show the relative suboptimality of MVC-WP-ER and MVC-WP-SF, respectively, compared with alternative algorithms. The y-axes show the number of benchmark instances for a range of relative suboptimality divided into bins of 1% (ranges beyond −10% and 10% are treated as single bins). Bars of different colors indicate different algorithms. Higher bars in the left half indicate that MVC-WP-ER and MVC-WP-SF, respectively, produce VCs of sizes smaller than the alternative algorithms. (Color figure online)

We used 4 sets of benchmark instances[2]. The first 3 sets of benchmark instances were selected from the "misc networks", "web networks", and "brain networks" categories in Network Repository[3] [23]. All instances with no less than 100,000 vertices as of July 8, 2017 were used. The fourth set of benchmark instances consists of the benchmark instances in the "street networks" category in the 10th DIMACS Implementation Challenge[4] [3], in which 7 out of 8 benchmark instances have more than 1 million vertices. To obviate the influence of the orders in which the edges are specified in the input files, we shuffled the edges for each benchmark instance before applying the algorithms.

---

[2] We compiled these benchmark instances in the DIMACS format and made them available online at http://files.hong.me/papers/xu2018b-data.

[3] http://networkrepository.com/.

[4] http://www.cc.gatech.edu/dimacs10/archive/streets.shtml.

Table 3. Shows the number of vertices and edges of benchmark instances in the web networks, street networks, and brain networks benchmark instance sets.

	Instance	|V|	|E|
Web Networks	web-wikipedia-link-it	1,051,219	25,199,339
	web-wikipedia-growth	898,367	4,468,005
	web-BerkStan	5,121	8,345
	web-italycnr-2000	176,055	2,336,551
	web-uk-2005	127,716	11,643,622
	web-Stanford	226,733	1,612,323
	web-BerkStan-dir	552,353	5,674,493
	web-google-dir	451,765	2,434,390
	web-wikipedia2009	154,344	302,990
	web-it-2004	424,893	6,440,816
	web-wikipedia-link-fr	1,098,517	12,683,034
	web-hudong	53,799	286,998
	web-arabic-2005	102,515	1,560,020
	web-baidu-baike	56,346	104,037
	web-NotreDame	99,557	631,931
	web-sk-2005	42,237	225,932
	web-wiki-ch-internal	21,101	47,480
	web-baidu-baike-related	126,607	710,562
Street Networks	asia	2,642,989	3,032,404
	germany	2,455,500	2,673,629
	great-britain	1,284,868	1,383,591
	luxembourg	23,196	24,710
	belgium	460,536	503,521
	netherlands	726,730	815,305
	italy	1,783,377	1,942,410
	europe	12,512,346	13,711,218

	Instance	|V|	|E|
Brain Networks	0025871-session-1-bg	738,598	168,617,323
	0025872-session-2-bg	759,626	147,761,328
	0025869-session-1-bg	679,760	134,979,814
	0025876-session-1-bg	778,074	140,293,764
	0025865-session-2-bg	705,588	155,118,679
	0025868-session-1-bg	717,428	150,383,991
	0025872-session-1-bg	746,316	166,528,410
	0025864-session-2-bg	682,197	133,656,879
	0025912-session-2	771,224	147,496,369
	0025868-session-2-bg	717,420	158,562,090
	0025869-session-2-bg	705,280	151,476,861
	0025873-session-1-bg	636,430	149,483,247
	0025870-session-2-bg	799,455	166,724,734
	0025865-session-1-bg	725,412	165,845,120
	0025889-session-2	731,931	131,860,075
	0025876-session-2-bg	766,763	139,801,374
	0025867-session-1-bg	735,513	145,208,968
	0025873-session-2-bg	758,757	163,448,904
	0025873-session-2-bg	682,580	140,044,477
	0025889-session-1	694,544	144,411,722
	0025870-session-1-bg	785,719	148,684,011
	0025871-session-2-bg	724,848	170,944,764
	0025864-session-1-bg	685,987	143,091,223
	0025867-session-2-bg	724,276	154,604,919
	0025878-session-1-bg	690,012	127,838,275
	0025886-session-1	769,878	158,111,887

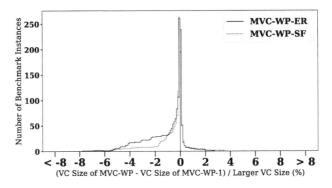

Fig. 3. Compares sizes of VCs produced by MVC-WP-ER and MVC-WP-SF with those produced by MVC-WP-1. The x-axis shows the relative suboptimality of MVC-WP compared with MVC-WP-1. The y-axis shows the number of benchmark instances. In the left half, for each point on the curve, its y coordinate shows the number of benchmark instances with relative suboptimality smaller than its x coordinate. In the right half, for each point on the curve, its y coordinate shows the number of benchmark instances with relative suboptimality larger than its x coordinate. Larger areas under the curves in the left half and smaller areas under the curves in the right half indicate that MVC-WP-ER and MVC-WP-SF, respectively, produce VCs of sizes smaller than MVC-WP-1.

To evaluate those algorithms that use random number generators, i.e., MVC-WP-ER, MVC-WP-SF, R, MVC-L and MVC-MPL, we ran them 10 times on each benchmark instance using different seeds. We recorded the average of the VC sizes produced by these 10 runs. For all algorithms compared in this section, we applied `Prune-Leaves` and `Remove-Redundancy` as preprocessing and postprocessing steps, respectively, since they are universally useful.

Table 4. Compares sizes of VCs produced by MVC-WP-ER, MVC-WP-SF, and alternative algorithms on the web networks, street networks, and brain networks benchmark instance sets. The numbers of vertex and edge of each benchmark instance are shown in Table 3. The smallest sizes of VCs produced for each benchmark instance are highlighted.

Instance	ConstructVC	MVC-2	R	MVC-L	MVC-MPL	MVC-WP-ER	MVC-WP-SF
web-wikipedia-link-it	991,272	987,621	1,039,011	1,018,672	1,020,827	972,275	972,670
web-wikipedia-growth	914,746	926,530	966,410	950,741	940,302	909,910	909,989
web-BerkStan	5,542	5,469	5,605	5,567	5,726	5,463	5,469
web-italycnr-2000	99,645	99,609	110,559	104,272	103,153	97,844	97,932
web-uk-2005	127,774	127,774	127,774	127,774	127,774	127,774	127,774
web-Stanford	126,603	126,960	136,048	130,248	128,296	123,540	123,890
web-BerkStan-dir	290,206	291,277	320,384	310,483	304,623	285,593	285,934
web-google-dir	350,676	355,930	387,462	379,508	365,427	351,241	351,343
web-wikipedia2009	650,888	654,152	657,343	656,131	655,813	652,241	652,363
web-it-2004	415,408	415,083	415,915	415,533	415,042	414,835	414,972
web-wikipedia-link-fr	1,574,973	1,558,998	1,626,052	1,598,021	1,597,887	1,538,658	1,538,960
web-hudong	503,373	504,025	506,598	505,903	504,839	503,335	503,359
web-arabic-2005	114,504	114,743	115,161	114,999	115,004	114,721	114,727
web-baidu-baike	637,805	638,538	640,537	639,811	639,935	637,796	637,815
web-NotreDame	76,468	76,257	80,341	79,013	77,893	75,735	75,953
web-sk-2005	58,238	58,300	58,669	58,443	58,370	58,347	58,349
web-wiki-ch-internal	260,354	260,476	261,571	261,244	260,927	260,213	260,231
web-baidu-baike-related	144,388	146,588	151,689	149,957	148,749	145,272	145,249
asia	6,087,218	6,099,227	6,130,265	6,104,699	6,018,875	6,053,077	6,049,489
germany	5,822,566	5,834,966	5,864,005	5,841,507	5,768,621	5,792,165	5,789,947
great-britain	3,837,647	3,843,986	3,857,098	3,844,972	3,804,317	3,821,741	3,820,618
luxembourg	58,168	58,267	58,456	58,230	57,417	57,823	57,810
belgium	739,185	741,647	747,374	742,968	729,190	733,264	732,877
netherlands	1,133,606	1,141,977	1,147,202	1,141,786	1,131,859	1,127,358	1,125,978
italy	3,425,723	3,434,538	3,452,907	3,434,830	3,374,512	3,401,824	3,400,005
europe	25,903,178	25,968,573	26,104,371	25,983,279	25,589,132	25,743,670	25,730,398
0025871-session-1-bg	688,391	695,636	701,228	698,818	699,234	694,616	694,625
0025872-session-2-bg	706,691	714,407	720,128	717,692	718,017	713,557	713,599
0025869-session-1-bg	629,715	637,159	642,647	640,327	641,147	636,057	636,050
0025876-session-1-bg	712,322	721,476	728,763	725,679	726,560	720,332	720,404
0025865-session-2-bg	656,483	663,359	669,205	666,846	666,996	662,446	662,564
0025868-session-2-bg	662,749	670,415	676,703	674,029	674,729	669,473	669,452
0025872-session-1-bg	692,476	700,076	705,979	703,495	704,039	699,263	699,284
0025864-session-2-bg	631,361	638,668	644,392	641,875	642,737	637,833	637,831
0025912-session-2	716,154	724,000	730,201	727,577	727,982	723,070	723,165
0025868-session-2-bg	662,025	669,661	676,133	673,415	673,972	668,845	668,784
0025869-session-2-bg	651,656	659,116	665,348	662,796	662,931	658,201	658,171
0025873-session-1-bg	595,166	601,378	605,916	604,042	604,270	600,488	600,467
0025870-session-2-bg	735,348	744,205	751,736	748,478	749,706	743,052	743,103
0025865-session-1-bg	676,416	683,459	689,108	686,788	687,193	682,528	682,526
0025889-session-2	674,887	683,034	689,378	686,679	687,462	682,117	682,056
0025876-session-2-bg	701,544	710,627	717,965	714,784	715,793	709,579	709,604
0025867-session-1-bg	682,118	689,846	695,806	693,294	693,749	688,861	688,829
0025874-session-2-bg	701,559	709,752	716,330	713,553	714,320	708,796	708,786
0025873-session-2-bg	635,363	641,971	647,566	645,325	645,449	641,349	641,441
0025889-session-1	644,858	651,812	657,483	655,226	655,525	651,102	651,113
0025870-session-1-bg	721,906	730,773	738,123	734,922	736,193	729,761	729,805
0025871-session-2-bg	674,478	681,551	687,309	684,955	685,328	680,690	680,686
0025864-session-1-bg	634,702	642,146	647,884	645,431	646,130	641,119	641,058
0025867-session-2-bg	673,075	680,577	686,239	683,853	684,641	679,654	679,647
0025878-session-1-bg	636,617	644,044	650,351	647,661	648,109	643,237	643,190
0025886-session-1	713,597	721,617	728,101	725,332	726,009	720,641	720,638

We evaluated MVC-WP-ER and MVC-WP-SF by comparing them with various other algorithms, namely ConstructVC, MVC-2, R, MVC-MPL, and MVC-L. We set $M = 3$ for both MVC-WP-ER and MVC-WP-SF, noting that $M = 3$ is a very small number of iterations of warning propagation.

Tables 2 and 4 and Fig. 2 compare these algorithms. In the misc networks and web networks benchmark instance sets, both MVC-WP-ER and MVC-WP-SF outperformed all other algorithms in terms of sizes of produced VCs. In the brain networks benchmark instance set, both MVC-WP-ER and MVC-WP-SF outperformed all other algorithms except ConstructVC. In the street networks benchmark instance set, both MVC-WP-ER and MVC-WP-SF outperformed

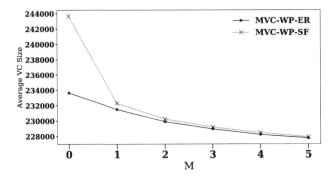

Fig. 4. Compares sizes of VCs produced by MVC-WP-ER and MVC-WP-SF for different values of M on the misc networks benchmark instance set.

all other algorithms except for MVC-MPL. The reason may be that street networks are always planar and thus in general cannot be well modeled as ER or SF graphs. Overall, MVC-WP-ER and MVC-WP-SF conclusively outperformed their competitors. We also conducted further experiments to demonstrate the usefulness of various individual steps of MVC-WP-ER and MVC-WP-SF.

To demonstrate the effectiveness of the message initialization step in MVC-WP-ER and MVC-WP-SF, i.e., assigning messages to be zero with probability p_0 computed from random graph models, we compared MVC-WP-ER and MVC-WP-SF with variants thereof in which p_0 is always set to 1 in order to mimic the message initialization in the standard warning propagation algorithm [20]. We refer to this variant as MVC-WP-1.

Figure 3 compares MVC-WP-ER and MVC-WP-SF with MVC-WP-1 on the misc networks benchmark instance set. Both MVC-WP-ER and MVC-WP-SF significantly outperformed MVC-WP-1 in terms of sizes of produced VCs. These results demonstrate the importance of our message initialization step.

To study the effect of M on MVC-WP-ER and MVC-WP-SF, we ran them for different values of M. For both MVC-WP-ER and MVC-WP-SF with $M \in \{0, 1, \ldots, 5\}$, Fig. 4 shows the sizes of the VC averaged over all benchmark instances in the misc networks benchmark instance set. The average VC size decreases with increasing M. The results demonstrate the usefulness of warning propagation iterations in MVC-WP-ER and MVC-WP-SF.

To demonstrate the effectiveness of Algorithm 2, we compared MVC-WP-ER and MVC-WP-SF with and without the use of it on the web networks benchmark instance set. MVC-WP-ER and MVC-WP-SF produced VCs of sizes that are on average 0.51% and 1.0% smaller than their counterparts without the use of Algorithm 2. These results demonstrate the importance of Algorithm 2.

Due to the fact that all algorithms are linear-time, all of them terminated very quickly. Despite that MVC-WP-ER and MWVC-WP-SF are slower than alternative algorithms, over 80% of their runs terminated within 300ms, which makes it difficult to measure the algorithms' running times on a single benchmark instance. In addition, it took much longer time (a few hundred times longer) to read input files from the hard disk than running these algorithms, which

Table 5. Compares sizes of MVCs and produced VCs by ConstructVC, MVC-2, MVC-R, MVC-L, MVC-MPL, MVC-WP-ER, and MVC-WP-SF, respectively, on benchmark instances with solutions from the Second DIMACS Implementation Challenge.

Graph			Algorithm						
Instance	$\|V\|$	MVC	ConstructVC	MVC-2	R	MVC-L	MVC-MPL	MVC-ER	MVC-SF
c-fat200-1	200	188	188	188	189	189	189	188	190
c-fat200-2	200	176	176	176	177	177	177	177	177
c-fat200-5	200	142	142	142	143	142	143	142	142
c-fat500-10	500	374	374	374	374	375	375	374	374
c-fat500-1	500	486	486	486	487	487	487	487	488
c-fat500-2	500	474	474	474	474	475	474	474	474
c-fat500-5	500	436	436	436	437	437	437	437	436
hamming10-2	1024	512	760	674	787	741	738	593	651
hamming6-2	64	32	45	40	44	40	39	34	41
hamming6-4	64	60	60	60	60	60	60	60	60
hamming8-2	256	128	155	185	188	182	178	141	160
hamming8-4	256	240	250	248	248	246	246	245	248
johnson16-2-4	120	112	112	112	112	112	112	112	112
johnson8-2-4	28	24	24	24	24	24	24	24	24
johnson8-4-4	70	56	56	60	61	60	60	58	56
keller4	171	160	162	164	163	163	163	163	161
p-hat300-1	300	292	295	294	295	294	294	295	295
p-hat300-2	300	275	279	285	286	286	284	280	282
p-hat300-3	300	264	270	273	278	278	277	273	271
p-hat500-1	500	491	494	494	494	494	494	493	495
p-hat500-2	500	464	470	479	482	480	481	472	473
san200-0	200	170	184	185	185	185	185	185	185
san200-0	200	130	155	155	159	156	154	154	155
san200-0	200	140	163	163	168	167	166	164	165
san200-0	200	156	169	169	173	172	171	171	172
sanr200-0	200	182	187	188	187	188	187	187	188

makes it difficult to reliably count the numbers of benchmark instances solved within a certain amount of time. For these reasons, it is difficult to have reliable comparisons of running times of all algorithms. Therefore, we skip the detailed comparison here, while this may be an interesting future work.

It is also interesting to compare the VCs produced by these linear-time-and-space algorithms with the sizes of MVCs. Since the MVC problem is NP-hard, it is elusive to find MVCs on the giant graphs in our previous used benchmark instances. Therefore, we ran all algorithms on the benchmark instances with provided solutions from the Second DIMACS Implementation Challenge[5] [17]. Since the given solutions are for the maximum clique problem, we ran all algorithms on the complements of the graphs in these benchmark instances, since the maximum clique problem on a graph is equivalent to the MVC problem on the

---

[5] http://dimacs.rutgers.edu/pub/challenge/graph/benchmarks/clique/.

complement of the graph. The solutions are shown in Table 5. From the table, we see that all linear-time-and-space algorithms produced VCs of similar results. We also see that, the produced VCs have sizes very close to the sizes of MVCs on all benchmark instances except hamming6-2 and hamming10-2.

6   Conclusions and Future Work

We developed MVC-WP, a warning propagation-based linear-time-and-space algorithm that finds small minimal VCs for giant graphs. We empirically showed that MVC-WP outperforms several other linear-time-and-space algorithms in terms of sizes of produced VCs. We also empirically showed that the theoretical underpinnings of MVC-WP significantly contribute to its success. These include both the way in which MVC-WP performs message initialization by computing p_0 and the iterations of warning propagation. We also made secondary contributions in computing various special functions efficiently with numerical accuracy sufficient for many AI applications. Future directions include applying similar techniques to solving other fundamental combinatorial problems on giant graphs.

Acknowledgment. The research at the University of Southern California was supported by the National Science Foundation (NSF) under grant numbers 1724392, 1409987, and 1319966. The views and conclusions contained in this document are those of the authors and should not be interpreted as representing the official policies, either expressed or implied, of the sponsoring organizations, agencies or the U.S. government.

References

1. Abu-Khzam, F.N., Collins, R.L., Fellows, M.R., Langston, M.A., Suters, W.H., Symons, C.T.: Kernelization algorithms for the vertex cover problem: theory and experiments. In: The Workshop on Algorithm Engineering and Experiments (2004)
2. Andrade, D.V., Resende, M.G.C., Werneck, R.F.: Fast local search for the maximum independent set problem. J. Heuristics **18**(4), 525–547 (2012). https://doi.org/10.1007/s10732-012-9196-4
3. Bader, D.A., Meyerhenke, H., Sanders, P., Wagner, D. (eds.): Graph Partitioning and Graph Clustering. Discrete Mathematics and Theoretical Computer Science. American Mathematical Society and Center, Providence (2013)
4. Barabási, A.L., Albert, R.: Emergence of scaling in random networks. Science **286**(5439), 509–512 (1999). https://doi.org/10.1126/science.286.5439.509
5. Cai, S.: Balance between complexity and quality: local search for minimum vertex cover in massive graphs. In: The International Joint Conference on Artificial Intelligence, pp. 747–753 (2015)
6. Cai, S., Su, K., Luo, C., Sattar, A.: NuMVC: an efficient local search algorithm for minimum vertex cover. J. Artif. Intell. Res. **46**(1), 687–716 (2013). https://doi.org/10.1613/jair.3907
7. Corless, R.M., Gonnet, G.H., Hare, D.E.G., Jeffrey, D.J., Knuth, D.E.: On the LambertW function. Adv. Comput. Math. **5**(1), 329–359 (1996). https://doi.org/10.1007/BF02124750

8. Dinur, I., Safra, S.: On the hardness of approximating minimum vertex cover. Ann. Math. **162**(1), 439–485 (2005). https://doi.org/10.4007/annals.2005.162.439
9. Erdős, P., Rényi, A.: On random graphs I. Publicationes Mathematicae **6**, 290–297 (1959)
10. Fang, Z., Li, C.M., Xu, K.: An exact algorithm based on MaxSAT reasoning for the maximum weight clique problem. J. Artif. Intell. Res. **55**, 799–833 (2016). https://doi.org/10.1613/jair.4953
11. Filiol, E., Franc, E., Gubbioli, A., Moquet, B., Roblot, G.: Combinatorial optimisation of worm propagation on an unknown network. Int. J. Comput. Electr. Autom. Control Inf. Eng. **1**(10), 2931–2937 (2007)
12. Finch, S.R.: Mathematical Constants, Encyclopedia of Mathematics and Its Applications, vol. 94. Cambridge University Press, Cambridge (2003)
13. Flum, J., Grohe, M.: Parameterized Complexity Theory. TTCSAES. Springer, Heidelberg (2006). https://doi.org/10.1007/3-540-29953-X
14. Goyal, A., Lu, W., Lakshmanan, L.V.S.: SIMPATH: an efficient algorithm for influence maximization under the linear threshold model. In: The IEEE International Conference on Data Mining, pp. 211–220 (2011). https://doi.org/10.1109/ICDM.2011.132
15. Haynsworth, E.V., Goldberg, K.: Bernoulli and Euler polynomials-Riemann zeta function. In: Abramowitz, M., Stegun, I.A. (eds.) Handbook of Mathematical Functions: With Formulas, Graphs, and Mathematical Tables, pp. 803–819. Dover Publications, Inc., Mineola (1965)
16. Hiary, G.A.: Fast methods to compute the Riemann zeta function. Ann. Math. **174**(2), 891–946 (2011). https://doi.org/10.4007/annals.2011.174.2.4
17. Johnson, D.J., Trick, M.A. (eds.): Cliques, Coloring, and Satisfiability: Second DIMACS Implementation Challenge. American Mathematical Society, Providence (1996)
18. Karp, R.M.: Reducibility among combinatorial problems. In: Complexity of Computer Computations, pp. 85–103. Plenum Press, New York (1972)
19. Korte, B., Vygen, J.: Combinatorial Optimization: Theory and Algorithms, 5th edn. Springer, Heidelberg (2012). https://doi.org/10.1007/978-3-642-24488-9
20. Mézard, M., Montanari, A.: Information, Physics, and Computation. Oxford University Press, Oxford (2009)
21. Niskanen, S., Östergård, P.R.J.: Cliquer user's guide, version 1.0. Technical report T48, Communications Laboratory, Helsinki University of Technology, Espoo, Finland (2003)
22. Pullan, W.: Optimisation of unweighted/weighted maximum independent sets and minimum vertex covers. Discret. Optim. **6**(2), 214–219 (2009). https://doi.org/10.1016/j.disopt.2008.12.001
23. Rossi, R.A., Ahmed, N.K.: The network data repository with interactive graph analytics and visualization. In: the AAAI Conference on Artificial Intelligence, pp. 4292–4293 (2015). http://networkrepository.com
24. Sherali, H.D., Rios, M.: An air force crew allocation and scheduling problem. J. Oper. Res. Soc. **35**(2), 91–103 (1984)
25. Vazirani, V.V.: Approximation Algorithms. Springer, Heidelberg (2003). https://doi.org/10.1007/978-3-662-04565-7
26. Weigt, M., Zhou, H.: Message passing for vertex covers. Phys. Rev. E **74**(4), 046110 (2006). https://doi.org/10.1103/PhysRevE.74.046110
27. Xu, H., Kumar, T.K.S., Koenig, S.: A new solver for the minimum weighted vertex cover problem. In: Quimper, C.-G. (ed.) CPAIOR 2016. LNCS, vol. 9676, pp. 392–405. Springer, Cham (2016). https://doi.org/10.1007/978-3-319-33954-2_28

28. Xu, H., Kumar, T.K.S., Koenig, S.: A linear-time and linear-space algorithm for the minimum vertex cover problem on giant graphs. In: The International Symposium on Combinatorial Search, pp. 173–174 (2017)
29. Yamaguchi, K., Masuda, S.: A new exact algorithm for the maximum weight clique problem. In: The International Technical Conference on Circuits/Systems, Computers and Communications. pp. 317–320 (2008)

Symbolic Bucket Elimination for Piecewise Continuous Constrained Optimization

Zhijiang Ye[✉], Buser Say, and Scott Sanner

University of Toronto, Toronto, Canada
tonyyezj@gmail.com, {bsay,ssanner}@mie.utoronto.ca

Abstract. Bucket elimination and its approximation extensions have proved to be effective techniques for discrete optimization. This paper addresses the extension of bucket elimination to continuous constrained optimization by leveraging the recent innovation of the extended algebraic decision diagram (XADD). XADDs support symbolic arithmetic and optimization operations on piecewise linear or univariate quadratic functions that permit the solution of continuous constrained optimization problems with a symbolic form of bucket elimination. The proposed framework is an efficient alternative for solving optimization problems with low tree-width constraint graphs *without* using a big-M formulation for piecewise, indicator, or conditional constraints. We apply this framework to difficult constrained optimization problems including XOR's of linear constraints and temporal constraint satisfaction problems with "repulsive" preferences, and show that this new approach significantly outperforms Gurobi. Our framework also enables symbolic parametric optimization where closed-form solutions cannot be computed with tools like Gurobi, where we demonstrate a final novel application to parametric optimization of learned Relu-based deep neural networks.

Keywords: Bucket elimination · Decision diagram
Constrained optimization · Symbolic dynamic programming

1 Introduction

Bucket elimination [2,7] is a generalized dynamic programming framework that has been widely applied to probabilistic reasoning problems on graphical models [8], including cost networks, constraint satisfaction [6], and propositional satisfiability [5]. The application of this framework to combinatorial optimization problems has been shown to be highly competitive against alternative techniques [14,16]. In this paper, we propose symbolic bucket elimination (SBE) as a novel method of solving mixed discrete and continuous constrained optimization problems (i.e., covering MILPs and a subclass of MIQPs). SBE critically leverages recent innovations in the extended algebraic decision diagram (XADD) that enable the exact representation and manipulation of piecewise linear and

© Springer International Publishing AG, part of Springer Nature 2018
W.-J. van Hoeve (Ed.): CPAIOR 2018, LNCS 10848, pp. 585–594, 2018.
https://doi.org/10.1007/978-3-319-93031-2_42

univariate quadratic functions [17]. We show that SBE can outperform Gurobi on low tree-width constrained optimization problems and that SBE can also perform symbolic *parameteric* optimization of learned Relu-based deep neural networks [15]—something tools like Gurobi cannot do exactly in closed-form.

2   Background

2.1   Case Representation and Operations

The case statement constitutes the foundational symbolic mathematical representation that is used throughout this paper and is presented below.

Case Statement. The *case* statement takes the following form:

$$
f = \begin{cases} \phi_1 : & f_1 \\ \vdots & \vdots \\ \phi_k : & f_k \end{cases}
$$

where ϕ_i is a logical formula over domain $(\boldsymbol{b}, \boldsymbol{x})$ with discrete[1] $\boldsymbol{b} \in \mathbb{B}^m$ and continuous variables $\boldsymbol{x} \in \mathbb{R}^n$, and is defined by arbitrary logical combinations $(\wedge, \vee, \neg)$ of (1) boolean variables in $\boldsymbol{b}$ and (2) linear inequality relations $(\geq, >, \leq, <)$ over continuous variables in $\boldsymbol{x}$. Each ϕ_i is disjoint from other ϕ_j $(j \neq i)$ and exhaustively covers the entire domain such that f is well defined. Each f_i is a linear or univariate quadratic function (LUQF) of $\boldsymbol{x}$, e.g. $f_1 = x_1 + 3x_2$ or $f_2 = 5x_3^2 - 2x_3 + 1$. Only one variable can be quadratic in a case statement and wherever it occurs it must be univariate, hence given the previous examples $f_3 = x_1^2 + 2x_3$ would be disallowed with f_1 and f_2 in the same case statement.

Binary Operations. For binary operations, the cross-product of logical partitions of each case is taken. For example, the "cross-sum" $\oplus$ is defined as:

$$
\begin{cases} \phi_1 : & f_1 \\ \phi_2 : & f_2 \end{cases} \oplus \begin{cases} \psi_1 : & g_1 \\ \psi_2 : & g_2 \end{cases} = \begin{cases} \phi_1 \wedge \psi_1 : & f_1 + g_1 \\ \phi_1 \wedge \psi_2 : & f_1 + g_2 \\ \phi_2 \wedge \psi_1 : & f_2 + g_1 \\ \phi_2 \wedge \psi_2 : & f_2 + g_2 \end{cases}
$$

Note that the case representation is closed under general conditions for $\oplus$.

Case Maximization. Maximization of two case statements is a piecewise operator that can be defined easily (e.g., consider the maximum of two hyperplanes):

$$
\mathrm{casemax}\left(\begin{cases} \phi_1 : & f_1 \\ \phi_2 : & f_2 \end{cases}, \begin{cases} \psi_1 : & g_1 \\ \psi_2 : & g_2 \end{cases} \right) = \begin{cases} \phi_1 \wedge \psi_1 \wedge f_1 > g_1 : & f_1 \\ \phi_1 \wedge \psi_1 \wedge f_1 \leq g_1 : & g_1 \\ \phi_1 \wedge \psi_2 \wedge f_1 > g_2 : & f_1 \\ \phi_1 \wedge \psi_2 \wedge f_1 \leq g_2 : & g_2 \\ \vdots & \vdots \end{cases}
$$

---

[1] For simplicity of exposition, we presume that non-binary discrete variables of cardinality k are encoded in binary with $\lceil \log_2(k) \rceil$ boolean variables.

The casemin operator is defined analogously. We remark that the case representation is closed for casemax and casemin for linear ϕ_i, ψ_j, f_i, and g_j. These operators are not necessarily closed for LUQF operands since the newly introduced constraints $f_i > g_j$ may become non-LUQF. However, we can often order eliminations to avoid application of casemax or casemin on LUQF statements.

Case Substitution. The case substitution operator defined as $\sigma = (y/g)$ triggers the substitution of a variable y with a case statement g. Similar to the $\oplus$ operation, $f\sigma$ results in a case with conditions as cross-products of case conditions between f and g, and value expressions in f with variable y replaced by the value corresponding to the case condition in g. As an illustrative example:

$$f\sigma = \begin{cases} z \leq 0 \wedge x \leq v : & x + v \\ z \leq 0 \wedge x > v : & z + v \\ z > 0 \wedge x \leq -2w : & x - 2w \\ z > 0 \wedge x > -2w : & z - 2w \end{cases}, f = \begin{cases} x \leq y : & x + y \\ x > y : & z + y \end{cases}, g = \begin{cases} z \leq 0 : & v \\ z > 0 : & -2w \end{cases}$$

Maximization/Minimization over a Variable. In symbolic optimization we will want to maximize over both boolean and continuous variables. For a boolean max over b_i,[2] we simply take the casemax over both instantiations $\{0, 1\}$ of b_i: $f(\boldsymbol{b}_{\backslash i}, \boldsymbol{x}) = \max_{b_i} g(\boldsymbol{b}, \boldsymbol{x}) = \text{casemax}(g(b_i = 0, \boldsymbol{b}_{\backslash i}, \boldsymbol{x}), g(b_i = 1, \boldsymbol{b}_{\backslash i}, \boldsymbol{x}))$. Symbolic maximization over a continuous variable x_i is a much more involved operation written as $f(\boldsymbol{b}, \boldsymbol{x}_{\backslash i}) = \max_{x_i} g(\boldsymbol{b}, \boldsymbol{x})$ and discussed in detail in [20]. This operation *is* closed-form for LUQF $g(\boldsymbol{b}, \boldsymbol{x})$ and results in a purely symbolic case statement for $f(\boldsymbol{b}, \boldsymbol{x}_{\backslash i})$. Minimization operators are defined analogously.

2.2 Extended Algebraic Decision Diagrams (XADDs)

Due to cross-product operations, a data structure such as decision diagrams are required to maintain a tractable case representation. Bryant [4] introduced the reduced ordered binary decision diagram (BDD) representing boolean functions; algebraic decision diagrams (ADD) [1] extended BDDs to non-boolean functions. The extended algebraic decision diagram (XADD) [18] shown in Fig. 1 extends the ADD to allow continuous variables with inequalities for decisions and LUQF expressions for leaves. As for ADDs, XADDs have a fixed order of decisions from root to leaf. The standard ADD operations to build a canonical ADD (REDUCE) and to perform a binary operation on two ADDs (APPLY) also apply

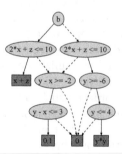

Fig. 1. Example XADD. The true branch is solid, the false branch is dashed.

---

[2] We use $\boldsymbol{b}_{\backslash i}$ to denote the set $\boldsymbol{b}$ with the variable b_i excluded. Similarly $\boldsymbol{x}_{\backslash i}$ denotes exclusion of x_i from $\boldsymbol{x}$.

for XADDs. The XADD can be exponentially smaller than the case representation (each path from root to leaf is a case partition) and all previous case operations can be implemented to exploit the DAG structure of XADDs [20].

3   Symbolic Bucket Elimination

In this section, we introduce our novel framework: symbolic bucket elimination for continuous constrained optimization. Problems can be specified as follows:

$$\max_{b,x} \sum_{i=1}^{n} R_i(\boldsymbol{b}, \boldsymbol{x}) \text{ subject to } C_j, \ \forall j \in \{1, \cdots, k\} \tag{1}$$

In our case, the R_i can be LUQF expressions and the C_j are linear constraints. We translate problems of this form into their symbolic equivalent:

$$\max_{b,x} \bigoplus_{i=1}^{m} F_i(\boldsymbol{b}, \boldsymbol{x}), \text{ where } F_i = \begin{cases} C_i : & 0 \\ \neg C_i : & -\infty \end{cases}, \forall i \in \{1, \ldots, k\}, \tag{2}$$

where $m = k + n$ and for each linear constraint C_j, there is a corresponding linear case statement $F_i(\boldsymbol{b}, \boldsymbol{x})$. A maximum value of $-\infty$ for the problem would indicate that the problem is overconstrained and infeasible. We also note that while a case representation would require a big-M formulation to handle piecewise, indicator, and conditional constraints possible in cases, this framework represents *all* of these logical constraints natively in the symbolic case form and thus as XADDs.

3.1   Symbolic Bucket Elimination Algorithm (SBE)

We can solve the general optimization problem of (2) by using a fully symbolic variant of bucket elimination [8] leveraging the case (XADD) representation and its efficient operations. In bucket elimination, each function F_i is placed into ordered, variable-specific buckets. Variable ordering is determined by heuristics that aim to minimize the induced tree-width of the underlying graph [8], with the restriction that variables appearing on the left hand side of equality constraints will be ordered such that their representative buckets will be eliminated first. The rule for bucket assignment is to identify the variable in each function that appears the latest in the ordering, and place the function in the bucket of the respective variable. The buckets are then eliminated sequentially in the forward step. In backtracking, optimal assignments are obtained with an arg max [20] on the summed function for each bucket. The SBE algorithm is presented in Algorithm 1. If the objective is minimization, (arg)max replaces (arg)min.

Computational Complexity. For bucket elimination over discrete domains, complexity is bounded by an exponential function of the tree-width of the constraint graph [8]. When we extend bucket elimination to continuous domains

---

Algorithm 1. Symbolic Bucket-Elimination (SBE)

Input: Input XADD functions $\{F_1, \ldots, F_m\}$, a variable ordering d
Output: The optimal objective value and variable assignments

1: *Initialization*: bucket assignment
2: Assign each input function to an ordered set of buckets: $bucket_1, \ldots, bucket_n$. Let $f_{j,p}$ denote a function residing in $bucket_p$ that is either an input function, or a resultant function from a bucket already eliminated
3: *Forward Elimination*: eliminate each bucket sequentially
4: **for** $p \leftarrow n$ down to 1 in ordering d **do**
5:     $g_p \leftarrow \bigoplus_j f_{j,p}$
6:     **if** x_p is on left hand side of equality constraint: $x_p = cons$ **then**
7:         $h_p \leftarrow g_p\sigma$, where $\sigma = (x_p/cons)$ // substitute variable for constraint
8:     **else** $h_p \leftarrow \max_{x_p} g_p$
9:     **if** $p > 1$ **then** assign h_p to a bucket according to rule ($bucket_1$ if constant)
10: *Backtracking*: recover optimal variable assignments, $\boldsymbol{x}^*$
11: **for** $p \leftarrow 1$ up to n in ordering d **do**
12:     $x_p^* \leftarrow \arg\max_{x_p} g_p(x_1 = x_1^*, \ldots, x_{p-1} = x_{p-1}^*)$, $\boldsymbol{x}^* \leftarrow \boldsymbol{x}^* \cup \{x_p^*\}$
13: **return** $h_1, \boldsymbol{x}^*$

---

using XADD, the complexity is not explicitly tree-width dependent. While a constraint with many decision variables may be represented compactly as a piecewise expression, one can generally only upper bound the number of pieces needed in a case statement as an exponential function of the number of primitive binary operations ($\oplus$, casemax) used by bucket elimination. Nevertheless, the XADD does maintain compact representations much smaller than the worst-case upper bound and proves to be particularly advantageous when the underlying constraint graph has low tree-width, as we show in the experimental results section.

4   Experimental Results

In this section, we demonstrate the computational efficiency and the expressiveness of symbolic bucket elimination framework on three distinct problems. First, we present two problems in which the symbolic bucket elimination framework outperformed the state of art solver AMPL-Gurobi (7.5.0) with default settings on a 2.20 GHz processor [10]. Second, we present a novel application of symbolic parametric optimization to Relu-based deep neural networks.

4.1   Problems with XOR Conditional Constraints

Following is a synthetic problem involving constraints combined with XOR ($\veebar$):

$$\max \sum_{i=1}^{n} r_i$$

where $r_i = \text{if}\,(x_i \geq x_{i+1} \veebar x_{i+1} \geq x_{i+2})$ then $\max(x_{i+1}, x_{i+2}) - \min(x_{i+1}, x_{i+2})$

else $\min(x_i, x_{i+1}) - \max(x_i, x_{i+1})$,

subject to $- 10 \leq x_i \leq 10, \forall i \in \{1, \ldots, n+2\}, -20 \leq r_j \leq 20, \forall j \in \{1, \ldots, n\}$

In this problem, the reward term in the objective r_i is determined by an XOR conditional expression involving the respective decision variables x_i, x_{i+1}, x_{i+2}.

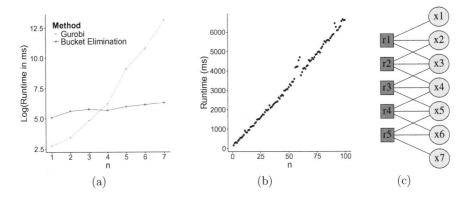

(a)   (b)   (c)

Fig. 2. (a) Comparison of log runtime of SBE vs. Gurobi, with SBE outperforming Gurobi for $n > 4$. (b) Non-log runtime for $1 \leq n \leq 100$ showing SBE is linear in n. (c) Constraint graph for $n = 5$, showing the low tree-width nature of this problem.

This problem structure is particularly advantageous for the proposed framework due to the small size of the decision diagram for each constraint term, as well as the sparsity of the constraints. This is illustrated through evaluations of its runtime performance, with comparisons to that of Gurobi in Fig. 2. Symbolic bucket elimination outperforms Gurobi, even at a very small n. The performance gap becomes significant as n increases—while the runtime for Gurobi scales *exponentially* in n, the bucket elimination framework scales *linearly* in n as evidenced in the non-log plot Fig. 2(b).

4.2   Temporal Constraint Satisfaction with Preferences

Temporal constraint problems with preferences deal with finding optimal assignments to time events based on preferences [12]. The objective is to optimize the total preference value, subject to a set of constraints such as ordering of certain time events, or time delays between events. This class of problems is a combination of Temporal Constraint Satisfaction Problems [9] with soft constraints [3].

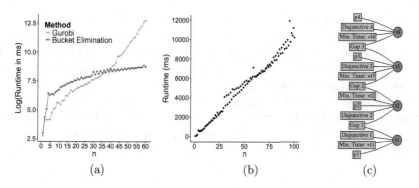

Fig. 3. (a) Comparison of log runtime of SBE vs. Gurobi, with SBE outperforming Gurobi for $n > 40$. (b) Non-log runtime for $1 \leq n \leq 100$ showing SBE is linear in n. (c) Constraint graph for $n = 4$, demonstrating the low tree-width structure.

The problem considered is:

$$\min \sum_{i=1}^{n} t_i + p_i$$

$$\text{where } p_i = \text{ if } (t_i \leq 10(i+1)) \text{ then } t_i{}^2 \text{ else } (t_i - 10n)^2, \forall i \in \{1, \ldots, n\}$$
$$\text{subject to } 10i \leq t_i \leq 10(i+1) \vee 10(i+2) \leq t_i \leq 10(i+3), \forall i \in \{1, \ldots, n\}$$
$$t_i + 10 \leq t_{i+1}, \forall i \in \{1, \ldots, n-1\}, \{t_i, p_i\} \in \mathbb{R}, \forall i \in \{1, \ldots, n\}$$

The definition of p_i is analogous to the mid-value preference constraint presented in [12]. The preference value p_i is dictated by whether the condition $t_i \leq 10(i+1)$ is true. If so, then it is preferred for time event t_i to occur as close to time 0 as possible, otherwise t_i should occur close to time $10n$ (i.e., the preferences are "repulsive" and prefer opposite ends of the timeline). The objective is to minimize the sum of time events t_i and preference values p_i. The first constraint is a disjunctive type constraint on time event t_i. The second constraint imposes a gap requirement between time events. We note that although quadratic terms appear in the leaves of the representative constraint XADD, decisions will remain linear throughout SBE as there are no explicit discrete variables to maximize and therefore no casemax operations to "promote" quadratic terms into decisions. The runtime evaluations and problem structure are visualized in Fig. 3. As in the previous example, the runtime for Gurobi scales *exponentially* in problem size while SBE scales *linearly*. SBE outperforms Gurobi for $n > 40$.

4.3 Symbolic Parametric Optimization of Deep Neural Networks

We demonstrate a novel application of SBE to perform symbolic parametric optimization of deep neural networks with rectified linear units (Relu) that are piecewise linear. Previous work has shown Relu-based deep neural networks can be compiled into linear constraint optimization programs and solved nonparametrically with applications in automated planning [19] and verification [11].

In this section, we show promising results for symbolic parametric optimization on the learned output units, $y_j(\boldsymbol{x})$ of a Relu-based deep network trained on $h(\boldsymbol{x})$ by first compiling the constraints and then maximizing $y_j(\boldsymbol{x})$ w.r.t. a subset of the input variables $\boldsymbol{x}_s$ using SBE. This results in new *symbolic* piecewise functions that represent the *maximal* deep network output values that can be achieved for the best $\boldsymbol{x}_s$ as a function of the remaining inputs. Such symbolic parametric (partial) optimization is not possible with Gurobi. The Relu-based deep network is represented by the following objective, piecewise linear case statements and constraints implementing the connections and Relu functions:

$$\max_{\boldsymbol{x}_s \subset \{x_1,\ldots,x_n\}} y_j, \forall j \in \{1,\ldots,m\}$$

$$y_j = \sum_{i \in \{1,\ldots,p\}} w_{i,j,l} r_{i,l} + b_{j,l+1}, \forall j \in \{1,\ldots,m\}$$

$$r_{j,k} = \max(\sum_{i \in \{1,\ldots,p\}} (w_{i,j,k-1} r_{i,k-1}) + b_{j,k}, 0), \forall j \in \{1,\ldots,p\}, k \in \{2,\ldots,l\}$$

$$r_{j,1} = \max(\sum_{i \in \{1,\ldots,n\}} (w_{i,j,0} x_i) + b_{j,1}, 0), \forall j \in \{1,\ldots,p\}$$

$$-10 \leq x_i \leq 10, \forall i \in \{1,\ldots,n\}, y_j \in \mathbb{R}, \forall j \in \{1,\ldots,m\}$$

$$0 \leq r_{i,k} \in \mathbb{R}, \forall i \in \{1,\ldots,p\}, k \in \{1,\ldots,l\}$$

where parameters n, m, p and l denote the number of input units, number of output units, width (units in a hidden layer) and depth (hidden layers) of the network, $w_{i,j,k}$ denotes the weight between unit i at layer k and unit j, and $b_{j,k}$ denotes the bias of unit j at layer k.

In Fig. 4, we show an example neural network structure and the runtime results of using SBE to parametrically optimize a network trained to learn $h(\boldsymbol{x}) = x_1^2 + x_2^2$ with $n = 2$, $m = 1$ for various width p and depth l. Runtimes are heavily width-dependent since tree-width grows with the width of a deep net, but not depth. The SBE eliminates the nodes in the hidden layers in a backward manner until it reaches the input layer, where the variable $\boldsymbol{x}_s = (x_1)$ is maximized out. We note that for networks with more than one output, it is possible with SBE to parametrically optimize on different sets of $\boldsymbol{x}_s$ for different outputs. Other types of activation functions (i.e., linear or step) are also possible, as long as each unit can be represented

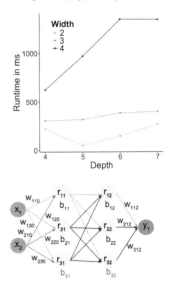

Fig. 4. Top: runtime for SBE for network widths 2–4, depths 4–7. Bottom: deep neural network structure with 2 input units, 1 output unit, width of 3, hidden layer depth of 2 ($n = 2, m = 1, p = 3, l = 2$).

as piecewise functions. SBE applied to deep nets as done here has potential applications in planning and verification: *i.e.*, what is achievable as a function of an input?

5   Conclusion and Future Work

We introduced a novel symbolic bucket elimination (SBE) framework for representing and solving constrained optimization problems symbolically (and even parametrically), that can *exponentially* outperform Gurobi when the underlying constraint graph has low tree-width. In terms of future work, we remark that previous investigations in the discrete domain using mini-buckets [7] and heuristic search have demonstrated excellent improvement over exact bucket elimination [13,16]. Hence, a promising direction for future work is mini-bucket extensions of SBE to allow it to scale to higher tree-width constrained optimization problems, vastly extending the scope of applicability of SBE.

References

1. Bahar, R.I., Frohm, E.A., Gaona, C.M., Hachtel, G.D., Macii, E., Pardo, A., Somenzi, F.: Algebraic decision diagrams and their applications. In: Proceedings of the 1993 IEEE/ACM International Conference on Computer-Aided Design, ICCAD 1993, pp. 188–191. IEEE Computer Society Press, Los Alamitos (1993). http://dl.acm.org/citation.cfm?id=259794.259826

2. Bertele, U., Brioschi, F.: Nonserial Dynamic Programming. Academic Press Inc., Orlando (1972)

3. Bistarelli, S., Montanari, U., Rossi, F.: Semiring-based constraint satisfaction and optimization. J. ACM **44**(2), 201–236 (1997). https://doi.org/10.1145/256303.256306

4. Bryant, R.E.: Graph-based algorithms for Boolean function manipulation. IEEE Trans. Comput. **35**(8), 677–691 (1986). https://doi.org/10.1109/TC.1986.1676819

5. Davis, M., Putnam, H.: A computing procedure for quantification theory. J. ACM **7**(3), 201–215 (1960). https://doi.org/10.1145/321033.321034

6. Dechter, R., Pearl, J.: Network-based heuristics for constraint satisfaction problems. In: Kanal, L., Kumar, V. (eds.) Search in Artificial Intelligence. SYMBOLIC, pp. 370–425. Springer, London (1988). https://doi.org/10.1007/978-1-4613-8788-6_11. http://dl.acm.org/citation.cfm?id=60727.60738

7. Dechter, R.: Bucket elimination: a unifying framework for reasoning. Artif. Intell. **113**(1), 41–85 (1999). http://www.sciencedirect.com/science/article/pii/S0004370299000594

8. Dechter, R.: Reasoning with Probabilistic and Deterministic Graphical Models: Exact Algorithms. Morgan & Claypool Publishers, San Rafael (2013)

9. Dechter, R., Meiri, I., Pearl, J.: Temporal constraint networks. Artif. Intell. **49**(1–3), 61–95 (1991). https://doi.org/10.1016/0004-3702(91)90006-6

10. Gurobi Optimization, Inc.: Gurobi optimizer reference manual (2016). http://www.gurobi.com

11. Katz, G., Barrett, C., Dill, D.L., Julian, K., Kochenderfer, M.J.: Reluplex: an efficient SMT solver for verifying deep neural networks. In: Majumdar, R., Kunčak, V. (eds.) CAV 2017. LNCS, vol. 10426, pp. 97–117. Springer, Cham (2017). https://doi.org/10.1007/978-3-319-63387-9_5

12. Khatib, L., Morris, P., Morris, R., Rossi, F.: Temporal constraint reasoning with preferences. In: Proceedings of the 17th International Joint Conference on Artificial Intelligence, IJCAI 2001, vol. 1, pp. 322–327. Morgan Kaufmann Publishers Inc., San Francisco (2001). http://dl.acm.org/citation.cfm?id=1642090.1642135

13. Larrosa, J., Dechter, R.: Boosting search with variable elimination in constraint optimization and constraint satisfaction problems. Constraints 8, 303–326 (2003)

14. Larrosa, J., Morancho, E., Niso, D.: On the practical use of variable elimination in constraint optimization problems: 'still-life' as a case study. J. Artif. Intell. Res. 23, 421–440 (2005)

15. Nair, V., Hinton, G.E.: Rectified linear units improve restricted Boltzmann machines. In: ICML, pp. 807–814 (2010). http://www.icml2010.org/papers/432.pdf

16. Rollón, E., Larrosa, J.: Bucket elimination for multiobjective optimization problems. J. Heuristics 12(4), 307–328 (2006). https://doi.org/10.1007/s10732-006-6726-y

17. Sanner, S., Abbasnejad, E.: Symbolic variable elimination for discrete and continuous graphical models. In: Proceedings of the Twenty-Sixth AAAI Conference on Artificial Intelligence, AAAI 2012, pp. 1954–1960. AAAI Press (2012). http://dl.acm.org/citation.cfm?id=2900929.2901004

18. Sanner, S., Delgado, K., Barros, L.: Symbolic dynamic programming for discrete and continuous state MDPs. In: UAI, pp. 643–652, January 2011

19. Say, B., Wu, G., Zhou, Y.Q., Sanner, S.: Nonlinear hybrid planning with deep net learned transition models and mixed-integer linear programming. In: Proceedings of the Twenty-Sixth International Joint Conference on Artificial Intelligence, IJCAI 2017, pp. 750–756 (2017). https://doi.org/10.24963/ijcai.2017/104

20. Zamani, Z., Sanner, S., Fang, C.: Symbolic dynamic programming for continuous state and action MDPs. In: Proceedings of the Twenty-Sixth AAAI Conference on Artificial Intelligence, AAAI 2012, pp. 1839–1845. AAAI Press (2012). http://dl.acm.org/citation.cfm?id=2900929.2900988

Learning a Classification of Mixed-Integer Quadratic Programming Problems

Pierre Bonami[1], Andrea Lodi[2], and Giulia Zarpellon[2(✉)]

[1] CPLEX Optimization, IBM Spain, Madrid, Spain
pierre.bonami@es.ibm.com
[2] CERC, École Polytechnique Montréal, Montreal, Canada
{andrea.lodi, giulia.zarpellon}@polymtl.ca

Abstract. Within state-of-the-art solvers such as IBM-CPLEX, the ability to solve both convex and nonconvex Mixed-Integer Quadratic Programming (MIQP) problems to proven optimality goes back few years, yet presents unclear aspects. We are interested in understanding whether for solving an MIQP it is favorable to linearize its quadratic part or not. Our approach exploits machine learning techniques to learn a classifier that predicts, for a given instance, the most suitable resolution method within CPLEX's framework. We aim as well at gaining first methodological insights about the instances' features leading this discrimination. We examine a new dataset and discuss different scenarios to integrate learning and optimization. By defining novel measures, we interpret and evaluate learning results from the optimization point of view.

1 Introduction

The tight integration of discrete optimization and machine learning (ML) is a recent but already fruitful research theme: while ML algorithms could profit of choices of discrete type, until now disregarded, various are the discrete optimization settings and situations that could benefit from a ML-based heuristic approach. Although a number of fresh applications is recently appearing in this latter direction, that one could call "learning for optimization" (e.g., [1,2]), two main topics in this thread of research involve ML-based approaches for the branch-and-bound scheme in Mixed-Integer Linear Programming (MILP) problems (see [3] for a survey on the theme) and the usage of predictions to deal with the solvers' computational aspects and configuration (see, e.g., [4,5]). We shift from those two main ideas and position ourselves somehow in between, to tackle a new application of ML in discrete optimization. We consider Mixed-Integer Quadratic Programming (MIQP) problems, which prove to be interesting for modeling diverse practical applications (e.g., [6,7]) as well as a theoretical ground for a first extension of MILP algorithms into nonlinear ones.

Within state-of-the-art solvers such as IBM-CPLEX [8], the ability to solve both convex and nonconvex MIQPs to proven optimality goes back few years (see, e.g., [9]), but theoretical and computational implications of the employed

© Springer International Publishing AG, part of Springer Nature 2018
W.-J. van Hoeve (Ed.): CPAIOR 2018, LNCS 10848, pp. 595–604, 2018.
https://doi.org/10.1007/978-3-319-93031-2_43

resolution methods do not seem fully understood yet. We are interested in learning whether it is favorable to linearize the quadratic part of an MIQP or not. As was firstly suggested in [10], we believe that MIQPs should be solved in an intelligent and adapted way in order to improve their resolution process; currently, the decision *linearize vs. not linearize* can be specified by CPLEX users via the linearization switch parameter. We interpret the question *linearize vs. not linearize* as a classification one: we learn a classifier predicting, for a given MIQP, the most suited resolution method within CPLEX, possibly gaining first methodological insights about the problems features leading to such prediction.

After a quick dive into the MIQPs algorithmic framework (Sect. 2), we motivate and state our research question (Sect. 3). Methodological details and learning-related aspects are presented in Sect. 4, while Sect. 5 is devoted to discuss results, new evaluation measures and different scenarios to integrate the learning and the optimization processes.

2   Solving MIQPs with CPLEX

We consider general MIQP problems of the form

$$\min \left\{ \frac{1}{2} x^T Q x + c^T x : Ax = b, \ l \le x \le u, \ x_j \in \{0,1\} \, \forall \, j \in I \right\} \qquad (1)$$

where $Q = \{q_{ij}\}_{i,j=1,\ldots,n} \in \mathbb{R}^{n \times n}$ is a symmetric matrix, $c \in \mathbb{R}^n$, $A \in \mathbb{R}^{m \times n}$ and $b \in \mathbb{R}^m$. Variables $x \in \mathbb{R}^n$ are bounded, and $I \subseteq N = \{1, \ldots, n\}$ is the set of indices of variables that are required to be binary. We say that a problem is *pure (binary)* when $I = N$, and *mixed* otherwise; we do not consider the continuous case of $I = \emptyset$. We refer to an MIQP *relaxation* as to the continuous version of (1), where integrality requirements are dropped.

Depending on its relaxation being convex or nonconvex, and on the types of variables involved, an MIQP can be tackled in different ways by CPLEX.

Convex Problems. A relaxed MIQP is *convex* if and only if the matrix Q is positive semi-definite ($Q \succeq 0$). In this setting, both pure and mixed MIQPs can be solved by the nonlinear programming-based branch and bound [11] (NLP B&B) (see also [12]), a natural extension of the integer linear B&B scheme [13] in which a QP is solved at each node of the tree. Another common resolution approach for convex problems is that of Outer Approximation algorithms [14], which are however not implemented in CPLEX for MIQPs.

Nonconvex Problems. When the relaxed MIQP is *not* convex ($Q \not\succeq 0$), variable types play an import role. A binary nonconvex MIQP can be transformed into a convex one by means of augmenting the main diagonal of Q: using $x_j = x_j^2$ for $x_j \in \{0,1\}$, $x^T Q x$ can be replaced by $x^T (Q + \rho \mathbb{I}_n) x - \rho e^T x$, where $Q + \rho \mathbb{I}_n \succeq 0$ for some suitable $\rho > 0$, $\mathbb{I}_n$ denotes the $n \times n$ identity matrix, and e the vector with all ones. Alternatively, a binary nonconvex MIQP can be linearized and transformed into a MILP. Without performing diagonal augmentation, nonzero terms $q_{ii}x_i^2$ are rewritten as $q_{ii}x_i$, while bilinear terms $q_{ij}x_ix_j$ are handled by the so-called McCormick inequalities [15]: a variable $y_{ij} \geq 0$ is added to represent x_ix_j, together with linear constraints

$$x_i + x_j - 1 \leq y_{ij} \text{ if } q_{ij} > 0, \text{ or } y_{ij} \leq x_i, y_{ij} \leq x_j \text{ if } q_{ij} < 0. \qquad (2)$$

In this way, the problem formulation grows in size, but the resulting model can be solved with standard MILP techniques.

For mixed nonconvex MIQPs, there is no straightforward way to convexify or linearize an instance, and CPLEX relies on the so-called Spatial B&B (see, e.g., [16]) to solve these problems to global optimality.

Although a number of possibilities can be explored to perform linearization and convexification, their discussion is not within the scope of the present paper. For more details, we refer the reader to [9] and the references therein.

3 Linearize vs. Not Linearize

One could assume that the linearization approach discussed for pure nonconvex MIQPs could be beneficial for the convex case as well: a binary problem would turn into a MILP, while in the mixed case one could linearize all bilinear products between a binary and a bounded continuous variable with generalized McCormick inequalities. However, nonzero products between two continuous variables would remain in the formulation, so that a mixed convex MIQP could still be quadratic after linearization, and hence solved with a NLP B&B.

We restrict our focus to *pure convex*, *mixed convex* and *pure nonconvex* problems; the *mixed nonconvex* case should be treated separately, due to the very different setup of Spatial B&B. Currently, in our three cases of interest, the solver provides the user the possibility to switch the linearization on or off by means of the preprocessing parameter `qtolin`, and the default strategy employed by CPLEX is to always perform linearization, although this approach does not dominate in theory the non-linearization one [9].

We aim at learning an offline classifier predicting the most suited resolution approach in a flexible and instance-specific way. We summarize in what follows the main steps undertaken in the development of our method, leaving the details for the next section.

1. **Dataset generation:** we implement a generator of MIQP instances, spanning across various combinations of structural and optimization parameters.
2. **Features design:** we identify a set of features describing an MIQP in its mathematical formulation and computational behavior.
3. **Labels definition:** we define rigorous procedures to discard troublesome instances, and assess a label depending on running times.
4. **Learning experiments:** we train and test traditional classifiers and interpretable algorithms such as ensemble methods based on Decision Trees.

4   Methodological Details

We now go through the development steps sketched above more in details, discussing how the dataset is generated, and features and labels defined.

4.1   Dataset Generation

To build complete MIQP instances, data generation is made of two separate steps. First, symmetric matrices Q are generated by the MATLAB function `sprandsym` [17], to which desired size n, density d and eigenvalues $\lambda_i, i = 1, \ldots, n$ are specified; (in)definiteness of Q is controlled by the spectrum. The second step is implemented with CPLEX Python API: quadratic data can be complemented with a linear vector c, uniform with density d; binary and continuous variables are added in various proportions; finally, a constraints set is defined. We monitor the addition of the following types of constraints, in different combinations:

- a single cardinality constraint $0 \leq \sum_{j \in I} x_j \leq r$, with $r < |I|$ varying;
- a standard simplex constraint $\sum_{j \notin I} x_j = 1, x_j \geq 0$;
- a set of η multi-dimensional knapsack constraints $\sum_{j \in I} w_{ij} x_j \leq f_i$, for $i = 1, \ldots, \eta$. We follow the procedure described in [18] to generate coefficients w_{ij} and f_i, without correlating them to the objective function.

4.2   Features Design

A (raw) formulation like (1) cannot be fed directly as input to a learning algorithm. We depict an MIQP by means of a set of 21 hand-crafted features, summarized in Table 1. *Static features* describe the instance in terms of variables, constraints and objective function, and are extracted before any solving (pre)process. Few *dynamic features* collect information on the early optimization stages, after the preprocessing and the resolution of the root node relaxation.

4.3   Labels Definition

To each MIQP we assign a label among L (*linearize*, i.e., `qtolin` on), NL (*not linearize*, i.e., `qtolin` off) and T, the latter to account for *tie* cases between L and NL. To deal with performance variability [19], each instance is run in both `qtolin` modes with 5 different random seeds; we enforce a timelimit of 1 h for each run. To monitor troublesome instances, we implement:

- solvability check: instances that cannot be solved within timelimit by any method (neither L nor NL) for any seed are discarded;
- seed consistency check: for each seed, unstable instances with respect to lower and upper bounds of L and NL are discarded;
- global consistency check: a global check on the best upper and lower bounds for the two methods is performed to discard further unstable instances.

Table 1. Overview and brief description of the complete features set.

#	Group name	Features general description
Static features		
2	Generic problem type	Size of the problem, variables proportions per type
2	Constraints matrix composition	Density w.r.t. different types of variables, magnitudes of nonzero (nnz) coefficients
5	Quadratic matrix composition	Magnitudes of coefficients, nnz diagonal and bilinear (*continuous · continuous*) and (*binary · continuous*)
7	Spectrum description	Shares of positive/negative eigenvalues, magnitude and value of the smallest one, trace and spectral norm
3	Other properties of Q	Density, rank, a measure of "diagonal dominance"
Dynamic features		
2	Root node information	Difference of lower bounds and resolution times at the root node, between *linearize* and *not linearize*

If an MIQP passes all these checks, we assign a label. When one mode is never able to solve an instance, the other wins. If both L and NL could solve the instance at least once, running times on each seed are compared and a "seed win" is assigned to one mode if at least 10% better. We assign L or NL only if their seed wins are consistent through the 5 runs, opting for a tie T otherwise.

5   Data, Experiments and Results

The generation procedure is run with MATLAB 9.1.0, Python 2.7 and CPLEX 12.6.3 on a Linux machine, Intel Xeon E5-2637 v4, 3.50 GHz, 16 threads and 128 GB. To label the dataset, we used a grid of 26 machines Intel Xeon X5675, 3.07 GHz (12 threads each) and 96 GB; each problem is restricted to one thread.

We generate 2640 different MIQPs, with size $n \in \{25, 50, \ldots, 200\}$ and density of Q $d \in \{0.2, 0.4, \ldots, 1\}$. For mixed convex MIQPs, the percentage of continuous variables is chosen from $\{0, 20, \ldots, 80\}$. We discard 340 instances due to solvability or consistency failures, ending up with a dataset $\mathcal{D}$ of 2300 problems.

We report in Table 2 the composition of dataset $\mathcal{D}$ with respect to problem types and assigned labels. The dataset is highly unbalanced: the majority of instances is tagged as NL, with a very small share of T. Also, the NL answer is strongly predominant for mixed convex instances, suggesting that there could be a clear winner method depending on the type of problem itself.

Table 2. Composition of dataset $\mathcal{D}$. For each type and label we report the total number of instances and their percentage.

	L	NL	T	Total (%)
0–1 convex	195	600	35	830 (**0.36**)
0–1 nonconvex	392	312	39	743 (**0.32**)
Mixed convex	11	701	15	727 (**0.32**)
Total (%)	598 (**0.26**)	1613 (**0.70**)	89 (**0.04**)	**2300**

Table 3. Classification measures for different learning settings. The best performing classifiers are boldfaced.

(a) Multiclass - All features

	SVM	**RF**	EXT	GB
Accuracy	85.22	**88.87**	84.00	87.65
Precision	81.91	**85.51**	81.26	84.79
Recall	85.22	**88.87**	84.00	87.65
F1-score	83.16	**87.11**	82.52	86.19

(b) Binary - Static features

	SVM	RF	EXT	GB
Accuracy	**86.80**	86.08	85.53	86.62
Precision	**86.48**	85.69	85.20	86.32
Recall	**86.80**	86.08	85.53	86.62
F1-score	**86.28**	85.53	85.30	86.03

5.1 Learning Experiments and Results

Learning experiments are implemented in Python with Scikit-learn [20], and run on a personal computer with Intel Core i5, 2.3 GHz and 8 GB of memory. We randomly split $\mathcal{D}$ into $\mathcal{D}_{\text{train}}$ and $\mathcal{D}_{\text{test}}$, a training and a test sets of, respectively, 1725 (75%) and 575 (25%) instances; data is normalized in $[-1, 1]$. We perform training with 5-fold cross validation to grid-search hyper-parameters, and test on the neutral $\mathcal{D}_{\text{test}}$. We try Support Vector Machine (SVM) with RBF kernel [21], together with Random Forests (RF) [22], Extremely Randomized Trees (EXT) [23] and Gradient Tree Boosting (GB) [24].

Our first experiment involves a multiclass scheme with labels {L, NL, T}, and exploits all features. Table 3a reports the standard measures for classification in this setting: for all classifiers we compare accuracy, precision, recall, f1-score (weighted by classes' supports, to account for unbalance). In this setting, RF is best performing in all measures. Features importance scores among RF, EXT and GB show that the subset of features that are more influential for the prediction comprises both dynamic features (difference of lower bounds and times at root node) and information on the convexity of the problem (e.g., value of the smallest nonzero eigenvalue and spectral norm of Q).

Examining the classifiers' confusion matrices, a major difficulty seems to be posed by the T class, which is (almost) always misclassified. Ultimately, we aim at providing a reliable classification of those "extreme" cases for which a change in the resolution approach produces a change in the instance being solved or not. Thus, we carry out further experiments in a binary setting: we remove all tie cases and rescale the data accordingly. All measures are overall improved for the new binary classifiers, and again RF performs as the best algorithm.

We also try classifiers trained without dynamic features: albeit this may sound in conflict with the features importance scores mentioned above, from the optimization solver's point of view is it useful to test a scenario in which a prediction is cast without the need of solving twice the root MIQP. All measures slightly deteriorate without dynamic features, and SVM becomes the best performing algorithm; nonetheless, "static" predictors and their original counterparts are coherent in their (mis)classifications on $\mathcal{D}_{\text{test}}$.

Results in a setting simplified in terms of both labels and features are reported in Table 3b: performance is balanced in the improvement brought by the removal of ties, and the degradation due to the absence of dynamic features, and again SVM performs better.

5.2 Complementary Optimization Measures

To determine the effectiveness of our learned approach with respect to the solver's strategy, we define "optimization measures" scoring and evaluating the classifiers in terms of resolution runtimes.

We run each instance i of $\mathcal{D}_{\text{test}}$ for three `qtolin` values - CPLEX default (DEF), L and NL. Each problem is run only once, with timelimit of 1h; we focus on the Multiclass and All features setting. We remove never-solved instances, to remain with 529 problems in $\mathcal{D}_{\text{test}}$. For each classifier clf, we associate to the vector of its predicted labels y_{clf} a vector of predicted times t_{clf} by selecting t_{L}^i or t_{NL}^i depending on y_{clf}^i for $i \in \mathcal{D}_{\text{test}}$ (we choose their average if a tie was predicted). We also build t_{best} (t_{worst}) selecting runtimes of the correct (wrong) labels for the samples. Note that t_{DEF} is directly available, without labels' vector.

Sum of Predicted Runtimes. We compare $\sigma_{\text{clf}} := \sum_{i \in \mathcal{D}_{\text{test}}} t_{\text{clf}}^i$ for clf $\in$ {SVM, RF, EXT, GB} with σ_{best}, σ_{worst} and σ_{DEF}. Results are in Table 4a: RF is the closest to *best* and the farthest from *worst*; also, DEF could take up to 4x more time to run MIQPs in $\mathcal{D}_{\text{test}}$ compared to a trained classifier. Note that the real gain in time could be even bigger than this, given the fact that we set a timelimit of 1 h.

Normalized Time Score. We then consider the shifted geometric mean of t_{clf} over $\mathcal{D}_{\text{test}}$, normalized between *best* and *worst* cases to get a score $N\sigma_{\text{clf}} \in [0, 1]$:

$$sgm_{\text{clf}} := \sqrt[|\mathcal{D}_{\text{test}}|]{\prod_{i \in \mathcal{D}_{\text{test}}} (t_{\text{clf}}^i + 0.01)} - 0.01, \quad N\sigma_{\text{clf}} := \frac{sgm_{\text{worst}} - sgm_{\text{clf}}}{sgm_{\text{worst}} - sgm_{\text{best}}}. \quad (3)$$

The measure is reported in Table 4a: all predictors are very close to 1 (this score highly reflects classification performance), while DEF is almost halfway between *best* and *worst*.

Table 4. Complementary optimization measures. Best classifiers are boldfaced.

(a) Multiclass - All features					
	SVM	**RF**	EXT	GB	DEF
$\sigma_{\text{clf}}/\sigma_{\text{best}}$	1.49	**1.31**	1.43	1.35	5.77
$\sigma_{\text{worst}}/\sigma_{\text{clf}}$	7.48	**8.49**	7.81	8.23	1.93
$\sigma_{\text{DEF}}/\sigma_{\text{clf}}$	3.88	**4.40**	4.04	4.26	-
$N\sigma_{\text{clf}}$	0.98	**0.99**	0.98	0.99	0.42

(b) Binary - Static features					
	SVM	RF	EXT	GB	DEF
$\sigma_{\text{clf}}/\sigma_{\text{best}}$	**1.80**	2.04	2.01	1.82	5.81
$\sigma_{\text{worst}}/\sigma_{\text{clf}}$	**6.23**	5.50	5.59	6.19	1.93
$\sigma_{\text{DEF}}/\sigma_{\text{clf}}$	**3.22**	2.85	2.89	3.20	-
$N\sigma_{\text{clf}}$	**0.98**	0.98	0.98	0.98	0.43

The presence of timelimiting cases in the test runs is also well reflected in σ_{clf} and $N\sigma_{\text{clf}}$, which are better for classifiers hitting timelimits less frequently (3 times only for RF, 39 for DEF). Note that both L and NL do reach the limit without finding a solution (38 and 55 times, respectively), and that due to variability even *best* hits the timelimit once. We compute σ_{clf} and $N\sigma_{\text{clf}}$ in the Binary - Static features setting as well. Results in Table 4b are in line with what previously discussed for this setup.

6   Conclusions and Ongoing Research

We propose a learning framework to investigate the question *linearize vs. not linearize* for MIQP problems. Results on a generated dataset are satisfactory in terms of classification performance and promising for their interpretability. Novel scoring measures positively evaluate the classifiers' performance from the optimization point of view, showing significant improvements with respect to CPLEX default strategy in terms of running times.

In ongoing and future research, we plan to focus on four main directions.

- Analyze other benchmark datasets: the analysis of public libraries containing MIQPs (e.g., [25]) is crucial to understand how representative the synthetic $\mathcal{D}$ is of commonly used instances, which can then be used to form a more meaningful and comprehensive final dataset.
  So far, we analyzed a share of CPLEX internal MIQP testbed $\mathcal{C}_{\text{test}}$ of 175 instances: the data is very different from $\mathcal{D}$ in features' distribution ($\mathcal{C}_{\text{test}}$ is dominated by the presence of very structured combinatorial MIQPs, like Max-Cut and Quadratic Assignment Problems). The majority class is that of ties, followed by L and with very few NL cases. Preliminary experiments on $\mathcal{C}_{\text{test}}$ used as a test set for classifiers trained on $\mathcal{D}_{\text{train}}$ produce very poor classification results, as most often misclassification happens in form of a T predicted as NL. In fact, complementary optimization measures are not discouraging: given that $\mathcal{C}_{\text{test}}$ contains mostly ties, albeit the high misclassification rate, the loss in terms of solver's performance is not dramatic.
- Deepen features importance analysis, to get and interpret methodological insights on the reasons behind the decision *linearize vs. not linearize*. As

we mentioned in Sect. 5, the problem type itself might draw an important line in establishing the winning method, which seems strongly linked to the information collected at the root node.

- Identify the best learning scenario, in order to successfully integrate the learning framework with the solver. We already considered the simplified Binary - Static features one; it could be interesting to perform static features selection based on their correlation with dynamic ones.
- Define a custom loss function: the complementary optimization measures that we propose showed effective in capturing the optimization performance as well as the classification one. We plan to use these and other intuitions to craft a custom loss/scoring function to train/validate the learning algorithm, in a way tailored to the solver's performance on MIQPs.

References

1. Kruber, M., Lübbecke, M.E., Parmentier, A.: Learning when to use a decomposition. In: Salvagnin, D., Lombardi, M. (eds.) CPAIOR 2017. LNCS, vol. 10335, pp. 202–210. Springer, Cham (2017). https://doi.org/10.1007/978-3-319-59776-8_16
2. Khalil, E.B., Dilkina, B., Nemhauser, G., Ahmed, S., Shao, Y.: Learning to run heuristics in tree search. In: 26th International Joint Conference on Artificial Intelligence (IJCAI) (2017)
3. Lodi, A., Zarpellon, G.: On learning and branching: a survey. TOP **25**(2), 207–236 (2017)
4. Hutter, F., Hoos, H.H., Leyton-Brown, K.: Automated configuration of mixed integer programming solvers. In: Lodi, A., Milano, M., Toth, P. (eds.) CPAIOR 2010. LNCS, vol. 6140, pp. 186–202. Springer, Heidelberg (2010). https://doi.org/10.1007/978-3-642-13520-0_23
5. Hutter, F., Xu, L., Hoos, H.H., Leyton-Brown, K.: Algorithm runtime prediction: methods and evaluation. Artif. Intell. **206**, 79–111 (2014)
6. Bienstock, D.: Computational study of a family of mixed-integer quadratic programming problems. Math. Program. **74**(2), 121–140 (1996)
7. Bonami, P., Lejeune, M.A.: An exact solution approach for portfolio optimization problems under stochastic and integer constraints. Oper. Res. **57**(3), 650–670 (2009)
8. CPLEX (2017). http://www-01.ibm.com/software/commerce/optimization/cplex-optimizer/index.html
9. Bliek, C., Bonami, P., Lodi, A.: Solving mixed-integer quadratic programming problems with IBM-CPLEX: a progress report. In: Proceedings of the Twenty-Sixth RAMP Symposium, pp. 16–17 (2014)
10. Fourer, R.: Quadratic optimization mysteries, part 1: two versions (2015). http://bob4er.blogspot.ca/2015/03/quadratic-optimization-mysteries-part-1.html
11. Gupta, O.K., Ravindran, A.: Branch and bound experiments in convex nonlinear integer programming. Manag. Sci. **31**(12), 1533–1546 (1985)
12. Bonami, P., Biegler, L.T., Conn, A.R., Cornuéjols, G., Grossmann, I.E., Laird, C.D., Lee, J., Lodi, A., Margot, F., Sawaya, N., et al.: An algorithmic framework for convex mixed integer nonlinear programs. Discret. Optim. **5**(2), 186–204 (2008)
13. Land, A., Doig, A.: An automatic method of solving discrete programming problems. Econometrica **28**, 497–520 (1960)

14. Duran, M.A., Grossmann, I.E.: An outer-approximation algorithm for a class of mixed-integer nonlinear programs. Math. Program. **36**(3), 307–339 (1986)
15. McCormick, G.P.: Computability of global solutions to factorable nonconvex programs: Part I—convex underestimating problems. Math. Program. **10**(1), 147–175 (1976)
16. Belotti, P., Kirches, C., Leyffer, S., Linderoth, J., Luedtke, J., Mahajan, A.: Mixed-integer nonlinear optimization. Acta Numerica **22**, 1–131 (2013)
17. MATLAB: Version 9.1.0 (2016). The MathWorks Inc., Natick, Massachusetts
18. Puchinger, J., Raidl, G.R., Pferschy, U.: The multidimensional Knapsack problem: structure and algorithms. INFORMS J. Comput. **22**(2), 250–265 (2010)
19. Lodi, A., Tramontani, A.: Performance variability in mixed-integer programming. In: Theory Driven by Influential Applications, INFORMS, pp. 1–12 (2013)
20. Pedregosa, F., Varoquaux, G., Gramfort, A., Michel, V., Thirion, B., Grisel, O., Blondel, M., Prettenhofer, P., Weiss, R., Dubourg, V., Vanderplas, J., Passos, A., Cournapeau, D., Brucher, M., Perrot, M., Duchesnay, E.: Scikit-learn: machine learning in Python. J. Mach. Learn. Res. **12**, 2825–2830 (2011)
21. Cortes, C., Vapnik, V.: Support-vector networks. Mach. Learn. **20**(3), 273–297 (1995)
22. Breiman, L.: Random forests. Mach. Learn. **45**(1), 5–32 (2001)
23. Geurts, P., Ernst, D., Wehenkel, L.: Extremely randomized trees. Mach. Learn. **63**(1), 3–42 (2006)
24. Friedman, J.H.: Stochastic gradient boosting. Comput. Stat. Data Anal. **38**(4), 367–378 (2002)
25. Furini, F., Traversi, E., Belotti, P., Frangioni, A., Gleixner, A., Gould, N., Liberti, L., Lodi, A., Misener, R., Mittelmann, H., Sahinidis, N., Vigerske, S., Wiegele, A.: QPLIB: a library of quadratic programming instances. Technical report (2017). Available at Optimization Online

Fleet Scheduling in Underground Mines Using Constraint Programming

Max Åstrand[1,2]([✉]), Mikael Johansson[2], and Alessandro Zanarini[3]

[1] ABB Corporate Research Center, Västerås, Sweden
max.astrand@se.abb.com
[2] KTH Royal Institute of Technology, Stockholm, Sweden
mikaelj@kth.se
[3] ABB Corporate Research Center, Baden-Dättwil, Switzerland
alessandro.zanarini@ch.abb.com

Abstract. The profitability of an underground mine is greatly affected by the scheduling of the mobile production fleet. Today, most mine operations are scheduled manually, which is a tedious and error-prone activity. In this contribution, we present and formalize the underground mine scheduling problem, and propose a CP-based model for solving it. The model is evaluated on instances generated from real data. The results are promising and show a potential for further extensions.

1 Introduction

Mining is the process of extracting minerals from the earth, commonly done either in open pit or underground mines. Underground mining is a cost-intensive industry, where the margins decrease as production goes deeper. The operational performance, and thus the profitability, of an underground mine is greatly affected by how the mobile machinery is coordinated. In most mines, the machine fleet is scheduled manually with methods and tools on the brim of what they can handle. In a survey [1] of more than 200 high level executives from mining companies all around the world, maximizing production effectiveness was identified as the top challenge for modern mines, even more than improving reliability of individual equipment. This fact highlights that system level coordination is critical for mines to remain profitable in the future.

Underground mining operation is planned on different levels with different horizons and task granularities. The longest planning horizon is in the *life-of-mine plan*, which contains a rough strategy for which year to extract what parts of the ore-body. The life-of-mine plan is further decomposed into *extraction plans* of various granularity that have time horizons of years or months. The extraction plans include more details about the amount of ore that is expected to be excavated under different parts of the plan. Lastly, the extraction plans are implemented as *short-term schedules* by allocating and time-tabling machines and personnel to the activities, producing a detailed schedule for roughly one week.

© Springer International Publishing AG, part of Springer Nature 2018
W.-J. van Hoeve (Ed.): CPAIOR 2018, LNCS 10848, pp. 605–613, 2018.
https://doi.org/10.1007/978-3-319-93031-2_44

To the best of the authors' knowledge, there is no previous work on using CP to schedule the mobile production fleet in underground mining. The most similar problem can be found in [2], where they study scheduling mobile machines in an underground potash mine using MIP-models and construction procedures. Another underground mine scheduling problem is described in [3] where schedules are created by enumerating all sequence permutations, and select the one with the shortest makespan. To limit the search space of all possible sequences (at worst factorial) the authors cluster the faces based on geographical distance. The authors in [4] study how to transport ore in an underground mine, via intermediate storages, to the mine hoist. They use a MIP-model to allocate machines to different work areas on a shift basis over a period of 2 months. The authors refine their work in [5] and propose several simplifications which decrease computation time.

A study of underground machine routing is introduced in [6], where they study the effect of using different heuristic dispatch strategies for routing machines in a diamond mine. The authors conclude that strategies that separate the machines geographically seem beneficial. This is due to the confined environment, in which avoiding deadlocks (e.g. machines meeting each other in a one-way tunnel) is crucial. More research on underground routing can be found in [7], where the authors continue on previous work in [8] by using dynamic programming.

In [9] the authors study scheduling open pit mines (a related but distinct problem) with MIP. Using a multiobjective approach, they study over 40 objectives, including maximizing the utilization of trucks and minimizing deviations from targeted production. Continuing on the topic of scheduling open pit mines, the authors in [10] develop a model for scheduling numerous autonomous drill rigs in an open pit mine. The problem naturally decomposes into subproblems, where a high-level CSP, linking the subproblems, is solved to find a solution in the joint search space.

Our contribution is the first study of production scheduling in underground mining using CP. Further, the problem introduced generalizes similar problems studied by other authors (using other methods). Most notably, we impose and exploit the presence of blast windows, allow for a mix of interruptible and uninterruptible tasks, and support tasks that have an after-lag.

The paper is organized as follows: Sect. 2 describes the problem and introduces the necessary notation. A CP-based model for solving the problem is developed in Sect. 3. Next, Sect. 4 reports experimental results on instances generated by data coming from a real mine. Conclusions are drawn in Sect. 5.

2   Problem Description

The mine operations that we consider are located at underground sites called *faces*, which denote the end of an underground tunnel. From here-on, the faces are labeled $F = \{1, \dots, n\}$. In order to extract ore from the mountain, a periodic sequence of activities takes place at each face: $C = (drilling, charging, blasting,$

ventilating, washing, loading, scaling, cleaning, shotcreting, bolting, face scaling, face cleaning). We refer to a full period of activities as a *cycle*. All the activities except blasting and ventilation require a specific (today) human-operated machine to be used, and we denote with $\hat{C}$ this subset of activity types. Specifically, drilling requires a drill rig in order to drill holes in the face processed; when charging, the holes are filled with explosives typically using a wheel loader with a platform; after blasting, the rock that has been separated from the mountain is sprayed with water to reduce the amount of airborne particles. The rock is then removed (loaded) from the face with an LHD (load, haul, dump machine); smaller rocks loosely attached to the mountain are later mechanically removed with a scale rig (scaling), and removed from the drift with an LHD (cleaning); the two successive steps are for ensuring safety, namely to secure the rock to the insides of the tunnel (bolting), and spraying the insides of the tunnel with concrete (shotcreting). Finally, the face is prepared for the next cycle by a scaling rig (face scaling), and the separated rock is removed by an LHD (face cleaning). For each activity type $c \in \hat{C}$, a non-empty set of machines M_c is available in the mine to perform that specific operation. In the general case we are studying, some machines can be employed to perform different activity types.

Blasting is a key activity in underground mine operations and it sets the overall pace of production for the entire mine. It is common that blasts occur in predetermined time windows during the day (typically twice or thrice); for safety reasons, no human operator is allowed in the mine during the blasting time window and subsequent ventilation of toxic blast fumes, independently of whether blasting occurs, and on which faces; in other words, no other activity can take place at any face across the whole mine.

As blasting and ventilation have the same properties, happen one after the other, and affect the scheduling in the same manner, in the following, whenever we refer to blasts we denote an activity that spans over the duration of the blast and the subsequent ventilation. We refer to the candidate blast time windows as $B = \{(s_{b_1}, e_{b_1}), (s_{b_2}, e_{b_2}), (s_{b_3}, e_{b_3}), \dots\}$ where the pair (s_{b_i}, e_{b_i}) defines the start and end of the time window; for simplicity, and without loss of generality, we assume that each time window has equal duration $d_b = e_b - s_b$ and that the blasting (and ventilation) can fit inside the time window (the problem would be trivially infeasible otherwise).

Most activity types are interruptible, i.e. they can start prior to a blast window, then be suspended during the blast, and resumed after the blast window. *Shotcreting* is however not interruptible; furthermore the subsequent *bolting* activity can only happen after a delay required for the concrete to cure; this delay, also referred to as after-lag, has a duration of d_{al}.

An instance of a problem is composed of a set of activities a_i^f indicating the i'th activity at face $f \in F$; the sequence of activities in each face is defined by $A^f = (a_1^f, \dots, a_{m_f}^f)$, which consists of a fixed number of production cycles that follow the sequence defined in C. We define by $d(a_i^f)$ the nominal duration of activity a_i^f, i.e. the duration in case no interruption takes place. Note that all the activity durations are in practice shorter than the time between two blasts, i.e. $d(a_i^f) < d_b$. Furthermore, let $c(a_i^f) \in C$ be a function indicating the activity type of activity a_i^f and let $I.(A^f) = \{i \mid c(a_i^f) = \cdot\}$ be the set of indices of the activities at face A^f of a given type; in this way, $I_{bolting}(A^f)$, for example, indicates the indices of all the bolting activities at face A^f.

Since the schedules are deployed in a rolling-horizon approach, where each face has a predefined number of cycles to be performed during the life of the mine, it makes sense to use an objective function that accounts for the state of all faces. Therefore, the scheduling problem consist of allocating the available mining machinery to the activities, and schedule them in order to minimize the sum of the makespans of all faces.

3 Model

The problem resembles a rich variant of the flow shop problem, with additional aspects such as unavailabilities due to blasts, after-lags, and a mix of interruptible and uninterruptible activities [11,12]. For each activity a_i^f, we employ $|M_{c(a_i^f)}|$ conditional interval variables [13] representing the potential execution of that activity on the candidate machines[1]. Specifically, each conditional interval variable consists of a tuple of four integer variables: $\mathsf{s}_{ir}^f, \mathsf{d}_{ir}^f, \mathsf{e}_{ir}^f, \mathsf{o}_{ir}^f$ indicating the start time, the duration, the end time $(\mathsf{s}_{ir}^f + \mathsf{d}_{ir}^f = \mathsf{e}_{ir}^f)$ and the execution status of activity a_i^f on machine $r \in M_{c(a_i^f)}$, respectively. This model allows for machine-dependent duration in case machines have different processing times.

The execution status takes value 0 if the activity is not executed with machine r, or 1 if it is executed with machine r. As each activity is executed using exactly one machine:

$$\sum_{r \in M_{c(a_i^f)}} \mathsf{o}_{ir}^f = 1 \quad \forall f \in F \quad \forall i \in 1, \ldots, |A^f| \tag{1}$$

The start times for blasts must be aligned with the blast time windows, therefore:

$$\mathsf{s}_{ir}^f \in \{s_{b_1}, s_{b_2}, s_{b_3}, \ldots\} \quad \forall f \in F \quad \forall i \in I_{blasting}(A^f) \quad \forall r \in M_{blasting} \tag{2}$$

Uninterruptible tasks (namely shotcreting) also must not overlap the blast time windows, independently of whether or not a blast occurs in that specific face:

$$\mathsf{s}_{ir}^f \in \mathcal{S}_{shotcreting} \quad \forall f \in F \quad \forall i \in I_{shotcreting}(A^f) \quad \forall r \in M_{shotcreting} \tag{3}$$

where $\mathcal{S}_{shotcreting} = \{0, \ldots, s_{b_1} - d(a_i^f), e_{b_1}, \ldots, s_{b_2} - d(a_i^f), \ldots\}$.

Both blast and shotcreting activities are uninterruptible therefore their respective durations are set to the nominal activity durations:

$$\mathsf{d}_{ir}^f = d(a_i^f) \quad \forall f \in F \quad \forall i \in I_{blasting}(A^f) \quad \forall r \in M_{blasting} \tag{4}$$

$$\mathsf{d}_{ir}^f = d(a_i^f) \quad \forall f \in F \quad \forall i \in I_{shotcreting}(A^f) \quad \forall r \in M_{shotcreting} \tag{5}$$

---

[1] In order to simplify the notation, we assume that for blasting activities we have a single machine $r \in M_{blasting}$ with infinite capacity.

Further, all activities, except blasting, cannot start during a blast window:

$$\mathbf{s}_{ir}^f \in \mathcal{S}_c \quad \forall f \in F \quad \forall c \in \tilde{C} \quad \forall i \in I_c(A^f) \quad \forall r \in M_c \tag{6}$$

where $\mathcal{S}_c = \{0, \ldots, s_{b_1}, e_{b_1}, \ldots, s_{b_2}, \ldots\}$ and $\tilde{C} = \hat{C} \setminus \{blasting\}$.

In order to model the interruptible activities, their associated intervals have variable durations. We introduce a variable $\mathbf{p}_{ir}^f$ indicating whether the interval of face f, index i, resource r has been interrupted. We use a sum of reified constraints to go over all the possible blast time windows, and verify if there is one that overlaps with the interval; if an interval starts before a blast time window and ends after the start of the blast time window then $p_{ir}^f = 1$. Note that the durations of the activities are such that no activity can span two blast time windows. Finally, whenever the interval gets interrupted, its duration needs to be extended by the duration of the blast time window.

$$\mathbf{p}_{ir}^f = \sum_k (\mathbf{s}_{ir}^f < s_{b_k}) * (\mathbf{s}_{ir}^f + d(a_i^f) > s_{b_k}) \tag{7}$$

$$\mathbf{d}_{ir}^f = d(a_i^f) + \mathbf{p}_{ir}^f * d_b \qquad \forall f \in F \quad \forall c \in \tilde{C} \quad \forall i \in I_c(A^f) \quad \forall r \in M_c$$

The full order of the cyclic activities is enforced by:

$$\mathbf{s}_{ir}^f + \mathbf{d}_{ir}^f < \mathbf{s}_{i+1r'}^f \quad \forall f \in F \quad \forall i \in 1, \ldots, |A^f| - 1 \quad \forall r, r' \in M_{c(a_i^f)} \tag{8}$$

and the after-lag of shotcreting as:

$$\mathbf{s}_{ir}^f + \mathbf{d}_{ir}^f + d_{al} < \mathbf{s}_{i+1r'}^f \quad \forall f \in F \quad \forall i \in I_{shotcreting}(A^f) \quad \forall r, r' \in M_{c(a_i^f)} \tag{9}$$

Unary constraints are used for all the faces and machines to enforce disjunctive execution:

$$\mathbf{unary}(\{[\mathbf{s}_{ir}^f, \mathbf{d}_{ir}^f, \mathbf{o}_{ir}^f] \mid i = 1, \ldots, |A^f|, \ r \in M_{c(a_i^f)}\}) \quad \forall f \in F \tag{10}$$

$$\mathbf{unary}(\{[\mathbf{s}_{ir}^f, \mathbf{d}_{ir}^f, \mathbf{o}_{ir}^f] \mid f = 1, \ldots, n, \ i \in \tilde{I}_r^f\}) \quad \forall c \in C \quad \forall r \in M_c \tag{11}$$

in which $\tilde{I}_r^f = \{i \mid i \in A^f \wedge r \in M_{c(a_i^f)}\}$ indicates the indices of all the activities of face f that can be performed by machine r. If all machines have identical processing times the machines can be modeled as a cumulative resource, which can provide significant speed-ups. However, in reality, the processing times are seldom identical due to the usage of a heterogeneous machine park, and due to different operator skills. Further, notice that the unary constraint for the faces is redundant, since a total ordering is enforced by constraints 8 and 9. However, adding a unary constraint for the faces enables scheduling auxiliary tasks that are not part of the production cycle.

Finally, the model minimizes the sum of the makespans across all faces by (recall that m_f is the index of the last activity of face f):

$$\sum_f \sum_{r \in M_{c(a_{m_f}^f)}} \mathbf{o}_{m_f r}^f * \mathbf{e}_{m_f r}^f \tag{12}$$

3.1 Search Strategy

Different generic and ad-hoc heuristics have been tested; for brevity, we present in the following what has been experimentally deemed the most effective one.

The search proceeds in two phases: the first phase considers machine allocation, the second is about scheduling the start times. In the former, tasks are ordered by their nominal durations, and for each task the least loaded machine is chosen among the compatible ones. As for the task scheduling, tasks are chosen based on an action-based heuristic and the value selection schedules the task as early as possible. Action-based heuristic learns during the search to branch on the variables that are likely to trigger the most propagation (also known as activity-based heuristic [14]).

Randomized Restarts. By systematically restarting the search procedure, a restart-based search samples a wider part of the search space. In order to not to end up on the same solution after each restart, some randomness needs to be included in the search heuristics. Therefore, the machine allocation chooses a random variable to branch on with probability p, otherwise it follows the heuristic described above. Similarly, the value selection is chosen at random with probability p, otherwise the least loaded heuristic is employed. Note that action-based branching tend to work well with restart-based search, since the restarts provide a lot of information of propagation.

4   Experimental Results

An underground mine scheduling scenario based on real data is here studied. All problems are solved using the model introduced in Sect. 3, where a baseline method without restarts is compared to a restart-based approach using the randomized search heuristics of Sect. 3.1. All problems are solved using Gecode 5.1 with 4 threads on a laptop with an i7-7500U 2.7 GHz processor.

The size of a problem is determined by (i) the number of faces, (ii) the number of cycles, and (iii) the number of machines of each type. In this work, we experimented with instances with 5 and 10 faces, with 1 or 2 cycles at each face, which results in a minimum of 55 tasks and a maximum of 220 tasks to be scheduled. The considered machine parks consist of 1, or 2 machines of each type, together with a non-uniform machine park inspired a real underground mine. The complex machine park consists of 4 drill rigs, 3 chargers, 1 water vehicle, 6 LHDs, 4 scaling rigs, 2 shotcreters and 3 bolters. The problem sizes are encoded as *(# faces)F(# cycles per face)C(# machines of each type)M*. For example, *5F2C2M* corresponds to a problem with 5 faces where each face has 2 cycles each, and the machine park consists of 2 machines of each type. The machine classification *CM* is used to encode the complex non-homogeneous machine park. In total, 8 different problem sizes are studied, where each problem size is studied in 5 instances where the task duration is varied. In each instance the task duration is based on nominal durations from an operational mine, however they are perturbed randomly by a factor between -25% and $+25\%$.

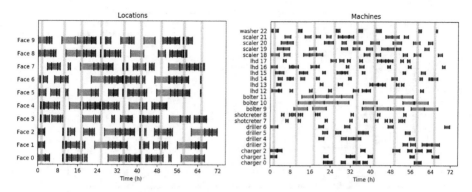

Fig. 1. The largest problem instance using 10 faces with 2 cycles at each face, together with a machine park that is based on a real underground mine.

A timeout of 12.5 min is used for the instances with 5 faces, and 25 min for the instances with 10 faces.

A solution to the largest problem studied, *10F2CCM*, using the restart-based search can be seen in Fig. 1. The blast windows are indicated by vertically aligned grey areas. In this solution we can see some features of the problem such as shotcreting tasks which are not split over blast windows whereas other activities are correctly interrupted and resumed after blast time windows. Evidently, manual scheduling quickly becomes a tedious and error-prone activity as the problem size increases.

In Table 1, we have aggregated the statistics of solving 40 instances using both the baseline and the restart-based method. The second to the fourth columns represent, respectively, the average, minimum and maximum objective across the instance set. The fifth to the seventh columns indicate, respectively, the

Table 1. Average, minimum, and maximum objective value after solving 5 instances of each problem size. The left column corresponds to the objective value using the search without restarts, while the right column corresponds to the reduction gained by using restart-based search.

	avg O_{BL}	min O_{BL}	max O_{BL}	avg $O^{\%}_{RES}$	min $O^{\%}_{RES}$	max $O^{\%}_{RES}$
5F1C1M	2082	1911	2214	0%	0%	1%
5F1C2M	1892	1751	1990	0%	0%	1%
5F2C1M	4200	3583	4575	−1%	−7%	1%
5F2C2M	4048	3560	4360	−7%	−11%	−3%
10F1C2M	4410	3973	4601	0%	−5 %	3 %
10F1CCM	3797	3606	4097	2 %	4%	1%
10F2C2M	11019	9803	12562	−15%	−11 %	−21%
10F2CCM	8222	7548	8825	−1%	−4%	2%

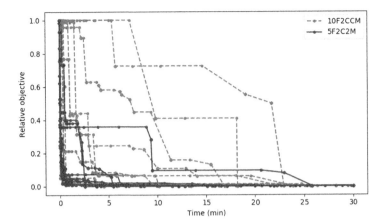

Fig. 2. The objective value using restart-based search on 5 samples of 10F2CCM and 5 samples of 5F2C2M. The plot shows the current objective divided by the value of the first found solution.

average, minimum, maximum reduction of the objective value obtained with randomized restarts w.r.t. the deterministic baseline. It is evident that using restart-based search can be advantageous, particularly for the larger instances. Note furthermore that, as shown in Fig. 1, the complex machine park problem is not very constrained, i.e. machines are not highly utilized, and the problem is therefore simpler than the instances with only 2 machines per face. This can also be seen in Table 1, where randomized restarts for *10F2CCM* does not bring significant improvements, whereas restarts seem beneficial for *10F2C2M*.

Finally, we are only able to solve the smallest instances, *5F1C1M* and *5F1C2M*, to optimality. Figure 2 shows how the objective function evolves with solution time for larger problem instances, and it motivates the timeouts chosen. No significant improvements were seen when extending the timeout up to an hour.

5   Concluding Remarks

In this paper, we presented, to the best of the authors' knowledge, the first CP-based model for underground mine scheduling with promising preliminary results. The model resembles a flow shop problem with the addition of periodic unavailabilities, after-lags, and a mix of interruptible and uninterruptible activities. Ongoing and future work includes: the integration of travel times for the machine park, replanning based on previous solutions using Large Neighborhood Search to minimize schedule disruptions, as well as personnel assignment and rostering. We would also like to explore decomposition approaches where the machine allocation and the scheduling are solved in a Logical Benders decomposition framework.

Acknowledgements. This work was partially supported by the Wallenberg AI, Autonomous Systems and Software Program (WASP).

References

1. Mincom: Annual Study: Mining Executive Insights 2011, Denver, CO (2011)
2. Schulze, M., Rieck, J., Seifi, C., Zimmermann, J.: Machine scheduling in underground mining: an application in the potash industry. OR Spectr. **38**(2), 365–403 (2016)
3. Song, Z., Schunnesson, H., Rinne, M., Sturgul, J.: Intelligent scheduling for underground mobile mining equipment. PloS One **10**(6), e0131003 (2015)
4. Nehring, M., Topal, E., Knights, P.: Dynamic short term production scheduling and machine allocation in underground mining using mathematical programming. Min. Technol. **119**(4), 212–220 (2010)
5. Nehring, M., Topal, E., Little, J.: A new mathematical programming model for production schedule optimization in underground mining operations. J. South. Afr. Inst. Min. Metall. **110**(8), 437–446 (2010)
6. Saayman, P., Craig, I.K., Camisani-Calzolari, F.R.: Optimization of an autonomous vehicle dispatch system in an underground mine. J. South. Afr. Inst. Min. Metall. **106**(2), 77 (2006)
7. Beaulieu, M., Gamache, M.: An enumeration algorithm for solving the fleet management problem in underground mines. Comput. Oper. Res. **33**(6), 1606–1624 (2006)
8. Gamache, M., Grimard, R., Cohen, P.: A shortest-path algorithm for solving the fleet management problem in underground mines. Eur. J. Oper. Res. **166**(2), 497–506 (2005)
9. Blom, M., Pearce, A.R., Stuckey, P.J.: Short-term scheduling of an open-pit mine with multiple objectives. Eng. Optim. **49**(5), 777–795 (2017)
10. Mansouri, M., Andreasson, H., Pecora, F.: Hybrid reasoning for multi-robot drill planning in open-pit mines. Acta Polytechnica **56**(1), 47–56 (2016)
11. Pinedo, M.: Scheduling. Springer, Heidelberg (2015). https://doi.org/10.1007/978-3-319-26580-3
12. Baptiste, P., Le Pape, C., Nuijten, W.: Constraint-Based Scheduling: Applying Constraint Programming to Scheduling Problems, vol. 39. Springer Science & Business Media, Heidelberg (2012). https://doi.org/10.1007/978-1-4615-1479-4
13. Laborie, P., Rogerie, J.: Reasoning with conditional time-intervals. In: FLAIRS Conference, pp. 555–560 (2008)
14. Michel, L., Van Hentenryck, P.: Activity-based search for black-box constraint programming solvers. In: Beldiceanu, N., Jussien, N., Pinson, É. (eds.) CPAIOR 2012. LNCS, vol. 7298, pp. 228–243. Springer, Heidelberg (2012). https://doi.org/10.1007/978-3-642-29828-8_15

Author Index

Printed in the United States
By Bookmasters